# Lecture Notes in Networks and Systems

Volume 170

The series "Lecture Notes in Networks and Systems" publishes the latest developments in Networks and Systems—quickly, informally and with high quality. Original research reported in proceedings and post-proceedings represents the core of LNNS.

Volumes published in LNNS embrace all aspects and subfields of, as well as new challenges in, Networks and Systems.

The series contains proceedings and edited volumes in systems and networks, spanning the areas of Cyber-Physical Systems, Autonomous Systems, Sensor Networks, Control Systems, Energy Systems, Automotive Systems, Biological Systems, Vehicular Networking and Connected Vehicles, Aerospace Systems, Automation, Manufacturing, Smart Grids, Nonlinear Systems, Power Systems, Robotics, Social Systems, Economic Systems and other. Of particular value to both the contributors and the readership are the short publication timeframe and the world-wide distribution and exposure which enable both a wide and rapid dissemination of research output.

The series covers the theory, applications, and perspectives on the state of the art and future developments relevant to systems and networks, decision making, control, complex processes and related areas, as embedded in the fields of interdisciplinary and applied sciences, engineering, computer science, physics, economics, social, and life sciences, as well as the paradigms and methodologies behind them.

Indexed by SCOPUS, INSPEC, WTI Frankfurt eG, zbMATH, SCImago.

All books published in the series are submitted for consideration in Web of Science.

More information about this series at http://www.springer.com/series/15179

Arnab Kumar Maji · Goutam Saha ·
Sufal Das · Subhadip Basu ·
João Manuel R. S. Tavares
Editors

# Proceedings of the International Conference on Computing and Communication Systems

## I3CS 2020, NEHU, Shillong, India

Springer

*Editors*
Arnab Kumar Maji
Department of Information Technology
North-Eastern Hill University
Shillong, Meghalaya, India

Goutam Saha
Department of Information Technology
North-Eastern Hill University
Shillong, Meghalaya, India

Sufal Das
Department of Information Technology
North-Eastern Hill University
Shillong, Meghalaya, India

Subhadip Basu
Department of Computer Science
and Engineering
Jadavpur University
Kolkata, West Bengal, India

João Manuel R. S. Tavares
Departamento de Engenharia Mecânica
Faculdade de Engenharia
Universidade do Porto
Porto, Portugal

ISSN 2367-3370 ISSN 2367-3389 (electronic)
Lecture Notes in Networks and Systems
ISBN 978-981-33-4083-1 ISBN 978-981-33-4084-8 (eBook)
https://doi.org/10.1007/978-981-33-4084-8



This Springer imprint is published by the registered company Springer Nature Singapore Pte Ltd.
The registered company address is: 152 Beach Road, #21-01/04 Gateway East, Singapore 189721, Singapore

# Organizing Committee

The Third International Conference on Computing and Communication Systems (I3CS-2020) was held during August 10–11, 2020, in the Department of Information Technology, North-Eastern Hill University, Shillong 793022, Meghalaya, India.

## Chief Patron

Prof. Sri Krishna Srivastava, V.C., NEHU

## Patron

Prof. L. J. Singh, Dean School of Technology, NEHU

## General Chair

Dr. D. Kandar, I.T., NEHU

## Organizing Chair and Co-Chair

Dr. A. K. Maji, I.T., NEHU
Dr. S. Das, I.T., NEHU

## Editorial Board

Dr. A. K. Maji, I.T., NEHU, India (Corr. Edtr.)
Dr. G. Saha, I.T., NEHU, India
Dr. S. Das, I.T., NEHU, India
Dr. S. Basu, C.S.E., Jadavpur University, India
Dr. J. Manuel R. S. Tavares, M.E., Universidade do Porto, Portugal

## Program Chairs

Prof. G. Saha, I.T., NEHU
Dr. A. K. Maji, I.T., NEHU
Dr. K. Amitab, I.T., NEHU

## Finance Chair

Dr. S. Das , I.T., NEHU

## Program Committee (Partial List)

Prof. Dr. G. Anjan Babu, Sri Venkateswara University, Tirupati
Dr. P. Pakray, NIT, Silchar, India
Prof. H. K. Kalita, CIT, Kokrajhar
Dr. A. Nag, CIT, Kokrajhar
Dr. S. Roy, CUS, Sikkim
Dr. A. Halder, NEHU, TURA
Dr. A. K. Khan, Assam University, Silchar
Dr. Somnath Mukhopadhyay, Assam University, Silchar
Dr. Pankaj Sarkar, NEHU, Shillong
Dr. Rupaban Subadar, NEHU, Shillong
Dr. Manashjit Gogoi, NEHU, Shillong
Prof. Suprativ Saha, Brainware University, Kolkata
Prof. Diptendu Sinha Roy, NIT, Meghalaya, Shillong
Prof. Rajat K. Pal, University of Calcutta, Kolkata
Prof. Palash Dey, Indian Institute of Technology, Kharagpur
Prof. P. Chakrabarty, Techno India NJR Institute of Technology, Udaipur
Dr. S. Padhye, Motilal Nehru National Institute of Technology, Allahabad
Dr. Ashis Maurya, Motilal Nehru National Institute of Technology, Allahabad

Dr. Sushanta Kabir Dutta, NEHU, Shillong
Dr. V. Dhilip Kumar, Vel Tech Rangarajan Dr.Sakunthala R&D Institute of Science and Technology, Chennai
Prof. Maad M. Mijwil, Baghdad College of Economics Science University, Baghdad
Dr. Yousef Daradkeh, Prince Sattam bin Abdulaziz University (PSAU), KSA

## Local Organizing Committee (Partial List)

Dr. Md. I. Hussain, I.T., NEHU
Dr. D. Kandar, I.T., NEHU
Dr. R. Subadar, E.C.E, NEHU
Mrs. S. Neog, I.T., NEHU
Dr. S. C. Sahana, I.T., NEHU
Dr. A. Nath, I.T., NEHU
Dr. K. Amitab, I.T., NEHU
Mr. N. R. Pal, I.T., NEHU

## Advisory Committee (Partial List)

Prof. Jugal K Kalita, University of CCS, Colorado, USA
Prof. Biplab Sikdar, National University of Singapore
Prof. Goutam Chakraborty, Iwate Prefectural University, Japan
Prof. Babusena Paul, University of Johannesburg, South Africa
Prof. Sanjoy Misra, Covenant University, Nigeria
Prof. Khalid Saeed, Bialystok University of Technology, Poland
Prof. Dipankar Dasgupta, University of Memphis, USA
Prof. Xiao Zhi Gao, School of Computing, University of Eastern Finland, Finland
Prof. Kiranmai Bellam, Prairie View A&M University, Prairie View, TX
Prof. Jemal H. Abawajy, Deakin University, Geelong, VIC 3220
Prof. Alexander Gelbukh National Polytechnic Institute, Mexico
Prof. Phalguni Gupta, Indian Institute of Technology, Kanpur, India
Prof. Jyotsna Kumar Mandal, Kalyani University, India
Prof. Nilanjan Dey, Techno India, Kolkata
Prof. Milli Pant, Indian Institute of Technology, Roorkee
Prof. Sukumar Nandi, Indian Institute of Technology, Guwahati
Prof. S. M. Hazarika, Indian Institute of Technology, Guwahati
Prof. Santi P. Maity, Indian Institute of Engineering Science and Technology, India
Prof. Rajat K. Pal University of Calcutta, India
Prof. Sugata Sanyal, Ex-Professor, TIFR, Mumbai, India
Prof. Paramartha Dutta, Visva-Bharati, India

Prof. Sudipta Roy, Assam University, Silchar, India
Prof. Debashis De, Maulana Abul Kalam Azad University of Technology, India
Prof. L. Lolit Kumar Singh, Mizoram University, India

# Preface

The Third **International Conference on Computing and Communication Systems** (I3CS-2020) was organized by the Department of Information Technology, North-Eastern Hill University, Shillong, India. The main objective of the conference was to provide an excellent interdisciplinary platform for international scientists, researchers, and engineers to exchange their innovative views and discuss many other emerging issues on several areas specifically in the area of computing, communication, and system developments. This conference desired to bring researchers, educators, professional engineers, and technologists under a single forum to discuss several path-breaking developments in the field of technology.

The earlier two editions of I3CS were held in the years of 2015 and 2016. After the successful conduction of first and second editions of I3CS, we got the courage to conduct the third edition I3CS-2020. The proposed conference was initially planned to be conducted in the month of April 2020. But unfortunately, due to global corona pandemic, we were forced to reschedule it in the month of August 2020, and ultimately with the help of everyone concerned, the I3CS-2020 was held successfully during August 10 and 11, 2020, in online mode. The reason of conduction of I3CS-2020 in online mode was mainly due to travel restrictions. We are proud to say that despite several restrictions, more than 100 researchers across the country and abroad participated in this unique event. Several keynote speakers from India and abroad had also shared their thoughtful ideas here. The objective of I3CS-2020 was sufficiently fulfilled by the nice participation of scientists and researchers all over the globe.

We are proud to say that despite all odds, we received 231 nos. of papers for I3CS-2020 Conference. All the papers were reviewed thoroughly by at least two independent reviewers. Based on primarily the quality of research and outcomes and plagiarism reports, reviewers accessed them and noted their learned comments. Finally on the basis of reviewers' comments, and their final presentation in the conference, 69 papers were considered for their further publication.

Shillong, India — Arnab Kumar Maji
Shillong, India — Goutam Saha
Shillong, India — Sufal Das
Kolkata, India — Subhadip Basu
Porto, Portugal — João Manuel R. S. Tavares
August 2020

# Acknowledgments

First of all, we would like to thank Prof. Sri Krishna Srivastava, Vice Chancellor, NEHU, for his continuous support and encouragement. We are also thankful to Prof. L. Joyprakash Singh, Dean, School of Technology, NEHU, for giving us the opportunity to arrange this international event. We also would like to thank Dr. Nilanjan Dey, Assistant Professor, Department of Information Technology, Techno International New Town, Kolkata, India, who has always been the guiding force behind this event's success. Dr. Debdatta Kandar, General Chair of I3CS-2020 and Associate Professor, Department of Information Technology, NEHU, deserves special thanks for successfully performing his role. It was his dream that we have strived to make a reality. Our special thanks to Springer and especially the editorial staff who were patient, meticulous, and friendly with their constructive criticism on the quality of papers and outright rejection at times without compromising on the quality of the papers as they are always known for publishing the best international papers.

Finally, we would like to express our heartfelt gratitude and warmest thanks to the I3CS-2020 organizing committee for all the hard work and outstanding contributions as a member of the organizing committee. We know how much time and energy this assignment demanded, and we deeply appreciate all your efforts to make it a grand success.

Our special thanks to all the authors who have contributed to publish their research work here in this conference and participated to make this conference a success, and thanks to all of you for your exceptional work. Last but not least, we would like to thank all the reviewers who took the trouble to review all the research works submitted to the conference. We would like to thank all the session chairs who spent their precious times to hold the sessions. We are thankful to our international keynote speakers for their valuable speech.

Regards,
Editorial Board
I3CS 2020

# About the Conference

This is the third edition of International Conference on Computing and Communication Systems (I3CS-2020). Earlier two versions were held successfully in the years 2015 and 2016. The present version of I3CS-2020 was initially planned to be organized during the month of April 2020. But due to global pandemic scenario, it could not be held in its scheduled time. Finally, with lot of efforts we were able to conduct it on August 10 and 11, 2020, in online mode. The conference was organized by the Department of Information Technology, North-Eastern Hill University, Shillong 793022. North-Eastern Hill University is a Central University situated in a picturesque city, Shillong, which is located at northeastern part of India. Shillong is also the capital of the state Meghalaya.

This conference was organized in an aim to provide an appropriate interdisciplinary forum for researchers, engineers, and scientists to discuss and promote research ideas and help in developing innovative technological skills in the thrust areas, specifically like computing, information science, communication engineering, and their relevant application domains. This conference helped successfully in bringing researchers, educators, professional engineers, and technologists on a single forum to discuss and debate on the emerging research areas in the above-mentioned domains.

We are delighted to announce that despite all the odds, in this I3CS-2020, 231 papers were submitted from across the globe. All the submitted papers had undergone strict peer-review process. The selection processes were based on the review outcome and similarity index score. The corresponding successful authors were notified about the outcome of the review process. Finally, corresponding authors of the selected papers presented their papers during the conference in front of distinguished session chairs and audience.

I3CS-2020 consisted of keynote talks, oral presentations, invited sessions on the applications and theory of computing, communication systems, and related areas. Nine distinguished learned figures of international and national fame enlightened us with their keynote speeches on the state-of-the-art domain of technical aspects. It is to mention that four of the keynote speakers are from foreign origin like Sweden, Portugal, Turkey, and Nigeria, and the remaining five speakers were from well-known institutes of India like Indian Institute of Technology, Jadavpur University, Calcutta University, and Tezpur University. The topics of the keynote speeches included the following: “Self Aware Strategies for Embedded and Cloud Security,” “A computational approach to drug repurposing for COVID-19 treatment,” “Machine Learning for Electronics Design Automation,” “Image Analysis by Computational Algorithms: Principles & Applications,” “ACE (Autonomous, Connected & Electric): Next Gen Transport Solution,” “Learning from Small Data,” “Analysis of Software Quality Assurance Implementations,” “Explainable AI for Biomedical Applications,” and “IoT Security from Machine Learning.” The conference succeeded in providing excellent opportunities for the successful paper presenter to present their interesting new research results and discuss its further upgradation or implementation, leading to knowledge transfer and helping in creating new ideas.

The conference proceedings is to be published by a renowned publishing house, Springer, in its one of the popular series LNNS.
In addition of the authors, this conference provided a wonderful knowledge exchange platform for the students and scholars of north east including NEHU, especially for them who are pursuing studies and researches in the area of communication and computing.

Being one of the premier institutes in north east, hosting this conference helped our university, North-Eastern Hill University, Shillong, to make an imprint in the global map as well as in the country as one of the excellent venues for knowledge exchange for technical domain.

# About This Book

The book consists of seven parts.

The first part is devoted on the research works on *Natural Language Processing*. Eight research papers have been presented in this part. Several investigations on native Indian languages like Bengali, Hindi, Assamese, Manipuri, and Khasi are accommodated here. Most of the works are based on low-resource Asian languages.

The second part covered the *Data Mining Techniques and Their Applications* domain. Thirteen research works have been published in this part. Several interesting data mining techniques are applied on various emerging or new application domains. One of the papers in this domain is based on text classifications. Few of the papers are on classification and clustering techniques, having some interesting applications. One of the interesting works on statistical analysis of emotional response is also presented in this part.

The third part is on research *Application of Machine Learning, Image Processing, and Pattern Recognition Techniques*. There are altogether 12 research papers in this part of which three papers are based on image processing algorithms. The remaining papers are based on application of machine learning and pattern recognition techniques. All the research works presented in this part consist of interesting applications.

The fourth part focuses on *Analysis of Biomedical Image, Data and Signal*. In this part, nine research articles have been presented. Some interesting investigations on MRI, cardiac ECHO, EEG, breast and lung cancer, melanoma, and hepatocellular carcinoma have been discussed in this part. One very interesting investigation on Yoga-Asana is also presented here.

The fifth part concentrates on *Networks and Communication*. A total of 11 papers have been discussed in this part. Few papers in this part are based on software-defined network (SDN)-based IoT system. Some papers are based on Cloud. Other papers are presented in the domain of antenna and optical communication.

The sixth part discusses on *Security and Cryptography*. Several interesting cryptographic algorithms along with secret sharing schemes have been discussed here. A total of eight papers are published in this part. Interesting malware detection techniques and HMAC-based authentication in fog-based Internet of Things are presented in this part.

The seventh part is on *Systems and Computing*. A total of eight papers have been presented in this part. Some of the papers in this part are control systems based, and the remaining are circuit and VLSI design based. One interesting paper based on quantum mechanics is also presented here. Other papers are on design of algorithm for several interesting problems.

# Contents

## Data Mining Techniques and Their Applications

## Networks and Communication

## Security and Cryptography



## Systems and Computing

# Editors and Contributors

## About the Editors

**Arnab Kumar Maji** received B.E. degree in Information Science and Engineering from Visvesvaraya Technological University (VTU) in 2003 and M.Tech in Information Technology from Bengal Engineering and Science University, Shibpur (Currently IIEST, Shibpur) in 2005. He received his Ph.D. from Assam University Silchar (A Central University of India) in the year 2016. He has 16 years of professional experience. He is currently working as Assistant Professor (Stage-3) in the Department of Information Technology, North Eastern Hill University, Shillong (A Central University of India). He has published more than 30 nos. of articles in different reputed International Journals and Conferences. He has published more than 20 nos. of articles as book chapter and co-authored 01 no. of book with several international Publishers like Elsevier, Springer, and IGI Global, McMilan International etc. 04 nos. of PhD scholars are currently pursuing Doctoral Degrees under his active supervision. He has also guided successfully 14 nos. of M.Tech thesis and 03 nos. of Ph.D. scholars. His Research interests include Machine Learning, Image Processing, and Natural Language Processing.

**Goutam Saha** received the B.E. degree in Electrical Engineering and the M.E. degree in Electronics and Telecommunication Engineering from the Bengal Engineering College, Shibpur, under the University of Calcutta, India in 1984 and 1989, respectively. He received the PhD degree from the Indian Institute of Technology, Kharagpur in 1999. He also has Post-Doctoral Research experience at the Indian Institute of Technology, Kharagpur and the Ben Gurion University, Israel between 2000 and 2001. Presently, he is working as a Professor in the Department of Information Technology, North-Eastern Hill University, Shillong, and is also holding the position of Head, Department of Information Technology at the same university. Dr. Saha's current research interests include computational biology, bioinformatics, systems biology, bioremediation, internet of things, software-defined radio, etc. Dr Saha has more than 50 research papers and has authored and co-authored two books. Dr. Saha also holds an international patent.

**Sufal Das** Ph.D. (Technology), M.Tech (Computer Science and Engineering), B.Tech (Computer Science and Engineering) is an Assistant Professor of Department of Information Technology, North-Eastern Hill University Shillong. He is having nearly 10 years of professional experience. He has guided around 08 nos. of M.Tech Students. He has currently having nearly 20 publications in several national and International Conferences and Journal. His research area includes Big Data and Data Mining.

**Subhadip Basu** received his PhD (Engineering) degree from Jadavpur University in 2006 and did his post-doctoral research from the University of Iowa, USA, in 2010 and from the University of Warsaw, Poland, in 2012. He joined the Computer Science and Engineering Department of Jadavpur University in 2006, where he is currently working as a Full Professor since 2017. He also holds an honorary position as a Research Scientist at the University of Iowa, USA, since 2016. Dr Basu has more than 200 research publications in notable International Journals and Conference proceedings in the areas of Pattern Recognition, Machine Learning, Bioinformatics, Biomedical Image Analysis etc. He has edited eight books and has two US patents on image binarization methods. Dr Basu has worked at reputed International Institutes, like, Hitachi Central Research Laboratory, Japan, University of Iowa, USA, Bournemouth University, UK, University of Lorraine, France, University of Warsaw, Nencki Institute of Experimental Biology, Poland and Hannover Medical School, Germany. He has supervised eight PhD students and received several major research grants from UGC, DST and DBT, Govt. of India. Dr. Basu is the recipient of the 'Research Award' from UGC, Govt. of India in 2016 and the DAAD fellowship from Germany in 2018. He has also received BOYSCAST and FASTTRACK fellowships from DST, Govt. of India, HIVIP fellowship from Hitachi, Japan, EMMA and CLINK fellowships from the European Union. He is the current Chairperson of the Computer Chapter, IEEE Kolkata Section, a senior member of IEEE and life member of IUPRAI.

**João Manuel R. S. Tavares** graduated in Mechanical Engineering at the Universidade do Porto, Portugal in 1992. He also earned his M.Sc. degree and Ph.D. degree in Electrical and Computer Engineering from the Universidade do Porto in 1995 and 2001 and attained his Habilitation in Mechanical Engineering in 2015. He is a senior researcher at the Instituto de Ciência e Inovação em Engenharia Mecânica e Engenharia Industrial (INEGI) and Associate Professor at the Department of Mechanical Engineering (DEMec) of the Faculdade de Engenharia da Universidade do Porto (FEUP). João Tavares is co-editor of more than 55 books, co-author of more than 50 book chapters, 650 articles in international and national journals and conferences, and 3 international and 3 national patents. He has been a committee member of several international and national journals and conferences, is co-founder and co-editor of the book series "Lecture Notes in Computational Vision and Biomechanics" published by Springer, founder and Editor-in-Chief of the journal "Computer Methods in Biomechanics and Biomedical Engineering: Imaging & Visualization" published by Taylor & Francis, Editor-in-Chief of the journal "Computer Methods in Biomechanics and Biomedical Engineering" published by Taylor & Francis, and co-founder and co-chair of the international conference series: CompIMAGE, ECCOMAS VipIMAGE, ICCEBS and BioDental. Additionally, he has been (co-)supervisor of several MSc and PhD thesis and supervisor of several post-doc projects and has participated in many scientific projects both as researcher and as scientific coordinator. His main research areas include computational vision, medical imaging, computational mechanics, scientific visualization, human-computer interaction, and new product development.

## Contributors

**Alpana Agarwal** Electronics and Communication Engineering Department, Thapar Institute of Engineering and Technology, Patiala, Punjab, India

**Anu Agarwal** Department of Computer Science, Delhi Technological University, Delhi, India

**Ahmed J. I. Alaff** Institute of Information Technology, Gazi University, Ankara, Turkey

**Ritu Anil kumar** North Eastern Space Applications Centre, Umiam, Meghalaya, India

**Subhash C. Arya** Department of Electronics and Communication, North-Eastern Hill University, Shillong, India

**Bunil Kumar Balabantaray** Department of Computer Science and Engineering, National Institute of Technology Meghalaya (NITM), Shillong, India;
Computer Vision and Machine Intelligence Laboratory, National Institute of Technology Meghalaya, Shillong, Meghalaya, India

**Sivaji Bandyopadhyay** Department of Computer Science and Engineering, National Institute of Technology Silchar, Silchar, Assam, India;
Department of Computer Science and Engineering and Center for Natural Language Processing (CNLP), National Institute of Technology Silchar, Silchar, India

**Arundhati Banerjee** Department of Biochemistry and Biophysics, University of Kalyani, Kalyani, West Bengal, India

**Manu Bansal** Electronics and Communication Engineering Department, Thapar Institute of Engineering and Technology, Patiala, Punjab, India

**Abhilasha Barla** Central Research Laboratory, Bharat Electronics Limited, Bengaluru, India

**Diganta Barman** North Eastern Space Applications Centre, Umiam, Meghalaya, India

**Golap Gunjan Barman** Department of Computer Science and Engineering, Girijananda Chowdhury Institute of Management and Technology, Guwahati, Assam, India

**Niharika Barman** North Eastern Hill University, Shillong, India

**Rajiv Barman** Department of Information Technology, Gauhati University, Guwahati, Assam, India

**Utpal Barman** Department of Computer Science and Engineering, Girijananda Chowdhury Institute of Management and Technology, Guwahati, Assam, India

**Barun Barua** Assam Don Bosco University, Guwahati, India

**Debina Basu** Amity Institute of Biotechnology, Amity University, Kolkata, India

**Saikat Basu** School of Medical Science & Technology, Indian Institute of Technology Kharagpur, Kharagpur, India;
Maulana Abul Kalam Azad University Of Technology, Kolkata, West Bengal, India

**Nayan Basumatary** Department of Computer Science and Engineering, Tezpur University, Tezpur, Assam, India

**K. B. Bhaskar** Adhi College of Engineering and Technology, Chennai, India

**Dinesh Bhatia** Department of Biomedical Engineering, North Eastern Hill University, Shillong, Meghalaya, India

**Rupak Bhattacharjee** Department of Mathematics, JIS College of Engineering, Kalyani, India

**Shreya Bhattacharya** Department of Biosciences and Bioengineering, IIT Guwahati, Guwahati, Assam, India

**Tanmay Bhattacharya** Department of Information Technology, Techno India, Saltlake, Kolkata, India

**Jaydeb Bhaumik** Department of ETCE, Jadavpur University, Kolkata, India

**Bishanka Brata Bhowmik** Tripura University, Suryamaninagar, Tripura, India

**Sanjukta Bhowmik** Tripura University, Suryamaninagar, Tripura, India

**Bubu Bhuyan** Department of Information Technology, North Eastern Hill University, Umshing Mawkynroh, Shillong, Meghalaya, India

**Kangkana Bora** Cotton University, Guwahati, India

**Navneet Kaur Brar** Electronics and Communication Engineering Department, Thapar Institute of Engineering and Technology, Patiala, Punjab, India

**Kisalaya Chakrabarti** Department of Electronics and Communication Engineering, Haldia Institute of Technology, Haldia, WB, India

**Susanta Chakraborty** Department of Computer Science and Technology, Indian Institute of Engineering Science and Technology, Shibpur, India

**Rommel Chakravarty** Computer Vision and Machine Intelligence Laboratory, National Institute of Technology Meghalaya, Shillong, Meghalaya, India

**Somdatta Chakravortty** Maulana Abul Kalam Azad University of Technology, Haringhata, West Bengal, India

**Kamlendra Chandra** Central Research Laboratory, Bharat Electronics Limited, Bangalore, India

**Hussain Ahmed Choudhury** Associate Professor, Department of IT, CVR College of Engineering, Hyderabad, India

**Genevieve Chyrmang** Assam Don Bosco University, Guwahati, India

**Pankaj Dadure** Department of Computer Science and Engineering, National Institute of Technology Silchar, Silchar, Assam, India

**Mrinmoy Dam** Mahishakati Netaji Subhash Chandra High School, Kahankia, India

**Anwesha Das** Department of Computer Science and Engineering, National Institute of Technology Silchar, Silchar, Assam, India

**Arnab Das** Department of Computer Science and Engineering, National Institute of Technology, Agartala, India

**Ringki Das** Department of Computer Science and Engineering, National Institute of Technology Silchar, Silchar, Assam, India

**Rohit Kumar Das** North-Eastern Hill University, Shillong, Meghalaya, India

**Sangeeta Das** Electronics and Communication Engineering Department, School of Technology, North Eastern Hill University, Shillong, India

**Srirupa Das** RCC Institute of Information Technology, Kolkata, India

**Sufal Das** North Eastern Hill Univeristy, Shillong, Meghalaya, India

**Tanuja Das** Department of Information Technology, GUIST, Guwahati, Assam, India

**Suman Deb** Department of Computer Science and Engineering, National Institute of Technology, Agartala, India

**Soumyabuddha Debnath** Maulana Abul Kalam Azad University Of Technology, Kolkata, West Bengal, India

**Swaraj Deep** Department of CSE, National Institute of Technology Durgapur, Durgapur, India

**Debasis Dhal** Department of Computer Science and Engineering, University of Calcutta, Kolkata, West Bengal, India

**N. Donald Jefferson Thabah** Department of Computer Science, Assam University, Silchar, India

**Amartya Dutta** Department of Computer Science and Engineering, B. P. Poddar Institute of Management and Technology, Kolkata, West Bengal, India

**C. Emmanuel** Global Hospitals and Health City, Chennai, India

**Chinmoy Ghosh** Department of Computer Science and Engineering, Jalpaiguri Government Engineering College, Jalpaiguri, West Bengal, India

**Debabrata Gogoi** North Eastern Hill University, Shillong, Meghalaya, India

**Rajlakshmi Guha** Centre of Education Technology, Indian Institute of Technology Kharagpur, Kharagpur, India

**Meena Gupta** Amity Institute of Physiotherapy, Amity University, Noida, Uttar Pradesh, India

**Krishnandu Hazra** Department of CSE, National Institute of Technology Durgapur, Durgapur, India

**Md. Iftekhar Hussain** Department of Information Technology, North Eastern Hill University, Shillong, Meghalaya, India

**Mir Wajahat Hussain** Department of Computer Science and Engineering, National Institute of Technology, Meghalaya, Laitumkhrah, Shillong, India

**Amit Prakash Jagtap** Central Research Laboratory, Bharat Electronics Limited, Bangalore, India

**Jhilam Jana** Department of ETCE, Jadavpur University, Kolkata, India

**Nabakumar Jana** Department of Mathematics, Indian Institute of Technology (ISM) Dhanbad, Dhanbad, India

**Sunanda Jana** Department of Computer Science and Engineering, Haldia Institute of Technology, Haldia, India

**A. Joshuva** Hindustan Institute of Technology and Science, Chennai, India

**Sushanta Kabir Dutta** North Eastern Hill University, Shillong, Meghalaya, India

**Anubha Kabra** Department of Computer Science, Delhi Technological University, Delhi, India

**Kamarujjaman** Department of Computer Application, Techno India University, Kolkata, West Bengal, India

**R. Kannan** B. S. Abdur Rahman Crescent Institute of Science and Technology, Chennai, India

**Balbindar Kaur** Department of Electronics and Communication, North-Eastern Hill University, Shillong, India

**Jaskaranbeer Kaur** Electronics and Communication Engineering Department, Thapar Institute of Engineering and Technology, Patiala, Punjab, India

**Ajoy Kumar Khan** National Institute if Technology Mizoram, Aizawl, Mizoram, India

**Utku Kose** Department of Computer Engineering, Suleyman Demirel University, Isparta, Turkey

**Vamshi Krishna** Department of Electrical Engineering, Indian Institute of Technology Kharagpur, Kharagpur, India

**Ajay Kumar** Electronics and Communication Engineering Department, Thapar Institute of Engineering and Technology, Patiala, Punjab, India

**Bhupender Kumar** Department of Information Technology, North Eastern Hill University, Umshing Mawkynroh, Shillong, Meghalaya, India

**Shubham Kumar** Central Research Laboratory, Bharat Electronics Limited, Bengaluru, India

**Somesh Kumar** Department of Mathematics, Indian Institute of Technology Kharagpur, Kharagpur, India

**Kuldeep Kushwah** NIT Silchar, Silchar, India

**Afsana Laskar** North Eastern Hill Univeristy, Shillong, Meghalaya, India

**Sahinur Rahman Laskar** Department of Computer Science and Engineering, National Institute of Technology Silchar, Silchar, Assam, India

**Saralin Lyngdoh** North-Eastern Hill University, Shillong, India

**Manjunatha Mahadevappa** School of Medical Science & Technology, Indian Institute of Technology Kharagpur, Kharagpur, India

**Lipi B. Mahanta** Institute of Advanced Study in Science and Technology, Guwahati, India

**Mausumi Maitra** Department of Information Technology, Govt. College of Engineering and Ceramic Technology, Kolkata, India

**Arnab Kumar Maji** North-Eastern Hill University, Shillong, Meghalaya, India; Department of Information Technology, North Eastern Hill University, Shillong, India

**Debasish Majumder** Department of Mathematics, JIS College of Engineering, Kalyani, India

**Riya Majumder** Department of Computer Science and Engineering, Supreme Knowledge Foundation Group of Institutions, Chandannagar, Hooghly, West Bengal, India

**Richa K. Makhijani** Indian Institute of Information Technology, Nagpur, India

**Saurav Malla** Department of Computer Science and Engineering, National Institute of Technology, Agartala, India

**Manjarini Mallik** Department of Computer Science and Engineering, University of Calcutta, Kolkata, India

**Loitongbam Sanayai Meetei** Department of Computer Science and Engineering and Center for Natural Language Processing (CNLP), National Institute of Technology Silchar, Silchar, India

**P. Mercy Nesa Rani** College of Post-Graduate Studies in Agricultural Sciences, Umiam, Meghalaya, India

**Maad M. Mijwil** Baghdad College of Economic Sciences University, Baghdad, Iraq

**M. Mohit Jaisingh** B. S. Abdur Rahman Crescent Institute of Science and Technology, Chennai, India

**Satyendranath Mondal** Department of Information Technology, Kalyani Government Engineering College, Kalyani, West Bengal, India

**Fisokuhle Mthethwa** Computer Science Department, Eswatini Medical Christian University, Mbabane, Eswatini

**Rajkishur Mudoi** Department of Electronics and Communication Engineering, North-Eastern Hill University, Shillong, India

**Hosam H. A. Mukhairez** Institute of Information Technology, Gazi University, Ankara, Turkey

**Jayanta Mukherjee** Department of Computer Science and Engineering, Indian Institute of Technology Kharagpur, Kharagpur, India

**Nirmalya Mukherjee** Department of Computer Science and Engineering, National Institute of Technology, Agartala, India

**Amitava Nag** Department of Computer Science and Engineering, Central Institute of Technology Kokrajhar, Btad, Assam, India

**D. Najumnissa Jamal** B. S. Abdur Rahman Crescent Institute of Science and Technology, Chennai, India

**Subrata Nandi** Department of CSE, National Institute of Technology Durgapur, Durgapur, India

**Ritika Nandy** Amity Institute of Biotechnology, Amity University, Kolkata, India

**Amitabha Nath** Information Technology Department, North-Eastern Hill University, Shillong, India;
Department of IT, NEHU, Shillong, Meghalaya, India

**Bhabesh Nath** Department of Computer Science and Engineering, Tezpur University, Tezpur, Assam, India

**Keshab Nath** Department of Computer Science and Engineering, Indian Institute of Information Technology Kottayam, Kottayam, Kerala, India

**Shaik Nawaz** Department of Electronics and Communication Engineering, Center for Advanced Computing and Research Laboratory (C-ACRL), Vardhaman College of Engineering, Shamshabad, Telangana, India

**Rajashree Nayak** Computer Vision and Machine Intelligence Laboratory, National Institute of Technology Meghalaya, Shillong, Meghalaya, India

**Vincent Kiplimo Ngetich** Department of Computer Science and Engineering, Maulana Abul Kalam Azad University of Technology, Kolkata, West Bengal, India

**Nilay Nishant** North Eastern Space Applications Centre, Umiam, Meghalaya, India

**Satyam Omar** Department of Mathematics, Motilal Nehru National Institute of Technology Allahabad, Prayagraj, India

**Sahadeo Padhye** Department of Mathematics, Motilal Nehru National Institute of Technology Allahabad, Prayagraj, India

**Partha Pakray** Department of Computer Science and Engineering, National Institute of Technology Silchar, Silchar, Assam, India

**Rajat Kumar Pal** Department of Computer Science and Engineering, University of Calcutta, Kolkata, West Bengal, India

**Akash Kumar Panda** Computer Vision and Machine Intelligence Laboratory, National Institute of Technology Meghalaya, Shillong, Meghalaya, India

**Alagumariappan Paramasivam** B. S. Abdur Rahman Crescent Institute of Science and Technology, Chennai, India

**Anil Singh Parihar** Department of Computer Science, Delhi Technological University, Delhi, India

**Parag Moni Patangia** North Eastern Hill University, Shillong, India

**Fabiola Hazel Pohrmen** North-Eastern Hill University, Shillong, Meghalaya, India

**Ashis Pradhan** Department of CSE, Sikkim Manipal Institute of Technology, SMU, Sikkim, India

**Mohan P. Pradhan** Department of Computer Application, Sikkim University, Sikkim, India

**Partha Pratim Kalita** Department of Information Technology, GUIST, Guwahati, Assam, India

**Bipul Syam Purkayastha** Department of Computer Science, Assam University, Silchar, India

**Md. Atiqur Rahman** Department of Electronics and Communication Engineering, NEHU, Shillong, India

**Ravinandan** Department of Information Technology, North Eastern Hill University, Shillong, Meghalaya, India

**J. V. R. Ravindra** Department of Electronics and Communication Engineering, Center for Advanced Computing and Rcscarch Laboratory (C-ACRL), Vardhaman College of Engineering, Shamshabad, Telangana, India

**Kaushik Ray** Tezpur University, Tezpur, Assam, India

**Sujay Ray** Amity Institute of Biotechnology, Amity University, Kolkata, India

**Amit Kumar Roy** National Institute of Technology Mizoram, Aizawl, Mizoram, India

**Swarup Roy** Network Reconstruction and Analysis (NetRA) Lab, Department of Computer Applications, Sikkim University, Gangtok, Sikkim, India

**Goutam Saha** North-Eastern Hill University, Shillong, Meghalaya, India;
Department of IT, NEHU, Shillong, Meghalaya, India;
Department of Information Technology, NEHU, Shillong, Meghalaya, India

**Rishiraj Saha** Department of Information Technology, JIS College of Engineering, Kalyani, India

**Soham Saha** Department of Information Technology, JIS College of Engineering, Kalyani, India

**Sujoy Saha** Department of CSE, National Institute of Technology Durgapur, Durgapur, India

**Suprativ Saha** Department of Computer Science and Engineering, Brainware University, Barasat, Kolkata, India

**Sarat Saharia** Tezpur University, Tezpur, Assam, India

**Lizia Sahkhar** Department of Computer Science and Engineering, National Institute of Technology Meghalaya (NITM), Shillong, India;
Department of Computer Science, Lady Keane College, Shillong, India

**Sudarsan Sahoo** NIT Silchar, Silchar, India

**Diganto Sahu** Department of Computer Science and Engineering, Girijananda Chowdhury Institute of Management and Technology, Guwahati, Assam, India

**R. Sai Roshan** Department of Electronics and Communication Engineering, Center for Advanced Computing and Research Laboratory (C-ACRL), Vardhaman College of Engineering, Shamshabad, Telangana, India

**Monjul Saikia** North Eastern Regional Institute of Science and Technology, Nirjuli, Arunachal Pradesh, India

**Shyam Sundar Santra** Department of Mathematics, JIS College of Engineering, Kalyani, India

**Valluri Sarimela** Central Research Laboratory, Bharat Electronics Limited, Bengaluru, India

**Pankaj Sarkar** Electronics and Communication Engineering Department, School of Technology, North Eastern Hill University, Shillong, India;
Department of Electronics and Communication Engineering, NEHU, Shillong, India

**Puja Sarkar** Department of Computer Science and Engineering, Central Institute of Technology Kokrajhar, Btad, Assam, India

**Saptaparni Sarkar** Department of Computer Science and Engineering, Brainware University, Barasat, Kolkata, India

**Kandarpa Kumar Sarma** Department of ECE, Gauhati University, Guwahati, Assam, India

**Nityananda Sarma** Department of Computer Science and Engineering, Tezpur University, Tezpur, Assam, India;
Tezpur University, Tezpur, Assam, India

**Parismita Sarma** Department of Information Technology, Gauhati University, Guwahati, Assam, India

**Animesh Sarmah** Department of Computer Science and Engineering, Girijananda Chowdhury Institute of Management and Technology, Guwahati, Assam, India

**G. Selvakumar** Central Research Laboratory, Bharat Electronics Limited, Bengaluru, India

**Mirinso Shadang** Department of Electronics and Communication Engineering, IIIT Senapati, Imphal, Manipur, India

**Mridusmita Sharma** Department of ECE, Gauhati University, Guwahati, Assam, India

**Sabna Sharma** School of Information Technology, SRM University, Sikkim, India

**Dhanapati Shougrakpam** Department of Linguistics, Central University of Rajasthan, Ajmer, Rajasthan, India

**Alok Singh** Department of Computer Science and Engineering and Center for Natural Language Processing (CNLP), National Institute of Technology Silchar, Silchar, India

**Telem Joyson Singh** Department of Computer Science and Engineering, IIIT Senapati, Imphal, Manipur, India

**Thoudam Doren Singh** Department of Computer Science and Engineering, National Institute of Technology Silchar, Silchar, Assam, India;
Department of Computer Science and Engineering and Center for Natural Language Processing (CNLP), National Institute of Technology Silchar, Silchar, India

**Satyabrata Singha** Tripura University, Suryamaninagar, Tripura, India

**Nidul Sinha** National Institute of Technology, Silchar, Assam, India

**Nitish Sinha** Tripura University, Suryamaninagar, Tripura, India

**Diptendu Sinha Roy** Department of Computer Science and Engineering, National Institute of Technology, Meghalaya, Laitumkhrah, Shillong, India

**Md. Amir Sohail** Department of Computer Science and Engineering, Jalpaiguri Government Engineering College, Jalpaiguri, West Bengal, India

**Sriraksha** Department of IT, NEHU, Shillong, India

**Sangeeta Srivastava** Central Research Laboratory, Bharat Electronics Limited, Bangalore, India

**Rupaban Subadar** North Eastern Hill University, Shillong, India

**R. Suresh** Central Forensic Science Laboratory, Guwahat, Assam, India

**Jaydeep Swarnakar** Department of Electronics and Communication Engineering, School of Technology, North-Eastern Hill University, Shillong, Meghalaya, India

**Upasana Talukdar** Department of Data Science and Analytics, Central University of Rajasthan, Ajmer, Rajasthan, India

**Surmila Thokchom** Department of Computer Science and Engineering, National Institute of Technology Meghalaya, Shillong, India

**Leki Chom Thungon** Department of Information Technology, North Eastern Hill University, Shillong, Meghalaya, India

**Pradeep Kumar Tiwari** Central Research Laboratory, Bharat Electronics Ltd., Bangalore, India

**Sayan Tripathi** Department of ETCE, Jadavpur University, Kolkata, India

**T. Velayutham** Central Research Laboratory, Bharat Electronics Ltd., Bangalore, India

**Akshitha Vuppala** Department of Electronics and Communication Engineering, Center for Advanced Computing and Research Laboratory (C-ACRL), Vardhaman College of Engineering, Shamshabad, Telangana, India

**Imayanmosha Wahlang** North-Eastern Hill University, Shillong, Meghalaya, India

**Sunita Warjri** North-Eastern Hill University, Shillong, India

# Natural Language Processing

# Low Resource Neural Machine Translation from English to Khasi: A Transformer-Based Approach

**N. Donald Jefferson Thabah and Bipul Syam Purkayastha**

**Abstract** With the increased amount of information that is readily available in digital platforms, the demand and significance of machine translation (MT) between languages have also risen. MT can bridge this gap where the need for a professional translator can be minimized. This paper focuses on low resource machine translation from English to Khasi and vice-versa which was encouraged from the recent development of the cross-lingual language model pretraining system developed by Facebook AI Research. The approach is well suited for a low resource language and the translation model was trained using a combination of monolingual and parallel corpus. When tested using a similar domain test corpus, a Blue score of 39.63 and 32.69 were achieved for translating English–Khasi and Khasi–English respectively. However, while testing by using a general test domain corpus, a Blue scores of 5.33 and 2.86 for translating English–Khasi and Khasi–English were obtained respectively. Considering the minimal corpus size, the technique shows promising results, especially when low resource language is considered.

**Keywords** Low resource MT · NMT for English–Khasi · XLM

## 1 Introduction

Language translation is an important aspect to bridge the communication gap among people of different geographical and cultural boundaries. With the vast availability of digital content, it has made this task more achievable. In this work, we have attempted to accomplish a translation from English to Khasi and vice-versa. Khasi is one of the main spoken languages in Meghalaya, a North Eastern State of India. Presently, the language is employed as a mode of instruction in lower and upper primary schools

---

N. Donald Jefferson Thabah · B. S. Purkayastha (✉)
Department of Computer Science, Assam University, Silchar 788011, India
e-mail: bipul_sh@hotmail.com

N. Donald Jefferson Thabah
e-mail: jefson08@gmail.com

 
A. Maji et al. (eds.), *Proceedings of the International Conference on Computing and Communication Systems*, Lecture Notes in Networks and Systems 170,
https://doi.org/10.1007/978-981-33-4084-8_1

and furthermore is taught as a subject at the high schools, colleges and university levels. Khasi is also an official language in the State of Meghalaya.

Recently, neural MT (NMT) has gained its popularity [1, 2]. It follows an end-to-end training approach for translating from a source language to the targeted language [2–4]. Even though the recurrent sequence to sequence neural network for MT have been a popular mechanism but recent development of the transformer model has shown improved performance in terms of computation speed and translation accuracy [5].

Transformer model and Bidirectional Encoder Representation from Transformer (BERT) have accomplished encouraging results in natural language processing task. The recent work by Lample et al. from Facebook AI Research [6] proposed an version of BERT for cross-lingual language model (XLM) which contributes to promising results in text classification and MT especially beneficial for low resource language pairs.

The outcome of XLM is fed as input to the transformer translation model. In turn, XLM uses a byte pair encoding (BPE) preprocessing technique [7] and a dual-language training method using BERT in learning the relationship among words in different languages.

This paper extends the existing work on XLM Pretraining by Lample and Conneau from Facebook AI Research [6]. We consider a language pair of Khasi and English. XLM stresses in solving language understandability, especially for low resource language which consequently improves machine translation accuracy. Therefore, Khasi being a low resource language fits perfectly and we are able to obtain a promising translation accuracy given the fact that we have used minimal dataset size to train the network.

## 2 Related Work

"Unsupervised translation system was based from the previous work on unsupervised embedding mappings" [8]; it has a modified attention encoder–decoder model which was trained using only monolingual corpora employing a combination of denoising and back-translation. "They obtained a translation accuracy (Blue score) of 15.56 for French–English and 10.21 for German–English using WMT'14 dataset" [9].

Facebook AI Research suggested two methods to learn cross-lingual language model (XLM) [6]. The first model relies only on monolingual data while the other uses parallel data with new XLM objectives. XLM pretraining has led to the improvement of natural language understanding which will be beneficial to the NLP task [10]. The author makes use of the XLM output and feeds it to the transformer model for language translation. They obtained an increase by more than 9 Blue scores when contrasted with the former state-of-the-art unsupervised translation using WMT'16 German–English. Similarly, they obtained an incremental of more than 4 Blue scores as that of the state-of-the-art supervised translation utilizing WMT'16 Romanian–English. The paper has shown promising encouragement in further exploration in

**Table 1** English and Khasi datasets

| Monolingual corpus | | Parallel corpus | | |
|---|---|---|---|---|
| Language | Size (MB) | Language | Size (MB) | No. of Sentence |
| English | 5.4 | English | 4.2 | 51,596 |
| Khasi | 6.3 | Khasi | 5.1 | |

this arena given the fact that it out performed the translation accuracy of the previous work [6]. Moreover, the technique incorporated in this work fit very well for low resource language translation.

## 3 Dataset

The dataset used in implementing this work was extracted from English–Khasi Bible [11, 12], Tatoeba project [13] and the remaining datasets were obtained from books, articles, newspapers, etc. Table 1 shows the corpus size and number of sentences [19–28].

Below is the sample example of parallel sentences in the dataset.

1. *it took Tom several hours to assemble a bunk bed that the instructions said would take less than an hour to assemble*
2. *ka shim ia u Tom katto katne kynta ban pyndait ia ka jingthiah bad ki jingbatai ki ong ba kan shim shi kynta duna ban pyndait.*

## 4 Methodology

### *4.1 Transformer*

Transformers [5] use a self-attention method to process the entire sequence of a sentence simultaneously. It consists of six stacks of encoder–decoder blocks. All the blocks are identical but they do not share any weight. The input words in a sentence are embedded into vectors using word2vec algorithm [5]. Positional encoding vectors of the same size as that of the embedded vector are added; this helps in imparting positioning information of a word in a sentence.

Referring from Fig. 1, in the self-attention sub-layer, the input matrix $X$ is multiplied with learnable matrices query ($Q$), key ($K$) and value ($V$).

$$X \times W_0^Q = Q_0 \tag{1}$$

$$X \times W_0^K = K_0 \tag{2}$$

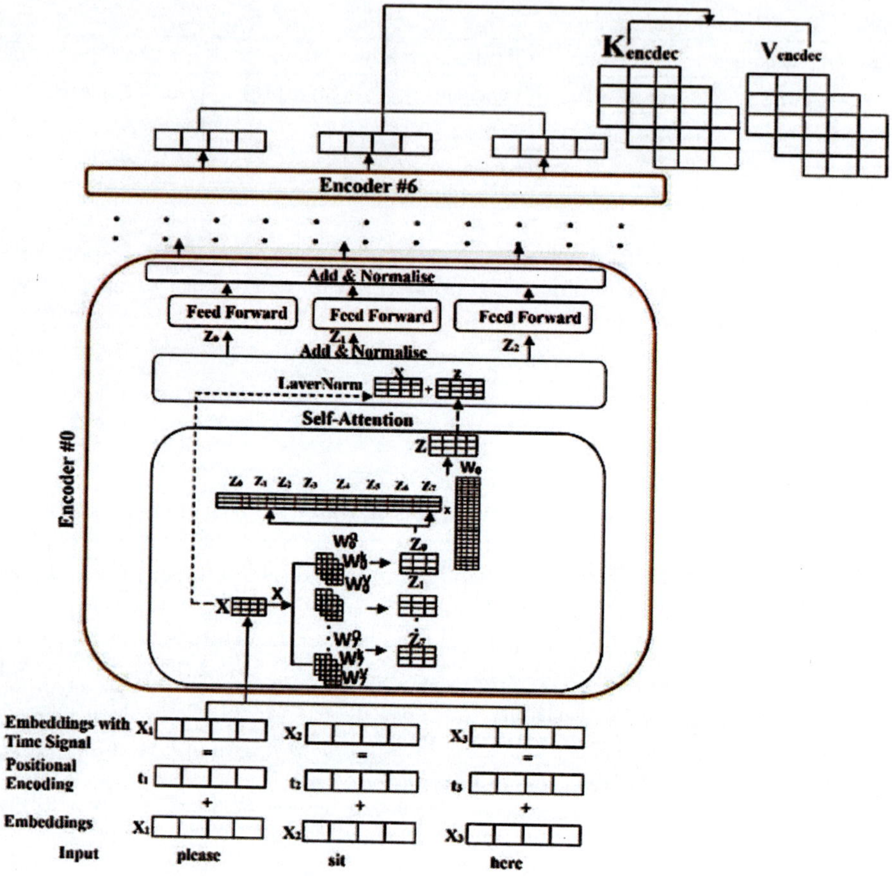


Fig. 1 Encoder of a transformer model [14]

$$X \times W_0^V = V_0 \tag{3}$$

$$= \text{Softmax}\left(\frac{Q_0 \times K_0^T}{\sqrt{d_{k_0}}}\right) \cdot V_0 \tag{4}$$

$$= Z_0 \tag{5}$$

where $K_0^T$ is $K_0$ Transpose.

$d_{k_0}$ is the dimension of $K_0$ vector.

In the softmax function, multiplying Q $\times K^T$ gives a score matrix which indicates a similarity between a given word and the remaining words of a sentence. This score tells us how much we need to focus on the other words of a sentence as we encode a particular word. Dividing the product $Q_0 \times K_0^T$ by the key vector dimension ($d_k$) gives a better stable gradient. Finally, multiplying by value vector ($V$), it stresses more on the current word and drops irrelevant words in a sentence.

To help focus on other words of a sentence, transformer uses multi-head attention. For example in (see Fig. 1), there are eight heads which produce $Z_0$–$Z_7$ attention matrices. Moreover, it gives the attention layer multiple representation space. Further, since the encoder layer needs only a single matrix Z, therefore $Z_0$–$Z_7$ are concatenated and multiplied by another learnable matrix $W_0$ to product Z. The add and normalize layer normalizes 0 mean and variance 1. The outcome matrix is then broken down into individual vectors and passed on to feed forward neural network which are processed in parallel. The final encoder layer will then be transformed into a set of attention matrices keys and values.

The steps in the decoder layer shown in Fig. 2 are very identical as that of the encoder layer except that the keys ($K$) and ($V$) attention outcome from the encoder layer will be fed into encoder–decoder attention sub-layer to decode and emphasis on the most significant places in the input sequence [14].

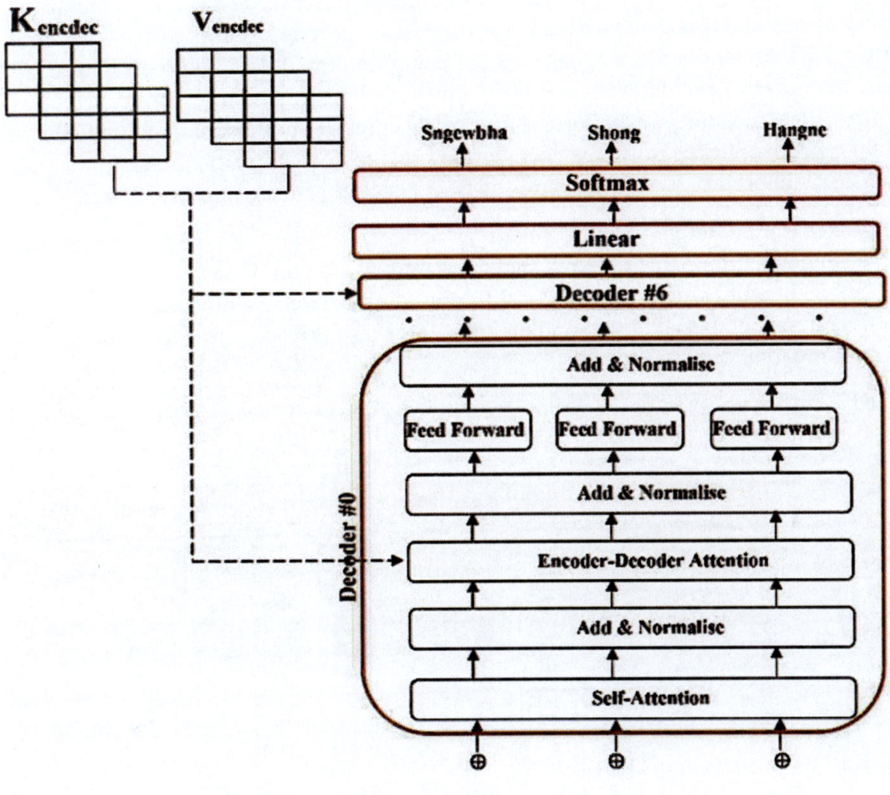


**Fig. 2** Decoder of a transformer model [14]

Training the transformer model is very similar to that of the sequence–sequence recurrent neural network. A source sentence is inputted to the encoder layer and a target sentence shifted by one place is fed to the decoder layer simultaneously.

During translation, the previous outputted sequence from the decoder is fed back as input into the current prediction till it reaches the end of a sequence. "Self-attention is only allowed to attend to earlier positions in the output sequence by masking future positions (setting them to $-\infty$) before the softmax step in the self-attention layer" [14].

The outcome from the decoder is generally a much smaller in dimension as compare to the target language vocabulary size. Hence, the Linear layer helps in projecting this smaller dimension to match the dimension of the target vocabulary language.

## *4.2 BERT*

The primary difficulties of natural language processing application are the lack of sufficient training datasets. Generally, enormous unprocessed or raw data are available but task-specific data like language translation corpus are limited. To overcome this, BERT pre-trained a language model using unaltered data and then fine-tuned it using task-specific datasets. BERT is a modified version of transformer which is bidirectional in nature. Learning and improving language understanding was BERT's main objective; hence, only the encoder part of transformer is utilized [10].

The main preprocessing measures of BERT model as shown in Fig. 3 are:

a. Token embeddings—[CLS] token is annexed toward the beginning of the sentence while [SEP] token is appended toward the end of each sentence.
b. Segment embedding—A marker differentiating between sentences.
c. Positional embeddings—Supplemented to every token to induce word positioning in an input sequence.

BERT does not predict the next word in the sentence. Training is as follows:

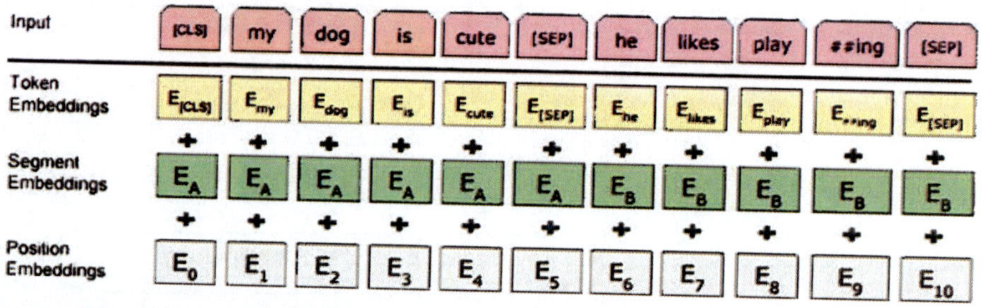


**Fig. 3** BERT language model [15]

Masked language model (MLM) randomly masks out (drop) 15% of the input token (words) in a sentence. Out of this 15% drop, 80% of the time the tokens are replaced with [MASK] token, 10% of the time are swapped by random token from the sequence and remaining 10% of the time the sequence is left unchanged. When the updated sequence is trained in BERT, it will be able to predict the [MASK] token.

### *4.3 Cross-Lingual Language Model (XLM)*

XLM gives a strong pretraining method for language understanding [6]. There are three XLM models namely causal language modeling (CLM), masked language modeling (MLM) and translation language modeling (TLM). CLM used a modified version of transformer model (BERT language model) to predict the outcome of a given word given its previous words given in the sentence. An MLM objective is to learn a language model by masking (dropping) some of the words and retry to predict them back [6]. TLM is an extension of MLM where parallel sentences are used to model the language in contrast to CLM and MLM which use only monolingual data. In TLM, the parallel sentences of both the language are concatenated where masking is performed randomly on both source and target sentences. This approach helps the model to align the parallel language representation.

### *4.4 Low Resource Machine Translation*

Lample et al. translation model used the cross-lingual language pre-trained model output as initial embedding to the translation model [6]. The translation model uses self-attention transformer encoder–decoder architecture [5]. The encoder–decoder, in turn, can utilize a combination of MLM/CLM, CLM/MLM, CLM/CLM and MLM/MLM, etc., which means, MLM output can be given as initial embedding input to the encoder to encode source sentence into a hidden attention vector representation and CLM output as input to the decoder to decode the hidden attention vector into the target sentence. It was observed that using MLM in both encoder and decoder gives a better translation accuracy in terms of Blue score. In fact, the use of cross-lingual word embedding as input in to the translation model has improved the performance of unsupervised machine translation [6].

## 5 Experiment Setups

The monolingual and parallel corpus is preprocessed (tokenization, true casing, etc.) using Statistical Moses Tools [16]. "Byte pair encoding (BPE) is learned from the source and target tokenizes corpus" [17]. BPE is a data compression technique which

**Table 2** Blue score for translating English to Khasi language and vice-versa

| Test corpus | Blue score English to Khasi | Blue score Khasi to English |
|---|---|---|
| Similar domain | 39.63 | 32.69 |
| General domain | 5.33 | 2.86 |

**Table 3** Model translation results from English to Khasi

| | |
|---|---|
| English (Source) | "People will come from the east and the west from the north and the south and sit down at the feast in the Kingdom of God" [11] |
| Khasi (Reference) | "Ki briew kin ïawan na mihngi bad na sepngi na shatei bad na shathie bad shong ha ka jingkhawai ha ka Hima U Blei" [12] |
| Khasi (Translated) | Ki briew kin wan na shaphang mihngi bad na sepngi kiba shaphang shatei bad shathie bad kiba shong ban sah ha ka jingkhawai ha ka Hima U Blei |

helps in solving rare word problem by splitting the word into more frequent sub-word. With the learned BPE codes, we applied (applybpe) to both source and target corpus. Source and target vocabulary are then extracted using the command getvocab and these are converted to binary files. Similarly, a full vocabulary of both source and target language is extracted into a common file.

The next step is to pretrain a language model with both MLM and TLM making use of the preprocessed binary files mentioned above. Finally, the pre-trained language model output is initialized in the transformer for training the model to perform machine translation from English to Khasi language and vice-versa. The detailed steps of processing are given in github by Facebook AI Research.[1]

## 6 Results and Discussion

To determine the translation accuracy, a Blue score using Moses tools was calculated and shown in Table 2. Blue is a measure to calculate the correctness of a translation by comparing a machine-translated sentence with a reference translation. Every $n$-gram in a deciphered sentence is checked against the entirety of the references words to acquire the precision [18]. In general, $n$ range from $1 \leq n \leq N$ and normally $N$ is chosen to be 4.

Table 2 shows the Blue score for translating English to Khasi and vice-versa. Conventionally, the main corpus is divided into train, test and validate sub-corpus. Hence, we use this test corpus in the first row of Table 2 and we characterize this as a similar domain test corpus. In the second row, we consider a new test corpus with less relation to that of the main corpus and we characterized this as a general domain test corpus.

[1]https://github.com/facebookresearch/XLM.

**Table 4** Model translation results from Khasi to English

| | |
|---|---|
| Khasi (Source) | "Ki briew kin ïawan na mihngi bad na sepngi na shatei bad na shathie bad shong ha ka jingkhawai ha ka Hima U Blei" [12] |
| English (Reference) | "People will come from the east and the west from the north and the south and sit down at the feast in the Kingdom of God" [11] |
| English (Translated) | People will come from the east and from the north and from the south and sit in the feast in the Kingdom of God |

Tables 3 and 4 show a sample output which are translated by the transformer model from English to Khasi and vice-versa. Tables 3 and 4 source refers to a source language sentence, reference is the manually translated target sentence and translated means a machine-translated sentence.

## 7 Conclusion and Future Work

The paper extends the work on XLM Pretraining by Lample and Conneau from Facebook AI Research [6]. The core foundation of the work was based on the self-attention transformer model and with a slight tweak on BERT model which is also a modified version of Transformer can achieve a better language understanding. Language understanding is helpful in building a cross-lingual language model, thereby enhancing a significant improvement in language translation.

We have considered a language pair of Khasi and English to perform translation from English to Khasi and Khasi to English, respectively. Since the described approach was intended to improve translation accuracy on low resource languages, Khasi language which is considered as low resource language will perfectly fit in this approach. In view of our minimal corpus size, we obtained a translation accuracy of 39.63 and 32.69 Blue scores for translating English to Khasi and Khasi to English, respectively. This comparatively high score was achieved because of the domain similarity of our test data. If the test data are from a generic domain, then the Blue score of 5.33 for English to Khasi and 2.86 for Khasi to English are obtained. Given the amount of corpus size and the type of test data, the result shows promising outcomes.

Therefore, one of the possible and future enhancements to ponder upon the translation accuracy is to accumulate more corpus dataset. Moreover, in the case of low resource language, a domain-specific corpus would further improve the translation accuracy.

**Acknowledgements** The author acknowledges the blind peer reviewers for their suggestions to improve upon the paper and the immense help rendered by the translation staff of Bible Society, Shillong, the Ri Khasi Press and the staff of Central Library of North Eastern Hill University, Shillong, in accessing the rich resources in their respective possessions without which the completion of this paper would not have been possible.

## References


1. Bahdanau D, Cho K, Bengio Y (2014) Neural machine translation by jointly learning to align and translate. arXiv preprint arXiv:1409.0473
2. Sutskever I, Vinyals O, Le QV (2014) Sequence to sequence learning with neural networks. In: Advances in neural information processing systems conference, pp 3104–3112
3. Kalchbrenner N, Blunsom P (2013) Recurrent continuous translation models. In: Proceedings of the conference on empirical methods in natural language processing, USA, pp 1700–1709
4. Cho K, Van Merriënboer B, Bahdanau D, Bengio Y (2014) On the properties of neural machine translation: encoder-decoder approaches. arXiv preprint arXiv:1409.1259
5. Vaswani A, Shazeer N, Parmar N, Uszkoreit J, Jones L, Gomez AN et al (2017) Attention is all you need. In: Proceedings of the 31st conference on neural information processing systems (NIPs), USA, pp 5998–6008
6. Lample G, Conneau A (2018) Cross-lingual language model pretraining. arXiv preprint arXiv:1901.07291
7. Sennrich R, Haddow B, Birch A (2015) Neural machine translation of rare words with subword units. arXiv preprint arXiv:1508.07909
8. Mikolov T, Sutskever I, Chen K, Corrado GS, Dean J (2013) Distributed representations of words and phrases and their compositionality. In: Advances in neural information processing systems, pp 3111–3119
9. Artetxe M, Labaka G, Agirre E, Cho K (2018) Unsupervised neural machine translation. In: Proceedings of the sixth international conference on learning representations
10. Devlin J, Chang MW, Lee K, Toutanova K (2018) Bert: pre-training of deep bidirectional transformers for language understanding. arXiv preprint arXiv:1810.04805
11. GNB Bible YouVersion homepage. https://www.bible.com/en-GB/bible/296/GEN.1.GNB
12. KHASICLBSI Bible YouVersion homepage. https://www.bible.com/en-GB/bible/1865/GEN.1.KHASICLBSI
13. Tatoeba Project Homepage. https://www.manythings.org/anki/
14. The Illustrated Transformer Homepage. https://jalammar.github.io/illustrated-transformer, last accessed 2020/03/06
15. BERT Explained Homepage. https://towardsml.com/2019/09/17/bert-explained-a-complete-guide-with-theory-and-tutorial, last accessed 2020/03/02
16. Koehn P, Hoang H, Birch A, Callison-Burch C, Federico M, Bertoldi N et al (2007) Moses: open source toolkit for statistical machine translation. In: Proceedings of the 45th annual meeting of the association for computational linguistics companion volume proceedings of the demo and poster sessions, pp 177–180
17. Artetxe M, Labaka G, Agirre E (2017) Learning bilingual word embeddings with (almost) no bilingual data. In: Proceedings of the 55th annual meeting of the association for computational linguistics (vol. 1: long papers), pp 451–462
18. Smith A, Hardmeier C, Tiedemann J (2016) Climbing Mont BLEU: the strange world of reachable high-BLEU translations. In: Proceedings of the 19th annual conference of the European association for machine translation, pp 269–281
19. Mawblei G (2017) The Khasis: culture and beliefs as confronted by the Gospel. Bilingual U Khasi: Ka Dustur bad ka Jingngeit Ba Lyngkhuh ia Ka Gospel. Syiem Offset Printers, Shillong
20. Roy J (2005) Shaphang U wei U Blei: About One God. A Translation by Bijoy Sawian. Ri Khasi Press, Shillong
21. Tham M (2015) The Golden Duitara: translation from Soso Tham's Duitara Ksiar Khasi to English. Print Xpress, Shillong
22. Singh N, Gurdon RRT (eds) (1904) Khasi English dictionary. Mittal Publication, New Delhi
23. Seng Khasi Seng Kmie (nd) Centenary Souvenir: Ka Shad Suk Mynsiem. Bhabani Print & Publication, Guwahati (1911–2011)
24. Syiem S (2006) The olden days of the seven hut. Translated from Soso Tham's Ki Sngi Ba Rim U Hynniewtrep. Walvens Computer System. Synod Complex, Shillong

25. Tham S (1972) Ka Duitara Ksiar Ne Ki Poetry Khasi. Dispur Print House, Guwahati
26. Tham S (1936) Ki Sngi Barim U Hynniew Trep. Ri Khsi Press, Shillong
27. Kharwaniang R (1978) Ka Jingsneng Tymmen Khasi. Book Stall, Shillong
28. Kharwaniang R (2016) The teaching of Khasi elders: Ka Jingsneng Tynmen Part I & II Translated by BijoyaSawian. Vivekananda Kendra Institute of Culture, Guwahati

# A Step Towards Sentiment Analysis of Assamese News Articles Using Lexical Features

**Ringki Das and Thoudam Doren Singh**

**Abstract** With the growth of the internet, more and more digital content gives rise to each day, resulting in an 'age of data'. It brings a trend for reading online news from various digitally available newspapers. Positive news spread positivity, and negative news spread negativity to our minds and our society. Since the last few decades, sentiment analysis has become a fascinating and salient area for researchers in natural language processing to understand the sentiment of the news. Therefore, it becomes necessary to classify into positive and negative polarity to measure the daily and overall news sentiment. In this paper, we aim to carry a sentiment polarity classification model by applying machine learning classifiers on low resource Assamese language using lexical features on the news domain. The baseline system works only with a bag of words without any feature-based polarity. But, our proposed model uses lexical features like adjectives, adverbs, and verbs. The proposed model has shown improvement over our baseline model in terms of F1-score on the standard data set.

**Keywords** Polarity identification · Lexical features · Low resource language · Machine learning based classifier

## 1 Introduction

Computationally processes to interpret and classify the emotions hidden in a text define as sentiment analysis. Sentiment polarity identification is a sentiment analysis task to detect the polarity from a labeled text data, which machine learning-based classifiers show hug success in polarity classification. The abundance of text data

R. Das (✉) · T. D. Singh (✉)
Department of Computer Science and Engineering,
National Institute of Technology Silchar, Silchar, Assam, India
e-mail: ringkidas@gmail.com

T. D. Singh
e-mail: thoudam.doren@gmail.com

 
A. Maji et al. (eds.), *Proceedings of the International Conference on Computing and Communication Systems*, Lecture Notes in Networks and Systems 170,
https://doi.org/10.1007/978-981-33-4084-8_2

in digital format and machine learning has empowered document classification and text analytics. The motto "Think positive be positive" has a strong bonding with the identification of positive or negative news. The aspect level, document level, paragraph level, and sentence level are the main classification level in sentiment analysis. Till today's date, most of the sentiment analysis related tasks have performed on some of the major languages such as English. Few works on some Indian languages like Bengali, Hindi, Punjabi, etc. reported. There are mainly three language families in Indian contexts, such as Indo-Aryan, Dravidian, and Tibeto-Burman. Bengali, Hindi, Marathi, Oriya, Punjabi, and Urdu come under the Indo-Aryan language family. Telugu comes under the Dravidian family, and Manipuri belongs to the Tibeto-Burman family. To the best of our knowledge, very less work on the Assamese language has reported in this field. So, there are interesting problems to explore in this language. The Assamese language belongs to the Indo-European language family. It is spoken mainly in Assam by approximately 15 million people as per the 2011 census. As a resource-constrained language like the Assamese, it has many challenges in NLP. In this paper, we propose a sentiment polarity identification model in a less privileged language like the Assamese. The present system classifies news texts into positive and negative polarity using different machine learning-based classifiers. A large manually annotated corpora require for supervised machine learning method. However, there is no such publicly available annotated data set for sentiment analysis of the Assamese. So, we started collecting and annotating the news domain data set. Sentiment analysis can be employed to news recommendation system, online question-answer system, and as an opinionated text summarization.

The main contribution of our research is:

1. Collection of news domain dataset for sentiment analysis and followed by manual annotation.
2. We collected small lexicon based on word class including, adjective, adverb, and verb as an additional feature set.
3. Development of sentiment polarity identification system using various machine learning algorithms and additional feature sets of the news text.

The rest of this paper is divided into four sections, with Sect. 2 discussing the related previous works. Sections 3 and 4 describe the system design used in the model and the analysis of experimental results respectively. Finally, summing up with the conclusion and future work in Sect. 5.

## 2 Related Work

Several works have been reported on sentiment classification using different machine learning approaches. In paper [1], Amitava et al. developed a sentiment polarity classification model on news corpus using the SVM classifier. The evaluation result of the proposed model is 70.04%. An online interactive gaming technology known as Dr. Sentiment has been reported by Amitava et al. [2]. In this paper by Gautam et al. [3]

discussed supervised learning which is a better way to understand sentiment analysis. An existing work carried out by Gogoi et al. [4] on a document classification using naïve Bayes classifier. Hu et al. [5] reported a very promising technique for feature-based summarization of the product reviews. Kaur et al. [6] reported on Punjabi news articles polarity detection using support vector machines. The paper by Kim et al. [7] suggested a model determine the opinion of the sentiment using part of speech (POS) tagger and WordNet to identify the opinion bearing words. A text classifier model learning from positive and unlabelled data (LPU) was suggested by Liu et al. [8]. A twitter sentiment analysis work was reported by Neethu et al. in [9] using different machine learning classifiers. To detect the subjectivity based on minimum-cut, a framework was developed by Pang et al. [10] on the movie review data set. The minimum-cut framework significantly improved the polarity classification accuracy in sentiment analysis. Pang et al. [11] raises a rating inference problem using a meta-algorithm based on metric labeling. A sentiment analysis model was proposed by Pang et al. [12] for document classification of movie reviews. Poria et al. [13] proposed to merge both WordNet and SenticNet for sentiment analysis with 85.12% accuracy.

## 3 System Design

A high-level graphical representation of the proposed model is shown in Fig. 1. The first and foremost step is data acquisition from local daily newspapers in the Assamese. The data collection step is followed by the data cleaning and annotation. After pre-processing the data, we divide the data set into the train data set and test data set. We then use the train data set to train the models and evaluate the trained models using the test data set. The approach of the model building is described in the following subsections.

### *3.1 Data Collection*

Data set acquisition is a very salient and time-consuming process. We prepare our data set by collecting the news articles from the local daily newspapers separately for Pratidin[1] and Ganaadhikar.[2] A separate scraper built in-house using Java and its Jsoup library[3] is used to collect the data set. The news articles on the Assamese language are collected from June 2018 to June 2019. The collected data set is used as a training data set for our model after pre-processing and annotation. A statistics of the Assamese corpus is mentioned in Table 1.

---

[1] https://www.asomiyapratidin.in/.

[2] http://ganaadhikar.com.

[3] https://jsoup.org/.

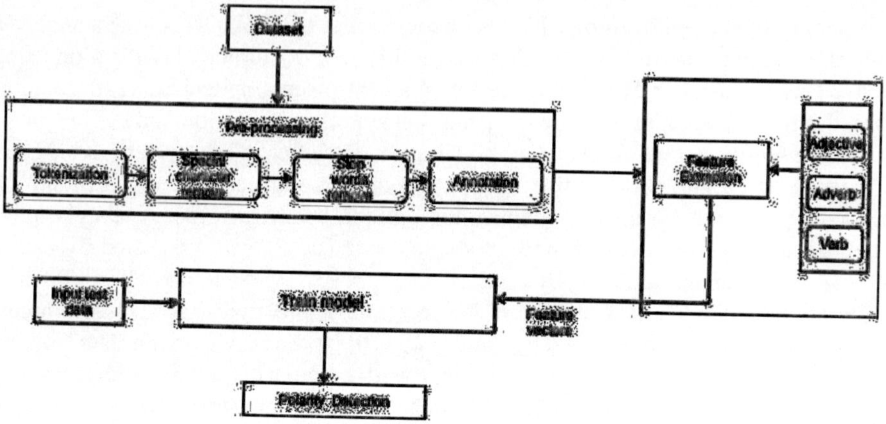


**Fig. 1** Our proposed model

**Table 1** Statistics of news corpus

| Category of articles | No. of documents | No. of sentences | No. of words |
|---|---|---|---|
| Positive | 1150 | 8958 | 169,701 |
| Negative | 1150 | 9748 | 179,579 |
| Total | 2300 | 18,806 | 349,280 |

## 3.2 *Pre-processing*

The news articles collected are noisy. To convert the data into a consumable format, a series of pre-processing steps are carried out. In sentiment analysis, along with data collection, data annotation is also a necessary pre-processing task for developing the NLP system. The news corpus is annotated into positive and negative polarity and used to train the model. The annotation is performed manually. Different annotator has a different point of view. For that reason, the annotation may vary from person to person based on the ambiguity of the sentence. The pre-processing phase includes the following steps separately:

1. Tokenization of a sentence into a token of words
2. Removing special characters and punctuation
3. Replacing multiple spaces into a single space
4. Removal of single-character tokens
5. Remove stop words.
6. Annotation.

### 3.3 Feature Extraction

All words can not be treated equally. Therefore, the sentiment bearing words such as adjectives, adverbs, and verbs are considered as relevant words for feature extraction. To extract these words, we have first prepared lexicons having 462 adjectives, 121 adverbs, and 577 verbs as it is not freely available in the Assamese. We have also created 116 numbers of stop words lexicon in the Assamese. As such, the stop words are not a significant feature. Therefore, it is better to remove the stop words which do not contribute any relevant information and have low discrimination value. After cleaning and annotating the data set, we extract the feature set and divide the data set into the train set and test set.

### 3.4 Machine Learning Classifier

The classifiers that we have trained are specifically naïve Bayes, multinomial naïve Bayes, Bernoulli naïve Bayes, logistic regression, support vector classifier (SVC), linear SVC, NuSVC, stochastic gradient descent, decision tree and random forest.

## 4 Experimental Result and Analysis

In this section, we will discuss the performance of different models. A brief description of the feature sets is shown in Table 2. The training and validation are carried out with a break of the dataset to 80% for training and 20% for validation.

### 4.1 Evaluation Metrics

The performance is compared based on four parameters namely: accuracy, precision, recall and F1-score. Equations (1), (2), (3), and (4) are used for calculating accuracy,

**Table 2** A description about feature set

| Feature set | Description |
|---|---|
| Adjective | Contains types of adjectives |
| Adverb | Contains types of adverbs |
| Verb | Contains types of verbs |
| Adjective + Adverb + Verb (AAV) | Contains combination of adjective, adverb and verb |

precision, recall and F1-score respectively. Accuracy is the percentage of correctly classified cases. Precision is the number of positive classes that are classified correctly as positive. Recall quantifies the number of positive predictions. F1 score is the harmonic mean of recall and precision.

$$\text{Accuracy} = \frac{\text{Tp} + \text{Tn}}{\text{Tp} + \text{Tn} + \text{Fp} + \text{Fn}} \tag{1}$$

$$\text{Precision (Pr)} = \frac{\text{Tp}}{\text{Tp} + \text{Fp}} \tag{2}$$

$$\text{Recall (Re)} = \frac{\text{Tp}}{\text{Tp} + \text{Fn}} \tag{3}$$

$$\text{F1-score} = \frac{2 * \text{Pr} * \text{Re}}{\text{Pr} + \text{Re}} \tag{4}$$

where, Tp = True positive, Fp = False positive, Tn = True negative, Fn = False negative.

## 4.2 Results

A feature wise machine learning classifier model F1- score is presented in Table 3.

We built an ensemble classifier model that takes votes from each classifier and to give the final decision with a certain confidence level. This model consists of naïve Bayes, logistic regression, support vector classifier, and decision tree. Ensemble model F1- score is shown in Table 4. In general, the ensemble model works better than an individual model. It is not necessary ensemble classifier accuracy is always better than any of its models. In our dataset, the ensemble model accuracy is comparatively less than the individual model train with respective features.

**Table 3** A feature wise system F1-score

| Model | F1-score |
|---|---|
| AAV + random forest | 0.67 |
| Adjective + decision tree | 0.64 |
| Verb + random forest | 0.63 |
| Adverb + decision tree | 0.61 |

**Table 4** An ensemble model F1-score

| Model | F1-score |
|---|---|
| AAV+ ensemble | 0.66 |
| Adjective+ ensemble | 0.63 |
| Verb+ ensemble | 0.60 |
| Adverb+ ensemble | 0.55 |

**Table 5** Sample 1 input-output

| Particulars | Sentence details |
|---|---|
| Input text 1 | ইণ্ডিগ' যাত্ৰীৰ বাবে সুখবৰ দেশীয় বিমান সেৱা সংস্থা ইণ্ডিগ'ই ইয়াৰ যাত্ৰীসকলৰ বাবে আগবঢ়াইছে বিশেষ ৰেহাই আঁচনি |
| Translation | Good news for indigo passengers, country's air service provider or company (institute) Indigo for it passengers has bought a special discount scheme |
| Ground truth | Positive |
| Predicted polarity | Positive with polarity probability score 0.8 |

### 4.3 *Analysis*

The proposed system can recognize the polarity from the news corpus to a certain extent. The model based on classical machine learning algorithms, feature extraction method i.e. bag of words. Sentiment lexicon such as an adjective, adverb, and verb are extracted as feature set which is not able to capture all the context of the features. A sample input-output using random forest classifier is shown in Tables 5 and 6. From Table 5, we can observe input text 1 shows the correct polarity. But in Table 6, input text 2 shows false polarity. It is a compound sentence, where the first part is negative but the second part is positive. Most of the time the word কৰায়ত্ব is considered as positive polarity although in example 2 the context is negative. Due to this reason, the system has misclassified. Tables 5 and 6 shows sample Input-Output for a correct and an incorrect classification of sentiment. There are some limitations of our model:

1. Inadequate feature set
2. Ambiguity in functional words.

## 5 Conclusion

Our experimental finding shows that a random forest-based model with the AAV feature set performs the best on news articles of the Assamese. About a mile to cross to develop a standard model in Assamese sentiment analysis. Our work is just a beginning that has to improve system performance. For regional languages like Assamese,

**Table 6** Sample 2 input-output

| Particulars | Sentence details |
|---|---|
| Input text 2 | বঙাইগাঁও জিলাৰ কীৰ্তনপাৰাত যুৱকক হত্যাৰ চেষ্টাঃ গুলিচালনা কৰা ডকাইত কৰায়ত্ব |
| Translation | Attempt to kill youth at Kirtanpaar in Bongaigaon district: bullet firing dacoit has been arrested |
| Ground truth | Negative |
| Predicted polarity | Positive with polarity probability score 0.6 |

polarity detection is a very arduous and daring task and suffers from the scarcity of resources. Tools like part of speech (POS) tagger and sentiment lexicon help in improving the performance in sentiment analysis. For a low resource-constrained language like Assamese, building quality language-specific tools POS tagger and sentiment lexicon is a labor-intensive task as it requires an annotated training corpus. Our model achieves a promising result in fewer data. But, it is unable to differentiate the correct sentence polarity for partial positive and partial negative compound sentences. In the future, we will propose a more enhanced polarity detection model with a large data set size along with lexical features.

## References


1. Das A, Bandyopadhyay S (2010) Opinion-polarity identification in bengali. In: International conference on computer processing of oriental languages, pp 169–182
2. Das A, Bandyopadhyay S (2011) Dr sentiment knows everything! In: Proceedings of the 49th annual meeting of the association for computational linguistics: human language technologies: systems demonstrations. Association for computational linguistics, pp 50–55
3. Gautam G, Yadav D (2014) Sentiment analysis of twitter data using machine learning approaches and semantic analysis. In: 2014 seventh international conference on contemporary computing (IC3). IEEE, pp 437–442
4. Gogoi M, Sarma S (2015) Document classification of assamese text using naïve bayes approach. Int J Comput Trends Technol 30:182–186. https://doi.org/10.14445/22312803/IJCTT-V30P132
5. Hu M, Liu B (2004) Mining opinion features in customer reviews. AAAI 4:755–760
6. Kaur G, Kaur K (2015) Sentiment analysis on punjabi news articles using svm. Int J Sci Res 6(8):414–421
7. Kim SM, Hovy E (2004) Determining the sentiment of opinions. In: Proceedings of the 20th international conference on Computational Linguistics. Association for Computational Linguistics, p 1367
8. Liu B, Dai Y, Li X, Lee WS, Yu PS (2003) Building text classifiers using positive and unlabeled examples. In: Third IEEE international conference on data mining. IEEE, pp 179–186
9. Neethu M, Rajasree R (2013) Sentiment analysis in twitter using machine learning techniques. In: 2013 fourth international conference on computing, communications and networking technologies (ICCCNT). IEEE, pp 1–5

10. Pang B, Lee L (2004) A sentimental education: Sentiment analysis using subjectivity summarization based on minimum cuts. In: Proceedings of the 42nd annual meeting on association for computational linguistics. Association for Computational Linguistics, p 271
11. Pang B, Lee L (2005) Seeing stars: exploiting class relationships for sentiment categorization with respect to rating scales. In: Proceedings of the 43rd annual meeting on association for computational linguistics. Association for Computational Linguistics, pp 115–124
12. Pang B, Lee L, Vaithyanathan S (2002) Thumbs up? Sentiment classification using machine learning techniques. In: Proceedings of the ACL-02 conference on Empirical methods in natural language processing, vol 10. Association for Computational Linguistics, pp 79–86
13. Poria S, Gelbukh A, Cambria E, Yang P, Hussain A, Durrani T (2012) Merging senticnet and wordnet-affect emotion lists for sentiment analysis. In: 2012 IEEE 11th international conference on signal processing, vol 2. IEEE, pp 1251–1255

# Efficient Assessment of Formula Representation in Embedded Vector

**Pankaj Dadure, Partha Pakray, and Sivaji Bandyopadhyay**

**Abstract** Mathematical information retrieval is a comparatively early stage of research, lying at the interaction of a text-based retrieval approach and mathematical knowledge management. There are a wide variety of approaches have been developed to cope-up with the challenge of formula representation, indexing, and relevance measurement. In this work, a vector-based formula embedding approach is proposed where a mathematical formula is transformed into the fixed-size vector of 202 bit. For relevance measurement, the minimum difference between the user query vector and the formula vector is reckoned. To estimate the proficiency, the Wikipedia dataset of NTCIR-12 is used. Finally, the benchmarking of the proposed approach is done against other published literature as it gives better results in terms of P_5 = 0.42, P_10 = 0.38, P_15 = 0.25, and P_20 = 0.19.

**Keywords** Mathematical information retrieval · Mathematical knowledge management · Indexing · Formula embedding

## 1 Introduction

Mathematical formula is an important mean of any scientific documents. Mathematical Information Retrieval (MIR), in other word searching for a mathematical formula, postulation or notation, is an observable task in Information Retrieval (IR). Retrieval of such information is a complex natural language processing task which requires to understand the meaning of a formula and the ability to reason over relevant facts.

P. Dadure · P. Pakray (✉) · S. Bandyopadhyay
Department of Computer Science and Engineering, National Institute of Technology Silchar, Silchar, Assam, India
e-mail: parthapakray@gmail.com

P. Dadure
e-mail: krdadure@gmail.com

S. Bandyopadhyay
e-mail: sivaji.cse.ju@gmail.com

 
A. Maji et al. (eds.), *Proceedings of the International Conference on Computing and Communication Systems*, Lecture Notes in Networks and Systems 170,
https://doi.org/10.1007/978-981-33-4084-8_3

Mathematical formulae inherent the predefined structure and script which are relatively difficult to handle by conventional search engine. According the prior work done by Ganesalingam [1], the idea of mathematical formula is delivered through the surrounding textual context and symbolic structure of formula. To interpret the mathematical formula is difficult for both human and machine because of large set symbols, ambiguities, semantic, alignment and context. In MIR, without knowing the precedence and associativity of operations lead to unclear query formation and irrelevant search. Evaluation of MIR system is an another challenge due to discrepancy in the information need of individual user. This challenge lead to definition of relevance is an considerable objective in MIR. In academic life, people gone through the articles, books, lecture notes etc. where mathematical formula are often significant part than other textual information and images. But to understand the meaning of formula, their surrounding text play an vital role. Therefore, many researcher have incorporate the textual contribution to accomplish the better retrieval efficiency in MIR system [2]. Unskillfully, many documents hold the math expression in image format such as .tiff or .png. This makes retrieval system helpless to retrieves such notations.

In this paper, the formula embedding approach is proposed where the mathematical formula is converted into the vector of size 202 bit, each bit demonstrates the occurrence count of a particular entity in a formula. To validate the performance of the proposed approach, the Wikipedia corpus of NTCIR-12 [3] is used.

The paper is structured as follows: Sect. 2 describes prior works in domain of MIR. Section 3 provides detailed description about the system architecture. Section 4 gives a detailed account of the dataset. Section 5 describes the experimental results and their comparative analysis. Section 6 conclude with summary and directions of further research and developments.

## 2 Related Work

This section provide the brief overview about the existing state-of-the-art formula retrieval approaches. Based on the notion of formula representation, existing work have classified into four category namely text-based, tree-based, vector-based and neural network based approach.

### 2.1 *Text-Based Approach*

The prior research study of text based approaches witness that the retrieval of textual information is different from MIR. To deal with mathematical information, ordinary text based approaches requires the significant improvement. At NTCIR-10, team MIRMU [4] used conventional full text retrieval approach to retrieve both text and math information. To make the formula compatible for text retrieval system, several preprocessing operations have been adopted. Moreover, team BRKLY [5] has adopt the keyword based approach to mutate the XML documents into the textual form. For indexing, conventional full text indexing approach has applied. The type embedding mechanism [6] uses the types (sequence of word or words) to communicate the idea of mathematical expression. In which, the type has created for the collection of documents where mathematical formulae have associated with concepts and object properties. This prior work imparts the use of semantic relationships in the retrieval of mathematical formulae.

### 2.2 *Tree-Based Approach*

In the retrieval of math information, tree-based approaches have divulged the considerable performance and set the benchmark for other approaches. To overcome the restraint of semantic enhancement, the ICST math retrieval system [7] has generates sub-tree from the semi-operator tree of formulas. In which, the layout presentation of tree transformed into a semantic presentation to enrich the retrieval of formula. Tangent-CFT [8] uses two distinct hierarchical representations of formulae i.e. Symbol Layout Tree (SLT) and Operator Tree (OPT). The path between the pair of symbols generates the tuples which have then embed using the n-gram embedding model. To achieve the stat-of-the-art results, structural matching has been deployed to retrieves partial similar formulas.

### 2.3 *Vector-Based Approach*

The BVTMF approach [9] generates the binary vector of size 202 where '1' represents the presence and '0' represent the absence of entity in a formula. In this approach, both textual and mathematical information have been incorporated to achieve the better retrieval efficiency. In a vector-based approach, the similarity calculation is one of the time taking task where the system mapped the similarities between the user query and indexed vectors. For instance, cosine similarity calculation between the query and indexed formulas have shows the valuable contribution and leading scope in similarity measurement [10].

### 2.4 *Neural Network-Based Approach*

Now a days, a neural network is an emerging field in the domain of IR and natural language processing. While the remarkable performance of the neural network in the retrieval of textual information, many researchers have attempted at formula retrieval and achieves the benchmark results. For instance, Formula2vec based approach [11] explores the proficiency of neural network based vector representation of mathematical formulas. It uses the Distributed Memory Model of Paragraph Vectors (PV-DM) to learn a different representation of formula. The influence of LSTM in sequence-to-sequence tasks, LSTM based formula entailment approach [12] has been proposed to formulate the entailment between the user entered math query and the formula contains in a scientific document.

### 2.5 Other Approaches

In the retrieval of mathematical information, some of the researchers have uses both text and math based approaches to achieves the best results. The Tangent math search system [13] uses two indices to attains their best retrieval results. The first index has based on the concept of TF-IDF textual search and second index based on the concept of query-by-expression. The query-by-expression strategy uses the bag-of-word approach to represent the formula in pairs of symbols which are computed from the symbol layout tree. The SMSG5 research team [14] from India introduced the math search engine which explores the co-occurrence of formula and keyword. For co-occurrence findings, LDA and doc2vec techniques have been used. Moreover, they have used common patterns between formula and keywords to achieves their best.

## 3 System Architecture

The system architecture of proposed approach is shown in Fig. 1, where each module have their own contribution to retrieve the relevant search.

In the proposed approach, the first operation is preprocessing where the documents have been preprocessed and clips the mathematical information tagged by <math> and </math> and also clean out the unnecessary MathML elements and attributes. In clipped mathematical information, the math formulas have presented in three different formats i.e. $L^AT_EX$, Presentation MathML, and Content MathML format. These three different representations of formula lead to huge index size and slow down the searching mechanism. For better retrieval and small index size, the proposed model uses only the Presentation MathML format and discards the remaining two formats i.e. $L^AT_EX$ and Content MathML.

After successful preprocessing, the proposed model moves towards the formula embedding module. In the formula embedding module, the formulas in Presentation MathML format have been transformed into an embedded vector of size 202 bit. Each bit of vector represents the count of a particular entity in a formula. For vector generation, the proposed system uses the Bit Position Information Table (BPIT) [15]. In the generated vector, "0" represents the absence and a decimal value represents the presence count of a particular entity in a formula. For 579,608 mathematical

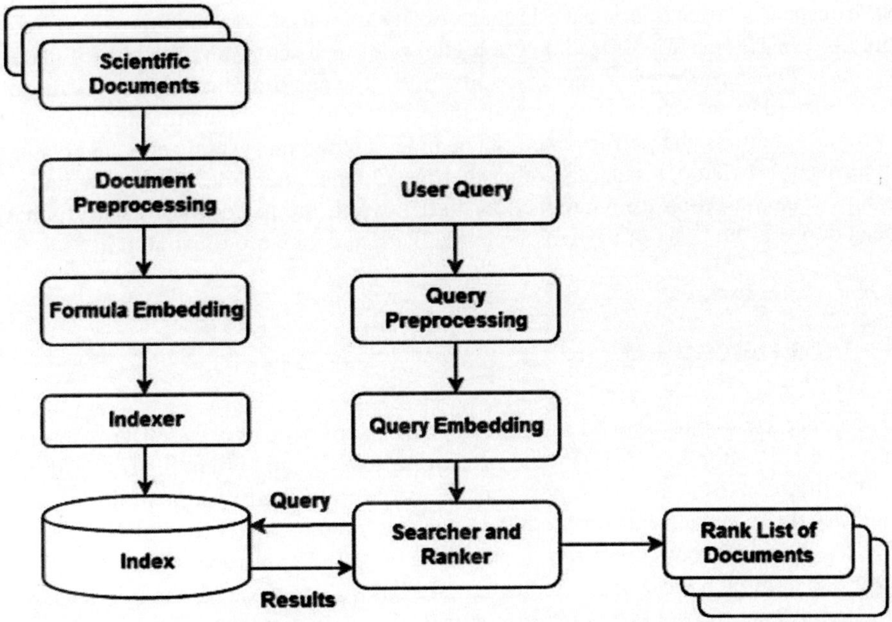


**Fig. 1** Proposed system architecture

formulas, the proposed system created the 579,608 vectors. The third operation of system architecture is indexing where vectors have been indexed by the indexer. Each index stored the three different fields namely embedded vector, the formula in Presentation MathML format, and document name from which the formula is originated. In the query preprocessing phase, user enters the query in the Presentation MathML format. After successful insertion, the query has been preprocessed by the query preprocessing module. The prime task of the query preprocessing module is to remove the irrelevant tags and noise from the user entered query. Afterward, the preprocessed query has been fed into a query embedding module. Query embedding module converts the preprocessed Presentation MathML user query into a vector of size 202. For query vector generation, the query embedding module considered the entity position format of BPIT [15]. In a generated query vector, '0' represents the absence and decimal number represents the count of a particular entity in a user entered query.

The searcher and ranker module proceeds to a similarity mapping function. Searcher module compared the query vector with each indexed formula vector where each bit position is compared and the minimum difference has been calculated as a relevance score. The maximum relevance score corresponds to the minimum difference between the query vector and the formula vector. After the completion of the searcher module task, the ranker module ranks the documents based on the higher relevance score. The documents which consist of more than one relevant formula with respect to user query have higher priority than those documents which hold only the single formula. Based on the higher relevance score, a ranked list of documents has been retrieved as search results. In IR, the performance of the system can be measured by automatic system evaluation and/or human evaluation. In automatic system evaluation, the performance can be judged based on system generated results in terms of precision, recall, and other metrics. On the other side, human evaluation can be done based on user satisfaction. In this work, the performance analysis has been done on the basis of system generated metrics which are illustrated in Sect. 5.

## 4 Data Description

Data collection can be defined as a crucial phase in any type of research. Any imprecise data collection may activate the inaccurate analysis and affect the result of the study. In order to measure the performance of the proposed approach, the experiment has been carried out on MathTagArticles of Wikipedia corpus of NTCIR-12 [3]. The metadata about the corpus is shown in Table 1. The retrieval efficiency of the proposed approach has been tested using 10 MathML queries. All the queries have been represented in the Presentation MathML format and maintain the symmetrical structure with indexed formula. For relevance measurement of retrieves documents, the Gold dataset has been used which contains the relevant (1) and irrelevant (0) documents with respect to 10 MathML queries. The Gold dataset strictly follows the structure guidelines of TREC qrel format [16].

## 5 Experimental Results and Analysis

This section demonstrates the extensive performance study of the proposed approach in comparison with the existing state-of-the-art retrieval approaches. Both tabular and graphical visualization of comparative analysis is shown in Table 2 and Fig. 2 which depicts the efficiency and effectiveness of the proposed approach. The experimental results have been reported using TREC tool [16] and analyzed to indicate the

**Table 1** Corpus description

| Corpus | MathTagArticles of wikipedia of NTCIR-12 |
|---|---|
| Size | 2.1 GB |
| No. of articles | 31,839 |
| Article format | HTML |
| No. of formulas | 579,608 |
| No. of test queries | 10 |

**Table 2** Evaluation report

| Parameters | Text_based | Formula embedding | Proposed approach |
|---|---|---|---|
| P_5 | 0.29 | 0.38 | 0.42 |
| P_10 | 0.2 | 0.31 | 0.38 |
| P_15 | 0.13 | 0.21 | 0.25 |
| P_20 | 0.1 | 0.16 | 0.19 |

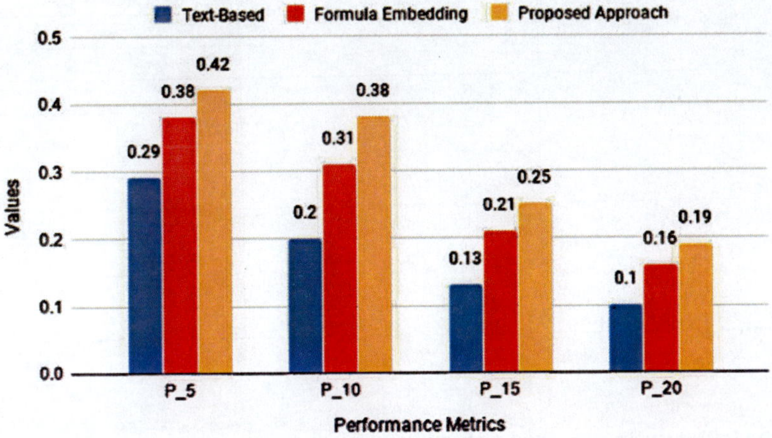


**Fig. 2** Result comparison

effectiveness of the proposed approach. The result set of retrieved documents and Gold dataset has been compared to obtains the evaluation metrics. The performance evaluation has been indicated in terms of precision at 5, 10, 15, and 20 as the primary measures respectively. Compared to text-based search engine implemented using Apache Nutch[1] platform and formula embedding [15], proposed approach achieves at least 30%, 47%, 48% & 47% and 10%, 18%, 38% and 37% improvement in terms of P_5, P_10, P_15 and P_20 respectively. This demonstrates that the count of a particular entity in a vector can significantly improve the retrieval effectiveness of the system while estimating the presence or absence of an entity.

Out of test 10 queries, the search results for a few queries are shown in Table 3. The outcomes for the first query consist of a different form of user need with different variables and scientific symbols. This retrieval characteristic of the proposed approach is a considerable achievement in MIR. The search results of the second query delineate that the proposed approach successfully retrieves the results for complex queries as well. The symbols, music notation, and scientific terms have a predefined structure and require special effort to deal with them. In the search results of the third query, the proposed approach smoothly retrieves such special notations. The retrieval outputs of the fourth query divulge that the proposed approach is pro-

[1] http://nutch.apache.org/.

**Table 3** Obtained search results

| No. of queries | User query | Search results |
|---|---|---|
| 1. | P = NP | P = NP |
| | | $\mu = NP$ |
| | | N = pt |
| | | n = pq |
| 2. | $ax^2 + bx + c = 0$ | $ax^2 + bx + c = 0$ |
| | | $ax^2 + bx + c = 0$ |
| | | $ax^2 + bx = c$ |
| | | $ax^3 + bx + c = 0$ |
| 3. | $A \Leftrightarrow B$ | $A \Leftrightarrow B$ |
| | | $aRb \Leftrightarrow bRa$ |
| | | $A \leftrightarrow B \leftrightarrow C$ |
| | | $A \cdot B$ |
| 4. | f*g | f*g |
| | | $f* = g$ |
| | | $f * g = g * f$ |
| | | E*F |

ductively retrieves the exact match formulas, subformula and parent formula as well. The noteworthy difference between the obtained results of the text-based approach and proposed approach disclose that the text-based approach requires significant improvement to deal with scientific information.

## 6 Conclusions and Future Scope

The objective of this paper is to retrieves the syntactically similar formula and magnify the retrieval efficiency. This work introduces the task of formula embedding where the count of particular entity in a formula is estimated. The experiments associated with the data set containing 31,839 scientific articles and near about 579,608 formula. The constructed model has formed the embedded vector to retrieves the documents which hold similar formula and shown the outstanding performance as compared to current state-of-the-art methods developed for similar tasks.

In future work, unification module will be deployed to boost the search score by retrieving the subformula. In addition, modification in the process of vector representation will be adopt to enhance the MIR performance.

**Acknowledgements** The authors would like to express their gratitude to the Department of Computer Science and Engineering and Center for Natural Language Processing, National Institute of Technology Silchar, India for providing the infrastructural facilities and support.

## References

1. Ganesalingam M (2013) The language of mathematics. Ph.D. thesis, Cambridge University Computer Laboratory, pp 1–277
2. Stathopoulos Y, Baker S, Rei M, Teufel S (2018) Variable typing: assigning meaning to variables in mathematical text. In: Proceedings of the North American chapter of the association for computational linguistics: human language technologies, vol 1, pp 303–312
3. Zanibbi R, Aizawa A, Kohlhase M, Ounis I, Goran T, Davilla K (2016) NTCIR-12 MathIR task overview. In: Proceedings of the 12th NTCIR conference on evaluation of information access technologies, Tokyo, Japan, pp 299–308
4. Líška M, Sojka P, Ružicka M (2013) Similarity search for mathematics: Masaryk University team at the NTCIR-10 Math task. In: Proceedings of the 10th NTCIR conference on evaluation of information access technologies, June 18–21. Tokyo, Japan, pp 686–691
5. Ray L, Reynolds C, Gey F (2013) The abject failure of keyword IR for mathematics search: Berkeley at NTCIR-10 Math. In: Proceedings of the 10th NTCIR conference on evaluation of information access technologies, June 18–21. Tokyo, Japan, pp 662–666
6. Stathopoulos Y, Teufel S (2016) Mathematical information retrieval based on type embeddings and query expansion. In: Proceedings of COLING, the 26th international conference on computational linguistics: technical papers, December 11–17. Osaka, Japan, pp 2344–2355
7. Gao L, Wang Y, Hao L, Tang Z (2014) ICST Math retrieval system for NTCIR-11 Math-2 task. In: Proceedings of the 11th NTCIR conference on evaluation of information access technologies, December 9–12. Tokyo, Japan, pp 99–102
8. Mansouri B, Rohatgi S, Oard D, Wu J, Giles C, Zanibbi R (2019) Tangent-CFT: an embedding model for mathematical formulas. In: Proceedings of the 2019 ACM SIGIR international conference on theory of information retrieval, Santa Clara, CA, USA, October 2–5, pp 11–18
9. Pathak A, Pakray P, Gelbukh A (2019) Binary vector transformation of math formula for mathematical information retrieval. J Intell Fuzzy Syst 36(5):4685–4695
10. Dadure P, Pakray P, Bandyopadhyay S (2019) An empirical analysis on retrieval of math information from the scientific documents. In: International conference on communication and intelligent systems, Springer, Singapore, pp 301–308
11. Gao L, Jiang Z, Yin Y, Yuan K, Yan Z, Tang Z (2017) Preliminary exploration of formula embedding for mathematical information retrieval: can mathematical formulae be embedded like a natural language? arXiv preprint arXiv:1707.05154. Washington
12. Pathak A, Pakray P, Das R (2019) LSTM neural network based math information retrieval. In: Second international conference on advanced computational and communication paradigms (ICACCP), SMIT, Sikkim, India, February 25–28, pp 1–6
13. Pattaniyil N, Zanibbi R (2014) Combining TF-IDF text retrieval with an inverted index over symbol pairs in math expressions: the tangent math search engine at NTCIR 2014. In: Proceedings of the 11th NTCIR conference on evaluation of information access technologies, December 9–12 , Tokyo, Japan, pp 135–142. In NTCIR
14. Thanda A, Agarwal A, Singla K, Prakash A, Gupta A (2016) A document retrieval system for math queries. In: Proceedings of the 12th NTCIR conference on evaluation of information access technologies, June 7–10. Tokyo, Japan, pp 346–353
15. Pathak A, Pakray P, Gelbukh A (2018) A formula embedding approach to math information retrieval. Computación y Sistemas 22(3):819–833
16. Voorhees E, Harman D (2005) TREC: experiment and evaluation in information retrieval. Computational linguistics, vol 32(4). MIT Press, Cambridge, 563–567

# Neural Machine Translation for Low Resource Assamese–English

**Sahinur Rahman Laskar, Partha Pakray, and Sivaji Bandyopadhyay**

**Abstract** Neural machine translation (NMT) attracts attention to the machine translation (MT) community because of its potential to persist sequential data over variable lengths of input and output sentences. With the attention mechanism, the NMT system achieves state-of-the-art technique which allows focusing on specific input vectors of the input sentence instead of learning a single vector for each input sentence. Although, the NMT system improves translation quality by handling long-term dependency issue and facilitate the ability of contextual analysis, and it requires adequate parallel train corpus, which is difficult in the context of low resource languages. This paper contributes an Assamese–English parallel corpus and builds two NMT systems, sequence-to-sequence recurrent neural network (RNN) with attention mechanism (NMT-1) and a transformer model having self-attention mechanism (NMT-2). The NMT-2 system has achieved a higher bilingual evaluation understudy (BLEU) score of 10.03 for English-to-Assamese and BLEU score 13.10 for Assamese-to-English translation, respectively.

**Keywords** Assamese–English · Neural machine translation · Attention · BLEU

## 1 Introduction

MT falls under the domain of natural language processing (NLP), where automatic translation occurs in between languages. MT attempts to reduce the language barrier in a multilingual country like India. English and Hindi are the official languages stipu-

S. R. Laskar (✉) · P. Pakray · S. Bandyopadhyay
Department of Computer Science and Engineering,
National Institute of Technology Silchar, Silchar, Assam, India
e-mail: sahinurlaskar.nits@gmail.com

P. Pakray
e-mail: parthapakray@gmail.com

S. Bandyopadhyay
e-mail: sivaji.cse.ju@gmail.com

 
A. Maji et al. (eds.), *Proceedings of the International Conference on Computing and Communication Systems*, Lecture Notes in Networks and Systems 170,
https://doi.org/10.1007/978-981-33-4084-8_4

lated by the Central Government of India for all means of communication. Moreover, there are 22 scheduled languages in India that are adopted as regional languages in different State Governments of India. Hence, there is a need for translating official documents like notices, orders, and messages issued by the Central Government into the respective regional languages of the State Governments to facilitate greater access among local people. Assamese language is one of the regional language of Indian state of Assam and widely accepted language in the entire northeastern region of India. It belongs to the Indo-European language family and the basic vocabulary originated from "Sanskrit." The native speakers of Assamese language are known as Assamese people (also known as Xomiya). In order to have a proper communication, there is a great demand for automatic translation of Assamese language from English and vice-versa. Although, Google provides automatic translation of 103 languages over the globe but lags behind in Assamese. MT system is shifted from rule-based to corpus-based approach that eliminated support for linguistic experts. In corpus-based MT system, statistical machine translation (SMT) relies on Bayesian inference that is used to predict translation probabilities of phrase pairs in parallel source–target sentences. Although SMT system overcomes problems of rule-based system but suffers from context analyzing ability, long-term dependency issue, less accuracy, and system complexity. To overcome such problems, NMT system has been introduced [2, 13]. The primary requirement of such a system is a sufficient parallel training corpus, which is a challenge in case of a low resource language. To address this challenge, we have also built an Assamese–English parallel corpus.

The main purpose of current research is to explore automatic translation for a low resource language, especially the Assamese–English language pair. The same has been investigated in both forward and backward directions using two different NMT approaches: the NMT-1 system with attention mechanism and the NMT-2 system using self-attention. The automated BLEU score is used to evaluate the predicted translations of the NMT systems [10].

The rest of the paper is categorized as follows: Section 2, briefly outlines related works. Corpus preparation and system description are described in Sects. 3 and 4. Section 5 presents results and error analysis, and finally, Sect. 6 concludes the paper with future directions.

## 2 Related Works

The NMT system aims to map natural languages using neural networks, which is capable of learning representations from the data. The system obtained promising outcome over the previously acclaimed SMT system in various languages [1, 2, 7, 13]. For low resource, Indian language translations like English to Hindi, English to Tamil, English to Mizo, English to Punjabi, Hindi ⇔ Nepali, Punjabi to Hindi, and NMT system have achieved remarkable results based on BLEU score [8, 11, 12]. The machine translation on Assamese–English language pair still remains in an outset stage though [5] obtained a BLEU score of 5.02 for English-to-Assamese translation and 9.72 for Assamese to English translation using SMT-based approaches. Current work aims to improve existing MT work on Assamese–English language pair using the NMT system.

## 3 Corpus Preparation

The corpus of parallel source–target sentence pairs has been developed through manual back-translation, where English and Assamese are the source and target language and vice-versa. The parallel sentences have been cross-verified by linguistic experts, who possess linguistic knowledge of the Assamese language. For our work, the linguistic expert is a native Assamese undergraduate student. The parallel sentences are self-curated from various online sources such as the Bible,[1] Multilingual Question Paper,[2] Parallel Dictionary Words,[3] Wikipedia[4], and other websites.[5] Since in the Wikipedia source, parallel sentences are not available, we have manually translated into Assamese sentences. We have also collected Assamese sentences from many online blogs and translated into the corresponding English sentences. The parallel source–target sentences were partitioned into train, validation, and test sets, and the details of which are summarized in Table 1. This work will be considered as a baseline paper of the English–Assamese low resource language pair. Because of this reason, we have considered only 100 test and 500 validation sentences. The validation data is required to verify the convergence of the training system.

## 4 System Description

The current work builds two NMT systems, the NMT-1 system for sequence-to-sequence RNN using the OpenNMT[6] toolkit and the NMT-2 system for the transformer model using Marian.[7] Both the toolkits are open source and publicly available. Data preprocessing, system training, and system testing are the major steps of operations, which are elaborated in the subsequent subsections.

---

[1] https://www.bible.com.

[2] https://sebaonline.org/.

[3] http://www.xobdo.org/dic.

[4] https://en.wikipedia.org/wiki/History_of_Assam.

[5] http://learn101.org/assamese_grammar.php.

[6] https://pypi.org/project/OpenNMT-py/.

[7] https://marian-nmt.github.io/.

**Table 1** Corpus statistics

| Corpus nature | Corpus name | Total instances |
|---|---|---|
| Train | Train set | 201,007 |
| Validation | Validation set | 500 |
| Test | Test set | 100 |

## 4.1 Data Preprocessing

The preprocessing step is used for the tokenization of the source and target sentences and creates a dictionary that indexes words while the system is training. In both the systems, the vocabulary dimension is 50,002, and 46,845 for English and Assamese sentences, respectively.

## 4.2 System Training

After data preprocessing, both NMT-1 and NMT-2 systems are trained separately up to 100,000 epochs and saved at an interval of 1000. The systems are trained on a single NVIDIA Quadro P2000 GPU.

**NMT System with Attention Mechanism** In the training process of NMT-1 system, parallel source–target sentences of Assamese–English language pair are fed into the sequence-to-sequence RNN with attention. In the architecture of the NMT system, the encoder and decoders are the key ingredients. The encoder is used to compress the entire source sentence into a context vector and the decoder is responsible for depicting the context vector. The encoder and decoder of the NMT-1 system consist of a two-layer network of long short term memory (LSTM) nodes with 500 units in each layer. The NMT-1 system is trained for Assamese-to-English and English-to-Assamese following default settings of [6]. The primary encoder–decoder model adopted LSTM to enhance the translation/prediction quality of longer sentences. However, the encoder fails to capture all the important information in the context vector when the sentence is too long. For this reason, the attention-based encoder–decoder model was developed. The attention mechanism aids the decoder to focus on different parts of the source sequence at different decoding steps in global as well as local terms [9].

For illustration, the source Assamese sentence "মই আপোনাক বুজোঁ" is shown in Fig. 1, NMT-1 system predicts the English target sentence "I understand you" where $< eos >$ indicates the termination of a sentence.

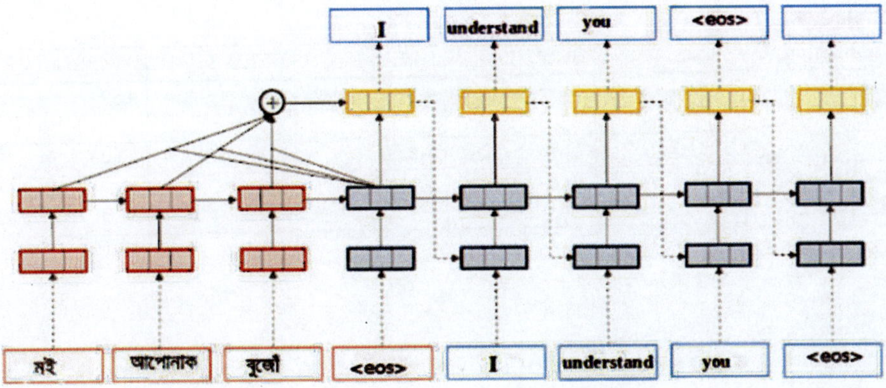


**Fig. 1** NMT-1 system architecture

**NMT System with Self-Attention Mechanism** Although LSTM-based sequence-to-sequence RNN improves translation quality of long sentences, accuracy decreases and there is limitation of parallelization. To handle these issues, the transformer model is introduced [14]. The attention mechanism that relies on the transformer model is called "self-attention." It is a state-of-the-art model for NMT. The encoder–decoder model of transformer consists of stacked self-attention and fully connected layers. The encoder and decoder contain two sub-layers: one multi-head self-attention mechanism and a feed-forward network. The decoder has an additional multi-head attention sub-layer. In the training process of the NMT-2 system, parallel sentences are fed into the transformer model, and the system is trained following the default settings of [4].

### *4.3 System Testing*

After the training processes is over, the NMT-1 and NMT-2 systems undergo the testing/translation on unseen test sentences following setting of [4, 6]. For optimum translation, the decoder uses a beam search algorithm to minimize the search space of possible translations.

## 5 Results and Error Analysis

The empirical results acquired from the NMT-1 and NMT-2 systems have been evaluated based on uni-gram (BLEU-1), bi-gram (BLEU-2), and tri-gram (BLEU-3) scores. We have considered BLEU score up to tri-gram because the BLEU score diminishes after passing over BLEU-3. Table 2 presents results for English-to-

**Table 2** BLEU scores on Assamese–English pair translation

| Translation | | BLEU-1 | BLEU-2 | BLEU-3 | AVG BLEU |
|---|---|---|---|---|---|
| English to Assamese | NMT-1 | 22.7 | 4.5 | 1.4 | 9.53 |
| | NMT-2 | 23.7 | 5.2 | 1.2 | 10.03 |
| Assamese to English | NMT-1 | 30.4 | 5.7 | 1.8 | 12.63 |
| | NMT-2 | 31.7 | 5.8 | 1.8 | 13.10 |

Assamese as well as Assamese-to-English translations, which reveal the optimum average BLEU score of 10.03 in English-to-Assamese translation and a score of 13.10 in Assamese-to-English translation. To capture the optimum BLEU score corresponding to different epochs for the predicted sentences on the NMT-1 and NMT-2 systems, these scores with respect to number of epochs are plotted. Figures 2 and 3 present such graphs for the NMT-2 system where $X$-axis represents 1000 epochs at each point from 0 to 100, i.e., up to 100,000 epochs, and $Y$-axis represents corresponding BLEU scores. From the graphs, it is observed that the computed average BLEU score up to BLEU-3 is close to the BLEU-2 score, which made evident the effect of the bi-gram model in the Assamese language [3, 7]. Figure 4 compares the BLEU scores of the various MT systems, where NMT-2 acquires a higher BLEU score than NMT-1 and the existing SMT system [5] in both directions.

To examine translation effectiveness of the NMT-1 and NMT-2 systems, sample predicted sentences are considered in Tables 3, 4, and 5. Table 3 presents an example of the best performance for both directions of translation where NMT-2 system correctly translates the source sentence unlike the NMT-1 system. From the predicted sentences, it is observed that both these system suffer various issues like under-translation, over-translation, and wrong translation of named entities and number of source words. In Table 4, NMT-2 suffers issues of under-translation and wrong translation of named entities while NMT-1 predicts mix words of source and target words for English-to-Assamese translation. Similarly, in Table 5, both NMT systems suffer problem of wrong translation of named entities. Also, in this example, NMT-2 suffers issue of under-translation unlike over-translation in case of NMT-1 system for English-to-Assamese translation.

## 6 Conclusion and Future Work

The present work investigates two different NMT systems for a low resource language pair, i.e., Assamese–English. In this work, we have achieved higher average BLEU score using the transformer model than sequence-to-sequence RNN model

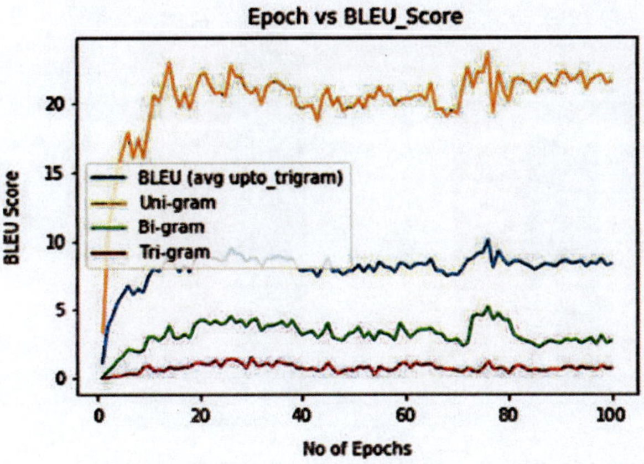


**Fig. 2** Number of epochs versus BLEU scores for English-to-Assamese language translation (NMT-2)

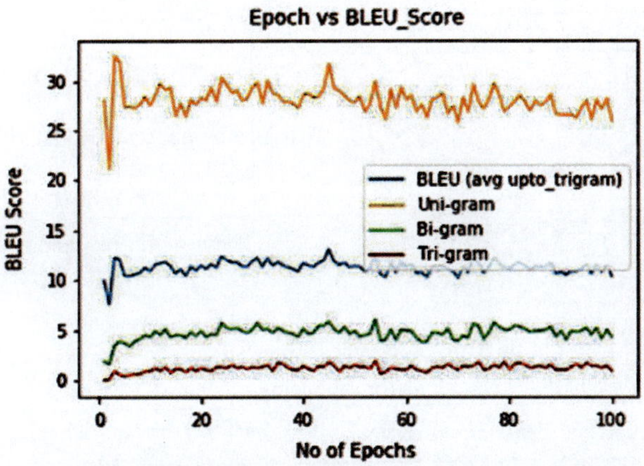


**Fig. 3** Number of epochs versus BLEU scores for Assamese-to-English language translation (NMT-2)

in both forward and backward directions of translation. The acquired BLEU scores are optimum for both Assamese-to-English and vice-versa, since much work has not been advanced for such a language pair. However, close observation of predicted sentences on the given test set remarks that the present NMT system needs improvement for appropriate translation. Hence, the conventional NMT system requires modification to resolve various issues for low resource languages, and also other evaluation metrics will be considered in future.

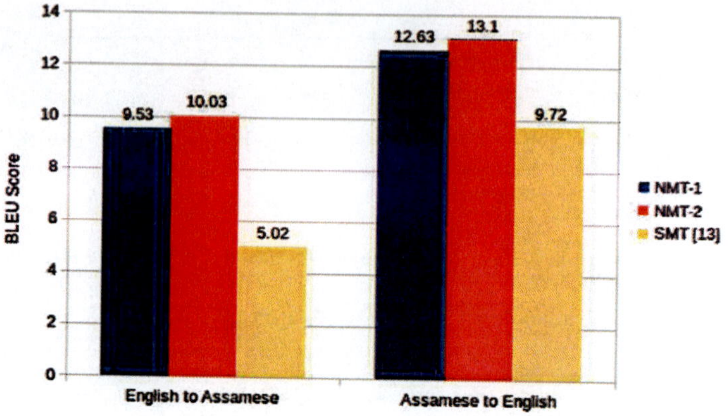


**Fig. 4** Comparison of BLEU scores among various MT systems

**Table 3** Example 1: Best performance of NMT-2 system on Assamese–English language pair translation

| Translation | English to Assamese | Assamese to English |
|---|---|---|
| Source | Patna is the capital city of Bihar | পাটনা বিহাৰৰ ৰাজধানী। |
| NMT-1 | ঔৰংগাবাদৰ ৰাজধানী চহৰ। | An important capital of India |
| NMT-2 | পাটনা বিহাৰৰ ৰাজধানী। | Patna is the capital city of Bihar |
| Target | পাটনা বিহাৰৰ ৰাজধানী। | Patna is the capital city of Bihar |

**Table 4** Example 2: Error in English-to-Assamese translation

| | |
|---|---|
| Source | Sualkuchi place is famous for producing golden silken thread |
| NMT-1 | Producing Sualkuchi Sualkuchi বাবে বিখ্যাত। |
| NMT-2 | স্বৰ্ণ মন্দিৰ বাবে বিখ্যাত। |
| Target | শুৱালকুছি সোণালী ৰেচম সূতা উৎপাদনৰ বাবে প্ৰসিদ্ধ । |

**Table 5** Example 3: Error in Assamese to English translation

| | |
|---|---|
| Source | শুৱালকুছি সোণালী ৰেচম সূতা উৎপাদনৰ বাবে প্ৰসিদ্ধ । |
| NMT-1 | The golden cottages is famous for its finely silk plantations |
| NMT-2 | It was famous for tourism |
| Target | Sualkuchi place is famous for producing golden silken thread |

**Acknowledgements** We would like to thank Department of Computer Science and Engineering and Center for Natural Language Processing (CNLP) at National Institute of Technology Silchar for providing the requisite support and infrastructure to execute this work.

## References

1. Bahdanau D, Cho K, Bengio Y (2015) Neural machine translation by jointly learning to align and translate. In: Bengio Y, LeCun Y (eds) 3rd international conference on learning representations, ICLR 2015, San Diego, CA, USA, May 7–9, 2015, conference track proceedings. http://arxiv.org/abs/1409.0473
2. Cho K, van Merriënboer B, Gulcehre C, Bahdanau D, Bougares F, Schwenk H, Bengio Y (2014) Learning phrase representations using RNN encoder–decoder for statistical machine translation. In: Proceedings of the 2014 conference on empirical methods in natural language processing (EMNLP). Association for computational linguistics, Doha, Qatar, pp 1724–1734. https://doi.org/10.3115/v1/D14-1179. https://www.aclweb.org/anthology/D14-1179
3. Ghosh S, Girish KVV, Sreenivas T (2011) Relationship between indian languages using long distance bi-gram language models. In: Proceedings of ICON 2011, 9th international conference on natural language processing. Macmillan, pp 104–113
4. Junczys-Dowmunt M, Grundkiewicz R, Dwojak T, Hoang H, Heafield K, Neckermann T, Seide F, Germann U, Aji AF, Bogoychev N, Martins AFT, Birch A (2018) Marian: fast neural machine translation in C++. In: Proceedings of ACL 2018, system demonstrations. Association for Computational Linguistics, Melbourne, Australia, pp 116–121. https://doi.org/10.18653/v1/P18-4020. https://www.aclweb.org/anthology/P18-4020
5. Kanchan Baruah K, Das P, Hannan A, Sarma SK (2014) Assamese-english bilingual machine translation. Int J Nat Lang Comput (IJNLC) 3: https://doi.org/10.5121/ijnlc.2014.3307
6. Klein G, Kim Y, Deng Y, Senellart J, Rush A (2017) OpenNMT: Open-source toolkit for neural machine translation. In: Proceedings of ACL 2017, system demonstrations. Association for Computational Linguistics, Vancouver, Canada, pp 67–72. https://www.aclweb.org/anthology/P17-4012
7. Laskar SR, Dutta A, Pakray P, Bandyopadhyay S (2019) Neural machine translation: English to hindi. In: 2019 IEEE conference on information and communication technology, pp 1–6. https://doi.org/10.1109/CICT48419.2019.9066238
8. Laskar SR, Pakray P, Bandyopadhyay S (2019) Neural machine translation: Hindi-Nepali. In: Proceedings of the fourth conference on machine translation (Volume 3: shared task papers, Day 2). Association for Computational Linguistics, Florence, Italy, pp 202–207. https://doi.org/10.18653/v1/W19-5427. https://www.aclweb.org/anthology/W19-5427
9. Luong T, Pham H, Manning CD (2015) Effective approaches to attention-based neural machine translation. In: Proceedings of the 2015 conference on empirical methods in natural language processing. Association for Computational Linguistics, Lisbon, Portugal, , pp 1412–1421. https://doi.org/10.18653/v1/D15-1166. https://www.aclweb.org/anthology/D15-1166
10. Papineni K, Roukos S, Ward T, Zhu WJ (2002) Bleu: a method for automatic evaluation of machine translation. In: Proceedings of the 40th annual meeting on association for computational linguistics, ACL '02. Association for Computational Linguistics, Stroudsburg, PA, USA, pp. 311–318. https://doi.org/10.3115/1073083.1073135. https://doi.org/10.3115/1073083.1073135
11. Pathak A, Pakray P (2018) Neural machine translation for indian languages. J Intell Syst, pp 1–13. https://doi.org/10.1515/jisys-2018-0065
12. Pathak A, Pakray P, Bentham J (2018) English-Mizo machine translation using neural and statistical approaches. Neural Comput Appl 30:1–17. 10.1007/s00521-018-3601-3. https://doi.org/10.1007/s00521-018-3601-3

13. Sutskever I, Vinyals O, Le QV (2014) Sequence to sequence learning with neural networks. In: Ghahramani Z, Welling M, Cortes C, Lawrence ND, Weinberger KQ (eds) Advances in neural information processing systems 27. Curran Associates, Inc., pp 3104–3112. http://papers.nips.cc/paper/5346-sequence-to-sequence-learning-with-neural-networks.pdf
14. Vaswani A, Shazeer N, Parmar N, Uszkoreit J, Jones L, Gomez AN, Kaiser LU, Polosukhin I (2017) Attention is all you need. In: Guyon I, Luxburg UV, Bengio S, Wallach H, Fergus R, Vishwanathan S, Garnett R (eds) Advances in neural information processing systems 30. Curran Associates, Inc., pp 5998–6008. http://papers.nips.cc/paper/7181-attention-is-all-you-need.pdf

# Review Comments of Manipuri Online Video: Good, Bad or Ugly

**Thoudam Doren Singh, Telem Joyson Singh, Mirinso Shadang, and Surmila Thokchom**

**Abstract** Sentiment analysis is a novel approach towards analyzing the sentiment of a given text or document. In particular, the sentiment analysis of short texts such as single sentences and social media tweets and comments are one of the main challenges of natural language processing (NLP) given the limited contextual information they usually convey. In recent years, deep learning application has shown promising result in natural language processing. This paper presents the comparative analysis of various deep learning, traditional machine learning and lexicon based methodologies of sentiment analysis. We present the importance of pre-processing and feature engineering for sentiment analysis on a resource constrained data-set of Manipuri comments gathered from social media platform using these approaches.

**Keywords** Manipuri language · Sentiment analysis · Social media comments · Machine learning · Neural networks

T. D. Singh
Department of Computer Science and Engineering,
National Institute of Technology Silchar, Silchar, Assam, India
e-mail: thoudam.doren@gmail.com

T. J. Singh (✉)
Department of Computer Science and Engineering, IIIT Senapati, Imphal, Manipur, India
e-mail: joyson@iiitmanipur.ac.in

M. Shadang
Department of Electronics and Communication Engineering,
IIIT Senapati, Imphal, Manipur, India
e-mail: mirinso@iiitmanipur.ac.in

S. Thokchom
Department of Computer Science and Engineering,
National Institute of Technology Meghalaya, Shillong, India
e-mail: surmila.th@gmail.com

 
A. Maji et al. (eds.), *Proceedings of the International Conference on Computing and Communication Systems*, Lecture Notes in Networks and Systems 170,
https://doi.org/10.1007/978-981-33-4084-8_5

## 1 Introduction

In the last few years, the use of social networking websites such as Facebook, Youtube, Twitter etc has been increasing significantly and has given every company or agency a good impulse to mine such sites in order to collect information from customer on their services or products review [1]. For instance, if someone wants to buy tickets for a film, a sentiment classifier can predict the overall sentiment of the film rather than going through all long reviews manually. Buying tickets can be decided on the basis of a positive or negative sentiment. This paper examines the performance of different sentiment analysis methods and methodologies. Using Manipuri language dataset as the main dataset, Naïve Bayes, support vector machine, lexicon based, deep dense, convolutional and recurrent neural networks (RNN) algorithms are implemented and the results are compared. This work also provides a dataset of 4000 properly labeled Manipuri comments with positive and negative sentiments as well as a pre-processing method for the dataset.

## 2 Motivation

Initially, sentiment analysis is used for the analysis of sentiment based on long texts like letters, emails etc. It is also used to analyze criminal activities in both the pre and post-crime areas. This field of studies became a major focus with the explosion of web applications such as microblogging, social networking sites and forums. People talk, comment, criticize and write reviews, recommendations, and more through these applications. A lot of valuable information about products, people, events, etc. are provided by user generated data. The modern world's cutting-edge competition has led to advanced mechanisms for feedback from customers. Automatic mining and test techniques are needed for the large user content because mining and analytics produced by the crowd are difficult to use. Work in this field originated in blogs [2] and product/movie reviews [3] early in their use. It is a challenge to apply this field to the fraternity of microblogging. Challenges are like short comments using improper words, use of emojis, etc. These challenges of the ill-formed comments and analysing their meaning for the sentiment analysis became our motivation.

## 3 Dataset Collection

A new dataset is built for our task based on the collection done in 2018. This dataset consists of several comments from social networking websites such as Youtube and Facebook. We identified a set of Youtube Manipuri video along with the respective comments from the subscribers of the respective channel. The main reason about selecting the Manipuri video comments are to identify the successful Manipuri video

**Table 1** Sample comment dataset

| Category | Manipuri comment and translation into English |
|---|---|
| Positive | Mashige song se tarubada eigonda kei hairtre ymna nungaiba ama phaojei ema leibk manipur ashigum eshei kayajina mai oirebne ym nungc esheise shigumba esei matungdasu puthokpirk u |
| Translation | I feel so happy after listening to such kind of patriotic songs and such songs are pride of motherland, please produce more such songs |
| Negative | asss actor si di loire hey . . . acting su heitey . . . yam waray . . . actress khufajsiga . . . . . |
| Translation | Alas, this actor is too much... don't know acting...can't take it.. but the actress is beautiful |

songs, actors and to identify the talents of the crew members in the production process as per the viewers and subscribers feedback. There are equal number of 2000 positive and 2000 negative comments in the dataset. The raw form of the dataset is noisy but cleaned before feeding into the sentiment analysis systems. All texts are lowercased. There are several Manipuri comments written in Roman script. White spaces are found around punctuation along with periods, commas and brackets. It is very common to use code-switched comments in many social media platforms. So, small percentage of code-switched English-Manipuri comments are also included on both positive and negative category. A sample comment is shown by Table 1. Current trend of using short form of words using Roman script for which a detailed analysis on this is very important to understand what kind of words often used in positive or negative or neutral comments. The spelling variations of the same word by different reviewers of the video is one of the key aspect on the how the machine learning systems can learn the orthographic variations in the social media comment.

### *3.1 Dataset Pre-processing*

Pre-processing of text for a specific task includes cleaning and preparation in the required format. Some of the important role of pre-processing in sentiment analysis are presented by Haddi et al. [4]. Usually, there are lot of noise in the online texts and uninformative entities. Not every word in the text has an impact on the tasks proposed. The dimensionality of the problem will be high if we retain these words and the task of classification becomes difficult for each word considering all the dimensions. Thus, the data is properly preprocessed so that text noise is minimized and speed up the classification task with higher accuracy. Several steps are involved in the process of transformation such as—cleaning of online text, removal of white space, abbreviation expansion, stemming, removal of stop words, handling of negation etc. This is followed by the selection of feature. All but the last step are transformations. This is followed by filtering to apply some functions to choose the patterns we need. In the opinion mining context, features may be terms, words or phrases having strong expression about opinion as negative or positive. These features have a higher impact on the text orientation than other words in same text. Several methods are used in feature selection based on the syntactic position of the word such as adjective, univariate and multivariate. A relevant work on comparison of pre-processing methods for Twitter sentiment analysis is presented by Zhao and Gui [5]. The methods assessed are—replacing negative mentions, removing URL links in the corpus, correct the spelling for the words that contain repeated letters to the respective original English form, removing stop words, expanding acronyms to their original words. In a separate work by Giulio et al. [6], the techniques of pre-processing for sentiment analysis discussed about handling emoticon, negation, dictionary and stemming. Table 2 shows the statistics of the dataset collected and used for our work.

Social media comments are noisy in many a cases. We exercise a careful selection of the comments from the reviewers of Manipuri music video, news and discussion channels. These comments include emoticons, short forms, ill-formed words and ill-formed sentences etc. For convenience, we choose the Manipuri comments written in Roman script only in order to gather a large collection. Table 3 shows a sample of the mis-spelled words and its corresponding corrections. Though, there is abundance of positive comments as compared to the negative comments, we try to balance the numbers for both the cases. There are situation where there is code-switching in the comments between English and Manipuri. But, the code-switched comments are less as compared to the positive and negative comments. When we create frequency distribution of words in the comments, we found that punctuation marks or stop-words are the most common words. Stop words are the common words that have no meaning in the analysis of texts. For example, *ei* (meaning I), *ng* (meaning You), *adu* (meaning That), *nggi* (meaning Yours), *masi* (meaning This), etc. Punctuation marks

**Table 2** Statistics of comment Corpus

| Category of comments | No. of comments | No. of tokens |
|---|---|---|
| Positive | 2000 | 16,512 |
| Negative | 2000 | 19,922 |
| Total | 4000 | 36,434 |

**Table 3** Expansion of mis-spelled words in comments

| Mis-spelled words | Expansion |
|---|---|
| ym | yam |
| ngw | ngao |
| khre | khare |
| cdi,cd | sidi |
| dna,dn | dana |

like comma, fullstop, inverted comma, etc. occur highly in any text data. By removing stop words and punctuation, we clean data. One interesting feature we found in the comments is that people uses vowel-less words instead of writing full words most of the time. It might be because of time and space saving. In order to reduce different spellings for same word, we introduce a dictionary which maps commonly used vowel-less words to full words.

## 4 Approaches of Classification

We attempted three approaches of classification to understand the underlying reviews of the online video.

- Traditional machine learning algorithms rely on the description of samples by common features, that are typically used for classification, regression, and/or clustering [7–9].
- Deep learning algorithms such as multi-layer perceptron (MLP), convolutional neural network (CNN) and recurrent neural network (RNN) use multiple layers to progressively extract higher level features from the raw input [10–12]. The big difference is ordinary machine learning usually use hand-crafted features whereas in deep learning the features are crafted in unsupervised manner by machine through its deep structure.
- Lexicon-based sentiment classification uses sentiment lexicon to find polarity of a text [13]. Sentiment of a text is defined by aggregate sum of the polarities of the words in the text.

## 5 Analysis of Results

We experimented the different methods using dataset of 2000 positive and 2000 negative comments. The dataset are collected from social medias like Youtube, Facebook, Twitter, etc. We split the dataset—75% as training dataset and 25% as testing set. The performance of our sentiment analysis system is calculated by using equation:

$$\text{Accuracy} = \frac{(\text{TP} + \text{TN})}{(\text{TP} + \text{TN} + \text{FP} + \text{FN})} \quad (1)$$

where TP, TN, FP and FN represent the number of true positives, true negatives, false positives and false negatives respectively. The results of the different approaches before pre-processing and after the pre-processing step in this work are shown in Table 7. Figures 1 and 2 show the confusion matrices of the two best models from traditional machine learning and deep learning based classifier. Referring to the

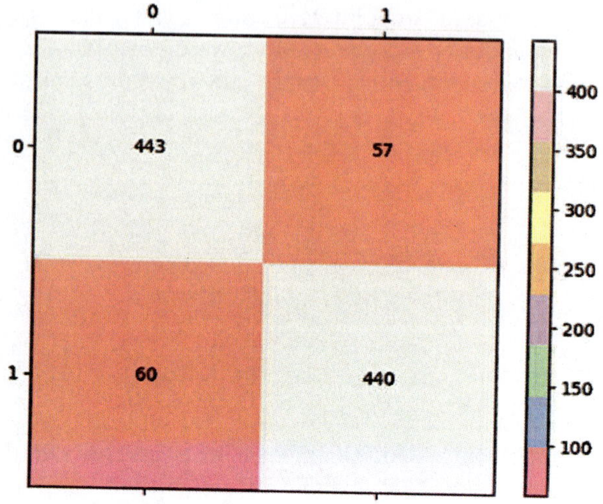


**Fig. 1** Confusion matrix: Naïve Bayes with bag-of-words

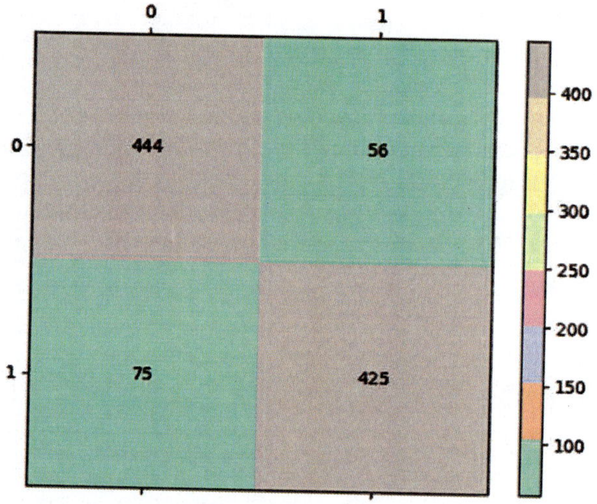


**Fig. 2** Confusion matrix: convolutional neural network

confusion matrix for Naïve Bayes and CNN, we observe that the correct predictions are distributed more evenly across different classes.

Naïve Bayes with top-n words feature achieves an accuracy of 86%, Naïve Bayes with bag-of-words features achieves 88.3% accuracy, and Naïve Bayes with n-gram (unigram plus bi-gram) achieves an accuracy of 87.1%. We train our linear Support Vector Machine with bag-of-words feature and we get an accuracy of 86.7%. Simple MLP achieves an accuracy of 86.6%. We have used two LSTM layers with dropout 0.4 and 0.5 respectively, before passing to the dense layer. The recurrent neural network gives a decent accuracy of 85.4% accuracy but convolutional networks outperforms it. The CNN based model trains very fast and it achieves an accuracy of 86.9%. Our lexicon based approach achieves a testing accuracy of 66.8%. One of the

**Table 4** Example of correct classification of polarity

| Particulars | Comments details |
|---|---|
| Comment | nang yam fhajei |
| Ground truth | Positive |
| Translation | You are beautiful |
| Predicted polarity | Positive (score: 0.67) |

**Table 5** Example of wrong classification of polarity

| Particulars | Comments details |
|---|---|
| Comment | nang yam fhajakhredako |
| Ground truth | Positive |
| Translation | You are too beautiful |
| Predicted polarity | Negative (score: 0.7) |

**Table 6** Example of correctly classified polarity of code-mixed comment

| Particulars | Comments details |
|---|---|
| Comment | Quality yam lei epa epuna thambirmb eralse . keep it up god bless u |
| Ground truth | Positive |
| Translation | The inherited properties of our forefather are quality one. Keep it up, God bless u |
| Predicted polarity | Positive (Score: 0.59) |

challenges we faced is the agglutinative nature of Manipuri language where words tend to be composed of many morphemes all glued together. On the contrary, the English language is analytical which have a very low (possibly 1) morpheme to word ratio. In our sentiment analysis system with Naïve Bayes(bag-of-words), we tested with two sentences—one with less agglutinated which is correctly classified and other with highly agglutinated which is wrongly classified as given by Tables 4 and 5 respectively. The Naïve Bayes (bag-of-words) classifier also correctly identified the code-mixed comment polarity.

We observed that the one with less agglutinated sentence returns correct result while the second highly agglutinated sentence gives incorrect result. One possible reason for this is the invisible feature complexity of the agglutinated words. Next we experimented with a sentence which is code-mixed comment of Manipuri and English words. We found the predicted class as well as its score to be satisfactorily correlating (Table 6).

Table 7 shows the result with and without data pre-processing. It shows that a sophisticated data processing techniques can improve accuracy. This attributes to the noise in the raw comments in the present day social media. The linear SVM

**Table 7** Performance of the classifiers

| Approach | Accuracy before preprocessing | Accuracy after preprocessing |
|---|---|---|
| Naïve Bayes (top-n) | 84.3 | 86.0 |
| Naïve Bayes (bag-of-words) | 85.6 | 88.3 |
| Naïve Bayes (n-gram) | 85.6 | 87.1 |
| Linear SVM (bag-of-words) | 83.8 | 86.7 |
| Logistic regression | 84.7 | 87.0 |
| MLP (word vector) | 85.4 | 86.6 |
| CNN (word vector) | 84.0 | 86.9 |
| RNN (word vector) | 84.5 | 85.4 |
| Lexicon based | 65.7 | 66.8 |

(bag-of-words) and CNN (word vector) showed the biggest jump from the raw data after pre-processing by a margin of 2.9. However, the RNN based classifier showed the least improvement over the raw data.

## 6 Conclusion and Future Works

After these experiments, some variants of Naïve Bayes and logistic regression show higher accuracy as compared to the deep learning based techniques. The MLP with word vector shows competitive accuracy as compared to the traditional machine learning techniques. Again, CNN outperforms RNN. So, CNN is good in capturing features from the short comments. Lexicon based approach is the least performer. We train our logistic regression model with bag-of-words feature and we get an accuracy of 87%. One big advantage of lexicon based approach is that it doesn't involve training. So it will be useful for languages with less dataset. Our lexicon based approach achieves a testing accuracy of 66.8% after the pre-processing. Some of the major hurdle in the whole work is to prepare a large dataset covering various common as well as important phenomena. Thereafter, a more elaborate pre-processing phase of the whole social media comment in order to embed all the important expressions and messages which are passed from one user to the other can be carried out. The best performer of the sentiment classification for Manipuri online video review comments is achieved using Naïve Bayes (bag-of-words) classifier. Thus, additional phases of the language processing such as syntax analysis, parsing, semantic analysis and modeling of topics can be carried out to enhance the performance of the classification tasks in the future with special emphasis on the processing of the social media comments.

## References

1. Feldman et al (2013) Techniques and applications for sentiment analysis. Commun ACM
2. Bautin M, Vijayrenu L, Skiena S (2008) International sentiment analysis for news and blogs. In: Second international conference on weblogs and social media ICWSM
3. Turney PD (2002) Thumbs up or thumbs down? Semantic orientation applied to unsupervised classification of reviews. In: Proceedings of the 40th annual meeting of the association for computational linguistics, pp 417–424
4. Haddi E, Liu X, Shi Y (2013) The role of text pre-processing in sentiment analysis. In: Proceedings of the first international conference on information technology and quantitative management (ITQM 2013). Suzhou, China
5. Zhao J, Gui X (2017) Comparison research on text pre-processing methods on twitter sentiment analysis. IEEE Access 5:2870–2879
6. Giulio A, Ferrari L, Fontanini T, Fornacciari P, Iotti E, Magliani F, Manicardi S (2016) A comparison between pre-processing techniques for sentiment analysis in Twitter. In: Proceedings of the 2nd international workshop on knowledge discovery on the WEB, Cagliari, Italy, 8–10 Sept 2016
7. Prasad S (2010) Micro-blogging sentiment analysis using Bayesian classification methods
8. Vapnik V (1995) The nature of statistical learning theory
9. Durant K, Smith M (2006) Mining sentiment classification from political web logs. In: Proceedings of workshop on web mining and web usage analysis of the 12th ACM SIGKDD international conference on knowledge discovery and data mining. Philadelphia
10. Zhang L, Wang S, Liu B (2018) Deep learning for sentiment analysis : a survey. Data mining and knowledge discovery. Wiley Interdisciplinary Reviews. https://doi.org/10.1002/widm.1253
11. Kim Y (2014) Convolutional neural networks for sentence classification. arXiv preprint arXiv:1408.5882
12. Hochreiter S, Schmidhuber J (1997) Long short-term memory. Neural Comput 9(8):1735–1780
13. Asghar, Dr. Muhammad, Kundi F, Khan A, Ahmad S (2014) Lexicon-based sentiment analysis in the social web. J Basic Appl Sci Res 4:238–248

# Development of English-to-Bengali Neural Machine Translation Systems

**Anwesha Das and Thoudam Doren Singh**

**Abstract** The development strategies of English-to-Bengali neural machine translation (NMT) systems are reported in this paper. Several attention mechanisms on the development of English-to-Bengali NMT system are reported transparently. Further, the translation system has been experimented with some preprocessing steps such as tokenization, truecasing to obtain a better translation. The automatic evaluation techniques are used to measure the performance of the translated systems followed by human judgment.

**Keywords** Neural machine translation · Self-attention · LSTM · Encoder–decoder

## 1 Introduction and Motivation

Machine translation is nothing but doing translation from a source language into another target language by utilizing the capability of the machine. The definition of machine translation was first proposed by Weaver et al. [13] which merged a new path to the existing translation mechanism. In that era, rule-based machine translation (RBMT) was prominent which was later overtaken by statistical methods [3] after a significant development in the statistical field. In 1990, Brown et al. [3] approached statistical-based machine translation which later on opens a new door of doing machine translation research works. Bengio et al. [2] developed a neural network-based language model and showed a promising result as compared to the n-gram-based language model. The key benefit of neural machine translation is its

A. Das (✉) · T. D. Singh (✉)
Department of Computer Science and Engineering, National Institute of Technology Silchar, Silchar, Assam 788010, India
e-mail: anwesha.21031995@gmail.com

T. D. Singh
e-mail: thoudam.doren@gmail.com



A. Maji et al. (eds.), *Proceedings of the International Conference on Computing and Communication Systems*, Lecture Notes in Networks and Systems 170,
https://doi.org/10.1007/978-981-33-4084-8_6

ability to do a direct translation from one language to another in an end-to-end way unlike the phase-based machine translation where different components were trained separately.

In this work, several experiments have been conducted on neural machine translation with different kinds of attention mechanism in English-to-Bengali machine translation domain. It is found that the model with self-attention gives the best outcome than the other attention mechanisms. BLEU score evaluation plays a crucial role to determine the efficiency of those procedures, whereas human judgments like adequacy and fluency are also performed in this case. This paper is organized as the following. Section 2 describes the related work, whereas Sect. 3 introduces the proposed methods. Sections 4 and 5 are corpus statistics and experiment, respectively. Result analysis and conclusion are drawn in Sects. 6 and 7.

## 2 Related Work

An application of RNN model is encoder–decoder approach which takes input from the encoder side and produces output at the decoder side. From the work by Cho et al. [4], authors introduce encoder–decoder model in NMT domain which showed a poor outcome for long sentence but indicated a new path to NMT research field. The problem of memory mentioned in the previous paper has been solved by introducing a new type of RNN unit known as long short-term memory (LSTM). Sutskever et al. [11] have given the idea of using LSTM in NMT domain. It not only has given a new idea but also including an efficient approach of NMT. Comparison between LSTM and attention mechanism has been done in the paper by Bahdanau et al. [1]. In this paper, authors have shown that attention mechanism has been proven as a better way of doing NMT. The paper "Effective Approaches to Attention Based Neural Machine Translation" by Luong et al. [8] has done an extension of the above-mentioned paper by adding dot product besides the additive method. The self-attention mechanism which is a great replacement of RNN unit has been introduced in the paper "Attention is all you Need" by Vaswani et al. [12]. Positional embedding has been experimented instead of vector embedding in this paper which has given a noticeable increment in outcome. The work of [9] made a comparison between statistical machine translation (SMT) and NMT based on simple and complex sentences. As a result, it has been shown that NMT worked better for small sentence, whereas SMT was good for whole corpus. Application of bidirectional LSTM on different corpus and spotting out of pair creating low accuracy has been experimented by Hasan et al. [5]. The work by Ojha et al. [10] determined both NMT and PBSMT (phrase-based SMT) for the available data sets. Both the automatic and manual evaluations have been done to measure the performance.

## 3 Proposed Techniques

**Long Short-Term Memory** Recognized as LSTM [6] is merely a recurrent unit for predicting the sequence more specifically. The big problem with seq2seq is that all the previous steps must be recalled by the encoder and decoder which has been recovered by LSTM to some extent due to the support of long-term dependencies. The following equation gives the working principle of LSTM model.

$$C_t = \mathrm{fr}_t \cdot C_{t-1} + \mathrm{in}_t \cdot \sigma_C(M_C x_t + N_C ht - 1 + B_c) \tag{1}$$

$$h_t = \mathrm{ot}_t \cdot \sigma_h(C_t) \tag{2}$$

Here, the input vector $x_t$ whereas $\mathrm{fr}_t$, $\mathrm{in}_t$, $\mathrm{ot}_t$ are forgetting, input and output gates, respectively. $h_t$ is a neural network hidden layer vector. $C_t$ represents a neural network cell vector. $M$ and $N$ are weight metrics, while $B$ indicates the bias (Figs. 1 and 2).

**Encoder–Decoder** Model [4] is used to predict word series. It is a combination of many recurrent units.

$$h_t = \mathrm{RNN}(h_{t-1}, x_{it}) \tag{3}$$

Equation (3) represents basic recurrent neural network equation where $h_t$ depends on the previous hidden step output that is $h_{t-1}$ and present input.

$$s_0 = h_T \tag{4}$$

$$s_t = \mathrm{RNN}(s_{t-1}, e(\hat{y}_{t-1})) \tag{5}$$

$$P(y_t|y^(t-1)_1, x) = \mathrm{softmax}(Vs_t + B) \tag{6}$$

**Fig. 1** Encoder–decoder model

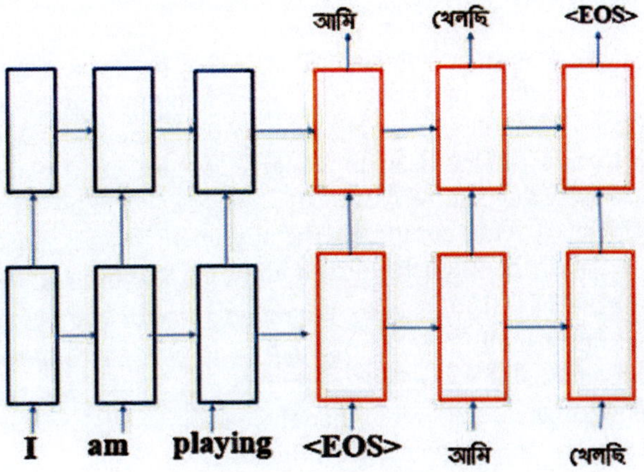

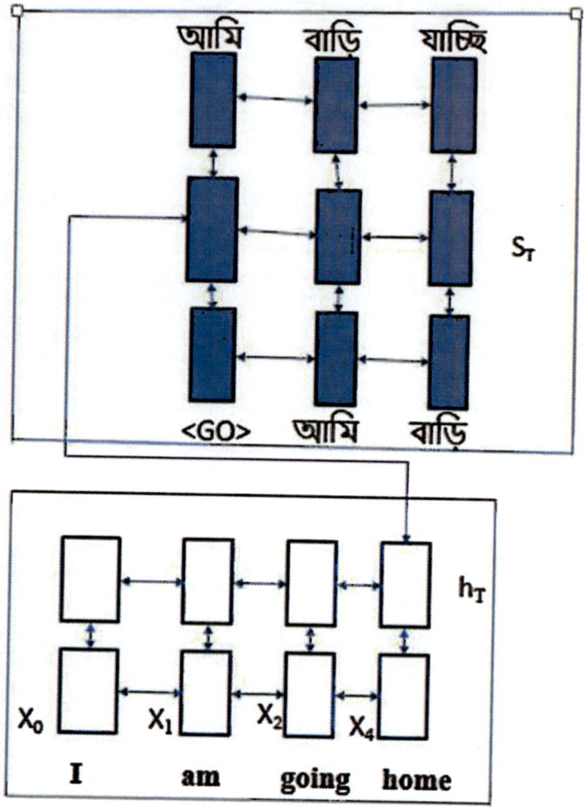


**Fig. 2** Working principle of encoder–decoder

Equation (4) is the initial input state of decoder. T is used as the length of the input sentence and B indicates bias. By giving the input "I am going home" in encoder part, decoder will generate আমি বাড়ি যাচ্ছি..

**Global Attention** Bahdanau et al. [1] has proposed a method for computing attention value of words in a sentence. This technique is implemented for the current work. Figure 3 elaborately explains the working method behind the MLP approach. Bidirectional RNN has been implemented to get a good result. The above figure describes that by giving source sentence "I am going home", all the alignment vector will be calculated. After comparing the alignment vectors, the best one will be chosen. For translating the English word "home", the highest alignment value will be for Bengali word "বাড়ি" . In this case, the context vector can be calculated by doing the summation of multiplication of alignment vector and hidden layer value produced by bidirectional RNN unit. By using this technique in this experiment, BLEU score of approximately 1.33 more than the baseline standard has been gained as this system managed long sentences well (Fig. 4).

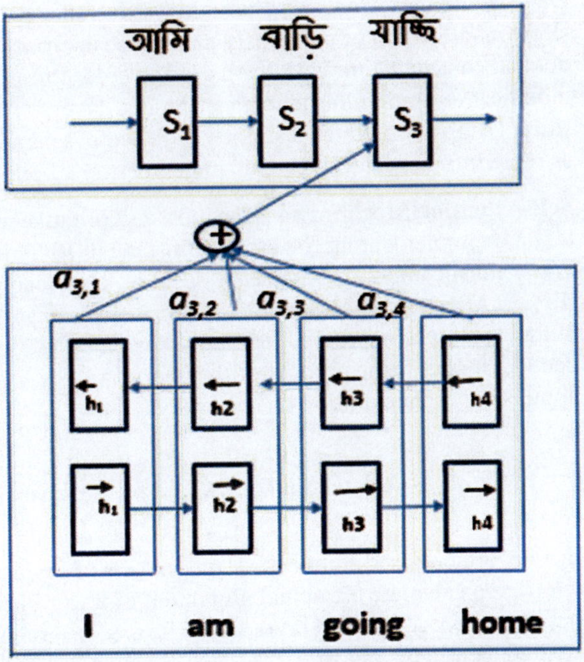


**Fig. 3** Global attention (MLP)

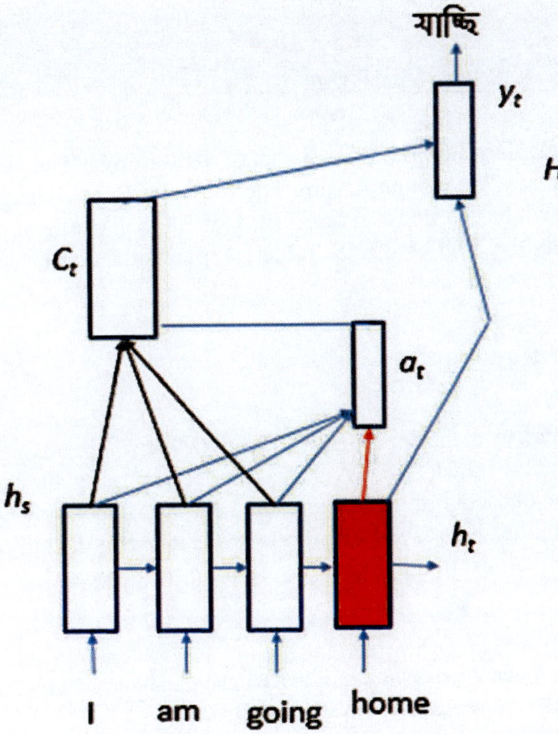


**Fig. 4** Global attention (dot)

**Dot Product Attention Mechanism** was introduced by Luong et al. [8] which was a slight modification of previously mentioned approach. In dot product attention, both dot and concatenation have been used in order to calculate attention value. Another modification is like introducing of hidden layer for both the encoder and decoder sides. Using dot product, a slightly high score is earned as opposed to concatenation as more information can be exchanged between both encoder and decoder sides.

**Self-Attention Mechanism** is the most authoritative way to do neural machine translation by implementing transformer instead of RNN which plays a very crucial role in the enhancement of efficiency. A set of key-value pairs $(K_y, V_l)$ of both dimensions of $N$ has been given as the transformer's input to the encoder part. The decoder output shall be compressed into a dimension of M vector known as query($Q_r$). The decoder output is the product of the query and key-value mapping. The transformer applies user concentration on scalar dots.

$$\text{Attention}(Q_r, K_y, V_l) = \text{softmax}(\frac{Q_r K_y^T}{\sqrt{N}})V_l \tag{7}$$

In transformer-based architecture, the concept of positional encoding was introduced. In order to calculate the actual alignment of a word in the target sentence, only one step is needed. Alignment tasks are performed in more prominent way than any other method mentioned above.

## 4 Corpus Statistics

Bengali or Bangla is the dialect of West Bengal, Bangladesh and a few parts of India such as Tripura and Assam. The source of Bengali language is Indo-Aryan language which has come from Sanskrit. Our English–Bengali parallel data set was obtained from the WAT[1] website. Table 1 explains the corpus statistics.

## 5 Experiment

To carry out the experiments, OpenNMT-py [7],[2] an open-source tool for performing machine translation task, based on deep learning and popular for simplification, has been chosen. For this project, Ubuntu 16.04 operating system is used. Google Colab Notebook has played an important role in executing our code. Data prepossessing jobs like tokenization and true casing have been applied to obtain more accurate results.

[1] http://lotus.kuee.kyoto-u.ac.jp/WAT/indic-multilingual/index.html.
[2] https://opennmt.net/OpenNMT-py/main.html.

**Table 1** Corpus statistics

| Type of data set | No. of sentences | No. of words in English data set | No. of words in Bengali data set |
|---|---|---|---|
| Train data set | 337,428 | 2,177,584 | 1,932,213 |
| Test data set | 1000 | 14,214 | 12,859 |
| Validation data set | 500 | 7762 | 6897 |

To obtain this, Moses script[3] and IndicNLP [4] are used for the prepossessing job. Dropout rate 0.3 with learning rate 0.01 has been applied for vanilla LSTM model, whereas 0.1 dropout rate has been chosen for attention mechanisms. The model has been trained after developing vocabulary followed by translation from source to target. This machine-generated text has been compared with the reference translation of test data by both automatic and human evaluation methods.

## 6 Results and Analysis

Without tokenization and true casing, the translated corpus has shown poor outcomes. Doing prepossessing with Moses scripts, highest BLEU score has been noticed. However, using IndicNLP has more adequacy and fluency than any other translated outputs as Moses script is not suitable for Indian language as Indian language has structural variation.

### *6.1 Without Text Normalization*

For short sentences, the LSTM model works well. The based model LSTM behaves badly when sentence length increases. By applying self-attention mechanism, 6.73 BLEU score has been noticed which is slightly higher than any other mechanism proposed here. As there is no idea of RNN in self-attention mechanism, the concept of positional embedding has been introduced. From the table, it can be clearly viewed that the translation is good for a small sentence in case of the baseline model, but when the length increases, the result has suffered. In terms of adequacy, the baseline model performs good but the fluency is not good because there is some grammatical mistake. From Table 2, it can be observed that MLP attention is not good because all the translated words at the output have not been translated as per the input sentence. Dot product attention and self-attention are also good in terms of both fluency and adequacy.

---

[3] http://www.statmt.org/moses/.

[4] https://indicnlp.org/.

**Table 2** Sample input and output without text normalization

| English Input | He will be at your funeral parlor in one hour. | Go back the way you came |
|---|---|---|
| Bengali Reference | তিনি আপনার জন্য আপনার ফিউনেরল পার্লারে পেক্ষা করবেন | এই, যেম্নে আসছিস সেভাবেই যাবি, বুঝছিস |
| Base Line Model | এক ঘন্টার মধ্যে সে তোমার funeral হয়ে যাবে। | যেখান থেকে ফিরে যা |
| Global Attention(MLP) | সে তোমার কাছে এক ঘন্টা পরেই থাকবোই | তুমি ফিরে যাও |
| Global Attention(dot) | তিনি আপনার funeral এক ঘন্টার মধ্যেই এখানে উপস্থিত হবেনই | তুমি ফিরে যাও |
| Self Attention | এক ঘন্টার মধ্যেই সে তোমার funeral কাছে আসবে। | যাও ফিরে যাও |

## 6.2 *Result Using Test Normalization*

**Using Moses Scripts** When Moses script has been applied for text normalization such as tokenization and true casing, a good increment in the result has been found. For the baseline model, the BLEU score of 14.5 was achieved whereas 15.7 is the best BLEU score obtained after applying the self-attention mechanism. From the sample table, it can be noticed that after applying standardization, the baseline model also performs the same as any other model attention in case of a long sentence. Table 3 stated that both self-attention based and baseline models perform well in terms of adequacy and fluency. But both the global attention mechanism-based modes give poor adequacy and fluency (Table 4).

**Table 3** Sample input and output using Moses scripts for text normalization

| English Input | He will be at your funeral parlor in one hour. | Go back the way you came |
|---|---|---|
| Base Line Model | এক ঘন ্ টার মধ ্ যে সে তোমার অন ্ ত ্ যেষ ্ টিক ্ রিয়ায় থাকবে | আপনি ফিরে আসেন |
| Global Attention(MLP) | তিনি এক ঘন ্ টার মধ ্ যে আপনার অন ্ তর ্ ভুক ্ ত হবে | তুই ফিরে যা |
| Global Attention(dot) | সে এক ঘন ্ টার মধ ্ যে তোমার অন ্ তরের সাথে দেখা করবে | তুই ফিরে যা |
| Self Attention | এক ঘণ ্ টার মধ ্ যে সে তোমার অন ্ ত ্ যেষ ্ টিক ্ রিয ় েন ্ টমেন ্ টে থাকবে । | তুই ফিরে যা |

**Table 4** BLEU score

| Attention Mechanisms | Without Normalization | Normalization Using Moses Script | Normalization Using IndicNLP |
|---|---|---|---|
| Base Line Model | 4.58 | 14.5 | 10.3 |
| Global Attention(MLP) | 5.75 | 14.8 | 10.4 |
| Global Attention(dot) | 5.91 | 15.5 | 10.4 |
| Self Attention | 6.73 | 15.7 | 10.56 |

Table 5 Sample input and output using IndicNLP for text normalization

| English Input | He will be at your funeral parlor in one hour. | Go back the way you came |
|---|---|---|
| Base Line Model | সে তোমাকে এক ঘন্টার মধ্যে পৌঁছে দিবে | আপনি ফিরে আসা উপায় ফিরে যান |
| Global Attention(MLP) | সে তোমাকে এক ঘন্টার মধ্যে পৌঁছে দিবে। | আপনি উপায় ফিরে আসেন |
| Global Attention(dot) | আসে তোমাকে এক ঘন্টার মধ্যে পৌঁছে দিবে। | আপনি উপায় ফিরে আসেন |
| Self Attention | এক ঘন্টা পরেই সে তোমার অন্ত্যেষ্টিক্রিয়ায় থাকবে। | আপনি উপায় ফিরে আসেন |

**Using IndicNLP** Slightly low BLEU score has been gained using IndicNLP than Moses scripts for doing prepossessing. In this case, only self-attention is good enough in respect of adequacy and fluency. But for any other techniques, the adequacy and fluency are very poor as the meaning of the input sentence has been changed (Table 5).

## 7 Conclusion

In this study, different kinds of attention mechanisms are applied to the data sets to compare their performance. Variety of attention and LSTM framework has been employed to carry out this research. According to our finding, the self-attention mechanism using Moses script achieve the highest BLEU score. IndicNLP has been shown better adequacy and fluency as compared to the other. Further, this study for enhancing current English-to-Bengali neural machine translation research using the role of attention mechanism will be useful in the other Indian languages contexts.

## References


1. Bahdanau D, Cho K, Bengio Y (2015) Neural machine translation by jointly learning to align and translate. In: 3rd international conference on learning representations, ICLR 2015, San Diego, CA, USA, May 7–9, 2015, Conference Track Proceedings
2. Bengio Y, Ducharme R, Vincent P, Jauvin C (2003) A neural probabilistic language model. J Mach Learn Res 3:1137–1155
3. Brown PF, Cocke J, Della Pietra SA, Della Pietra VJ, Jelinek F, Lafferty J, Mercer RL, Roossin PS (1990) A statistical approach to machine translation. Comput Linguist 16(2):79–85
4. Cho K, Van Merriënboer B, Bahdanau D, Bengio Y (2014) On the properties of neural machine translation: encoder-decoder approaches. arXiv preprint arXiv:1409.1259
5. Hasan MA, Alam F, Chowdhury SA, Khan N (2019) Neural vs statistical machine translation: Revisiting the bangla-english language pair. In: International conference on Bangla speech and language processing (ICBSLP), vol 27, p 28
6. Hochreiter S, Schmidhuber J (1997) Long short-term memory. Neural Comput 9(8):1735–1780
7. Klein G, Kim Y, Deng Y, Senellart J, Rush AM (2017) Opennmt: open-source toolkit for neural machine translation. arXiv preprint arXiv:1701.02810
8. Luong MT, Pham H, Manning CD (2015) Effective approaches to attention-based neural machine translation. arXiv preprint arXiv:1508.04025
9. Mahata SK, Mandal S, Das D, Bandyopadhyay S (2018) Smt vs nmt: a comparison over Hindi & bengali simple sentences. arXiv preprint arXiv:1812.04898

10. Ojha AK, Chowdhury KD, Liu CH, Saxena K (2018) The rgnlp machine translation systems for wat 2018. arXiv preprint arXiv:1812.00798
11. Sutskever I, Vinyals O, Le QV (2014) Sequence to sequence learning with neural networks. In: Advances in neural information processing systems, pp 3104–3112
12. Vaswani A, Shazeer N, Parmar N, Uszkoreit J, Jones L, Gomez AN, Kaiser Ł, Polosukhin I (2017) Attention is all you need. In: Advances in neural information processing systems, pp 5998–6008
13. Weaver W (1949) The mathematics of communication. Sci Am 181(1):11–15

# Generation and Evaluation of Hindi Image Captions of Visual Genome

**Alok Singh, Loitongbam Sanayai Meetei, Thoudam Doren Singh, and Sivaji Bandyopadhyay**

**Abstract** The automatic image caption generation with proper fluency and expressiveness is an emerging area of research. A lot of research has been done on image caption generation for English, but very few work has been done in the area of generating and evaluating captions in Hindi. In this paper, the problem of generation and evaluation of captions in Hindi is addressed by using a framework based on convolutional neural network (CNN) and long short-term memory (LSTM). This model maximizes the likelihood of the target caption for an input image. The framework is experimented over Hindi Visual Genome dataset. Human evaluation and pre-defined automatic evaluation metrics are used for the evaluation of generated output. The experimental results of the framework manifest that the model is generating reasonably impressive Hindi captions.

**Keywords** Image captioning · CNN · LSTM · Hindi · Fluency · Adequacy

## 1 Introduction

The generation of natural language description for an image is grasping the interest of many researchers due to an emerging application-based area. It is an interdisciplinary problem of computer vision (CV) and natural language processing (NLP). The effective implementation of image caption generation (ICG) can be used as a

A. Singh (✉) · L. S. Meetei · T. D. Singh · S. Bandyopadhyay
Department of Computer Science and Engineering and Center for Natural Language Processing (CNLP), National Institute of Technology Silchar, Silchar, India
e-mail: alok.rawat478@gmail.com

L. S. Meetei
e-mail: loisanayai@gmail.com

T. D. Singh
e-mail: thoudam.doren@gmail.com

S. Bandyopadhyay
e-mail: sivaji.cse.ju@gmail.com



A. Maji et al. (eds.), *Proceedings of the International Conference on Computing and Communication Systems*, Lecture Notes in Networks and Systems 170,
https://doi.org/10.1007/978-981-33-4084-8_7

basic stage for several application-based tasks such as searching and retrieval of an image from a database based on the text and multimodal machine translation. This is relevant to help the people who are visually impaired since we are addressing the problem of generating captions in Hindi, so it would be more beneficial for the visually impaired people whose native language is Hindi. Capturing, analyzing and generating some inference from the analysis in the form of linguistics are the core steps of every ICG framework. Just recognizing objects, their attributes and their relationships in an image are not sufficient for generating a caption of good quality. The system should be capable of generating some logical inferences from them.

The ICG approaches used in the past are mainly categorized into two classes, viz., template-based matching approach and retrieval-based approach [5]. In the template-based approach, the captions are generated by extracting the objects, actions and other attributes in an image and fill them into a pre-defined template. Due to the rigid sentence template, the generated captions lack fluency and expressiveness most of the time which is one of the main drawbacks of template-based approach, whereas, in retrieval-based approach, a large database of images is used for extracting the visually similar image and generating a caption for a query image by using the caption of the retrieved image. This approach slightly helps in generating a flexible and relevant caption for an input image. Some of the deep learning (DL)-based approaches are proposed for caption generation with the recent success of DL in the field of CV. From the study, it is found that the caption generated using DL networks is more fluent and expressive than the traditional caption generation approaches. Some end-to-end frameworks are also proposed for ICG in which image is encoded using CNN and captions are encoded using some encoding scheme followed by mapping them into a similar space and fed to an RNN for training a model.

Furthermore, the evaluation of the generated caption is also important to the same extent apart from generating a caption for an image. Unlike other problems where the generated output is validated by comparing it with the ground truth, it is not easy in the image captioning task because a single image can have multiple inferences. So, along with the automatic evaluation metrics such as Bilingual Evaluation Understudy (BLEU) [10] and Metric for Evaluation of Translation with Explicit ORdering (METEOR) [1], human evaluation of generated output is necessary.

In this paper, we adopt a framework proposed by Vinyals et al. [15] for generating Hindi captions followed by human evaluation by native Hindi speakers. The contribution of the proposed paper is:

1. Generating captions for an image in Hindi which is a challenging task due to the unavailability of images with their corresponding reference in Hindi; in this paper, we address the problem of Hindi image captioning.
2. For ensuring the correctness of the generated output, we carry out human evaluation on the generated output by using the criterion given in [2] by Linguistic Data Consortium (LDC) used in various National Institute of Standards and Technology (NIST) tasks.

## 2 Related Work

In this section, a brief literature review on the previous work of ICG is done. Till now, various approaches have been proposed for an automatic image captioning, but the success of deep CNNs-based frameworks in image classification tasks attracts the attention of researchers for using CNN for extracting the rich representation of the input image. In [15], CNN is used for encoding the input image into a visual feature vector, and then, that feature vector is passed to a decoder for generating a caption. Kiros et al. [8] proposed the first neural network-based approach for caption generation by employing a multimodal log bi-linear model. Xu et al. [17] used an attention-based mechanism for caption generation. In this approach, firstly, CNN is used as an encoder for generating visual feature vector of whole and sub-regions of an image, and, then it is passed to a decoder for a caption generation. This work is extended by Yang et al. [18] and proposed a review module for improving the attention mechanism. There are various approaches proposed for caption generation [4] using pre-trained object detectors and generate caption using template matching. Most of the time, the result of template-based matching approaches is not fluent and expressive [5]. Jia et al. [7] proposed an extension of LSTM named gLSTM (guided LSTM). Ren et al. [12] proposed a reinforcement learning-based method which employs two networks that together predict the upcoming word at every time step. In unidirectional LSTM-based caption generation techniques, the prediction of the upcoming text is based on the previous textual context and visual context. The caption generated by unidirectional LSTM is not contextually well [6]. To address the drawback of unidirectional LSTM, [16] proposed an approach based on deep bidirectional LSTM for image captioning. In this paper, we use CNN and RNN with a beam search for caption generation over the Hindi Visual Genome dataset. A related work on caption generation for the Hindi is reported by Meetei et al. [9].

## 3 Image Caption Generation Using CNN and LSTM with Beam Search

In this paper, a CNN-LSTM-based framework with beam search proposed in [15] is tested over the Hindi Visual Genome dataset. The detail of the dataset can be found in Sect. 4.1. In this framework, the probability of generated output is directly maximized using Eq. (1)

$$P^* = \underset{w}{\operatorname{argmax}} \sum_{I,S} \log p(S|I; w) \tag{1}$$

where $I$ is an image, $S$ is a reference caption, and $w$ is a model parameter.

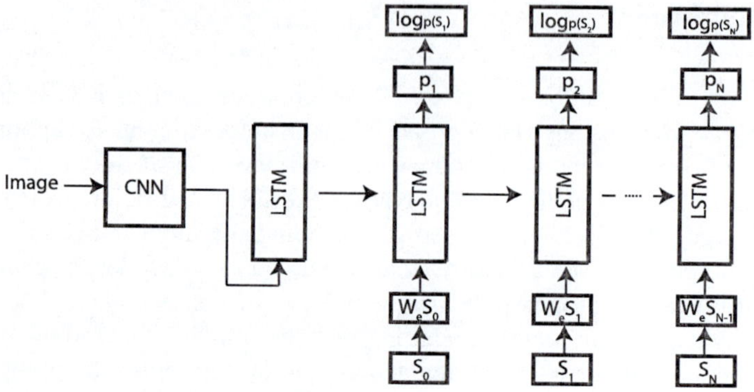


**Fig. 1** Architecture of CNN-LSTM with word embedding

CNN outperforms other classical methods for various image tasks [15]. So, a CNN is used in this paper for encoding the visual feature vector, and LSTM is used for caption generation. The architecture of the framework used in this paper is shown in Fig. 1. After encoding the visual features using CNN, the decoder LSTM is trained in such a way that it predicts the next word in the caption after seeing the image and previously generated output words. During training, LSTM is framed in such a way that the parameters shared by LSTMs are same and the output word generated at time step $t-1$ is passed to LSTM at instance $t$ as shown in Fig. 1. Further, if $I$ and $S = (S_0, \ldots, S_N)$ correspond to an input image and a caption describing the image, respectively, then Eq. (2) shows the unrolled procedure of the model.

$$\begin{aligned} m_{-1} &= \mathrm{CNN}(I) \\ y_t &= W_e S_t, \quad t \in \{0 \ldots N-1\} \\ p_{t+1} &= \mathrm{LSTM}(x_t, m_{-1}), \quad t \in \{0 \ldots N-1\} \end{aligned} \tag{2}$$

where $S_t$ is the dense embedding of words in input caption. Here, $S_0$ and $S_N$ are the special start and stop markers, respectively. The image $I$ is fed only once at $t = -1$. For the caption generation, beam search is used in which the best $k$ caption up to time step $t$ is considered as candidates for generating caption of size $t+1$ and keeping only the best $k$ captions.

## 4 Experimental Results and Discussion

This section contains a detail about the dataset which is used for Hindi caption generation. Further, a discussion about the evaluation of CNN- and LSTM-based model with beam search is performed.

(a) (b)

**Fig. 2** **a** Is a sample image and **b** is the cropped image from (**a**) with caption यातायात रोशनी पीले हैं। **(Yatayat roshni pili hain.)** *The traffic lights are yellow*

## 4.1 Dataset

In this paper, the performance of CNN-LSTM with beam search framework is evaluated over Hindi Visual Genome dataset[1] [11]. In this dataset, each image is associated with its caption in both Hind and English. The dataset consists of $29K$ images for training the model and reserved $1K$ and $1.6K$ images for validation and test, respectively. Instead of considering the overall description of an image, a rectangular portion is cropped out and the caption is generated for the cropped part. The details of the image segment defined in the caption can be seen in Fig. 2.

## 4.2 Evaluation Metrics

An automatic measure that can copy human judgments capability for evaluating the generated captions is the most urgent requirement of the image captioning field [3]. But, some automatic metrics are proposed for evaluating the generated caption are BLEU, METEOR and CIDEr [14]. BLEU evaluates the score for the output based on the part of n-grams (up to four) that are matched. For the evaluation of our model, we use BLEU overall, unigram matching score (BLEU-1) and soft cosine similarity [13] between referenced captions and generated captions as shown in Table 1. Table 2 shows output generated by the framework.

[1] https://ufal.mff.cuni.cz/Hindi-visual-genome.

**Table 1** Performance of the framework on Hindi caption generation

| Methods | Score |
|---|---|
| BLEU (overall) | 2.35 |
| BLEU-1 | 0.18 |
| Cosine | 0.19 |

**Table 2** Sample Hindi output and reference captions

| Input Image | Reference Caption | Generated Caption |
|---|---|---|
|  | टेनिस रैकेट पकडे हुए सफेद कपडे में आदमी<br>**(Tennis Recket pakde hue safed kapde mai aadmi)**<br>*Man in white holding tennis racket* | यह एक टेनिस खिलाड़ी है<br>**(Yeh ek Tennis khiladi hai)**<br>*This is a tennis player* |
|  | काला और सफेद चिह्न<br>**(Kala or safed chinh )**<br>*Black and white square sign* | यह एक चिह्न है<br>**( Yeh ek chinh hai)**<br>*This is a mark* |
|  | एक व्यक्ति जो एक स्केटबोर्ड पर निष्पादित करता है<br>**(Ek vayakti jo ek skate-board par nispadit karta hai)**<br>*a man performing a stunt on a skateboard* | यह एक व्यक्ति है<br>**(Yeh ek vayakti hai )**<br>*This is a person* |

### *4.3 Model Performance*

In this section, the overall performance of the CNN-LSTM model with beam search is discussed as shown in Table 1 using automatic evaluation metrics. The human evaluation is performed for evaluating the adequacy and fluency of the generated captions discussed in Sect. 4.3. The results shown in Table 1 are not so high which is due to some unavoidable challenges such as lack of sufficient training data in Hindi and lack of relevant evaluation parameter for evaluating Hindi captions.

**Training Details** The task of image captioning is slightly harder than the task of image classification and object recognition. It required a large sized dataset that

**Table 3** Fluency and adequacy score evaluation criterion

| Score | Fluency evaluation criterion | Adequacy evaluation criterion |
|---|---|---|
| 1 | Incomprehensible | None of the meaning is delivered |
| 2 | Inarticulate Hindi and ungrammatical overall | Little meaning is delivered |
| 3 | Non-native Hindi, with possibly a few minor grammatical errors | Much of the meaning is delivered |
| 4 | Fluent Hindi without any grammatical error | Every meaning is delivered |

**Table 4** Validation of model using human evaluation metrics on some example

| Reference Caption | Generated Caption | Adequacy Score | Fluency Score |
|---|---|---|---|
| स्त्री के चेहरे पर चश्मा | मनुष्य के चेहरे पर चश्मा लगाना | 4 | 2 |
| एक इमारत में घड़ी | एक घड़ी का हिस्सा | 2 | 4 |
| एक तट पर हाथियों का झुंड | दो हाथी | 1 | 2 |
| एक घोड़े पर एक महिला | यह एक घोड़ा है | 2 | 4 |
| Average Score | | 2.25 | 3.00 |

will contain a sufficient number of images for training a model effectively because it is a data-driven approach. In this paper, we have trained the CNN and LSTM framework with beam search [15] over the Hindi Visual Genome dataset and analyzed its performance. The visual features of dimension 4096 are extracted from pre-final layer of VGG16. The dimension of embedding layer is set to 300.

**Human Evaluation** BLEU is not a perfect metric for evaluation [15] because it is unable to capture the difference between Neural Image Caption (NIC) and caption generated by a human. For more full-proof evaluation of the model, we use some rating for both fluency and adequacy of the generated caption as shown in Table 3. For convenience, we randomly select 1000 images from the dataset and calculated fluency and adequacy scores according to Table 3. For the selected 1000 samples, the average adequacy and fluency scores are 1.5 and 2.9, respectively. Table 4 shows some examples of human evaluation in which both adequacy and fluency score is given.

## 5 Conclusion

The captions generated from an image may not be a well-formed sentence but also an ambiguous one. The overall issue of evaluation of the automatically generated caption using deep learning techniques can be attempted twofold, viz automatic approach and subjective approach. We observed that the automatic scoring lacks capturing the relevant n-gram orders of the model. So, the alternative approach of human evaluation techniques of adequacy and fluency is devised. We found that the overall adequacy is lesser as compared to the given reference. However, the fluency of the generated captions is found to be better albeit the lack of adequacy.

**Acknowledgements** This work is supported by Scheme for Promotion of Academic and Research Collaboration (SPARC) Project Code: P995 of No: SPARC/2018–2019/119/SL(IN) under MHRD, Govt of India.

## References

1. Banerjee S, Lavie A (2005) Meteor: an automatic metric for mt evaluation with improved correlation with human judgments. In: Proceedings of the ACL workshop on intrinsic and extrinsic evaluation measures for machine translation and/or summarization, pp 65–72
2. Denkowski M, Lavie A (2010) Choosing the right evaluation for machine translation: an examination of annotator and automatic metric performance on human judgment tasks
3. Elliott D, Keller F (2014) Comparing automatic evaluation measures for image description. In: ACL
4. Farhadi A, Hejrati M, Sadeghi MA, Young P, Rashtchian C, Hockenmaier J, Forsyth D (2010) Every picture tells a story: generating sentences from images. In: European conference on computer vision. Springer, pp 15–29
5. He X, Deng L (2017) Deep learning for image-to-text generation: a technical overview. IEEE Sig Process Mag 34(6):109–116
6. Hossain M, Sohel F, Shiratuddin MF, Laga H (2019) A comprehensive survey of deep learning for image captioning. ACM Comput Surv (CSUR) 51(6):118
7. Jia X, Gavves E, Fernando B, Tuytelaars T (2015) Guiding the long-short term memory model for image caption generation. In: The IEEE international conference on computer vision (ICCV)
8. Kiros R, Salakhutdinov R, Zemel R (2014) Multimodal neural language models. In: Proceedings of the 31st international conference on international conference on machine learning, volume 32, ICML'14, pp II–595–II–603. JMLR.org
9. Meetei LS, Singh TD, Bandyopadhyay S (2019) Wat2019: English-hindi translation on hindi visual genome dataset. In: Proceedings of the 6th workshop on Asian translation, pp 181–188
10. Papineni K, Roukos S, Ward T, Zhu WJ (2002) Bleu: a method for automatic evaluation of machine translation. In: Proceedings of the 40th annual meeting on association for computational linguistics, ACL '02. Association for Computational Linguistics, Stroudsburg, pp 311–318
11. Parida S, Bojar O, Dash SR (2019) Hindi visual genome: a dataset for multimodal English-to-Hindi machine translation. Computación y Sistemas. In print. Presented at CICLing 2019, La Rochelle, France
12. Ren Z, Wang X, Zhang N, Lv X, Li LJ Deep reinforcement learning-based image captioning with embedding reward. In: Proceedings of the IEEE conference on computer vision and pattern recognition, pp 290–298
13. Sidorov G, Gelbukh A, Gómez-Adorno H, Pinto D (2014) Soft similarity and soft cosine measure: Similarity of features in vector space model. Computación y Sistemas 18(3):491–504
14. Vedantam R, Lawrence Zitnick C, Parikh D (2015) Cider: Consensus-based image description evaluation. In: Proceedings of the IEEE conference on computer vision and pattern recognition, pp 4566–4575
15. Vinyals O, Toshev A, Bengio S, Erhan D (2015) Show and tell: a neural image caption generator. In: Proceedings of the IEEE conference on computer vision and pattern recognition, pp 3156–3164

16. Wang C, Yang H, Meinel C (2018) Image captioning with deep bidirectional lstms and multi-task learning. ACM Trans Multimedia Comput Commun Appl 14(2s):40:1–40:20
17. Xu K, Ba J, Kiros R, Cho K, Courville A, Salakhudinov R, Zemel R, Bengio Y (2015) Show, attend and tell: neural image caption generation with visual attention. In: International conference on machine learning, pp 2048–2057
18. Yang Z, Yuan Y, Wu Y, Cohen WW, Salakhutdinov RR (2016) Review networks for caption generation. In: Advances in neural information processing systems, pp 2361–2369

# Adopting Conditional Random Field (CRF) for Khasi Part-of-Speech Tagging (KPOST)

**Sunita Warjri, Partha Pakray, Saralin Lyngdoh, and Arnab Kumar Maji**

**Abstract** The main purpose of this paper is to develop part-of-speech (PoS) tagging for the Khasi language based on conditional random field (CRF) approaches. Khasi belongs to a Mon–Khmer language of the Austroasiatic language family that is spoken by the native people of the state Meghalaya, Northeastern Part of India. This work is the first instance of a CRF approach in Khasi language for PoS tagging. A grammatically tagged corpus is the main requirement to work on part-of-speech tagging. But, for Khasi language, we do not have a standard corpus for PoS tagging. Therefore, in this paper, Khasi lexicon or corpus is developed for training and testing the system. The designed Khasi corpus consists of around 41,000 tokens. Using the conditional random field (CRF) model for learning the created corpus, an F1-score of 0.92 is the yield for testing data and 0.95 is obtained on training data.

**Keywords** Natural language processing (NLP) · Computational linguistics · Parts of speech (PoS) · PoS tagging · Khasi language · Lexical morphology · Conditional random field

---

S. Warjri (✉) · S. Lyngdoh · A. K. Maji
North-Eastern Hill University, Shillong, India
e-mail: sunitawarjri@gmail.com

S. Lyngdoh
e-mail: saralyngdoh@gmail.com

A. K. Maji
e-mail: arnab.maji@gmail.com

P. Pakray
National Institute of Technology, Silchar, Assam, India
e-mail: parthapakray@gmail.com

 
A. Maji et al. (eds.), *Proceedings of the International Conference on Computing and Communication Systems*, Lecture Notes in Networks and Systems 170,
https://doi.org/10.1007/978-981-33-4084-8_8

## 1 Introduction

Natural language processing (NLP) is a field of artificial intelligence (AI) that involves the interaction of machine and human language. NLP has expanded to many languages with its different applications. Natural languages are employed in the fields of NLP for making the machine or system understand human language or text. Many kinds of research have been taken place in many different fields of NLP, to accomplish the task of making the machine understand human language. Different fields of NLP are like: named-entity recognition (NER), part-of-speech (POS) tagging, information extraction, machine translation, question answering, and speech recognition.

In this present paper, we are using the Khasi language for interacting with the machine. Here, we are discussing on part-of-speech (POS) tagging. This field of POS tagging is one of the initial stages for working toward NLP or computational linguistics (CL). POS tagging deals with the grammatical class categories. This field helps in identifying the grammatical category of particular words with respect to its context. The POS tagging process requires many stages for the system to identify or determine each tag for each word. Different procedures that are followed in POS tagging are: creating tagsets, preprocessing, creating a corpus, taking into consideration the context rules and also bearing in mind the inflections, dependent irregularity of the particular language [12].

Khasi language belongs to Mon–Khmer languages which are a division of Austroasiatic. Khasi is the language spoken in the state of Meghalaya, Northeast India. One of the tribes that live in the state of Meghalaya is the Khasi tribe. The word Khasi itself implies the tribe and also the language. The Khasi language is also found to be spoken in the borderland of India–Bangladesh and also in the border area of Meghalaya–Assam.

There have been very limited research works in computational linguistics concerning Khasi language. Therefore, in this paper, POS tagging is discussed for Khasi language based on the conditional random field (CRF). This work is the first instance of a CRF approach in Khasi language for PoS tagging. A grammatically tagged corpus is the main requirement to work on part-of-speech tagging. But, for Khasi language, we do not have a standard PoS tagged corpus. Therefore, we have created a Khasi lexicon for this research work as discussed below.

The paper is organized as follows: Sect. 2 describes related works on POS tagging; Sect. 3 describes the methods which are followed for Khasi POS tagging using CRF model; Sect. 4 shows the experimental results; Sect. 5 consists of conclusions and some future perspectives of the work.

## 2 The Literature Review

In this section, some of the existing research works in POS tagging based on conditional random field (CRF) are presented. There are several existing research works using CRF on POS tagging for different languages as discussed below.

The paper [1] discussed the Assamese language using transformation-based learning (TBL) model and the CRF model for POS tagging. Using the designed corpus, it is reported to achieve an accuracy of 67.73% for CRF and an accuracy of 87.17% for TBL.

In paper [9], parsing and POS tagging were discussed using the CRF model in the Kannada language (Suraksha, shma, Kumar, 2017). For this work, tagset consisting 19 tags was used to build the corpus. Around 3000 sentences consist of the corpus, out of which for training data consists of 2500 sentences. To test the system, 500 sentences were used from the designed corpus. It is reported to achieve an accuracy of 96.86% by the system using the corpus.

The paper [8] discussed Manipuri language for POS tagging using support vector machine (SVM) and CRF. Annotation of the corpus was done using 26 tags, and 63,200 tokens were annotated. For training, 39,449 tokens were used and 8672 tokens were used for testing. It has been reported that the system yields an accuracy of 74.38% for SVM and an accuracy of 72.04% for CRF.

In 2015, the paper [5], Bhojpuri, Hindi, and Odia language was discussed using CRF++ and SVM. In this work, the corpus consists of 92k tokens, and for annotation, BIS tags were used. For training, 90k were used, and for testing, 2k tokens were used. Results for SVM vary from 88 to 93.70%, whereas for CRF, it varies from 82 to 86.70%.

The paper [6] discussed POS tagging for the Indonesian language using the maximum entropy and CRF method. For this work, two different types of the corpus were used. The corpus one was annotated using 37 and also with 25 POS tags, and similarly, for the second corpus, annotation was done with 37 and 25 POS tags. The comparison of both the corpus has been conducted in this work. The accuracy of the maximum entropy and CRF methods using the Corpus I is: (1) for CRF-10 fold, an accuracy of 69.48% achieved for annotation with 37 tags and accuracy of 82.89% for annotation with 25 tag, and (2) for Max Entropy-10, an accuracy of 77.36% was achieved for annotation with 37 tags and an accuracy of 85.02% was achieved for annotation with 25 tags. Similarly, the accuracy of the maximum entropy and CRF methods using the Corpus II is: (1) for CRF-10 fold, an accuracy of 83.72% achieved for annotation with 37 tags and accuracy of 91.15% for annotation with 25 tag, and (2) for Max Entropy-10, an accuracy of 95.19% was achieved for annotation with 37 tags and an accuracy of 97.57% was achieved for annotation with 25 tags.

In the paper [4], part-of-speech tagging is discussed for French and also for English based on CRF. In this research work, linear-chain CRF is employed along with many configuration features in the system. Different configurations are B: Basic feature templates, C1: Brown word-clustering templates feature, and C2: MKCLS word-clustering feature templates. Therefore, the performance of this proposed tagger is

91.9% for token accuracy, and sentence accuracy is 51.1%. An evaluation of Twitter data with the Penn Treebank (PTB) tagset gives an accuracy of 90.1%. Similarly, an evaluation of Chat data with Penn Treebank (PTB) tagset yields an accuracy of 92.7%.

Apart from these languages We also have some of the works for POS tagging done on the Khasi language. The paper [10] discussed POS for Khasi using hidden HMM and NLTK tools. In this work, corpus of 86,087 tokens was used. Using the NLTK tool, an accuracy of 89.7% was achieved, and using the HMM approach, 95.68% accuracy was achieved. In paper [12], POS tagger using HMM was discussed for the Khasi language. Khasi corpus was annotated using 53 tags, and train data consists of 7500 tokens and test data 312 words. As a result, the system produces 76.70% of accuracy.

## 3 Probabilistic Part-of-Speech Tagging for Khasi

For Pos tagging, many processes involve such as knowing a particular language, designing the tagset, checking the lexical features of the language, collecting raw data, preprocessing, annotation, and using different POS tagging algorithms, and so on. In this section, we are discussing different steps followed for the proposed Khasi part-of-speech tagging. In this paper, the investigation is made using conditional random field (CRF) methods.

### *3.1 Conditional Random Fields*

Using conditional probabilities, one has the facilities to feed the feature of the language based on its context for determining the POS tags. The corpus that has been manually annotated helps to obtain the probability values. For the tag labeling or sequencing problem, the conditional random field (CRF) is a system that makes distinctions with accuracy for building probabilistic models [2, 6, 7]. The CRF is a chain model with an undirected graph having $X$ and $Y$ nodes. The vertex $y_i$ for a random variable with x sequence as a condition of observation and edge is used for a dependency across the random variable. The CRF conditional probability chain structured model is represented as follows:

$$P(Y|X) = \frac{1}{Z_x} \exp[\sum i \sum j \lambda j f_j(y_{i-1}, y_i, X, i)] \tag{1}$$

The Eq. (1) represents the probability of $Y$ as the tag sequence with an observation of $X$ as a word sequence. The condition of each $Y$ that depends on $X$ is determined using features function as $f_j(y_{i-1}, y_i, X, i)$, whereas $Z_x$ is a factor for normalization and the $\lambda j$ is the values of weights that are received by learning the train data.

### 3.2 *Tagset for Khasi Language*

For creating any POS tagged corpus to have tagsets is important. Tagset is a collection of grammatical tags or labels. For this work, a tagset for Khasi language has already been designed in [11]. The tags are used for annotating each collected words accordingly concerning its context. Though many already existing tagsets are available, to follow tagset which have been designed for its particular language is important. We have used a tagset that consists of 53 tags for annotating the Khasi corpus. Different tags that have been used for annotation in this work are shown in Table 1 along with its description. For more details regarding the Khasi tagset, it can be found in the paper [11].

### 3.3 *Khasi Corpus or Lexicon*

Corpus is basically a collection of written texts on a particular language, mainly the complete written works of an appropriate author or a variety of writing on a particular topic. In this research work, building a corpus for Khasi POS tagging is done by collecting the raw Khasi text data from [3]. The collected data consists of mainly political news. The collected data are then pre-processed manually, and also, they are manually tagged using the tagset. The corpus consists of around 41,000 tokens. Table 2 shows some sample Khasi text that is tagged manually.

## 4 Experimental Results Evaluation

This subsection presents a brief discussion on the experimental result achieved that has been conducted using the CRF model on the Khasi lexicon or corpus.

### 4.1 *Results*

The designed Khasi POS tagged corpus is fed to the conditional random field (CRF) model. In this experiment from the corpus, a ration of 80:20 is used for the training and testing data, respectively. As a result, the system yields an F1 score of 0.92 for the test data and 0.95 as an F1 score for the train data. Table 3 shows the average precision, recall, and F1-score yield by the system.

The Table 4 presents the precision, recall, and F-score of each tag.

**Table 1** POS tagset for Khasi Language [11]

| No. | Tag | Description | No. | Tag | Description |
|---|---|---|---|---|---|
| 1 | PPN | Proper nouns | 28 | 2PG | 2nd person singular/plural common gender |
| 2 | CLN | Collective nouns | 29 | 2PF | 2nd person singular/plural Feminine gender |
| 3 | CMN | Common nouns | 30 | 2PM | 2nd person singular/plural Masculine gender |
| 4 | MTN | Material nouns | 31 | 3PSF | 3rd person singular Feminine gender |
| 5 | ABN | Abstract nouns | 32 | 3PSM | 3rd person singular Masculine gender |
| 6 | RFP | Reflexive pronoun | 33 | 3PPG | 3rd person plural common Gender |
| 7 | EM | Emphatic pronoun | 34 | 3PSG | 3rd person singular common Gender |
| 8 | RLP | Relative pronouns | 35 | VPT | Verb, present tense |
| 9 | INP | Interrogative pronouns | 36 | VPP | Verb, present progressive participle |
| 10 | DMP | Demonstrative pronouns | 37 | VST | Verb, past tense |
| 11 | POP | Possessive pronoun | 38 | VSP | Verb, past perfective participle |
| 12 | CAV | Causative verb | 39 | VFT | Verb, future tense |
| 13 | TRV | Transitive verb | 40 | Mod | Modalities |
| 14 | ITV | Intransitive verb | 41 | Neg | Negation |
| 15 | DTV | Ditransitive verb | 42 | CLF | Classifier |
| 16 | ADJ | Adjective | 43 | COC | Coordinating conjunction |
| 17 | CMA | Comparative adjective marker | 44 | SUC | Subordinating conjunction |
| 18 | SPA | Superlative adjective marker | 45 | CRC | Correlative conjunction |
| 19 | AD | Adverb | 46 | CN | Cardinal number |
| 20 | ADT | Adverb of time | 47 | ON | Ordinal number |
| 21 | ADM | Adverb of manner | 48 | QNT | Quantifiers |
| 22 | ADP | Adverb of place | 49 | CO | Copula |
| 23 | ADF | Adverb of frequency | 50 | PaV | Passive voice |
| 24 | ADD | Adverb of degree | 51 | COM | Complementizer |
| 25 | IN | Preposition | 52 | FR | Foreign words |
| 26 | 1PSG | 1st person singular common gender | 53 | SYM | Symbols |
| 27 | 1PPG | 1st person plural common gender | | | |

**Table 2** Khasi Lexicon: manually tagged dataset for training

| Sentences | Khasi Lexicon |
|---|---|
| 1 | U/3PSM nongïalam/CMN ba/COM dang/VPP samla/CMN jong/POP ka/3PSF All/FR India/PPN Congress/FR Committee/FR ,/SYM u/3PSM Gaurav/PPN Gagoi/PPN ,/SYM u/3PSM la/VST ong/ITV ba/COM u/3PSM Myntri/CMN Rangbah/ADJ ka/3PSF Meghalaya/PPN ruh/COC u/3PSM la/VST bud/ITV ïa/InP ka/3PSF nuksa/CMN jong/POP u/3PSM Myntri/CMN Rangbah/ADJ duh/SPA ka/3PSF ri/CMN ,/SYM u/3PSM Narendra/PPN ,/SYM u/3PSM ba/COM shu/ADJ heh/ADJ tang/ADD ka/3PSF nia/CMN ,/SYM Katba/IN Ka/3PSF kam/CMN pat/AD um/Neg ler/ITV satia/AD ./ |
| 2 | Katba/IN na/IN ka/3PSFliang/CMN u/3PSM Zenith/PPN Sangma/PPN ,/SYM u/3PSM wei/CN na/IN ki/3PPG MLA/FR ka/3PSF Congress/PPN ,/SYM u/3PSM la/VST ong/ITV ba/COM ki/3PPG paidbah/CLN ki/3PPG m/Neg shym/Neg la/VST ai/DTV jingjop/ABN tang/ADD ïa/InP ka/3PSF wei/CN ka/3PSF Seng/CMN ban/VFT synshar/CMN ha/IN jylla/CMN ,/SYM bad/COC wat/SUC ka/3PSF NPP/FR ruh/COC ka/3PSF donkam/TRV ïa/InP ki/3PPG paralok/CMN ban/VFT ïa/InP trei/ITV lang/ADD ./ |
| 3 | Da/EM Ka/3PSF ba/COM ther/TRV tyngeh/ADM ïa/InP u/3PSM Myntri/CMN Rangbah/ADJ ka/3PSF Meghalaya/PPN ,/SYM u/3PSM Gogoi/PPN ,/SYM u/3PSM la/VST ong/ITV ba/COM lah/VSP hynriew/CN bnia/CMN ka/3PSF sorkar/CMN MDA/FR ka/3PSF la/VST don/CO ha/IN ka/3PS bor/CMN ,/SYM hynrei/COC kaei/INP ka/3PSF ba/COM ka/3PSF la/VST leh/TRV ,/SYM ym/Neg don/CO eiei/AD ,/SYM lai/ITV tang/ADD ki/3PPG jingkren/ABN bad/COC ki/3PPG jingkop/ABN ./ |

**Table 3** Average precision, recall, and F-measure of the proposed CRF for Khasi Lexicon

| Sl. no. | Khasi Corpus | Precision | Recall | F-measure |
|---|---|---|---|---|
| 1. | Training data of 32800 tokens<br>Testing data of 8200 words | 0.922 | 0.922 | 0.921 |

## 5 Conclusion and Future Works

In this paper, work for POS tagging on the Khasi language using conditional random field (CRF) has been discussed. The accuracy results of the system are done by training using Khasi corpus data and testing the system. Different scores of testing data are presented. Therefore, for future works, more raw data will be collected to increase the size of the token on the Khasi lexicon or corpus for POS tagging. A thorough investigation of the corpus to account for the problems of ambiguities and orthography will also be conducted. It will help in accomplishing the main aim to have a standard Khasi gold corpus. This work is indeed a resource work building for Khasi language toward the NLP perspective.

**Table 4** Average precision, recall and F-measure of each POS tags

| Sl.no. | Tags | Precision | Recall | F-score |
|---|---|---|---|---|
| 1 | SUC | 0.732 | 0.652 | 0.690 |
| 2 | 3PSF | 0.987 | 0.987 | 0.987 |
| 3 | PPN | 0.881 | 0.899 | 0.890 |
| 4 | POP | 0.932 | 0.946 | 0.939 |
| 5 | SYM | 0.989 | 0.995 | 0.992 |
| 6 | VST | 0.963 | 0.967 | 0.965 |
| 7 | CO | 0.905 | 0.930 | 0.917 |
| 8 | COM | 0.986 | 0.990 | 0.988 |
| 9 | VFT | 0.978 | 0.981 | 0.980 |
| 10 | CMN | 0.873 | 0.898 | 0.885 |
| 11 | AD | 0.743 | 0.694 | 0.718 |
| 12 | EM | 0.985 | 1.000 | 0.992 |
| 13 | CN | 1.000 | 0.882 | 0.938 |
| 14 | CAV | 1.000 | 0.971 | 0.985 |
| 15 | IN | 0.957 | 0.956 | 0.956 |
| 16 | TRV | 0.791 | 0.851 | 0.820 |
| 17 | 3PPG | 0.981 | 0.981 | 0.981 |
| 18 | ABN | 0.958 | 0.995 | 0.976 |
| 19 | ADP | 0.660 | 0.633 | 0.646 |
| 20 | COC | 0.943 | 0.963 | 0.953 |
| 21 | NEG | 0.952 | 0.967 | 0.959 |
| 22 | ADD | 0.721 | 0.766 | 0.742 |
| 23 | ITV | 0.789 | 0.678 | 0.729 |
| 24 | DMP | 0.855 | 0.933 | 0.892 |
| 25 | 3PSM | 0.986 | 0.982 | 0.984 |
| 26 | ADT | 0.862 | 0.700 | 0.772 |
| 27 | . | 1.000 | 0.994 | 0.997 |
| 28 | 1PSG | 1.000 | 1.000 | 1.000 |
| 29 | 1PPG | 1.000 | 1.000 | 1.000 |
| 30 | ADJ | 0.789 | 0.667 | 0.723 |
| 31 | MOD | 0.889 | 1.000 | 0.941 |
| 32 | VPP | 0.421 | 0.667 | 0.516 |
| 33 | QNT | 0.744 | 0.762 | 0.753 |
| 34 | FR | 0.933 | 0.932 | 0.932 |
| 35 | PAV | 1.000 | 0.875 | 0.933 |
| 36 | CLN | 0.969 | 0.969 | 0.969 |
| 37 | MTN | 0.500 | 0.250 | 0.333 |
| 38 | CLF | 1.000 | 0.929 | 0.963 |

(continued)

**Table 4** (continued)

| Sl.no. | Tags | Precision | Recall | F-score |
|---|---|---|---|---|
| 39 | DTV | 0.818 | 0.900 | 0.857 |
| 40 | CRC | 0.950 | 0.950 | 0.950 |
| 41 | INP | 0.500 | 0.200 | 0.286 |
| 42 | ON | 1.000 | 1.000 | 1.000 |
| 43 | SPA | 0.714 | 0.909 | 0.800 |
| 44 | VPT | 0.000 | 0.000 | 0.000 |
| 45 | 3PSG | 0.750 | 0.750 | 0.750 |
| 46 | RFP | 1.000 | 0.333 | 0.500 |
| 47 | XX | 0.500 | 1.000 | 0.667 |
| 48 | CMA | 0.875 | 0.700 | 0.778 |
| 49 | 2PG | 1.000 | 1.000 | 1.000 |
| 50 | VSP | 0.000 | 0.000 | 0.000 |
| 51 | RLP | 0.000 | 0.000 | 0.000 |
| 52 | ADF | 0.000 | 0.000 | 0.000 |
| 53 | ADM | 0.000 | 0.000 | 0.000 |

**Acknowledgements** Authors would like to thanks and acknowledge the Government of India, Ministry of Science and Technology, Department of Science and Technology (DST), KIRAN Division, Technology Bhavan, New Delhi for the financial assistance (Grant: DST/WOS-B/2018/1216/ETD/Sunita(G)) during the study.

## References

1. Barman AK, Sarmah J, Sarma SK (2013) Pos tagging of Assamese language and performance analysis of CRF++ and FNTBL approaches. In: 2013 UKSim 15th international conference on computer modelling and simulation. IEEE, pp 476–479
2. Lafferty J, McCallum A, Pereira FC (2001) Conditional random fields: probabilistic models for segmenting and labeling sequence data
3. Mawphor: Mawphor (2017). https://www.mawphor.com/index.php/, [Online; accessed Nov-2017 to June-2019]
4. Nooralahzadeh F, Brun C, Roux C (2014) Part of speech tagging for French social media data. In: Proceedings of COLING 2014, the 25th international conference on computational linguistics: technical papers, pp 1764–1772
5. Ojha AK, Behera P, Singh S, Jha GN (2015) Training & evaluation of pos taggers in indo-aryan languages: a case of hindi, odia and bhojpuri. In: the proceedings of 7th language & technology conference: human language technologies as a challenge for computer science and linguistics, pp 524–529
6. Pisceldo F, Manurung R, Adriani M (2009) Probabilistic part-of-speech tagging for bahasa indonesia. In: Third international MALINDO workshop, pp 1–6
7. Pvs A, Karthik G (2007) Part-of-speech tagging and chunking using conditional random fields and transformation based learning. Shallow Pars South Asian Lang 21:21–24

8. Singh TD, Ekbal A, Bandyopadhyay S (2008) Manipuri pos tagging using CRF and SVM: a language independent approach. In: proceeding of 6th international conference on natural language processing (ICON-2008), pp 240–245
9. Suraksha N, Reshma K, Kumar KS (2017) Part-of-speech tagging and parsing of Kannada text using conditional random fields (CRFS). In: 2017 international conference on intelligent computing and control (I2C2). IEEE, pp 1–5
10. Tham MJ (2018) Challenges and issues in developing an annotated corpus and hmm pos tagger for Khasi. In: The 15th international conference on natural language processing
11. Warjri S, Pakray P, Lyngdoh S, Kumar Maji A (2018) Khasi language as dominant part-of-speech (pos) ascendant in NLP. Int J Comput Intell IoT 1(1)
12. Warjri S, Pakray P, Lyngdoh S, Maji AK (2019) Identification of pos tag for khasi language based on hidden Markov model pos tagger. Computación y Sistemas 23(3)

# Data Mining Techniques and Their Applications

# An Explainable Artificial Intelligence Model for Detecting COVID-19 with Twitter Text Classification: Turkey Case

**Ahmed J. I. Alaff, Hosam H. A. Mukhairez, and Utku Kose**

**Abstract** In December 2019, the first case of the coronavirus was reported, specifically in Wuhan, in China, and the virus began to spread very quickly until it reached more than 3 million cases around the world. But with the lack of technology and medical equipment and the existence of low health awareness in many countries, there is an intense research to combat with that massive problem. In this context, objective of this paper is to discover the spread of the virus according to the countries, by following what users around the world publish on social networking sites, especially on Twitter. In detail, we proposed an explainable artificial intelligence (XAI)-based text classification model that depends on three main steps to discover approximate numbers of infects, by checking the symptoms that published within Twitter posts. A dataset publishing cases of infected and the accompanying symptoms of more than 112 k cases was considered and a model with Naïve Bayes was trained through a comparative work with eight different classifiers. Naïve Bayes is a transparent machine learning technique so it is easier to use probabilistic relations between inputs and the outputs to explain results. Eventually, that XAI-based Naïve Bayes model reached to the highest accuracy rate with 93.6%.

**Keywords** COVID-19 · Machine learning · Twitter feeds · Text classification · Explainable artificial intelligence

---

A. J. I. Alaff · H. H. A. Mukhairez
Institute of Information Technology, Gazi University, Ankara, Turkey
e-mail: aelaff@gmail.com

H. H. A. Mukhairez
e-mail: h.mukhairez@outlook.com

U. Kose (✉)
Department of Computer Engineering, Suleyman Demirel University, Isparta, Turkey
e-mail: utkukose@sdu.edu.tr



A. Maji et al. (eds.), *Proceedings of the International Conference on Computing and Communication Systems*, Lecture Notes in Networks and Systems 170,
https://doi.org/10.1007/978-981-33-4084-8_9

## 1 Introduction

Social media is one of the global effective new technologies in the twenty-first century, besides it became one of the sources for natural language processing and big data analysis. Thanks to interactive environments by social media platforms, people often indicate what they are thinking instantly or expressing something about their situations particularly. Twitter is one of the strongest platforms for instant feeding and it provides a strong developer search kit used for searching and getting some statistics.

The world passed through a lot of diseases crises in the history of humankind. Critical diseases like Spanish flu [1], SARS [2], MERS [3], and recently COVID-19, which was reported officially in Wuhan city of China in December 2019 [4], have all threatened the humankind massively. Currently, COVID-19 spreads very quickly to cover all countries in all over the world, with infected estimated at 3 million or more, according to the statistics by World Health Organization (WHO) [4].

One of the most important challenges faced within the health sector of especially developing countries is the difficulty of knowing true numbers of infected people. That is actually because of the lack of health awareness among some people in addition to not very well-organized health measurements done by governmental units. Although all of these are disadvantages, it is still possible to see accurate numbers through the world-wide metrics Web site [5], which is updated instantly.

In the context of artificial intelligence research, explainable models are gaining more popularity day by day. In the literature, there are lots of different techniques to ensure explainable aspects in traditional transparent machine learning techniques as well as blackbox techniques such as artificial neural networks, support vector machines, and deep learning solutions like CNN, DBN, and LSTM [6]. That efforts have been combined under the research topic of explainable artificial intelligence (XAI). As transparent traditional machine learning techniques have the advantages of being faster and easier to be interpretable (at least mathematically and logically adjustable) for ensuring objectives of the XAI, it can be a good way to employ such techniques in detecting COVID-19 cases. At this point, Naïve Bayes is a strong technique as based on Bayes probability so it will be a good candidate for such application.

Moving from the explanations so far, this paper focused on the Twitter as the social media environment to detect the outbreak of COVID-19 all around the world, by following a methodology which has three main steps: (1) building the classification model, (2) working on Twitter daily feeds, and (3) determining the results of the COVID-19 cases/outbreak, a simple but effective enough solution was designed. That model gives the specialists the ability to mine tweeters' datasets and give some indicators of COVID-19 cases, according to country, sex, age, etc., depending on the available Twitter Search API [7] and more detailed adjustments in future works. Additionally, a XAI infrastructure was employed in order to give some brief explanations regarding how each output was detected by the developed model. It is also remarkable that this study has focused on Turkey as a case perspective, which will give insight for further studies.

The importance of the research in this paper is to determine places where possible the outbreak of the virus happens, as helping governments, workers of health sectors, and responsible health institutions to control outbreaks and take all necessary precautions, as dictated by the WHO. The model developed in this study can be considered as an assistive component for specialists as it actually does not guarantee accurate results because human expressions may not always reflect the real health situation or wrong words may be used by Twitter users. However, XAI approach may give some clues for further applications.

In terms of research over social media, the associated literature has been an intense effort in terms of similar studies. A lot of researches use social media as a source of information to be followed for knowing diseases outbreaks. As examples, [8] and [9] built a general research that used all social media to follow the diseases outbreaks customization. The authors of the [10] used Twitter as an effective tool to predict crimes before it may be happening.[11], and [12] employed Twitter for diseases breakouts, as predicting influenza epidemics by analyzing Twitter messages. Additionally, [13] used Twitter for improving the tracking of Zika virus outbreaks. The [14] negotiates the content analysis of tweets during 2009, for H1N1 virus [15] talked about the early warning of outbreaks system, especially for swine flu. In detail they used the Twitter as a case study. In [8] and [9], the authors talked about the use of social media to detect disease outbreaks and they did not study a specific disease or even used an effective tool such as Twitter. However, [10] used the Twitter platform to detect and track crimes before they occurred. As compared to these works, that study differs as it focuses on the outbreak of COVID-19 disease. In the studies [11–15], the authors have studied outbreaks of various diseases such as swine flu, regular flu, sway flu and Zika virus by tracking what people post on Twitter, but none of them has discussed about COVID-19 as done in this study.

## 2 Methodology

The solution methodology followed in this study employs three main steps as mentioned in Fig. 1. These main steps are, respectively, building the classification model with XAI support, working on the daily feeds over Twitter, and finally evaluation of the results regarding the COVID-19 outbreak. The next paragraphs explain main details about each of steps.

### *2.1 Building Classification Model*

In order to ensure a classification approach/model, the infected people were considered based on: cough, fever, sore throat, shortness of breath, and headache symptoms.

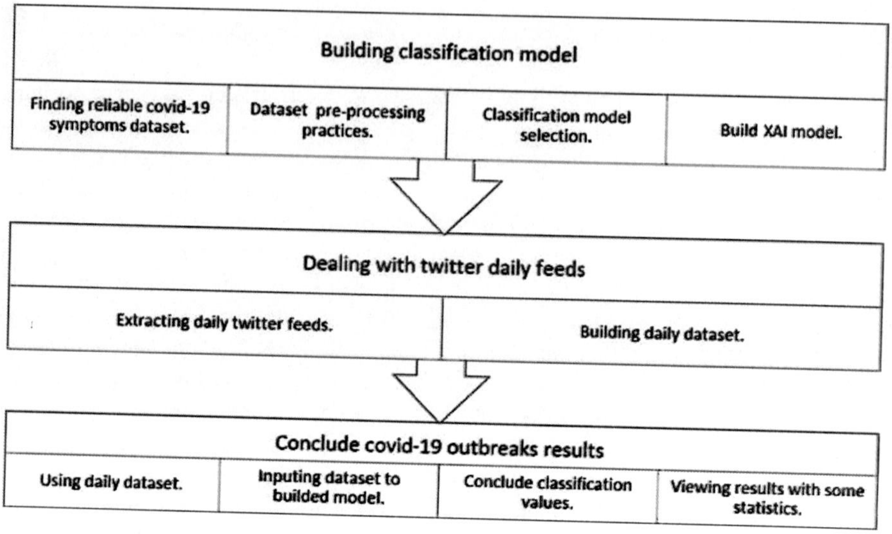


**Fig. 1** Text classification model structure

Here, especially Naïve Bayes was chosen as a remarkable, fast technique, which is also transparent in terms of defining explainable outputs. In the context of the related process, the following substeps were taken:

**Determining reliable COVID-19 symptoms dataset**. After searching for a formal dataset from Health Ministry (of Turkey), a dataset with 112 k records (as shown in Table 1) was obtained. The related dataset columns correspond to the symptoms of: "cough, fever, sore throat, shortness of breath, and headache" with binomial datatype:

**Table 1** COVID-19 symptoms dataset with each of parameters/properties

| Columns | Values | Explanation | Count |
|---|---|---|---|
| Cough | 0 | Negative symptom | 94,264 |
| | 1 | Positive symptom | 17,840 |
| Fever | 0 | Negative symptom | 102,609 |
| | 1 | Positive symptom | 9494 |
| Sore_throat | 0 | Negative symptom | 110,976 |
| | 1 | Positive symptom | 1369 |
| Shortness_of_breath | 0 | Positive symptom | 111,351 |
| | 1 | Positive symptom | 994 |
| Headache | 0 | Negative symptom | 110,401 |
| | 1 | Positive symptom | 1944 |
| Corona_result | Negative | COVID-19 Negative | 101,841 |
| | Positive | COVID-19 Positive | 9158 |
| | Other | Has another disease | 1346 |

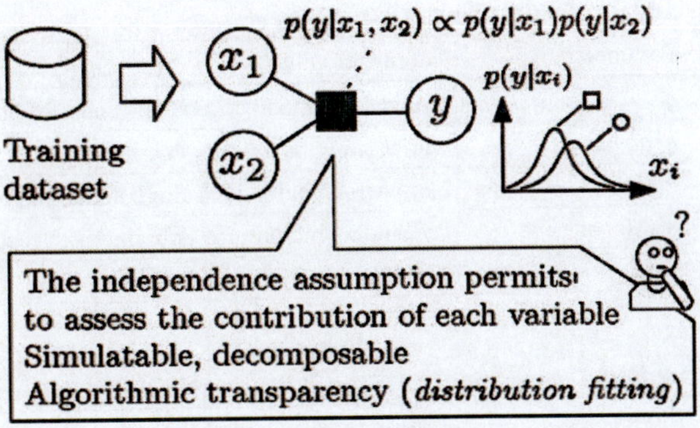


**Fig. 2** Ensuring explainable aspects in Naïve Bayes classification technique [6]

"0" for "negative symptom" and "1" value for "positive symptom." The final/output column was "Corona result" which determine if the diagnosis was "negative" or "positive."

**Dataset Preprocessing**. After getting the dataset, the next step included a preprocessing in order to ensure a data cleansing approach for adjusting/eliminating missing values and getting a proper dataset for classification.

**Selecting the model for classification**. A widely known software environment: RapidMiner [16] was employed as a tool for machine learning/classification process for ensuring a wide training/test application period. As the main classification technique was chosen as Naïve Bayes, a wider evaluation planning was done by considering a total of eight different models for the classification process. As mentioned in Table 3, it was found that the Naïve Bayes [17] was already the best one in terms of accuracy as well as the fastest score and training time.

**Building the explainable model**. After the previous training process and model selection step, the exact classification model was developed by using the Naïve Bayes classifier, which will be also used in the last process for obtaining findings from classification processes based on the symptoms. Additionally, probabilistic parameters inside that classifier were connected with the outputs and a simple enough interface was prepared in order to ensure an explaining interface for the outputs. In this way, inputs (symptoms) were connected to the output, as an explaining approach followed in the literature, as explained briefly in Fig. 2 [6].

## 2.2 *Working on Twitter Feeds*

**Extracting Daily Tweets**. In order to get some statistics like longitude and latitude attributes, Twitter Search API (TSA) [7], which is a powerful API provided by Twitter Developer gate, was used in this study. As general, the TSA fetches the feeds based on some parameters such as geocode, lang, result type and until, with up to

**Table 2** Twitter search API parameters

| Parameter | Parameter value |
|---|---|
| *q* | Cough, fever, sore throat, shortness of breath, and headache |
| *geocode* | Variable by the world's countries |
| *lang* | "en": which belongs to English language |
| *result type* | "recent": which return only the most recent results in the response |
| *until* | Variable day by day date "30-04-2020" |

**Table 3** Comparative evaluation of the classification techniques/models

| Model | Error rate (%) | Total time | Training time (1000 row) | Scoring time (1000 row) |
|---|---|---|---|---|
| Naïve Bayes | 6.4 | 4653.0 | 2.2 | 92.5 |
| Generalized linear model | 9.3 | 18,672.0 | 77.9 | 94.7 |
| Logistic regression | 6.7 | 14,087.0 | 38.5 | 198.6 |
| Fast large margin | 6.7 | 261,967.0 | 10.6 | 229.1 |
| Deep learning | 6.7 | 36,984.0 | 160.1 | 155.8 |
| Decision tree | 6.7 | 5530.0 | 2.5 | 93.2 |
| Random forest | 6.8 | 39,689.0 | 12.6 | 1117.8 |
| Gradient boosted trees | 6.7 | 156,180.0 | 43.4 | 195.0 |

100 results per connection. A day by day search was used by using the parameters mentioned in Table 2. By using the "*q*" attribute, it was aimed to find some keywords from the gathered daily tweet feedings, which exactly include symptoms words like cough, fever, sore throat, shortness of breath and headache (or some related words to every symptom like "shortness of breath" may have other synonymous like "trouble breathing, difficulty breathing and suffocation" words).

**Building daily dataset**. According to the previous step, a small C# desktop application, which takes all the symptoms and fill them in excel CSV sheet as mentioned in Table 1, was developed accordingly. That step has been automatically taken every day, with multiple times.

## 2.3 *Final Results*

After using the XAI-based Naïve Bayes model, which was developed in the first step, the research was directed to use of the daily extracted dataset as a model input to get final results, which can be negative, positive, or other diseases (as diagnosis regarding to COVID-19). Following to that, by following some reserving statistics (like location) from the extracted tweets, a small local database was built in order to store the increases for the infected people, as belonging to their location. Finally, the model/system developed here gives the final results as numbers of infected related to their locations.

## 3 Evaluation

As mentioned before in the first step of the methodology, the RapidMiner tool was used for testing the created dataset, through eight different classifiers including Naïve Bayes. Table 3 briefly gives comparative findings for each of techniques, which are, respectively: Naive Bayes [17], generalized linear model [18], logistic regression [19], fast large margin [20], deep learning [21], decision tree [22], random forest [23], gradient boosted trees [24], and support vector machine [25].

The related classifiers were tested in the same dataset with 30 independent runs. All the techniques were set with default values, as reported optimum resulting ones in the literature. In order to evaluate the findings, three main properties, error rate, scoring, and training time were considered. The best classifier should have the least rate of these properties, in order to be the overcoming classifier among all these solutions.

As seen from Table 3 providing average values for each property, the Naïve Bayes classifier has the least error rate with 6.4%, as also being the fastest scoring, and getting the best training time with, respectively, 92.5 and 2.2 for the 1000-raw of the dataset. It may be possible to update other alternative techniques for better findings but that may cause negative results in terms of performance and also, XAI aspects need more efforts for especially blackbox models. So, this study has given more consideration for the Naïve Bayes technique, which gave desired findings for the organized comparative evaluation.

In order to test the Naïve Bayes system, the RapidMiner simulator was used for feeding the system with different inputs. Figure 3 shows a screenshot from the system as when it was evaluated that the patient has all the symptoms, it concludes that the COVID-19 may be positive with the rate of 69% (In that example the input values are sore throat: 1, headache: 1, fever: 1, cough: 1, and shortness of breath: 1, respectively).

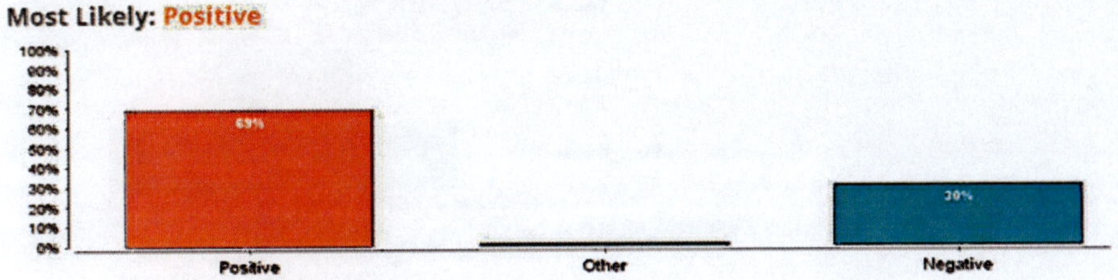


**Fig. 3** Naive Bayes simulator results for a COVID-19 positive case

Regarding that result the XAI interface explains the output by considering the rates of inputs, as provided in Fig. 4. As it is seen, rates are in different colors according to changing levels.

In order to ensure a negative case evaluation for COVID-19, a sample can give given with the input values as: sore throat: 0, headache: 0, fever: 1, cough: 1, and shortness of breath: 0, respectively. In that case, the Naïve Bayes model gives the diagnosis result as the negative case with the rate of 62% (Fig. 5). Figure 6 represents the XAI interface explaining that output, according to the rates of inputs.

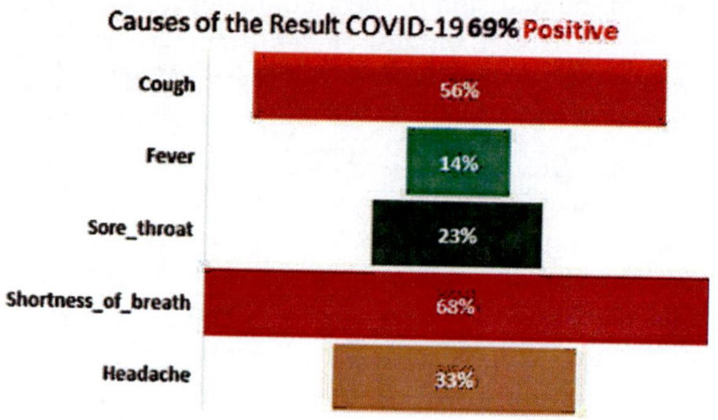


**Fig. 4** XAI explanation for the COVID-19 positive case

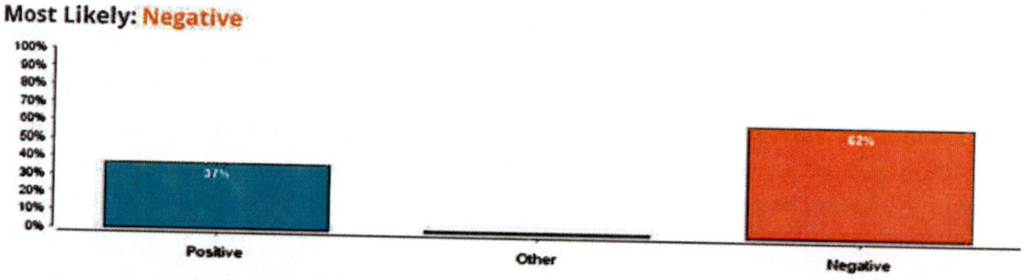


**Fig. 5** Naive Bayes simulator results for a COVID-19 negative case

**Fig. 6** XAI explanation for the COVID-19 negative case

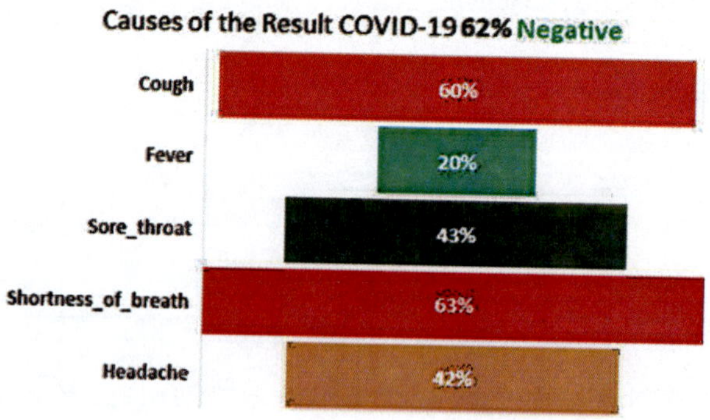

It is important to explain that the XAI results indicate the importance of the inputs with the rates explaining how intense the decision was taken according to a corresponding input. That is done by evaluating the probability metrics that can be read from the context of Naïve Bayes classification model.

## 4 Conclusions and Future Work

In this study, an XAI-based machine learning model for detecting COVID-19 cases was developed. The developed model consists of Naïve Bayes, which is a simple but fast, effective classification model, which uses Bayes probability calculations. The model in this study has been also supported with an XAI interface as Naïve Bayes classifier is one of transparent traditional machine learning models, making explainable aspects real by using probability measures connecting inputs to the output. In order to evaluate COVID-19 cases, an approach for text classification over Twitter environment was planned. In order to ensure a unique and instant solution, tweets in Turkey were considered for creating a reliable dataset. After some processes, the system can be ready for evaluating COVID-19 outbreak.

In order to understand if the developed system is a successful tool for the related purposes, a comparative evaluation work was done by considering eight classifiers including the Naïve Bayes. The finding showed that the Naïve Bayes classifier achieved the least error rate with 6.4%, as also being the fastest scoring, and providing the best training time with, respectively, values of 92.5 and 2.2. According to the XAI interface designed for the Naïve Bayes, some inputs like cough, and shortness of breath are important factors to determine whether the diagnosis is negative or positive (i.e., appropriate to the reported symptoms of the COVID-19). As general, the system has been an effective tool for diagnosis as it hits a massive target and ensures some important information such as location of the tweet with COVID-19 positive.

According to the first findings/results, the developed system seems successful for combating COVID-19. The system can be used by governmental units as well as health institutions. As following the positive findings, the authors have encouraged for further developments. These further works include improving capabilities of the system by running hybrid machine learning techniques like genetic algorithm trained artificial neural networks for evaluating the tweets in a wider scope. Another future work planning has been directed to advanced XAI developments according to any model-based updates in the system. Finally, more COVID-19 case evaluations will be done by considering tweets from different regions of the world.

## References


1. Trilla A, Trilla G, Daer C (2008) The 1918 "Spanish flu" in Spain. Clin Infect Dis 47(5):668–673
2. Fouchier RA, Kuiken T, Schutten M, Van Amerongen G, Van Doornum GJ, Van Den Hoogen BG et al (2003) Koch's postulates fulfilled for SARS virus. Nature 423(6937):240–240
3. Wong G, Liu W, Liu Y, Zhou B, Bi Y, Gao GF (2015) MERS, SARS, and Ebola: the role of super-spreaders in infectious disease. Cell Host Microbe 18(4):398–401
4. World Health Organization (2020) Coronavirus disease 2019 [COVID-19]: Situation report, 70
5. Worldometers Coronavirus Update [Live]: 3,085,120 Cases and 212,546 deaths from COVID-19 virus pandemic—worldometer. Available from: https://www.worldometers.info/coronavirus/
6. Arrieta AB, Díaz-Rodríguez N, Del Ser J, Bennetot A, Tabik S, Barbado A et al (2020) Explainable artificial intelligence (XAI): concepts, taxonomies, opportunities and challenges toward responsible AI. Inf Fusion 58:82–115
7. Twitter Standard Search API. Available from: https://developer.twitter.com/en/docs/tweets/search/api-reference/get-search-tweets
8. Schmidt CW (2012) Trending now: using social media to predict and track disease outbreaks. National Institute of Environmental Health Sciences
9. Seo DW, Shin SY (2017) Methods using social media and search queries to predict infectious disease outbreaks. Healthcare Inf Res 23(4):343–348
10. Gerber MS (2014) Predicting crime using Twitter and kernel density estimation. Decis Support Syst 61:115–125
11. Culotta A (2010) Towards detecting influenza epidemics by analyzing Twitter messages. In: Proceedings of the first workshop on social media analytics, pp 115–122
12. Lee K, Agrawal A, Choudhary A (2013) Real-time digital flu surveillance using twitter data. In: The 2nd workshop on data mining for medicine and healthcare
13. Masri S, Jia J, Li C, Zhou G, Lee MC, Yan G, Wu J (2019) Use of Twitter data to improve Zika virus surveillance in the United States during the 2016 epidemic. BMC Public Health 19(1):761
14. Chew C, Eysenbach G (2010) Pandemics in the age of Twitter: content analysis of Tweets during the 2009 H1N1 outbreak. PLoS ONE 5(11):e14118
15. Kostkova P, De Quincey E, Jawaheer G (2010) The potential of social networks for early warning nad outbreak detection systems: the swine flu Twitter study. Int J Inf Dis 14:e384–e385
16. RapidMiner RapidMinerldata science and machine learning platform. Available from: https://rapidminer.com/
17. Rish I (2001) An empirical study of the naive Bayes classifier. In: IJCAI 2001 workshop on empirical methods in artificial intelligence, vol 3, issue 22, pp 41–46
18. Chou W, Li L (2006) Method and apparatus for text classification using minimum classification error to train generalized linear classifier. Google Patents
19. Greiner R, Su X, Shen B, Zhou W (2005) Structural extension to logistic regression: discriminative parameter learning of belief net classifiers. Mach Learn 59(3):297–322
20. Hu W, Chung FL, Wang S (2012) The maximum vector-angular margin classifier and its fast training on large datasets using a core vector machine. Neural Netw 27:60–73
21. LeCun Y, Ranzato M (2013) Deep learning tutorial. In: Tutorials in international conference on machine learning (ICML'13), pp 1–29
22. Safavian SR, Landgrebe D (1991) A survey of decision tree classifier methodology. IEEE Trans Syst Man, Cybern 21(3):660–674
23. Rodriguez-Galiano VF, Ghimire B, Rogan J, Chica-Olmo M, Rigol-Sanchez JP (2012) An assessment of the effectiveness of a random forest classifier for land-cover classification. ISPRS J Photogram Remote Sens 67:93–104

24. Ganjisaffar Y, Caruana R, Lopes CV (2011) Bagging gradient-boosted trees for high precision, low variance ranking models. In: Proceedings of the 34th international ACM SIGIR conference on research and development in information retrieval, pp 85–94
25. Chen HL, Yang B, Liu J, Liu DY (2011) A support vector machine classifier with rough set-based feature selection for breast cancer diagnosis. Expert Syst Appl 38(7):9014–9022

# Cluster Analysis Using Improved Bio-inspired Fuzzy C-Means Algorithm

**Fisokuhle Mthethwa and Amitabha Nath**

**Abstract** Optimization is a vital constituent of machine learning as it brings about efficiency of the analysis process. An optimizer is used to minimize an objective function, which in turn leverages efficiency of any given model. Many metaheuristic algorithms have been applied on many models for optimization purposes. The mostly favoured metaheuristic algorithms are the population-based ones as they initialize a population of candidate solutions, rather than a single-solution strategy. This paper takes off from the original research made on fuzzy C-means (FCM), where the weighting component ($m$) was set at 2. Setting the weighting component at 2 may not ideally work well for all applications. Then, the point of this examination is to optimize the weighting component $m$ using bio-inspired algorithms that are population-based in nature. The idea is to compare performance of conventional FCM with improved version of FCM by augmenting it with different optimization techniques. Three comparatively new and popular bio-inspired optimization algorithms such as moth-flame Optimizer (MFO), gray wolf optimizer (GWO) and whale optimization algorithm (WOA) are selected to optimize tunable parameters of conventional FCM. For experimentation purpose, five different data sets namely seeds, iris, glass identification, mall customer segmentation and BuddyMove are selected. The experimental results show considerable improvements in the overall performance of conventional FCM as a consequence of parameter optimization. A new approach for optimizing FCM parameters using bio-inspired algorithms is presented.

**Keywords** Fuzzy C-means · MFO · GWO · WOA · Clustering · Optimization

F. Mthethwa (✉)
Computer Science Department, Eswatini Medical Christian University, Mbabane H101, Eswatini
e-mail: fisokuhle90@gmail.com

A. Nath
Information Technology Department, North-Eastern Hill University, Shillong 793022, India
e-mail: amitabha.me@gmail.com

 
A. Maji et al. (eds.), *Proceedings of the International Conference on Computing and Communication Systems*, Lecture Notes in Networks and Systems 170,
https://doi.org/10.1007/978-981-33-4084-8_10

## 1 Introduction

Machine learning paradigms can be exercised to leverage knowledge discovery in many aspects of research such as biology, medicine, marketing and so on. Scalability problems have been addressed in clustering by sampling data to ensure low computational complexity and better performance in general [1]. The procedure for applying the optimization algorithms is to identify the parameters that are possible to optimize in order to reduce the value of the objective function for the identified method. For the case of fuzzy C-means (FCM), a single parameter $m$ referred to as the weighting component of the objective function $J_m$ is the parameter meant to be altered to achieve the best results. Originally, Bezdek (1981) set the weighting component $m = 2$, which is not ideal for all situations [2, 3]. Another drawback of FCM is that, FCM employs a single-solution strategy. The problem with such strategies is that the solutions are formulated around the local search space.

In a paper by Singh et al., authors have highlighted that the drawback with fuzzy clustering techniques such as FCM is the time complexity when compared to hard clustering [4]. Furthermore, FCM has tendency to be stuck on the local minima which may not bring about desirable results [4, 5]. The logic behind FCM is fuzzy, where a data point can be a constituent of one or more clusters and the focus being on reducing the value of the objective function that looks into the partitioning quality. Dissimilar to K-means, FCM is finely built for large databases [4]. It grants finest results for overlapped data sets and relatively better than the K-means algorithm.

There have been attempts to improve FCM using various optimization techniques. Niu and Huang used particle swarm optimization (PSO) algorithm to improve FCM by optimizing its parameters [6]. Their results showed a great improvement of PSO-FCM over the performance of conventional FCM. In another work by Alata et al., conventional FCM is optimized using the genetic algorithm (GA) to optimize FCM parameters where subtractive clustering was employed [2]. Their work also showed a better performance achieved by the optimized FCM. Similarly, there are numerous instances available from various domains where better performance of optimized FCM algorithm has been reported [7–9].

In this experiment, a similar attempt of improving capability of conventional FCM algorithm has been carried out. Essentially, optimal value for $m$ actually determines the effectiveness of FCM. Prevalent metaheuristic-based optimization algorithms such as moth-flame optimizer (MFO), gray wolf optimizer (GWO) and whale optimization algorithm (WOA) are therefore chosen for optimization in this regard. The main idea of this research is to validate the impact of parameter optimization of FCM and compare the performance with conventional FCM with respect to different data sets.

## 2 Models Used

### 2.1 *Fuzzy C-Means Algorithm*

Algorithms that can represent data in detailed precision are of a great necessity in the data realm. As FCM is part of soft clustering, objects are not compelled to belong to one cluster. Here, objects are not forced to belong to a single cluster but any number of clusters through partial membership, which makes FCM a more natural method than hard clustering algorithms. The membership degrees in FCM are between 0 and 1. FCM makes use of the kind of partitioning that is fuzzy where an object can be a part of all or some groups with membership ranks that lie between 0 and 1 [4, 5].

Fuzzy C-means (FCM) is a system for grouping data objects which grants one bit of a data object of information to have a place with at least two clusters [10]. In FCM calculation, most researchers fix weighting exponent ($m$) to a customary estimation of two which probably will not be the fitting for all applications. This research utilizes $J_m$ at the target parameter intended to be reduced.

$$J_m = \sum_{i=1}^{P} \sum_{j=i}^{R} U_{ij}^{m} |Q_i - R_i| \tag{1}$$

where;
$m$ : A genuine number more prominent than 1, originally set to 2 by Bezdek (1981)
$U_{ij}$ : The grade of membership of $Q_i$ in the cluster $j$
$Q_i$ : The $i$th of $x$-dimensional calculated data
$R_i$ : The $x$-dimension focal point of the cluster
$| * |$ : A standard communicating the similitude between any calculated object and the focal point (center).

Fuzzy partitioning is carried out by iteratively optimizing the objective function presented above, with the update of membership $u_{ij}$ and the $R_j$ cluster centers by:

$$U_{ij} = \frac{1}{\sum_{k=1}^{R} \left( \frac{|Q_i - R_j|}{|Q_i - R_k|} \right)^{\frac{2}{m-1}}} \tag{2}$$

$$R = \frac{\sum_{k=1}^{R} U_{ij} . Q_i}{\sum_{k=1}^{R} U_{ij}} \tag{3}$$

The iteration will be halted by the following condition:

$$\max_{i \quad j} = | \, U_{ij}{}^{(Z+1)} - U_{ij}{}^{(Z)} \, | < \gamma \tag{4}$$

where $Z$ signifies the steps of the iteration and $\gamma$ is the termination condition which stands between 0 and 1.

The algorithm is composed of the following steps:

1. Choose cluster focal point (center), arbitrarily
2. reset $U = [u_{i_j}]$ matrix to $U^{(0)}$ and Calculate the $u_{i_j}$.
3. At k-step: calculate the centers vectors $R^{(Z)} = [R_j]$ with $U^{(Z)}$

$$C = \frac{\sum_{k=1}^{R} U_{ij}.X_i}{\sum_{k=1}^{R} U_{ij}} \tag{5}$$

4. Update $U^{(Z)}$, $U^{(Z+1)}$
5. If $\max_i < \gamma$ or the minimum $J$ is conceived, then HALT; otherwise return to second step.

## 2.2 *Optimization Methods Employed*

**Moth-Flame Optimization Algorithm** The moth-flame optimization (MFO) algorithm is an algorithm inspired by nature, and it is a population-based optimization algorithm inspired by the behavior of moths [11]. The fundamental motivation of this optimization agent is the route technique for moths in nature called transverse direction; moths fly during the night by keeping up a fixed verge concerning the moon, an exceptionally viable component for going in a straight queue for prolonged separations [11].

Since the MFO algorithm is a population-based algorithm, it should be noted here that moths and flames are mutually solutions. The difference between them is the method in which they are treated and reorganized in each repetition. The moths are real pursuit specialists that move around the hunt space, though flames are the best position of moths that acquires up until a certain point. Each moth scrutinizes around a flag and updates it in case of discovering a better solution. By this setup, a moth never loses its finest solution [11].

Figure 1 illustrates the process of MFO is finding an optimal solution out of a population in each iteration. In every iteration, the flame number is updated by considering the distance of the moth from the the flame. There can be many numbers of flames, as well the number of moths.

**Grey Wolf Optimization Algorithm** Gray wolf optimization is a swarm intelligent technique that mimics the hunting practices of gray wolves [12]. This algorithm has a few number of parameters merely and not difficult to implement, which makes it greater than prior ones.

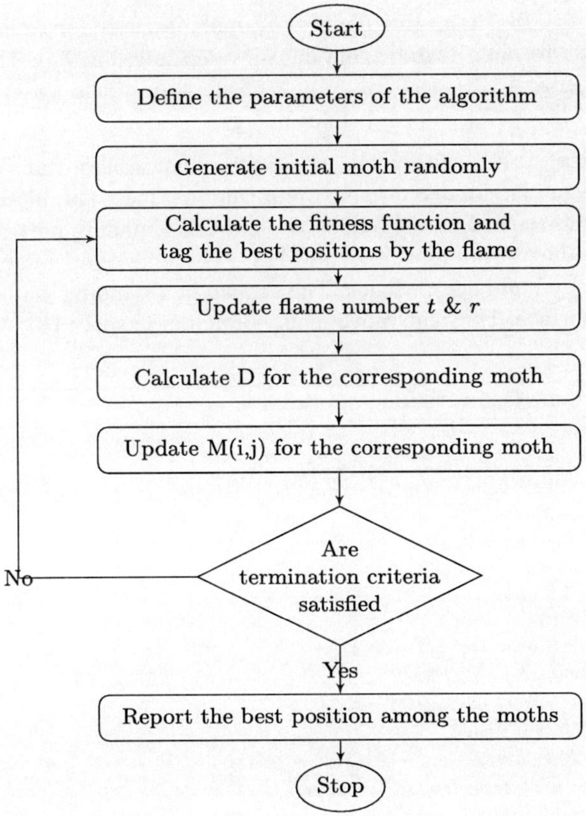


**Fig. 1** MFO flow diagram

---

**Algorithm 1** Gray Wolf Optimization Algorithm

---

**procedure** GWO($problemSize$, $populationSize$, $PgBest$)
    Initialize the population of gray wolves $X_i$ (i=1, 2,.., n)
    Initialize a, A and C ▷ A and C are vectors
    Calculation of the fitness values of the search agents and grading agents.
($X_\alpha$ = the best solution in the search agent, $X_\beta$ = the second best solution in the search agent, and $X_\delta$ = the third best solution in the search agent.)
    t = 0
    **while** t < Maximum number of iterations **do**
        **for** each search agent **do**
            Updating the position of the current search
    Updating a, A and C
    Calculation of the fitness values of all search agents and grading them
    Updating the positions $X_\alpha$, $X_\beta$ and $X_\delta$
    t = t + 1

---

The algorithm 1 illustrates the role of the population that is initialized from unity to n. The best performing search agent has its value stored in $X_\alpha$ . The value of the alpha might change to the best value by each iteration. Each agent has its fitness calculated and classified as alpha, beta or omega.

**Whale Optimization Algorithm** The whale optimization algorithm (WOA) is a new optimization technique for deciphering optimization issues. This algorithm includes three operators to impersonate the quest for prey, surrounding prey and bubble-net rummaging conduct of humpback whales [13]. These three operators are activities that are shown by humpback whales. They move in a winding way to achieve the prey from the surface. They can move in any direction in space [14, 15].

**Algorithm 2** Whale Optimization Algorithm

**procedure** WOA()<br>
    Randomly initialize the *whale* population<br>
    Evaluate the fitness values of *whales* and find out the best search agent $X^*$<br>
    **while** $t < t_{max}$ **do**<br>
        Calculate the values of $a$ +<br>
        **for** each search agent **do**<br>
            **if** $h < 0.5$ **then**<br>
                **if** $\mid S \mid < 1$ **then**<br>
                    $Q(r+1) = q^*(r) - S \cdot T$<br>
                **else if** $\mid S \mid \geq 1$ **then**<br>
                    $Q(t+1) = q_{rand}(t) - S \cdot T''$<br>
            **else if** $h \geq 0.5$ **then**<br>
                $Q(R+1) = T^x \cdot e^{bl} \cdot cos(2\pi l) + Q^*(t)$<br>
        Evaluate the fitness of $Q(t+1)$ and update $Q^*$<br>
    End of **WHO**

## 3 Methodology

The improved FCM differs a little bit from general FCM. In the case of improved FCM, subsequent to the calculation the euclidean distance, the optimal solution for $m$ of $U_{i_j}$ is sought through incorporating optimization mechanism provided by the optimization algorithms in the improved FCM. In case of FCM-MFO, the distance $T_i$ must be calculated and updated by adjusting the parameters $M_i$ and $E_j$ which in turn finds the optimal value for weighting component $m$ of $U_{i_j}$ in an iteration.

In the case of FCM-GWO, there is some major difference in the manner at which parameter $m$ is updated. In this case, the optimal value is found by updating vector coefficients $\overrightarrow{S}$ and $\overrightarrow{R}$ for weighting $m$ of $U_{i_j}$ in an iteration. It is to be noted that the value of the optimal solution must be greater than 1.

In the case of FCM-WOA, coefficients $S$ and $R$ must be updated by updating the spiral $D(t+1)$ (see equation. The change in parameters of the spiral ($S$ and $R$) either

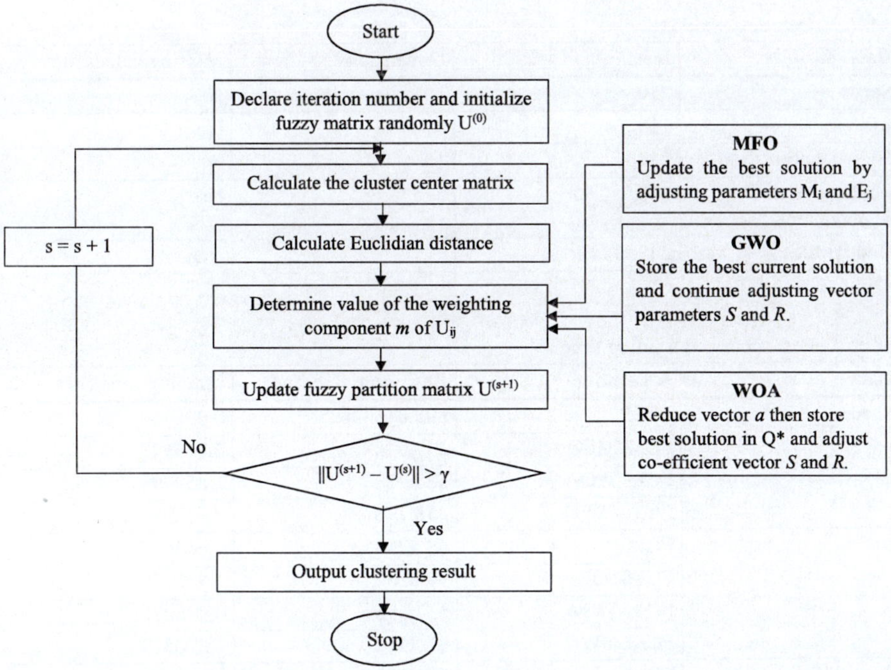


**Fig. 2** Flowchart of improved FCM technique

gives a better solution which is updated or gives a non-desirable solution which is discarded. The flowchart on Fig. 2 shows the procedure of the improved FCM using the optimization technique.

# 4 Results and Discussion

The number of iterations considered in this study is 20 across all optimization algorithms. Additionally, the number of candidate solutions (population) is 30. The candidate solutions have their fitness calculated and ranked from 1 to 3. The strategy is to adjust the parameters of the algorithms in each iteration to find the best solution. The test data sets were adapted from UCI and Kaggle. The data sets had the following characteristics in terms of number of instances, dimensions and missing values:

Table 1 illustrates the performances of the algorithms on different data sets. The considered validation parameters are the value of the objection function and the training time experienced by each algorithm on each data set.

The first section of Table 2 shows the performance of the algorithms on the seeds data set. The results exhibit a great improvement by the improved methods. All improved well by reducing the value of the objective function. It can be noted that

**Table 1** Data sets used

| Data set | Instances | Dimensions | Missing values |
|---|---|---|---|
| Seeds | 210 | 7 | N/A |
| Iris | 150 | 4 | No |
| Glass identification | 214 | 10 | No |
| Mall customer segmentation data | 200 | 5 | No |
| BuddyMove | 249 | 7 | N/A |

**Table 2** Best score and training time

| Data | Algorithm | Best score (Obj fun) | Training time (in s) |
|---|---|---|---|
| Seeds | FCM | 895.616496 | 0.036444 |
| | FCM-MFO | 457.8448 | 2.775923 |
| | FCM-WOA | 535.7786 | 1.859793 |
| | FCM-GWO | 458.7987 | 2.026988 |
| Iris | FCM | 60.575962 | 0.014655 |
| | FCM-MFO | 99.8264 | 1.501758 |
| | FCM-WOA | 137.4567 | 0.996033 |
| | FCM-GWO | 97.8413 | 1.070529 |
| Glass identification | FCM | 353.112171 | 0.106332 |
| | FCM-MFO | 350.8714 | 1.330237 |
| | FCM-WOA | 443.6656 | 1.042561 |
| | FCM-GWO | 305.8017 | 1.037286 |
| Mall customer segmentation data | FCM | 88,191.914035 | 0.051424 |
| | FCM-MFO | 4769.7327 | 1.412173 |
| | FCM-WOA | 5227.3578 | 1.020324 |
| | FCM-GWO | 4757.7641 | 1.053621 |
| BuddyMove | FCM | 491,522.745184 | 0.051598 |
| | FCM-MFO | 13,958.5492 | 1.394046 |
| | FCM-WOA | 16,672.974 | 1.047565 |
| | FCM-GWO | 13,337.5056 | 0.881009 |

the value of the objective function is inversely proportional to the training time. When it comes to training time, FCM without optimization comes first than all improved methods. Figure 3 graphically shows the results of the algorithms on seeds data set. The parameter considered on the data set is the best score of the objective function. Conventional FCM starts with the highest value then becomes constant. FCM-MFO starts with the lowest objective function and ends up with the lowest objective function. A probable explanation to the behavior of conventional FCM is the fact that it has no more options of solutions in an iteration.

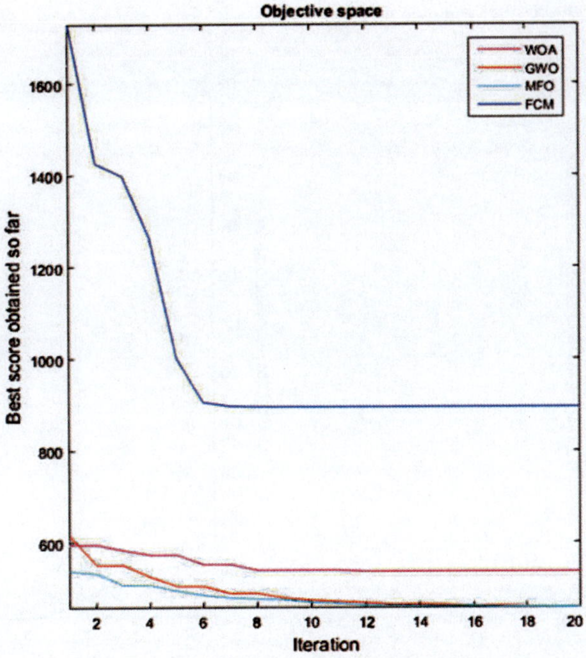


**Fig. 3** Performance on seeds data sets

Meanwhile, the second section of Table 2 shows the performance of the algorithms on the iris data set. The results show no improvement by the optimization algorithms. FCM has the lowest value of the objective function, while FCM-GWO comes in second. There is no consistent variation between the training time and the value of the objective function. FCM without optimization still has better performance in terms of training time. Figure 4 graphically shows a rapid decline from the first iteration to the fourth by conventional FCM.

The third section of Table 2 shows the performance of the algorithms on the glass data set. There is an improvement by two of the optimization algorithm over FCM. FCM-GWO shows a significant difference between FCM and itself. Also, FCM-MFO also performed better than FCM. FCM-WOA had the worst performance compared to all the algorithms on the glass data sets. Figure 5 shows the results of the algorithms on glass data set.

The fourth section of Table 2 shows the performance of the algorithms on the mall customer segmentation data. The results show a significant improvement of the objective function by the improved algorithms. FCM has the highest value of the objective function in this regard, and FCM-GWO has the lowest score, which is the best score. There is a pattern in the variation between training time and the score of the objective function. The objective function score varies inversely with the training time. It can be noticed that FCM has the best training time and FCM-GWO has the worst training time. Figure 6 shows the results of the algorithms on mall customer segmentation data set.

**Fig. 4** Performance on iris data set

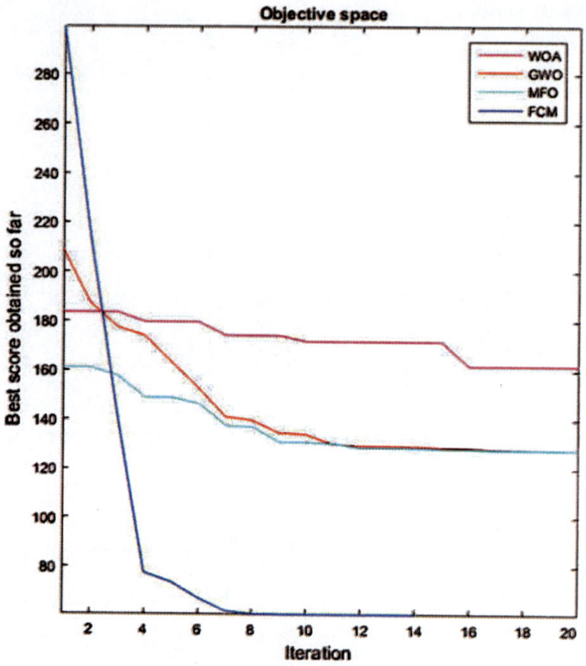


**Fig. 5** Performance on glass data set

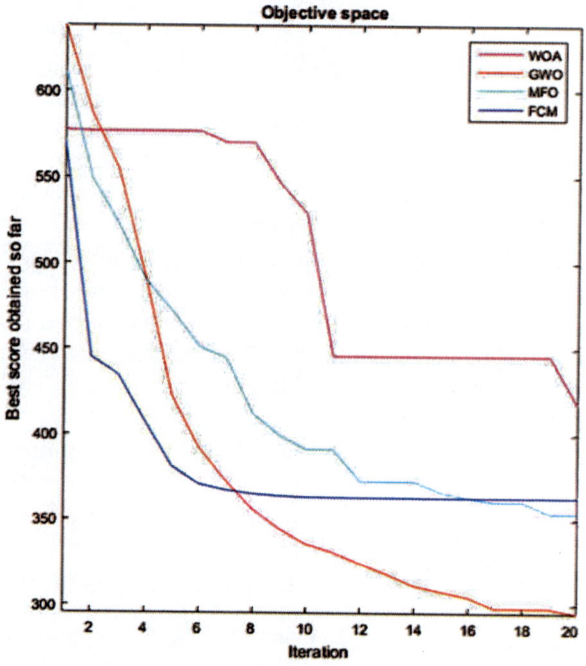

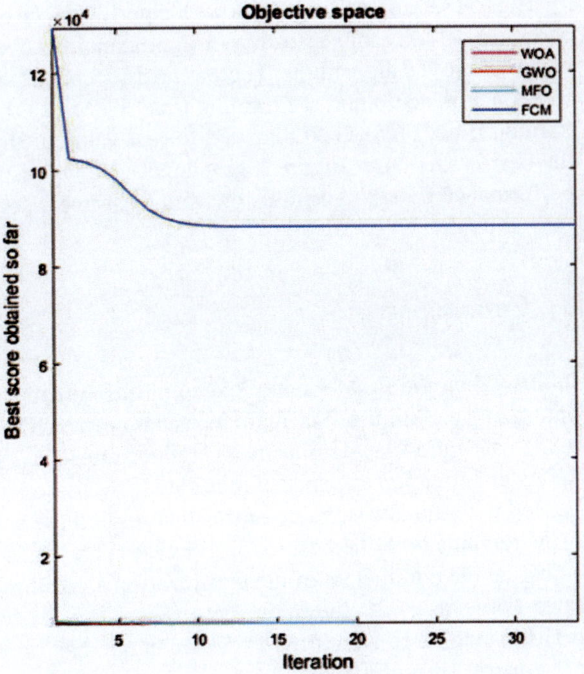


**Fig. 6** Performance on mall customer data set

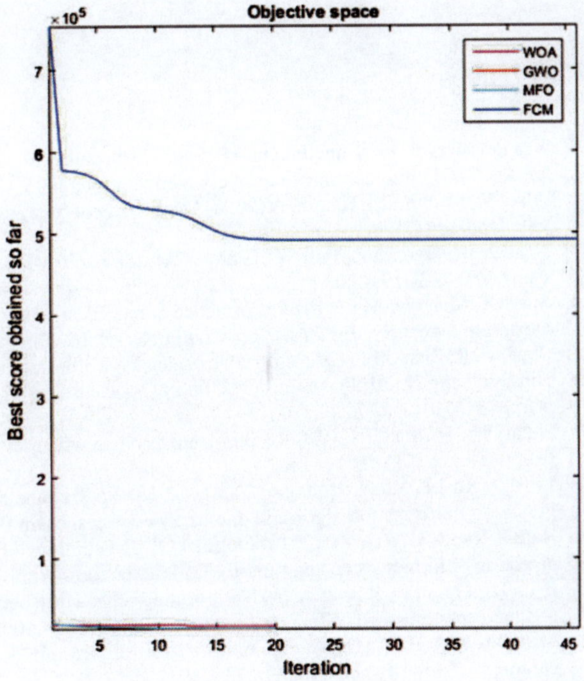


**Fig. 7** Performance on BuddyMove data set

The last section of Table 2 shows the performance of the algorithms on the Buddy-Move data. The results also show a significant improvement of the objective function by the improved algorithms. There is no direct pattern in variation between the objective function score and the training time. FCM has the worst score and the shortest training time. FCM-GWO has the lowest value of the objective function, which is the best is this case. Figure 7 graphically shows results that strongly resemble the performance by all algorithms on Mall Customers segmentation data.

## 5 Conclusion

In this paper, the FCM's membership partition matrix is updated by using different optimizing techniques. The main focus has been on minimizing the objective function to increase intra-class similarity and reduce inter-class similarity. The achieved result indicates better performance of modified FCM in 4 out of 5 test data sets. FCM-GWO has led with the lowest value of the data set in three cases. In terms of training time, FCM remains with the best in all five cases. The FCM parameter was optimized by adjusting the parameters of the optimization algorithm. FCM-MFO has led with the lowest objective function in one case. Therefore, it can be safely concluded that the performance of conventional FCM technique has improved with the incorporation of heuristic algorithms.

## References

1. Van der Merwe DW, Engelbrecht AP (2003) Data clustering using particle swarm optimization. In: The 2003 congress on evolutionary computation, 2003. CEC'03, vol. 1. IEEE, pp 215–220
2. Alata M, Molhim M, Ramini A (2008) Optimizing of fuzzy c-means clustering algorithm using GA. Update 1(5)
3. Bezdek JC, Ehrlich R, Full W (1984) FCM: the fuzzy c-means clustering algorithm. Comput Geosci 10(2–3):191–203
4. Singh T, Mahajan M (2014) Performance comparison of fuzzy C means with respect to other clustering algorithm. Int J Adv Res Comput Sci Softw Eng 4(5):89–93
5. Suganya R, Shanthi R (2012) Fuzzy c-means algorithm-a review. Int J Sci Res Publ 2(11):1
6. Niu Q, Huang X (2011) An improved fuzzy C-means clustering algorithm based on PSO. JSW 6(5):873–879
7. Katarya R, Verma OP (2018) Recommender system with grey wolf optimizer and FCM. Neural Comput Appl 30(5):1679–1687
8. Li MQ, Xu LP, Xu N, Huang T, Yan B (2018) SAR image segmentation based on improved grey wolf optimization algorithm and fuzzy c-means. Math Probl Eng
9. Raju P, Rao VM, Rao BP (2019) Optimal GLCM combined FCM segmentation algorithm for detection of kidney cysts and tumor. Multimedia Tools Appl 78(13):18419–18441
10. Bora DJ, Gupta D, Kumar A (2014) A comparative study between fuzzy clustering algorithm and hard clustering algorithm. arXiv preprint arXiv:1404.6059
11. Mirjalili S (2015) Moth-flame optimization algorithm: a novel nature-inspired heuristic paradigm. Knowl-Based Syst 89:228–249
12. Mirjalili S, Mirjalili SM, Lewis A (2014) Grey wolf optimizer. Adv Eng softw 69:46–61

13. Mirjalili S, Lewis A (2016) The whale optimization algorithm. Adv Eng Softw 95:51–67
14. Abdel-Basset M, Manogaran G, El-Shahat D, Mirjalili S (2018) A hybrid whale optimization algorithm based on local search strategy for the permutation flow shop scheduling problem. Future Gener Comput Sys 85:129–145
15. Reddy PDP, Reddy VV, Manohar TG (2017) Whale optimization algorithm for optimal sizing of renewable resources for loss reduction in distribution systems. Renewables Wind Water Solar 4(1):3

# Overlapping Community Detection in Static and Dynamic Networks: A Qualitative Assessment

**Keshab Nath and Swarup Roy**

**Abstract** Community detection (or graph clustering) is one of the fundamental graph analytic techniques for numerous graph-theoretic applications. Many community detection methods have been developed during the last few decades to reveal the concealed properties of complex networks. There are a good number of methods are available for detecting overlapping communities too. It is therefore important to access the performance of the different overlapping community detectors, empirically. More specifically, how good they are in detecting overlapping communities in both static and dynamic networks. In this paper, we present a study on a range of different overlapping community detection methods and access them experimentally to understand better the performance and biases of such methods. We employ six (06) real-world networks to test five (05) state-of-the-art algorithms. We quantify the performances using four (04) different statistical assessment parameters considering both static and dynamic networks. Our experiments reveal that SLPA and SLPAD are the best performers in static and dynamic scenarios, respectively, based on various assessment parameters.

**Keywords** Overlapping community · Dynamic network · Comparative analysis · Community detection.

K. Nath (✉)
Department of Computer Science and Engineering, Indian Institute of Information Technology Kottayam, Kottayam, Kerala 686635, India
e-mail: keshabnath@iiitkottayam.ac.in

S. Roy
Network Reconstruction and Analysis (NetRA) Lab, Department of Computer Applications, Sikkim University, Gangtok, Sikkim 737102, India
e-mail: sroy01@cus.ac.in

 
A. Maji et al. (eds.), *Proceedings of the International Conference on Computing and Communication Systems*, Lecture Notes in Networks and Systems 170,
https://doi.org/10.1007/978-981-33-4084-8_11

## 1 Introduction

Graph theoretically, a network is a collection of nodes, where the mutual interaction among them is represented by the links. Each node and links of a network possess certain characteristics and functions. Nodes with similar functionality or properties can be accumulated to a single module (sub-graph). A coherent sub-graph consists of nodes having high density intra-connections among them and low inter-connections among the nodes from other modules. These sub-graphs are called *clusters* or *communities* of a network. Formally, community may be defined as follows.

**Definition 1** (*Network Community*) For a given network $\mathcal{G}(\mathcal{V}, \mathcal{E})$, where $\mathcal{V}$ represent the set of nodes and $\mathcal{E}$ represent the set of edges. A community $C_i = \{V_i, E_i\}$ $(V_i \subseteq \mathcal{V}$ and $E_i \subseteq \mathcal{E})$ can be represented as a sub-module of the network $\mathcal{G}$, i.e., $C_i \subseteq \mathcal{G}$, where affinity score of each $v_i \in V_i$ w.r.t $C_i$ is higher than the other community $C_j$.

The classical definition of disjoint clusters is crisp in nature, where the members can not be a part of multiple communities simultaneously. However, in the case of real networks, the user interest or properties are imprecise, it alters over time, as a result, it leads to shared participation of a user in more than one group, simultaneously. Communities that share common members across different communities are frequently termed as *overlapping* communities. Mathematically, overlapping communities can be defined as follows.

**Definition 2** (*Overlapping Community*) For any two given communities $C_i = \{V_i, E_i\}$ and $C_j = \{V_j, E_j\}$ are said to be overlapping, if there exist some nodes $v_x$, which have shared membership concerning both the communities $C_i$ and $C_j$, i.e., $V_i \cap V_j = v_x$, whereas $E_i \cap E_j = \phi$.

For the last few decades, an impressive number of methods has been proposed for effective overlapping community detection in static along with growing networks [1–4]. However, it remains unclear which methods are effective and can be used for the effective graph or network analysis. Understanding the structural properties of the communities detected, its effectiveness and the computation time required by different methods are of great interest while applying for large-scale real-life graphs.

As a case study, we consider social networks for the current study. However, the same may be extended to any large and complex networks. In our current study, we select a few popularly used community detectors that are known to be performed well on synthetic (benchmark) and real-world social graphs. In this study, we try to evaluate and compare the qualitative performance of these methods based on different validity measures, which are particularly designed for testing the quality of the overlapping communities.

Our major contributions of this paper are: Firstly, we examine which is the most effective algorithm based on various statistical assessment. Secondly, the reliability of the various methods is tested with varying mixing parameters. Finally, we inspect the effectiveness and computing time of algorithms concerning the network size.

The paper is structured as follows. In Sect. 2, we discuss various overlapping community detection methods applicable to both static and evolving networks. In Sect. 3, various assessment parameters and performance evaluation of the different community detectors are reported in Sect. 3. Eventually, conclusions and future work are given in Sect. 4.

## 2 Overlapping Community Detection Methods

We have considered a few spectrum of overlapping community detectors applicable for both static and evolving networks. Next, we address a few available overlapping community detectors work in static and dynamic networks.

### *2.1 Overlapping Community Detection in Static Networks*

SLPA [4] is a label propagation-based overlapping community detection method. SLPA assigns separate memory for each node such that it can store label information obtained after each iteration. COPRA [5] is proposed to detect overlapping communities by label propagation. COPRA extend the vertex label and the propagation step of RAK algorithm [6] and also incorporate the synchronous updating process. Agglomerativ**E** hierarchic**A**l clusterin**G** based on maxima**L** cliqu**E** (EAGLE) [7] is able to detect both hierarchical and overlapping communities. This algorithm deals with the set of maximal cliques and adopts an agglomerative framework. CONGA [8] is an algorithm for discovering overlapping communities in complex networks based on betweenness centrality measure. CONGA is an extended version of Girvan and Newmans algorithm. LPANNI [9] is a label propagation-based approach to detect overlapping community structures in large-scale complex networks. LPANNI improves the label update strategy by computing the neighbor nodes influence concerning the similarity between node pairs and node significance.

### *2.2 Overlapping Community Detection in Dynamic Networks*

SLPA Dynamic (SLPAD) [2] is the extended version of SLPA [4] to detect overlapping communities in dynamic environment. OSLOM [10] is an optimized fitness function-based approach to detect overlapping and hierarchical community structures in diverse types of networks. iLCD [1] is an overlapping community detection approach in dynamic networks. iLCD states that communities are self-reliant of its background and it transforms from one state (origin) to other (current). **A**daptive **F**inding **O**verlapping **C**ommunity **S**tructure (AFOCS) [11] is an adaptive framework-based approach to detect, update and tracking the overlapping communities in an incremental network. InOvIn [12] is a rough-fuzzy based method for detecting disjoint, overlapping and hierarchical (intrinsic) communities in evolving networks. In this approach, the author adopts a rough-fuzzy and mean-standard deviation model to uncover overlapping and intrinsic community structures, respectively.

## 3 Empirical Assessments

We use both synthetic (benchmark) and real-world networks to compare the performance of ten (10) (both static and dynamic) contemporary community detection methods (*see* Sect. 2). We assess the effectiveness of the candidate methods in terms of four (04) distinct assessment parameters.

### 3.1 Assessment Parameters

Over the recent years, numerous measures have been proposed to evaluate the quality of disjoint community detection. However, a few validity indices are available to evaluate the quality of overlapping community structure. Next, we discuss a few statistical assessment parameters available for evaluating the overlapping community.

NFMI [13] measure is appropriate to evaluate the overlapping community structures. The high value of NFMI represents better quality. Overlapping Normalized Mutual Information (ONMI) [14] compute the fraction of shared links among the modules. A high ONMI value indicates the higher precision of the identified communities. Omega Index (OI) [15] is a variant of the original Rand Index [16]. A high Omega value signifies greater precision. Overlapping modularity ($Q_{ov}$) [17] is an extended version of original modularity index [18]. Higher value of $Q_{ov}$ means better community results.

### 3.2 Performance Evaluation: Synthetic Networks

We investigate the performance of different community detectors on the LFR benchmark [19, 20] networks. For this experiment, the number of nodes ($N$) ranges from 1000 to 5000 and the mixing coefficient ($\mu$) from 0:10 to 0:80. We consider the average degree ($\langle k \rangle$) is 20 and maximum degree ($k_{max}$) is 50. The community sizes diverge from $S = (10, 50)$ and $B = (20, 100)$ are categorized as small range and large range communities, respectively.

The degree of overlap is measured concerning the number of overlapping nodes ($O_n$) and the respective overlapping communities ($O_m$). For our experiment, we vary the value of $O_n$ from 10% (low overlapping density) to 50% (high overlapping

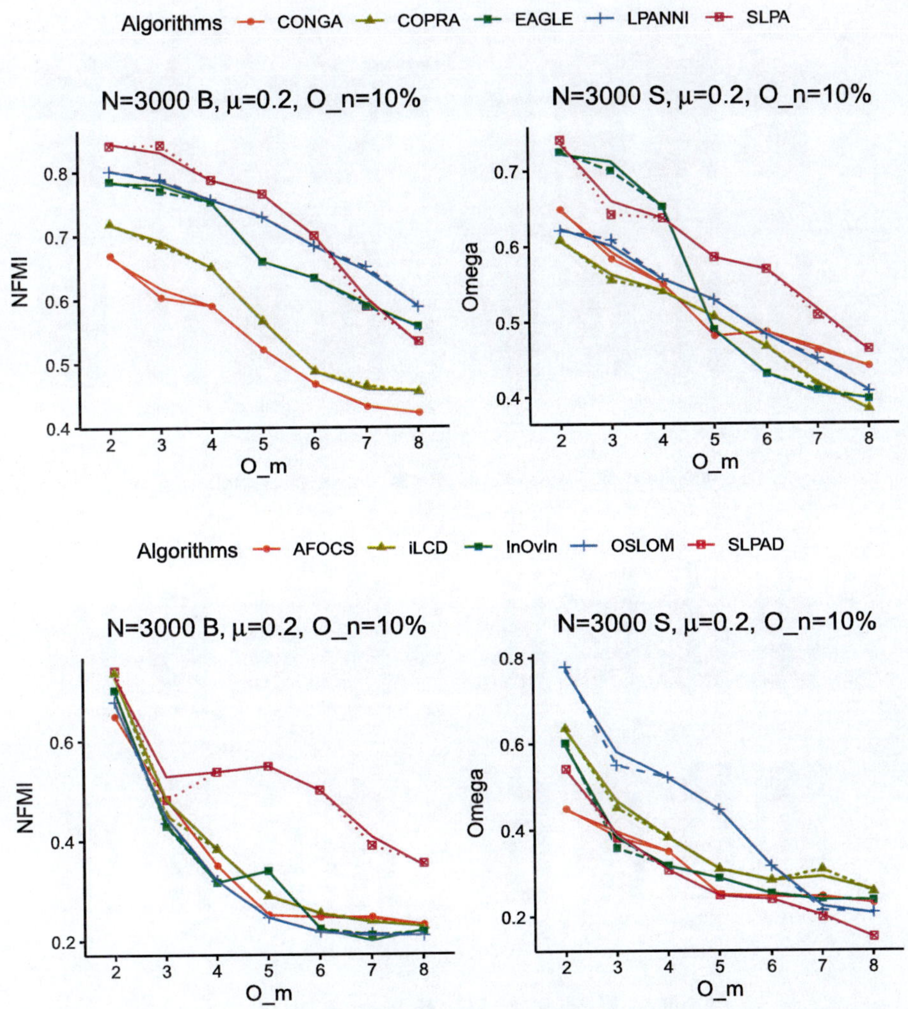


**Fig. 1** Evaluations of both static and dynamic overlapping community detectors on LFR networks with $N$ = 3000 and $\mu$= 0.2. **Left column** Performance of candidate methods in terms of NFMI ($y$-axis) and $O_m$ ($x$-axis), concerning large community size range with low overlap density $O_n$ = 10%; **Right column** Performance of candidate methods with respect to Omega Index ($y$-axis) as a function of the number of memberships $O_m$ ($x$-axis) and small community size range with low overlap density $O_n$ = 10%.

density) of $N$. Similarly, we also consider the overlapping assortment of overlapping nodes by taking distinct values of $O_m$ from 2 to 8.

Figure 1 represents the performance of static methods, measured by NFMI and Omega with respect to different values of $O_m$ (2–8), $N$ (1000–5000) and $\mu$ (0.10–0.80).

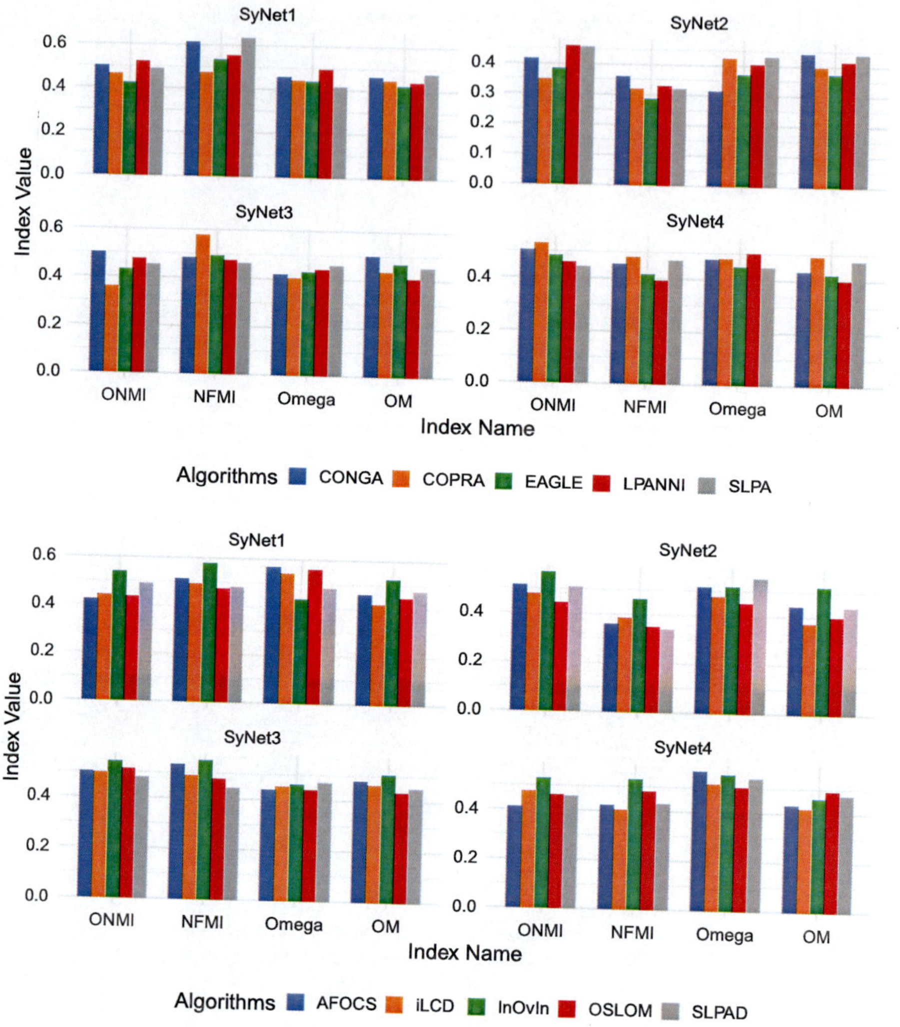


**Fig. 2** Evaluations of static (top) and dynamic (bottom) overlapping community detectors on **SyNet** (1 to 4) (LFR networks). All results are from networks with high overlap density $O_n = 50\%$

We create artificial Networks **SyNet**(1 to 4) with different parameters values ($N$, $\mu$, $O_m$). For **SyNet (1 to 4)** the parameters values are as (2000, 0.1, 4), (3000, 0.2, 4), (4000, 0.4, 6) and (5000, 0.3, 6), respectively. Results obtained from both the static and dynamic algorithms on **SyNet**(1 to 4) networks are presented in Fig. 2.

**Table 1** Real-world networks used for experiments

| Dataset | Available | #Nodes | #Edges |
|---|---|---|---|
| American college football | UCI[3] | 115 | 616 |
| Books about US politics | | 105 | 441 |
| Word adjacencies | | 112 | 425 |
| ca-GrQc | SNAP[3] | 5242 | 14,496 |
| ca-HepTh | | 9877 | 25,998 |
| Social circles from Facebook | | 4039 | 88,234 |

### 3.3 *Performance Evaluation: Real-World Networks*

We also consider a few well-known real networks, which are available at UCI[1] machine repository and SNAP.[2] Table 1 summarized both synthetic and real-world networks used in the experiment. The performance of both static and dynamic methods on real networks with respect to validity measures like ONMI, NFMI, Omega and OM is presented in Fig. 3.

### 3.4 *Ranking for Community Detectors*

A comprehensive analysis has been carried out over distinct overlapping densities and community size ranges. Performance measured by NFMI, Omega, ONMI and OM is shown in Figs. 1 and 2. Individual algorithms are tested with respect to the assessment parameters for the given $N$, $\mu$, $O_m$ and $O_n$ value. The final ranking among algorithms is obtained by sorting the individual assessment score in decreasing order. Table 2 summarized the final ranking for the community detection approaches for $N$ = 1000–5000, $\mu$ =0.1 to 0.8, $O_n$ = 10–50% and $O_m$ = 2–8.

## 4 Conclusions

In this research work, we have reported a comparative analysis of overlapping community detectors for social networks. Experiments we performed on the LFR benchmarks and real-world networks. We created several synthetic LFR networks with diverse network structures and a distinct degree of overlapping. Based on our exper-

[1] http://archive.ics.uci.edu/ml/datasets.html.

[2] https://snap.stanford.edu/data/#citnets.

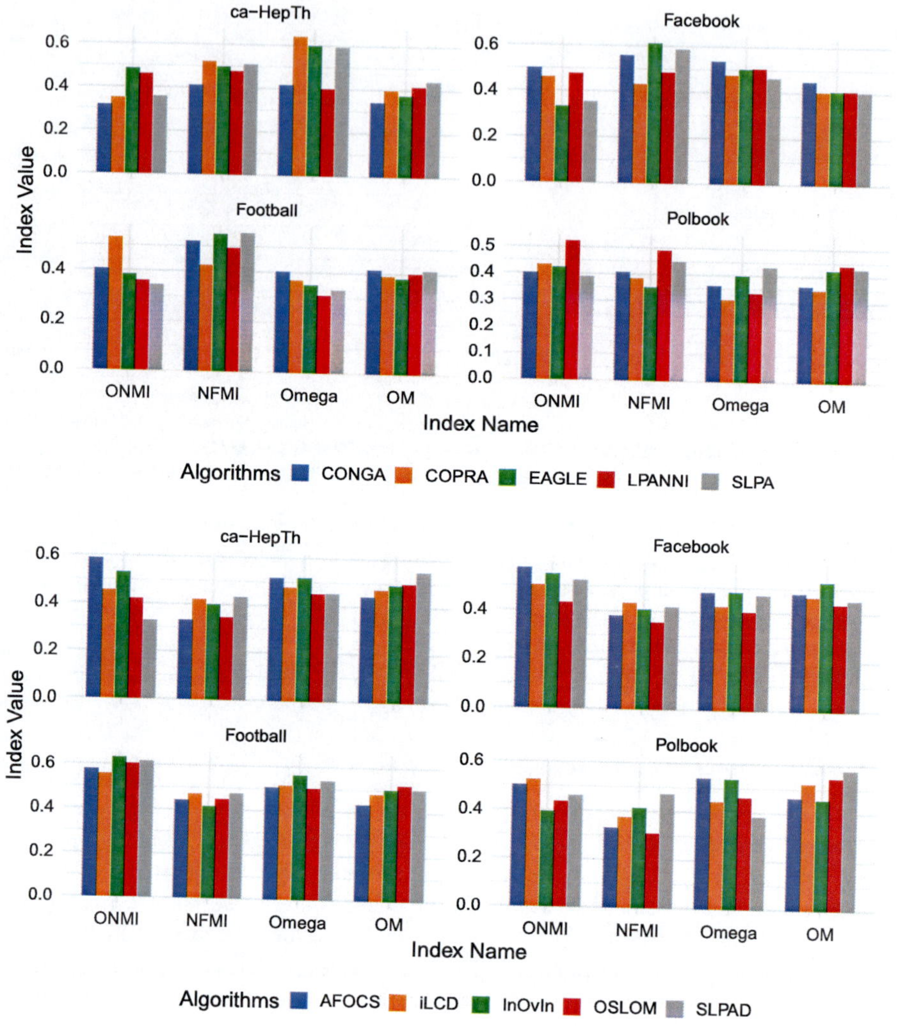


**Fig. 3** Effectiveness of static (top) and dynamic (bottom) overlapping community detection algorithms on real-world networks

iments, we conclude that the overlapping community detector SLPA (for static) and SLPAD (for dynamic) are the best performers on the set of benchmarks and real-world networks we used. Results imply that community detection still left considerable scope for betterment in the presence of both high overlapping assortment and overlapping density.

**Table 2** Ranking for social networks community detectors

| Network | Rank | Low overlapping density | | | | High overlapping density | | | | Final rank |
|---|---|---|---|---|---|---|---|---|---|---|
| | | NFMI $N = 4000$, $\mu = 0.2$, $O_n = 10\%$, $O_m$=5 | ONMI $N = 5000$, $\mu = 0.3$, $O_n = 10\%$, $O_m$=4 | Omega $N = 4000$, $\mu = 0.4$, $O_n = 10\%$, $O_m$=3 | OM $N = 5000$, $\mu = 0.5$, $O_n = 10\%$, $O_m$=2 | NFMI $N = 1000$, $\mu = 0.2$, $O_n = 50\%$, $O_m$=5 | ONMI $N = 3000$, $\mu = 0.3$, $O_n = 50\%$, $O_m$=4 | Omega $N = 4000$, $\mu = 0.4$, $O_n = 50\%$, $O_m$=3 | OM $N = 5000$, $\mu = 0.5$, $O_n = 50\%$, $O_m$=2 | |
| Static | 1 | SLPA | CONGA | SLPA | SLPA | COPRA | CONGA | SLPA | SLPA | SLPA |
| | 2 | CONGA | SLPA | CONGA | CONGA | CONGA | LPANNI | COPRA | CONGA | CONGA |
| | 3 | LPANNI | EAGLE | COPRA | EAGLE | EAGLE | SLPA | LPANNI | COPRA | EAGLE |
| | 4 | COPRA | LPANNI | EAGLE | COPRA | SLPA | COPRA | CONGA | EAGLE | COPRA |
| | 5 | EAGLE | COPRA | LPANNI | LPANNI | LPANNI | EAGLE | EAGLE | LPANNI | LPANNI |
| Dynamic | 1 | SLPAD | InOvIn | SLPAD | SLPAD | SLPAD | OSLOM | SLPAD | iLCD | SLPAD |
| | 2 | InOvIn | iLCD | InOvIn | AFOCS | InOvIn | SLPAD | OSLOM | InOvIn | InOvIn |
| | 3 | OSLOM | AFOCS | OSLOM | InOvIn | OSLOM | InOvIn | iLCD | AFOCS | OSLOM |
| | 4 | iLCD | SLPAD | AFOCS | OSLOM | iLCD | AFOCS | AFOCS | SLPAD | AFOCS |
| | 5 | AFOCS | OSLOM | iLCD | iLCD | AFOCS | iLCD | InOvIn | OSLOM | iLCD |

**Acknowledgements** This research is partially supported by the Department of Science & Technology (DST), Govt. of India under DST-ICPS Data Science program [DST/ICPS/Cluster/Data Science/General].

## References

1. Cazabet R, Amblard F, Hanachi C (2010) Detection of overlapping communities in dynamical social networks. In: 2010 IEEE second international conference on social computing (SocialCom). IEEE, pp 309–314
2. Aston N, Hertzler J, Hu W et al (2014) Overlapping community detection in dynamic networks. J Softw Eng Appl 7(10):872
3. Agarwal P, Verma R, Agarwal A, Chakraborty T, Dyperm: maximizing permanence for dynamic community detection. In: 22nd Pacific-Asia conference advances in knowledge discovery and data mining, PAKDD 2018, Melbourne, VIC, Australia, June 3–6, 2018, Proceedings, Part I, 2018, pp 437–449. https://doi.org/10.1007/978-3-319-93034-3_35
4. Xie J, Szymanski BK, Liu X (2011) Slpa: uncovering overlapping communities in social networks via a speaker-listener interaction dynamic process. In: 2011 IEEE 11th international conference on data mining workshops (ICDMW). IEEE, pp 344–349
5. Gregory S (2010) Finding overlapping communities in networks by label propagation. New J Phys 12(10):103018
6. Raghavan UN, Albert R, Kumara S (2007) Near linear time algorithm to detect community structures in large-scale networks. Phys Rev E 76(3):036106
7. Shen H, Cheng X, Cai K, Hu M-B (2009) Detect overlapping and hierarchical community structure in networks. Phys Stat Mech Appl 388(8):1706–1712
8. Gregory S (2007) An algorithm to find overlapping community structure in networks. In: European conference on principles of data mining and knowledge discovery. Springer, pp 91–102
9. Lu M, Zhang Z, Qu Z, Kang Y (2018) Lpanni: overlapping community detection using label propagation in large-scale complex networks. IEEE Trans Knowl Data Eng 31(9):1736–1749
10. Lancichinetti A, Radicchi F, Ramasco JJ, Fortunato S (2011) Finding statistically significant communities in networks. PloS one 6(4):e18961
11. Nguyen NP, Dinh TN, Tokala S, Thai MT (2011) Overlapping communities in dynamic networks: their detection and mobile applications. In: Proceedings of the 17th annual international conference on Mobile computing and networking. ACM, pp 85–96
12. Nath K, Roy S, Nandi S (2020) Inovin: a fuzzy-rough approach for detecting overlapping communities with intrinsic structures in evolving networks. Appl Soft Comput 89:106096. https://doi.org/10.1016/j.asoc.2020.106096, http://www.sciencedirect.com/science/article/pii/S1568494620300363
13. Kundu S, Pal SK (2015) Fuzzy-rough community in social networks. Patt Recognit Lett 67:145–152
14. McDaid AF, Greene D, Hurley NJ, Normalized mutual information to evaluate overlapping community finding algorithms, CoRR abs/1110.2515. http://arxiv.org/abs/1110.2515
15. Collins LM, Dent CW (1988) Omega: a general formulation of the rand index of cluster recovery suitable for non-disjoint solutions. Multivariate Behav Res 23(2):231–242
16. Rand WM (1971) Objective criteria for the evaluation of clustering methods. J Am Stat Assoc 66(336):846–850
17. Nicosia V, Mangioni G, Carchiolo V (2009) Malgeri M (2009) Extending the definition of modularity to directed graphs with overlapping communities. J Stat Mech Theory Exp 03:P03024
18. Newman ME (2006) Modularity and community structure in networks. Proc Nat Acad Sci 103(23):8577–8582

19. Lancichinetti A, Fortunato S, Radicchi F (2008) Benchmark graphs for testing community detection algorithms. Phys Rev E 78(4):046110
20. Lancichinetti A, Fortunato S (2009) Benchmarks for testing community detection algorithms on directed and weighted graphs with overlapping communities. Phys Rev E 80(1):016118

# Cluster-Based Deep Contextual Reinforcement Learning for Top-k Recommendations

**Anubha Kabra, Anu Agarwal, and Anil Singh Parihar**

**Abstract** Rapid advancements in the E-commerce sector over the last few decades have led to an imminent need for personalized, efficient and dynamic recommendation systems. To sufficiently cater to this need, we propose a novel method for generating top-k recommendations by creating an ensemble of clustering with reinforcement learning. We have incorporated DB scan clustering to tackle vast item space, hence increasing the efficiency multi-fold. Moreover, by using deep contextual reinforcement learning, our proposed work leverages the user features to its full potential. With partial updates and batch updates, the model learns user patterns continuously. The dueling bandit-based exploration provides robust exploration as compared to the state-of-the-art strategies due to its adaptive nature. Detailed experiments conducted on a public dataset verify our claims about the efficiency of our technique as compared to existing techniques.

**Keywords** Recommendation system · Deep neural networks · Deep reinforcement learning · Clustering

## 1 Introduction

With the ever-growing availability of commodities and services online, the consumer community has conveniently started shifting to online mediums for fulfilment of their needs. To capture the interest of consumers online, prompt and real-time recommendation systems are needed. The current recommendation systems are non-dynamic, i.e., they do not update themselves with time. This discards the trend that might uplift

---

A. Kabra · A. Agarwal · A. S. Parihar (✉)
Department of Computer Science, Delhi Technological University, Delhi, India
e-mail: parihar.anil@gmail.com

A. Kabra
e-mail: anubhakabradtu@gmail.com

A. Agarwal
e-mail: anuag7699@gmail.com

 
A. Maji et al. (eds.), *Proceedings of the International Conference on Computing and Communication Systems*, Lecture Notes in Networks and Systems 170,
https://doi.org/10.1007/978-981-33-4084-8_12

the recommendation which is more relevant to the consumer. To keep the consumer retention intact, it is necessary that the recommendation systems update themselves from time to time to learn the changing user behaviour. These changing trends can be captured using the upcoming reinforcement learning techniques. Reinforcement learning focuses on updating the policy in batches rather than splitting the data into training and testing. When such split is made, the policy does not generally learn anything from the testing data. Following are the challenges that the system faces with static recommendation systems:

1. **Ignoring trend**: Since the system is trained on the entire training set all at once, the trends present in the consumer behaviour data get mixed up and averaged out. This leads to non-relevant recommendation for consumers. For industry, tackling the non-stationary environment is very important or it may lead to consumers looing interest. Reinforcement learning techniques update the system frequently, hence learning the changes in the environment with time.
2. **Disregarding features**: Most of the existing recommendation systems rely only on the consumer and item frequency or rating. They usually do not have a provision of utilizing additional features about the consumers that may be available, that may help in recommendation. Currently, the recommendation systems rely on the narrow scope of how frequently a consumer or item is seen, and explicit user feedback in terms of ratings [1]. With reinforcement-based learning, these problems can be curbed by using contextual RL, which considers all features available in the data set before making valid recommendations.
3. **No accommodation for new consumers**: Due to the training and testing steps being discrete, if a new user joins the system, the entire system needs to be retrained. This is a very cumbersome process. Reinforcement learning-based strategies can easily handle this situation by updating the system with each new trial. So, if a new user joins the system, the agent will learn about the user with time without re-training.

This paper focuses on exploration using dueling bandit gradient descent for reinforcement learning-based recommendation strategy. The proposed work also narrows the item set to be considered for recommendation by using DB scan clustering (Fig. 1). The novelty lies in combining both strategies to overcome the drawbacks of current recommendation strategies and make it fit for continuously changing online environment. The combined model meets the industry standards due to its swift working, while exhibiting recommendations highly applicable to users.

## 2 Related Work

Existing recommendation systems can be broadly divided into major categories. The first category consists of the traditional recommendation algorithms which are either popularity-based or collaborative. Popularity-based systems always recommend the

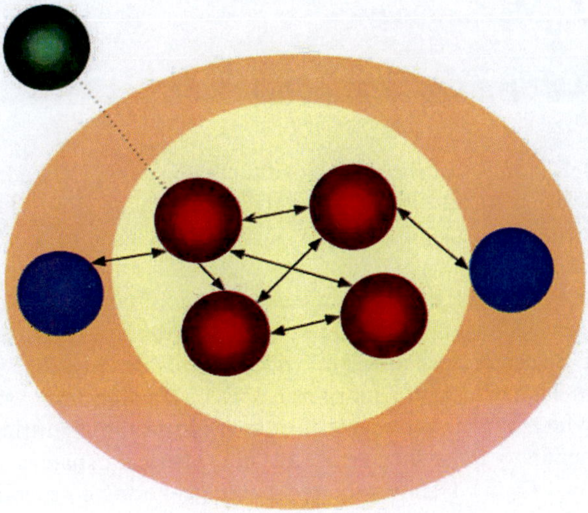

**Fig. 1** DB scan

same items at given time, irrespective of the user. On the other hand, collaborative-filtering (CF)-based systems [2, 3] look for other users which are similar to the given user and recommend him/her the items already purchased by them. It is still one of the most popular and successful techniques. Content-based filtering [4, 5] forms another category of recommenders which generates recommendations based on items already purchased by the user. However, this technique generally does not work well for platforms serving items of multiple categories as this method does not generalize well across different categories. Yet another traditional algorithm is hybrid filtering [6, 7] which uses a combination of above two techniques. Matrix factorization [8] extends CF as it attempts to identify latent features from user-ratings data by expressing the user-item interaction matrix as a product of multiple lower-dimensional matrices. Further attempts [9] have been made to incorporate the use of neural networks with the aforementioned techniques to improve the results.

The second major and relatively modern category includes recommendation systems which use reinforcement learning [10–12] and are inherently dynamic in nature. The recommendation scenario maps well to a usual Reinforcement Learning problem where an agent learns to behave in an environment via a series of sequential observations. These sequential observations can be mapped to the user-item interaction logs generated over time. Further, this task has been modelled by some as Markov decision problem [13, 14] and others as a multi-armed bandit problem [15]. Attempts have also been made to incorporate session context into the system to provide more relevant recommendations [16–18]. Our proposed work combines contextual multi-armed bandits strategy with DB-SCAN clustering dueling bandits exploration strategy to provide highly personalized recommendations in a more efficient manner.

# 3 Methodology

## 3.1 DB Scan Clustering

DB scan is a clustering algorithm that is density-based. The distance between points is used to find the neighbours (points with close vicinity). This measurement is generally done on the basis of Euclidean distance. Furthermore, it specifies the out-of-bound points and low-density regions. DB scan has two parameters: **minPts**: For the formation of a cluster, least number of points to be considered. **eps**: The distance between two points is to be called neighbours. This distance measures the closeness. DB scan corresponds well with the real-life data. Due to the unpredictable structure in such data with a lot of noise, DB scan algorithm can fit these irregularities well, which is generally denied in other clustering algorithms. E.g., K-means clustering generally fails when the data is not spherical in nature, whereas DB scan can capture this. Figure 1 shows the red points to be the core points which satisfy the minPts criteria and are in the eps vicinity. The blue points satisfy the eps distance but exceed the minPts criteria and hence are not included in the cluster (Yellow Area). The green point does not satisfy any of these conditions and hence is referred to as noise.

**Algorithm 1** DB Scan Algorithm(eps, minPts)

**Initialization**
D : dataset
$p_u$ : Un-visited Points in the dataset
$p_v$ : Visited points in dataset C : cluster index
**for** *non-visited* $p_u$ *in* $D$ **do**
    $p_{u_i}$ changes to $p_{v_i}$
    N : neighbours($p_i$,eps)
    **if** $N > minPts$ **then**
        N = N U N'
    **if** *found p' does not belong to any cluster* **then**
        p' $\in C$
        **end**

## 3.2 Dueling Bandit Gradient Descent

Dueling bandit gradient descent is an adaptive exploration strategy unlike epsilon greedy, UCB, TS, etc. It attempts to present the best recommendations to the user by combining exploration and exploitation in appropriate proportions. As the name suggests, in this technique, two models work simultaneously. First model $M$ takes decision based on patterns learnt from the data, whereas the second model $M^t$ is exploratory in nature. For every row in the dataset, both models generate a list of k items each. These lists are then combined by the reinforcement learning agent using probabilistic interleaving [19] to generate a final list of k items. This combined list is presented to the user and partial feedback is obtained for each item on the list. If more items from the list generated by $M^t$ get positive, then the parameters ($P$) of model $M$ are updated to incline it towards $M^t$ as shown in Eq. 1, else the model is kept unchanged.

$$P_{\text{new}} = P_{\text{old}} + \beta P^t \quad (1)$$

where other exploration strategies tend to lower the performance of the recommender in the long run because of being predominantly random in nature, this technique strikes a good balance between exploration and exploitation and adapts as and when required.

## 3.3 Model Pipeline

**STEP 1: Clustering** Due to the vast number of applications which can be recommended to user, DB scan clustering can help reduce the number of items to be taken into consideration before recommendation. The initial batch size does not implement DB scan clustering. Collection of correctly predicted user context and item is done. Clusters of items are made using context features of the first batch. For the next batch, when a trial sample is passed through this clustering mechanism, it first clusters applications relatable to the user and its features, from which best k items are selected to be shown to user with that particular features. The re-clustering is done after every batch is processes with the building collection of trail and corresponding correctly predicted arm. This technique works very well to improve efficiency since the pool of applications considered is not the entire item-set but a smaller chunk of it. Moreover it helps to understand the user behaviour with applications better.

**STEP 2: Exploration or Exploitation** After clustering, decision is made on whether to pursue exploration (selecting an item for user with some form of randomized way) or exploitation (selecting best item for the user); however, this is not as straightforward as epsilon greedy. The strategy used for this decision is dueling bandits gradient descent which generates a list combined from both exploration and exploitation models. The exploit network adapts itself further in the process if required based on the user feedback.

**STEP 3: Reward Formulation** The k selected items are passed through the base model to predict if user has chosen any one of them or not. There are two base models experimented with. First one is a basic logistic regression (LR)-based model while the other is a deep neural network (DNN). DNN gives higher performance than LR. If any of the k items matches with the actually selected item, the reward is 1, else the reward is 0. This reward, along with the trial features of that particular user is fed back into the model for continuous learning.

**STEP 4: Partial Update** The model is updated in random batches on default size 1000 selected from all the trials experienced till now. The trial and reward information is stored over time in memory. Random mini batch of 1000 from this set after every

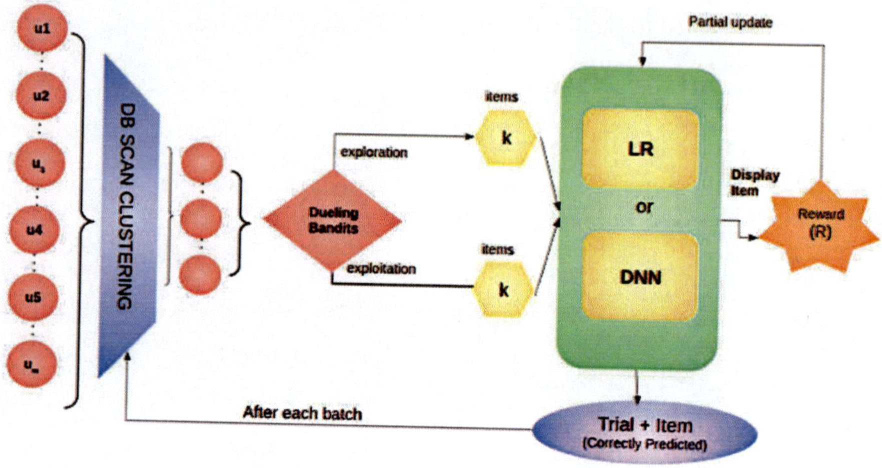


**Fig. 2** Proposed model pipeline

5000 trials updated back to the base model. This helps to keep the trend of the market intact and learn a pattern similarity between context features of different users. The model pipeline is illustrated diagrametically in Fig. 2.

# 4 Experimentation

## 4.1 Dataset

The dataset used consists of user and mobile application interaction logs. The data is collected from TalkingData SDK[1] integrated within mobile apps TalkingData serves under the service term between TalkingData and mobile app developers. Here, the apps are considered as items to be recommended and user and their information as the context used for the same. The categorical features are one hot encoded and continuous features are scaled before being given as input.

## 4.2 Baselines Static Techniques

- *LR:* Logistic regression is widely used in the industry as a baseline method due to its easy implementation and high efficiency. It is implemented using Keras.
- *FM:* Factorization machines is a feature handy recommendation system which takes in all contextual features and predicts clicks/non-click.

[1]https://www.kaggle.com/c/talkingdata-mobile-user-demographics/data.

- *W&D:* Wide and deep combines dependent learning using textbfwide: a normal representation of the features and **deep**: embedding representation of feature to predict the click label.

**Dynamic Techniques**

- *Contextual Bandits* These are reinforcement learning techniques with base classifier as logistic regression. The following exploration strategies are explored.
- *Epsilon greedy:*The idea here is to explore with certain fixed possibility. The exploration threshold is generally kept very small. Here default is set to 0.2. This has many variations for improved exploration.
- *First explore then exploit:* The exploration in this case is done initially, and later only exploitation is taken into consideration. The system assumes that initial exploration is good enough for system to learn.
- *Bootstrapped UCB* [16]*:* UCB works on upper confidence bound strategy for exploration. The bootstrapping ensures small pool of trails to be taken at a single time.
- *Bootstrapped TS* [16]*:* Thompson sampling creates distribution for each item, which changes according to the reward updation for the selected trial. This strategy involves forming a prior distribution for each arm.
- *Active explorer* [16]*:* This strategy focuses on weighted label-based probabilities.

## *4.3 Evaluation Metrics*

**CTR**: Click behaviour can be used to analyse how well a recommendation system is performing. CTR helps to capture the user click behaviour.

$$\mathrm{CTR}_i = \frac{\text{Total number of clicks}}{\text{Total number of occurrences}} \tag{2}$$

The average of CTR values of all the items is as follows:

$$\mathrm{CTR} = \frac{\sum_{i=0}^{N} \mathrm{CTR}_i}{N} \tag{3}$$

**Precision@k**: Precision@k for top-k recommendations is calculated as follows:

$$\text{Precision@}k = \frac{\text{number of clicks in top} - \text{k recommended items}}{k} \tag{4}$$

**Table 1** Abbreviation and code number for each policy

| Code | Abbreviation | Policy |
|---|---|---|
| 1 | LR | Logistic regression |
| 2 | W&D | Wide and deep |
| 3 | FM | Factorization machines |
| 4 | B-UCB | Bootstrapped upper confidence bound |
| 5 | B-TS | Bootstrapped Thompson sampling |
| 6 | E-Greedy | Epsilon greedy |
| 7 | FEE | First explore then exploit |
| 8 | AE | Active explorer |
| 9 | **DB + LR** | **Dueling bandits + LR** |
| 10 | **DB + DNN** | **Dueling bandits + DNN** |
| 11 | **DB_Scan + DB + DNN** | **DB scan + dueling bandits + DNN** |

Proposed policies are given in bold

### 4.4 *Experimentation Settings*

Three simulations were used for experimentation as given in Table 1. The techniques **9** and **10** use different base model. Technique **9** has used logistic regression to simulate [16] strategy. **10** uses deep neural network (DNN) as the base model. Since DNN gives significantly better results, technique **11** uses DNN as base model with DB scan clustering. Experiments have been conducted for top-1 and top-3 recommendations. While static strategies have a train–test split of 70:30. The dynamic techniques rely on learning throughout the process and no split has been done. To evaluate the results, however, 30% of the data is considered, which is kept identical for all experiments.

## 5 Results and Analysis

It can be clearly deduced from the CTR and Precision@k values mentioned in Table 2 that the methods which rely on static (one-time) model training such as LR, W&D and FM, perform poorly in comparison with the rest of the reinforcement learning based techniques. Thus, it can be concluded that dynamic model training is pivotal when it comes to applications such as recommender systems, because the user's underlying behaviour pattern keeps changing with time. Further, on comparing the CTR values for different exploration strategies, it can be seen that the adaptive techniques such

**Table 2** Result for each policy mentioned with its respective CTR and accuracy values

| Abbreviation | CTR | | Precision@k | |
|---|---|---|---|---|
| | $k = 1$ | $k = 3$ | $k = 1$ | $k = 3$ |
| LR | 0.0022 | 0.0042 | 0.28 | 0.28 |
| W&D | 0.0032 | 0.0010 | 0.33 | 0.36 |
| FM | 0.0036 | 0.0051 | 0.35 | 0.32 |
| B-UCB | 0.0243 | 0.0276 | 0.54 | 0.59 |
| B-TS | 0.0243 | 0.0276 | 0.67 | 0.70 |
| E-Greedy | 0.0288 | 0.0323 | 0.44 | 0.48 |
| FEE | 0.0294 | 0.0388 | 0.69 | 0.68 |
| AE | 0.0285 | 0.3117 | 0.48 | 0.49 |
| **DB + LR** | **0.0432** | **0.0566** | **0.75** | **0.77** |
| **DB + DNN** | **0.0965** | **0.0972** | **0.82** | **0.85** |
| **DB_Scan + DB + DNN** | **0.1342** | **0.0157** | **0.89** | **0.92** |

as active explorer perform better than purely random exploration strategies like *E*-greedy. However, the best performance is given by dueling bandit gradient descent strategy as it diversifies the network and inclines it towards exploration if favourable user feedback is obtained. The relative CTR graphs in Fig. 4 for different values of k show that inspite of a slight decrease in the overall values from $k = 1$ to $k = 3$, the general trend still remains the same. Non-RL based techniques show just about no change in results with changing k values. Since RL based techniques incorporate feedback for every recommended item, the change in values is comparatively more significant for them. Figure 3 shows a direct performance comparison between different variations of deep contextual bandits proposed in this paper. Even though DB + DNN seems to be the best strategy as it give the highest values throughout, DB_SCAN + DB + DNN is the most efficient and practical alternative. Inclusion of clustering straightaway reduces the number of comparisons by a factor of $x/N$ where $x$ is the average size of a cluster.

## 6 Conclusion

The proposed work focuses on improving recommendations of items to users in real-time scenario. Using DB scan clustering on the data reduces the processing cost making the recommender efficient. Moreover, Reinforcement learning techniques with dueling bandit gradient descent exploration and partial feedback tends to learn the user pattern and adapts to it quickly. Dueling bandit exploration gives this work an edge as it performs adaptive exploration with lesser number of parameters requiring manual tuning. This leads to relevant item suggestions to users without cumbersome

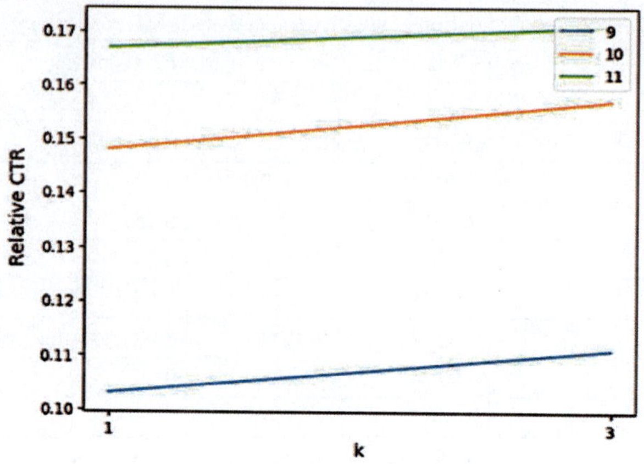


**Fig. 3** Comparative analysis of proposed methods for 1 and 3 item recommendations

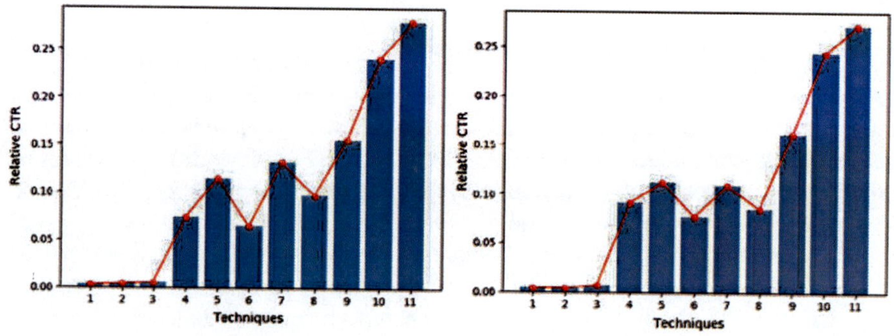


(a) $k = 1$ (b) $k = 3$

**Fig. 4** Relative CTR trends for different values of *k*

re-training. For future work, various other clustering methods could be experimented with for item space reduction. Ensemble of exploration strategies can also be tried.

## References


1. Harper FM, Konstan JA (2015) The movielens datasets: History and context. ACM Trans Inter Intell Syst (Tiis) 5(4):1–19
2. Das AS, Datar M, Garg A, Rajaram S (2007) Google news personalization: scalable online collaborative filtering. In: Proceedings of the 16th international conference on world wide web, pp 271–280
3. Marlin B, Zemel RS (2004)Themultiplemultiplicativefactormodelforcollaborative filtering. In: Proceedings of the twenty-first international conference on Machine learning, pp 73
4. IJntema W,Goossen F,Frasincar F,Hogenboom F (2010)Ontology-basednews recommendation. In: Proceedings of the 2010 EDBT/ICDTworkshops, pp 1–6
5. Mooney RJ, Roy L (2000)Content-basedbookrecommendingusinglearningfor text categorization. In: Proceedings of the fifth ACM conference on digital libraries, pp 195–204
6. De Francisci Morales G, Gionis A, Lucchese C (2012) From chatter to headlines: harnessingthereal-timewebforpersonalizednewsrecommendation.In:Proceedings of the fifth ACM international conference on Web search and data mining, pp 153–162
7. Liu J,Dolan P,Pedersen ER (2010)Personalizednewsrecommendationbasedon click behavior. In: Proceedings of the 15th international conference on Intelligent user interfaces, pp 31–40
8. Koren Y, Bell R, Volinsky C (2009) Matrix factorization techniques for recommender systems. Computer 42(8):30–37
9. Zhang S, Yao L, Sun A, Tay Y (2019) Deep learning based recommender system: a survey and new perspectives. ACM Comput Surv (CSUR) 52(1):1–38
10. Mahmood T, Ricci F (2009) Improving recommender systems with adaptive conversational strategies. In: Proceedings of the 20th ACM conference on Hypertext and hypermedia, pp 73–82
11. Taghipour N, Kardan A (2008) A hybrid web recommender system based on q-learning. In: Proceedings of the 2008 ACM symposium on Applied computing, pp 1164–1168
12. Taghipour N,Kardan A,Ghidary SS (2007)Usage-basedwebrecommendations: a reinforcement learning approach. In: Proceedings of the 2007 ACM conference on recommender systems, pp 113–120
13. Lu Z, Yang Q (2016)Partiallyobservablemarkovdecisionprocessforrecommender systems. arXiv preprint arXiv:1608.07793
14. Rojanavasu P, Srinil P, Pinngern O (2005) New recommendation system using reinforcement learning. Special Issue IntJ Comput Internet Manag 13(SP 3)
15. Li L, Chu W, Langford J, Schapire RE (2010) A contextual-bandit approach to personalized-newsarticlerecommendation.In:Proceedingsofthe19thinternational conference on World wide web, pp 661–670
16. Cortes D (2018) Adapting multi-armed bandits policies to contextual banditsscenarios. arXiv preprint arXiv:1811.04383
17. Wang H,Wu Q,Wang H (2016)Learninghiddenfeaturesforcontextualbandits. In: Proceedings of the 25th ACM international on conference on information and knowledge management, pp 1633–1642
18. Zeng C,Wang Q,Mokhtari S,Li T (2016)Onlinecontext-awarerecommendation with time varying multi-armed bandit. In: Proceedings of the 22nd ACM SIGKDD international conference on Knowledge discovery and data mining, pp 2025–2034
19. Grotov A, deRijke M (2016)Onlinelearningtorankforinformationretrieval:Sigir 2016 tutorial. In: Proceedings of the 39th international ACM SIGIRconference on research and development in informationretrieval, pp 1215–1218

# A Novel Local Consensus Inspired Spatial Fuzzy Method for Classification and Spectral Unmixing of Hyperspectral Data

**Srirupa Das and Somdatta Chakravortty**

**Abstract** In this paper, a novel local consensus inspired spatial fuzzy method has been introduced to perform classification and spectral unmixing of hyperspectral image pixels. Here, a local consensus index inspired spatial membership has been combined with the global membership to deal with non-homogeneity and noise present in the hyperspectral image scene. The modified membership function has been used to produce the fractional abundance for end members present in the considered image scene, which serves the purpose of spectral unmixing. To analyze the superiority of the proposed method, partition coefficient and partition entropy of the proposed method have been thoroughly compared with the fuzzy-based state of the art methods and some recently published novel works. The qualitative result and fractional abundance are observed and thoroughly analyzed.

**Keywords** Hyperspectral image · Spectral unmixing · Global membership · Local membership · Fractional abundance

## 1 Introduction

Hyperspectral image pixel covers a huge land cover because of the sensors properties, i.e., low spatial resolution. The classification and spectral unmixing of the hyperspectral pixels have been considered as a complex and next to impossible task by using manual segmentation techniques because of the spatial resolution and artifacts. There are several unmixing techniques available for the spectral unmixing [1, 2]. Among them, linear mixing model [3, 4] has been widely used for estimation of abundance values, where endmembers are residing in a linear fashion and single interactions between endmembers occur. But, in most cases, the endmembers reside in very close

S. Das
RCC Institute of Information Technology, Kolkata, India
e-mail: srirupadas000@gmail.com

S. Chakravortty (✉)
Maulana Abul Kalam Azad University of Technology, Haringhata, West Bengal, India
e-mail: csomdatta@rediffmail.com

 
A. Maji et al. (eds.), *Proceedings of the International Conference on Computing and Communication Systems*, Lecture Notes in Networks and Systems 170,
https://doi.org/10.1007/978-981-33-4084-8_13

association among each other, and in such cases, non-linear interactions among them take place. To deal with non-linearity many non-linear and bilinear models [5–8] such as Nascimento's model [8], Fan's model [6], Support Vector Machine (SVM) [9, 10], and Gaussian mixture model (GMM) [11] are also being used widely. Conventional fuzzy-based methods [12, 13] are also popular to estimate the fractional abundance values, where fuzzy membership function plays an important role. It is very common for hyperspectral pixels to have non-homogeneity and uncertainty into it due to the presence of large land cover within a single pixel and artifacts. FCM-based methods work better if the considered pixel is free from noise and uncertainty [14, 15]. To deal with these issues, influences of the neighbors are considered in the FCM-based methods [14]. It is possible to get rid of noise and non-homogeneity but blurring effect arises in the edges of the image. To overcome these issues, authors in [14] (FLICM) have defined fuzzy local information-based c-means technique. Recently [16] has proposed another approach on entropy-based FCM technique to segment image data.

To overcome the above mentioned issues and for the betterment of classification in the presence of noise and high degree of non-homogeneity, a novel local consensus index [17] inspired spatial fuzzy method has been proposed in this paper. Here, a $5 \times 5$ neighborhood is considered for incorporating their spatial influence on the considered hyperspectral pixel by means of Gaussian-based distance instead of intensity-based distance. The proposed algorithm has been simulated and examined for Samson dataset, and it has shown it's superiority in terms of visual output, partition coefficient, and entropy over considered techniques.

Rest of the paper is arranged as follows: Sect. 2 clarifies the proposed method, Sect. 3 elaborates the result analysis, and Sect. 4 states the conclusion.

## 2 Methodology

The proposed method deals with the issues raised when the state of the art fuzzy-based methods try to classify and segment the hyperspectral images which are having pixels with high degree of non-homogeneity and noise and as well as low spatial resolution. Here, a $5 \times 5$ neighborhood is considered to incorporate the local spatial information along with the global membership for each pixel $(x_i)$ of the hyperspectral dataset, $X = \{x_i \vee i = 1, 2, \ldots (\mathrm{nr} * \mathrm{nc} * \mathrm{nz})\}$, where $(\mathrm{nr} * \mathrm{nc} * \mathrm{nz})$ is the size of the hyperspectral dataset $X$. At the very beginning, the number of clusters $(C)$ and initial cluster centers $(v_j)$ are determined by histogram pick associative rule [18], and the standard deviation has been calculated considering the cluster centers as sample mean. Then, the global membership has been calculated for the entire image based on the intensity difference between the centers of the cluster and pixels of the input image, described in (Eq. 1).

$$u_{ij} = \frac{1}{\sum_{l=1}^{C}\left(\frac{d_{ij}^2}{d_{il}^2}\right)^{\frac{2}{(m-1)}}} \quad (1)$$

Here, $u_{ij}$ is the membership of $i$th element to $j$th cluster. $d_{ij}^2$ is the Euclidean distance, defined as intensity difference between pixel $x_i$ and cluster mean $v_j$. $m$ is the degree of fuzziness for the processing pixel $(x_i)$.

A $5 \times 5$ mask $(M_i)$ has been used to identify the neighborhood (Eq. 2), and a local distance matrix $(D_{ij})$ is constructed based on the difference between the neighbors and the cluster centers for each processing pixel, defined by (Eq. 3).

$$M_i = \{x_k \vee M_i \subset X, \forall k\} \quad (2)$$

$$D_{ij} = \{||M_i - V|| \vee V = \{v_j\}\forall j\} \quad (3)$$

Again Gaussian-based distances $(G_{dij})$ between each member of the local distance matrix $(D_{ij})$ and it's center $(D_c)$ have been calculated using (Eq. 4).

$$G_{dij} = \mathrm{e}^{-\left(\frac{\|D_{ij} - D_c\|^2}{2\sigma_1^2}\right)} \quad (4)$$

Now, a spatial distance matrix $(S_{di})$ has been constructed (Eq. 5) by computing the Gaussian based geometric distance of the neighbors $\overrightarrow{(s,t)}$ from the processing pixel $\overrightarrow{(i,j)}$.

$$S_{di} = \mathrm{e}^{-\left(\frac{\|\overrightarrow{(i,j)} - \overrightarrow{(s,t)}\|^2}{2\sigma_2^2}\right)} \quad (5)$$

Now, both the matrices generated from the above two stages are convolved (Eq. 6) to produce the resultant distance matrix $(\hat{L}_{ij})$ for Eq. 7, which is defined to generate the local membership. The local membership has been generated by normalizing the $(\hat{L}_{ij})$ to satisfy the constraints of fuzzy membership function.

$$\hat{L}_{ij} = \sum_{\forall k, j \vee x_k \subset M_i} \left(S_{di} * G_{dij}\right) \quad (6)$$

$$L_{ij} = \frac{\hat{L}_{ij}}{\sum_{\forall j} \hat{L}_{ij}} | \sum_{\forall j} L_{ij} = 1 \quad (7)$$

After the generation of the local membership, the modified membership function $(u'_{ij})$ (Eq. 8) has been defined by combining both local and global membership functions. New cluster centers (Eq. 9) are generated based on the modified membership, and new standard deviation (Eq. 10) has been calculated for the image based on new cluster centers. This iterative process optimizes by optimizing the objective function defined in (Eq. 11).

$$u'_{ij} = \frac{u_{ij}^{p} \times L_{ij}^{q}}{\sum_{j=1}^{C}\left(u_{ij}^{p} \times L_{ij}^{q}\right)} \Big| \sum_{j=1}^{C} u'_{ij} = 1 \tag{8}$$

$$V = \left\{ v_j \,\middle|\, v_j = \left\{ \frac{\sum_{i=1}^{\mathrm{nr}\times\mathrm{nc}}\left(u_{ij}^{\prime m} \times x_i\right)}{\sum_{i=1}^{\mathrm{nr}\times\mathrm{nc}}\left(u_{ij}^{\prime m}\right)} \right\} \middle| x_i \in X \right\} \tag{9}$$

$$\mathrm{SD}_j = \frac{\sum_{\forall i}\left\|X - v_j\right\|^2}{\mathrm{nr} \times \mathrm{nc}} | v_j \in V \tag{10}$$

$$J = \sum_{j=1}^{C} \sum_{i=1}^{nr \times nc} u_{ij}^{m} d_{ij} + \sum_{j=1}^{C} \sum_{i=1}^{nr \times nc} L_{ij}^{m} G_{dij} \tag{11}$$

Figure 1 describes the flowchart for the working principle of the proposed method.

## 3 Result and Discussion

This study has been done on a scene of 95 × 95 pixels of Samson dataset [19], which is a hyperspectral image of 156 bands, covers visible range, i.e., 401 nm to some part of NIR region, i.e., 889 nm. The spectral resolution is 3.13 nm. This image contains three major endmembers, i.e., bare soil, vegetated land cover, and water body. In this section, the results of the proposed method are thoroughly compared with the results of the considered state of the art and recently published fuzzy-based classification methods by means of visual result, partition coefficient (Vpc), and partition entropy (Vpe) [20]. The simulation is carried out on MATLAB (v 2015a) at the environment of Windows 7 with 4 GB RAM and 3.10 GHz Intel(R) i3 processor. Here, $\sigma_1$ and $\sigma_2$ are considered to be 10 and 2, respectively. The results are thoroughly analyzed upon the whole spectral range of the image and compared accordingly.

Between range 680–730 nm, the change in reflectance of vegetation is the maximum, and this range is referred to as red edge region. Keeping this in mind, the visual results of band number 96 (706.2478 nm) have been projected in this paper, depicted in Fig. 2, from which it can be clearly identified that the clustering has been done in a better manner, and the clarity is increasing for the proposed technique over the other considered methods.

Also the most significant part is the edges between two homogeneous land cover types, where the demarcation is clearly visible in the proposed methods.

Table 1 shows the comparative result of Vpc and Vpe of the proposed algorithm with the other considered methods for visible and NIR region.

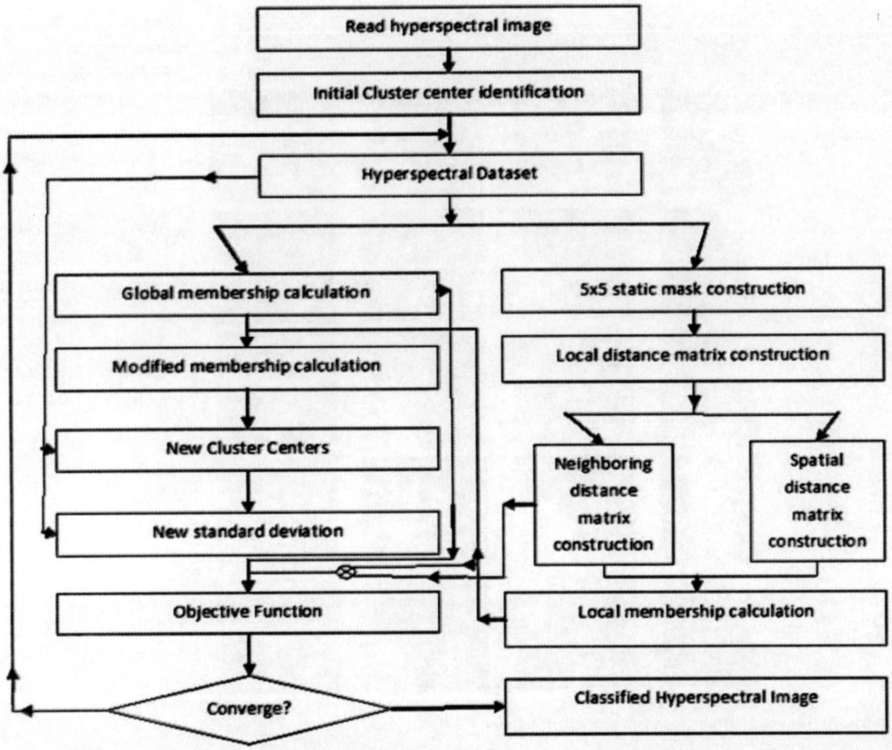


**Fig. 1** Flowchart of the proposed method

It has been seen that the Vpc and Vpe are much superior for the proposed method in comparison with the other considered methods. For better realization of the superiority of the proposed local consensus inspired spatial fuzzy method, the comparison of Vpc and Vpe for the whole spectral region has been shown in Fig. 3.

## 4 Conclusion

A novel local consensus inspired spatial fuzzy method has been introduced in this paper, to perform classification and spectral unmixing of hyperspectral image pixels. To deal with non-homogeneity and noise present in the hyperspectral image scene, here, a local consensus index inspired spatial membership has been introduced along with the global membership to produce the modified membership for each pixel. This modified membership function has been utilized to produce the fractional abundance for endmembers present in the considered image scene. To analyze the superiority of the proposed method, partition coefficient and partition entropy have been thoroughly compared with the fuzzy-based state of the art methods and some recently

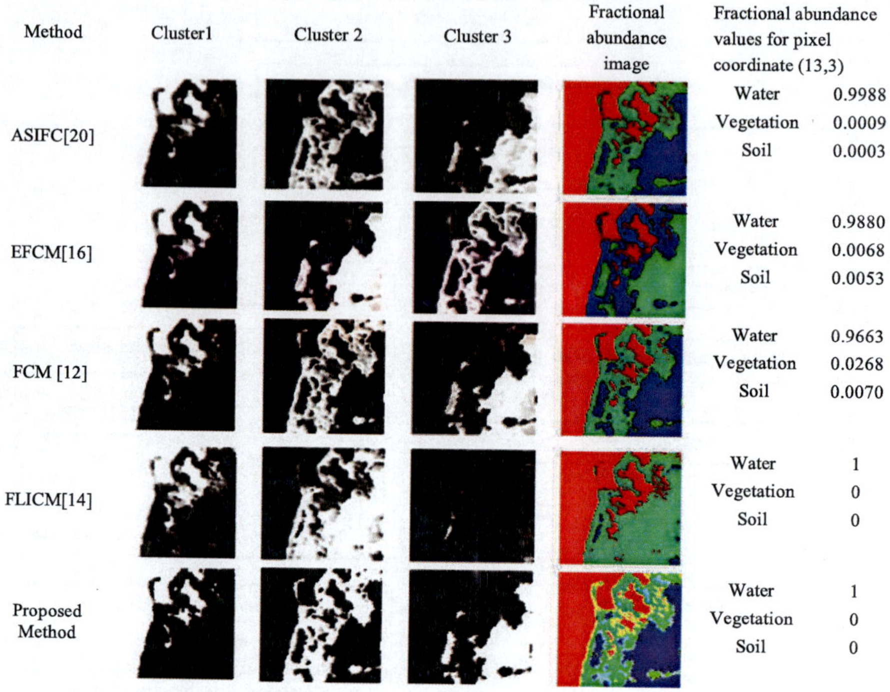


**Fig. 2** Visual result of considered methods

**Table 1** Comparative results of Vpc and Vpe of the methods

| Band | Method | Vpc | Vpe |
|---|---|---|---|
| 48 (551.9864 nm) | ASIFC [20] | 0.8102 | 0.3554 |
| | EFCM [16] | 0.7832 | 0.3866 |
| | FCM [12] | 0.7800 | 0.3832 |
| | FLICM [14] | 0.7774 | 0.3557 |
| | **Proposed method** | **0.9733** | **0.0463** |
| 110 (751.2407 nm) | ASIFC [20] | 0.8493 | 0.2759 |
| | EFCM [16] | 0.8248 | 0.3374 |
| | FCM [12] | 0.8150 | 0.3359 |
| | FLICM [14] | 0.9414 | 0.1004 |
| | **Proposed method** | **0.9726** | **0.0452** |

published novel works. The qualitative result and fractional abundance are observed and thoroughly analyzed to confirm the superiority of the proposed LCI inspired spatial fuzzy-based method.

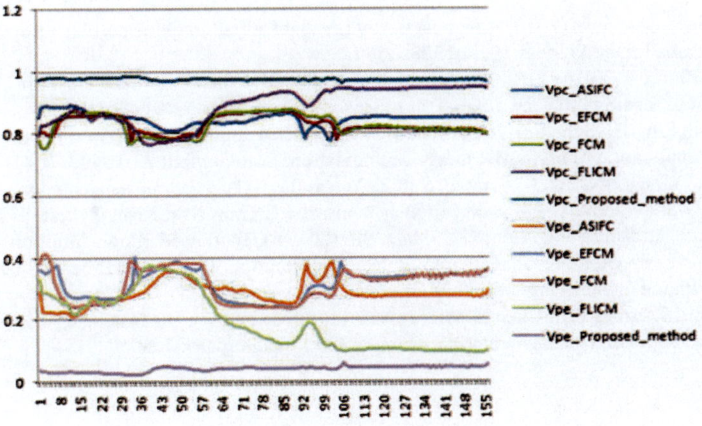


**Fig. 3** Vpc and Vpe plot for proposed and considered methods

## References


1. Keshava N, Mustard JF (2002) Spectral unmixing. IEEE Signal Process Mag 19(1):44–57
2. Meer FVD (1999) Iterative spectral unmixing (isu). Int J Remote Sens 20(17):3431–3436
3. Heinz DC, Chang C-I (2001) Fully constrained least squares linear spectral mixture analysis method for material quantification in hyperspectral imagery. IEEE Trans Geosci Remote Sens 39(3):529–545
4. Zanotta DC, Haertel V, Shimabukuro YE, Renno CD (2014) Linear spectral mixing model for identifying potential missing endmembers in spectral mixture analysis. IEEE Trans Geosci Remote Sens 52(5):3005–3012
5. Dobigeon N, Tits L, Somers B, Altmann Y, Coppin P (2014) A comparison of nonlinear mixing models for vegetated areas using simulated and real hyperspectral data. IEEE J Select Top Appl Earth Observ Remote Sens 7(6):1869–1878
6. Fan W, Baoxin H, Miller J, Mingze L (2009) Comparative study between a new nonlinear model and common linear model for analysing laboratory simulated—forest hyperspectral data. Int J Remote Sens 30(11):2951–2962
7. Halimi A, Altmann Y, Dobigeon N, Tourneret J-Y (2011) Nonlinear unmixing of hyperspectral images using a generalized bilinear model. IEEE Trans Geosci Remote Sens 49(11):4153–4162
8. Nascimento JMP, Bioucas-Dias JM (2012) Nonlinear mixture model for hyperspectral unmixing. In: Bruzzone L, Notarnicola C, Posa F (eds) Proceedings of the SPIE image and signal processing for remote sensing XV, Berlin, Germany, pp 7477
9. Ping-Xiang L, Wu B, Zhang L (2005) Abundance estimation from hyperspectral image based on probabilistic outputs of multi-class support vector machines. Paper presented at the IEEE international geoscience and remote sensing symposium, Seoul, pp 4315–4318
10. Tang Y, Krasser S, Yuanchen H, Yang W, Alperovitch D (2008) Support vector machines and random forests modeling for spam senders behavior analysis. Paper presented at the IEEE global telecommunications conference, New Orleans
11. Cheng B, Zhao C, Wang Y (2010) Algorithm to unmixing hyperspectral images based on APSO-GMM. Paper presented at the 2010 first international conference on pervasive computing, signal processing and applications signal processing and applications, Harbin, pp 964–967
12. Foody GM (1996) Approaches for the production and evaluation of fuzzy land cover classifications from remotely-sensed data. Int J Remote Sens 17:1317–1340
13. Bastin L (1997) Comparison of fuzzy c-means classification, linear mixture modelling and mlc probabilities as tools for unmixing coarse pixels. Int J Remote Sens 18(17):3629–3648

14. Krinidis S, Chatzis V (2010) A Robust Fuzzy Local information C-means clustering algorithm. IEEE Trans Image Process 19(5):1328–1337
15. Ma A, Zhong Y, Zhang L (2015) Adaptive multiobjective memetic fuzzy clustering algorithm for remote sensing imagery. IEEE Trans Geosci Remote Sens 53(8):4202–4217
16. Kahali S, Sing JK, Saha PK (2019) A new entropy-based approach for fuzzy c-means clustering and its application to brain MR image segmentation. Soft Comput 23:10407–10414
17. Xiao X, Xiong NN, Lai J, Wang C, Sun Z, Yan J (2019) A local consensus index scheme for random-valued impulse noise detection systems. IEEE Trans Syst Man, Cybern Syst 1–17
18. Namburu A, Samayamantula SK, Edara SR (2017) Generalised rough intuitionistic fuzzy c-means for magnetic resonance brain image segmentation. IET Image Proc 11(9):777–785
19. https://www.escience.cn/people/feiyunZHU/Dataset_GT.html
20. Chuang KS, Tzeng HL, Chen S, Wu J, Chen TJ (2006) Fuzzy c-means clustering with spatial information for image segmentation. Comput Med Imaging Graph 30(1):915

# Gated Recurrent Unit: An Effective Tool for Runoff Estimation

**Amitabha Nath, Diganta Barman, and Goutam Saha**

**Abstract** Runoff estimation has been an active area of research in the field of hydrology. It is often considered to be one of the most complicated processes owing to its spatio-temporal distribution, inadequate data and uncertainties of the contributing factors. In recent years, soft computing techniques have evolved to be effective in modeling such complex phenomena. Particularly with the success of artificial neural networks and its variants such as recurrent neural network, computational modeling techniques have achieved satisfactory acceptance level. In this paper, we present a sophisticated gated variant of recurrent neural network known as gated recurrent unit (GRU) to forecast runoff values. The performance of GRU is then compared with another famous gated variant of RNN known as long short-term memory (LSTM). RMSE value and model training time are used as a performance evaluation criteria for the selected models. Experimental result suggests that GRU's result is at par with its compatriot LSTM model, in fact it is better in some cases and that too in significantly reduced computational time.

**Keywords** Rainfall-runoff · Time series estimation · GRU · LSTM · Deep learning

## 1 Introduction

Runoff estimation is an important parameter in hydrology [1]. It is a very complex process which requires several direct and indirect input data parameters like rainfall,

---

A. Nath (✉) · G. Saha
Department of IT, NEHU, Shillong, Meghalaya , India
e-mail: amitabha.me@gmail.com

G. Saha
e-mail: dr.goutamsaha@gmail.com

D. Barman
North Eastern Space Applications Centre, Umiam, Meghalaya, India
e-mail: diganta_isro1@yahoo.co.in

 
A. Maji et al. (eds.), *Proceedings of the International Conference on Computing and Communication Systems*, Lecture Notes in Networks and Systems 170,
https://doi.org/10.1007/978-981-33-4084-8_14

temperature, topography, vegetation, soil characteristics and other physical parameters. Modeling this phenomenon has been an active field of research in hydrology domain. In general, hydrological model can be classified into three major categories, i.e., (i) Empirical model, (ii) Conceptual model and (iii) Physically based model. Empirical model considers only the existing data for modeling purpose without considering actual process of hydrological system and hence gives a black box view of the whole process. These are also known as data-driven model, e.g., ANN, unit hydrograph. Conceptual models are considered as parametric or gray box model as semi-empirical equations are used in this process, and model parameters are acquired from field data as well as through model calibration. Examples of this type of modeling are HBV model, TOPMODEL. Physical-based model on the other hand is considered as the white box model as it tries to represents the real phenomenon mathematically. Examples of such modeling are SHE, SWAT. These types of models are quite accurate because of the ability to overcome many shortcomings present in the other two models. However, physically based models are very complex and require voluminous data, human expertise and computational capability for its operation. In recent years, data-driven empirical models have gained popularity to attract everyone's attention in this field. Often this popularity is attributed to the advent of various soft computing techniques.

Among various soft computing techniques, artificial neural network (ANN) has been the preferred choice of modeling technique over the years. It is a computational model inspired by the working of human brain and its nervous system. It is well known for its ability to mimic complex nonlinear system [1–3]. In almost every aspect of hydrological modeling including rainfall-runoff, ANNs have been used successfully [4–7]. However, the main drawback of commonly used feed forward ANNs is its inability to retain any information as per the sequential order of input data. Recurrent neural network (RNN), a variant of ANN is been known to overcome this limitation and has proven to be very efficient in dealing with time series data.

RNN has shown promising result in different machine learning task. It is a very powerful modeling tool known for its ability to capture time varying nonlinearity very well. It has a chain of information processing system known as cell which makes it possible to capture temporal dynamics of the input data. It was first proposed by David Rumelhart in 1986 [8]. RNN has the capability of handling spatio-temporal relationship in the input data. This makes it a suitable candidate for modeling in hydrology. Some of the areas where RNN has performed reasonably well include river flow forecasting, flood forecasting, landslide prediction, etc. [9, 10]. However, it is found that conventional RNN suffers from inability to handle long-term dependencies due to the presence of exploding or vanishing gradient problem [11].

One way to deal with this problem is to use learning algorithms that perform better than a stochastic gradient descent algorithm or by changing the method of calculating derivatives such that it can avoid getting trapped in either of such problems. Hochreiter and Schmidhuber proposed one such RNN variant known as long short-term memory (LSTM) and claimed to have alleviated the problem of exploding or vanishing gradient problem [12]. Often this ability is credited to its upgraded activation function consisting of affine transformation followed by array of gating units. There are many instances available in the literature describing the applications of LSTM in modeling rainfall-runoff process [13–15]. In addition to rainfall-runoff, LSTM networks have successfully been used in other fields of hydrology such as flood prediction [16], stream-flow forecasting [17] as well.

Recently one more variant of RNN, known as gated recurrent unit (GRU) was proposed by Cho et al. [18]. It is very much similar to the structure of LSTM but differs by the number of gates that is used in the construction of LSTM. The advantage of the GRU lies in the fact that, it is simple when compared with other RNN variants and has the similar ability to learn long-term dependencies in a time series data. In hydrology, application of GRU network can be traced in flood forecasting, simulating reservoir operation, forecasting water table fluctuations, etc. [19–21]. However, applicability of GRU networks in rainfall-runoff modeling is yet to be explored. No such instances exist in the literature to the best of our knowledge. It will be particularly interesting to analyze its performance in modeling rainfall-runoff relation. The main purpose of this study is to analyze performance of GRU in rainfall-runoff modeling and present a comparison with that of LSTM. For experimental purpose, model configuration has been kept same. Also, to understand the effect of different optimization algorithms on the model performance, usefulness of RMSProp, ADAM and SGD functions are explored. Root mean square error (RMSE) value and training time are used to evaluate the model performance.

Organization of this paper is as follows. Section 2 introduces the profile of the study area and the data preprocessing steps used for this experimentation purpose. Experimentation and results are discussed in Sect. 3, and finally, Sect. 4 provides the concluding remarks.

## 2 Study Area and Data Preparation

For experimentation purpose, we have concentrated on Dikhow river basin (94.879399, 26.591392 Decimal Degrees), spanning over the state of Assam and Nagaland, two of North–Eastern states of India as shown in Fig. 1.

The monsoon period for this part of the world spreads from July to September. The daily runoff values for this monsoon period from 2006 to 2018 and corresponding runoff values are considered as the input data. Rainfall-runoff essentially exhibits time series behavior. In time series data, summary statistics like mean, variance autocorrelation, etc., of the observations keeps changing over time. It provides additional information that can be used in the process of model building along with existing data. Modeling process needs to incorporate not only the input features but also the changes occurring over time. Formally, it can be defined as a sequence of vectors, depending on time $t$.

$$\vec{d}(t), \quad t = 0, 1, .... \tag{1}$$

where vector $\vec{d}$ represents any variable changing over time $t$. In most cases of time series analysis, data preprocessing plays an important role in getting the reliable

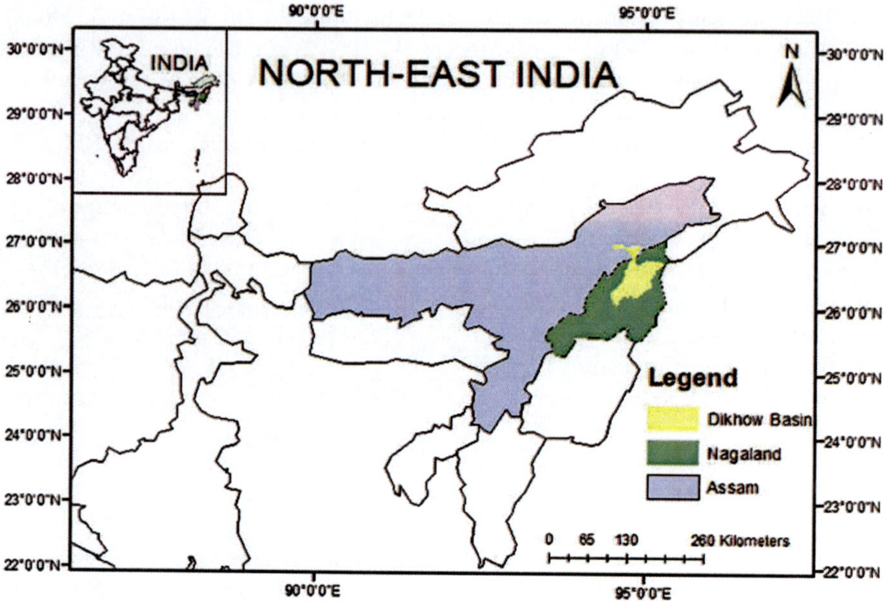


**Fig. 1** Dikhow watershed

result. Ideally, we want our time series estimation model to extract any kind of systematicities present in the time variant data. However, we also do not want known or obvious systematicities to affect its performance. That is why classical time series analysis techniques convert non-stationary time series data into stationary by identifying and removing trends and seasonal effects. This is done by using a technique named differencing process. It basically finds the difference between two successive values as given in Eq. (2).

$$\vec{a}'^{(t)} = \vec{a}(t) - \vec{a}(t-1), \quad t = 0, 1, .... \tag{2}$$

In this regard, input data is prepared with three different time lags such as $Q_t$, $Q_{t-1}$ and $Q_{t-2}$ where $t$, $t-1$ and $t-2$ represent previous one day, two day and three days of runoff values, respectively.

Moreover, as a preprocessing step, linear interpolation method is used to incorporate missing values in the input data. Thereafter, input data is normalized as it improves performance of deep learning models. This also helps the model to be trained a bit faster. The available dataset is then divided to 70–30 ratio as training and testing purpose, respectively.

## 3 Experimentation and Result

The main focus of the present investigation is on establishing rainfall-runoff relation by using GRU network and compares its performance with LSTM. For the comparison purpose, statistical measure known as root mean square error (RMSE) value has been used. RMSE is an index that determines the closeness of predicted and actual observations. Mathematically, it can be represented by Eq. (3).

$$\text{RMSE} = \sqrt{\frac{1}{n}\sum_{1}^{n}(\text{Target}_i - \text{Output}_i)^2} \tag{3}$$

In model building purpose, necessary parameters like number of epochs, batch size are deliberately kept same for both GRU and LSTM network evaluation. Looking at the trend of epochs vs. RMSE score (Fig. 2), total 100 number of epochs are selected for carrying out the experimentation.

The proposed working mechanism is shown in Fig. 3. The input data, consisting of historical rainfall-runoff values, is first processed to be fed to the GRU network. Neural networks are prone to biasness when variance in the input data is high [22, 23]. Therefore, preprocessing steps such as scaling often help to improve overall performance. After performing several rounds of experiment and tracking of network performance, total of four GRU layers have been used for this experiment. Finally, the output of the last GRU layer is the input to a feed forward neural network which then maps the output of GRU layer to corresponding runoff forecast.

The proposed prediction method can be summarized in following steps. Step-1: Defines the process of scaling the input data. Step-2: Explains network construction process. Step-3: Network compilation. Step-4: Network training and Step-5: Defines the runoff forecast. The partial codes for completing this task are included in algorithm 1.

Algorithm 1: GRU Network - Partial code

```
procedure GRU
    Step-1: Data_Preprocessing
    sc = MinMaxScaler(feature_range = (0, 1))
    training_set_scaled = sc.fit_transform(training_set)
    training_set_scaled.shape

    Step-2: Define_Network
    regressorGRU = Sequential()
    GRU(units = 50, return_sequences = True, input_shape =
  (X_train.shape[10], 1), activation =' tanh'))
    regressorGRU.add(Dropout(0.2))
    #Theoutputlayer
    regressorGRU.add(Dense(units = 1))
```

```
Step-3: Network Compilation
    regressor.compile(optimizer = SGD(lr = 0.01, decay =
  1e − 7, momentum = 0.9), loss =' mean_squared_error', metrics = [rmse])
    regressor.compile(loss =' mse', optimizer =' adam', metrics = [rmse])
    regressor.compile(loss =' mse', optimizer =' rmsprop', metrics = [rmse])

Step-4: Network training
    start_time = datetime.now()

  history = regressorGRU.fit(trainX, trainY, epochs = 100, batch_size = 50)
    time_elapsed = datetime.now() − start_time

Step-5: Performance evaluation
    Runoff = regressorGRU.predict(testX)
    plt.plot(history.history['rmse'])
    print('Timerequired(hh : mm : ss.ms) : '.format(time_elapsed))
End procedure
```

Table 1 represents the RMSE score of GRU and LSTM techniques using different learning techniques such as RMSProp, ADAM and SGD and input data with $t$ time lags where $t = 1,\ 2,\ 3$ days, respectively. It is clear that variation of $t$ (duration of past time period) reflects significant difference on performance of the model. For example, the best RMSE value of 1.54 and 1.7 is evaluated for GRU and LSTM, respectively, with input $Q_t$, $Q_{t-1}$ and $Q_{t-2}$. RMSE trend hints towards the fact that, greater lag in input window actually helps recurrent model to learn and capture the nonlinearity better.

The impact of different learning algorithms (RMSProp, ADAM and SGD) on the performance of GRU or LSTM is also investigated. A significant difference in terms of computational time requirement for its convergence is observed because of the choice of optimization algorithm used. Three well-known techniques namely stochastic gradiant descent (SGD) with momentum, Adam and root mean square propagation (RMSProp) are selected on the basis of performance as reported by various researchers. SGD with momentum is a modified approach where not only the gradient of the current step but also of the past step is used to determine the direction to follow for reaching the optimal minima. Therefore, it converges faster compared to the conventional SGD. RMSProp adapts different learning rates for different parameters, based on the average of recent changes in the gradient of their weight. Adam or adaptive moment optimization algorithm on the other hand combines the idea of both momentum and RMSProp together in it. The impact of different optimizing techniques is also visible when we take into account the amount of time GRU takes to complete its training. A summary of the time taken by GRU and LSTM in minutes for different values of $t$ and with different optimizing algorithm is presented as a chart in Fig 4.

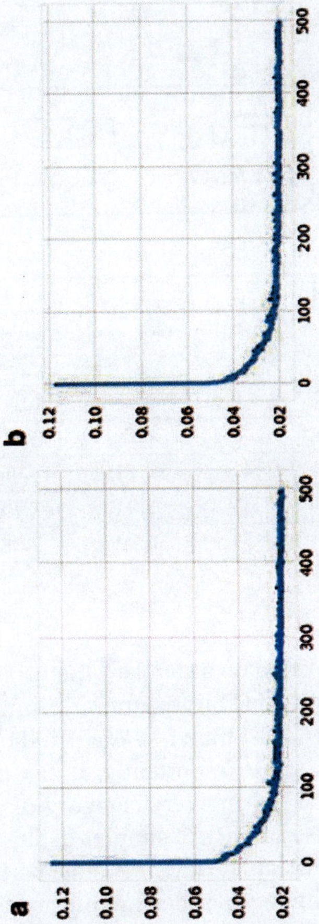


**Fig. 2** Epoch versus RMSE of **a** GRU and **b** LSTM

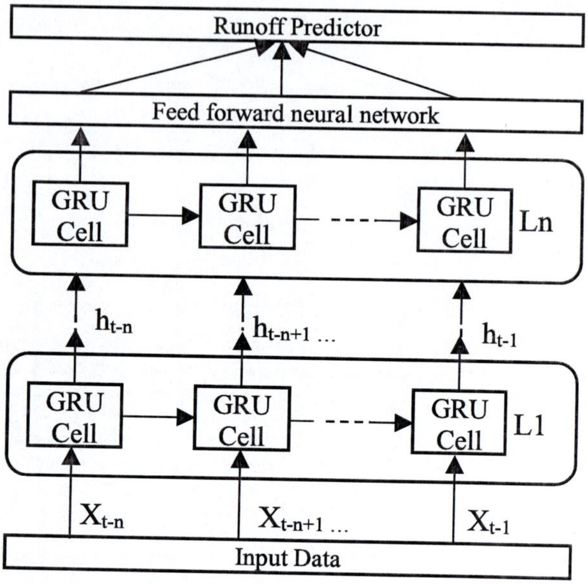


**Fig. 3** Application of GRU network for runoff forecasting process

**Table 1** Test RMSE comparison

| | GRU | | | LSTM | | |
|---|---|---|---|---|---|---|
| Lag | RMSProp | ADAM | SGD | RMSProp | ADAM | SGD |
| $Q_t$ | 15.36 | 15.44 | 15.71 | 14.96 | 15.36 | 16.18 |
| $Q_t$ and $Q_{t-1}$ | 3.71 | 1.65 | 3.04 | 1.75 | 3.1 | 4.48 |
| $Q_t$, $Q_{t-1}$ and $Q_{t-2}$ | 2.72 | 1.54 | 3.19 | 2.08 | 1.7 | 4.7 |

This experiment has been carried on a standard Intel core i5 machine without any GPU. It is apparent from the figure that highest training time is incurred by SGD, then comes RMSProp and finally the least time is taken by ADAM or at least by ADAM and RMSProp both. However, most importantly, it can be observed that training time of GRU is far less than LSTM in every category of $t$ irrespective of learning techniques used. This is because of GRU's simpler gating structure. It is important to mention that LSTM needs to calculate memory state with hidden state while GRU can avoid calculating these using its upgraded architecture structure. Thereby, GRU does the same with less overhead than that of LSTM does. Naturally, GRU takes lesser time in training compared to that of LSTM. In all the three possible learning method implementations, GRU's training time is far lesser than that of LSTM (Fig. 4). In fact, in some cases, GRU takes almost half the time required by that of LSTM. This is possible because GRU cell consists of smaller number of states compared to LSTM. Hence, overall computational time is much less in case of GRU compared to that of LSTM. Overall, performance of both GRU and LSTM in runoff estimation is presented in Fig. 5.

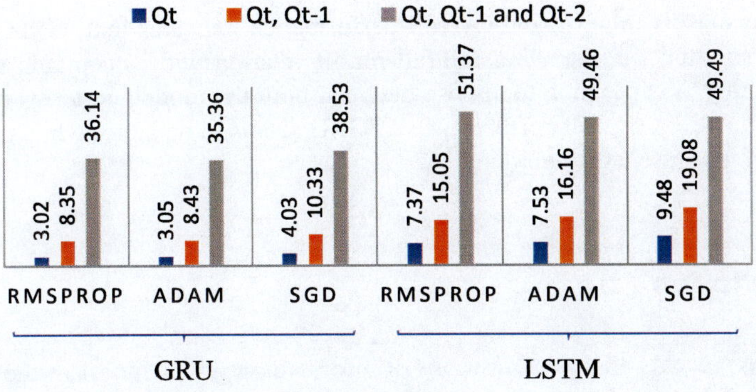


**Fig. 4** Training time experienced by GRU and LSTM

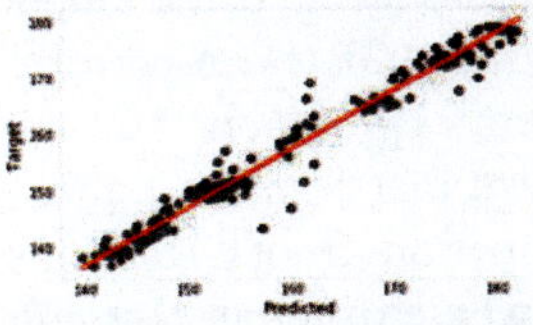

(a) GRU with lr method= Adam & $Q_t$, $Q_{t-1}$& $Q_{t-2}$

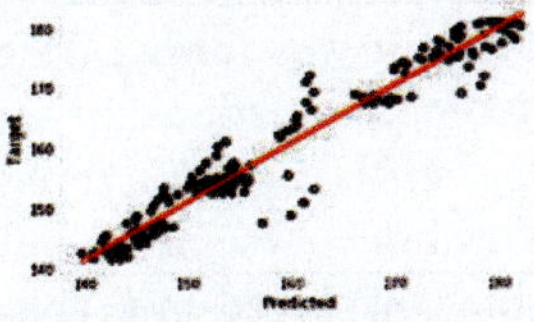

(b) GRU with lr method= RMSProp & $Q_t$, $Q_{t-1}$& $Q_{t-2}$

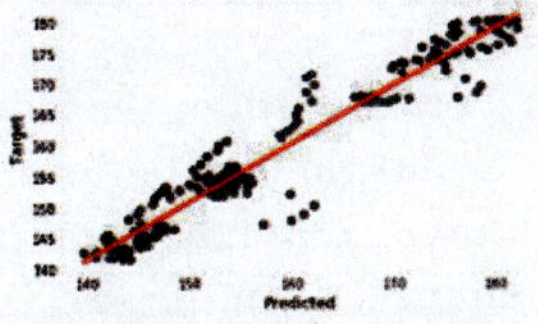

(c) GRU with lr method= ADAM & $Q_t$, $Q_{t-1}$& $Q_{t-2}$

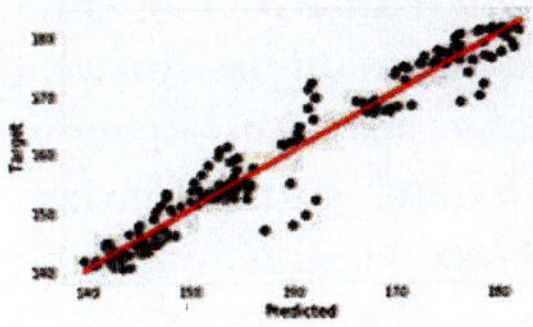

(d) LSTM with lr method= ADAM & $Q_t$, $Q_{t-1}$& $Q_{t-2}$

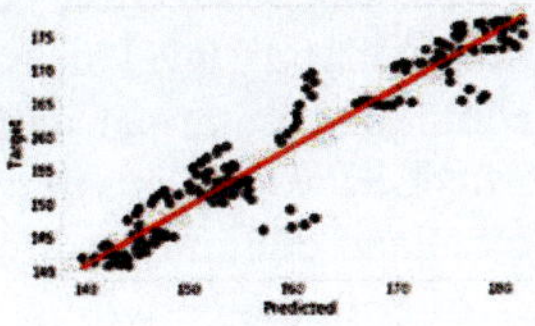

(e) LSTM with lr method= RMSProp & $Q_t$, $Q_{t-1}$& $Q_{t-2}$

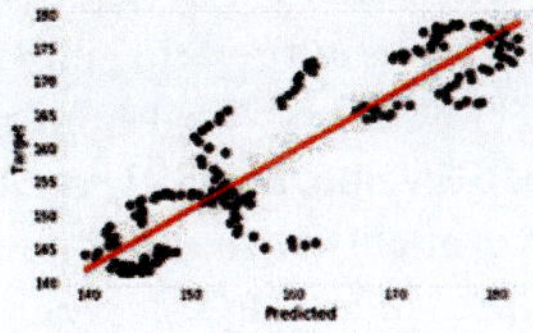

(f) LSTM with lr method= SGD & $Q_t$, $Q_{t-1}$& $Q_{t-2}$

**Fig. 5** Comparison of GRU and LSTM performance when used with Adam, RMSProp, SGD learning algorithm

From the Fig. 5, performance of GRU and LSTM can be observed when applied with three different learning algorithms. The scatterplot presented in Fig. 5a–c shows the observed vs. predicted plot of GRU, and Fig. 5d–f shows LSTM with respect to lag $Q_t$, $Q_{t-1}$ *and* $Q_{t-2}$ (i.e., previous two-day runoff data). The best RMSE score of 1.54 and 1.7 is measured for the case of GRU and LSTM, respectively, when these models were compiled using ADAM learning algorithm. Inclusion of all three algorithms into model building is important because there are no established theory to suggest which technique to use at what situation. Therefore, findings of this experiments are particularly important to understand the applicability of appropriate learning algorithms and its importance. Overall performance indicates that, GRU can be a potential solution for modeling rainfall-runoff relation with acceptable accuracy. Although there is not much to choose between both the models in terms of RMSE measure, however, overall performance of GRU seems to be better when the model training time is also considered.

## 4 Conclusion

The article presents an application of GRU for estimation of runoff in the Dikhow river basin. The simulation results as shown in Fig. 5 and RMSE value provided in Table 1 suggest that GRU has performed well and outperformed LSTM in this regard. It is also observed that with increase in the size of input sequence, the model performance improves accordingly. It was further observed that GRU can perform better in unearthing the hidden temporal relationship or dynamics present in the data.

Comparing the performance of GRU and LSTM, it is clear that both are comparable with different learning algorithm but the best result comes out from GRU. This is eminent from the RMSE values displayed in (Table 1) where different learning approaches were used both for GRU and LSTM. The differences are not that prominent to clearly distinguish which one is better. This is so because, the functionality of both the models is almost same, only differing by number of states.

Thus, it can be concluded that GRU can be used as a better alternative for modeling rainfall-runoff relationship with a satisfactory level of accuracy. The comparison with LSTM suggests that both the techniques are at par with each other with respect to RMSE value, but GRU outperforms LSTM in terms of overall training time. Further, experiments with multi-step look ahead condition and increased lag time might be helpful in testing full potentiality of both the networks.

## References

1. Chattopadhyay PB, Rangarajan R (2014) Application of ann in sketching spatial nonlinearity of unconfined aquifer in agricultural basin. Agric Water Manage 133:81–91
2. McCulluch W, Pitts W (1943) A logical calculus of the ideas imminent in nervous activity. Bull Mathe Biophys 5:115–133
3. Somers MJ, Casal JC (2009) Using artificial neural networks to model nonlinearity: the case of the job satisfaction-job performance relationship. Organ Res Methods 12(3):403–417
4. Ghose DK, Samantaray S (2019) Estimating runoff using feed-forward neural networks in scarce rainfall region. In: Satapathy SC, Bhateja V, Das S (eds) Smart intelligent computing and applications. Springer Singapore, Singapore, pp 53–64
5. Asadi H, Shahedi K, Jarihani B, Sidle RC (2019) Rainfall-runoff modelling using hydrological connectivity index and artificial neural network approach. Water 11(2):212
6. Kashani MH, Soltangeys R (2018) Comparison of three intelligent techniques for runoff simulation. Civil Eng J 4(5):1095–1103

7. Jimeno-Sáez P, Senent-Aparicio J, Pérez-Sánchez J, Pulido-Velazquez D (2018) A comparison of swat and ann models for daily runoff simulation in different climatic zones of peninsular spain. Water 10(2):192
8. Rumelhart DE, Hinton GE, Williams RJ et al (1988) Learning representations by back-propagating errors. Cognit Model 5(3):1
9. Kumar DN, Raju KS, Sathish T (2004) River flow forecasting using recurrent neural networks. Water Resourc Manage 18(2):143–161
10. Chiang YM, Hsu KL, Chang FJ, Hong Y, Sorooshian S (2007) Merging multiple precipitation sources for flash flood forecasting. J Hydrol 340(3–4):183–196
11. Bengio Y, Simard P, Frasconi P et al (1994) Learning long-term dependencies with gradient descent is difficult. IEEE Trans Neural Netw 5(2):157–166
12. Hochreiter S, Schmidhuber J (1997) Long short-term memory. Neural Comput 9(8):1735–1780
13. Kratzert F, Klotz D, Brenner C, Schulz K et al (2018) Rainfall-runoff modelling using long short-term memory (lstm) networks
14. Demir I, Xiang Z (2018) Runoff prediction using long-short term memory model. In: AGU fall meeting abstracts
15. Hu C, Wu Q, Li H, Jian S, Li N, Lou Z (2018) Deep learning with a long short-term memory networks approach for rainfall-runoff simulation. Water 10(11):1543
16. Liu Z, Xu W, Feng J, Palaiahnakote S, Lu T et al (2018) Context-aware attention lstm network for flood prediction. In: 2018 24th international conference on pattern recognition (ICPR). IEEE, pp 1301–1306
17. Yan L, Feng J, Hang T (2019) Small watershed stream-flow forecasting based on lstm. In: International conference on ubiquitous information management and communication. Springer, pp 1006–1014
18. Cho K, Van Merriënboer B, Bahdanau D, Bengio Y (2014) On the properties of neural machine translation: Encoder-decoder approaches. arXiv preprint arXiv:1409.1259
19. Jeong J, Park E (2019) Comparative applications of data-driven models representing water table fluctuations. J Hydrol
20. Zhang D, Peng Q, Lin J, Wang D, Liu X, Zhuang J (2019) Simulating reservoir operation using a recurrent neural network algorithm. Water 11(4):865
21. Sit M, Demir I (2019) Decentralized flood forecasting using deep neural networks. arXiv preprint arXiv:1902.02308
22. Caruana R, Lawrence S, Giles CL (2001) Overfitting in neural nets: Backpropagation, conjugate gradient, and early stopping. In: Advances in neural information processing systems, pp 402–408
23. Geman S, Bienenstock E, Doursat R (1992) Neural networks and the bias/variance dilemma. Neural Comput 4(1):1–58

# A Proposal for Predicting Missing Values in a Spectrum Dataset Using Supervised Learning

**Nayan Basumatary, Bhabesh Nath, and Nityananda Sarma**

**Abstract** Missing values which occurred frequently in our test bed spectrum dataset posed the challenge of knowing how to predict those values which can impact the learning of classification process for detection of primary user in a Cognitive Radio Network. Here, in our work we propose an Imputation method to predict the missing values based on supervised learning classification scheme. The proposed method first maps the missing value problem into a classification problem by discretization of the known available values. Further,we make use of C 4.5 decision tree algorithm for prediction of the discrete nominal values corresponding to the missing values. Finally,we predict the numeric values for the missing places using Local Closet Fit algorithm where the term local is defined by the discretization of the known values of the attribute with missing values. The performance of the proposed method is compared with the existing schemes for data imputation where the results show that the proposed method has higher prediction accuracy.

**Keywords** Cognitive radio network · Spectrum sensing · USRP · GNU radio · Spectrum dataset · Missing value · Data imputation · Supervised learning

## 1 Introduction

Missing data in a data set refers to an instance wherein no data value is stored for the variable in the observation of interest. Various problems are faced with missing data while mining the dataset [1]. Firstly, the absence of data reduces the probability that

N. Basumatary (✉) · B. Nath · N. Sarma
Department of Computer Science and Engineering, Tezpur University,
Tezpur, Assam, India 784028
e-mail: nayanb@tezu.ernet.in

B. Nath
e-mail: bnath@tezu.ernet.in

N. Sarma
e-mail: nitya@tezu.ernet.in

 
A. Maji et al. (eds.), *Proceedings of the International Conference on Computing and Communication Systems*, Lecture Notes in Networks and Systems 170,
https://doi.org/10.1007/978-981-33-4084-8_15

the test will reject the null hypothesis when it is false since it does not have the complete data. Secondly the lost data causes bias in the estimation of parameters. Also missing values reduces the significance of the samples obtained. Hence, missing values makes the analysis of dataset highly complicated and leads to invalid conclusions several times.Due to frequent occurrence of missing values in the training observation, prediction of the missing data has always remained at the center of attention of knowledge discovery in databases and data mining research community [2]. One could also think of discarding the instances with missing value but this would lead to loss of important information and inaccurate inference from about the data [3]. Hence prediction of missing data is a better choice than eliminating the instance as whole. A number of approaches for prediction of missing values have been devised over time. Some of these methods include concept mean method, k-means clustering [4], unsupervised learning [5], event covering, LEM2 [6] etc.

Missing data are inherent problem and common in data collection especially when dealing with real world dataset [7]. Missing values in the dataset make the analysis very unreliable which affects the decisions [8, 9]. Traditional methods cannot deal with missing values [10] as they require numeric values to be used in mathematical equation. Hence, if an observation has a missing value in any of the required variables, the whole observation (data record) needs to be omitted from the analysis. So, knowing how to predict those missing values is very important. Most of the existing learning methods are not well adapted to some application domains due to the difficulty with missing attributes values [11]. In case of classification, if missing attribute values appear in the training set, implying that in the training phase, a reliable method for dealing with those missing values is necessary, so that the classifier can be designed properly [12]. In, general incomplete data greatly affects the performance of classification algorithms [13–15]. The problem of missing values has been investigated in the last two decades and a number of strategies have been devised for dealing with missing values [16]. The simplest solution is to discard the data instances with missing values [17] but it may lead to loss of information [18]. The authors in [19] presents an empirical evaluation of missing values. One of the simplest method is to handle missing attribute values is to use the most common value of the attribute which has been implemented in [2]. An approach to handle missing data problem was presented in [1] which states that every instance worth missing attribute values can be replace by all possible know values in all the instances for that particular attribute. Another approach is obtained by combining replacing by attribute mean method and replacing by most common value [7]. A method known as the Global Closest Fit method [7] replaces a missing attribute by the known value in another instances that resembles as much as possible to the instances with the missing attribute value. An extension to [7] is known as Concept closest fit or Local Closest fit. Another method is the Event-Covering Method which is described in [20, 21], is a probabilistic approach to fill in the unknown attribute values. An another method discussed in [9] is based on the idea of assigning almost all possible values and then use K-means.Each new value is assigned and the data set is clustered using k-means.In [22] the authors address this issue with a novel deep learning technique to impute missing values in health data. Here, we propose a method for prediction of missing attribute values of our spectrum data in our sensing test bed. Our attribute in study having missing values in our dataset is normally distributed. The objective of this paper is to predict the missing value of an attribute using a supervised classification scheme for our test bed data which is very crucial for detection of a primary user in a cognitive radio network. The rest of the paper is organized as follows: In Sect. 2, the system model and the assumptions are presented. The spectrum dataset missing value problem is discussed in length along with different process involved in Sect. 3. In Sect. 4 the proposed method for prediction of missing values for an attribute in a dataset is explained. Section 5 shows the results and analysis of the proposed algorithm on weka tool. Section 6 finally concludes the paper along with the future scope of research in the present study.

## 2 System Model and Assumptions

The system model proposed is a homogeneous Cognitive Radio network which consist of both the Primary and Secondary users. The primary network is formed by '*N*' number of primary users denoted as $PU_1, PU_2, \ldots, PU_N$. There exists a set of licensed channels $C = 1 \ldots L$ used for primary transmissions.It is assumed that the channel occupancy states are independent, and the primary user can use any channel. The secondary network consists of a set of '*K*' users named secondary users (SU) denoted by $SU_1, SU_2, \ldots, SU_K$ are also present in the CR network which senses these primary channels in an opportunistic way in order to access the channels. The secondary users are equipped with certain learning techniques to detect if the channels are accessible and then for their own transmission to control such primary channels. Each user independently searches for a spectrum resource inside this secondary network. It is assumed that each secondary user could only sense one channel at a time, and that they would not communicate with each other. It is believed that both the main and secondary networks are synchronized with a global clock, and that transmission is performed in a time slotted way. In Cognitive Radio, spectrum sensing essentially decides if there is a signal present or absent between two channel states. To determine between two hypotheses, the features extracted from the samples obtained from the signal can be analysed:

$$H_0 : x(i) = n(i) \quad \text{(Signal is absent.)}$$
$$H_1 : x(i) = s(i) + n(i) \quad \text{(Signal is present.)}$$

The signal samples received are represented by $x(i)$, the primary user signal $s(i)$, and the $n(i)$ noise, where i denote the sample *i*th. The transmitting of the signal and the received signal is achieved in a continuous way. Here , in this paper the signal is treated as discrete , since the receiver or the sensor takes separate samples of the signal. Here, in case of indoor environment, the primary transmission in Line of sight propagation (LOS) between the transmitter antenna and the receiver antenna, hence the Rician distribution can describe the probability density function of the rapidly varying amplitude of the received instant signal. The environ-mental noise is assumed to be AWGN (Additive white Gaussian Noise). The average path loss PL($d$) for a transmitter and receiver with separation distance $d$ is

$$PL(d) = PL(d_0) + 10n\log(d/d_0)$$

where '$n$' is path loss component and indicates the rate at which path loss increases with distance $d$, with close in reference distance $d_0$.

# 3 Spectrum Dataset Missing Value Problem

## 3.1 *Cognitive Radio*

Cognitive radio [23] have been widely accepted as the emerging technology in the domain of new age wireless communication [24]. A cognitive radio (CR) is defined as a radio [25] that can change its transmission parameters based on the changes in environmental factors [26]. An important process involved with cognitive radio is spectrum sensing which is very important for opportunistic spectrum access [27]. Spectrum sensing has been identified as a key enabling functionality to ensure that cognitive radios would not interfere with primary users, by reliably detecting primary user signals.

The spectrum dataset which is generated is a real raw data generated from our CRN (Cognitive Radio Network) test bed used for implementing spectrum sensing which USRP based using Energy detection method in low SNR. In Fig. 1 the experimental setup is shown

Spectrum Data collection experiment is performed using GNU Radio and USRP1 device [28, 29]. Two USRP, one laptop and one PC was used for performing this experiment. The USRP are connected to the PC and Laptop through the USB connectors. This spectrum data experiment is performed indoor to sense the spectrum band with center frequency 2.426 GHz using different modulations. The band selected is the ISM band (2.4–2.5 GHz (WLAN /IEEE 802.11 b/g/n)) which is having 14 Wi-fi channels with channel 1, 6, 11 being non overlapping channels. Here, each channel is of 22 MHz bandwidth with lower, upper and center frequency [30]. During the spectrum sensing experiment a lot of dataset is collected by making the channel free and busy. The main attribute under observation is the power of the signal available in channel under low SNR condition.

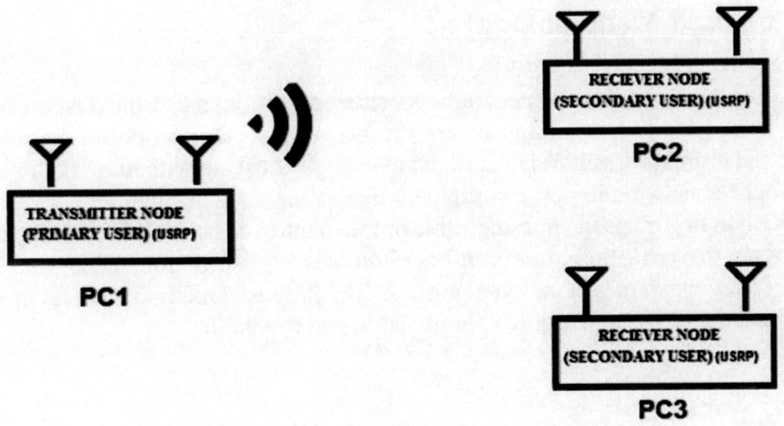


**Fig. 1** One pair transmitter and receiver spectrum sensing experimental setup

### *3.2 Dataset Creation*

In our sensing experimental setup for real time data set comprise of GNU Radio and USRP devices. In our CRN testbed we consider six PCs named as $PC_1$, $PC_2$, $PC_3$, $PC_4$, $PC_5$, $PC_6$ connected with respective USRP devices. Each PC (Transmitter/Receiver/Sensing Node) comprises of USRP device with inbuilt RFX 2400 daughter board. RFX 2400 daughter board can sweep across 8 MHz channel with 4 MHz each side from the Center Frequency. We performed the experiment in ISM band consisting of 14 Wifi channels. Out of 14 Wifi channels, 1 and 6 are non-overlapping channels. So we considered these two channels for sensing purpose. Since our USRP device can sweep only 8 MHz, so the channel bandwidth is defined as in Table 1.

During the preparation of data set from the spectrum sensing experiments conducted in our CRN test bed, we could see the certain values of attributes changing drastically at some interval of time. This may could be due to environmental factors involved while performing the experiments. Earlier, we discarded the values simply to perform our classification. But this makes the dataset to be incomplete and low rate of classification accuracy. Thus,it become important to know how to anticipate certain missing values.

**Table 1** Frequency range of Channel 1 and 6

| Channel No. | Low cut frequency (MHz) | Centre frequency (MHz) | High cut frequency (MHz) |
|---|---|---|---|
| Channel 1 | 2408 | 2412 | 2416 |
| Channel 6 | 2433 | 2437 | 2441 |

# 4 Proposed Methodology

Predicting missing values is generally considered to be a part of the data cleansing phase done before data mining or any further analysis.Our proposed method for prediction of missing values is restricted to a single attribute with numerical values. Here, our focus is in the power attribute which is the main attribute in our data set. This method first maps the missing value problem into a classification problem using a proposed discretization algorithm based on normal distribution. Then the values are predicted using classification algorithm. The proposed method consists of three modules namely Discretization, Classification and Prediction.

## 4.1 *Discretization*

The proposed Normal Distribution based discretization method consists of following steps:
Step 1: Take all the available instance of the attribute with the missing value.
Step 2: Find the maximum and minimum values for it.
Step 3: Compute the Mean ($\mu$) and Standard Deviation ($\sigma$).
Step 4: Partition into k number of classes based on ($\mu$) and ($\sigma$).

## 4.2 *Classification*

The classification in our proposed method is done using the C 4.5 classification method that generates classifiers expressed as decision tree [8]. It is one of the best decision tree algorithms that can be easily interpreted and can deal with noise.

## 4.3 *Prediction*

We use Local Closest Fit (LCF) approach [10] for performing the prediction of numerical value from the interval predicted by the trained classifier as output. The LCF algorithm works as follows.

Suppose a dataset $D_{\text{old}}$ having missing value in attribute $a_i$ is separated in two datasets $F$ and $M$ where in $A$ is the class label then the pseudo code for LCF algorithm is given as:
For each instance X in M with X.class = A do
Compute the distance (X, Y) between X and every instance
X such that Y.class = A
MinInstance = the instance with the minimum value of

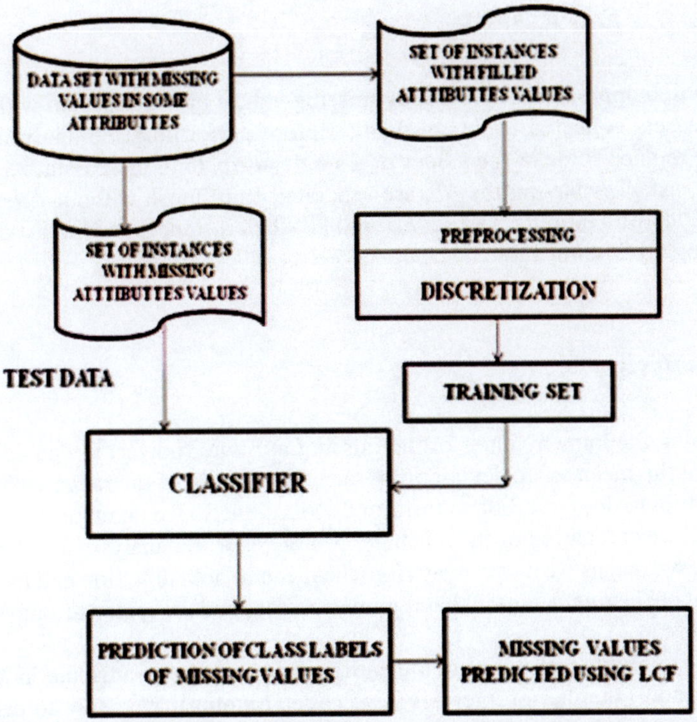


**Fig. 2** Flow chart of proposed prediction scheme

distance(X, Y)
X.$a_i$ = MinInstance.$a_i$
End For

Where the value of distance (*X*, *Y*) is calculated as $X_i - Y_i$ divided by the difference of maximum and minimum values in that class (*r*).

Figure 2 represents the proposed prediction algorithm in terms of a flowchart as follows.

We can briefly say that the proposed algorithm consists of following steps:
Step 1: Take the supplied data set $D_{\text{old}}$ with missing values in the attribute $a_i$.
Step 2: Split $D_{\text{old}}$ into two datasets *F* (containing all filled instances) and *M* (containing all instances with missing attribute values).
Step 3: Discretize the attribute $a_i$ in *F* using normal distribution based discretization.
Step 4: Build a C 4.5 classifier by training the dataset *F* with nominal values of $a_i$ as the target class.
Step 5: Test the dataset *M* on the above classifier to predict the nominal values corresponding to the missing values.
Step 6: Use the Local Closet Fit algorithm to predict the numeric value corresponding to the nominal value of $a_i$.

## 5 Analysis and Results

The proposed approach for predicting missing values has been tested with two different datasets collected in our testbed . Before performing the analysis we first manually replace some of the values of a single attribute in these datasets with "?". Then the instances having the "?" are separated from those without it resulting in the formation of a training data set $F$ (with all filled instances) and a test data set M (containing all missing value instances).

### 5.1 *Analysis*

By analysing the known values of the missing attribute (power) in consideration it is seen that the mean value deviate from the central value of the range which means that the attribute does not follow uniform distribution. So we do not use equal width binning for discretization of the attribute. Moreover while analysis the attribute we could observe that as we move away from the mean in both direction and frequencies of the attribute having values in that region decreases and the region around the mean is very dense.

First we apply discretization to the entire range for a given attribute in $K$ (where $K = 5, 7$ and 9) number of intervals (also called binning). We chose to discretized for the attribute into odd number of intervals. This is done because for any two values say ( $\mu - \sigma$) and ( $\mu + \sigma$) which is very close to each other, where $\sigma$ corresponds to very small change, if we chose the cut point as $\mu$ then this two values will deviate from each other. Based on this observation the choice of discretization interval is done. Table 1. shows for a 5 Bin Discretization the cut points, intervals and class label. The value of cut points are based on $\mu$ and $\sigma$

The 7 and 9 bins discretization are carried out in the similar way. After discretizing the known values of the missing power attribute, we get a new class label which corresponds to the discrete nominal values of the attribute and for the unknown values (?) we do not have the class label. Then removing the old class label we built the classifier model using C4.5 algorithm in WEKA using the new training set with the new class label, the classifier model is built which is then tested upon by the new test set.Hence, the new class label is predicted testing sample is predicted. Next we replace the predicted class label of the testing set by a continuous numerical value. Then the prediction of this continuous numerical value can be done by the attribute mean belonging to the same class, the most common value belonging to the same class and the local closest fit (Table 2).

**Table 2** Cut points, Intervals and class labels for 5 bin discretization

| Cut point (4) | Symbol (4) | Interval (5) | And class label (5) |
|---|---|---|---|
| $\mu - 3\sigma$ | A1 | [min – A1 ] | A |
| $\mu - \sigma/2$ | A2 | [A1, A2] | B |
| $\mu + \sigma/2$ | A3 | [A2, A3] | C |
| $\mu + 3\sigma$ | A3 | [A3, A4] | D |
| – | A4 | [A4, max] | E |

## 5.2 *Tool Used: Weka*

Weka stands for Waikato Environment for Knowledge Analysis which is a collection of many state of the art machine learning algorithms and data pre-processing tools [14]. It was developed at the University of Waikato in New Zealand. It provides extensive support for the whole process of experimental data mining, evaluating learning schemes statistically and visualizing results of learning algorithms.

## 5.3 *Results*

The results of the prediction of the missing value for out power attribute using the old class label as well as the new class label in the three prediction approaches is compared below for the datasets. Table 3 shows the comparison of the actual value and the predicted value for a value of each attribute using the old class label and new class label (5 bins discretization). Table 4 shows the results for 7 Bins discretization and Table 5 shows the results for 9 Bins discretization.

**Table 3** Actual values versus predicted value For different dataset (5 bin discretization)

| Power attribute value actual value | Concept mean predicted value (CM) | Proposed method with CM (Predicted value) | Most common value (MCV) predicted value | Proposed method with MCV (Predicted value) | Local closet fit (LCF) Predicted value | Proposed method with LCF (Predicted value) |
|---|---|---|---|---|---|---|
| 48 | 50 | 51 | 51 | 51 | 46 | 49 |
| 34 | 30 | 35 | 30 | 34 | 31 | 34 |
| 50 | 52 | 51 | 55 | 55 | 51 | 50 |
| 25 | 20 | 24 | 18 | 24 | 23 | 24 |
| 16 | 14 | 15 | 13 | 15 | 15 | 16 |

**Table 4** Actual values versus predicted value for different dataset (7 bin discretization).

| Power attribute value actual value | Concept mean predicted value (CM) | Proposed method with CM (Predicted value) | Most common value (MCV) predicted value | Proposed method with MCV (Predicted value) | Local closet fit (LCF) predicted value | Proposed method with LCF (Predicted Value) |
|---|---|---|---|---|---|---|
| 48 | 50 | 51 | 51 | 51 | 46 | 47 |
| 34 | 30 | 36 | 30 | 34 | 31 | 35 |
| 50 | 52 | 52 | 55 | 55 | 51 | 51 |
| 25 | 20 | 24 | 18 | 24 | 22 | 23 |
| 16 | 14 | 15 | 13 | 15 | 15 | 17 |

**Table 5** Actual values versus predicted value for different dataset (9 bin discretization).

| Power attribute value actual value | Concept mean predicted value (CM) | Proposed method with CM (Predicted value) | Most common value (MCV) predicted value | Proposed method with MCV (Predicted value) | Local closet fit (LCF) Predicted value | Proposed method with LCF (Predicted value) |
|---|---|---|---|---|---|---|
| 48 | 50 | 51 | 51 | 51 | 46 | 50 |
| 34 | 30 | 35 | 30 | 34 | 31 | 35 |
| 50 | 52 | 52 | 55 | 55 | 51 | 51 |
| 25 | 20 | 24 | 18 | 26 | 23 | 26 |
| 16 | 14 | 15 | 13 | 15 | 15 | 17 |

## 6 Conclusion and Future Work

In this work, we proposed a method for prediction of missing values in our spectrum data set collected in test bed based on classification scheme. The proposed method first maps the missing value problem into a classification problem by performing normal distribution based discretization of the known values of the missing attribute. Then it performs the prediction of the nominal value corresponding to the missing values using classification. Finally known prediction approaches are employed on the new data set to predict the values. In our case of missing value prediction of the spectrum data the 5 bin discretization based LCF prediction accuracy is more than the other methods. Our Future work would be to check whether the proposed method is also applicable for other attributes who does not follow a normal distribution. Also there would be attempt to detect whether the data is missing data or not in the Spectrum Data set.

## References


1. Grzymala-Busse JW (1991) On the unknown attribute values in learning from examples. Proc of the ISMIS-91. In: 6th International symposium on methodologies for intelligient systems, Charlotte, North Carolina, 16–19 Oct 1991
2. Clark P, Nblett T (1989) The CN2 induction algorithm, Mach. Learn. 3:261–283
3. Nikam SS (2015) A comparative study of classification techniques in data mining algorithms. Oriental J Comput Sci Technol 8(1):13–19. ISSN 0974-6471
4. Prajapati DJ, Prajapati JH (2011) Handling missing values: application to university data set, vol 1, issue 1. ISSN 2249-6149
5. Ssali G, Marwala T (2007) Estimation of missing data using computational intelligence and decision trees. In: Proceeding of IEEE international joint conference on neural networks, Hong Kong
6. Sugana S, Thanushkodi KG (2011) Predicting missing values using K-means clustering. J Comput Sci 7(2):216–224
7. Grzymala-Busse JW, Goodwin LK (2002) A comparison of three closest fit approaches to missing attribute values in preterm birth data. Int J Intell Syst 17:125–134
8. Kaiser J (2012) Algorithm For missing values imputation in categorical data with use of association rules. In: The ninth international conference on web-age information management. IEEE
9. Enders CK (2010) Applied missing data analysis. Guildford Press, Guidford
10. Luengo J, Garca S, Herrera F (2012) On the choice of the best imputation methods for missing values considering three groups of classification methods. Knowl Inf Syst 32:77–108
11. Sumathi A (2012) Missing value imputation techniques depth survey and an imputation algorithm to improve the efficiency of imputation
12. Patil DV (2010) Multiple imputation of missing data with genetic algorithm based techniques. IJCA special issue on "Evolutionary Computation for Optimization Techniques" (ECOT)
13. Rajnikanth TV (2013) An enhanced approach on handling missing values using bagging K-NN imputation
14. Poolsawad N, Moore L, Kambhampati C, Cleland JGF (2012) Handling missing values in data mining—A case study of heart failure dataset. In: 9th International conference on fuzzy systems and knowledge discovery (FSKD)
15. Fujikawa Y (2001) Efficient algorithms for dealing with missing values in knowledge discovery. Master degree thesis. Japan Advanced Institute of Science and Technology
16. Petrozillo A, Jordanov I (2017) Coulmn wise guided data imputation. Elsevier B.V
17. Lam D, Wei M, Wunsh D (2015) Clustering data of mixed categorical and numerical type with unsupervised feature learning. IEEE
18. Acuna E, Rodriguez C (2004) The treatment of missing values and its effect in the classifier accuracy. Springer, Berlin, Heidelberg, pp 639–648
19. Singh SK, Purwar A (2014) Empirical evaluation of algorithms to impute missing values for financial dataset. In: IEEE International conference
20. Chiu DK, Wong AKC (1986) Synthesizing knowledge: a cluster analysis approach using even covering. IEEE Trans Syst Man Cybern SMC-16:251–259
21. Wong KC, Chiu KY (1987) Synthesizing statistical knowledge for incomplete mixed-mode data. IEEE Trans Pattern Anal Mach Intell 9:796–805
22. Phung S, Kumar A, Kim J (2019) A deep learning technique for imputing missing healthcare data. In: 2019 41st Annual international conference of the IEEE engineering in medicine and biology society (EMBC), Berlin, Germany, pp 6513–6516. https://doi.org/10.1109/EMBC.2019.8856760
23. Haykin S (2005) Cognitive radio: brain-empowered wireless communications IEEE J Sel Areas Commun 23:201–220 (Stockholm)
24. Fette BA (2006) Cognitive radio technology. Newnes 656 p. ISBN-10:0750679522, ISBN-13:978-0750679527

25. Mitola J (2000) An integrated agent architecture for software defined radio. Ph.D. dissertation. KTH Royal Institute of Technology, Stockholm, Sweden
26. I. F. Akyildiz, W.Y. Lee, M. C. Vuran S. Mohanty, "NeXt generation/ dynamic spectrum access/cognitive radio wireless networks: A survey" Computer Networks", vol.50, no.13, pp. 2127–2159",2006
27. Software communications architecture specification (version 2.2.2). Joint Program Executive Office (JPEO), Joint Tactical Radio System (JTRS), 15 May 2006
28. Sarijari M et al (2009) Energy detection sensing based on GNU radio and USRP: an analysis study. In: IEEE 9th Malaysia international conference on communications
29. Dhar R et al (2006) Supporting integrated MAC and PHY software development for the USRP SDR. In: 1st IEEE workshop on networking technologies for software defined radio networks
30. Nafkha A et al (2014) Experimental spectrum sensing measurements using USRP software radio platform and GNU radio. In: 9th International conference on cognitive radio oriented wireless networks (CROWNCOM)

# Clustering-Based Techniques for Big Data Analysis of Gene Expression

**Tanuja Das, Partha Pratim Kalita, and Goutam Saha**

**Abstract** Proper investigation of cancer has always been of foremost importance for its accurate forecasting, thereby aiding the correct cure. Microarray-based gene expression profiling is being practised for this purpose making it one of the leading research interests for discovering gene clusters accountable for a particular behavior. Big data analytics provides an efficient way to seek facts about the biological processes inherent from this microarray data. Previously, many attempts have been made to achieve this using numerous clustering approaches, but the results were quite deviating from the reality. In this work, we have attempted to discover potential and accurate gene indicators from the gene expression data by using a well-known quantitative measure called quantum clustering. The characteristic feature of this concept is that the total estimate of clusters formed is not predetermined but is determined depending on the nature of the data. As the concept is established on the grounds that a cluster is formed by density wise spaces, where the center is formed based on the density maxima point, this motivated us to detect those clusters which may be engaged in a certain biological process. The clustering approach becomes privileged in that extremely dense spaces are inherently detected and combined to produce arbitrarily shaped clusters without regarding the dimension of the space. For the purpose of comparing the results obtained, we have also applied a non-parametric measure, namely, the mean shift clustering on the gene expression data. For validation purpose, we used DAVID to check the significance of the clusters created. Results show that the genes so discovered are highly indicative in the pursuit of rare diseases.

**Keywords** Quantum clustering · Mean shift clustering · Microarray data · Big data · Gene expression data

T. Das (✉) · P. Pratim Kalita
Department of Information Technology, GUIST, Guwahati, Assam, India
e-mail: tanujadas55@gmail.com

P. Pratim Kalita
e-mail: parthakalita142@gmail.com

G. Saha
Department of Information Technology, NEHU, Shillong, Meghalaya, India
e-mail: dr.goutamsaha@gmail.com



A. Maji et al. (eds.), *Proceedings of the International Conference on Computing and Communication Systems*, Lecture Notes in Networks and Systems 170,
https://doi.org/10.1007/978-981-33-4084-8_16

# 1 Introduction

Due to the rapid progress of DNA microarray technology, huge amount of data is being produced which is categorized as big data [1]. The easy access to these big datasets contributes to unparalleled possibilities, but it also introduces novel challenges in the analytics process. All the data about a cell like its nature, state, stage and environment can be deduced from the highly dynamic gene expression data [2]. Also, information about a pathological situation or a homeostasis reaction can be latent in these data, which in turn can contribute to synthesis of novel drugs, investigation of diseases [3] or modeling biological pathways [4]. A limited number of genes from this microarray data is effective for extracting biologically relevant information. Thus, by identifying those marker genes, the cost of decisions of biological experiments can be significantly reduced [5].

Many clustering techniques have been used previously to find correlations amongst the genes [6], like the k-means clustering [7], hierarchical clustering [8], etc., from the gene expression data. These conventional clustering techniques are generally established on probabilistic views [9], and the clustering is done depending only on conditions. Then, many bi-clustering approaches [10] were developed which aim to cluster gene expression data along with the conditions. The most notable ones are being the plaid model approach [11], xMotif [12], BiMax [13], OPSM [14], Bicluster [15], BCC [16] and ROCC [17]. The issue with these methods is that the total estimate of clusters has to be determined previously and so the locations of the samples in the data-space may not be defined clearly, leading to incorrect results.

In this paper, we have applied two variants of density-based clustering algorithms, namely quantum clustering and mean shift clustering on the gene expression data. The aim is to detect the genes which are responsible for a particular behavior. The significance of these algorithms is that the clusters are estimated automatically depending on the density of the dataset. Also, we have validated the results obtained from the techniques using DAVID [18], which infers that the results obtained are biologically relevant.

# 2 Background

This section consists of a brief summary of the background of our research. Sect. 2.1 discusses the basics of quantum clustering. In Sect. 2.2, a brief idea about mean shift clustering is given.

## 2.1 *Quantum Clustering*

The quantum clustering algorithm is a density-based clustering approach build on the physical insight developed from quantum mechanics [19]. The algorithm starts by assigning a Gaussian [19],

$$\psi(x_i) = \sum_{j=1}^{N} e^{-\frac{(x_i - x_j)^2}{2\sigma^2}} \tag{1}$$

with width $\sigma$ to each data points.

The potential function $V(x)$ associated with the data points is calculated [17] as :

$$V(x) = \frac{1}{2\sigma^2} \sum_i (x - x_i)^2 e^{-\frac{(x-x_i)^2}{2\sigma^2}} \tag{2}$$

In Eq. 2, the potential $V(x)$, for each data points, attracts the distribution $\psi$, whereas the parameter $\sigma$ is the source of diffusion of the data points. This clustering approach determines the minimum values of $V$, which corresponds to the cluster centers [19]. From the minimum values of $V$ obtained, the data points can be assigned to the respective clusters using the gradient descent algorithm [20]. Data converging to the same minimum points belongs to the same cluster.

## 2.2 *Mean Shift Clustering*

The mean shift clustering method [21] is a non-parametric iterative algorithm used for clustering purpose based on kernel density estimation.

In machine learning, kernel is simply a weighing metric which is usually used in convolution [8]. The kernel density estimate [22] of $n$ datapoints $x_i$, $i = 1, 2, \ldots, n$), with kernel $k(x)$ and window radius $h$, is given by the following equation:

$$f(x) = \frac{1}{nh^d} \sum_{i=1}^{n} k^{\frac{(x-x_i)^2}{h}} \tag{3}$$

The values of the density function occurring most frequently are calculated from the gradient function $\nabla f(x) = 0$ which are positioned at the zeroes. So, Eq. 3 on algebraic manipulation becomes

$$\nabla f(x) = \frac{2c_{k,d}}{nh^{d+2}} \left[ \sum_{i=1}^{n} g\left( ||\frac{x - x_i}{h}||^2 \right) \right] \left[ \frac{\sum_{i=1}^{n} x_i g(\frac{y_t - x_i}{h})^2}{\sum_{i=1}^{n} g(\frac{y_t - x_i}{h})} - x \right] \tag{4}$$

The first term in Eq. 4 denotes the density evaluation of the data point $x$, and the corresponding term denotes the mean shift vector, which gives the direction of data points shifting towards the maximum.

The mean shift vector computes a series of estimation points $y_{t\,(t=1,2,\ldots)}$ the equation for which is

$$m_h(x) = \frac{\sum_{i=1}^{n} x_i g\left(\left\|\frac{x-x_i}{h}\right\|\right)^2}{\sum_{i=1}^{n} g\left(\left\|\frac{x-x_i}{h}\right\|\right)} - x \tag{5}$$

Points which converge towards the same estimation points are assembled into the same cluster.

## 3 Methodology

This section describes the various procedure used in our work.

Procedure A is the nonlinear clustering approach of quantum clustering used for clustering data points belonging to the same minima as cluster centers.

Procedure A: QC
Input: instances $instance(i)$
width parameter: $\sigma$
Output: clusters $v(k)$
$n$ = the number of instances
for $i = 1$ to $n$
    $x = instance(i)$
    $V(i) = \frac{1}{2\sigma^2} \sum_i (x - x_i)^2 e^{-\frac{(x-x_i)^2}{2\sigma^2}}$
    Density of a point where the minima is zero
        $\nabla V(i) = 0$
Return limiting values of $V(i)$
End Procedure

Procedure B depicts the mean shift clustering approach which assigns the data points to the clusters iteratively by shifting the data based on kernel density estimation.

Procedure B: MSC
Input: Data points $x(i)$
Output: cluster centers $y(i)$
Repeat until convergence, i.e., $\nabla f(x) = 0$
    Calculate the mean shift vector $m_h(x^t)$ from Eq. 5

$$m_h(x^t) = \frac{\sum_{i=1}^{n} x_i g\left(\left\|\frac{x^t-x_i}{h}\right\|\right)^2}{\sum_{i=1}^{n} g\left(\left\|\frac{x^t-x_i}{h}\right\|\right)} - x^t$$

    Calculate the density estimation points $x_{t+1}$

$$x_{t+1} = \frac{\sum_{i=1}^{n} x_i g\left(\left\|\frac{x^t-x_i}{h}\right\|\right)^2}{\sum_{i=1}^{n} g\left(\left\|\frac{x^t-x_i}{h}\right\|\right)}$$

Return limiting values of $x(t)$
End Procedure

# 4 Results and Discussion

The two density-based approaches, viz, quantum clustering and mean shift algorithm were used on four gene expression datasets, the details of which are given in Table 1. The objective was to detect genes accountable for a particular disease from the respective datasets.

Clusters with similar co-expression values are found by the two clustering algorithms. Table 2 demonstrates the effectiveness of quantum clustering (QC) and mean shift clustering(MSC) based on normalized mutual information [27].

We try to verify how this genes may be responsible for a particular disease by using a tool named DAVID [18]. DAVID assists researchers to comprehend the inherent interpretation of the huge number of genes [18]. Using the GO finder function [18] of DAVID, we find the effect of the genes for a particular disease. The details of the most relevant GO terms determined from the quantum clustering approach are given in Table 3.

Also, when searched against the OMIM and GAD list, those marker genes were present there as well. Online Mendelian Inheritance in Man (OMIM) is a continuously updated database of genes and genetic abnormalities and attributes of organisms, with a specific interest in the gene-phenotype relationship [28]. The Genetic Association Database (GAD) on the other hand is a database for associating gene and complicated diseases [29].

**Table 1** Dataset description

| Dataset | No. of genes | No. of samples | Classification of samples | |
|---|---|---|---|---|
| Breast cancer [23] | 7129 | 49 | 25 Positive | 24 Negative |
| Leukemia [24] | 7070 | 72 | 47 Acute | 25 Myeloid |
| Colon [25] | 2000 | 62 | 40 Tumor | 22 Normal |
| RAOA [26] | 18000 | 30 | 21 RA | 9 OA |

**Table 2** Effectiveness of the two density-based algorithms on the four gene expression datasets based on NMI

| Dataset | Mean shift clustering | Quantum clustering |
|---|---|---|
| Breast cancer | 96.4 | 98.0 |
| Leukemia | 97.6 | 98.6 |
| Colon | 98.5 | 98.8 |
| RAOA | 97.1 | 98.1 |

**Table 3** Significant GO terms for genes in the clusters obtained by quantum clustering

| Dataset | Biological processes |
|---|---|
| Breast cancer | Expression of pAkt, GLUT1 and TKTL1 was higher in breast cancer and DCIS than in normal tissue |
| | M2PK expression was highest in normal breast tissue |
| Leukemia | The expression level of SPINK2 is significantly elevated in most leukemia ell lines except B-lymphoblast TK-6 cells |
| | The inability of neutrophils and phagocytes to kill microbes that they have ingested |
| Colon | Hypermethylation of the spastic paraplegia-20 promoter is a highly specific and sensitive biomarker for screening colorectal cancer in stool samples as a non-invasive method |
| | Up-regulation of SPG20 expression, brought about by trisomy 13 in colon cancer cells trisomy 13 amniocytes, is sufficient for the cytokinesis failure phenotype |
| RAOA | Recombinant transgene facilitates the transcription of pro-inflammatory cytokines in fibroblast like synoviocytes in rheumatoid arthritis |
| | Recurrent mutation of the gene encoding sequestosome 1 (SQSTM1/p62) in paget disease of bone |

The genes which pass the *p*-value are plotted in the form of heat maps as shown in Fig. 1. A matrix of genes has been shown plotted against the samples, showing the gene expression levels or ratios. It represents the clusters of genes and acts as a mechanism to detect the co-expressed genes.

## 5 Conclusion

Huge amount of gene expression data is being produced using microarray technology. The proper analysis of this big data remains an important research aspect of the day as it consists of priceless genetic information. Through this work, we have tried to achieve clustering of the gene expression data using the quantum clustering and mean shift clustering algorithms. The essential characteristic of these two algorithms is that the number of clusters need not be predetermined but is determined by the density of the data.

The clusters computed by the algorithms, on introducing to a tool named DAVID, which is basically used for annotation, are validated in terms of accurately uncovering the genes relevant for the particular disease. For future work, we would like to incorporate some additional clustering constraints to these algorithms, so that the results of clustering can directly be inferred without using any supplementary tools.

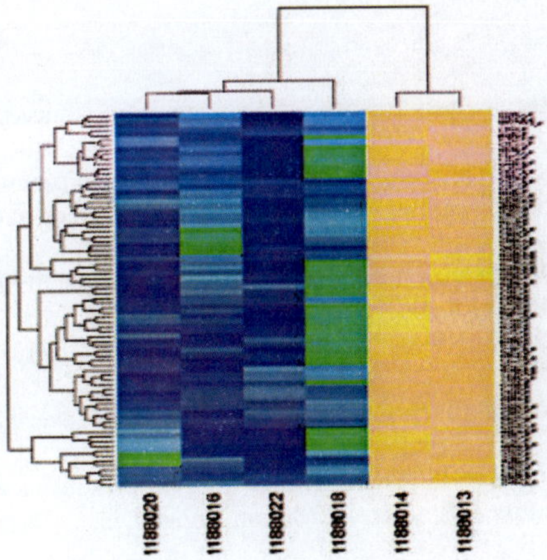


(a) Heatmap for Breast Cancer Dataset

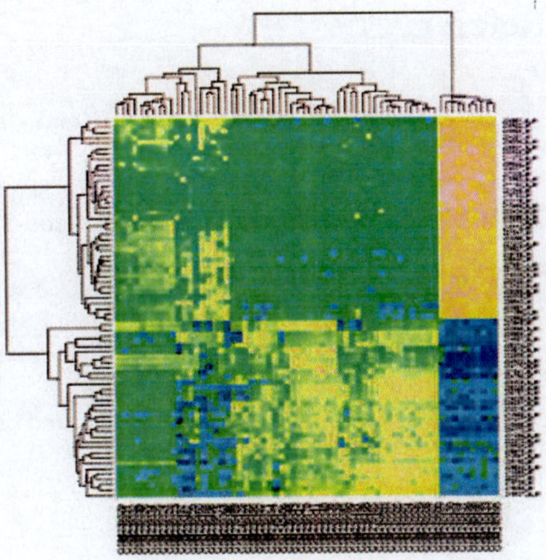

(b) Heatmap for Leukaemia Dataset

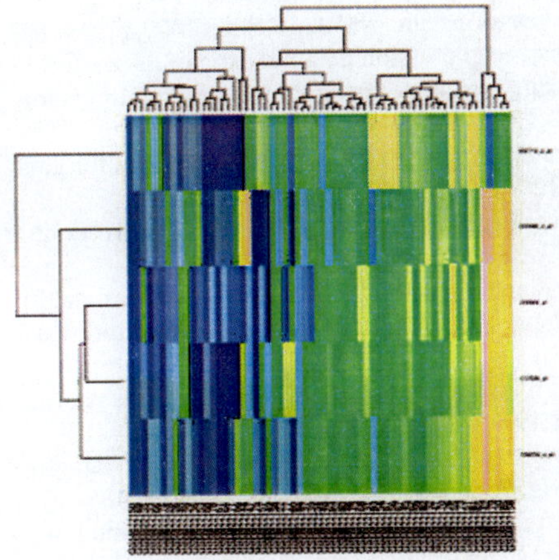

(c) Heatmap for Colon Cancer Dataset

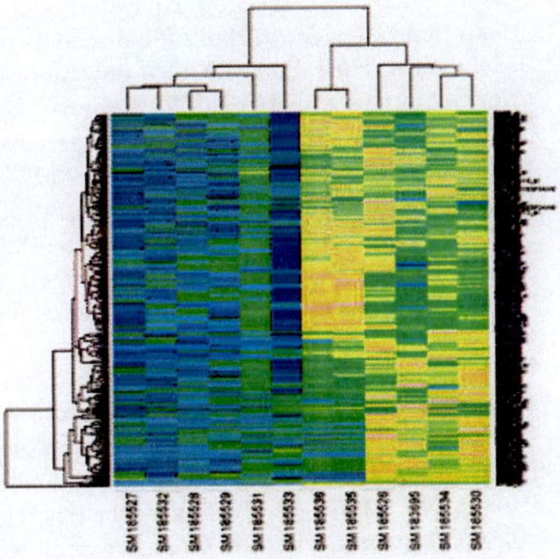

(d) Heatmap for RAOA Dataset

**Fig. 1** Heatmaps of the expression values for the four gene expression datasets

## References

1. Jiang D, Tang C, Zhang A (Nov 2004) Cluster analysis for gene expression data: a survey. IEEE Trans Knowl Data Eng 16(11):1370–1386
2. Board FS (2017) Artificial intelligence and machine learning in financial services. http://www.fsb.org/2017/11/artificialintelligence-and-machine-learning-in-financialservice/. Accessed 30 Jan 2018
3. Maji P (2012) Mutual-information-based supervised attribute clustering for microarray sample classification. IEEE Trans Knowl Data Eng 24(1):127–140
4. Pita-Juarez et al (2018) The pathway coexpression network: revealing pathway relationships. PLoS Comput Bifol 14(3):e1006042
5. Kim J, Shin M (2017) Inferring genes and biological functions that are sensitive to the severity of toxicity symptoms. Int J Mol Sci 18(4):755
6. Kriegel HP, Kroger P, Zimek A (2009) Clustering high-dimensional data: a survey on subspace clustering, pattern-based clustering, and correlation clustering. ACM Tran Knowl Discov Data (TKDD) 3(1):1
7. Huang Z (1998) Extensions to the k-means algorithm for clustering large data sets with categorical values. Data Mining Knowl Discov 2(3):283–304
8. Breyne P, Zabeau M (2001) Genome-wide expression analysis of plant cell cycle modulated genes. Current Opin Plant Biol 4(2):136–142
9. Fukunaga K (2013) Introduction to statistical pattern recognition. Academic Press
10. Cheng Y, Church GM (2000) Biclustering of expression data. ISMB, vol 8, no 2000, pp 93–103
11. Lazzeroni L, Owen A (2002) Plaid models for gene expression data. Statistica Sinica 61–86
12. Abdullah A, Hussain A (2006) A new biclustering technique based on crossing minimization. Neurocomputing 69(16):1882–1896
13. Preli A et al (2006) A systematic comparison and evaluation of biclustering methods for gene expression data. Bioinformatics 22(9):1122–1129
14. Ben-Dor A et al (2003) Discovering local structure in gene expression data: the order-preserving submatrix problem. J Comput Biol 10(3–4):373–384
15. Cho H et al (2004) Minimum sum-squared residue co-clustering of gene expression data. In: Proceedings of the 2004 SIAM international conference on data mining, society for industrial and applied mathematics, pp 114–125
16. Banerjee A et al (2007) A generalized maximum entropy approach to Bregman co-clustering and matrix approximation. J Mach Learn Res 8:1919–1986
17. Deodhar M et al (2008) Hunting for coherent co-clusters in high dimensional and noisy datasets. In: IEEE international conference on data mining workshops ICDMW08. IEEE, pp 654–663
18. Huang DW (2007) DAVID bioinformatics resources: expanded annotation database and novel algorithms to better extract biology from large gene lists. Nucl Acids Res 35(suppl 2) W169–W175
19. Horn D, Gottlieb A (2002) The method of quantum clustering. In: Advances in neural information processing systems, pp 769–776
20. Sebastian R (2016) An overview of gradient descent optimization algorithms. vol 1609, no 04747
21. Fukunaga K, Hostetler L (1975) The estimation of the gradient of a density function, with applications in pattern recognition. IEEE Trans Inf Theory 21(1):32–40
22. Comaniciu D, Meer P (2002) Mean shift: a robust approach toward feature space analysis. IEEE Trans Pattern Anal Mach Intell 5:603–619
23. West M et al (2001) Predicting the clinical status of human breast cancer by using gene expression profiles. Proc Natl Acad Sci 98(20):11462–11467
24. Golub TR, Slonim DK, Tamayo P, Huard C, Gaasenbeek M, Mesirov JP, Coller H et al (1999) Molecular classification of cancer: class discovery and class prediction by gene expression monitoring. Science 286(5439):531–537

25. Alon U, Barkai N, Notterman DA, Gish K, Ybarra S, Mack D, Levine AJ (1999) Broad patterns of gene expression revealed by clustering analysis of tumor and normal colon tissues probed by oligonucleotide arrays. Proceedi Natl Acad Sci 96(12):6745–6750
26. Van der Pouw Kraan TCTM et al (2007) Rheumatoid arthritis subtypes identified by genomic profiling of peripheral blood cells: assignment of a type I interferon signature in a subpopulation of patients. Ann Rheum Dis 66(8):1008–1014
27. Liu X, Cheng HM, Zhang ZY (2019) Evaluation of community detection methods. IEEE Trans Knowl Data Eng
28. Hamosh A et al (2005) Online Mendelian Inheritance in Man (OMIM), a knowledgebase of human genes and genetic disorders. Nucl Acids Res 33(suppl 1):D514–D517
29. Becker KG et al (2004) The genetic association database. Nature Gen 36(5):431–432

# A Review on Protein Structure Classification along with a Proposed Classifier Using Data Mining Techniques

**Suprativ Saha, Saptaparni Sarkar, and Tanmay Bhattacharya**

**Abstract** Protein structure classification is an approach to classify unknown protein into its class using the structural properties of the protein. As new protein structures are rapidly increasing, the need for efficient and automated data mining techniques for classifying proteins into classes with high accuracy and low computational time is also increasing. There are numerous techniques which are available for Protein structural classification, some of them have been reviewed in this paper. To execute the same, various researchers have proposed many efficient data mining approaches like representing protein structures in a hierarchy tree structure, random forest approach, amino acid residue's averaged chemical shifts (ACS), Hidden Markov Model (HMM), etc. A comparative analysis is also performed with the limitations of above approaches to find the gap of research in this field. Finally, a new classification model is proposed based on the limitation of the above approaches, which can classify the protein with better accuracy and low computational time.

**Keywords** PSIMAP · CASP · MCC · NMR · ACS · HMM

## 1 Introduction

Protein structure refers to the 3-D ordering of particles or atoms in a chain of amino acid molecules. Proteins are polymers, precisely polypeptides that are generated from arrangements of amino acids in a sequence. Four levels of protein structures

S. Saha (✉) · S. Sarkar
Department of Computer Science and Engineering, Brainware University, Barasat, Kolkata 700125, India
e-mail: reach2suprativ@yahoo.co.in

S. Sarkar
e-mail: saptaparni.sarkar91@gmail.com

T. Bhattacharya
Department of Information Technology, Techno India, Saltlake, Kolkata 700091, India
e-mail: dr.tb1029@gmail.com



A. Maji et al. (eds.), *Proceedings of the International Conference on Computing and Communication Systems*, Lecture Notes in Networks and Systems 170,
https://doi.org/10.1007/978-981-33-4084-8_17

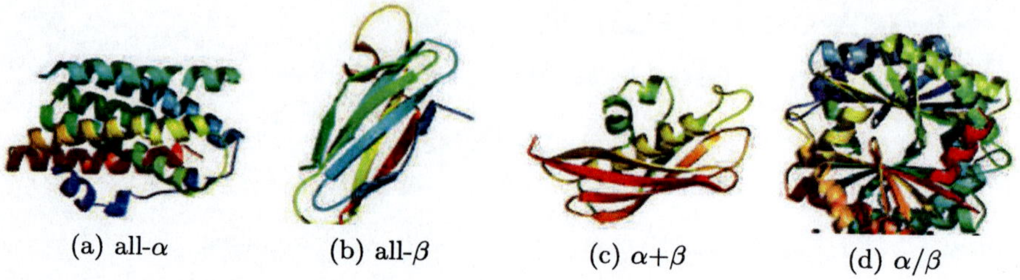

Fig. 1 Pictorial representation of structural fold classes of protein

are defined which are (1)'**Primary Structure**' alludes to the ordering of amino acids, held by peptide in a chain of polypeptide. (2) **Secondary Structure** alludes to extremely structured local sub-structures on the original polypeptide backbone, that were defined by ordering of hydrogen bonds between the groups of main chain peptide. Two main secondary structures are '$\alpha$-helix' and the '$\beta$-strand' or '$\beta$-sheets'. (3) **Tertiary Structure** alludes to the 3-D structure made of monomeric and multimeric protein molecules. (4) **Quaternary structure** also alludes to 3-D structure made of two or more individual polypeptide chains (subunits) which work as a single functional unit (multimer).

The protein structural classes are highly related with the composition of its amino acid. The four structural fold classes are defined as (1) '**all-$\alpha$**' (Fig. 1a) proteins represent a part of structural domains where the secondary structure made of '$\alpha$-helices' completely, except the periphery where few isolated '$\beta$-sheets' are present. (2) '**all-$\beta$**' (Fig. 1b) proteins represent a part of structural domains where the secondary structure made of '$\beta$-sheets' completely, except the periphery where few isolated '$\alpha$-helices' on the periphery. (3) '$\alpha + \beta$' (Fig. 1c) proteins represent a part of structural domains where the secondary structure made of '$\alpha$-helices' and '$\beta$-strands' that appear independently on the backbone of the chain. Hence the '$\beta$-strands' are mainly antiparallel. (4) '$\alpha/\beta$' (Fig. 1d) proteins represents a part of structural domains where the secondary structure made of alternating '$\alpha$-helices' and '$\beta$-strands' on the backbone and therefore the '$\beta$-strands' are mainly parallel.

Here, a literature review of different ways of protein structure classification is elaborated in Sect. 2, then a comparative analysis with the limitation of those models is done in Sect. 3. Section 4 proposes a new classification approaches and lastly a conclusion is presented in Sect. 5.

## 2 Literature Review

### *2.1 Hierarchy Tree Structure Model*

In paper [1], Rahman et al. had presented a model based on the hierarchy tree structure that classify unknown protein into its proper structure class. This proposed approach selected six major features of a protein that were (1) 'Sequence Comparison' (2)'Structure comparison' (3) 'Cluster Index' (4)'Connectivity' (5) 'Taxonomic' (6) 'Interactivity' [2, 3] for classifying the protein taken from the databank. The classification is done by creating a hierarchical tree structure based on the calculation of probability or degree of likeness of each new protein against the proteins which are already existing in the system, found by applying probability theorem on all the above mentioned six properties. At first, the general probability function was applied to calculate by $P(X) = \text{Total Outcome}/\text{Expected Outcome}$, where $P(X)$ is the probabilistic function of event $X$. After that, Bond factor is calculated using the probabilistic function of all the events. i.e. $P(\text{Total}) = P(X_1) + P(X_2) + P(X_3) + P(X_4) + \cdots + P(X^n)$. Then the above mentioned six properties were used on unknown proteins like $P(z_1.\,z_2.\text{Structure})$ = Similarity between $z_1$ and $z_2$ with respect to structure/expected similarity of $z_1$ and $z_2$ with respect to structure. In the same way others properties i.e. $P(z_1.z_2.\text{Connectivity})$, $P(z_1.z_2.\text{Cluster index})$, $P(z_1.z_2.\text{Interactivity})$, $P(z_1.z_2.\text{Taxonomic and age diversity})$ were also calculated. Finally total probability of $z_1$ with respect to $z_2$ i.e. $P(z_1.z_2) = P(z_1.z_2.\text{Structure}) + P(z_1.z_2.\text{Sequence}) + P(z_1.z_2.\text{Connectivity}) + P(z_1.z_2.\text{Cluster index}) + P(z_1.z_2.\text{Interactivity}) + P(z_1.z_2.\text{Taxonomic and age diversity})$ was identified. Then, the combined probabilistic value was compared with the probabilistic value of that protein that belonged to the root for arranging them in a hierarchy tree. Finally this model can claim the 98% accuracy of the classification but problem is that, to implement this model data by data analysis is required. In this case, knowledge-based implementation is missing which could enhance the accuracy level of the classification. Additionally, hand computational time is also very high of this model which should be reduced.

## 2.2 *ProteinNet: A Standard Data Set for Machine Learning Models of Protein Structure*

In Paper [4], AlQuraishi had found a standard data set for machine learning models of protein structure named as ProteinNet. ProteinNet is a type of data sets that was created for providing a structured method for machine learning to assess the relationship between Protein sequence and structure. This approach had three desirable properties which includes, firstly influencing the selection of structures by using 'CASP (Critical Assessment of protein Structure Prediction)' [5] structures for data set that solemnly vary from the 'PDB' structures [6] at a given point of time. Secondly, multiple validation sets with different sequence identity levels were used by 'ProteinNet' to provide intermediaries to both 'TBM (20′40% seq. id.)' and 'FM (<10% seq. id.)' CASP proteins. Thirdly, this approach encourages CASP to appreciate the involvement of all vital predictors as it is considered as standard for evaluating structure prediction methods. Many versions of ProteinNET were released with more improvement in the maximum percentage sequence identity to training set with highest percentage as 90%. There are few drawbacks in this approach which we can conclude is that only one type of file format mmCIF is used for assessment which is a huge limitation and, also many other algorithms or models might not be able to support the data sets ProteinNet for their analysis purposes. This approach might also have higher computational time as it includes involvement of third-party tools.

## 2.3 Random Forest Model

In paper [7], Jain et al. had proposed a classifier using random forest for automated structural classification of small proteins. Random forest, a group based supervised machine learning algorithm, was used to find the SCOP [8] structural classification for a target protein structure on the basis of likeness of its structural features with a sample structure having an equivalent number of 'secondary structure elements (SSEs)'. It was considered that 'precision', 'recall' and 'Matthew's correlation coefficient (MCC)' had the ability to differentiate Protein pairs and predict the shared structural level accordingly [9]. This approach achieves up to 94% of predictive accuracy on a non-overlapping and independent test set taken from 'SCOP version 1.73'. There is one drawback in this approach which is, constructing a complete decision tree has time complexity of $O(w * r\log(r))$, where r denotes number of records and w denotes number of variables/attributes. So, with the increase in the number of Proteins samples and in the number of SSEs (attributes), the time complexity of this algorithm also increases. Hence the execution time of this algorithm on larger Protein domains is highly increased which is not favourable.

## 2.4 NMR Chemical Shifts Model Using Protein Residues

In Paper [10], Arun et al. had proposed an efficient method for classification of protein structure using averaged 'NMR chemical shift's based on protein residues. Two databases BioMagResBank (BMRB) [11] and RefDB [12] have been used to obtain the structure files. These chemical shifts were reduced to per-residue 'averaged chemical shift (ACS)' with the formula $AA^{K}_{\mathrm{ACS}(j)} = \frac{1}{\mathrm{N}x} \sum_{n=1}^{\mathrm{N}x} \mathrm{CS}(j, n)$ for nuclear species "*j*". It was stated in the paper that in this formula $j$ = $^{12}$CO, $^{12}$C$\alpha$, $^{13}$C$\beta$ , $^{1}$HN, $^{1}$H$\alpha$ or $^{15}$N; $N^{x}$ denotes the total number of residues of type '*x*'. CS(*j*, *n*) denotes the CS value of the *j*th nucleus at the *n*th residue of type '*x*'. Using this approach, each protein had been categorized into one of the six structural classes of SCOP: (1) 'all alpha' proteins ($\alpha$), (2) 'all beta' proteins ($\beta$), (3) 'mixed $\alpha\beta$' proteins ($\alpha\beta$), (4) 'coiled coil', (5) 'small proteins' and (6) 'membrane/cell surface proteins and multi-domain proteins' [13–15]. This approach gives different accuracy level for different algorithms with highest as 94% when used in Random Forest algorithm. So, the major drawback of this approach is that it cannot give constant accuracy level for all the algorithms used for implementing average NMR chemical shifts as it differs with different algorithms. It is also observed that this method is used only for proteins having 50 or more amino acid residues and having chemical shifts assigned to at least 70% of their residues. Hence there is a scope of improvement here so as to include all the available proteins for structural classifications.

### 2.5 *Hidden Markov Model Based Classifier*

In paper [16], Mirceva et al. had built a separate Hidden Markov Model (HMM) for each Protein fold class. The HMM generated a protein sequence by producing symbols while going through a series of states having probabilities like the '*transition probability*' $T_{mn}$ where a state $s_m$ will move to another state $s_n$ and the '*emission probability*' $E(y|k)$ where a state $s^k$ will produce a symbol $y$ [17]. It was implemented using the Viterbi [18] or forward algorithm [19] to find the most probable path of a Protein sequence. The primary idea of the approach was to use HMM to structure the protein backbone folding around its center of mass. Hence the alignment of tertiary structures of protein molecules was done in same way as 3-D HMM model [20]. But, it was observed that this methodology was more accurate than 3-D HMM approach. This model claims 92% accuracy of the classification, but here the problem is data by data analysis is required to implement this mode and, also the accuracy % needs to be increased so that this model can be considered as base model for Protein structure classifications.

## 3 Comparative Analysis

See Table 1.

## 4 Proposed Model

A model has been proposed for classifying an unknown protein sequence to protein class based on the structural classification of protein with low computational time and higher accuracy. This proposed model (Fig. 2) consists of primarily several feature extraction procedures which are applied on the neural network base classifier.

Firstly, this model is used to create the training dataset which is required for the structural classification for an unknown protein sequence. The first aim is to reduce the input dataset from a given number of known protein sequences and then the focus is to build an efficient training dataset which will be able to classify any unknown protein sequence with higher accuracy. Later, this model is used for an unknown protein to classify into one of the four structural classes. The reduced dataset is achieved by the mean over distance in Di-sulphide bond of protein, a feature extraction procedure of the protein sequence and then comparing it with the

**Table 1** Comparative analysis of existing classification models with accuracy percentage and limitation

| Name of model | Database used | Accuracy | Limitation |
|---|---|---|---|
| Hierarchy tree structure model | SCOP, PDB | 98% | 1. To implement this model, data by data analysis is required and hence computational time is also very high<br>2. Knowledge-based implementation is missing which can improve the accuracy of the classification |
| ProteinNet: a standard data set for machine learning models of protein structure | CASP, PDB, UniParc | 90% | 1. Only one type of file format mmCIF is used for assessment<br>2. Many other algorithms or models might not support ProteinNet dataset for their analysis purposes<br>3. Higher computational time as it includes involvement of third-party tools |
| Random forest model | SCOP | 94% | The increase in the number of Proteins samples and SSEs (attributes), results in the increase of the time complexity of the algorithm |
| NMR chemical shifts model using protein residues | BRMB, RefDB | 85–94% | 1. It cannot give constant accuracy level for all the algorithms used for implementing average NMR chemical shifts<br>2. This method is used only for proteins having 50 or more amino acid residues and having chemical shifts assigned to at least 70% of their residues |
| Hidden Markov model | PDB | 92% | 1. Data by data analysis is required to implement this model and hence computational time high<br>2. The accuracy % of his model is quite low |

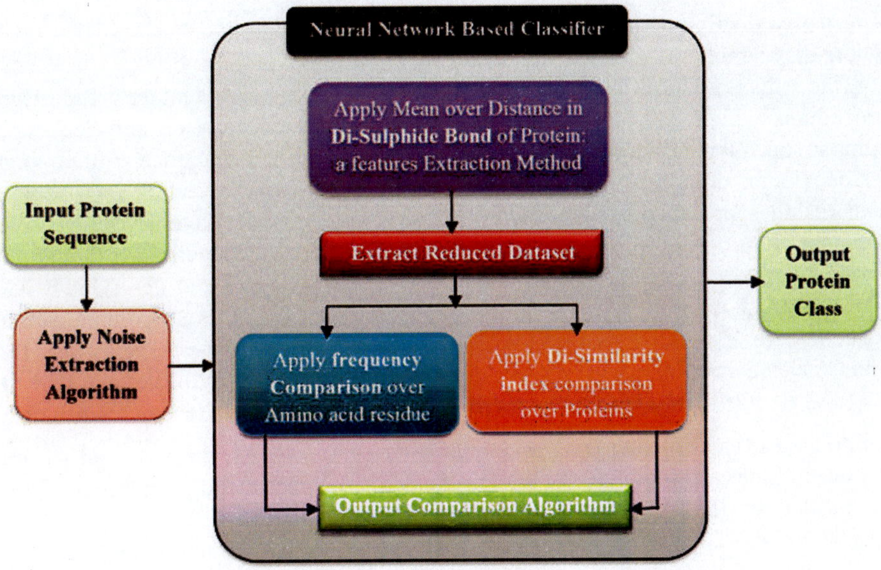


**Fig. 2** Proposed model of protein structural classifier

existing known values. Protein sequences whose values are more similar to the existing values are taken forward for the next processing. Disulphide connectivity is a covalent bond that is found between two cysteine residues in a Protein chain, which is classified into three connectivity patterns: (1) **Parallel bond** between adjacent cysteine residues, (2) **Alternative bond** between alternative cysteine residues and (3) **Quadrative bond** between quadrative separated cysteine residues. For better understanding, an example protein sequence is taken to calculate the mean over distance of these connectivity patterns i.e. QVIECQPQQS CTASLTGLNV CAPFLVPGSP TASTECCNAV QSINHDCMCN TMRIAAQIAC. The positions of cysteine residues are 5, 11, 21, 36, 37, 47, 49 and 60. For parallel bond, mean cysteine separation distance is $((11 - 5) + (36 - 21) + (47 - 37) + (60 - 49))/4$. Hence Mean Distance for parallel disulphide bond can be calculated as $\text{dis}_\text{p} = \frac{\sum_{j=1}^{m} X_{j+1} - X_j}{c}$, where *j*' indices the number of the 'C' Amino Acid in the Protein Sequence under consideration; *m* denotes the index of the last Cysteine molecule present in the sequence. $X^j$ denotes the position of the *j*th 'C' amino acid, corresponding to the bond and $X^{j+1}$ denotes the position of the next 'C' in the same bond. '*c*' represents the total number of bonds encountered in the sequence of the particular type. The values of $\text{dis}_\text{p}$ denotes the mean of the distance between the parallel bond in the protein. Similarly for alternative bond, mean cysteine separation Distance is $((21 - 5)+(36 - 11) + (49 - 37) + (60 - 47))/4$ using the formula $\text{dis}_\text{a} = \frac{\sum_{j=1}^{m} ((X_{j+2} - X_j) - (X_{j+3} - X_{j+1}))}{c/2}$ where $\text{dis}_a$ denote the mean of the distance between the alternative bond. Lastly for Quadrative bond, Mean Cysteine separation Distance is $((36 - 5) + (21 - 11) + (60 - 37) + (49 - 47))/4$. Hence Mean Distance for

Alternative disulphide bond can be calculated as $\mathrm{dis_q} = \frac{\sum_{j=1}^{m}((X_{j+3}-X_j)-(X_{j+2}-X_{j+1}))}{c/2}$ where $\mathrm{dis}_q$ denote the mean of the distance between the qurd-bond.

This reduced dataset is then forwarded to protein sequence feature comparison procedures which includes Amino acid residue index comparison and Sequence comparison for parallel processing of the unknown protein sequence. Amino acid residue index comparison includes comparing the index values of each amino acid residues present in the protein sequence with another protein sequence for the same amino acid. A more elaborate explanation is given with an example of two protein sequences (1) QVIECQPQAS CTQSLTGLNVQ and (2) TQSTECC-NQV QSINQDCMCN. For the amino acid residue **Q**, the index values in first sequence are 1, 6, 8, 13 and 21 and in the second sequence are 2, 9, 11 and 15. Then mean over distance between the two set of indexes for the amino acid **Q** is $((2-1)+(9-6)+(11-8)+(15-13)+(21-0))/5$. Hence, for a single amino acid it can be calculated as $\mathrm{SA_{aa}} = \frac{\sum_{i=1}^{n}\sum_{j=1}^{m}\mathrm{dis}(A_1^i, A_2^j)}{\mathrm{MaxLen}(A_1, A_2)}$ where $A_1$ and $A_2$ are the sets of index values of an Amino acid for the sequences 1 and 2 respectively and '$n$' and '$m$' are the respective length of the sets. Here $\mathrm{dis}(A_1^i, A_2^j) = |i - j|$ for $i = 1, 2...n$ and $j = 1, 2...\mathrm{m}$. This mean distance is found for each amino acid residues of both the protein sequences and then finally a mean sequence alignment distance between the two sequences is calculated by $\mathrm{SA_{Protein\,Sequences}} = \frac{\sum_{i=1}^{20}\mathrm{SA}_i}{20}$ where $\mathrm{SA}^i$ is the mean index distance between two sequence for an amino acid '$i$'. Here, 20% the available 20 amino acid residues.

Protein sequence comparison includes the ratio calculation of frequency of an amino acid residue in both the protein sequences. For example, in the above mentioned two sequences, the rate of occurrences of amino acid residue Q is 5 and 4 in sequence 1 and 2 respectively. Hence, we calculate the ratio as 5/4 for the amino acid residue $Q$. This ration can be represented by $\mathrm{SC}^i$ = (frequency of amino acid i in sequence 1)/(frequency of amino acid i in sequence 2). This ratio is calculated for each amino acid residues present in both the protein sequences and then finally the mean sequence ratio is calculated as $\mathrm{SC_{Protein\,Sequences}} = \frac{\sum_{i=1}^{20}\mathrm{SC}_i}{20}$ where $\mathrm{SC}^i$ is the frequency ratio between two sequence for an amino acid '$i$'. Here, 20% the available 20 amino acid residues.

Two matrices are formed using the values generated in the amino acid residue index comparison and protein sequence comparison respectively. These matrices show mean index distance values and mean sequence ratio values between any two protein sequences respectively. These matrices are used for comparing the values of an unknown protein sequence with the values of the known protein sequences to extract the association between the protein sequences and its structural classes. At the final stage, a neural network classifier is build using training datasets and it is then used for classifying the unknown protein sequence into its respective structural class.

## 5 Conclusion

Protein structure classification of all the uncharacterized proteins has become very important as thousands of new proteins are getting introduced every day and manual classification of them into protein structural classes is quite impossible. The 'Structural Classification of Proteins (SCOP)' database is mostly a manual protein structure classification into different domains depending on likenesses of the structures and sequences of amino acid. Hence the need for introduction of structured and enhanced data mining approaches to classify proteins into one of the protein structural fold groups has never been more than now. This may profoundly help in Protein and its structure analysis by the specialist. Many data mining techniques are used for classifying the unknown protein structures, but most of those approaches are having drawbacks like low accuracy level or high computational time. Hence this dissertation comprises a detailed review of few of the research paper work that involves five different techniques for classifying protein structures. It was perceived that knowledge-based data analysis and few models like Hierarchical tree structure, Random forest, Hidden Markov Model and Amino acid residue's Chemical shifts (ACS) were used for introducing new data mining techniques for protein structure classification. So, it can conclude that more research studies will be done in future to introduce new classification models with better accuracy and computation speed than the existing ones. Finally a new classification approach is introduced here for better classification of protein structure with low computational time, which will be implemented in future.

## References

1. Rahman MM, Ul Alam A, Al-Mamun A, Mursalin TE (2011) A more appropriate protein classification using data mining. JATIT 33–43
2. Bolser DM, Dafas P, Harrington R, Park J, Schroeder M (2003) Visualization and graph-theoretic analysis of a large-scale protein structural interactome. BMC Bioinf 4(45):1471-2105, 1–11
3. Watts DJ, Strogatz SH (1998) Collective dynamics of 'small-world' networks. Nature 393(6684):440–2
4. Al Quraishi M (2019) ProteinNet: a standardized data set for machine learning of protein structure. BMC Bioinf 20(Article number):311
5. John M et al (2018) Critical assessment of methods of protein structure prediction (CASP) round XII. Proteins Struct Funct Bioinf 86(S1):7–15
6. Berman HM et al (2000) The protein data bank. Nucleic Acids Res 28:235–242 https://doi.org/10.1093/nar/28.1.235
7. Jain P, Hirst JD (2010) Automatic structure classification of small proteins using random forest. BMC Bioinf 11( Article number):364
8. Murzin AG et al (1995) SCOP: a structural classification of proteins database for the investigation of sequences and structures. J Mol Biol 247:536–540
9. Jain P et al (2009) Supervised machine learning algorithms for protein structure classification. Comput Biol Chem 33:216–223

10. Kumar AV, Ali RFM, Cao Y, Krishnan VV (2015) Application of data mining tools for classification of protein structural class from residue based averaged NMR chemical shifts. Biochim Biophys Acta 1854(1000):1545–1552
11. Seavey BR, Farr EA, Westler WM, Markley JL (1991) A relational database for sequence-specific protein NMR data. J Biomol NMR 1:217–236
12. Zhang HY, Neal S, Wishart DS (2003) RefDB: A database of uniformly referenced protein chemical shifts. J Biomol NMR 25:173–195
13. Gewehr JE et al (2007) AutoSCOP: automated prediction of SCOP classifications using unique pattern-class mappings. Bioinformatics 23:1203–1210
14. Dubchak I et al (1999) Recognition of a protein fold in the context of the Structural Classification of Proteins (SCOP) classification. Proteins 35:401–407
15. Dubchak I et al (1997) Protein folding class predictor for SCOP: approach based on global descriptors. Proc Int Conf Intell Syst Mol Biol 5:104–107
16. Mirceva G et al (2009) HMM based approach for classifying protein structures. Int J Bio-Sci Bio-Technol 1(1)
17. Karchin R (1999 ) Hidden Markov models and protein sequence analysis. In: Seventh international conference on intelligent systems for molecular biology (ISMB)
18. Viterbi AJ (1967) Error bounds for convolutional codes and an asymptotically optimum decoding algorithm. IEEE Trans Inf Theory 13(2):260–269
19. Durbin R, Edy S, Krogh A, Mitchison G (1998) Biological sequence analysis: Probabilistic models of proteins and nucleic acids. Cambridge University Press
20. Alexandrov V, Gerstein M (2004) Using 3D hidden Markov models that explicitly represent spatial coordinates to model and compare protein structures. BMC Bioinf 5(2)

# Statistical Analysis of Emotional Response through Physiological Signals

**Saikat Basu, Nabakumar Jana, Vamshi Krishna, Manjunatha Mahadevappa, Jayanta Mukherjee, Rajlakshmi Guha, and Somesh Kumar**

**Abstract** Emotion plays an essential role in human interaction. This paper reports a new technique for accessing the effects of human emotional responses from physiological signals with the help of statistical analysis. For emotional elicitation, a two-dimensional valence-arousal model is used for emotion elicitation. Four types of Physiological signals, namely, Heart Rate (HR), Skin Temperatures (ST), Respiration Rate (RR) and Galvanic Skin Responses (GSR) were employed for emotional

S. Basu (✉) · M. Mahadevappa
School of Medical Science & Technology, Indian Institute of Technology Kharagpur, Kharagpur, India
e-mail: saikat.basu@smst.iitkgp.ernet.in; saikatbasu@gmail.com

M. Mahadevappa
e-mail: mmaha2@smst.iitkgp.ernet.in

N. Jana
Department of Mathematics, Indian Institute of Technology (ISM) Dhanbad, Dhanbad, India
e-mail: nabakumarjana@iitism.ac.in

V. Krishna
Department of Electrical Engineering, Indian Institute of Technology Kharagpur, Kharagpur, India
e-mail: vamshimuga304@gmail.com

J. Mukherjee
Department of Computer Science and Engineering, Indian Institute of Technology Kharagpur, Kharagpur, India
e-mail: jay@cse.iitkgp.ac.in

R. Guha
Centre of Education Technology, Indian Institute of Technology Kharagpur, Kharagpur, India
e-mail: rajg@cet.iitkgp.ac.in

S. Kumar
Department of Mathematics, Indian Institute of Technology Kharagpur, Kharagpur, India
e-mail: smsh@maths.iitkgp.ernet.in

S. Basu
Department of Computer Science and Engineering, Maulana Abul Kalam Azad University of Technology, Kolkata, West Bengal, India



A. Maji et al. (eds.), *Proceedings of the International Conference on Computing and Communication Systems*, Lecture Notes in Networks and Systems 170,
https://doi.org/10.1007/978-981-33-4084-8_18

analysis. A healthy group of volunteers were exhibited a few sets of images received from International Affective Picture System (IAPS) having four different states of emotion. The experiment is non-invasive. This paper aims to identify the emotional effects expressed in human physiological signals with the help of pictorial stimuli. We conducted a series of Analysis of Variance (ANOVA) tests using the following physiological signals, namely, HR, ST, GSR, and ST, to understand the emotional effect in each of the signals. We have observed significant differences between different emotional classes in terms of four bio-medical signals like HR, GSR, RR, and ST. Significant interactions among the subjects and the stimuli are observed. The ordering among the mean values of each of the physiological signals for various emotional classes has been established. It is observed that mean response values corresponding to physiological signal ST show a complete ordering, whereas, for the signal RR, they are homogeneous.

**Keywords** Emotion recognition · Physiological signal · Autonomic nervous system · Valence-arousal · ANOVA · International affective picture systems

## 1 Introduction

Emotion [24] expresses the psychological condition of a person. It is a tool of communication without any verbal and language hurdles. It is represented using physiological changes like changes in heartbeat, blood pressure, perspiration rate, gestures, facial expression, voice, respiration rate, skin temperature, etc. Automated or computer-based understanding of emotional states has drawn attention among researchers in recent times. Understanding emotional states are quite tricky, even for an intelligent person. So automatic apprehending of the emotional state is a challenging task. There could be several potential applications of automated interpretation of emotions. For example, the intelligent emotion recognition system may suggest a suitable movie as per the mood of the user or reducing the volume of music if a user is emotionally stressed. Emotion recognition helps in understanding the condition of the children who have autism, very aged persons who have paralysis, new-born who are unable to express their emotions explicitly [2, 20]. It may be used to understand the mood of a student interacting through online tutorials or webinars. A significant amount of research has been done in measuring emotion through different modalities like facial expression, gestures and speech [1, 4, 6, 11]. However, a person can suppress facial expressions or gestures. The measurements vary due to illumination, auditory noise, presence of spectacles, etc. These methods are not universal methods and dependent upon various factors, such as age, culture, gender, tempo, and modulation of voice, etc. Physiological signals [25] originate from an autonomous nervous system. Skin Temperature (ST), Respiration Rate (RR), Heart Rate (HR), and Galvanic Skin Response (GSR) [21] have been used previously for recognizing emotional states [3, 18, 19, 22, 27]. In this work, we present the effects of different kinds of emotions through different physiological signals, which are induced by visual stimuli. We have observed that the physiological signals carry enough information in determining emotional states. Emotion elicitation has been done by different techniques such as using picture, sound, video, etc. [10, 12, 28]. In our work, we used pictures to elicit emotion in a controlled atmosphere. The IAPS [17] data set was used for emotion elicitation. These data were evaluated with a large sample size over a long tenure having high accuracy of emotion elicitation [15]. In the past, many researchers also used IAPS [13, 26] for emotional elicitation. We group the labelled pictures under four categories of emotions, namely, (1) High Valence High Arousal (HVHA), (2) High Valence Low Arousal (HVLA), (3) Low Valence High Arousal (LVHA), and (4) Low Valence Low Arousal (LVLA). The emotional effects in terms of physiological signals in these four groups of emotions have been observed in a healthy group of volunteers. Analysis of Variance (ANOVA) was used to judge the effects on different physiological signals when different groups of emotional pictures were shown.

## 2 Mathematical Modeling of Emotion

In the nineteenth century, James [14] and Darwin and Prodger [8] proposed the theory of emotion. A large amount of research has taken place in the theory of emotions since then. Picard [24], said that emotion plays a pivotal role in perception, learning, and rational decision-making. The emotional effect is expressed by physiological changes like heart rate, blood pressure, degrees of the sweat secretion, skin color, voice, etc. Various theoretical models of emotion are proposed by neuroscientists and psychologists. Among them, the two most popular models of emotions are a discrete model by Ekman [9] and a two-dimensional valence arousal model by Lang [16]. The discrete model consists of a fixed number of emotions, which are claimed to be fundamental entities among all cultural backgrounds. There are considerable ambiguities in determining the number of basic emotions. There was a consensus between psychologists to agree with sadness, anger, happiness, fear, disgust, and surprise as basic six emotions. Valence and arousal are the two axes of a two-dimensional model. Valence depicts attractiveness or pleasantness, and varies from negative to positive, whereas arousal represents reactiveness, and varies from low to high. Figure 1 depicts a 2-D space of valence-arousal place indicating six basic emotions. For example, anger has low valence and high arousal.

## 3 Experimental Protocol

We have chosen visual stimuli from the set of IAPS images for the elicitation of emotion through physiological signals. Pictures from the International Affective Picture System (IAPS) were chosen for emotion elicitation. "Center for the Study of Emotion and Attention", of University Florida has permitted to use the IAPS

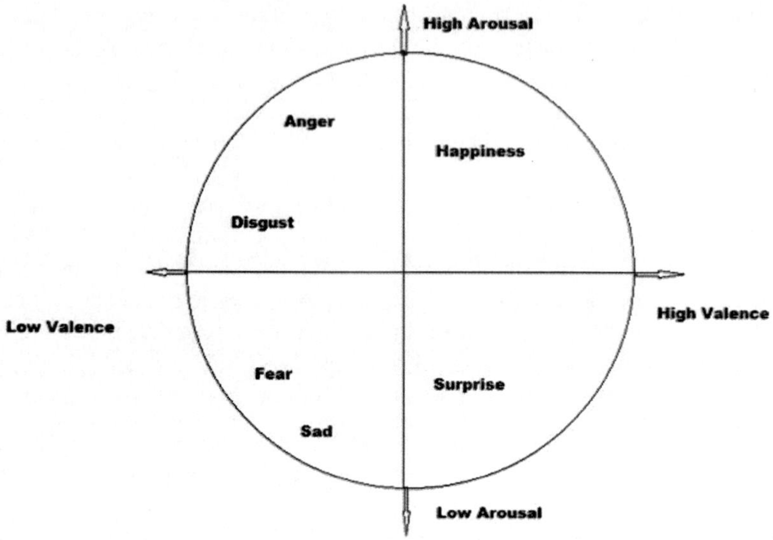


**Fig. 1** Experimental protocol

datasets. These data sets provide values for valence, arousal, and dominance for each item averaged over a large number of subjects, who rated the items using the Self-Assessment Manikin (SAM) [5]. We have categorized these pictures by their normative behavior. Dominance represents the level of control produced by the stimulus. We have used a two-dimensional model consisting of valence and arousal. Valence and arousal both vary in the 1–9 scale. The competent authority of the Indian Institute of Technology Kharagpur issued an Institutional Ethical Clearance Certificate to conduct the experiment. A group of thirty healthy students voluntarily participated in the study. Eighteen subjects were girls, and twelve were boys. The details are discussed in the following sections

### *3.1 Design of Experiments*

Consent forms were signed before the experiment. At first, a small briefing was done about the experiment. The experiments were conducted in an isolated room. The room is equipped with a proper lighting facility. A cozy chair was placed in front of the monitor for the participants' for the experiment. The subject was shown four sets of emotional pictures through a computer equipped with a 22-inch display screen. The subject acted as a passive watcher. Different sensors like GSR, ECG, RR electrode, and ST sensor have been attached. Initially, the person was told to take a rest for one minute. Soothing music was played to minimize the arousal from foreign sources in the environment. Four different slides consisting of seven emotional photographs of

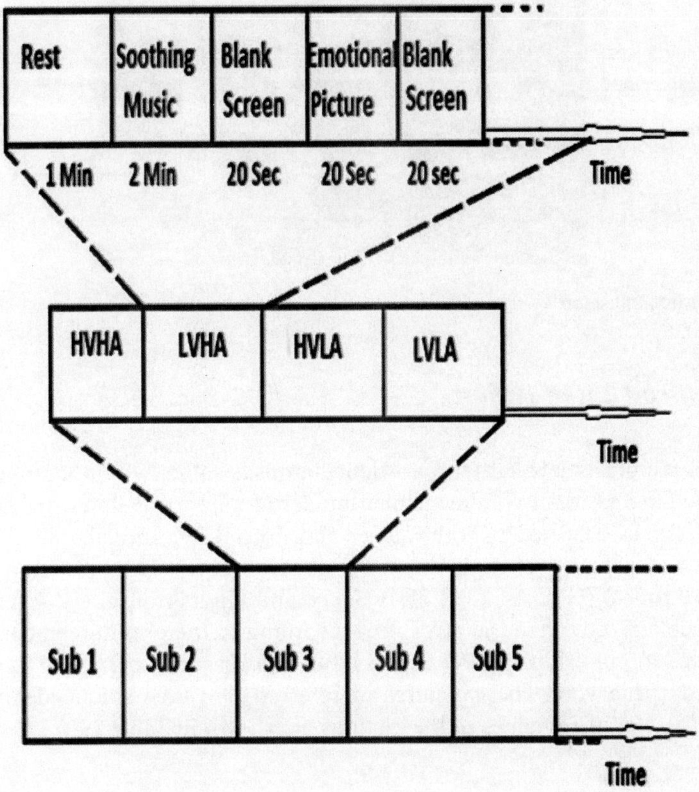


**Fig. 2** Experimental protocol

two and a half minutes were shown. In a slide show, IAPS pictures and a blank screen with a cross in the middle were shown one after another Each IAPS picture, and the blank screen was shown for ten seconds. This procedure was continued. Each slide show consists of a total of eight blanks with a cross in the middle and seven emotional pictures. Blank with a cross in the middle treated as a neutral picture. The schematic diagram of the experimental procedure was shown in Figure 2. Here sub1, sub2, etc. denote thirty different subjects appearing for the experiment. HVHA, HVLA, LVHA, and LVLA denote different slide-show shown to various subjects. Data have been captured with the help of a PowerLab data collection unit of ADInstruments. After a slide-show is over, a break of two minutes is followed. In between the break, anagrams were given to the subjects to minimize the residual effect of the emotional arousal. All the parameters are non-invasive. Emotional effects in the subject after seeing the pictures were calculated with respect to the biological signals like GSR, ECG, RR, and ST. Debriefing has been done after the experiments. Figure 3. shows the experimental setup.

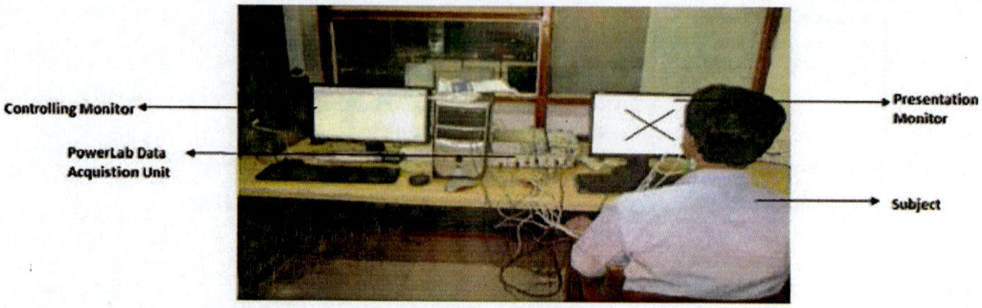


**Fig. 3** Experimental setup

## 3.2 *Picture Classification*

Each IAPS picture is labeled with a valence-arousal value, within a range of one to nine. We have used a two dimensional model of emotional states. Each type of emotional state is identified by both valence and arousal. We have classified emotion into four states, namely, (1) High Valence High Arousal (HVHA), (2) High Valence High Low Arousal (HVLA), (3) Low Valence and High Arousal (HVLA), and d. Low Valence and Low Arousal (LVLA). According to the user agreement, we are not permitted to publish the IAPS pictures. For that reason, the original images are not reported in this work. These pictures are referred to by their unique identification numbers. The classified values of the pictures are shown in Table 1.

## 3.3 *Inclusion Criteria for Experiment*

We have chosen healthy students in our institute as subjects. The mean and standard deviations of the volunteers are 24.24 years and ±3.03, respectively.

## 3.4 *Exclusion Criteria for Experiment*

The study excludes persons outside the age range of 18 and 45 years, pregnant women, persons having substance addiction, and individuals on psychiatric medications.

**Table 1** IAPS stimuli classifications

| IAPS picture categorisation | Classification based on IAPS rating |
|---|---|
| LVLA | Valence < 4, arousal < 5 |
| LVHA | Valence < 2, arousal > 6.5 |
| HVLA | Valence > 6.5, arousal < 4 |
| HVHA | Valence > 7 and arousal > 6 |

## 3.5 Signals Captured

The experiment proposes four physiological parameters such as GSR, HR, RR, and ST, to analyze human emotion. GSR, HR, and RR are produced by electro-dermal activity, electrocardiogram (ECG) and respiration signals, respectively. All these captured signals are of non-invasive types. GSR reflects skin conductance.

The sympathetic nervous system controls the secretion of the eccrine sweat glands present on the palmar surfaces of the hands and bottom of the feet. As sweat production increases, skin conductance also increases. This is the direct effect of the sympathetic nervous system. The unit of skin conductance is micro-Siemens ($\mu S$). GSR depicts the arousal component of the involved emotion. HR is defined by the numbers of heartbeats per unit time, which is usually expressed beats per minute. This is controlled by the Autonomic Nervous System (ANS), HR is measured by the number of QRS complexes in an ECG signal in a minute. GSR and HR behave differently for different persons. RR is defined as a number of respiration peaks per minute. Skin temperature is also controlled by the ANS. Emotional change may be reflected by the variation in the skin temperature. GSR of the subjects has been captured from the index and middle fingers of the left hand using the Ag/AgCl GSR electrode of ADInstruments attached by a Velcro strap. Proper care has been taken to fix the electrodes. Velcro belt provided by Zephyr BioHarness was set at the sternum to collect HR and RR. Different physiological signals have been monitored by LabChart 7 software by ADInstruments at the administrator's desktop. All of the physiological signal components are non-invasive. The sampling frequency of all the signals was kept at 1 kHz.

## 3.6 Instruments and Software Used

ADInstruments ML870 PowerLab 8/30 was used as a main data acquisition unit with a physiological sensor like bio amplifier ML-135, temperature sensor ML-309 and GSR amp FE-116. Zephyr BioHarness is a wireless data acquisition unit compatible with PowerLab, used for capturing respiration signal and the ECG. Both PowerLab and BioHarness can work simultaneously. LabChart7 by ADInstruments, MATLAB 2013a, Microsoft Excel and R-3.2 software was used for data analysis. Figure 4 represents BioHarness fixed on the chest to measure ECG and RR.

## 3.7 Pre-processing of Data

Pre-processing of physiological signals was required to get rid of noise and baseline wander. A cutoff frequency of the low pass filter was set for each of the physiological signals to remove noise. High pass filters with cut-off frequencies of 4 and 0.1 Hz

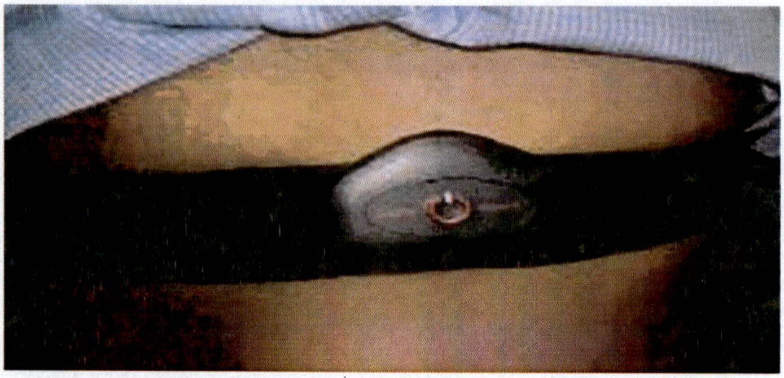

**Fig. 4** Experimental protocol

were used to separate baseline wanders for ECG and HR, respectively. The algorithm developed by Pan and Tompkins [23] was used for HR calculations. We have measured the duration between two $R$ ($R$-$R$ interval) waves in the QRS complex. This is expressed by $t_{R-R}$. The following formula denotes HR in beats per minutes: $\text{HR} = 60/t_{R-R}$.

## 4 Statistical Analysis

When the IAPS pictures were presented, that time, the calculations were also initiated. Values associated with the blank screen with a cross mark in the middle have been ignored while computing the mean. We have taken mid-five seconds of the display pictures into the calculations. In our experiment, sampling rate was 1 kHz. There were 5000 observations in 5 s. The mean of all 5000 observations has been calculated. Seven emotional pictures were shown to the healthy thirty subjects. We have done soft synchronization between PowerLab and BioHarness using LabChart software. Data capturing by BioHarness get over a few seconds earlier than the PowerLab. Because of this time gap, we have ignored the values obtained during the display of the last picture in the sequence and considered only the first six pictures out of seven pictures in the calculation. Averages of the first six pictures of individual biological markers associated with four emotional classes (HVHA, HVLA, LVHA, and LVLA) have been performed. Since the average values of a vast number of samples (5000 in this case) have been considered here, it assumed that the samples follow a normal distribution according to the central limit theorem [7]. For each of the physiological responses, the homogeneity of variances is tested using an $F$-test, and we observed that the variances of the populations are equal. For example, in the case of ST, the computed value of the test statistic is 1.105, and the $F$-critical value is 1.279 for testing the equality of the variances between the HVHA and HVLA classes. Hence the two way ANOVA tests are applied to analyze the primary relationships among responses of four classes (HVHA, HVLA, LVHA, and LVLA) and also among the subjects. The two way ANOVA model is briefly described here. Analysis Of Variance (ANOVA) is a statistical test to identify which parameters are significant. The two way ANOVA model is briefly described here. Let Y be the response variable. Levels $i = 1, \ldots, a$ exist in factor A while levels $j = 1, \ldots, b$ exist in factor B. A treatment is a particular combination of levels. So, there are $ab$ treatments. Let $Y_{i,j,k}$ be the $k$th observation for treatment $(i, j)$, $k = 1, \ldots, n$. These mean values follow a normal distribution. Also, the response variable observations are independent. The model is described by $Y_{i,j,k} = \mu_{i,j} + \epsilon_{i,j,k}$. Here, $\mu_{i,j}$ represents expected value or theoretical mean for all observations in cell $(i, j)$. The effect of both factors is explicitly written as $\mu_{i,j} = \mu + \alpha_i + \beta_j + \gamma_{i,j}$ , where is the grand mean, $\alpha_i$ is the additive main effect of level $i$ from the first factor, $\beta_j$ is the additive main effect of level $j$ of the second factor, and $\gamma_{i,j}$ is the interaction effect of the treatment $(i, j)$. The errors $\epsilon_{i,j,k}$ are identically and independently normally distributed with mean zero and variances $\sigma^2$ . Then $\gamma_{i,j,k}$ follow a normal distribution with mean $\mu_{i,j}$ for $i = 1, \ldots, a$ and $j = 1, \ldots b$.

For each of the physiological responses, the homogeneity of variances is tested using an $F$-test. It is observed that the variances of the populations are equal. For example, in the case of ST, the computed value of the test statistic is 1.105, and the $F$-critical value is 1.279 for testing the equality of the variances between the HVHA and HVLA classes. Hence the two way ANOVA tests are applied to analyze the primary relationships among responses of four classes (HVHA, HVLA, LVHA, and LVLA) and also among the subjects. In this study, it is considered the subjects (1 to 30) as the first factor and the emotional classes (HVHA, HVLA, LVHA, LVLA) as the second factor. So $a = 30$ and $b = 4$. Also $k = 6$. The following three steps explain the analysis of each of the biological markers.

1. Four types of emotional classes (HVHA, HVLA, LVHA, and LVLA) are present for every bio-physiological responses such as RR, GSR, HR, and ST. First, two way ANOVA test for each of these four emotional classes on thirty subjects were applied.
2. Assuming the six pictures as six replications, two way ANOVA tests with replications are used to test whether the mean values corresponding to the emotional classes are equal and whether any interaction is present among the emotions and the subjects.
3. If significant differences between average values related to the emotional classes HVHA, LVHA, HVLA, and LVLA are present, two way ANOVA tests were applied to select any two groups from HVHA, HVLA, LVHA, and LVLA at a time. Pairwise means between the considered groups are also compared. Finally, an ordering of the expected values amongst these four classes is also determined.

The ANOVA tables with replications for each of the physiological signals are given. The other ANOVA tables are not included here due to space constraints. In each of the ANOVA tables, the mean sum of square due to errors (MSS), the degrees of freedom (df), and the corresponding computed value of the test statistic (the $F$-computed value) are given.

**Table 2** Two way ANOVA with replication for GSR and HR

| | GSR | | | | HR | | | |
|---|---|---|---|---|---|---|---|---|
| Source | df | MSS | *F*-computed | *p*-value | df | MSS | *F*-computed | *p*-value |
| Subjects | 29 | 402.62 | 285.71 | 0 | 29 | 1525.04 | 260.32 | 0 |
| Emotions | 3 | 26.91 | 19.10 | 7.65E–12 | 3 | 95.12 | 16.24 | 3.69E–10 |
| Interaction | 87 | 26.08 | 18.50 | 7.9E–12 | 87 | 27.54 | 4.70 | 2.02E–30 |
| Error | 600 | 1.41 | – | – | 600 | 5.86 | – | – |

**Galvanic Skin Response (GSR)**: The results of the two way ANOVA tests are discussed. For each of the emotional classes HVHA, HVLA, LVLA, it is observed that there are significant differences among the pictures. Also, significant differences are observed between the effects on subjects. The *p*-values for all these cases were nearly zero. However, for the LVHA class, although there is a significant difference in the effect among the subjects, no significant difference is present among the pictures. The results of the ANOVA tests with replication are shown in Table 2. The effects of the pictures are different for each of the emotions. Also, significant interaction is present between the subjects and the emotions. Therefore different subjects show different effects on seeing the same picture. The emotional class HVHA has a higher GSR than that of the LVLA. Again, the emotional class HVLA has a higher GSR than that of HVHA. Both LVHA and HVLA classes have approximately equal GSR, and therefore they are comparable. GSR values corresponding to the classes LVHA and HVLA are usually higher compared to those observed in classes HVHA and LVLA.

**Heart Rate (HR)**: Significant differences between the subjects are found among emotional classes HVHA, LVHA, and LVLA. The corresponding *p*-value is nearly zero. But there is no significant difference between the pictures with corresponding *p*-values as 0.0176, 0.118, 0.8811, respectively. For the emotional class HVLA, significant variation is present between the subjects (*p*-value zero) as well as between the pictures (*p*-value 0). In the case of heart rate, the effects of the pictures are different for each of the emotions, and the data show a significant interaction between the subjects and the emotions (see Table 2). Therefore different subject shows a different effect on the same picture. The heart rates corresponding to the emotional classes HVHA, HVLA, and LVLA are similar and are higher than the corresponding rate of LVHA.

**Skin Temperature (ST)**: Application of two way ANOVA tests without replications for each of the emotions shows that, for the emotional class HVHA, there are significant differences among the pictures. Also, significantly different effect among the subjects is observed. The *p*-values for all these cases were nearly zero. Similar conclusions are obtained for each of the four emotional classes. It has been observed from Table 3, the effects of the pictures are different for each of the emotional classes. Significant interaction is also observed between the subjects and emotional responses. The emotional class LVHA has the highest skin temperature than others.

**Table 3** Two way ANOVA with replication for ST and RR

| | ST | | | | RR | | | |
|---|---|---|---|---|---|---|---|---|
| Source | df | MSS | $F$-computed | $p$-value | df | MSS | $F$-computed | $p$-value |
| Subjects | 29 | 20.31 | 5113.09 | 0 | 29 | 208.59 | 90.80 | 3.8E–198 |
| Emotions | 3 | 20.35 | 5125.45 | 0 | 3 | 1.45 | 0.63 | 0.59 |
| Interaction | 87 | 0.31 | 77.54 | 4.5E–274 | 87 | 6.98 | 3.04 | 2.07E–15 |
| Error | 600 | 0.003972 | – | – | 600 | 2.30 | – | – |

Therefore different subjects show different values for the same picture. The emotional class HVHA has the lowest skin temperature among all the emotions. The value of skin temperature for classes LVLA and HVLA are moderate, and it usually lies in between the values observed for the classes LVHA and HVHA. Also, the observed values in the emotional class LVLA are usually higher than those observed in the class HVLA.

**Respiration Rate (RR)**: The two way ANOVA tests without replications for each of the emotional classes shows that different subjects have a different effect on visual stimuli ($p$-value is close to zero). But for the emotional class HVHA, variation among the pictures is not significant ($p$-value = 0.7150). Similarly, for other emotional classes such as HVLA, LVHA, and LVLA, different pictures have different effects on a subject (corresponding $p$-value almost zero). A significant variation was not observed among the pictures. In case of respiration rate, the $p$-value corresponding to the emotional classes is 0.5951 (see Table 3) and ($F(29,600) = 1{:}4864$)). Therefore the ANOVA with replication shows no significant effect of emotional categories under RR. This implies that there is no significant variation of the mean rates corresponding to the emotional classes HVHA, HVLA, LVHA, and LVLA.

**Relationship between mean values of the emotional Classes**: Among the subjects, significant differences were observed and also significant interaction is present between subjects and emotional classes ($F(87,600) = 1{:}2873$). Let the mean value corresponding to the emotional class a in case of the physiological signal s be denoted by $\mu_a^s$ . Thee emotional classes like HVHA, HVLA, LVHA, and LVLA are denoted by a, b, c, and d, respectively. Table 4 presents the ordering among the mean values of the physiological signals of the considered emotional classes. Clearly, the mean

**Table 4** Relationship between mean values of the emotional classes

| Physiological signal | Ordering of the mean value |
|---|---|
| GSR | $\mu_b^{GSR} = \mu_c^{GSR} > \mu_a^{GSR} > \mu_d^{GSR}$ |
| HR | $\mu_a^{HR} = \mu_b^{HR} = \mu_d^{HR} > \mu_c^{HR}$ |
| ST | $\mu_c^{ST} > \mu_d^{ST} > \mu_b^{ST} > \mu_a^{ST}$ |
| RR | $\mu_a^{RR} = \mu_b^{RR} = \mu_c^{RR} = \mu_d^{RR}$ |

values corresponding to the skin temperatures are all different, whereas the values corresponding to RR are all equal. In the case of HR, $\mu_c^{\mathrm{HR}}$ is the lowest among all mean heart rates.

## 5 Conclusion

In this paper, we have proposed a new method for emotion elicitation using IAPS pictures. We have categorized IAPS stimuli into four groups, namely, HVHA, HVLA, LVHA, and LVLA. Physiological signals such as GSR, HR, RR, and ST have been used for emotion elicitation. We have recruited thirty healthy students as volunteers from the Institute for data collection. The nature of the study was non-invasive. Usually, each experiment has been performed for thirty minutes, including fixing the sensors. Subjects participated as passive watchers in the experiment. We have analyzed the experimental data using a series of ANOVA tests. Another unique aim to investigate the variations of physiological signal responses related to different groups of emotion elicitating pictures. For each of the physiological responses and emotions, two way ANOVA tests without replication as well as two way ANOVA tests with replications are used to study the significant differences among pictures, subjects, and emotional classes. In all the cases, a significantly different effect between the pictures was observed. For each of the physiological responses, we have observed a significant interaction between the emotional responses and the subjects. In the case of RR, no significant difference was found, whereas other physiological responses show significant differences among different emotional states. We have shown how different biological signals like RR, GSR, HR and ST behave in different emotional classes, such as HVHA, HVLA, LVHA, and LVLA.

## References

1. Adolphs R (2002) Recognizing emotion from facial expressions: psychological and neurological mechanisms. Behav Cogn Neurosci Rev 1(1):21–62
2. Bal E, Harden E, Lamb D, Van Hecke AV, Denver JW, Porges SW (2010) Emotion recognition in children with autism spectrum disorders: Relations to eye gaze and autonomic state. J Autism Dev Disord 40(3):358–370
3. Basu S, Bag A, Aftabuddin M, Mahadevappa M, Mukherjee J, Guha R (2016) Effects of emotion on physiological signals. In: IEEE annual India conference (INDICON), pp 1–6
4. Basu S, Chakraborty J, Aftabuddin M (2017) Affect detection from speech using deep convolutional neural network architecture. In: 2017 14th IEEE India council international conference (INDICON), pp 1–5
5. Bradley MM, Lang PJ (1994) Measuring emotion: the self-assessment manikin and the semantic differential. J Behav Ther Exp Psychiatry 25(1):49–59
6. Castellano G, Kessous L, Caridakis G (2008) Emotion recognition through multiple modalities: face, body gesture, speech. Springer
7. Chung KL, Zhong K (2001) A course in probability theory. Academic Press

8. Darwin C, Prodger P (1998) The expression of the emotions in man and animals. Oxford University Press, USA
9. Ekman P (1993) Facial expression and emotion. Am Psychol 48(4):384
10. Gross JJ, Levenson RW (1995) Emotion elicitation using films. Cognit Emotion 9(1):87–108
11. Gunes H, Piccardi M (2007) Bi-modal emotion recognition from expressive face and body gestures. J Netw Comput Appl 30(4):1334–1345
12. Harper D (2002) Talking about pictures: a case for photo elicitation. Vis Stud 17(1):13–26
13. Hempel RJ, Tulen JH, van Beveren NJ, van Steenis HG, Mulder PG, Hengeveld MW (2005) Physiological responsivity to emotional pictures in schizophrenia. J Psychiat Res 39(5):509–518
14. James W (1884) What is an emotion? Mind 9(34):188–205
15. Lang P, Bradley MM (2007) The international affective picture system (IAPS) in the study of emotion and attention. In: Handbook of emotion elicitation and assessment, vol 29
16. Lang PJ (1995) The emotion probe: studies of motivation and attention. Am Psychol 50(5):372
17. Lang P, Bradley M, Cuthbert B (2008) International affective picture system (IAPS): affective ratings of pictures and instruction manual. University of Florida, Gainesville
18. Li L, Chen J-H (2006) Emotion recognition using physiological signals. In: International conference on artificial reality and telexistence, pp 437–446
19. Lisetti CL, Nasoz F (2004) Using noninvasive wearable computers to recognize human emotions from physiological signals. EURASIP J Adv Sig PR
20. Martínez R, de Ipiña KL, Irigoyen E, Asla N, Garay N, Ezeiza A, Fajardo I (2010) Emotion elicitation oriented to the development of a human emotion management system for people with intellectual disabilities. Springer
21. Montagu J, Coles EM (1966) Mechanism and measurement of the galvanic skin response. Psychol Bull 65(5):261
22. Nasoz F, Alvarez K, Lisetti CL, Finkelstein N (2004) Emotion recognition from physiological signals using wireless sensors for presence technologies. Cognit Technol Work 6(1):4–14
23. Pan J, Tompkins WJ (1985) A real-time QRS detection algorithm. IEEE Trans Biomed Eng
24. Picard RW (2003) Affective computing: challenges. Int J Hum Comput Stud 59(1–2):55–64
25. Shu L, Xie J, Yang M, Li Z, Li Z, Liao D, Xu X, Yang X (2018) A review of emotion recognition using physiological signals. Sensors 18(7):2074
26. Valenza G, Lanata A, Scilingo EP (2011) The role of nonlinear dynamics in affective valence and arousal recognition. IEEE Trans Affect Comput 3(2):237–249
27. Wagner J, Kim J, André E (2005) From physiological signals to emotions: implementing and comparing selected methods for feature extraction and classification. In: IEEE international conference on multimedia and expo, pp 940–943
28. Zeng Z, Pantic M, Roisman GI, Huang TS (2008) A survey of affect recognition methods: audio, visual, and spontaneous expressions. IEEE Trans Pattern Anal Mach Intell 31(1):39–58

# Discernment of Human Stress Using Generic Computer-Aided Tasking—A Review

**Soumyabuddha Debnath and Saikat Basu**

**Abstract** When an undermining situation is seen, a course of physiological techniques happens setting up the body. A tangible framework is there to confront the approaching peril and fruitful assurance adjustment. It is known as stress. Stress has become a critical reason for mental illnesses in the cutting edge society. In this study, we will analyse the ongoing research on stress identification in everyday life, which is utilizing distinctive digital tasks. Along with digital tasks, various sensors estimating pulse, respiratory action, pressure sensors, galvanic skin reaction sensors and so forth are utilized. The data acquired in the experiments are analysed with various existing classifier models. Even though there are a few works identified with stress identification in controlled laboratory conditions, the number of studies analysing stress discovery in the day by day life is restricted.

**Keywords** Stress identification · Digital tasks · K-nearest neighbor · Logistic regression · Support vector machine · Linear discriminant analysis · Naive Bayes · Random forest

## 1 Introduction

Recently it has been comprehended that there is a significant association between the physical quality of every person and their emotional state. This understanding has prompted expanding interest for affective computing (AC) [1]. From that point forward, it becomes a branch under the human–computer correlative area. It utilizes corporal and physical indications of a person to distinguish the present feeling. Conspicuous rudimentary feelings identified in affective computing are bliss, outrage, shock, sicken, misery and dread [1]. Stress has been as of late added to

S. Debnath · S. Basu (✉)
Maulana Abul Kalam Azad University Of Technology, Kolkata, West Bengal 700064, India
e-mail: saikatbasu@gmail.com

S. Debnath
e-mail: neeldebnath99@gmail.com

 
A. Maji et al. (eds.), *Proceedings of the International Conference on Computing and Communication Systems*, Lecture Notes in Networks and Systems 170,
https://doi.org/10.1007/978-981-33-4084-8_19

previously mentioned feelings that can be resolved. It tends to be characterized as the response that individuals may have when they are liable to requests and weights which do not relate as far as anyone is concerned and capacities and that can challenge their dealing with abilities. Stress is a noteworthy issue in the present world. It is a developing snag and gradually it becomes a certain piece of our day by day lives. The soon we identify it, will diminish the harm it costs and keeps it resistible. The steps for building a model for stress detection are selecting a digital task for invoking stress conditions and physiological response storing devices, then extracting information from the experiment and analysing the data using classifiers.

## 2 Literature Review

The investigation of the related work on stress identification using software engineering shows that the focus moves from stress recognition in an obliged situation utilizing sensors to push location in an unconstrained domain utilizing better sensors. Different investigations were led to actualize stress identification utilizing a blend of sign handling and artificial intelligence. The greater part of them utilized information from a respiratory sensor [2–4], ECG sensor [5–7], pulse sensor [4, 8–10], electrodermal activity (EDA) sensor [6, 8, 11], blood volume pressure (BVP) sensor [2, 10] and electromyogram (EMG) sensor [5, 12]. Some are increasingly constrained, either physically (brain activity analyser [2, 13–15]) or as for privacy (analysing the participants sound or video [16, 17]). In [18], the researchers portrayed the sorts of physiological signs without referencing related researches. They just gave the kinds of corporal signs and a few highlights of them. In [19, 20], researchers distinguished stress at work environments and workplaces. The ongoing investigation for stress identification is to perceive stress in working environments as well as in the daily life of people.

Distinctive computerized tasks are being utilized as stressors, for example, the Stroop task [6], Montreal imaging task [11], stimulated stress talk [21], etc. These are demonstrated to be productive in instigating stress on condition. In [22], they utilized the mental arithmetic task as a stressor. It is additionally a viable technique under constrained time conditions. Berg card arranging test [23], which is utilized to administer brain damage or disorders, was additionally used as recognize stress-related conditions. Alongside the tasks, biosensors [2] are likewise demonstrated to be valuable to distinguish stress conditions in individuals. For estimating anxiety, particular sensors have been utilized. It incorporated various sensors to access diverse physiological reactions of our body when it is under pressure. For example, rear weight sensor, floor pressure sensor, seat pressure sensor, eye movement sensor and so on. The precision of the test was reliant on the blends of the various modules [10].

## 3 Review of Digital Tasks

The traditional Stroop test, otherwise called colour word test (CWT), is the most settled and extensively used stress acknowledgement tests. Changing the tasks, type, time for endeavour scoring frameworks, different assortments of the task were made. The typical rendition of the job involves an individual to peruse the label of the shaded terms. This task is classified as "word perusing". The next task is determining the shade of term. The final task is determining the hue of the word appearing. CWT is executed in this work as a game to challenge the test individually in various difficulties. Each level broadens the previous one by another part proposed to build the perplexity of the trial and thus the apparent pressure. The initial level speaks to a standard CWT with no alterations, as portrayed previously. In the following level, the CWT is stretched out with a pacing clock of five seconds to put pressure on the partaker. At that point in the next level of trails, the potential answers, the test individual can pick, are exhibited in various hues rather than only dark. This causes other interruption and along these lines more pressure. On the last level of trials, an errand variety is included by haphazardly asking the test individual to either name the text style shade of the word or the word itself. At long last on the last level, the time pressure is expanded by diminishing the utmost for a response to three seconds. In [12], Stroop test was utilized alongside electromyogram. Electromyogram was utilized to measure muscle working and strain by recording muscle activity. During continuous performance tests, for example, the Stroop test, electromyogram action expanded in a progressively obvious way. The proficiency of the investigation was around 90.7%. In [14] also, the Stroop test was utilized. Here, four degrees of stress were incited, and electroencephalogram information was recorded during the trial. Their experiment indicated expanded precision. There are different varieties of trials where the Stroop test has been utilized with other computerized assignments, for example, Stroop task and auditory stimulus [7], Stroop task and mental maths test, stress talk [9], Stroop task and information recalling test. Other variations were also made, such as Stroop task also has been tested with a different set of physiological signals. And the accuracy of each model was different. In [5], Stroop test was used with electrocardiogram and electroencephalogram, and the accuracy of this model was 90.35%. When Ren et al. [13] experimented with Stroop task, electrodermal sensors and pupil diameter measuring sensors, the accuracy was 85.53%. Best accuracy was found when electrodermal activity and electrocardiogram sensors were used along with Stroop task [6].

The Montreal Imaging Stress Task (MIST) incorporates modernized mental numbers juggling difficulties and social danger provoking components with the assistance of a computer application. An agent applies it. It can be divided into three stages relax, control and exploratory. In the first stage of the test, the individual undergoing the test is watched through functional magnetic resonance imaging (fMRI) or positron discharge tomography. In the relaxing stage, the individual takes a look at the PC screen. During this stage, individuals are not approached to play out any assignments. In the next step, a progression of mental number juggling undertakings appears to the individual and individual marks a response against each task by utilizing the application. In the final stage, the trial stage, the level of difficulty of the task is expanded and the time is diminished. The point is to develop the stress of the job over the individual's psychological capacities. To extend the pressure of the individual additionally, the normal execution of members and assessed culmination time is shown to the individual. After the finish of each assignment, the exhibition assessment is shown to additionally expand the pressure of the individual. In [11], MIST was utilized alongside the electrodermal movement. They were effective in finishing up whether the individual was stressed or not. In [13], additionally, MIST was being utilized alongside electroencephalogram. The accuracy of their model was 94%. The analyses clearly show MIST is a powerful stressor.

The International Affective Picture System (IAPS) is a picture database made for giving a standard arrangement of pictures for stress, emotion-related experiments. IAPS task was invented by the National Institute of Mental Health Center at the University of Florida [2]. It is actively used in measuring physiological changes due to emotion [24]. It comprises images fluctuating from straightforward family unit articles to extraordinary pictures, which cause excitement on people. In [2], IAPS task was utilized with physiological sensors like galvanic skin reaction, blood volume pressure, respiratory rate and electroencephalogram. The efficiency discovered was 82.60%. In [25], similar level of accuracy of 82.70% was found while measuring Respiration, skin conductance, photoplethysmograph, heart rate variability and EEG signals. In [16], IAPS task was utilized alongside the Stroop task. Their model had a proficiency of 91.68%. Stress detection by demonstrating a progression of photos from the IAPS is one more different test, which is quite popular. In mental arithmetic task, participants were asked to do sums [26]. The arithmetic test is seen as a standard moderate extreme stressor, and it is routinely used in physiology to recognize changes in the autonomic tactile framework. In [15], mental arithmetic test and Stroop test were used. They discovered diverse precision for the two tests. For the Stroop test, it was 88% and for the mental arithmetic task, it was 96%. In different investigations, the mental arithmetic test was utilized with various physiological sensors. In [8], mental arithmetic test was utilized with a pulse rate sensor and electrodermal action sensor. They found an accuracy of 96.19%. In [22] and [3], they utilized a mental arithmetic test with electrocardiogram, electroencephalogram, heart rate variability sensor and breathing rate sensor individually. In [3], they found 90.9% exactness; however, in [22], precision was 79.54%.

The standard Berg Card Sorting Task (BCST) comprises of deck cards showed on a PC screen. Each of the cards contains an alternate blend of four shapes, colours and amounts. Four example cards are shown at the highest point of the screen as a manual for help figure out the arrangement. The deck is uncovered each card in turn, and the obvious card is coordinated to scratch cards relying upon the specific standard (unknown to the individual) for a given arrangement. When ten cards are effectively coordinated, the set is finished and the arranging rule changes (additionally unknown to the individual). The new standard must be found utilizing experimentation through input got after each card is arranged. After a card is arranged, the member is furnished with information in regard to whether it was arranged expertly (i.e. as per the present standard). This procedure proceeds until the member either sort every one of the 128 cards or until the member effectively finishes nine sets, whichever starts things out. The test can hypothetically be finished in 90 trails [23]. In [4], researchers used BCST and obtained an efficiency of 86%. So, it is also an effective way to induce stress (Table 1).

Different emotions can be detected from speech also [27]. So for invoking stress, stimulated stress talk can be used [21], which is also proved to be effective.

## 4 Review of Classifiers

In [7, 15], support vector machine (SVM) was used. SVM amasses a hyperplane or set of hyperplanes in a high- or unending dimensional space, which can be utilized for classification, regression or various tasks like special case identification. Instinctively, an incredible division is done by the hyperplane that has the best expel to the closest preparing data point of any course, since in like manner the more prominent the edge, the lower the speculation blunder of SVM. In support vector machine, we can use the function describing hinge loss for data which cannot be separable linearly.

$$\left[\frac{1}{n}\sum_{i=1}^{n}\max(0, 1 - y_i(w \cdot x_i - b))\right] + \lambda\|w\|^2 \tag{1}$$

Equation 1 represents soft margin of SVM. Here, max(0, 1 − $y_i$ ($w \cdot x_i - b$)) is known as hinge loss and $\lambda kwk^2$ is the penalty. In [11], electroencephalogram and MIST were used. Parameter obtained from electroencephalogram was used as a feature in support vector machine classifier. The achieved percentage of accuracy was 82.80%. *K*-nearest neighbour (KNN) algorithm is widely used learning algorithm. It uses a database where data stored into it are divided into classes for determining the class of the new input. It does not make any assumptions. KNN model structure is derived from the data available. It is when we do not have much information about the data obtained. It is based on the similarities of the feature of the new sample. It uses the general formula for calculating Euclidean distance. Equation 2 describes the formula.

$$d(p, q) = d(q, p) = \sqrt{(q_1 - p_1)^2 + \cdots + (q_n - p_n)^2} \tag{2}$$

$p$ and $q$ describes the Euclidean vectors where $p = (p_1, p_2, \ldots, p_n)$ and $q = (q_1, q_2, \ldots, q_n)$. Logistic regression is a supervised classification model. In [9], using logistic regression 79% accuracy was found. Logistic regression turns into a classification procedure just when a choice limit is brought. The setting of the limit esteem is a significant part of logistic regression and is subject to the classification issue itself. The cost capacity of logistic regression is denoted by Eqs. 3 and 4 is the probability

**Table 1** Summary table of digital tasks used

| SL | Author | Physiological signals | Digital tasks |
|---|---|---|---|
| 1 | Elena Smets et al. (2018) | Galvanic skin response, heart rate, skin temperature | Stroop test, math test, stress talk |
| 2 | Zhang (2018) | Galvanic skin response, electrocardiogram, electromyography | Stroop test and auditory stimuli |
| 3 | Khosrowabadi (2018) | Electroencephalogram | AudioVisualemotional stimuli |
| 4 | Xia et al. (2018) | Electroencephalogram (EEG) and electrocardiogram (ECG) | Mental arithmetic task |
| 5 | Airij (2018) | Heart rate, electrodermal activity, skin temperature | Stroop test and mental Arithmetic test |
| 6 | Giannakakis et al. (2017) | Facial expression | Emotion recall, International Affective Picture System, Stroop test, videos |
| 7 | McDuff et al. (2016) | Heart rate, heart rate variability (HRV), breathing rate | Berg Card Sorting Task |
| 8 | Simantiraki et al. (2016) | Spectral slope (speech recognition) | SUSAS (simulated stress speech) |
| 9 | Jun and Smitha (2016) | Electroencephalogram | Stroop test and mental Arithmetic test |
| 10 | Hou et al. (2015) | Electroencephalogram | Stroop test |
| 11 | Al-shargie et al. (2015) | Electroencephalogram | Montreal Imaging Stress Task |
| 12 | Yoshiki Nakashima et al. (2015) | Blood volume pressure, heart rate (HR), back pressure sensor, floor pressure sensor | Stroop test and mental Arithmetic test |
| 13 | McDuff (2014) | Heart rate variability, breathing rate | Mental arithmetic test |
| 14 | Korbinian Frank et al.(2013) | ECG, heart rate, breathing rate, temperature | Stroop test |
| 15 | Ren et al. (2013) | Pupil, electrodermal activity | Stroop test |
| 16 | Giakoumis et al. (2012) | Electrodermal activity, electrocardiogram, body activity | Stroop test |
| 17 | Karthikeyan et al. (2012) | Electromyography | Stroop test |
| 18 | Hosseini and Khalilzadeh (2010) | Galvanic skin response, blood volume pressure, respiratory rate, electroencephalogram | International Affective Picture System |
| 19 | Setz et al.(2010) | Electrodermal activity | Montreal Imaging Stress Task |

of estimating $y = 1$ where $h_\theta$ $(x)$ is hypothesis. Here, $\theta$ denotes the parameter. The value of the $\theta$ such that cost function $J(\theta)$ need to be minimized.

$$J(\theta) = \frac{1}{m}\left[\sum_{i=1}^{m} y^{(i)} \log h_\theta\left(x^{(i)}\right) + \left(1 - y^{(i)}\right)\log\left(1 - h_\theta\left(x^{(i)}\right)\right)\right] \tag{3}$$

$$P(y = 1|x; \theta) = h_\theta(x) = \frac{1}{1 + e^{-\theta T_x}} \tag{4}$$

In [10], linear discriminant analysis (LDA) was used. LDA is utilized when the fluctuation covariance frameworks for all populaces are homogeneous. In LDA, our choice principle depends on the linear score function, a component of the populace implies for every one of our populaces, *i*, and the pooled difference covariance network. LDA score function is decision rule. In Eq. 5, score function is represented.

$$S_i^L(x) = -\frac{1}{2}\mu_i' \Sigma^{-1} \mu_i + \mu_i' \Sigma^{-1} X + \log P(\Pi_i) \tag{5}$$

$$S_i^L(x) = -d_i^L(X) + \log P(\Pi_i) \tag{6}$$

Here in Eq. (6) $d^L{}_i$ $(X)$ is known as the LDA function (Table 2).

In [17] and [5], Naive Bayes classifiers were used. These are a group of basic probabilistic classifier subject to applying Bayes' hypothesis with strong independent presumptions between the feature. They are among the easiest Bayesian system models. According to Bayes theorem, we know if feature vector $X = (x_1, x_2, x_3, \ldots, x_n)$ and a class variable $C_k$. Then,

$$P(C_k|X) = \frac{P(X|C_k)P(C_k)}{P(X)} \text{ for } k = 1, 2, 3 \ldots k. \tag{7}$$

In Eq. 7 $P$ $(C_k|X)$ is posterior probability. According to Naive independent assumption, it can also be written as described in Eq. 8.

$$P(C_k|X) = \frac{P(C_k)\prod_{i=1}^{n} P(x_i|C_k)}{P(X)} \tag{8}$$

For different class values, Naive Bayes classifier can be formulated as in Eq. 9.

$$\widehat{C} = \underset{C_k}{\arg\max}\, P(C_k)\prod_{i=1}^{n} P(x_i|C_K) \tag{9}$$

Random forest (RF) classifier built numerous individual choice trees at preparing. Expectations from all trees are pooled to make the last forecast. As they utilize an assortment of results to settle on an ultimate choice, they are alluded to as ensemble strategies.

**Table 2** Summary table of classifiers used

| SL | Author | Sample size | Classifiers | Accuracy (%) | Data set |
|---|---|---|---|---|---|
| 1 | Zhang (2018) | 12 | Support vector machine | 88.50 | EDA, EMG, heart rate volume, response time |
| 2 | Airij (2018) | 35 | Fuzzy logic, K-nearest neighbour | 96.19 | Heart rate, skin conductance, skin temperature |
| 3 | Xia et al. (2018) | 22 | Principal component analysis, support vector machine | 79.54 | EEG and ECG features |
| 4 | Song and Kim (2017) | 1280 | Deep belief network | 66.23 | pulse rate, systolic and diastolic blood pressure (SBP, DBP), height, heaviness, body mass index (BMI), averages of age, sleep time, smoking and drinking behaviours |
| 5 | Simantiraki et al. (2016) | 9 | Random forest | 92.06 | SUSAS data set |
| 6 | Yoshiki Nakashima et al. (2015) | 10 | K-nearest neighbour, support vector machine, linear discriminant analysis | 84.10 | EDA, BVP, HR, floor, backrest and seat pressure distributions |
| 7 | Hou et al. (2015) | 9 | K-nearest neighbour, support vector machine | 85.17 | EEG |
| 8 | Korbinian Frank et al. (2013) | 15 | Bayesian network | 90.85 | ECG, heart rate, breathing rate, temperature |
| 9 | Karthikeyan et al. (2012) | 10 | K-nearest neighbour | 90.70 | EMG |
| 10 | Hosseini and Khalilzadeh (2010) | 9 | Support vector machine | 82.60 | Respiration, skin conductance, photoplethysmograph, heart rate variability, EEG signals |

In [21], using random forest model, accuracy found was 92.06%. In [8], accuracy found using fuzzy logic was 96.19%. In a fuzzy classification framework, a case or an item can be ordered by applying a lot of fuzzy guidelines dependent on the semantic estimations of its characteristics. Each standard has a weight, which is a number somewhere in the range of 0 and 1, and this is used to the number presented by the antecedent. It includes two particular parts. The initial segment includes assessing the antecedent, fuzzifying the data and applying important fuzzy set operations. Some of them are given by Eqs. 10 and 11.

$$\text{Intersection}\quad \mu_{A\cap B}(x) = \text{Min}[\mu_A(x), \mu_B(x)] \tag{10}$$

$$\text{Complement}\quad \mu_A - (x) = 1 - \mu_A(x) \tag{11}$$

$\mu_A$ and $\mu_B$ are membership functions. They represent fuzzy sets $A$ and $B$, respectively. Equation 10 defines the intersection fuzzy set $A$ and $B$. Equation 11 defines complement of fuzzy set $A$.

Another interesting model is the deep belief network (DBN). It is a deep neural network with several layers of restricted Boltzmann machine network (RBM). It is isolated in two separate phases. The main phase is unaided prepreparing that learns traits just with input esteems without marks. The input learns the principal concealed layer x, which, thus, learns the next shrouded layer. In the subsequent phase, an error backpropagation assessment is done utilizing a supervised refinement [28]. In [28], pulse rate and other parameters like systolic and diastolic blood pressure (SBP, DBP), height, heaviness, body mass index (BMI), averages of age, sleep time, smoking and drinking behaviours of different low-stress and high-stress individuals were used as input nodes. A joint arrangement ($v$, $h$) of the observable nodes $v$ and concealed nodes h can be expressed by the following equation:

$$E(v, h) = -\sum_{i,j} v_i h_j w_{ij} - \sum_i b_i v_i - \sum_j b_j h_j \tag{12}$$

Here, $v_i$ is visible node, $h_j$ is concealed node, $w_{ij}$ is weight between nodes and $b_i$, $b_j$ are bias term of observable and concealed nodes, respectively.

## 5 Discussion

For understanding stress, it is important to understand its cause, detection and analysis. This survey reviews digital tasks that were been used along with the physiological signal sensing sensors to detect stress. In all of the research, biosignal features were used along with generic digital tasks. But there is no collective information on significant differentiation of role of each biosignal's features during detecting stress. We found that Stroop colour word test, International Affective Picture System, Montreal Imaging Stress Task, Berg Card Sorting Task, mental arithmetic task, stress inflicting videos or combination of these tasks were used to detect stress. These tasks were performed within a fixed time limit. Stroop task is a cognitive task, i.e. it is automatic. If the time limit is lesser than the standard response time of the individual while performing the Stroop task then the accuracy level decreases [29]. When it instantaneously reflects on the score it induces stress. Similarly, for mental arithmetic task, each individual has different processing time. If the time limit is lesser than response time, it can induce stress. In [12], when the Stroop task was used, the accuracy in detecting stress was 90.7%. And when in [15] Stroop task and mental arithmetic task were used, the accuracy level was 88% and 96%, respectively, provided in [12] and in [15] different physiological signals were measured. Over several more kinds of research, these two tasks have been found effective. For these reasons, the Stroop colour word test and mental arithmetic test were the most used tasks for invoking stress. Another stress inducer Montreal Imaging Stress Task which also involves solving mental arithmetic problems includes a social threat component. Social threat component shows the participant his or her performance and expected performance. It is equally effective as a mental arithmetic task. In the International Affective Picture System, emotional responses seeing the images can vary. For some individuals, surprise can be perceived as positive stimuli and for some individuals, it can be perceived as negative stimuli [30]. So it can be ambiguous. Berg Card Sorting Task which is derived from the Wisconsin Card Sorting Test has few limitations. It can help to find out specific aspects, specifically the mechanism of perseveration which causes the shifting from old rule to the new rule. But [24] suggests it is the learning of irrelevance. Learned irrelevance is the understanding that helps to find out the irrelevance of the old rule. So there is ambiguity regarding these two factors.

## 6 Conclusion

Depending on the social, contextual and cognitive-emotional parameters, the same stressor can invoke different stress responses in different individuals. So there is complexity in determining the stressed and relaxed condition. But the methods discussed here can provide a good basis of determining the initial stressed and relaxed conditions. The accuracy levels of the tests discussed here varied from 66% [28] to 96.6% [6]. This review aims to give an understanding of the underlying mechanisms of digital tasks and biosignals in determining the stress. Most of the studies that have been discussed were performed in a controlled environment. The tasks used were effective to determine intricate stress conditions but in real life, scenarios are more intricate that involve many aspects of human psychology and characteristic traits that involve in the impact of stress on a particular individual. Different types of tasks can be introduced but it is not always possible to map the experiences to real-life experiences. For that purpose, more refined tasks need to be developed that will help map the real-life experiences.

**Acknowledgements** I am thankful to my project guide Mr Saikat Basu for giving his time and insights for this review paper. We are also very grateful to honourable Vice-Chancellor of our university, Prof. Saikat Maitra for his support and encouragement was given to us in the idea of making the review paper.

## References

1. Picard RW (2016) Automating the recognition of stress and emotion: from labto real-world impact. IEEE Multim 23(3):3–7
2. Hosseini SA, Naghibi-Sistani MB (2011) Classification of emotional stress using brain activity. Appl Biomed Eng 313–336
3. McDuff D, Gontarek S, Picard R (2014) Remote measurement of cognitive stress via heart rate variability. In: 2014 36th annual international conference of the IEEE engineering in medicine and biology society. IEEE, pp 2957–2960
4. McDuff DJ, Hernandez J, Gontarek S, Picard RW (2016) Cogcam: contact-free measurement of cognitive stress during computer tasks with a digital camera. In: Proceedings of the 2016 CHI conference on human factors in computing systems, pp 4000–4004
5. Frank K, Robertson P, Gross M, Wiesner K (2013) Sensorbased identification of human stress levels. In: 2013 IEEE international conference on pervasive computing and communications workshops (PERCOM Workshops). IEEE, pp 127–132
6. Giakoumis D, Drosou A, Cipresso P, Tzovaras S, Hassapis G, Gaggioli A, Riva G (2012) Using activity-related behavioural features towards more effective automatic stress detection. PloS one 7(9)
7. Zhang B (2017) Stress recognition from heterogeneous data
8. Airij AG, Sudirman R, Sheikh UU (2018) Gsm and gps basedreal-time remote physiological signals monitoring and stress levels classification. In: 2018 2nd international conference on biosignal analysis, processing and systems (ICBAPS). IEEE, pp 130–135
9. Huysmans D, Smets E, De Raedt W, Van Hoof C, Bogaerts L, Van Diest I, Helic D (2018) Unsupervised learning for mental stress detection exploration of self-organizing maps. In: BIOSIGNALS 2018—11th international conference on bio-inspired systems and signal processing, proceedings; part of 11th international joint conference on biomedical engineering systems and technologies, BIOSTEC 2018, vol 4. SciTePress, pp 26–35
10. Nakashima Y, Kim J, Flutura S, Seiderer A, Andre E (2015) Stress recognition in daily work. In: International symposium on pervasive computing paradigms for mental health. Springer, pp 23–33
11. Setz C, Arnrich B, Schumm J, La Marca R, Troster G, Ehlert U (2009) Discriminating stress from cognitive load using a wearable eda device. IEEE Trans Inf Technol Biomed 14(2):410–417
12. Karthikeyan P, Murugappan M, Yaacob S (2012) Emg signal based human stresslevel classification using wavelet packet transform. In: International conference on intelligent robotics, automation, and manufacturing. Springer, pp 236–243
13. Al-Shargie FM, Tang TB, Badruddin N, Kiguchi M (2015) Mental stress quantification using eeg signals. In: International conference for innovation in biomedical engineering and life sciences. Springer, pp 15–19
14. Hou X, Liu Y, Sourina O, Eileen Tan YR, Wang L, Mueller-Wittig W (2015) Eeg based stress monitoring. In: 2015 IEEE international conference on systems, man, and cybernetics. IEEE, pp 3110–3115
15. Jun G, Gopi Smitha K (2016) Eeg based stress level identification. In: 2016 IEEE international conference on systems, man, and cybernetics (SMC). IEEE, pp 003270–003274.
16. Giannakakis G, Pediaditis M, Manousos D, Kazantzaki E, Chiarugi F, Simos PG, Marias K, Tsiknakis M (2017) Stress and anxiety detection using facial cues from videos. Biomed Signal Process Control 31:89–101
17. Ren P, Barreto A, Gao Y, Adjouadi M (2012) Affective assessment by digital processing of the pupil diameter. IEEE Trans Affect Comput 4(1):2–14
18. Sioni R, Chittaro L (2015) Stress detection using physiological sensors. Computer 48(10):26–33
19. Dedovic K, Renwick R, Khalili Mahani N, Engert V, Lupien SJ, Pruessner JC (2005) The montreal imaging stress task: using functional imaging to investigate the effects of perceiving and processing psychosocial stress in the human brain. J Psychiatry Neurosci 30(5):319

20. Sano A, Taylor S, McHill AW, Phillips AJK, Barger LK, Klerman E, Picard R (2018) Identifying objective physiological markers and modifiable behaviors for self-reported stress and mental health status using wearable sensors and mobile phones: observational study. J Med Internet Res 20(6):e210
21. Simantiraki O, Giannakakis G, Pampouchidou A, Tsiknakis M (2016) Stress detection from speech using spectral slope measurements. In: Pervasive computing paradigms for mental health. Springer, pp 41–50
22. Xia L, Malik AS, Subhani AR (2018) A physiological signalbased method for early mental-stress detection. Biomed Signal Process Control 46:18–32
23. Fox CJ, Mueller ST, Gray HM, Raber J, Piper BJ (2013) Evaluation of a short-form of the berg card sorting test. PloS one 8(5)
24. Basu S, Bag A, Aftabuddin M, Mahadevappa M, Mukherjee J, Guha R (2016) Effects of emotion on physiological signals. In: 2016 IEEE annual India conference (INDICON). IEEE, pp 1–6
25. Hosseini SA, Khalilzadeh MA (2010) Emotional stress recognition system using eeg and psychophysiological signals: using new labelling process of eeg signals in emotional stress state. In: 2010 international conference on biomedical engineering and computer science. IEEE, pp 1–6
26. Weinstein AA, Deuster PA, Francis JL, Bonsall RW, Tracy RP, Kop WJ (2010) Neurohormonal and inflammatory hyperresponsiveness to acute mental stress in depression. Biol Psychol 84(2):228–234
27. Basu S, Chakraborty J, Aftabuddin M (2017) Affect detection fromspeech using deep convolutional neural network architecture. In: 2017 14th IEEE India council international conference (INDICON). IEEE, pp 1–5
28. Song S-H, Kim DK (2017) Development of a stress classification model using deep belief networks for stress monitoring. Healthcare Inf Res 23(4):285–292
29. Kyllonen PC, Zu J (2016) Use of response time for measuring cognitiveability. J Intell 4(4):14
30. Neta M, Kelley WM, Whalen PJ (2013) Neural responses to ambiguity involve domain-general and domain-specific emotion processing systems. J Cognitive Neurosci 25(4):547–557

# Design a New Learning based Method for Smart Semantic Data Management System

**Afsana Laskar and Sufal Das**

**Abstract** Now a days, enormous quantity of data is being generated everyday from various sources such as online data, social media, scientific data, text data, images etc. When data becomes large in Volume, generated in high Velocity and Variety of data in the form of structured data or unstructured data is known as Big data. Traditional database tool such as RDBMs cannot deal with Big data. Challenges in big data are information storage, search and retrieval of data. Since most of the data are in different formats, we need tools and technology for transforming, analyzing, visualising data and extracting information for decision making. Collection of large data set is a significant issue and to make sense from that big data is a problem. We need ways to find semantic information from the data to help users make better decisions making. To facilitate better data management, Semantics can be used. Semantic can be efficiently use for management of large data sets and better information retrieval process. Semantic annotation of heterogeneous data sets could facilitate the search and retrieval using the concepts contained in them. This paper proposes a new methodology for designing a Learning based Method for Smart Semantic Data Management System.

**Keywords** Semantic · Semantic annotation · Resource description framework · Ontology · Sparql

## 1 Introduction

Big data describes huge amount of data set,that may compromise of petabytes or exabyte of data which consists of millions of records generated from different sources such as web, online data, social media, senors, scientific data, mobile data,

A. Laskar (✉) · S. Das
North Eastern Hill Univeristy, Shillong, Meghalaya, India
e-mail: laskar.afsana24@gmail.com

S. Das
e-mail: sufal.das@gmail.com



A. Maji et al. (eds.), *Proceedings of the International Conference on Computing and Communication Systems*, Lecture Notes in Networks and Systems 170,
https://doi.org/10.1007/978-981-33-4084-8_20

e-commerce,health care systems etc. Big Data systems are capable of storing, ingesting and processing massive amount of heterogeneous data from various sources [1]. Big Data consists of structured and unstructured data and are characterised by its Volume, Velocity, Variety, Veracity and Value [2].

- **Volume**: When data are generated in vast amount in a short period of time say seconds, minutes, hours or day, it is referred to as Volume.
- **Variety**: It consists data in structured,semi-structured or unstructured form. They are heterogeneous in nature. Data can be text data, log files, audio, video, images etc.
- **Velocity**: It refers to the speed in which new facts and figures are generated.
- **Value**: It is an essential *V* in big data. Among all the data available, we need to store only relevant and correct data.
- **Veracity**: It refers to data in doubt. With large volume of data, some data are going to be inaccurate, others may contain missing value or noisy data. So, we need ways to filter such data and store correct value.

With large volume of data, it is difficult to store and process using traditional database such as RDBMS. Due to heterogeneous nature of big data, decision making is difficult. We need tools and technology for storing, analysing, transforming and visualising data to make it understandable to the user for decision making. Metadata standard has not been defined for big data. With heterogeneous data, schema design in advance is also difficult. An improper data representation may reduce the accuracy or may invalidate the result.

Semantics can help in solving such problems. Semantic is the study of meaning. It explains how we derive meaning or symbols. Semantic data is data, that is organised so it can be understandable by machine. It refers to machine to see and discover relationship in the data.

Organization of the paper: Sect. 2 presents the background study. In the next section, four separate topic are presented. Each topic represents one particular aspect of management of big data using semantic. Sect. 4 provides an overview of the proposed method and finally we conclude the paper.

## 2 Background Study

### *2.1 Semantic Annotation*

Annotation is defined as the system of including semantic metadata to resources to make the data more meaningful. If we are searching for something from a huge pile of files. How do we find things.In the web, we do it by search engine. We search by keywords. If we search by its meaning, then data can be retrieved easily. We need metadata, that describes our data. Resources can be described via metadata. Metadata is data that describes other data, contains information about other data. According to the W3C, metadata is machine understandable information about web resources.

The semantics of semantic metadata is explicitly and is a formally defined (ontology) and therefore machine understandable (i.e. it can be automatically read and interpreted by machine).

Annotation implies to attach data to some other pieces of data, it establishes a typed relation between the annotated data and the annotating data. Annotation can denote both the process of annotating and the result of that data.

An annotation $A$ is defined by tuple $a_s, a_p, a_o, a_c)$
where
$a_s$ = subject of the annotation
$a_o$ = object of the annotation
$a_p$ = predicate
$a_c$ = context in which annotation is made.

Annotation is useful due to its capacity to deal with large collection of information/textual content. The annotation process is essential to reduce the burden of annotating new documents. It is beneficial in providing the scalability needed to annotate existing documents. It can also be used in information extraction techniques, among which named entity recognition (NER) is used to pick out ideas to annotate. Semantic annotation is the process of assigning metadata tags and or ontology classes to text segments as a further basis for knowledge access and information retrieval mechanisms. It relies on the meaning of data. The Semantic annotation of a document is composed of locating mappings between text chunks of a document and the instances or individuals in ontology [3]. Annotation is useful in applications such as extraction of open information, generation of linked data, ontology matching and semantic search.

## 2.2 *Ontology*

Ontology is a formal, explicit specification of a shared conceptualisation.An ontology is a formal illustration of knowledge,which plays a very vital role in semantic web [4], due to its functionality to express meanings and relationships among data. Ontology is beneficial in information extraction such as in the aggregation of knowledge from unstructured documents. Ontologies are useful for semantic annotation of the identified concepts. Ontology describes syntax and semantics to define complicated domain vocabularies. It leverages interoperability and offers path for semantic annotation. Knowledge can be harvested from ontologies for semantic annotation [5].

## 2.3 Resource Description Framework

RDF is proposed by W3C for representing semantic Web data It is a model for data interchange on the web [6]. To present resources on the web with structured machine understandable descriptions,RDF is used. It is also used as a method for conceptual description or modeling of knowledge about web resources so that information can be exchanged between web applications without loss of meaning. It is used to represent facts and relationships. It is a technique for describing information about a resource in a graphical shape,where resources are represented by nodes and relationship between them by edges. RDF triples are declared as Subject-Predicate-Object. Example of Data expressed in RDF triple is given in Fig. 1.

Example of RDF framework for semantics is given below in Fig. 2.

RDF feature allow us to merge data. Inference is about discovering new facts and information in database. New Facts and triples can be generated from triples. If we add more triples to the graph in the Fig. 2, we get new information. We can infer from Fig. 2, that Alvin lives at a latitude of 51.507.

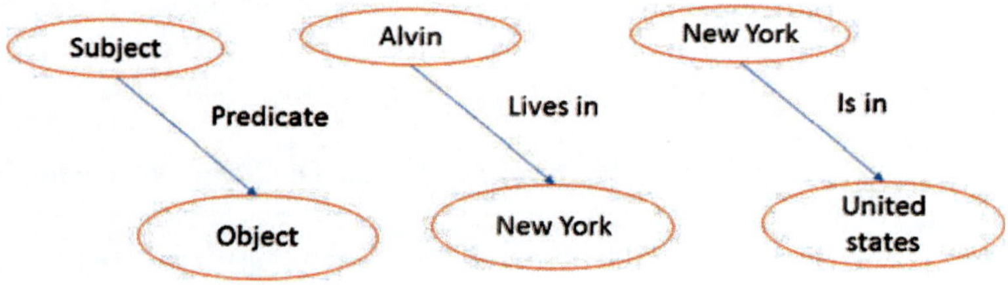


**Fig. 1** Data expressed in RDF

**Fig. 2** RDF: Framework for semantic

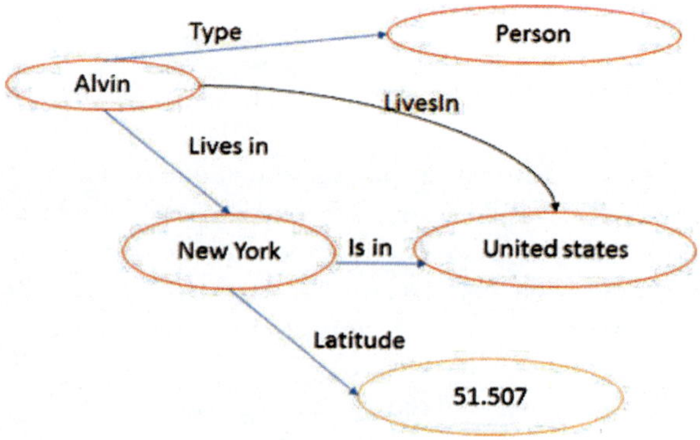

### 2.4 Sparql

SPARQL is a web query language. It is a semantic query language. It is used for retrieval mechanism in databases that store data in RDF format [7].

### 2.5 Protege

Protege is a a free,open-source ontology editor [8]. It is a framework for building intelligent systems. Protege is a java based application. It is a GUI to help in the creation, modification, reasoning and debugging of the ontology [9].

## 3 Related Works

Here,we discuss the use of semantics in Big Data.

### 3.1 Semantic Annotation for News Feed

**Introduction**
In the work done by Hendi et al. [10] have developed semantic framework to annotate RSS News Feed. It has used semantic metadata for searching and adding of web content. Here, methodology used is searching on the knowledge base instead of keyword search. Metadata is created for different instances and ontology is designed using protege.
**Dataset Description**
Input is in the form of CNN RSS feeds. It is loaded line by line in the system and annotation is done between the mapped instances of the loaded news feed and its ontology.
**Result**
Recall, Precision and $F$-measure [11, 12] are used for performance evaluation. It is applied on 8 different RSS URLs of the news domain,the precision is between 90 and 100%. Recall is between 63 and 93% and the value of the $F$-measure range is between 77 and 96% and average system accuracy is 87.8%.
**Advantage**
Semantic annotation is done using the ontology domain. It is effective and accurate for searching of web content semantically.
**Disadvantage**
Supports on English language only.

### 3.2 *Method to Improve Quality of Big RDF Data Using Semantic Based and Entity—Resolution Fusion [13]*

**Introduction**
Benbernou et al. [13] have used Resource Description Framework representation for querying Big Data for improving the quality of Big data and to fetch more relevant and complete query result. Here heterogeneous data format like csv, text, database are converted to RDF format for semantic annotation. Then data are identified with URI primarily based on the idea of Ontology. Hadoop HDFS is used for storing the RDF. To query the data, a MapReduce process is used along with SPARQL query. It is used for better evaluation and inference mechanism.
**Dataset Description**
Three open datasets of French insurance company is used. The first with housing contracts data set contains insurance contracts, housing properties, geographic information. Second data set used is Statistic open data set of INSEEE is used. Third data set used is open data from ONDRP using Dbpedia. The data sets are in the form of *N*-Triples.
**Result**
Distribution of MapReduce Jobs over different clusters,helps in reducing the execution time. When the size of the data is increased,the response time on the cluster of machine increases more slowly (linear) in comparison to using standalone machine (polynomial).
**Advantage**
The above framework improves the data quality of the result. It handles variety of data.
**Disadvantage**
It takes more Execution time.

### 3.3 *Framework for Big Data Integration Using Semantic Extract-Transform-Load(ETL) [14]*

**Introduction**
In the paper presented by Bansal [14], have proposed use of semantic technology for integrating and publishing data from multiple sources as open linked data. A semantic model using ontology is created. It provides a basis for data integration and understanding of knowledge from different sources. RDF is used as the graph data model and information is extracted using SPARQL.
**Dataset Description**
Open datasets from National Household Travel survey data and EPA's fuel economy data are used.

**Result**
The integrated data is query by using RDF Query language SPARQL. Accuracy is good.
**Advantage**
Semantic technology is used to produce rich and meaningful knowledge.
**Disadvantage**
Ontology generation from different data sources is a difficult task and requires a good understanding of the data. It requires more time for processing.

### 3.4 *Bolster Architecture for Big Data [15]*

**Introduction**
Nadal et al. [15] have developed architecture for Big Data System using Semantics.Basic architecture used is $\lambda$ architecture, which is a base model for Big data systems and is used to create *Bolster*. Bolster is an Software Reference Architecture for semantic aware big data system. $\lambda$ architecture includes three layers namely Batch layer, Serving layer and Speed layer.

In Bolster, an additional layer is added known as Semantic layer. One of the advantage is, it uses semantic and metadata to aid the user in decision making process. Bolster is suitable for handling large volume of data and variety of data. It can also handle the velocity and veracity of data.
**Dataset Description**
Real time data set is used.
**Result**
Good performance for real time data.
**Advantage**
Successfully handles the $V$'s of Big Data such as Volume, Velocity, Variety Variability and Veracity.
**Disadvantage**
Compiling all the components of *Bolster* together is a challenging task.

## 4 Proposed Method

Various methods have been developed for information retrieval and search using semantics and ontology. This paper proposes a method for querying RDF data using Ontology and updating of the Ontology terms to fetch more relevant and accurate result. Here, entities in the ontology are represented in the form of tree and with the help of learning rule,the terms are updated periodically. Agents can be used as system

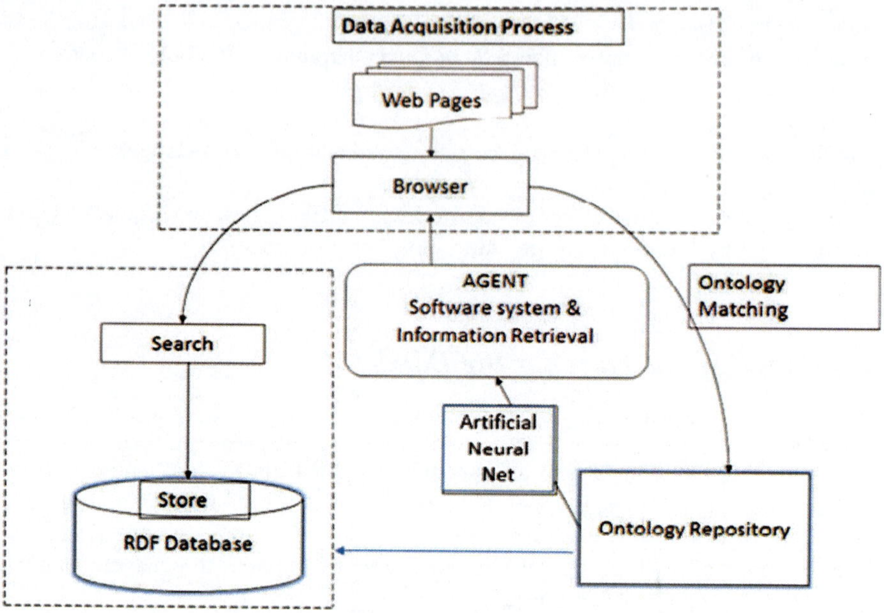


**Fig. 3** Agent based semantic management of data

software and information retrieval. For the purpose of analysis and learning of large amount of raw and unsupervised data,deep learning can be used.Neural network learning can be used for addressing issues in Big Data Analytic, such as extracting complex patterns from massive volumes of data, semantic indexing, data tagging and fast information retrieval [16]. As data is becoming larger,neural network learning is providing some solutions for big data analytics [17]. Ayadi et al. [18] have developed Ontology population using Deep Learning based NLP.
The architecture of the proposed system is depicted in Fig. 3.

**Data Acquisition Process**

This is the initial step in the process. It consists of searching and collecting web documents and files related to the domain/interest of our ontology.

**Agent**

Agents are used as persistent entity for a specific purpose. Agent are used along with learning rule, to perform task received from user. Here, agents are used as information (data or knowledge) retrieval from the Ontology.

**RDF Database**

It consist of set of resources with their descriptions. Resources are identified by their properties or characteristics. URI is used to uniquely identify an object in the RDF database.

**Ontology Repository**

It provides a shared and common understanding of the domain and is used to model the structure of the system. Here, the Ontology repository can communicate with

Agent. Learning rule is applied to update the Ontology, so that more relevant information can be achieved. Ontology matching corresponds to finding semantically related entities of ontology.

**Input Dataset** First dataset is the tourism dataset. The different entities of the dataset are hills, rivers, mountaintains, places, weather, temperature etc. Second dataset used is 20newsgroup open-source dataset. It is a collection of 11K newsgroup documents, divided into 20 different newsgroup.

**Result Analysis** Three parameters are used for performance evaluation.Precision, Recall and $F$-measure. Our precision obtained is between 8 and 100%. Recall is from 81 to 95% and $F$-measure is between 90 and 96%. If we look at the comparison between the SANF method and Our method using neural network, the precision obtained is better for large no of data (Figs. 4, 5 and 6).

$$\text{Precision} = \frac{\text{accurate}}{\text{accurate} + \text{inaccurate}}$$

$$\text{Recall} = \frac{\text{accurate}}{\text{all}}$$

$$F\text{-measure} = 2 * \frac{\text{Precision} * \text{Recall}}{\text{Precision} + \text{Recall}}$$

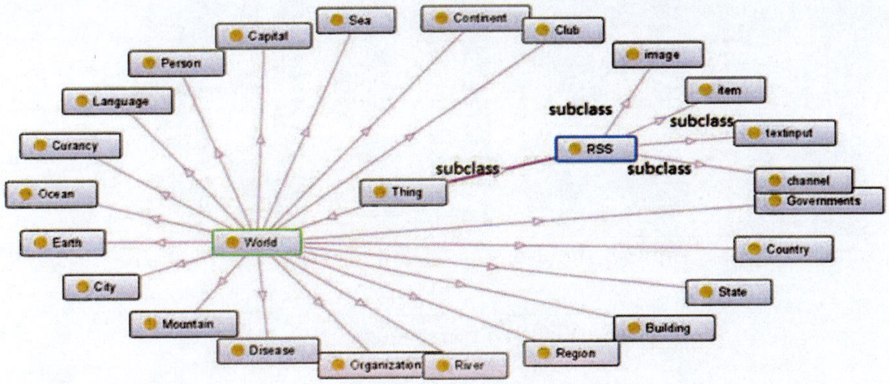


**Fig. 4** News ontology

**Fig. 5** Search result for 'Computer Science'

| index | Subject |
|---|---|
| 2231 | 2231 Computer Engr vs Computer Science |
| 10340 | 10340 Science and Methodology |
| 4173 | 4173 Rawlins debunks creationism |
| 4326 | 4326 Computer Engr vs Computer Science |
| 6921 | 6921 Automatic layout of state diagrams |
| 7618 | 7618 Solution Why do I need xrdb m when Xdefau... |
| 5741 | 5741 Rawlins debunks creationism |
| 8464 | 8464 Date is stuck |
| 3177 | 3177 Homeopathy a respectable medical tradition |
| 4366 | 4366 Computer Engr vs Computer Science |

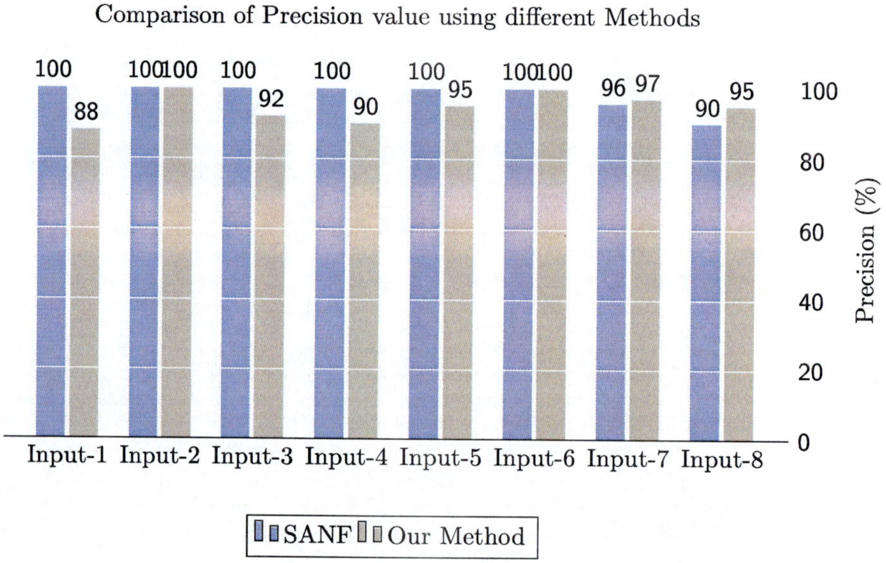


## 5 Conclusion

The above-discussed cases explain the use of semantic technology,ontology to integrate data from different sources.One of the challenges is to efficiently design Ontology for different entities. Applying Artificial Neural Network in large volume of unsupervised data, helps in data analysis. It also helps in extracting complex data representations from large volumes of data.

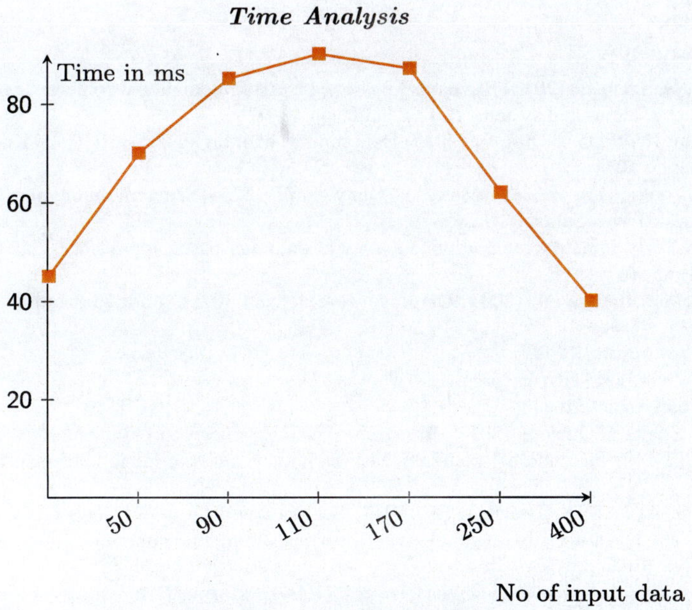
Time Analysis
Time in ms
80
60
40
20
50
90
110
170
250
400
No of input data

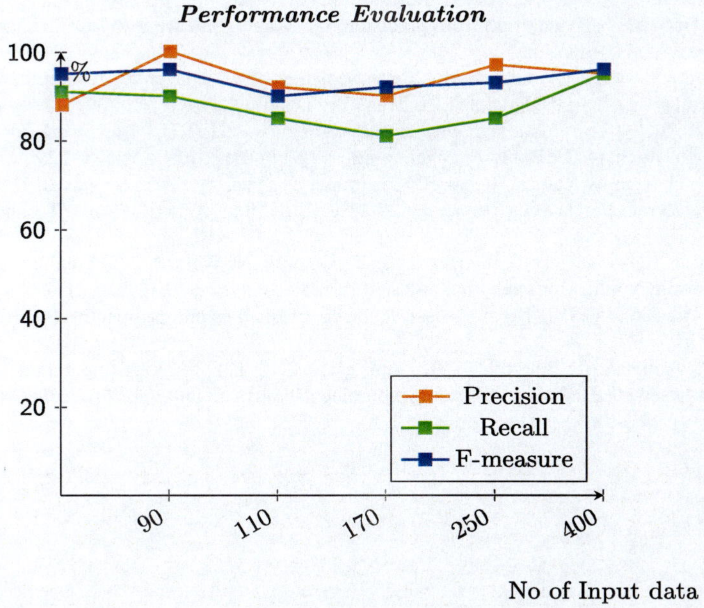
Performance Evaluation
100
%
80
60
40
20
Precision
Recall
F-measure
90
110
170
250
400
No of Input data

**Fig. 6** Result

# References


1. Sagiroglu S, Sinanc D (2013) Big data: A review. In: 2013 international conference on collaboration technologies and systems (CTS). IEEE, pp 42–47
2. Wu X, Zhu X, Wu GQ, Ding W (2013) Data mining with big data. IEEE Trans Knowl Data Eng 26(1):97–107
3. Kiryakov A, Popov B, Terziev I, Manov D, Ognyanoff D (2004) Semantic annotation, indexing, and retrieval. J Web Semant 2(1):49–79
4. Noy NF (2004) Semantic integration: a survey of ontology-based approaches. ACM Sigmod Rec 33(4):65–70
5. Noy NF, McGuinness DL (2001) Ontology development 101: a guide to creating your first ontology
6. https://www.w3.org/RDF/
7. https://www.w3.org/TR/rdf-sparql-query/
8. https://protege.stanford.edu/
9. Zhao H, Zhang S, Zhao J (2012) Research of using protege to build ontology. In: 2012 IEEE/ACIS 11th international conference on computer and information science. IEEE, pp 697–700
10. Hendi HG, Al-Feel H, Hassanein EE (2017) Semantic annotation for news feeds. In: 2017 8th IEEE annual information technology, electronics and mobile communication conference (IEMCON). IEEE, pp 185–190
11. Makhoul J, Kubala F, Schwartz R, Weischedel R (1999) Performance measures for information extraction. In: Proceedings of DARPA broadcast news workshop, pp 249–252
12. Powers DM (2011) Evaluation: from precision, recall and $F$-measure to ROC, informedness, markedness and correlation
13. Benbernou S, Huang X, Ouziri M (2017) Semantic-based and entity-resolution fusion to enhance quality of big RDF data. IEEE Trans Big Data
14. Bansal K (2014) Towards a semantic extract-transform-load (ETL) framework for big data integration. In: 2014 IEEE international congress on big data. IEEE, pp 522–529
15. Nadal S, Herrero V, Romero O, Abelló A, Franch X, Vansummeren S, Valerio D (2017) A software reference architecture for semantic-aware Big Data systems. Inf Softw Technol 90:75–92
16. Najafabadi M, Villanustre F, Khoshgoftaar TM, Seliya N, Wald R, Muharemagic E (2015) Deep learning applications and challenges in big data analytics. J Big Data 2(1):1
17. Chen X-W, Lin X (2014) Big data deep learning: challenges and perspectives. IEEE Access 2:514–525
18. Ayadi A, Samet A, de Beuvron FDB, Zanni Merk C (2019) Ontology population with deep learning-based NLP: a case study on the biomolecular network ontology. Procedia Comput Sci 159 (2019): 572–581

# Real-Time Data Analysis using Fog Intelligence

Sriraksha and Amitabha Nath

**Abstract** Internet is experiencing an unusual growth primarily because of geometric increase in the number of Internet of things (IoT) devices getting connected every day. The increase in the number of communicating devices is causing a stiff surge in overall data getting generated. In a typical cloud computing environment, this huge voluminous data are transmitted directly to the cloud for storage and computation purpose. This is causing serious pressure on the computing environment and network capacity as well. Thus, there is a need to perform the processing of these huge data closer to the source network so that only the relevant or meaningful data flow to the cloud. This resulted in the adoption of a new computing paradigm known as fog computing which aims at storage and computation nearer to the edge of the network thereby helps in overcoming various cloud computing drawbacks. The paper presents the taxonomy of fog computing and an application where different artificial neural network (ANN)-based models are developed and tested on various datasets such as room occupancy dataset, blood sugar percentage, and light and humidity dataset. The functional effectiveness of the proposed fog architecture and its performance in these adopted scenarios are presented through different experimental results.

**Keywords** Fog · Internet-of-Things (IoT) · Data analytics · Artificial Neural Network (ANN)

## 1 Introduction

The Internet is expanding because of various innovations. Internet of things (IoT) is one such innovation facilitating communication among various devices around the world. In the cloud computing scenario, collected data from these devices are

Sriraksha (✉) · A. Nath
Department of IT,NEHU, Shillong, India
e-mail: sriraksha98@gmail.com

A. Nath
e-mail: amitabha.me@gmail.com



A. Maji et al. (eds.), *Proceedings of the International Conference on Computing and Communication Systems*, Lecture Notes in Networks and Systems 170,
https://doi.org/10.1007/978-981-33-4084-8_21

transferred to the cloud for computation and storage purposes. As a result of the development in IoT applications, a huge volume of data is getting generated. Transmission of the large amount of data over the network is resulting in wastage of bandwidth, congestion in the network, and other similar problems. Recent developments in fog computing have provided a means to control the problems up to a certain extent. The fog is used to perform computation and storage just as a cloud, but it resides near to the end user or the sensor network. As a result, fog can facilitate filtering of unwanted or irrelevant data, and only the meaningful (novelty) data are allowed to flow to the cloud. The majority of the studies are based on analytic techniques employed at the cloud level. However, the trend is changing, and focus is shifting from cloud level analysis to fog level. The shift can be attributed to the need for preserving network bandwidth, transmission cost, security, privacy issues, etc. Some of the relevant and important references to the existing works are presented below.

Al-khafajiy et al. [1] introduced the concept of cloud computing for remote real-time health monitoring for elderly people via wearable devices. The authors discussed the working of the system with the help of three different layers. The model works by setting the threshold value (normal value) at cloud level, and the user will be intimated if the signal received by the sensor exceeds the threshold value.

Raafat et al. [2] discussed the data mining techniques suitable for the IoT environment. They introduced homoscedasticity and statistical feature extraction-based techniques for novelty detection and successfully built the room occupancy detection application. The authors also provided a discussion about the applicability of various classifiers such as random forest (RF), k nearest neighbors (KNN), feed forward neural network (FFNN), decision tree (DT), which can be used for this purpose.

Tortonesi et al. [3] introduced a platform called fog as a service, which proposes a utility-based service model and information-centric that can be executed either at fog level or cloud level.

He et al. [4] proposed large-scale data analytics service based on multitier fog computing for smart city applications, which contain both dedicated fogs and ad hoc fogs.

Perera et al. [5] used module-based data analytical components at fog level for building sustainable smart cities.

Mohammadian et al. [6] used deep normative modeling as a probabilistic novelty detection method at fog level to diagnose neuro-developmental and neurodegenerative disorders.

Lyu et al. [7] proposed a method to detect anomaly at fog level, by using an algorithm called hyperellipsoidal clustering and by harnessing the capacity of the fog computing paradigm in terms of processing.

Verma and Sood [8] proposed remote patient health monitoring in homes. The authors concluded that the proposed Bayesian belief network classifier-based model has high accuracy and response time in novelty detection of an event when compared with other classification algorithms.

Verma et al. [9] proposed a cloud computed IoT-based student health care monitoring system. The architecture computes the health severity in students by computing the data collected from different sensors. Results show that the decision tree(C4.5) and K nearest neighbor algorithms perform better as compared to other classifiers.

Baucas et al. [10] introduced a feasible and deployable real-time sound classification system. Three different configurations are tested in this paper for the neural network-based model.

Dhingra et al. [11] proposed a smart traffic monitoring system (STMS) to manage traffic light and monitor congestion. The idea was to overcome the limitations of latency, real-time analytic, and network congestion of cloud-based services for traffic monitoring. The author adopted the integration of fog and cloud computing paradigm.

Lin et al. [12] proposed a fog computing-based hybrid deep learning framework (FC-HDLF) that can help to find possible defective products in a factory environment. In this paper, the convolutional neural network (CNN) model is adapted to the fog computing environment.

Marković et al. [13] describe the IoT system for collecting and analyzing data for room occupancy detection. The authors used an artificial neural network (ANN) model to make real-time decisions. The accuracy was reported to be close to 99% in occupancy detection.

Parise et al. [14] developed a non-intrusive sensing method for predicting occupancy to reduce building emission and also promoting a productive and comfortable working environment. Data collected from sensors are merged in the gateway and then stored in the cloud. An occupancy detection model is then deployed at the cloud level by applying support vector machine (SVM) to predict occupancy patterns.

Welhenge et al. [15] proposed a long short-term memory (LSTM) network to determine a person's activities of daily living (ADL). MobiAct dataset was used to train and test the network. An accuracy of 90% was achieved using the LSTM network. In a similar experiment, Zhou et al. [16] proposed deep learning enhanced human activity recognition (HAR) in IoT environments. The authors proposed an LSTM-based classification model for the purpose.

In this article, neural network-based classifiers are used for analysis at fog level. The motivation has been ANN's ability to use sensor data as input features and perform classification of input signals into either novelty or non-novelty data. This will help in filtering unwanted data closer to the origin and allowing only the data with novelty to flow towards the cloud. Once the data is received at the cloud, appropriate processing can take place, and necessary action can be taken. Here in this experiment, we present real-time traffic management using fog level intelligence. Four different time-sensitive case studies such as room occupancy detection, diabetes diagnosis, human activity detection, and detecting lighting conditions of a room are considered. Five popular machine learning techniques such as support vector machine (SVM), decision tree (DT), random forest (RF), linear regression (LR), Naive Bayes (NB) have been used to compare the model performance. The organization of this paper is divided into four different sections. Section 2 presents the discussion about the various input dataset used in the experiment. The methodology part and the discussion about the neural network classifier are included in Sect. 3. The experimental results and conclusion are subsequently provided in Sects. 4 and 5, respectively.

## 2 Input Dataset

To run the models efficiently, we used different datasets from different sources for each test case as described in Table 1. For the diabetes diagnosis application, the neural network model is evaluated on the pima diabetes Indian dataset from the UCI repository, and out of 537 records, 30% is used for testing. For room occupancy detection and human activity detection, the dataset is collected from the UCI machine learning repository, where 8143 records are used for training, and 2665 records are used for testing, and 7352 records are used for training and 2947 records are used for testing, respectively. For room light detection, the dataset is simulated from cooja. A total of 394 records are simulated where 50% of it is used for testing.

## 3 Model Description

### *3.1 Artificial Neural Network*

The artificial neural network (ANN) is the information processing system. It receives data through input layer neurons, processes data in hidden layers, and finally gives out the output through the output layer neuron. In this paper, a neural network is used to perform binary classification of the sensor data as a novelty and no novelty. Figure 1 shows the architecture of the three hidden layers of a neural network.

### *3.2 ANN Model for Different Test Cases*

The four different ANN models to evaluate four different time-sensitive test cases are illustrated below.

**Table 1** Dataset used for four different test cases

| Test case | Dataset Source | No. of dataset for training | No. of dataset for testing |
|---|---|---|---|
| Diabetes diagnosis | Github | 161 | 376 |
| Room occupancy detection | UCI machine learning repository | 8143 | 2665 |
| Human activity detection | UCI machine learning repository | 7352 | 2947 |
| Room light detection | Simulated through cooja simulation | 197 | 197 |

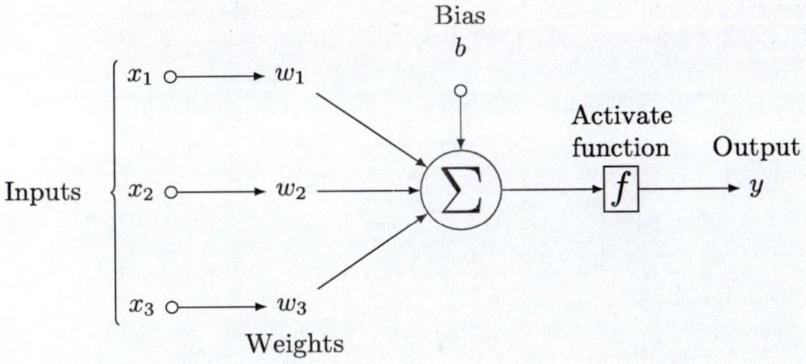


**Fig. 1** Internal structure of ANN network

In case of diabetes diagnosis, the model receives the input features such as body mass index, triceps trifold thickness, plasma glucose, number of times pregnant, age, blood pressure (BP), serum insulin, and diabetes pedigree function. It processes dataset and classifies it into a person with diabetes (novelty) and a person without diabetes (no novelty). In the case of room occupancy detection, the model receives input features such as light, temperature, humidity, carbon dioxide, and humidity ratio. It processes the dataset and classifies it into occupied (novelty) and not occupied (no novelty). In the case of human activity detection, the model receives the input features such as tBodyAcc, tGravityAcc, tBodyAccJerk, tBodyGyro, tBodyGyroJerk, tBodyAccMag, tGravityAccMag, tBodyAccJerkMag, tBodyGyroMag, tBodyGyroJerkMag, fBodyAcc, fBodyAccJerk, fBodyGyro, fBodyAccMag, fBodyAccJerkMag, fBody, GyroMag, fBodyGyroJerkMag, and activity. It processes the dataset and classifies it into motion (novelty) and rest (no novelty), and in case of room light detection, the model receives light as the input feature. It processes the dataset and classifies it into light-on (novelty) and light-off (no novelty). The neural network model in the case of room light detection is evaluated for the dataset simulated from cooja. Cooja is the application of the contiki operating system, acts as a network simulator that is mainly used to emulate the real hardware platforms [17]. In this paper, simulation is performed on two motes such as border router (gateway) and sky websense (sensors) as shown in Fig. 2a (a mote in green color is border router, motes in pink color is sky websense). The light values generated by any of the sky websense motes are used by the model as shown in Fig. 2b.

The detailed architecture of the model for four different test cases is shown in Table 2. Taking into consideration, the number of epochs and the hidden layer neurons, for diabetes diagnosis, room occupancy detection, human activity detection and room light detection, model gives 81.39% accuracy with 40.73% error, 97.86% accuracy with 6.07% error, 99.05% accuracy with 29.83% error and 98.98% accuracy with 16.48% error, respectively. The factors behind the obtained results are the number of datasets used and the selection of input features to train the model.

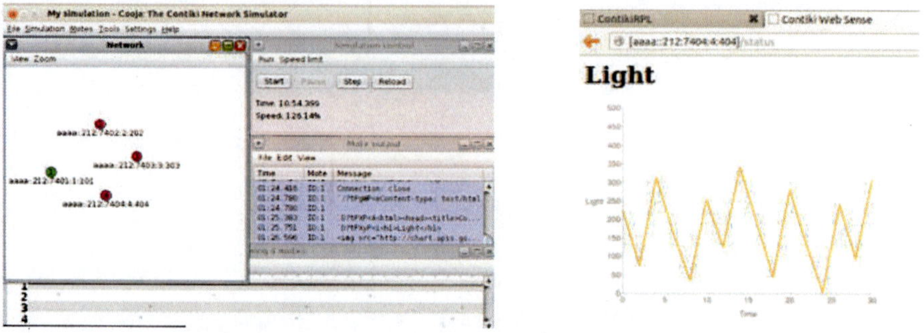


(a) Arrangement of sensor nodes (b) Light Sensor output

**Fig. 2** COOJA simulation result

**Table 2** Architecture of the model for four different test cases

| Test cases | No. of hidden layers | Neurons in each layer | Optimizer | Loss function (LF) | Activation function (AF) |
|---|---|---|---|---|---|
| Diabetes diagnosis | 3 | 4-4-4 | Adam | Binary cross entropy | ReLu at hidden layers, Sigmoid at output layer |
| Room occupancy detection | 2 | 6-7 | Adam | Binary cross entropy | ReLu at hidden layers, Sigmoid at output layer |
| Human activity detection | 4 | 5-7-4-9 | Adam | Binary cross entropy | ReLu at hidden layers, Sigmoid at output layer |
| Room light detection | 2 | 3-2 | Adam | Binary cross entropy | ReLu at hidden layers, Sigmoid at output layer |

## 4 Results and Discussions

The result of ANN-based model in classifying an input signal containing novelty or not in different scenarios has been listed in the previous sections. However, to understand how well ANN has performed, we have compared its performance with other states of the art modeling techniques that are commonly used in this field. For this purpose, we have considered five different techniques such as support vector machine (SVM), random forest (RF), Gaussian Naive Bayesian (GNB), logistic regression (LR), and decision tree (DT). In case of diabetes diagnosis test case, the ANN model gave 81% accuracy (60% precision, 78% recall, 78% F1-score) with

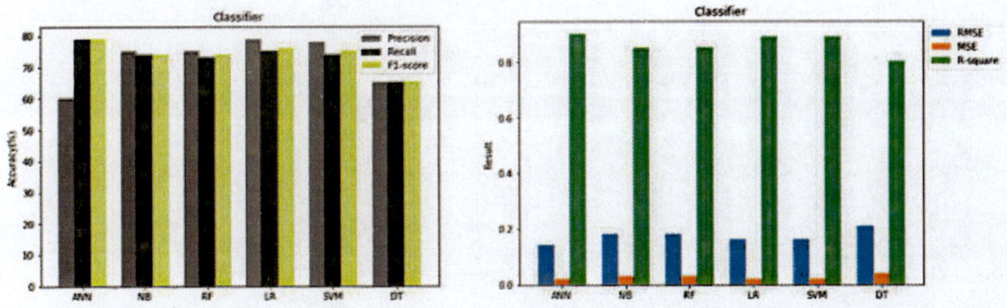


(a) In terms of accuracy (b) In terms of RMSE, MSE, R-square

**Fig. 3** Performance analysis for diabetes diagnosis

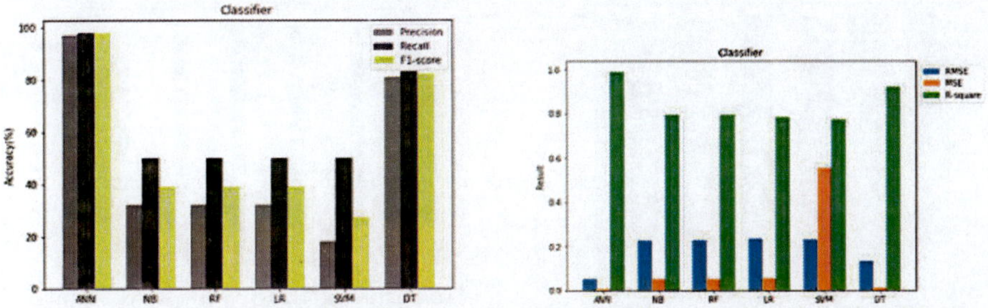


(a) In terms of accuracy (b) In terms of RMSE, MSE, R-square

**Fig. 4** Performance analysis for room occupancy detection

RMSE, MSE, *R*-square of 0.14, 0.02 and 0.90, respectively, which is considered to be slightly better than the other modeling techniques as shown in Fig. 3. Similarly in case of room occupancy detection, human activity detection and room light detection test cases, the ANN model performed better compared to other modeling techniques, i.e., the accuracy, RMSE, MSE and *R*-square of ANN model for room occupancy detection is 98%, 0.05, 0, 0.98, respectively, for human activity detection is 99%, 0.22, 0, 0.99, respectively, and for room light detection 99%, 0.02, 0, 0.98, respectively, as shown in Figs. 4, 5 and 6. The comparison result of these techniques for four different test cases is calculated and furnished in Table 3. The result suggests that, for each of these test cases, ANN scored a better performance in terms of accuracy, RMSE, MSE, *R*-Square, compared to other compatriot techniques. Therefore, it can be inferred that ANN proves to be a potentially viable option to be used as a technique for augmenting fog level intelligence.

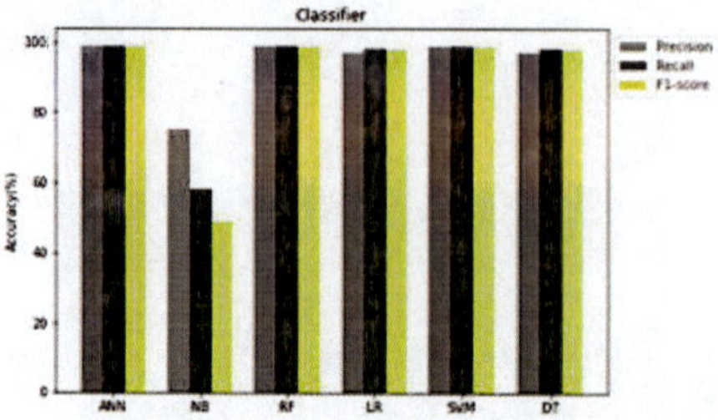


(a) In terms of accuracy

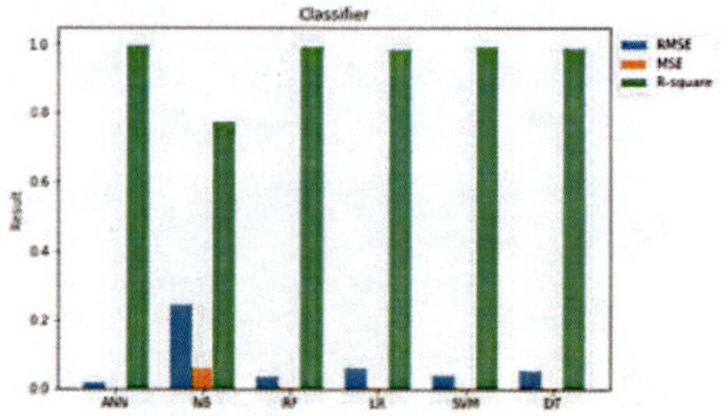


(b) In terms of RMSE, MSE, R-square

**Fig. 5** Performance analysis for human activity detection

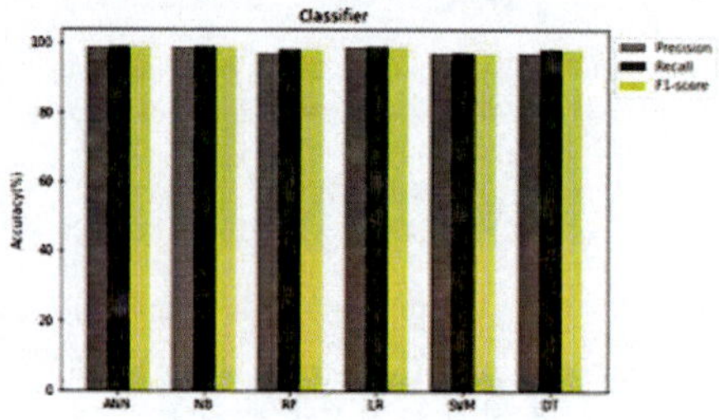


(a) In terms of accuracy

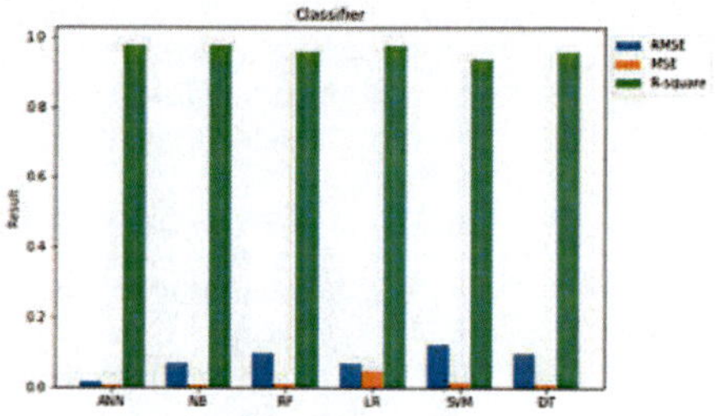


(b) In terms of RMSE, MSE, R-square

**Fig. 6** Performance analysis for room light detection

## 5 Conclusions and Future Work

This article presents a case study concerning experimentation with fog intelligence using ANN as the backbone computational model. Four different ANN architectures were built for each of the different time-sensitive test cases namely diabetes diagnosis, room occupancy detection, human activity detection, and room light detection. The performance of ANN was then compared with various other machine learning techniques such as SVM, LR, DT, RF, and GNB. The experimental result suggests that ANN performed better than other considered models in terms of classification accuracy. Therefore, from the achieved results, it can be safely be inferred that equipping fog level devices with smarter intelligence can help us to reduce wastage of network bandwidth and save cloud space and cloud processing time. This paper presents one such case where it showed that ANN posses a promising capability to provide the required fog intelligence in dealing with real-time traffic.

However, this work can further be extended in the future by incorporating other decision-making parameters such as latency and the proximity of the cloud. Various other states of the art machine learning techniques such as RNN, ANFIS may also be incorporated to augment this study.

**Table 3** Comparison results on six different classifiers for four different test cases

| Classifier | Performance matrix | Test cases | | | |
|---|---|---|---|---|---|
| | | Diabetes diagnosis | Room occupancy detection | Human activity detection | Room light detection |
| GNB | Accuracy | 0.7600 | 0.6400 | 0.5600 | 0.9900 |
| | RMSE | 0.1800 | 0.2270 | 0.2464 | 0.0710 |
| | MSE | 0.0300 | 0.0510 | 0.0600 | 0.0050 |
| | *R*-square | 0.8500 | 0.7935 | 0.7755 | 0.9798 |
| RF | Accuracy | 0.7600 | 0.6400 | 0.9900 | 0.9800 |
| | RMSE | 0.1800 | 0.2270 | 0.03680 | 0.1000 |
| | MSE | 0.0300 | 0.0510 | 0.0013 | 0.0100 |
| | *R*-square | 0.8500 | 0.7935 | 0.9945 | 0.9601 |
| LR | Accuracy | 0.7900 | 0.6300 | 0.9800 | 0.9900 |
| | RMSE | 0.1600 | 0.2340 | 0.0610 | 0.0710 |
| | MSE | 0.0200 | 0.0550 | 0.0037 | 0.0500 |
| | *R*-square | 0.8900 | 0.7809 | 0.9851 | 0.9798 |
| SVM | Accuracy | 0.7900 | 0.5600 | 0.9900 | 0.9700 |
| | RMSE | 0.1600 | 0.2348 | 0.0411 | 0.1230 |
| | MSE | 0.0200 | 0.5500 | 0.0016 | 0.0150 |
| | *R*-square | 0.8900 | 0.7700 | 0.9932 | 0.9391 |
| DT | Accuracy | 0.6800 | 0.8300 | 0.9800 | 0.9800 |
| | RMSE | 0.2100 | 0.1396 | 0.0552 | 0.1000 |
| | MSE | 0.0400 | 0.0190 | 0.0030 | 0.0100 |
| | *R*-square | 0.8000 | 0.9200 | 0.9897 | 0.9600 |
| ANN | Accuracy | 0.8100 | 0.9800 | 0.9900 | 0.9900 |
| | RMSE | 0.1400 | 0.0512 | 0.0184 | 0.0170 |
| | MSE | 0.0200 | 0.0026 | 0.0003 | 0.0050 |
| | *R*-square | 0.9008 | 0.9887 | 0.9986 | 0.9798 |

# References


1. Al-khafajiy M, Baker T, Chalmers C, Asim M, Kolivand H, Fahim M, Waraich A (2019) Remote health monitoring of elderly through wearable sensors. Multimedia Tools Appl 78(17):24681–24706
2. Raafat HM, Hossain MS, Essa E, Elmougy S, Tolba AS, Muhammad G, Ghoneim A (2017) Fog intelligence for real-time IoT sensor data analytics. IEEE Access 5:24062–24069
3. Tortonesi M, Govoni M, Morelli A, Riberto G, Stefanelli C, Suri N (2019) Taming the IoT data deluge: an innovative information-centric service model for fog computing applications. Future Gener Comput Syst 93:888–902
4. He J, Wei J, Chen K, Tang Z, Zhou Y, Zhang Y (2017) Multitier fog computing with large-scale iot data analytics for smart cities. IEEE Internet of Things J 5(2):677–686
5. Perera C, Qin Y, Estrella JC, Reiff-Marganiec S, Vasilakos AV (2017) Fog computing for sustainable smart cities: a survey. ACM Comput Surv (CSUR) 50(3):1–43

6. Mohammadian Rad N, Van Laarhoven T, Furlanello C, Marchiori E (2018) Novelty detection using deep normative modeling for IMU-based abnormal movement monitoring in Parkinson's disease and autism spectrum disorders. Sensors 18(10):3533
7. Lyu L, Jin J, Rajasegarar S, He X, Palaniswami M (2017) Fog-empowered anomaly detection in iot using hyperellipsoidal clustering. IEEE Internet of Things J 4(5):1174–1184
8. Verma P, Sood SK (2018) Fog assisted-IoT enabled patient health monitoring in smart homes. IEEE Internet of Things J 5(3):1789–1796
9. Verma P, Sood SK, Kalra S (2018) Cloud-centric IoT based student healthcare monitoring framework. J Ambient Intell Hum Comput 9(5):1293–1309
10. Baucas MJ, Spachos P (2020) Using cloud and fog computing for large scale IoT-based urban sound classification. Simul Model Pract Theory 101:102013
11. Dhingra S, Madda RB, Patan R, Jiao P, Barri K, Alavi AH (2020) Internet of things-based fog and cloud computing technology for smart traffic monitoring. Internet of Things 100175
12. Lin S-Y, Du Y, Ko P-C, Wu T-J, Ho P-T, Sivakumar V et al (2020) Fog computing based hybrid deep learning framework in effective inspection system for smart manufacturing. Comput Commun
13. Marković D, Vujičić D, Stamenkovič Z, Randič S (2020) IoT based occupancy detection system with data stream processing and artificial neural networks. In: 2020 23rd International symposium on design and diagnostics of electronic circuits & systems (DDECS). IEEE, pp 1–4
14. Parise A, Manso-Callejo MA, Cao H, Mendonca M, Kohli H, Wachowicz M (2019) Indoor occupancy prediction using an IoT platform. In: 2019 Sixth international conference on Internet of Things: systems, management and security (IOTSMS). IEEE, pp 26–31
15. Welhenge AM, Taparugssanagorn (A 2019) Human activity classification using long short-term memory network. Signal Image Video Process 13(4):651–656
16. Zhou X, Liang W, Kevin I, Wang K, Wang H, Yang LT, Jin Q (2020) Deep learning enhanced human activity recognition for internet of healthcare things. IEEE Internet of Things J
17. Mehmood T (2017) Cooja network simulator: exploring the infinite possible ways to compute the performance metrics of IoT based smart devices to understand the working of IoT based compression & routing protocols. arXiv:1712.08303

# Application of Machine Learning, Image Processing, and Pattern Recognition Techniques

# Health Monitoring of Wind Turbine Blades Through Vibration Signal Using Machine Learning Techniques

**Kuldeep Kushwah, Sudarsan Sahoo, and A. Joshuva**

**Abstract** Wind turbines are used for the transformation of wind energy into electrical energy with the help of rotating blades connected to the generator. It is an emerging source for the production of energy because of its dependence on advancement in technology and economics. The wind turbines have massive structures and are often located in remote areas near the sea or at the places, where the speed of the wind is sufficient enough to rotate the blades. The blades of the wind turbine play an important role since these are used for the conversion of kinetic energy into electrical energy. The blades of the wind turbine experience severe vibrations due to adverse environmental conditions, huge size, variation in wind speeds, and continuous operation throughout. These vibrations lead to serious damage to the turbine resulting in the reduction of its productivity and may lead to failure in the future. The faults must be recognized at the early stage so that energy conversion is not affected. Therefore, effective health monitoring and fault diagnosis are important for evading severe damage to the wind turbines. The identification of faults in any system requires physical knowledge of the parts which are not readily available every time. Hence, data-driven models are used for classification and diagnosis of faults. The present work aims to diagnose the faults in a wind turbine blade using vibration signals as the measured signal which is acquired from the hardware setup and classify the faults using different machine learning techniques and then compare the performance of the classifiers. The experimental work is carried out using a wind turbine setup by introducing the different conditions of the blades, i.e., healthy and defective blades. To find the suitability of the proposed method, the signal acquisition is done using a suitable instrumentation system. Various statistical features are extracted from the acquired vibration signals. Then, three different machine learning

---

K. Kushwah · S. Sahoo (✉)
NIT Silchar, Silchar, India
e-mail: sudarsan_iisc@yahoo.in

K. Kushwah
e-mail: kushwahkuldeepkd@gmail.com

A. Joshuva
Hindustan Institute of Technology and Science, Chennai, India
e-mail: joshuva1991@gmail.com



A. Maji et al. (eds.), *Proceedings of the International Conference on Computing and Communication Systems*, Lecture Notes in Networks and Systems 170,
https://doi.org/10.1007/978-981-33-4084-8_22

classifiers are applied for the classification of faults. The performances of the classifiers are evaluated in terms of the percentage of accuracy from the confusion matrix. The proposed work shows that the machine learning technique is a good approach for wind turbine fault classification, and the vibration signal is a good choice as a measuring signal for the detection and diagnosis the faults in wind turbine blades.

**Keywords** Wind turbine · Health monitoring · Fault diagnosis · Machine learning · Classifiers · Statistical features · Confusion matrix

## 1 Introduction

In the modern era, the development of wind turbines had been very fast due to advancements in technology, and at the same time, it has given a few obstacles. It is essential to sustain the healthful state of the operating turbine to maintain its efficiency. There is a continuous requirement to lessen the operation and maintenance costs [1, 2]. Hence, identification of defects/malfunctions at an early stage is important to decrease the shutdown, and productivity can also be increased. Wind turbines are used for the transformation of wind energy into electrical energy with the help of rotating blades connected to the generator. The blades are considered to be an important part of the wind turbine because of its huge structure and play a crucial role in the energy conversion mechanism [3]. The price of blades in the wind turbine can reach up to 15–20% of the whole setup. The blades are exposed to the environment; therefore, blades are subjected to various faults due to different environmental conditions and huge structures. The vibration caused in blades due to different wind speed, interaction with foreign particles, and various environmental changes (rain, snowfall, etc.) results in slow rotation or even failure of the turbine which affects the overall production and may cause a shutdown. The vibration of the blades is tough to evaluate on-line due to the huge structure, various operating conditions, and the remote location of the wind turbines [4, 5].

Every rotating element carries its vibration and acoustic signatures in its healthy condition. But when a defect occurs in it, then a variation appears in its vibration and acoustic signatures. This variation can be observed in time and frequency analysis [6, 7]. Vibration and acoustic signal processing are some of the vividly used ways to detect and diagnose the different machine faults. In some literature, it is found the use of wavelet transform or time–frequency analysis and adaptive wavelet transform to diagnose the bearing fault [8, 9].

Some literature introduces the signal strength improvement techniques using an active noise cancelation (ANC) technique [10]. Andrew Kusiak et al. show blade pitch fault monitoring in wind turbines [11].

In current scenario, machine learning techniques are widely used to diagnose the different types of machine faults. Therefore, a proper signature of faults is essential which allows us to move forward in the fault classification, results in a better and accurate fault diagnosis of the machine elements [12, 13].

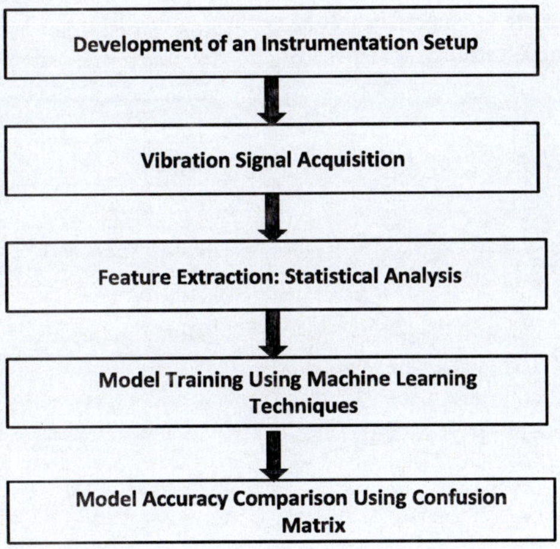


**Fig. 1** Flow of work

In the proposed work, one healthy blade and three defective blades are used for the experimental work, and vibration signal is used as a measuring parameter for the diagnosis of faults in wind turbine blades. Firstly, the instrumentation setup of the wind turbine is developed. Then, from that instrumentation setup, vibration signals are acquired for the different conditions of the blades, and from those vibration signals, various statistical features are extracted. The machine learning classifiers are trained by the evaluated statistical features. The machine learning classifiers are used for the classification of faults. The machine learning classifiers used in this work are k-nearest neighbor (K-NN), support vector machine (SVM), and decision tree. The performance comparisons of different classifiers are presented in terms of accuracy depicted from the confusion matrix. The objective of the proposed experimental work is to determine whether the blades of the wind turbine are in a good mode of operation or in a defective condition and then diagnose the faults in blades if it is defective. Figure 1 shows the flow of the proposed work.

## 2 Experimental Setup

The proposed work was performed using a wind turbine generator. DM 50 W wind turbine is used in the experiment. The technical specification of the model DM 50 W is listed in Table 1. A hardened galvanized iron platform is used to install the turbine against the wind source having an exit of around 1000 mm. To acquire the data, the turbine was set to run at 700 rpm. The experimental setup is shown in Fig. 2. For the acquisition of vibration signals, a piezoelectric type accelerometer is used. The accelerometer was fixed on the turbine near the hub using an adhesive solution. The accelerometer used was DYTRAN 3055B1 having a range of 500 g, resonating frequency of 40 Hz and sensitivity of 100 mV/g. A data acquisition system (model:

**Table 1** Specification of wind turbine

| | |
|---|---|
| Model | DM 50 W |
| Rated power | 50 W |
| Rated DC voltage | 12 V |
| Rated speed | 1050 r/m |
| Max power | 65 W |
| Rotor diameter | 1050 mm |
| Engine | 3 Ø |
| Blade material | Carbon fiber reinforced plastics |

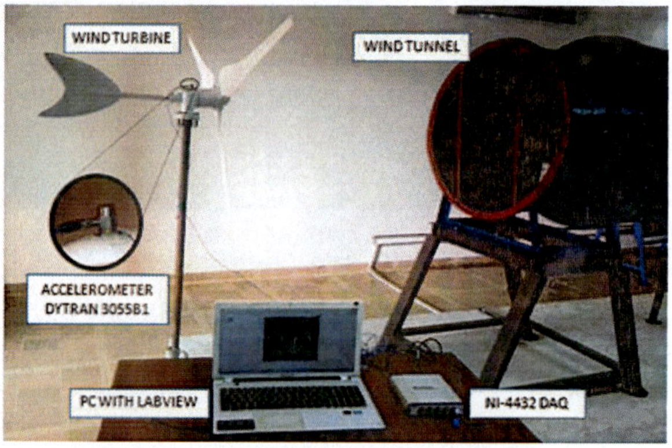


**Fig. 2** Experimental setup

NI USB 4432) is connected to the accelerometer through a cable. The software tool used for the experimental work is LabVIEW and MATLAB.

## 3 Experimental Procedure

A wind turbine having three blades was used for carrying out the experiment. At the initial stage, the vibration signals were acquired from the wind turbine when it is in its normal condition, i.e., all the blades are in healthy condition. The signals are acquired with the sampling frequency of 12 kHz. A sample length of 8192 was selected. For each condition of the blade, minimum 100 samples were taken.

In the next stage, the defective blades were used one at a time for acquiring the corresponding vibration signals as per the specifications mentioned above. The healthy and defective blades used in the experiment are shown in Fig. 3.

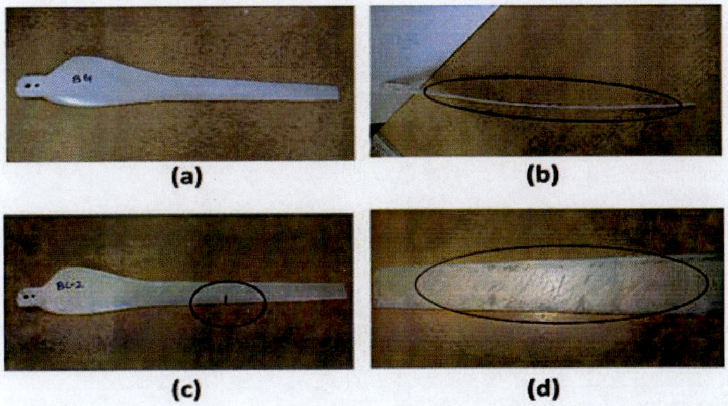


**Fig. 3** Different blades used in the experiment: **a** healthy blade (type-0 fault), **b** blade bend (type-1 fault), **c** blade crack (type-2 fault), **d** blade erosion (type-3 fault)

## 4 Statistical Analysis

The vibration signals are acquired from the hardware setup in four different conditions of the blades data acquisition which is done at sampling frequency of 12 kHz. The acquired vibration signal of the different blade faults and the variation in vibration is shown in Fig. 4. Type-0 blade is a healthy blade. Type-1 blade is a defective blade which is having bending. Type-2 blade is a defective blade having crack. Type-3 blade is a defective blade having erosion effect.

The statistical parameters used to identify the defects in blades are computed from vibration signals which are acquired from the healthy and defective blades by using

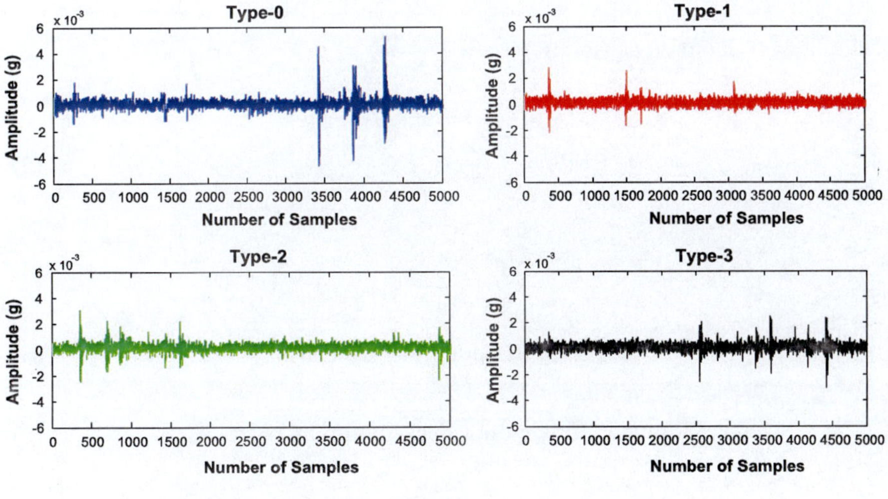


**Fig. 4** Vibration response of different type of faults

**Table 2** Statistical parameter comparison for vibration signal

| S. no. | Static parameters | Healthy blade (type-0) | Blade bend (type-1) | Blade crack (type-2) | Blade erosion (type- 3) |
|---|---|---|---|---|---|
| 1. | Mean | 4.5849 | 8.6502 | 1.0173 | 1.0085 |
| 2. | Standard deviation | 3.8211 | 3.4509 | 3.1012 | 3.1784 |
| 3. | Root mean square | 3.8483 | 3.5575 | 3.2636 | 3.3344 |
| 4. | Sum | 0.4585 | 0.8650 | 1.0173 | 1.0085 |
| 5. | Mode | 1.1200 | 1.4100 | 9.7156 | 1.6000 |
| 6. | Range | 0.0100 | 0.0089 | 0.0085 | 0.0056 |
| 7. | Minimum | −0.0047 | −0.0037 | −0.0041 | −0.0031 |
| 8. | Maximum | 0.0053 | 0.0052 | 0.0044 | 0.0025 |
| 9. | Skewness | 0.4655 | 1.0083 | 0.0737 | 0.0418 |
| 10. | Kurtosis | 31.6013 | 33.7253 | 22.1594 | 11.6717 |
| 11. | Variance | 1.4601 | 1.1908 | 9.6174 | 1.0102 |
| 12. | Crest factor | 13.6709 | 14.7071 | 13.4421 | 7.5306 |

the following mathematical formulas. The computed parameter values are shown in Table 2. The change in statistical parameter values can be clearly observed from this table. These changes show the presence of faults in blades. To train the model, these statistical parameter values are used as the input data for the classifiers.

$$\text{Mean} = \mu = \frac{1}{N}\sum_{i=1}^{n} X_i \tag{1}$$

$$\text{Standard Deviation} = \sigma = \left(\frac{1}{N-1}\sum_{i=1}^{n}(X_i - \mu)^2\right)^{\frac{1}{2}} \tag{2}$$

$$\text{RMS} = \sqrt{\frac{1}{N}\sum_{i=1}^{n}(X_i - \mu)^2} \tag{3}$$

$$\text{Sum} = \sum_{i=1}^{n} X_i \tag{4}$$

$$\text{Mode} = 3(\text{median}) - 2(\text{mean}) \tag{5}$$

$$\text{Range} = \text{maximum value} - \text{minimum value} \tag{6}$$

$$\text{Minimum} = \text{smallest value} \tag{7}$$

$$\text{Maximum} = \text{largest value} \tag{8}$$

$$\text{Skewness} = \frac{\left[\frac{1}{N}\sum_{i=1}^{n}(X_i - \mu)^3\right]}{\text{RMS}^3} \tag{9}$$

$$\text{Kurtosis} = K = \frac{\left[\frac{1}{N}\sum_{i=1}^{n}(X_i - \mu)^4\right]}{\text{RMS}^4} \tag{10}$$

$$\text{Variance} = \sigma^2 = \frac{\left[\sum_{i=1}^{n}(X_i - \mu)^2\right]}{N} \tag{11}$$

$$\text{Crest Factor} = C = \frac{P_v}{\text{RMS}} \tag{12}$$

## 5 Results

The scatter plot of the vibration signal acquired from different condition of the blade is shown in Fig. 5. In this, the blue, red, yellow, and violet show the type-0, type-1, type-2, and type-3 faults, respectively. Figure 5a shows the original dataset with full features, and Fig. 5b shows predicted model. In the predicted model, the circle represents the different condition of the blades that are being correctly classified by the algorithm, and cross represents those conditions of the blades that are incorrectly classified by the algorithm. The confusion matrix for the K-NN, decision tree, and SVM is shown in Fig. 6. The true positive and negative rate are reflected from these figures. The percentage model accuracy is calculated from these true positive and negative rate. The model formed using SVM classifier gives the highest accuracy of 87.00% for fault classification among all the classifiers. The performance comparison between the different classifiers is tabulated in Table 3.

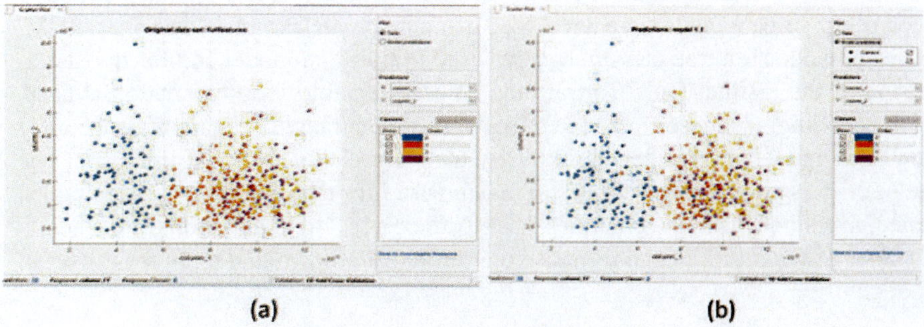


**Fig. 5** Scatter plot of the vibration signal: **a** original dataset, **b** predicted model

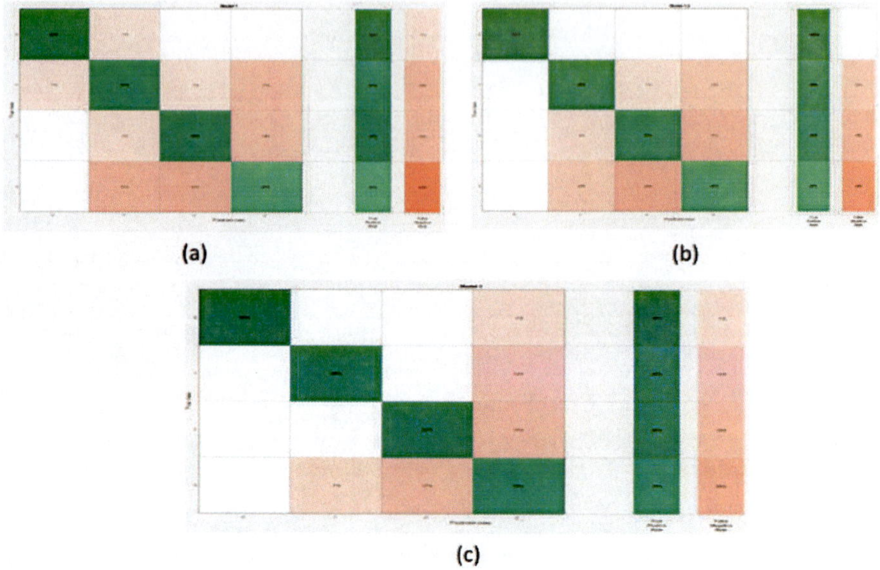

Fig. 6 Confusion matrix for classifiers: **a** accuracy opted using K-NN classifier, **b** accuracy opted using decision tree classifier, **c** accuracy opted using SVM classifier

**Table 3** Comparison of percentage model accuracy offered by the different classifier algorithm

| S. no | Classifier algorithm used | % model accuracy |
|---|---|---|
| 1. | K-nearest neighbor | 79.80 |
| 2. | Decision tree | 84.00 |
| 3. | Linear SVM | 87.00 |

## 6 Conclusion

This paper presented the diagnosis and classification of faults in wind turbine blades using machine learning-based approach. From the data acquired from the instrumentation setup, three models were developed using machine learning techniques. SVM, K-NN, and decision tree classifiers were used to train the model and for the classification of the distinct fault. To train the dataset, ten-fold cross-validation is used. All the classifiers were compared in terms of their accuracy. The maximum accuracy found among all the classifiers is 87.00% which is given by the SVM algorithm. The rate of error is comparatively less and may be used for the diagnosis of blade faults. Hence, machine learning techniques can be efficiently adopted to monitor the health of blades of wind turbines to decrease the turbine shutdown and to increase overall productivity.

## References


1. Zhang P, Lu D (2019) A survey of condition monitoring and fault diagnosis toward integrated o&m for wind turbines. Energies 12:2801
2. Ellabban O, Abu-Rub H, Blaabjerg F (2014) Renewable energy resources: current status, future prospects and their enabling technology. Renew Sustain Energy Rev 39:748–764
3. Liu WY, Tang BP, Han JG, Lu XN, Hu NN, He ZZ (2015) The structure healthy condition monitoring and fault diagnosis methods in wind turbines: a review. Renew Sustain Energy Rev 44:466–472
4. Joshuva A, Sugumaran V (2017) A data driven approach for condition monitoring of wind turbine blade using vibration signals through best-first tree algorithm and functional trees algorithm: a comparative study. ISA (Instrum Soc Am) Trans 67:160–172
5. Joshuva A, Sugumaran V (2017) Fault diagnosis for wind turbine blade through vibration signals using statistical features and random forest algorithm. IJPT 9
6. Jena DP, Sahoo S, Panigrahi SN (2014) Gear fault diagnosis using active noise cancellation and adaptive wavelet transform. Measurement 47:356–372
7. Kankar PK, Sharma SC, Harsha SP (2011) Fault diagnosis of ball bearings using continuous wavelet transform. Appl Soft Comput 11:2300–2312
8. Abraham SK, Sugumaran V, Amarnath M (2016) Acoustic signal based condition monitoring of gearbox using wavelets and decision tree classifier. Indian J Sci Technol 9(33)
9. Sahoo S, Das JK (2018) Bearing health monitoring and diagnosis using ANC based filtered vibration signal. J Eng Appl Sci 03:3587–3593
10. Sahoo S, Das JK, Bapi D (2018) Fault diagnosis in rolling element bearing using filtered vibration and acoustic signal. Int J Pure Appl Math 118:95–101
11. Kusiak A, Verma A (2011) A data-driven approach for monitoring blade pitch faults in wind turbines. IEEE Trans Sustain Energy 2(1):87–96
12. Gupta DL, Malviya AK, Singh S (2012) Performance analysis of classification tree learning algorithms. Int J Comput Appl 55(6):0975–8887
13. Amarnath M, Sugumaran V, Kumar H (2013) Exploiting sound signals for fault diagnosis of bearings using decision tree. Measurement 46:1250–1256

# Iraqi Food Image Detection Using Convolutional Neural Network Classification Method

**Maad M. Mijwil**

**Abstract** A restaurant is a public place, and it is open to everyone to take a drinks and food against money. The restaurant offers service of food and drinks desires to satisfy the customer. At the moment, the payment process at restaurants is still manual and ineffective because it uses a cash record. A cashier will check what foods are required, then calculate them with the cash record, and this process will take a long time and be ineffective. In this paper, design application for Iraqi food detection and also reduces long queues at certain hours in restaurants and estimate food costs automatically applying the convolutional neural network (CNN) classification method. The results of the application are can detect six types of food on 80% data partition training data and 20% test data, with the image taken in storage. The accuracy of the appley of CNN reaches 100% with a detection time of 3.59 s.

**Keywords** Image detection · Deep learning · Convolution neural network

## 1 Introduction

Food is a vital activity for all humanity, with food being connected to countless aspects and situations in our day. With the growth of modern technologies, such as computer vision and smartphones, food-related apps have increased [1]. Image detection is a method that processes the image and detects objects in it with different dimensions and size. People usually confuse image detection with image classification. Although the difference is very obvious [2]. Where image classification uses to classify image items, while image detection uses to get out the number of objects in the picture [3]. Image detection is a type of pattern detection technology that identifies features from images by identifying features [4]. Specifically, for example, the computer determines which object is in the image and makes the computer recognize the image [5]. By looking at images such as photographs, humans can infer from their own experiences what has moved and where. However, computers have no human

M. M. Mijwil (✉)
Baghdad College of Economic Sciences University, Baghdad, Iraq
e-mail: mr.maad.alnaimiy@baghdadcollege.edu.iq

 
A. Maji et al. (eds.), *Proceedings of the International Conference on Computing and Communication Systems*, Lecture Notes in Networks and Systems 170,
https://doi.org/10.1007/978-981-33-4084-8_23

memory and experience, so if they give for a computer an image of a "Wolf," then this computer will not recognize it as a "Wolf" [6]. Therefore, in image detection, a computer is given a large no. of images from a database, and this computer is automatically "learns" from the characteristics of the object. Then, the computer can "learn" the characteristics of the fox from the image data, and if an image with the same characteristics is given, then it can infer that it is a "Wolf." In the last years, this area of image detection has been quickly grown due to the improvement of deep learning technology [7].

On the other hand, a computer will extract features from any image in the database, and based on these features, the computer can determine what the image is moving to [8]. However, the way computers make decisions is different from humans. It takes a mathematical process of performing some type of operation on the pixel data describing the image and calculating the features. Hence, increasing the image detection ability of a computer means changing the calculation method and parameters applied in the calculation [9].

Fortunately, in image detection, machine learning (ML) is performed [10]; it is a system in which computers learn by themselves. In this method, humans just determine how the computer performs calculations and according to what commands to correct parameters, and the machine automatically corrects parameters based on a large amount of data and optimize it [11].

Furthermore, in image detection, a network model called a convolutional neural network (CNN) is often applied. This model is a developed version of a network model called a neural network (NN) that represents a neural network in the human brain. The feature of this model is that humans do not convert pixel data of an image into an abstract vector and allow the computer to perform feature extraction as it is on the image data [12, 13]. In a CNN, a part of the image data is filtered and calculated, and the convolution is performed by sliding the area and repeating to generate a feature map [14]. By this model, local features of the image can be extracted, and also the computer repeatedly extracts the features of the image, estimates the target, and matches the correct data to learn and increase the accuracy of image detection [15].

The article is organized as follows. Section 2 introduces the CNN model. Then, display the proposed scheme in detail in Sect. 3. In Sect. 4, present the experimental results followed by the conclusion in Sect. 5.

## 2 CNN Model

Deep learning is a part of ML that is motivated by artificial neural networks (ANNs). A special type of such deep learning is the CNN [16]. Nowadays, CNN can achieve high performance in detecting objects. Thus, accurate positions detected by neural networks can be used to track the movement of objects and count the number of line crossings. CNNs are a type of ANNs that are used, among other things, for analysis and image detection. The concentrate of CNNs on the rule that the input will be

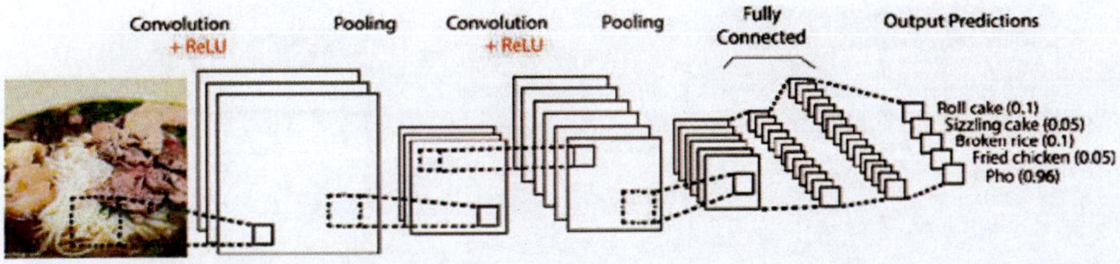


**Fig. 1** A simple CNN architecture

composed of images. This concentrates on the architecture to be installed in a way to optimal suit the requirement for dealing with the special type of data [17]. One of the most major differences is that the neurons that the layers within the CNN are composed of neurons classified into 3D, width, height, and depth. As for the depth is the third dimension of an activation volume, it did not refer to the total no. of layers within the ANN. Unlike standard ANNs, the neurons will only connect to a small area of the layer preceding it.

CNNs are categorized into "three" main layers (convolutional layers, pooling layers, fully connected layers). When these layers are collected, the architecture of CNN has been created. The example of simple CNN architecture is presented in Fig. 1 [18]

Moreover, there are two gains for the architecture of CNN; the first gain is created to use the 3D structure of an input image. This is done by local connections and tied weights supported by some form of pooling which results in translation-invariant characteristics [19]. The second gain is that they are easier to train and have many fewer parameters than fully connected nets with the same no. of hidden units [20].

## *2.1 Convolution Layer*

This layer applied to get the pattern/texture in the image by sending the convolution process between the filter and the image. In this layer placed on the first layer in the architecture of CNN, a 5 $\times$ 5 dimension image is made between the input image and also the 3 $\times$ 3 filter, with a 1-pixel shift filter (Fig. 2).

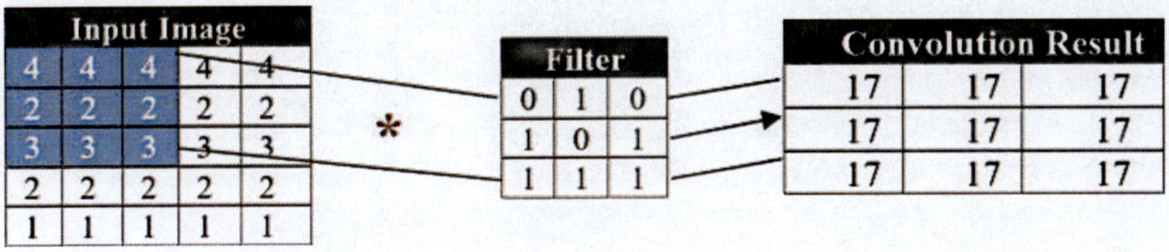


**Fig. 2** Convolution layer process

| Convolution Result | | |
|---|---|---|
| 17 | 17 | 17 |
| 17 | 17 | 17 |
| 17 | 17 | 17 |

Max Pooling filter 2*2 and stride 1 →

| Pooling Result | |
|---|---|
| 17 | 17 |
| 13 | 13 |

**Fig. 3** Pooling layer process

The input image is multiplied by the filter. Then, the results of this layer can be observed on the right. Example of the input image: $[(4 * 0) + (4 * 1) + (4 * 0) + (2 * 1) + (2 * 0) + (2 * 1) + (3 * 1) + (3 * 1) + (3 * 1)] = 17$. After that, all results from this layer will enter to the next layer (pooling layer).

## 2.2 *Pooling Layer*

This layer does the method of dividing several parts in a square shape. Next, this layer will receive the max. value of the pixel part from the results of the convolution layer. In the pooling, the max. value is explored by using the max-pooling filter $2 \times 2$ and stride 1 or shifting the filter by 1 pixel. The results of this layer reached the highest values. Next, go to the fully connected layer (Fig. 3).

## 2.3 *Fully Connected Layer*

This layer uses to save the results of the whole layer by connecting other neurons to one neuron and saving it in shape. At this layer is the arrangement of the pooling results. The results of pooling are 17 and 13; then, 2 vectors are worth 17 and 13. The following step is SoftMax is defined as activation function which is responsible for predicting the final output; in this step, the probability reached from every vector will then be matched with the class that has been created by the system. By using the following equation, the calculation can be viewed as follows:

$$M_1 = \frac{e^{17}}{e^{17} + e^{13}} = 0.982$$

$$M_2 = \frac{e^{13}}{e^{17} + e^{13}} = 0.017$$

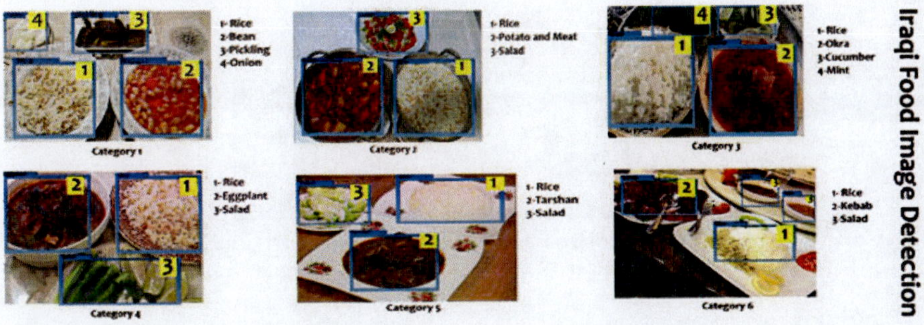


**Fig. 4** Categories of food used in database

## 3 The Proposed Scheme

The first step is the pre-processing of images. All images that have been taken by camera or smartphone, before will go into pre-processing which consists of resizing and changing the original image that has the JPEG format converted to the XML data file. Before the image is processed more, the image will be resized to resize the image < 200 KB with dimensions of 720 × 1280. The next step is training data using the CNN method, which from the results of pre-processing data will be input to the label terms. CNN will continue to detect food and match the label of each food. Training will continue until getting a loss value of 0.04, with no. steps of 8482 steps. After the training process, the image will be examined and will be seen how much accuracy on food objects are in one picture (Fig. 4).

## 4 Experimental Result

Tests carried out with a total of 521 combined images of bean, potato with meat, okra, eggplant, tarshan, and kebab. Where there are 521 pictures of rice and salad, which means that rice and salad are a food object that is always in all pictures. There are 260 pictures of bean, potato and meat, okra, eggplant, tarshan, and kebab, with dimensions of 720 × 1280 with image size < 200 Kb; to get a confident value of more than 50% in testing, it is important to segmentation data between training and test data. The division of training data and test data is divided into two tests, 90% (training data) 10% (test data), and also 80% (training data) 20% (test data) (Fig. 5).

**First**: Partitioning 90% training data and 10% test data.

In the training, data division as much as 90% with 466 images and 10% test data sharing with 55 images have 8482 steps with a loss of 0.04 and a learning rate of 0.0001. Tests using test data totaling 55 images, the following results are obtained (Table 1).

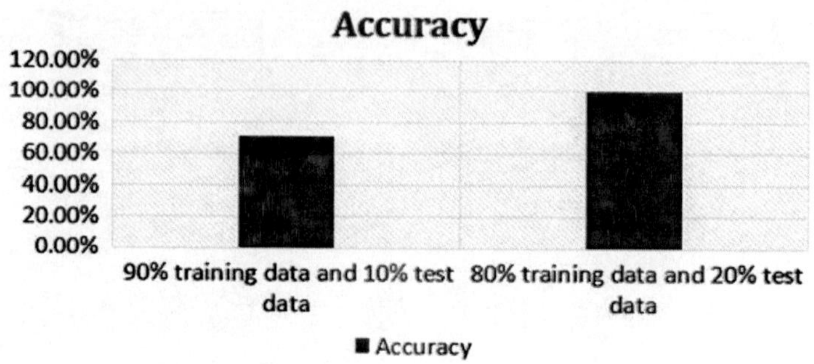


**Fig. 5** Result of best partition data

**Table 1** Partition testing table 90% training data and 10% test data

| Category | Food dish | True | False | Accuracy (%) |
|---|---|---|---|---|
| Category 1 | Rice | 5 | 1 | 86.13 |
| | Salad | 0 | 6 | 0 |
| | Bean | 5 | 1 | 86.13 |
| Category 2 | Rice | 5 | 1 | 86.13 |
| | Salad | 4 | 2 | 68.26 |
| | Potato with meat | 6 | 0 | 100 |
| Category 3 | Rice | 5 | 0 | 100 |
| | Salad | 5 | 0 | 100 |
| | Okra | 6 | 1 | 86.13 |
| Category 4 | Rice | 6 | 0 | 100 |
| | Salad | 2 | 4 | 37.43 |
| | Eggplant | 5 | 1 | 86.13 |
| Category 5 | Rice | 5 | 1 | 86.13 |
| | Salad | 0 | 6 | 100 |
| | Tarshan | 3 | 3 | 50 |
| Category 6 | Rice | 6 | 0 | 100 |
| | Salad | 1 | 5 | 17.71 |
| | Kebab | 4 | 2 | 68.12 |

**Second**: Partitioning 80% training data and 20% test data

In the training, data division of 80% with 431 pictures and 20% of the test data distribution with 90 images have 8482 steps with a loss of 0.04 and learning rate 0.0001. Then, the following results are obtained (Tables 2 and 3).

**Table 2** Partition testing table 80% training data and 20% test data

| Category | Food dish | True | False | Accuracy (%) |
|---|---|---|---|---|
| Category 1 | Rice | 6 | 0 | 100 |
| | Salad | 6 | 0 | 100 |
| | Bean | 6 | 0 | 100 |
| Category 2 | Rice | 6 | 0 | 100 |
| | Salad | 6 | 0 | 100 |
| | Potato with meat | 6 | 0 | 100 |
| Category 3 | Rice | 6 | 0 | 100 |
| | Salad | 6 | 0 | 100 |
| | Okra | 6 | 0 | 100 |
| Category 4 | Rice | 6 | 0 | 100 |
| | Salad | 6 | 0 | 100 |
| | Eggplant | 6 | 0 | 100 |
| Category 5 | Rice | 6 | 0 | 100 |
| | Salad | 6 | 0 | 100 |
| | Tarshan | 6 | 0 | 100 |
| Category 6 | Rice | 6 | 0 | 100 |
| | Salad | 6 | 0 | 100 |
| | Kebab | 6 | 0 | 100 |

**Table 3** Show average detection speed for each category

| Category | Average detection speed (seconds) | Accuracy (%) |
|---|---|---|
| Category 1 | 3.20 | 100 |
| Category 2 | 2.88 | 100 |
| Category 3 | 1.7 | 100 |
| Category 4 | 2.10 | 100 |
| Category 5 | 3.22 | 100 |
| Category 6 | 2.64 | 100 |

## 5 Conclusion

Generally, food detection is more challenging and harder to manage in all restaurants, due to various cooking styles, cuisines and the intrinsic complexity of modeling food from its visual manifestation. Image detection technology has evolved significantly with the introduction of AI. Object detection AI has the same correct answer rate as humans and is already being put to practical use in various industries. It is expected that the use of image detection technology will further expand with the development of IoT technology and the use of big data. Furthermore, the application of Iraqi food detection system can classify six types of food using CNN algorithm well on existing image data or stored in storage device with 80% data partition for training data and 20% for test data and get an accuracy of 100% and can detect food with an average speed of 3.59 s which has 8482 steps with a loss of 0.04 and learning rate 0.0001.

## References

1. Herranz L, Jiang S, Xu R (2016) Modeling restaurant context for food recognition. IEEE Trans Multimedia 1–12
2. Pouladzadeh P, Shirmohammadi S, Al-Maghrabi R (2014) Measuring calorie and nutrition from food image. IEEE Trans Instrum Meas 63:1947–1956
3. He Y, Xu C, Khanna N, Boushey CJ, Delp EJ (2014) Analysis of food images: features and classification. In: IEEE international conference on image processing, pp 2744–2748
4. Hassaballah M, Abdelmgeid AA, Alshazly HA (2016) Image features detection, description and matching. Springer International Publishing (Verlag), pp 11–45
5. Khurana K, Awasthi R (2013) Techniques for object recognition in images and multi-object detection. Int J Adv Res Comput Eng Technol (IJARCET) 2(4):1383–1388
6. Krizhevsky A, Sutskever I, Hinton GE (2017) ImageNet classification with deep convolutional neural networks. Commun ACM 60(6):84–90
7. Ciresgan D, Meier U, Schmidhuber J (2012) Multi-column deep neural networks for image classification. In: IEEE conference on computer vision and pattern recognition, pp 1–19
8. Ciocca G, Napoletano P, Schettini R (2017) Food recognition: a new dataset, experiments and results. IEEE J Biomed Health Inf 21(3):588–598
9. Beijbom O, Joshi N, Morris D, Saponas S, Khullar S (2015) Menu-match: restaurant-specific food logging from images. In: 2015 IEEE winter conference on applications of computer vision (WACV). IEEE, pp 844–851
10. Nichols JA, Chan HWH, Baker MAB (2018) Machine learning: applications of artificial intelligence to imaging and diagnosis. Biophys Rev 11:111–118
11. Staniewicz L, Midgley PA (2015) Machine learning as a tool for classifying electron tomographic reconstructions. Adv Struct Chem Imag 01(09):1–15
12. Hossain T, Shishir FS, Ashraf M, Al Nasim MA, Shah FM (2019) Brain tumor detection using convolutional neural network. In: 1st international conference on advances in science, engineering and robotics technology (ICASERT-2019), pp 1–6
13. Zhang W, Zhao D, Gong W, Li Z, Lu Q, Yang S (2015) Food image recognition with convolutional neural networks. In: 2015 IEEE 12th international conference on ubiquitous intelligence and computing and 2015 IEEE 12th international conference on autonomic and trusted computing and 2015 IEEE 15th international conference on scalable computing and communications and its associated workshops (UIC-ATC-ScalCom), pp 690–693
14. Yanai K, Kawano Y (2017) Food image recognition using deep convolutional network with pre-training and fine-tuning. In: 2015 IEEE international conference on multimedia and expo workshops (ICMEW), pp 1–7
15. Termritthikun C, Muneesawang P, Kanprachar S (2019) NU-InNet: Thai food image recognition using convolutional neural networks on smartphone. J Telecommun Electron Comput Eng 9(2–6):63–67
16. He K, Zhang X, Ren S, Sun J (2015) Deep residual learning for image recognition. arXiv preprint, ID:1512.03385
17. O'Shea K, Ryan Nash R (2015) An introduction to convolutional neural networks. arXiv preprint, ID: 1511.08458v2
18. Phat TV, Tien DX, Pham Q, Pham N, Nguyen BT (2017) Vietnamese food recognition using convolutional neural networks. In: 2017 9th international conference on knowledge and systems engineering (KSE), pp 11–18

19. Mittal S (2016) Unsupervised learning models of invariant features in images: recent developments in multistage architecture approach for object detection. Int J Soft Comput Artif Intell Appl (IJSCAI) 5(2):63–70
20. Yiğit GÖ, Özyildirim BM (2018) Comparison of convolutional neural network models for food image classification. J Inf Telecommun 2(3):347–357

# Performance Measurement Model for Wind Power Project Using Mamdani Fuzzy Inference System

**Debasish Majumder, Mrinmoy Dam, Rupak Bhattacharjee, Shyam Sundar Santra, Rishiraj Saha, and Soham Saha**

**Abstract** This paper presents a comprehensive performance measurement model for wind power projects based on fuzzy logic. Performance evaluation process suffers from uncertainties and/or vagueness, as it frequently involves subjective judgments of the experts at different stages due to the presence of numerous qualitative performance indicators from various aspects and their inexact causal relationship. In the proposed model, a hierarchical structure has been suggested for the metrics derived to measure the performance attributes and then the performance scores of the derived lagging (output) metrics have been estimated from the input values (leading metrics) based on their causal relationship. Mamdani fuzzy reasoning approach has been employed in order to capture the inexact causal relationship between the leading and lagging metrics. Finally, the behavior of the responses produced by the proposed model has been analyzed using random values, generated to simulate experimental data, which seems to be a feasible practice for the mangers in decision making for managing performance of wind power projects.

D. Majumder (✉) · R. Bhattacharjee · S. S. Santra
Department of Mathematics, JIS College of Engineering, Kalyani 741235, India
e-mail: debamath@rediffmail.com

R. Bhattacharjee
e-mail: rupakb13@yahoo.co.in

S. S. Santra
e-mail: shyam01.math@gmail.com

M. Dam
Mahishakati Netaji Subhash Chandra High School, Kahankia 743245, India
e-mail: dam.mrinmoy1@gmail.com

R. Saha · S. Saha
Department of Information Technology, JIS College of Engineering, Kalyani 741235, India
e-mail: saha.rishiraj1999@gmail.com

S. Saha
e-mail: sohamsaha999@gmail.com



A. Maji et al. (eds.), *Proceedings of the International Conference on Computing and Communication Systems*, Lecture Notes in Networks and Systems 170,
https://doi.org/10.1007/978-981-33-4084-8_24

**Keywords** Performance measurement · Wind power projects · Mamdani fuzzy inference system

## 1 Introduction

Performance measurement (PM) [1] and analysis of wind power projects (WPPs) have been identified as significant mechanism that all WPPs must possess for their sustainability [2, 3]. As wind energy is becoming more economically competitive and demand is changing rapidly, wind farm operators need to become more responsive to stakeholders and market needs. Despite the rapid development in the wind turbine technology, there are several challenges, like land acquisition and evacuation, high production cost, some negative impacts such as noise and aesthetic pollution, which remains the major obstacle in order to optimize its functionality [4]. Wind farm operators must understand the requirements of performance analysis of WPPs by means of current and accurate performance information of WPPs, in order to respond to all challenges proactively and to achieve the desired production and revenue goals.

A distinctive characteristic of PM of WPP is that the WPP should be spanned by the measurement system, including metrics derived to measure the performance attributes from various aspects, like technological factors, infrastructure, social issues, economic aspects, and government's perception along with their interdependencies [3, 5]. These multi-dimensional characteristics of WPP make it complex to measure performance of the WPP.

Several researches have been carried out over the years to deal with such multiple attribute decision-making complex issues in PM [6–12]. However, from these works, it has been revealed that some of the performance attributes are qualitative and cannot be expressed quantitatively (by crisp numbers), rather expressed in words and hence highly depend on expert(s)'s knowledge and experience [13, 14]. Due to frequent involvement of subjective judgment by the expert(s), vegness/fuzziness is inherent in PM systems. Over the years, it has been revealed that fuzzy set [15] theoretic approach has proficiency, over the other approaches, in capturing vegness/fuzziness present in performance evaluation process. Moreover, fuzzy logic [16] is identified as a very useful tool in dealing with causal relationship exists between input and output parameters of the PM systems [3, 5].

In this paper, performance indicators for WPP have been selected from the perspective of sustainability. In the context of REs, sustainability primarily refers to the performance of the systems in social, economic, and environmental aspects [17]. In order to measure these three performance indicators, a set of metrics has been considered as proposed in [18]. However, social and environmental impacts of WPPs, due to the qualitative nature, cannot be expressed quantitatively by crisp numbers. A hierarchical PM model has been proposed to determine the performance attributes, and then the performance scores of the derived lagging (output) metrics have been estimated from the input values (leading metrics) based on their causal relationship. Mamdani fuzzy reasoning approach has been employed in order to capture the inexact causal relationship between the leading and lagging metrics.

## 2 Sustainability Performance Indicators for WPPs

A typical performance evaluation system involves two types of components: performance attributes and metrics [5]. Metrics are derived from the strategic objectives and performance attributes to control the performance on multi-dimensions in a hierarchical structure. The level-1 metrics, also known as strategic metrics and key performance indicators (KPI), are diagnostics for the overall performance; the level-2 metrics provide the diagnostics for the level-2 metrics and so on. The analysis of performance of metrics from top level to bottom level is known as metrics decomposition, performance diagnosis or metrics root cause analysis.

The identification of key performance indicators (KPIs) for performance evaluation of WPP is a critical issue. In this paper, the KPIs for performance evaluation of WPP have been selected from the perspective of sustainability. The proposed PM model has been developed based on three performance attributes, viz. economic aspect, environmental aspect, and social aspect and several metrics to measure them as proposed in [18] (see Table 1).

## 3 Fuzzy Set Theory

The fuzzy set theory [15] has been recognized as a means for representation of vague, imprecise, and uncertain information mainly evolved from judgment of decision makers. Subjective judgments involved in decision-making problems are expressed qualitatively by linguistic terms, each of which represented by a fuzzy set, and characterized by membership function on the universe of discourse. In the present study, trapezoidal type fuzzy set, as a universal form of linear fuzzy set, has been employed due to its simple characteristic. A trapezoidal number $\tilde{A}$ is defined by the membership function

$$\mu_{\tilde{A}}(x; a, b, c, d) = \begin{cases} \frac{x-a}{b-a}, & \text{for } a \leq x \leq b \\ 1, & \text{for } b \leq x \leq c \\ \frac{d-x}{d-c}, & \text{for } c \leq x \leq d \\ 0, & \text{otherwise} \end{cases} \tag{1}$$

Moreover, the fuzzy inference system [19] allows the practitioners to mimic the human-like thinking in computers-aided modeling using a set of fuzzy IF-THEN rules, which form the knowledge base of a fuzzy system. In the present study, Mamdani fuzzy inference system (MFIS), introduced by Mamdani [19], has been

**Table 1** Membership functions corresponding to linguistic terms used for the leading and lagging metrics

| | Linguistic terms and their trapezoidal membership functions | | | | | | | | | | | | | |
|---|---|---|---|---|---|---|---|---|---|---|---|---|---|---|
| MFISs | Metrics | | Low | Medium | High | | | | | | | | | |
| | | | *a* | *b* | *c* | *d* | *a* | *b* | *c* | *d* | *a* | *b* | *c* | *d* |
| MFIS 1 | Leading metrics | Investment cost | 0 | 0 | 0.2 | 0.45 | 0.2 | 0.45 | 0.45 | 0.7 | 0.45 | 0.7 | 1 | 1 |
| | | Operation and maintenance cost | 0 | 0 | 0.25 | 0.55 | 0.25 | 0.55 | 0.55 | 0.85 | 0.55 | 0.85 | 1 | 1 |
| | | Levelized cost of energy | 0 | 0 | 0.2 | 0.5 | 0.2 | 0.5 | 0.5 | 0.8 | 0.5 | 0.8 | 1 | 1 |
| | Lagging metrics | Payback period | 0 | 0 | 0.15 | 0.45 | 0.15 | 0.45 | 0.45 | 0.75 | 0.45 | 0.75 | 1 | 1 |
| | | Economic impact | 0 | 0 | 0.3 | 0.5 | 0.3 | 0.5 | 0.5 | 0.7 | 0.5 | 0.7 | 1 | 1 |
| MFIS 2 | Leading metrics | Carbon dioxide emission reduction | 0 | 0 | 0.25 | 0.5 | 0.25 | 0.5 | 0.5 | 0.75 | 0.5 | 0.75 | 1 | 1 |
| | | PM 2.5 emission reduction | 0 | 0 | 0.3 | 0.55 | 0.3 | 0.55 | 0.55 | 0.8 | 0.55 | 0.8 | 1 | 1 |
| | Lagging metrics | Ecological impact | 0 | 0 | 0.35 | 0.55 | 0.35 | 0.55 | 0.55 | 0.75 | 0.55 | 0.75 | 1 | 1 |
| | | Land occupation | 0 | 0 | 0.35 | 0.5 | 0.35 | 0.5 | 0.5 | 0.65 | 0.5 | 0.65 | 1 | 1 |
| MFIS 3 | Leading metrics | Environmental impact | 0 | 0 | 0.3 | 0.45 | 0.3 | 0.45 | 0.45 | 0.6 | 0.45 | 0.6 | 1 | 1 |
| | | Poverty alleviation promotion | 0 | 0 | 0.15 | 0.5 | 0.15 | 0.5 | 0.5 | 0.85 | 0.5 | 0.85 | 1 | 1 |
| | | Technology innovative promotion | 0 | 0 | 0.3 | 0.55 | 0.3 | 0.55 | 0.55 | 0.8 | 0.55 | 0.8 | 1 | 1 |
| | Lagging metrics | Power quality improvement | 0 | 0 | 0.15 | 0.4 | 0.15 | 0.4 | 0.4 | 0.65 | 0.4 | 0.65 | 1 | 1 |
| | | Residential satisfaction | 0 | 0 | 0.1 | 0.4 | 0.1 | 0.4 | 0.4 | 0.7 | 0.4 | 0.7 | 1 | 1 |
| | | Social impact | 0 | 0 | 0.2 | 0.45 | 0.1 | 0.45 | 0.45 | 0.7 | 0.45 | 0.7 | 1 | 1 |
| MFIS 4 | Leading metrics | Economic impact | 0 | 0 | 0.3 | 0.5 | 0.3 | 0.5 | 0.5 | 0.7 | 0.5 | 0.7 | 1 | 1 |
| | | Environmental impact | 0 | 0 | 0.3 | 0.45 | 0.3 | 0.45 | 0.45 | 0.6 | 0.45 | 0.6 | 1 | 1 |
| | Lagging metrics | Social impact | 0 | 0 | 0.2 | 0.45 | 0.2 | 0.45 | 0.45 | 0.7 | 0.45 | 0.7 | 1 | 1 |
| | | Overall performance | 0 | 0 | 0.25 | 0.5 | 0.25 | 0.5 | 0.5 | 0.75 | 0.5 | 0.75 | 1 | 1 |

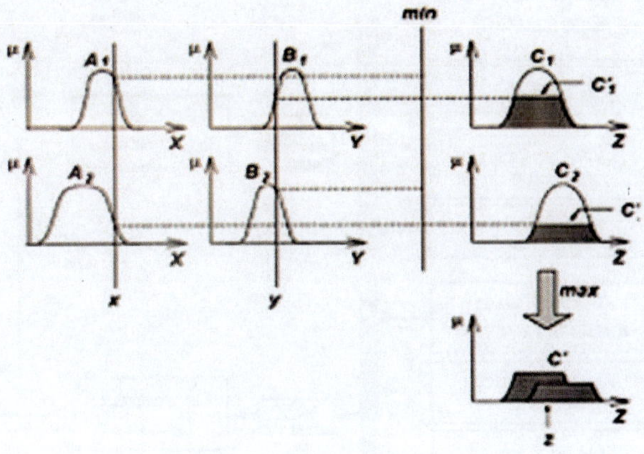

**Fig. 1** Pictorial description of max-min inference method

employed as it uses the linguistic terms in both antecedents and consequent of IF-THEN fuzzy rules as Eq. (2) based on expert's opinions making this method more suitable for supporting decisions in the PM process.

A typical fuzzy rule in MFIS with $n$ input linguistic variables $x_1, x_2, \ldots, x_n$ and one output linguistic variable $y$ has the form

$$\text{IF } x_1 \text{ is } \tilde{A}_1 \text{ and } x_2 \text{ is} \tilde{A}_2, \ldots, x_n \text{ is } \tilde{A}_n, \text{ THEN } y \text{ is } \tilde{B} \tag{2}$$

where $\tilde{A}_i, i = 1, 2, \ldots, n$ are linguistic terms in the antecedent and $\tilde{B}$ is a linguistic term in the consequent.

In the present study, the max-min inference method is used for fuzzy reasoning of the MFIS. The mechanism of such inference in an MFIS with two IF-THEN rules has been schematically shown in Fig. 1.

## 4 Methodology

This section presents the process of systematic development of the proposed PM model for WPP. The proposed model has been developed based on a hierarchical structure of the performance attributes and the metrics (see Fig. 2), derived from sustainability aspect. Finally, the MFISs have been used hierarchically to address the causal relations among the leading and lagging metrics. The entire process has been described in the following steps.

**Step1: Selection of Performance Metrics**. Identification of performance attributes and defining the appropriate metrics (leading and lagging) in order to measure them is a critical issue in any PM model. It primarily depends on

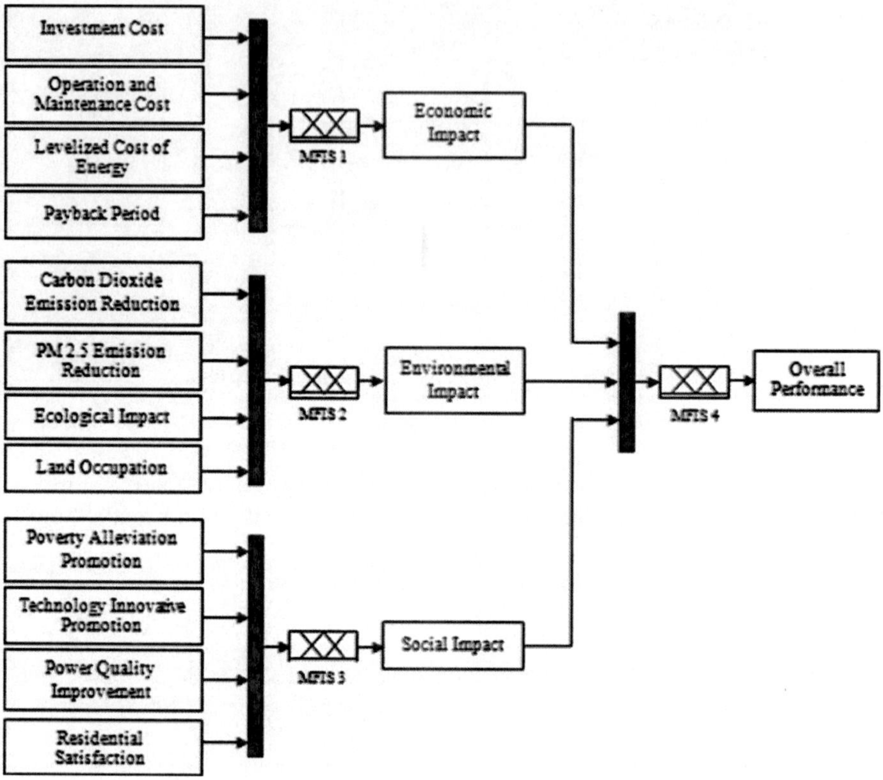


**Fig. 2** Pictorial description of hierarchical structure of the proposed PM model

the purpose of study. In the present study, metrics have been selected in accordance with the sustainable criteria, viz. economic aspect, social aspect, environmental aspect [18]. Three KPIs economic impact, social impact, and environmental impact have been considered to measure the overall performance of the WPP, whereas the three KPIs have been measured through the derived metrics. Based on the cause–effect relationship between leading and lagging metrics, a hierarchical structure has been developed.

**Step 2: Defining Universe of Discourse**. The universe of discourse for the leading or lagging metrics has been identified based on collected data and available benchmarking information. For the present study, universe of discourse for each leading or lagging metric has been normalized in order to obtain a common universe of discourse [0, 1] for all leading or lagging metrics.

**Step 3: Development of Membership Functions**. The universe of discourse of each leading or lagging metrics has been covered by qualitatively descriptors (fuzzy sets), each of which has been characterized by membership function (MF) based on experts' judgment and knowledge.

**Step 4: Generation of Fuzzy Rule-base**. Lagging metrics have been inferred from leading metrics based on their causal relationship defined by a set of fuzzy if-then rules. In the present study, all possible combinations of the predetermined linguistic terms (qualitatively descriptors) have been considered and corresponding to each combination a consequent linguistic term has been assigned based on expert's judgment.

**Step 5: Development of Inference Engine**. Inference engine involves two operations: implication and aggregation. In the present study, *min* operator has been used as an implication operator which determines the fuzzy output of each rule by the composition between a fuzzified value of crisp inputs and the implication relation. Finally, *max* operator has been used as an aggregation operation which develops the overall fuzzy output.

**Step 5: Development of Defuzzification Interface**. Defuzzification interface converts the overall fuzzy output, resulting from the inference process, into a crisp number. In the present study, center of area (COA) operator [19] for defuzzification process has been used for defuzzification process which computes the final output of MFISs.

The architecture of the PM model proposed for WPP has been shown in Fig. 2.

## 5 Simulation Results

In this section, the simulation results by measuring the performance of a WPP using random numbers have been presented to show the relevance of the proposed PM Model for WPP. The linguistic terms and the corresponding fuzzy sets those are implemented in the proposed model have been presented in Table 1. The proposed system has been developed using the fuzzy toolbox from the MATLAB® software.

The results of PM for WPP, obtained by the proposed model, have been presented in Table 2.

The sensitivity analysis of the four MFISs has been carried out using surface viewer of MATLAB® software, which allows the evaluation of interaction effects among the leading and lagging metrics of different levels. As a case example, the interaction effects between the leading metrics "power quality improvement" and "residential satisfaction" and the lagging metric "social impact" are shown in Fig. 3. This figure shows that increasing values of leading metrics "power quality improvement" and "residential satisfaction" will always produces a positive and nonlinear effect in the lagging metric "social impact."

**Table 2** Results of performance measurement for WPP

| Level 1 metrics | Input data | Level 2 metrics | Output data | | Output data |
|---|---|---|---|---|---|
| Investment cost | 0.23 | Economic impact | 0.535 | Overall performance | 0.561 |
| Operation and maintenance cost | 0.47 | | | | |
| Levelized cost of energy | 0.55 | | | | |
| Payback period | 0.67 | | | | |
| Carbon dioxide emission reduction | 0.88 | Environmental impact | 0.721 | | |
| PM 2.5 emission reduction | 0.75 | | | | |
| Ecological impact | 0.48 | | | | |
| Land occupation | 0.32 | | | | |
| Poverty alleviation promotion | 0.20 | Social impact | 0.485 | | |
| Technology innovative promotion | 0.72 | | | | |
| Power quality improvement | 0.46 | | | | |
| Residential satisfaction | 0.33 | | | | |

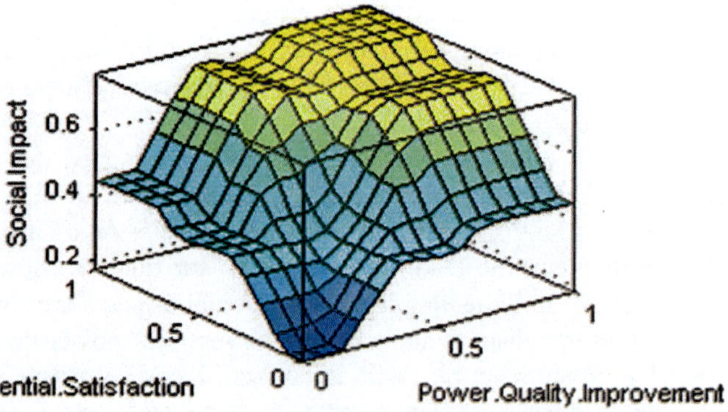


**Fig. 3** Pictorial description of interaction effect between the leading metrics "power quality improvement" and "residential satisfaction" and the lagging metrics "social impact"

## 6 Conclusion

This paper presents a comprehensive PM model for assessing the performance WPP from the point of view of sustainability. The sustainability perspective has been decomposed from three dimensions: economy, environment, and society; and a hierarchical assessment criteria system has been proposed. The employment of MFIS in evaluation process is useful to capture the vagueness and uncertainty and also to quantify the subjective judgment inherently presents in PM of WPP. The proposed model allows the practitioners to analyze the effects of interaction among the leading and lagging metrics. However, employment of weighted fuzzy reasoning approach to improve the proposed model is the further scope of study.

**Acknowledgements** The authors remain very much grateful to the anonymous reviewers for their insightful comments and valuable suggestions to improve the quality of the manuscript.

## References

1. Nudurupati SS, Bititci US, Kumar V, Chan FTS (2011) State of the art literature review on performance measurement. Comput Ind Eng 60(2):279–290. https://doi.org/10.1016/j.cie.2010.11.010
2. Buyukozkan G, Karabulut Y (2017) Energy project performance evaluation with sustainability perspective. Energy 119:549–560
3. Rathi R, Prakash C, Singh S, Krolczyk G, Pruncu CI (2020) Measurement and analysis of wind energy potential using fuzzy based hybrid MADM approach. Energy Rep 6:228–237. https://doi.org/10.1016/j.egyr.2019.12.026
4. Rohrig K et al (2019) Powering the 21st century by wind energy—options, facts, figures. Appl Phys Rev 6(3):031303. https://doi.org/10.1063/1.5089877
5. Ganga GMD, Carpinetti LCR (2011) A fuzzy logic approach to supply chain performance management. Int J Prod Econ 134(1):177–187. https://doi.org/10.1016/j.ijpe.2011.06.011
6. Vasileiou M, Loukogeorgaki E, Vagiona DG (2017) GIS-based multi-criteria decision analysis for site selection of hybrid offshore wind and wave energy systems in Greece. Renew Sustain Energy Rev 73:745–757. https://doi.org/10.1016/j.rser.2017.01.161
7. Xu F, Liu J, Lin S, Yuan J (2017) A VIKOR-based approach for assessing the service performance of electric vehicle sharing programs: a case study in Beijing. J Clean Prod 148:254–267. https://doi.org/10.1016/j.jclepro.2017.10.098
8. Xu D, Lv L, Ren J, Shen W, Wei S, Dong L (2017) Life cycle sustainability assessment of chemical processes: a vector based three-dimensional algorithm coupled with AHP. Ind Eng Chem Res 56(39):11216–11227. https://doi.org/10.1021/acs.iecr.7b02041
9. Tian Z, Wang J, Wang J, Zhang H (2018) A multi-phase QFD-based hybrid fuzzy MCDM approach for performance evaluation: a case of smart bike-sharing programs in Changsha. J Clean Prod 171(Supplement C):1068–1083. https://doi.org/10.1016/j.jclepro.2017.10.098
10. Xu D, Li W, Shen W, Dong L (2019) Decision-making for sustainability enhancement of chemical systems under uncertainties: combining the vector-based multiattribute decision-making method with weighted multiobjective optimization technique. Ind Eng Chem Res 58(27):12066–12079. https://doi.org/10.1021/acs.iecr.9b01531
11. Ghenai C, Albawab M, Bettayeb M (2020) Sustainability indicators for renewable energy systems using multi-criteria decision-making model and extended SWARA/ARAS hybrid method. Renew Energy 146:580–597. https://doi.org/10.1016/j.renene.2019.06.157

12. Majumder D, Bhattacharjee R, Dam M (2020) Fuzzy supply chain performance measurement model based on SCOR 12.0. In: Solanki V, Hoang M, Lu Z, Pattnaik P (eds) Intelligent computing in engineering. Advances in intelligent systems and computing, vol 1125. Springer, Singapore, pp 1129–1139
13. Charles Rajesh Kumar J, Vinod Kumar D, Majid, M (2019) Wind energy programme in India: emerging energy alternatives for sustainable growth. Energy Environ 30(7):1135–1189. https://doi.org/10.1177/0958305x19841297
14. Sellak H, Ouhbi B, Frikh B (2017) Energy planning under uncertain decision-making environment: an evidential reasoning approach to prioritize renewable energy sources. Inteligencia Artif 20(59):31–31. https://doi.org/10.4114/intartif
15. Zadeh LA (1965) Fuzzy sets. Inf Control 8(3):338–353. https://doi.org/10.1016/s0019-9958(65)90241-X
16. Zedeh LA (1989) Knowledge representation in fuzzy logic. IEEE Trans Knowl Data Eng 1(1):89–100. https://doi.org/10.1109/69.43406
17. Campos-Guzmán V, García-Cáscales MS, Espinosa N, Urbina A (2019) Life cycle analysis with multi-criteria decision making: a review of approaches for the sustainability evaluation of renewable energy technologies. Renew Sustain Energy Rev 104:343–366. https://doi.org/10.1016/j.rser.2019.01.031
18. Wu YN, Xu CB, Zhang BY, Tao Y, Li XY, Chu H et al (2019) Sustainability performance assessment of wind power coupling hydrogen storage projects using a hybrid evaluation technique based on interval type-2 fuzzy set. Energy 179:1176–1190. https://doi.org/10.1016/j.energy.2019.04.195
19. Mamdani EH (1974) Applications of fuzzy algorithm for control a simple dynamic plant. Proc IEEE 121(12):1585–1588. https://doi.org/10.1049/piee.1974.0328

# Automated Detection of Road Potholes Under Shadowed Condition Using Machine Learning

**Debabrata Gogoi, Nilay Nishant, Ritu Anil kumar, and Sushanta Kabir Dutta**

**Abstract** Efficient methodology for monitoring the condition of road is essential for the road departments and public safety. Advancements in computer vision technology and machine learning techniques present the possibilities of providing accurate information on road condition. Present methodologies include the use of UAV images from thermal bands as well as labor-intensive ground survey. The current study evaluates inter-comparison between the use of different segmentation techniques. Fusion of spectral difference and multiresolution segmentation-based study outperform among other segmentation techniques. Segmented output is processed with k-nearest neighbor classifier to build a model for automated detection of road potholes on the orthomosaic image captured through UAV platform. The performance is evaluated by $F$-score accuracy assessment technique.

**Keywords** Remote sensing (RS) · Multiresolution segmentation (MRS) · k-nearest neighbor (k-NN) · Unmanned aerial vehicle (UAV)

## 1 Introduction

Road network is the basic component for the transportation system, it is not only about building road but also maintenance and health monitoring of road is another challenging task. The goal of this paper is to build a model that detects road pot-

D. Gogoi (✉) · S. Kabir Dutta
North Eastern Hill University, Shillong, Meghalaya, India
e-mail: debabratagogoi@gmail.com

S. Kabir Dutta
e-mail: sushantatzp@gmail.com

N. Nishant · R. Anil kumar
North Eastern Space Applications Centre, Umiam, Meghalaya, India
e-mail: nilaynishant@gmail.com

R. Anil kumar
e-mail: ritu.anilkumar@gmail.com



A. Maji et al. (eds.), *Proceedings of the International Conference on Computing and Communication Systems*, Lecture Notes in Networks and Systems 170,
https://doi.org/10.1007/978-981-33-4084-8_25

holes based on machine learning algorithms for monitoring and updating the status of road structure automatically. Many factors need to be taken into account while classification of remote sensing data. The major steps of image classification include selection appropriate of data for classification, determination of a suitable classification algorithm, selection of training samples, image preprocessing, feature extraction, post-classification processing and evaluation of classification results [1].

## 2 Literature Review

Remote sensing technology is trending nowadays in research field for object detection that can capture larger swath in a single shot by either UAV or satellite. Many researchers use UAV as a platform to capture the road asphalt pavement and detect road cracks and potholes using machine learning algorithm. Saar and Talvik [2] uses grayscale images and equalized to balance the intensity value of cracks depend on location of the crack in the picture and divide the image in to smaller sub-matrix and detect the probable defects depending on the mean value of the sub-matrix. Pan et al. [3] use UAV to capture digital pavement images and perform multiresolution segmentation, and algorithm is run using spectral, geometrical and texture features. A subtype of a neural network called convolutional neural network is used in image-related tasks [4–6]. Wang et al. [7] uses convolution neural networks for identifying cracks on images of roads. Pan et al. [8] come up with multispectral sensor fixed in UAV to acquire road pavement images and detects road damages using machine learning technique. As researchers use UAV to capture road images from a height of more than 20 m which is not possible in case of areas where road surface is covered with tree branches and leaves again if we change the height of UAV captured images, then orthophoto preparation becomes difficult. Vigneshwar and Hema Kumar [9] detection of crack done by image preprocessing is based on difference of Gaussian filtering and clustering-based image segmentation.

## 3 Data Base Description

### *3.1 Study Area*

Study area is a section of road located in Umiam, Ri-Bhoi District, Meghalaya. The total length of the road captured was about 1200, and 500 m was taken for the processing. During the field investigation, it was observed that the maximum diameter of potholes was about 50 cm and maximum depth being 10 cm and in the shape elliptical, round and more in a polygonal shape. There are trees on the roadside that covers direct line of sight from the UAV flying above 15 m.

# 4 Experimental Setup

## 4.1 Preprocessing

Data preprocessing is an important step in the machine learning. If data have not been carefully analyzed before training, then that would produce misleading results. So before running an analysis of the representation, checking quality of data and filtering of data is very important. A digital elevation model (DEM) is built using images captured from the UAV. It contains the elevation information of the terrain over a specific area, and ortho rectification is carried out in order to ensure the planimetric position of the images. In a georeferenced image, only the Nadir point lies in its proper planimetric position. The error in the image increases as we move away from the Nadir. The off nadir errors are corrected by using the elevation information from the DEM. The AgiSoft processed orthophoto output and actual captured image is shown in Fig. 1.

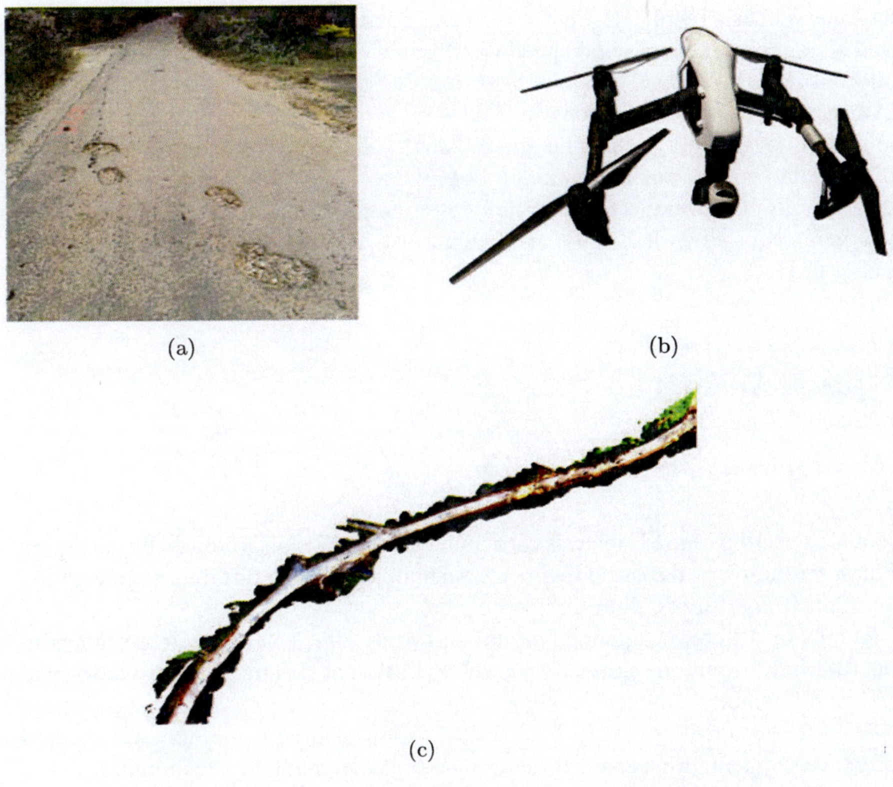

**Fig. 1** **a** Road image **b** UAV with integrated RGB camera and **c** Agisoft processed orthomosaic image

### 4.2 k-Nearest-Neighbor

k-nearest neighbor classification, or kNN, is a "Euclidian Distance"-based machine learning algorithm that is widely used for classification and clustering problem statements. The algorithm localizes a group of objects in training dataset and assigns appropriate label based on the proximity with reference to Eucledian distance [10, 11]. Three important components are needed for this algorithm: a group of labeled objects; a proximity metric and the number k-nearest neighbors [12]. The algorithm can be explained by the formula [13]:

$$D(x, y) = \left( \sum_{i=1}^{m} |w_i - y_i|^2 \right)^{1/2} \tag{1}$$

The RGB zenmuse Z3 camera integrated in UAV also integrated GPS receiver that will provide locations latitude, longitude and other information while capturing the images. The same sensor has been used to assessment of agricultural research [14]. Collected image is in .jpg format, and the survey was conducted in the 15–20 days interval basis regularly. Field survey was conducted mostly during Saturday or Sunday at noon time and wind speed (m/s), wind direction (°), humidity, etc., data collected from MOSDAC Web site to ensure the success of the planned flight. The UAV survey was carried out along with the road at the height of 8 m also 40 m above the ground level, while at 8 m height, the angle tilted 45° from the ground, and at 40 m height, camera position is nadir, but images taken from 40 m height are not useful due to the coverage of the road by the trees. Two hundred image tiles of the road were acquired with 70% of overlapping between two sequential images were also ensured.

## 5 Result Analysis

### 5.1 Segmentation

Manual extraction of road area was carried out using GIS tools followed by clustering of high resolution of the ortho image to create objects using different segmentation algorithm with different algorithm parameters.

Results of different segmentation are shown in Fig. 2: (a) Quadtree segmentation, (b) Multiresolution segmentation and (c) Fusion of spectral multiresolution and spectral difference segmentation algorithm.

From the result, it is clearly visible that fusion of multiresolution and spectral difference segmentation performs better than other segmentation techniques.

**Fig. 2** **a** Quadtree-based segmentation, **b** multiresolution segmentation and **c** fusion of spectral difference and multiresolution segmentation

## 5.2 Classification

After performing fusion of multiresolution segmentation and spectral difference algorithm for the targeted image, samples of "Potholes" and "Non Distress Road" are selected. Rather, detectors are programmed to extract optimum features from the data that provides the best output result and, these features are used as input to the machine learning algorithm. Started from seventy three different features and finalized by shrinking down to eleven feature that gives the optimum result using Feature Selection Optimization (FSO) techniques in e-Cognition software (Figs. 3 and 4; Table 1).

## 5.3 Accuracy Assessment

Accuracy assessment is very important in classified images to verify the performance of classification models. In accuracy assessment, system classified or detected map data compared with reference or ground truth data to identify and quantify error. In this project, confusion matrix is used to represent the test's accuracy of classification models. Here compared the classified "Potholes" and "Non Distress Road" layer map with ground truth reference layer and calculate the different parameters. The

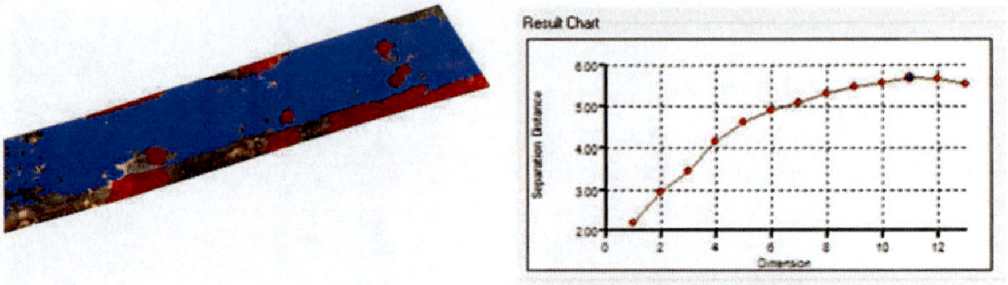


(a) (b)

**Fig. 3** **a** Sample selection image and **b** class separation distance

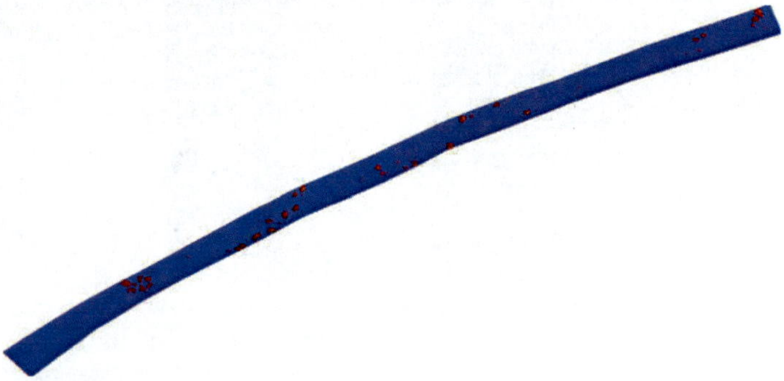

**Fig. 4** Nearest neighbor classified output

**Table 1** Features selection to run k-nearest neighbor

| Category | Name | Category | Name |
|---|---|---|---|
| Spectral | Mean of each band<br>STD of each band | Geometric | Elliptic fit Roundness |
| Textural | GLCM dissimilarity<br>GLCM homogeneity<br>Brightness | Layer values | Pixel based edge contrast of neighbor pixels |

below figure shows the k-NN classified image referencing with ground truth point shape potholes layer in ArcGIS (Fig. 5).

Confusion matrix is cross-tabulates predicted and observed with four options:
True Positive(TP): Correctly predicting a "Potholes",
True Negative(TN): Correctly predicting as a "Non Distress Road",
False Positive(FP): Falsely Predicting as a "Potholes" and
False Negative (FN): Predicting as "Non Distress Road" for "Potholes" (Table 2).

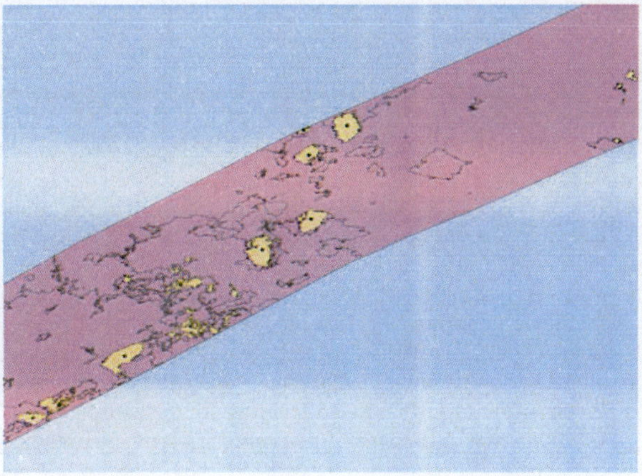

**Fig. 5** Classified image is referencing with the ground truth potholes

**Table 2** Accuracy table for k-NN

| Description | Result | Method |
|---|---|---|
| Precision | 0.8684 | Precision = TP / (FP + TP) |
| Recall | 0.9706 | Recall = TP / (FN + TP) |
| Negative Prediction (NP) | 0.0000 | NP = TN / (FN + TN) |
| False Detection Rate (FDR) | 0.1316 | FDR = FP / (TP + FP) |
| False Negative Rate (FNR) | 0.0294 | FNR = FN / (TP + FN) |
| Accuracy | 0.8462 | Accuracy = (TP + TN) / (TP + FP + TN + FN) |
| $F_1$ Score (FS) | 0.9167 | FS = 2 (Precision × Recall) / (Precision + Recall) |

## 6 Conclusion and Future Direction

This Study represents an approach for automated detection of potholes from the images taken from a RGB sensor that uses drone as a platform. Different object detection segmentation process used, and out of all, fusion of spectral multiresolution and spectral difference segmentation algorithm provide the best grouping of potholes. In this study, object-based features are applied, and these models provide relatively good results in detecting road potholes. k-nearest neighbor classifier detects the potholes with accuracy of 0.8462. Also, there are more algorithms available which can be implemented for further comparison of the output accuracy. There are also higher resolution sensors available that can detect more details about road surfaces and damages.

## References


1. Geetha P, Chitradevi B (2014) A review on image processing techniques for synthetic aperture radar (SAR) images. Int J Innov Res Comput 2(4):3823–3829
2. Saar T, Talvik O (2010) Automatic asphalt pavement crack detection and classification using neural networks. In: 2010 12th Biennial baltic electronics conference. IEEE
3. Pan Y et al (2017) Object-based and supervised detection of potholes and cracks from the pavement images acquired by UAV. Int Arch Photogram Remote Sens Spatial Inf Sci 42
4. Stenroos O (2017) Object detection from images using convolutional neural networks
5. Maeda H et al (2018) Road damage detection using deep neural networks with images captured through a smartphone. arXiv:1801.09454
6. Maeda H et al (2018) Road damage detection and classification using deep neural networks with smartphone images. Comput-Aided Civil Infrastruct Eng 33(12):1127–1141
7. Wang KC et al (2017) Deep learning for asphalt pavement cracking recognition using convolutional neural network. In: Proceedings of international conference on airfield highway pavements
8. Pan Y et al (2018) Detection of asphalt pavement potholes and cracks based on the unmanned aerial vehicle multispectral imagery. IEEE J Sel Top Appl Earth Observ Remote Sens 11(10):3701–3712
9. Vigneshwar K, Hema Kumar B (2016) Detection and counting of pothole using image processing techniques. In: 2016 IEEE international conference on computational intelligence and computing research (ICCIC). IEEE
10. Hoffmann M, Mock M, May M (2013) Road-quality classification and bump detection with bicycle-mounted smartphones. In: Proceedings of the 3rd international conference on ubiquitous data mining, vol 1088. CEUR-WS.org
11. Zhang K, Hutter M, Jin H (2009) A new local distance-based outlier detection approach for scattered real-world data. In: Pacific-Asia conference on knowledge discovery and data mining. Springer, Berlin, Heidelberg
12. Wu X, Kumar V, Ross Quinlan J, Ghosh J, Yang Q, Motoda H, McLachlan GJ et al (2008) Top 10 algorithms in data mining. Knowl Inf Syst 14(1): 1–37
13. Kotsiantis SB, Zaharakis I, Pintelas P (2007) Supervised machine learning: a review of classification techniques. Emerging Artif Intell Appl Comput Eng 160:3–24
14. Handique BK et al (2016) Rapid assessment of boro paddy infestation by brown planthopper in Morigaon district, Assam, India using unmanned aerial vehicle. Current Sci 111(10):1604–1606

# Combination of Cepstral Features and Temporal Learning Approach for Emotion Recognition

**Rajiv Barman, Mridusmita Sharma, Kandarpa Kumar Sarma, and Parismita Sarma**

**Abstract** Humans express their feelings through different behaviours in their day to day life. Facial emotion is one of them and can be recognized by various digital image processing techniques. Emotion can also be recognized from utterance of a speaker. MFCC, LPCC, MFSC are some of the cepstral features that can be used for the purpose of emotion recognition. We have used RNN and NARX for classifying the features according to the emotions contained in them. The network structures of RNN and the NARX help in capturing the temporal information present in the speech signal. The combination of cepstral features and temporal learning carried out using RNN and NARX provides efficient performance which is the key focus of the present work. After completion of the whole experiment, we observe that NARX exhibits better result than the RNN in case of emotion recognition by an increase of 3.3% performance.

**Keywords** Neural network · Recurrent · Autoregressive · Emotion recognition · Temporal · Filter

## 1 Introduction

Speech is quasi-periodic in nature, and besides containing verbal information that contains non-verbal information like, emotion, dialectal contents, etc.

R. Barman · P. Sarma (✉)
Department of Information Technology, Gauhati University, Guwahati, Assam, India
e-mail: parismita.sarma@gmail.com

R. Barman
e-mail: rajivbarman123@gmail.com

M. Sharma · K. K. Sarma
Department of ECE, Gauhati University, Guwahati, Assam, India
e-mail: mriduzb@gmail.com

K. K. Sarma
e-mail: kandarpaks@gauhati.ac.in

 
A. Maji et al. (eds.), *Proceedings of the International Conference on Computing and Communication Systems*, Lecture Notes in Networks and Systems 170,
https://doi.org/10.1007/978-981-33-4084-8_26

When the speech signal is interfered by noise signal, it becomes a challenging task to send the accurate data to the intended place and analysis of various parameters like emotion recognition in the main speech signal. For accurate, effective analysis and recognition of speech, it is important to make the signal completely noise free. Performance of any speech processing system may be degraded due to the presence of noise. Here, the need of speech enhancement technique or noise cancellation method arises [1].

The common filtering process is by using configuring fixed digital filters. From study, it is known that this type of filter is suitable for narrow band stationary noise. But in general, noise is not stationary by nature, and it occurs with wide frequency in speech signal. So the above mention filter underperforms for such situation and gives unsatisfactory result. In that case, adaptive filter works. Adaptive filter is capable of readjusting different constraints with a view to decrease error signal. Adaptive filter is suitable for unknown environment so it can play an important role in noise cancellation [2].

In our experiment, we have considered FIR-based adaptive filter. We have used three adaptive filters in our experiment. They are Least Mean Square (LMS), Filter-x Least Mean Square (FxLMS) and Recursive Least Square (RLS). After filtering the samples with these filters, we have used Artificial Neural Network (ANN) to classify the samples. Certain ANNs don't have the ability to learn temporal variations in a signal, for example, Multi Layer Perceptron (MLP). RNN, NARX, etc. are proper examples of ANNs that have the ability to track temporal variations. Here, we have considered the RNN and NARX for classifying human emotions from speech.

RNNs are suitable for machine learning which can remember inputs because of its internal memory. RNN entails sequential data such as video, speech, audio and many more. This network outlines a very profound learning of a sequence and content compared to others. Actually, RNN enables past information to keep intact and determines temporal correlation between far away data [3].

NARX is a non-linear dynamic Neural Network (NN) which is suitable to non-linear system and time series modelling. It consists of recurrent responses from some other network layers to input layer.

## 1.1 Features Used

We have used some commonly used features in our experiment. Here, a brief discussion about them is given:

MFCC: This is a popular and mostly used feature extraction technique for speech recognition. MFCC is famous for its ability to perform some operations, namely pre-emphasis, framing, windowing, DFT, Mel Filter Bank (MFB) and DCT. The expression (1) is used for computation of MFCC [4]:

$$\text{MFCC}_i = \sqrt{\frac{2}{N}} \sum_{j=1}^{N} \log[E_{\text{mel}}(j)] \cos\left(\frac{\pi i}{N}(j - 0.5)\right) \tag{1}$$

LPCC: Linear Predictive Cepstral Coefficient (LPCC) can be used to confine emotional features from speech signal that is manifested at vocal tract. The theory behind linear predictive analysis is the *n*th sample of a speech signal which can be predicted from linear combinations of *p* predecessor signals. The LPCC is derived from the Linear Predictive Coding (LPC) as shown in expression (2), [4]:

$$\text{LPCC}_i = \text{LPC}_i + \sum_{k=1}^{i-1} \frac{k - i}{i} \text{LPCC}_{i-k} \text{LPC}_k \tag{2}$$

MFSC: DCT transformation is done in MFCC. The features extorted from speech signal before the DCT transformation are called MFSC. It is Mel-frequency spectral coefficient, in other words, Log Mel-Frequency Spectral Coefficients (LMFSC) [5]. The features extracted using MFSC has many useful applications in speech recognition.

### *1.2 Classifiers*

As already mentioned, we have used learning-based classifiers for emotion recognition using speech samples. The classifiers used are RNN and NARX. Both of these classifiers have temporal ability and can track time variations and hence are preferred for speech applications. The dynamic processing ability of the said classifiers in combination with the above-mentioned features is crucial in making the performance satisfactory.

## 2 Literature Review

We have studied some recent journal papers to work with the system. Some of them are mentioned below.

A recent research paper we discussed is by Boussaada et al. [6], where a clear conception on NARX recurrent neural network is presented. Actually, this network encloses several layers of feedback connections. NARX is very effective for its memory utilization capability that uses real-time or predicted past values. They had used NARX for prediction of direct solar radiation on horizontal plane. Their finding is quite interesting; mainly, they were able to perform periodical training.

Kumar and Goel [7] assumed a range of Least Mean Squares (LMS) noise cancellation algorithmic techniques. Due to generally used digital filter and their unknown statistical characteristics, researchers proposed use of adaptive filter beyond fixed digital filter for noise cancellation purpose. They have used Hindi speech dataset. Through this paper, they have shown a comparative analysis of algorithms like BLMS, normalized NLMS, block VSLMS, etc.

Banda et al. [8] describe a NARX-RNN-based continuous emotion recognition using PSO optimization. Since the recognition of continuous emotion is a difficult task, they trained the NARX to detect the emotional patterns of the dataset which are in the condition of temporal process.

A new technique called transfer learning was introduced by Feng et al. [9] which tries to transfer knowledge among the datasets. For speech dataset, the networks may be generative and adversarial. In case of physiological signals like speech or brain wave, transfer learning endorses adapted models all the way through statistical methods. With the help of this technique, video as well as speech signals are allocated architecture of any refined system that is capable of detecting maximum seven emotions.

Sharma and Sarma [10] experimented on a dialect supported emotion recognition system. They had used a telephonic-based Assamese speech dataset for three learning-based classifiers. They are RNN, SVM and Distributed Time Delay Neural Network (DTDNN) sequentially. They were able to get an improved accuracy of 92% by using MFCC features for mood recognition.

## 3 Methodology

Figure 1 is general block diagram of our proposed system:

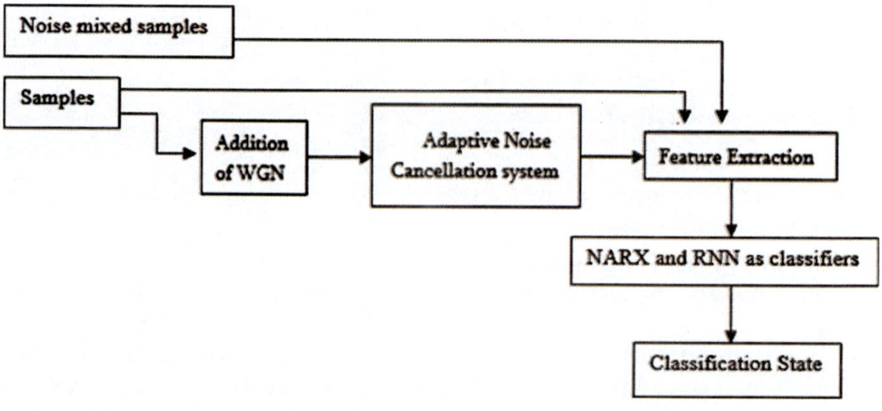


**Fig. 1** The proposed block diagram

The working of the system has two parts. First training and then testing. We have used about 540 samples for training which includes the dialectal variations of Central Assamese speech with background noise. Another set of samples derived using Central Assamese dialect with noise and mood variations containing two, three, four, five and six different sentence lengths. Both the networks are further subjected to train using original, noise mixed and filtered signals. Networks are also trained using Levenberg–Marquardt (LM), quasi-Newton method, resilient back propagation and scaled conjugate gradient training followed by testing.

The filtering of the corrupted signals are done using LMS, FxLMS and RLS methods. Moreover features of original noise corrupted and filtered signals are extracted. Three different feature sets of all the samples are generated and used for training on the classifiers as already discussed. In our experiment, we have chosen Elman Neural Network (ENN) which is a common form of RNN.

### 3.1 Experimental Analysis

The dataset is an important component. It is created by recording Assamese telephonic speech having two, three, four, five and six word sentences. These samples are of different utterances and moods. We have considered three mood variations viz. regular/normal, loud and irritated/angry. The audio clips have been recorded over telephone in a noise free environment.

Two important parameters for our analysis are discussed here. They are Mean Squared Error (MSE) and Signal to Noise Ratio (SNR).

MSE is measured by taking the difference between result acquired and target (output). A smaller value indicates enhanced performance with negligible fault. Its expression is shown in (3).

$$\text{MSE} = \frac{1}{N}\sum_{i=1}^{N} (\text{desired signal} - \text{error})^2 \tag{3}$$

SNR can be defined as rapport between speech signal strength and noise modulation. It is measured in decibels (dB). In general, SNR is found to be greater than 0 dB. It implies that whatever we speak is louder than background noise. Equation (4) is the expression for SNR.

$$\text{SNR}_{\text{dB}} = 10\log_{10}\left(\frac{\text{desired signal}}{\text{error}}\right)^2 \tag{4}$$

Figure 2 is the snapshot of example comparisons among noise free speech signal, speech signal with noise and signal of LMS adaptive filter. It is observed that LMS adaptive filter has positive impact on noised speech signal, and a clean signal can be achieved as a result. But little deviation is also seen sometimes.

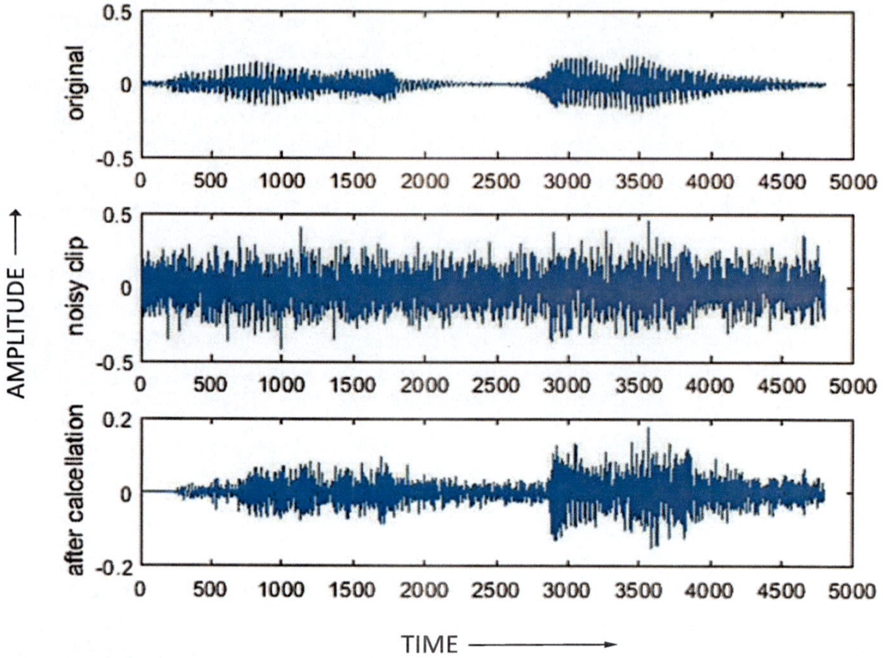


**Fig. 2** Comparison of signals after filtering by LMS

Figure 3 shows the significant noise reduction by FxLMS filter. The convergence times are much better in FxLMS with variable step size algorithm. The results show that the FxLMS converges to an absolute minimum at a slower rate than LMS.

LMS works effectively for small inputs on the other hand for huge size vector computation complexity increases. Big computation has to fit in small steps. Due to this difficulty, computation becomes lengthy, hence cannot be recommended for real-time application. The neural network specification and structure used in our experiment can be read from Table 1.

We have used the following Optimization techniques while training the data with NARX.

1. Gradient Descent (GD) or Steepest Descent (SD) algorithm
2. Resilient Back Propagation algorithm (RBP)
3. Scaled Conjugate Gradient Algorithm (SCGA)
4. Quasi–Newton Method (QNM)
5. Levenberg–Marquardt (LM) algorithm or Damped Least Square (DLS) method.

After experiments, in terms of ANN's speed and memory consumption found that above-mentioned LM algorithm gives better output. The scenario can be depicted in Fig. 4.

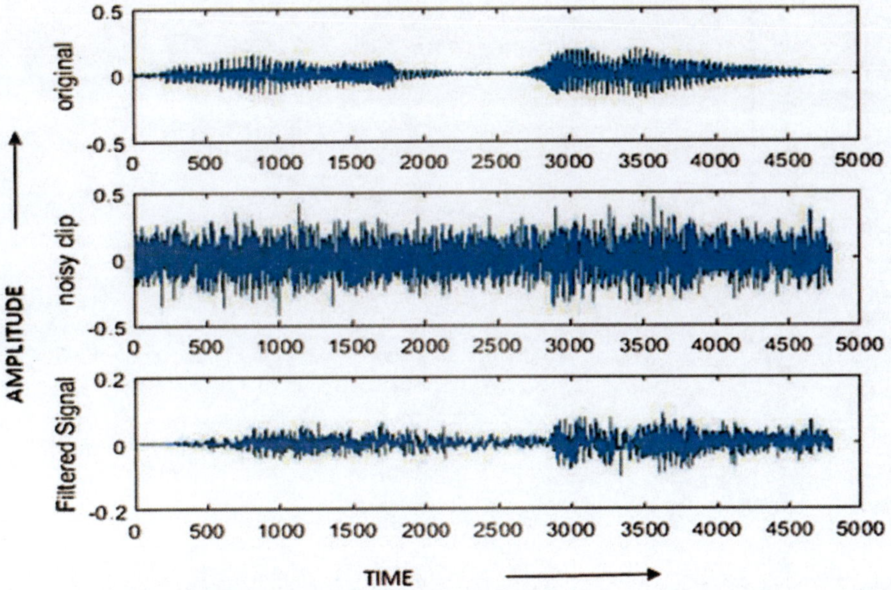


**Fig. 3** Comparison of signals after filtering by FxLMS

**Table 1** Specification of NN

| | |
|---|---|
| Name of the classifier | NARX, RNN |
| No. of hidden layers | 03 |
| Training functions in hidden layers | tansig, tansig, tansig |
| No. of neuron in input layer | 50 |
| No. of neuron in hidden layer—1 | 30 |
| No. of neuron in hidden layer—2 | 25 |
| No. of neuron in hidden layer—3 | 10 |
| No. of neuron in output layer | 03 |

We have evaluated performance of NARX algorithm for recognizing the emotions with the help of confusion matrix. Table 2 shows the performance of NARX with Levenberg–Marquardt algorithm using MFCC feature vector.

In the above confusion matrix, rows are for different moods that are predicted, whereas columns are the real or true classes of moods. The cells on diagonal line gives the correct values for correct classes. Other values in the table are erroneous. All the values in the cell represent percentage of observations. Hence, the recognition rate for correctly identified sample is $33.3 + 33.3 + 30.0 = 96.6\%$.

Similarly, we have built confusion matrices to measure performances of different training algorithms on feature vectors by using MFCC, LPCC and MFSC methods. Below mentioned confusion matrices, we have constructed for regular, loud and angry moods recognition.

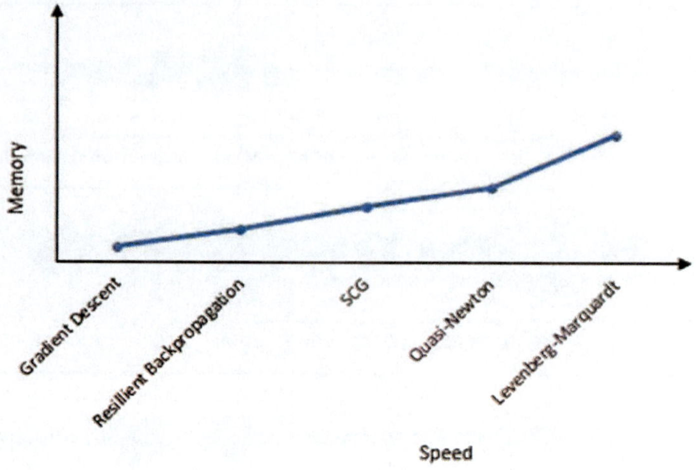


**Fig. 4** Performance comparison of optimization techniques

**Table 2** Use of MFCC on unfiltered speech signal using NARX with Levenberg–Marquardt algorithm: confusion matrix

| Moods | Regular | Loud | Irritated/angry |
|---|---|---|---|
| Regular | 33.3 | 0.0 | 0.0 |
| Loud | 0.0 | 33.3 | 3.3 |
| Irritated/angry | 0.0 | 0.0 | 30.0 |

(1) MFCC vector value of unfiltered output by quasi-Newton algorithm (NARX),
(2) MFCC vector value of FxLMS filtered output by NARX with Levenberg–Marquardt algorithm,
(3) MFCC vector value of FxLMS filtered output by NARX with quasi-Newton algorithm,
(4) LPCC vector value of RLS filtered output by NARX with Levenberg–Marquardt algorithm,
(5) LPCC vector value of RLS filtered output by NARX with quasi-Newton algorithm,
(6) MFSC vector value of FxLMS filtered output by NARX with Levenberg–Marquardt algorithm,
(7) MFSC vector value of FxLMS filtered output by NARX with quasi-Newton algorithm.

We have also carried out experiments for different feature vectors of filtered as well as unfiltered samples and obtained recognition rate with respect to times elapsed. From the above-mentioned seven tables, we achieved accuracy and computational time as shown in Table 3.

We have observed from Table 3 that original speech signals as well as noise filtered signals extracted with MFCC feature vector have offered the highest recognition rate of 97.9% average with computational time 1.2950 s in the NARX classifier. Then,

**Table 3** Features for NARX and RNN classifiers with accuracy and computational time

| Types of classifiers | Features used | Accuracy (%) | Computational time (s) |
|---|---|---|---|
| NARX | MFCC feature vectors | 97.9 | 1.2950 |
| | LPCC feature vectors | 88.7 | 2.7100 |
| | MFSC feature vectors | 95.0 | 1.5285 |
| RNN | MFCC feature vectors | 94.6 | 1.2448 |
| | LPCC feature vectors | 89.2 | 1.2776 |
| | MFSC feature vectors | 89.6 | 1.1109 |

**Table 4** Comparison with other existing approaches

| Authors | Types of classifiers | Features used | Accuracy (%) | Computational time (s) | Remarks |
|---|---|---|---|---|---|
| Rami and Lee 2012 | RNN + NARX | PHOG | 91.5 | – | Proposed model is capable of giving better accuracy |
| Sharma and Sarma (2016) | RNN | MFCC | 92.5 | 24.5 | |
| Present work | NARX | MFCC | 97.9 | 1.2950 | |
| | RNN | MFCC | 94.6 | 1.2448 | |

signals extracted by MFSC provide better recognition rate of 95.0% as compared to LPCC that offers 88.7% recognition rate in the NARX. The RNN classifier also provides highest recognition in case of MFCC feature with 94.6% for computational time 1.2448 s. Similarly, better accuracy was observed in case of MFSC feature with 89.6% than LPCC feature that provided 89.2% accuracy. Our results are also compared with previously reported works. Table 4 shows the comparative performance. Hence, the advantage of the proposed approach is obvious.

## 4 Conclusion

In this paper, an approach based on NARX has been applied for capturing the emotional content as well as the temporal attributes of Assamese telephonic speech after cancelling the Added White Gaussian Noise (AWGN) by three robust adaptive filter algorithms. The cepstral features used in our experiment are capable of representing the emotional contents in a satisfactory manner as found from the experimental results. There are two popular RNNs Jordan Network and Elman Network available for experiment [11]. The network structures of RNN and NARX classifiers are found to be capable of capturing the temporal aspects present in the speech signals. We have achieved 97.9% recognition rate with NARX, whereas with RNN, we achieved 94.6% accuracy rate. We also observed better performance of FxLMS filter than LMS and RLS filter algorithms.

The accuracy of the present work can be further dealt with a large as well as dynamic dataset and deep learning structures. Adoption of more robust adaptive filter for noise cancellation mechanisms is expected to boost the performance of our proposed system.

## References

1. Ram R, Panda S, Palo HK, Mohanty MN (2016) Impulsive noise cancellation for speech enhancement using state space adaptive algorithm. In: 2016 international conference on signal processing, communication, power and embedded system (SCOPES). IEEE, pp 1393–1396
2. Awachat PM, Godbole SS (2012) A design approach for noise cancellation in adaptive Lms predictor using matlab. Int J Eng Res Appl (IJERA) 2(4):2388–2391
3. Sharma M, Sarma M, Sarma KK (2013) Recurrent neural network based approach to recognize assamese vowels using experimentally derived acoustic-phonetic features. In: 2013 1st international conference on emerging trends and applications in computer science. IEEE, pp 140–143
4. Wong E, Sridharan S (2001) Comparison of linear prediction cepstrum coefficients and mel-frequency cepstrum coefficients for language identification. In: Proceedings of 2001 international symposium on intelligent multimedia, video and speech processing. ISIMP 2001 (IEEE Cat. No. 01EX489). IEEE, pp 95–98
5. Suniya VS, Mathew D (2015) Acoustic modelling using auditory model features and convolutional neural network. In: 2015 international conference on power, instrumentation, control and computing (PICC). IEEE, pp 1–4
6. Boussaada Z, Curea O, Remaci A, Camblong H, Mrabet Bellaaj N (2018) A nonlinear autoregressive exogenous (NARX) neural network model for the prediction of the daily direct solar radiation. Energies
7. Kumar A, Goel P, Gupta VK, Chandra M (2016) Comparative research of various adaptive algorithms for noise cancellation in speech signals. In: 2016 international conference on control, computing, communication and materials (ICCCCM), Allahabad, pp 1–5
8. Banda N, Engelbrecht A, Robinson P (2015) Continuous emotion recognition using a particle swarm optimized NARX neural network. In: 2015 international conference on affective computing and intelligent interaction (ACII). IEEE, pp 380–386
9. Feng K, Chaspari T (2020) A review of generalizable transfer learning in automatic emotion recognition. Front Comput Sci 2:9
10. Sharma M, Sarma KK (2016) Learning aided mood and dialect recognition using telephonic speech. In: 2016 international conference on accessibility to digital world (ICADW). IEEE, pp 163–167
11. Barman R, Sharma M, Sarma KK, Sarma P (2018) Content capture and noise cancellation aided mood recognition using assamese speech. In: 5th international conference on signal processing and integrated networks (SPIN), Noida. IEEE, pp 811–815

# Estimation of Tea Leaf Chlorophyll Using MLR, ANN, SVR, and KNN in Natural Light Condition

Utpal Barman, Animesh Sarmah, Diganto Sahu, and Golap Gunjan Barman

**Abstract** Traditionally, the tea leaf chlorophyll is estimated using spectrophotometer and soil plant analysis device (SPAD). The SPAD and spectrophotometer are costly and time-consuming methods. The present work proposed a new method to detect the chlorophyll of Assam tea leaf using multiple linear regression (MLR), artificial neural network (ANN), support vector regression (SVR) and K nearest neighbor (KNN). The tea leaf images are captured using a Nikon DSLR camera in a natural light condition. The color features of the tea leaves are extracted and fitted with the values of SPAD (Tea actual Chlorophyll). The results of the MLR, ANN, SVR, and KNN are validated with the actual chlorophyll value of the tea leaves. The models are evaluated using regression score ($R^2$) mean absolute error (MAE), mean squared error (MSE), and root mean squared error (RMSE). The MLP model outperforms as compared to the other model and it can be treated as a replica of SPAD.

**Keywords** Linear regression · SVR · KNN · ANN · SPAD · Image processing · Chlorophyll

---

U. Barman (✉) · A. Sarmah · D. Sahu · G. G. Barman
Department of Computer Science and Engineering, Girijananda Chowdhury Institute of Management and Technology, Guwahati, Assam, India
e-mail: utpalbelsor@gmail.com

A. Sarmah
e-mail: sarmahanimesh@gmail.com

D. Sahu
e-mail: digantosahu@gmail.com

G. G. Barman
e-mail: barmangolap15@gmail.com



A. Maji et al. (eds.), *Proceedings of the International Conference on Computing and Communication Systems*, Lecture Notes in Networks and Systems 170,
https://doi.org/10.1007/978-981-33-4084-8_27

## 1 Introduction

Agriculture is said to be the main livelihood of India. All agriculture plants require enough supplies of macronutrients and *N* (nitrogen). The chlorophyll content of leaf indicates the content of nitrogen along with some other additional information related to the plant's nutrient status [1]. The SPAD and spectrophotometer are there to calculate the chlorophyll of plants. But these processes are expensive. For this reason, the researchers developed digital chlorophyll meter using image analysis of leaf color. The rapidly growing technology empowers researchers to furnish with advanced tools that help in both image processing techniques and digital image acquisition, including high throughput, and gainful real-time quantification of leaf spectral properties by utilizing only a computer and a digital camera [2]. In this system, we have proposed such kind of system to predict the chlorophyll of tea leaves using MLR, ANN, SVR, and KNN.

In the year 2020, an algorithm [3] was designed to calculate the value of citrus leaves chlorophyll. The authors used the LR and ANN to compare and estimate the values of citrus leaf chlorophyll. They reported that the chlorophyll of citrus leaf had been changed with the leaf color and age of the leaf. They selected the ANN model as the best model with an average accuracy of 96%. They used a spectrophotometer to validate the result of the citrus chlorophyll. Apart from spectrophotometer, Mahmoodi et al. [4] used a SPAD meter to estimate the chlorophyll content of walnuts variants. They estimated the chlorophyll digitally using LR with a regression score of 0.87. In a study [5], presented a modest and easily operatable technique that estimates the chlorophyll using a smartphone. They captured the images using a contact imaging. The chlorophyll values were calculated manually using a spectrophotometer and fitted with the color features of the leaves using LR (84% accuracy). Vesali et al. [6] presented an android application to calculate the chlorophyll content in a corn leaf. They captured 480 contact images and extracted the RGB of the leaf images. The digitized values of chlorophyll were calculated using LR and ANN with an accuracy of 74% and 82%, respectively. Mohan and Gupta [1] described a new method to obtain the chlorophyll of rice leaf images under the field environment using a smartphone. They used SPAD for actual chlorophyll of rice leaf. They extracted the RGB values from the rice leaf and train the ANN model with the actual chlorophyll value. The model predicted the chlorophyll of the leaf with an accuracy of 77%. Peng and Wang [7] also used the SAPD meter for actual chlorophyll. They used the stacked spare auto encoder and ANN to predict the chlorophyll. Agarwal and Gupta [2] showed the application of principal component analysis and agglomerative hierarchical clustering analysis for chlorophyll prediction (85% accuracy). Ali et al. [8] proposed a chlorophyll prediction system to measure the chlorophyll of three different species of lettuce, broccoli, and tomato using image processing techniques. They used SPAD-502 chlorophyll meter to determine total chlorophyll in the leaf. The LR was used as a predictive algorithm. They found 0.96 in tomato, 0.89 in lettuce, and 0.91 in broccoli to be the best fit. A new method of chlorophyll prediction was reported [9] for betel vine's leaf using LR. They used MATLAB 2010 to acquire the values of RGB of leaf images and fitted these values with the actual chlorophyll values to estimate the chlorophyll content of the leaf.

The application of LR and ANN was reported in most of the research on chlorophyll prediction. The authors follow the same research methodology in their studies of chlorophyll prediction such as image acquisition, actual chlorophyll prediction using SPAD or spectrophotometer, feature extraction of the leaf, and prediction of chlorophyll using machine learning techniques. In this study also, we have been used the same methodology but compare the result of LR, ANN, SVR, and KNN. The proposed research contributes to the following for the digital chlorophyll prediction.

(i) We have predicted the tea leaf chlorophyll using machine learning using LR, ANN, KNN, and SVR.
(ii) The paper proposed a new image acquisition method using Nikon DLSR 3500 in environmental light conditions.
(iii) The paper compares the results of LR, ANN, SVR, and KNN.
(iv) The results of the models are compared using MAE, MSE, RMSE, and $R^2$.

## 2 Materials and Methods

### *2.1 Image Acquisition and SPAD Estimation*

The Assam tea leaf is selected for the research. The tea leaves are collected from Kokrajhar tea estate, Assam. A total of 750 Assam tea leaves are collected from the tea estate. The actual chlorophyll of tea leaf is estimated using SPAD-502 m and the 750 values of the tea leaf chlorophyll are saved in.csv file for further processing (Fig. 1). The chlorophyll values of Assam tea leaves are in the range of 9–73.7 with an average value of 35.221. The low chlorophyll shows low greenness in the leaf [3] and they may consider as immature tea leaves. But the tea leaf with high chlorophyll may consider an as mature leaf.

The images of the tea leaves are captured using NIKON D3500, placing the leaf on a white sheet with an exposure of 1/60, fixing the aperture at 2.8, a focal length of 50.0 mm, and ISO speed of 800. The images are captured in September in the

(a) (b) (c) (d)

**Fig. 1** **a–b** Sample of tea leaf, **c** image acquisition process, **d** SPAD estimation afternoon. A total of 750 images in RGB color space are captured by the camera and are transferred to a desktop computer for further analysis and model development [8]. All the images are in the high dimension of 6000 × 4000 with 72 dpi and stored in.jpg format (Fig. 1).

## 2.2 *Image Pre-processing and Features Extraction of the Tea Leaf Images*

The previous research showed the relation of leaf color and chlorophyll values [3]. In this paper, we have been extracted the color features of the tea leaf and fitted these values with the actual chlorophyll of tea leaf for the new prediction. Before the feature extraction, the images are resized into 500 × 500 for CPU processing. Since we deal with the chlorophyll prediction of tea leaves in natural environmental light conditions, we have not applied other pre-processing steps in our research. In the proposed system, we have extracted the different color features of the image because the color purity indicates the amount of chlorophyll present in the tea leaf. For calculating the color features, the mean of RGB, standard deviation (SD) of RGB and HSV, variance, kurtosis, and skewness of the leaf image are determined. A total of 12 different color features are extracted from the tea leaf image (Table 1). The average values of mean_R, mean_G, mean_B, SD_R, SD_G, SD_B, SD_H, SD_S, SD_V, variance, kurtosis, and skewness are 163.178, 167.1203, 157.0623, 43.7784, 35.6811, 57.1744, 48.1189, 69.6882, 36.4821, 2364.989, 7.1189, and –2.180, respectively. The mean values of all features show that the values of features are in different ranges. The value of variance is very high as compared to the other values of the features. So, there may be a possibility of influence by the variance in

**Table 1** Data without normalization

| Parameters | Count | Mean | Std. | Min. | Max. |
|---|---|---|---|---|---|
| mean_R | 750 | 163.17 | 36.16 | 2.25 | 208.4 |
| mean_G | 750 | 167.12 | 36.56 | 2.84 | 212.27 |
| mean_B | 750 | 157.06 | 35.8 | 1.24 | 202.98 |
| SD_R | 750 | 43.77 | 12.85 | 16.04 | 75.52 |
| SD_G_ | 750 | 35.68 | 10.54 | 12.75 | 65.00 |
| SD_B | 750 | 57.17 | 17.05 | 15.7 | 89.81 |
| Variance | 750 | 2364.98 | 1209.4 | 251.7 | 5790.7 |
| SD_H | 750 | 48.11 | 10.79 | 5.94 | 83.72 |
| SD_S | 750 | 69.68 | 14.55 | 2.95 | 103.07 |
| SD_V | 750 | 36.48 | 11.23 | 13.09 | 75.38 |
| Kurtosis | 750 | 7.11 | 10.24 | −0.9 | 156.22 |
| Skewness | 750 | −2.18 | 1.69 | −5.33 | 11.84 |

the prediction result. We have done the data normalization to overcome this issue before the prediction using the MinMax method. For each feature value, the MinMax method subtracts the value of the minimum value of the respective feature from the individual value of the feature and then divide the result by the difference between the maximum value and minimum value of the feature. In the previous study, authors [3, 6, 8] calculated only the RGB values of the images and then combined these values to form different color indexes to relate with the actual chlorophyll values. But in this paper, we have calculated the different color features without combining the RGB and HSV values. We also relate the variance, skewness, and kurtosis of the tea image with the tea chlorophyll.

## 2.3 Tea Chlorophyll Prediction Using MLR, ANN, SVR and KNN

In Sect. 2.2, we have already presented the data normalization of tea leaf features. In this section, we have explained the chlorophyll prediction using MLR, ANN, SVR, and KNN. Before the implementation of the prediction algorithm, we have divided the normalized dataset into two training and testing set as a ratio of 80:20. It means that 80% of the 750 leaf features (600 values) are used for training and 20% of the 750 (150 values) leaf features are used for testing.

At first, the prediction is done using MLR. We have fitted the 12 features of the tea leaf image with the actual SPAD estimation. The regression algorithm is reported between the chlorophyll value of the tea leaves and the features of the tea leaf images using the Eq. 1 where $x1$, $x2$, $x3$, $x4$, $x5$, $x6$, $x7$, $x8$, $x9$, $x10$, $x11$, and $x12$ denotes the 12 extracted features of the tea leaf image. The different regression coefficient of mean_R, mean_G, mean_B, SD_R, SD_G, SD_B, SD_H, SD_S, SD_V, variance, kurtosis and skewness are 45.45, −80.06, 3.41, 75.02, −8.39, 46.86, −4.06, −53.04, 40.81, −70.62, 11.63, and −43.60, respectively. The proposed MLR method predicts tea leaf chlorophyll based on the best fit line (Fig. 2 and Table 2).

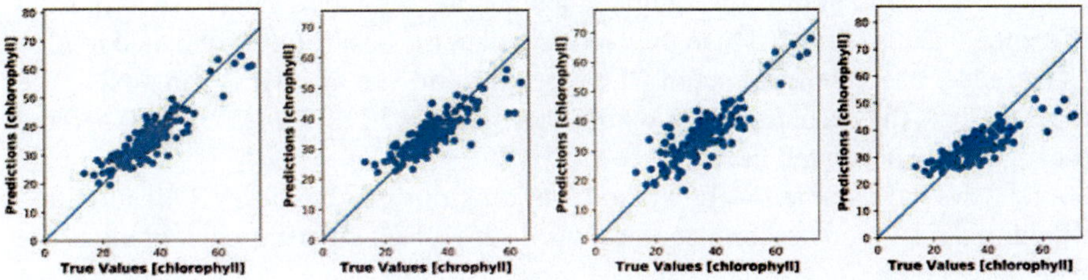


(a) MLR (b) ANN (c) KNN (d) SVR

**Fig. 2** Regression analysis of the MLR, ANN, KNN, and SVR

**Table 2** Actual chlorophyll versus predicted values (random 5 values)

| S. No. | Actual | MLR predicted | SVR predicted | KNN predicted | ANN predicted |
|---|---|---|---|---|---|
| 0 | 36.5 | 37.855538 | 34.713873 | 35.366667 | 37.18 |
| 1 | 27.3 | 25.616577 | 26.174367 | 27.9 | 26.01 |
| 2 | 29.8 | 29.342354 | 32.361575 | 30.6 | 26.05 |
| 3 | 31.5 | 33.71819 | 31.425221 | 28.366667 | 35.08 |
| 4 | 35.8 | 34.232223 | 34.799888 | 31.066667 | 33.25 |

**Table 3** Model evaluation for tea leaf prediction

| Algorithms | MAE | MSE | RMSE | $R^2$ |
|---|---|---|---|---|
| MLR | 3.375 | 19.133 | 4.374 | 0.801 |
| ANN | 3.75 | 30.03 | 5.54 | 0.502 |
| SVR | 4.86 | 44.58 | 6.67 | 0.538 |
| KNN | 4.05 | 28.83 | 5.36 | 0.701 |

$$\text{Tea Chlorohyll} = ax1 + bx2 + cx3 + \cdots + nx12 \tag{1}$$

Secondly, the prediction is done using ANN. The ANN is a good supervised technique to predict the unknown parameter. In the ANN model also, the data is divided into a ratio of 80:20. The 12 calculated tea leaf features are passed as an input to the model. The ANN model has two hidden layers with 64 hidden neurons in the first hidden layer and 32 hidden neurons in the second hidden layer. The activation function of both layers is Relu. The output layer contains only one hidden neuron to predict the tea leaf chlorophyll with a linear activation function. A total of 2945 parameters are trained during the training. The model is optimized with an RMSprop optimizer with a learning rate of 0.0001. The regression line and output of the proposed ANN model are presented in Fig. 2 and Tables 2, 3.

Thirdly, the tea chlorophyll is predicted using KNN. The KNN predicts the tea chlorophyll based on the distance metric. In this paper, we have used the Minkowski distance metric and uniform weight to calculate the prediction value. We have tried different $K$ values ($K = 2$–$10$) in the prediction but the best result is obtained in $K = 3$. The model used the same set of training and testing as like MLR and ANN. The model predicts the tea chlorophyll with an accuracy of 70% (Tables 2 and 3) and the best fine line is presented in Fig. 2.

At last, we have applied SVR to predict the tea chlorophyll. The SVR is a modified version of the support vector machine. It is a supervised algorithm that predicts the output based on a hyperplane [10]. The model is implemented with a linear kernel function by considering the $C = 1$, cache size $= 200$, epsilon $= 0.1$, and gamma $=$ auto. The model accuracy of the SVR is 53.8% and it is presented in Fig. 2 and Table 3.

## 3 Discussion

The fitting curve of the regression (Fig. 2) indicates that the model predicts the value of the chlorophyll efficiently only in MLR and KNN. The models are evaluated using the mean absolute error (MAE), mean squared error (MSE), and root mean squared error (RMSE) using the Eq. 2–4 (Table 3). The model accuracy is done in terms of regression score. The value of MAE, MSE, and RMSE and $R^2$ for MLR are 3.375, 19.133, 4.37, and 0.801, respectively. For ANN, 3.75, 30.03, 5.54, and 0.502 are reported for MAE, MSE, RMSE, and $R^2$. The value MAE, MSE, RMSE and $R^2$ for SVR and ANN are 4.86, 44.58, 6.67, 0.538, 4.05, 28.83, 5.36, and 0.701, respectively. The values presented in Table 3 show that the MLR model outperforms as compared to the other algorithm in tea leaf chlorophyll prediction. The regression score of MLR and KNN is acceptable and it efficiently predicts the chlorophyll. The KNN algorithm is tried for different $K$ values ($K = 2$–$10$). But for $K = 3$, the model performs well as compared to the other $K$ values. The regression score of the KNN with different $K$ values are 0.65 for $K = 2$, 0.70 for $K = 3$, 0.695 for $K = 4$, 0.694 for $K = 5$, 0.693 for $K = 6$, 0.692 for $K = 7$, 0.692 for $K = 9$, and 0.698 for $K = 10$. These values indicate that the KNN model outperforms at $K = 3$.

$$\text{MAE} = \frac{1}{n}\sum |\text{actual chlorophyll} - \text{predicted chlorophyll}| \tag{2}$$

$$\text{MSE} = \frac{1}{n}\sum (\text{actual chlorophyll} - \text{predicted chlorophyll})^2 \tag{3}$$

$$\text{RMSE} = \sqrt{\frac{1}{n}\sum (\text{actual chlorophyll} - \text{predicted chlorophyll})^2} \tag{4}$$

In the previous study, Ali et al. [8] used LR techniques to predict the chlorophyll of lettuce, broccoli, and tomato and reported an accuracy of 96, 89, and 91% for tomato, lettuce, and broccoli, respectively. Again an accuracy of 68% and 74% were reported by Vesali et al. [6] using LR. In this paper, we have reported the use of MLR with an MAE of 3.375, MSE of 19.13, and RMSE of 4.37. It means that the model is considered as the best model with an accuracy of 80.18. Again, the authors reported accuracy of more than 80 and 90% with LR and ANN [3] because they had captured the images of the citrus using a smartphone and a contact imaging method. In the contact imaging method, no variation and interference is found in between the image and camera. But in this paper, we have used a digital camera to capture the tea leaf images. Due to the light noise, the result of the MLR lies in 80% but it predicts the chlorophyll well as compared to the other one. It can be considered an easy method from a farmer's perspective and can deploy the model as an android app for the same. Most of the previous studies are based on RGB color analysis [11, 2]. In the proposed system, we have focused on different color features of an image. The feature vector of the proposed system is not only based on the RGB values but also depends on the other color parameters such as HSV, variance, skewness, and kurtosis. An Accuracy of 91% was reported by Ali et al. [8] for the RGB color-based prediction of broccoli chlorophyll. The results of the present study proved that LR is a good technique to predict the chlorophyll of tea leaves.

## 4 Conclusions

The present study of chlorophyll estimation is done in Assam tea. The result of the proposed system shows the relation of different color features with the tea chlorophyll. The proposed system proved that the chlorophyll of any plant may be obtained from different color features of the leaf image including RGB. The result of the proposed system can be increased with ample numbers of tea leaf images. The MLR model can be treated as a replica of SPAD with an accuracy of 80%. The model can further be improved using the deep neural network model and a convolutional neural network model.

**Acknowledgements** This research project is supported by All India Council for Technical Education under RPS (NER), vide Ref. No.: File No. 8-131/RIFD/RPS-NER/POLICY-1/2018-19 dated March 14, 2019.

## References

1. Mohan PJ, Gupta SD (2019) Intelligent image analysis for retrieval of leaf chlorophyll content of rice from digital images of smartphone under natural light. Photosynthetica 57:388–398
2. Agarwal A, Gupta SD (2018) Assessment of spinach seedling health status and chlorophyll content by multivariate data analysis and multiple linear regression of leaf image features. Comput Electron Agric 152:281–289
3. Barman U, Choudhury RD (2020) Smartphone image based digital chlorophyll meter to estimate the value of citrus leaves chlorophyll using linear regression, LMBP-ANN, and SCGBP-ANN. J King Saud Univ Comput Inf Sci
4. Mahmoodi M, Khazaei J, Vahdati K, Mohamadi N, Javanmardi Z (2013) Chlorophyll content estimation using image processing technique. World Appl Sci J
5. Barman U, Choudhury RD, Saud A, Dey S, Pratim MB, Gunjan BG (2018) Estimation of chlorophyll using image processing. Int J Recent Sci Res 9(3):24850–24853
6. Vesali F, Omid M, Kaleita A, Mobli H (2015) Development of an android app to estimate chlorophyll content of corn leaves based on contact imaging. Comput Electron Agric 116:211–220
7. Peng Y, Wang Y (2019) Prediction of the chlorophyll content in pomegranate leaves based on digital image processing technology and stacked sparse autoencoder. Int J Food Prop 22(1):1720–1732
8. Ali MM, Al-Ani A, Eamus D, Tan DK (2012) A new image processing based technique to determine chlorophyll in plants. Am Eurasian J Agric Environ Sci 12(10):1323–1328
9. Dey AK, Sharma M, Meshram MR (2016) An analysis of leaf chlorophyll measurement method using chlorophyll meter and image processing technique. Procedia Comput Sci 85:286–292

10. Barman U, Choudhury RD (2019) Soil texture classification using multi class support vector machine. In: Information processing in agriculture
11. Vesali F, Omid M, Mobli H, Kaleita A (2017) Feasibility of using smart phones to estimate chlorophyll content in corn plants. Photosynthetica 55(4):603–610

# Comparison of LR, SVR, and DNN for the Rainfall Forecast of Guwahati, Assam

**Utpal Barman, Diganto Sahu, and Golap Gunjan Barman**

**Abstract** Forecasting of rainfall is an important research area from a farmer's perspective. This paper presents the comparative assessment of the linear regression (LR), support vector regression (SVR), and deep neural network (DNN) for the forecasting of rainfall for Guwahati. For the research, the daily rainfall of the Guwahati region is collected from Regional Meteorological Center, Guwahati, Assam, India. The daily rainfall data is predicted based on the value of time, minimum temperature, maximum temperature, humidity, STP, mean sea level, STN, and wind speed. The algorithms are compiled with the different ratios of the training and testing set of rainfall data. The mean absolute error (MAE), mean squared error (MSE), and root mean squared error (RMSE) are reported to compare the LR, SVR, and DNN. The DNN model performs well as compared to the LR and SVR based on MSE and RMSE, but the SVR model outperforms based on the value of MAE. Both models can be considered as the best model irrespective of the ratio of the train and test set of rainfall data.

**Keywords** Linear regression · Support vector regression · Deep neural network · Rainfall · Forecast

## 1 Introduction

Assam is an agriculture-based state. The farmers of Assam always faced problems due to the artificial floods in Assam. Artificial floods are due to the cause of heavy

---

U. Barman (✉) · D. Sahu · G. G. Barman
Department of Computer Science and Engineering, Girijananda Chowdhury Institute of Management and Technology, Guwahati, Assam, India
e-mail: utpalbelsor@gmail.com

D. Sahu
e-mail: digantosahu@gmail.com

G. G. Barman
e-mail: barmangolap15@gmail.com

 
A. Maji et al. (eds.), *Proceedings of the International Conference on Computing and Communication Systems*, Lecture Notes in Networks and Systems 170,
https://doi.org/10.1007/978-981-33-4084-8_28

rainfall. It degrades the quality of the and quantity of farming. So, it is necessary to forecast the rainfall with an efficient method. In this paper, we have presented the comparative assessment of linear regression (LR), support vector regression (SVR), and deep neural network (DNN) for the forecasting of rainfall for Guwahati.

In the previous works, authors did valuable works in the rainfall forecast. Dounia et al. [1] investigated the application ANN for rainfall forecast in Algeria and reported 0.95 regression score (RC) for the correlation. The paper [2] reported the application of an adaptive neuro–fuzzy inference system (ANFIS) in rainfall prediction. They reported a 4.64% mean square percentage error (MAPE) and 0.1277 RC for the forecast. The author [3] used SVR for rainfall forecasting in Bangladesh. They demand that the model predicts the rainfall well with an accuracy of 99.92%. They evaluated the model with RMSE (18.60) and MAE (0.18). The RMSE (24.95) and MAE (4.99) are also reported for the rainfall prediction before seven days. In paper [4], the authors reported the application of optimized SVR. They recorded 0.3426 mean absolute percentage error (MAPE) for the rainfall prediction in China. The authors [5] investigated the rainfall data of Japan using ANN. They applied the multi-layered perceptron of ANN for the forecasting. The application of the ANN model is also reported by Abhishek et al. [6] for weather forecasting. The authors [7] reported the application of SVR and ANN for weather prediction using the past six years of data. They reported the SVR as the best model for forecasting. The authors [8] also reported the application of SVR fort the rainfall forecasting in Changchun, China. They evaluated the model based on the MAPE (14.90). The investigators of paper [9] investigated on LR for rainfall forecasting. They examined the model on 48 stations by taking the rainfall data of South Australia from 1959 to 1995.

In the above literature, it is observed that the application of SVR, LR, and ANN is more in the rainfall forecast. The result of SVR and ANN is more applicable and accurate than the result of LR [6, 8]. In this paper, we have presented the comparative assessment of LR, SVR, and ANN depending on the 14 different input parameters of the rainfall. The paper forwards the assessment based on the different ratios of the training and testing set of rainfall data.

## 2 Materials and Methods

### 2.1 About Dataset

The dataset of this paper is collected from the Regional Meteorological Center, Guwahati, Assam, India. It contains the daily rainfall data from January 1, 2017, to December 31, 2018. Along with the daily rainfall, the others rainfall dependent factors such as time, minimum temperature, maximum temperature, humidity at 8.30 am, humidity at 5.50 pm, STP_12, STP_48, MSL_12, MSl_48, STN_12, STN_48, and wind speed are also present in the dataset for the period. In the dataset, we separated the time parameters as three other parameters like year, date, and month.

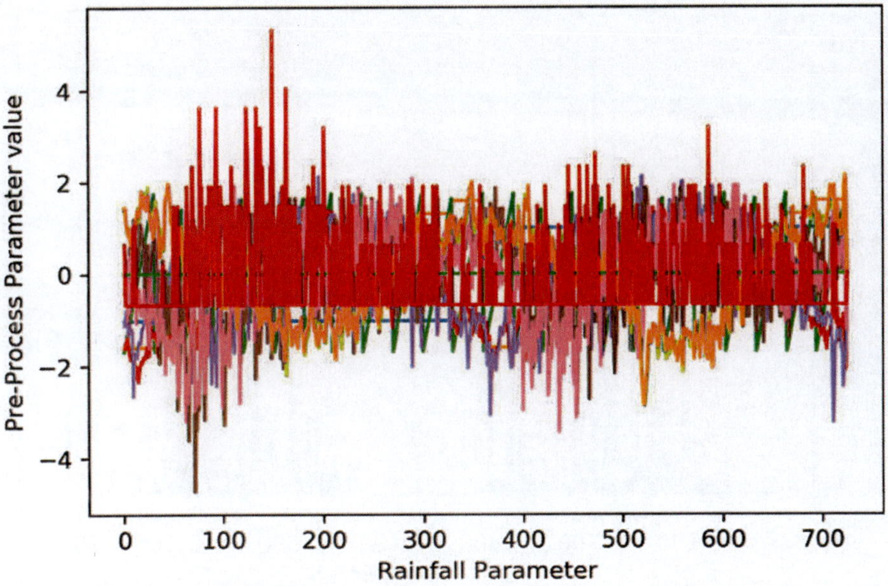


**Fig. 1** Preprocessed rainfall parameter

Here, rainfall is considered as the target value, and the other 14 parameters are considered as input variables.

## 2.2 *Preprocessing of Rainfall Data Dataset*

The dataset does not contain any missing values, but the individual value of the STN, STP, and year is more as compared to the individual value of the other parameters. It may sometimes influence the prediction result. So, to reduce sparsity, we have applied the preprocessing. Scale () algorithm of sklearn to put the data in one scale. This process makes the zero mean and unit variance to the input parameters. The preprocessed data of rainfall is presented in Figs. 1 and 2.

## 2.3 *Rainfall Prediction Using LR, SVR, and DNN*

After the preprocessing of the data, the rainfall is predicated using LR, SVR, and DNN. The predicated algorithms are applied to the different ratio of the rainfall dataset, and their performance is evaluated based on different evaluating parameters. The division of the dataset is presented in Table 1.

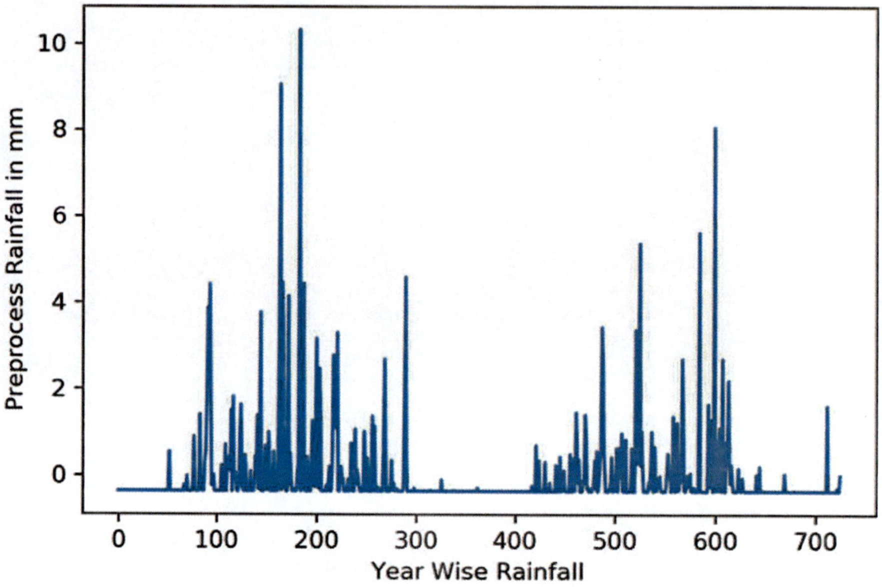


**Fig. 2** Year-wise preprocessed rainfall data

**Table 1** Different ratio of training and testing set of the rainfall dataset

| Rainfall dataset | Training (%) | Validation (%) |
|---|---|---|
| Set 1 | 60 | 40 |
| Set 2 | 70 | 30 |
| Set 2 | 80 | 20 |
| Set 4 | 90 | 10 |

The LR is a simple technique to predict the unknown value depending on the dependent value. Here, we have applied the multiple LR as the dataset contains 14 different independent parameters and 1 dependent parameter (Eq. 1). In Eq. 1, the value ×1, ×2, ×3, ×4, ×5, ×6, ×7, ×8, ×9, ×10, ×11, ×12, ×13, and ×14 denote the 14 different independent parameters of the rainfall. Here, the LR is applied using the sklearn of python.

$$\text{Rainfall} = a \times 1 + b \times 2 + c \times 3 + \cdots + n \times 14 \quad (1)$$

Like the LR, the SVR is also applied to forecast the rainfall. The SVR is applied using the sklearn package of Python. The SVR algorithm takes the concept of support vector machine, but it demands to set the value of margin of tolerance (Epsilon). Here, we have considered the value of epsilon as 0.2. The other parameters of the SVM such as gamma and C are kept as 'auto' and 2, respectively, for the experiment. The exact value of gamma and C determines the performance of the SVM, otherwise,

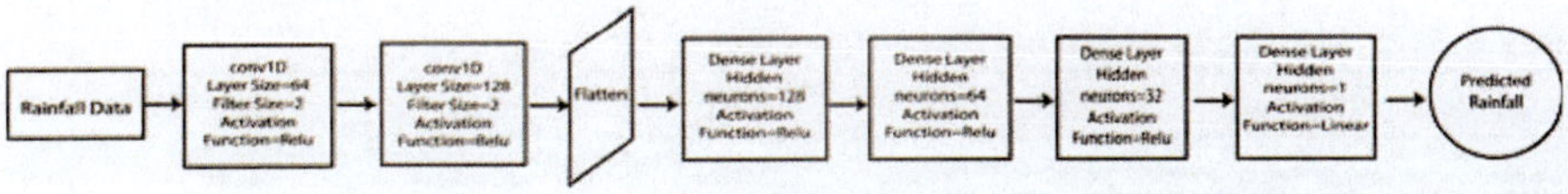


**Fig. 3** Block diagram of DNN

it leads to overfitting and underfitting of the model. So, keeping in mind, we have analyzed the SVR with the different values of gamma and C, but it gives the best performance with gamma = 'auto' and $C = 2$.

Finally, the forecasting is done using DNN (Fig. 3). The input shape of the model is (14, 1). In the first step, a one-dimensional convolution layer (Conv1D) with 64-layer size, 2-filter size, and ReLu activation function is applied in the model. In the next step, another Conv1D is applied in the model with 64-layer size, 2-filter size, and ReLu activation function. Then, the model is flattened. In the next step, three hidden (Dense) layers are applied in the model continuously with 128, 64, and 32 hidden neurons, respectively. The activation function of the hidden layer is ReLu. Finally, the output layer is defined with a single hidden neuron and linear activation function to forecast the rainfall. The architectural step of the DNN is presented below.

**Architectural Steps of DNN for rainfall Forecast**
**Input**: 14 different input parameters (input shape)
**Output**: Rainfall forecast

Step 1: Add first Conv1D (Layer size = 64, Filter size = 2, AF = ReLu)
Step 2: Add second Conv1D (Layer size = 128, Filter size = 2, AF = ReLu)
Step 3: Do the model flatten
Step 4: Apply the dense layer (Hidden neurons = 128, AF = ReLu)
Step 5: Apply the dense layer (Hidden neurons = 64, AF = ReLu)
Step 6: Apply the dense layer (Hidden neurons = 32, AF = ReLu)
Step 7: Final layer (Hidden neurons = 1, AF = linear).

After the implementation of the model in the Python environment, the models are evaluated using the mean absolute error (MAE), mean squared error (MSE), and root mean squared error (RMSE) using Eqs. (2)–(4).

$$\text{MAE} = \frac{1}{n}\sum |\text{actual rainfall} - \text{predicted rainfall}| \tag{2}$$

$$\text{MSE} = \frac{1}{n}\sum (\text{actual rainfall} - \text{predicted rainfall})^2 \tag{3}$$

$$\text{RMSE} = \sqrt{\frac{1}{n}\sum (\text{actual rainfall} - \text{predicted rainfall})^2} \tag{4}$$

The result of the three prediction algorithms is presented in Table 2. Table 2 shows the evaluating parameters of the model for the different sets of training and testing.

**Table 2** MAE, MSE, and RMSE of LR, SVR, and DNN

| Training/testing ratio | Algorithm | MAE | MSE | RMSE |
|---|---|---|---|---|
| 50:50 | LR | 0.5428 | 1.2315 | 1.109 |
| | SVR | 0.4637 | 1.156 | 1.0756 |
| | DNN | 0.457 | 0.870 | 0.932 |
| 60:40 | LR | 0.523 | 0.973 | 0.986 |
| | SVR | 0.404 | 0.923 | 0.962 |
| | DNN | 0.442 | 0.807 | 0.8983 |
| 70:30 | LR | 0.538 | 1.000 | 1.001 |
| | SVR | 0.385 | 0.925 | 0.961 |
| | DNN | 0.457 | 0.912 | 0.955 |
| 80:20 | LR | 0.536 | 1.213 | 1.101 |
| | SVR | 0.358 | 1.068 | 1.033 |
| | DNN | 0.396 | 1.032 | 1.016 |
| 90:10 | LR | 0.418 | 0.342 | 0.585 |
| | SVR | 0.238 | 0.215 | 0.464 |
| | DNN | 0.238 | 0.217 | 0.466 |

## 3 Discussion

From the evaluating parameters, it is found that the models are performed well. The application of LR is reported for rainfall prediction with wavelet transformation [9]. Here, the result of multiple LR is given in Table 2. In Table 2, it is shown that the MAE, MSE, and RMSE of LR at 50% testing size is 0.542, 1.231, and 1.109, respectively. The MAE, MSE, and RMSE values of multiple LR are 0.523, 0.973, and 0.986, respectively, at a 40% testing set of Guwahati rainfall data. The values of MSE are 1.00, 1.213, 0.342 at the testing test of 30%, 20%, and 10%, respectively (Table 2). The RMSE values of multiple LR at 10% of the testing set (0.585) is less as compared to the RMSE values of the 20% (1.101), 30% (1.001), and 40% (0.986) of test data. Here, it is found that the result of SVR is more efficient than the result of LR. The previous researches [3, 4, 10] also showed good accuracy with SVR in their investigation of rainfall forecast. They achieved very less error in their research such as 0.3426 (MAPE) [4] and 0.18 (MAE) [3]. In this research, we have achieved less MAE, MSE, and RMSE for all the sets of training and testing. The MAEs of SVR are 0.4637 at 50% of testing set, 0.404 at 40% of testing set, 0.385 at 30% of testing set, 0.358 at 20% of testing set, and 0.238 at 10% of the testing set of rainfall data (Table 2). The MSEs of SVR are reported as 1.156 for 50%, 0.923 for 40%, 0.925 for 30%, 1.068 for 20%, 0.215 for 10% of testing test (Table 2). The values 1.075 for 50%, 0.962 for 40%, 0.961 for 30%, 1.033 for 20%, 0.464 for 10% of testing set are reported in the investigation of RMSE (Table 2).

The result of DNN is the same as the result of SVR with a 90% training test. Since we have taken more samples in training (90%), it shows a good result with an MAE value of 0.238, an MSE value of 0.217, and an RMSE value of 0.466. With the training set of 80, 70, 60, and 50%, the MAE value of the DNN model is 0.396, 0.457, 0.442, and 0.457, respectively. From Table 2, it is shown that the DNN model has done well only with 90% and 50% training data as compared to SVR and LR. The DNN model also shows the good result in 80% training data as compared to LR with an MAE value of 0.396, MSE value of 1.032, and RMSE value of 1.016. The MSE values of the DNN model are 0.870, 0.807, 0.912, 1.023, and 0.217 at 50%, 40%, 30%, 20%, and 10% of testing set, respectively. The result shows that the DNN model can be considered as the best model with 10% of the testing set. The RMSE value (0.466) of the DNN model also shows it as the best model at 10% of testing data.

From Table 2, the reported values of MAE (0.457), MSE (0.870), and RMSE (0.932) of DNN show that the DNN model outperforms than LR and SVR at 50% of the testing set. The MAE (0.404) of SVR at 40% of the testing set choose the model as best model as compared to the LR and DNN, but MSE (0.80) and RMSE (0.89) of DNN show that the model DNN is best as compared to the LR and SVR. The MAE (0.385) value of SVR gain proved that the SVR is the best model for rainfall prediction at 30% of testing data, but the values of MSE (0.91) and RMSE (0.95) select the DNN as the best model. At 20% testing set, again the reported MAE (0.35) of SVR is less as compared to the MAE (0.39) of DNN. The less value of MSE (1.03) and RMSE (1.01) of DNN is reported as compared to SVR and LR. Both the models (SVR and DNN) perform the same at 10 of testing data of rainfall.

## 4 Conclusion

In this research, we have presented the performance of LR, SVR, and DNN for the parameters-based rainfall prediction. The rainfall is forecasted using 14 different parameters, and the results are compared based on three evaluating parameters. The model DNN can be considered as the best model irrespective of the training and testing set based on the values of MSE and RMSE. These two parameters can be considered as good evaluating parameters for rainfall prediction. But the evaluating parameter MAE shows the SVR model as the best model for the rainfall prediction in Guwahati, Assam. All the model outperforms, but the DNN can be considered as the best model for rainfall prediction.

**Acknowledgements** This research project is supported by Assam Science and Technology University, Guwahati, Assam under TEQIP-III, vide Ref. No.: ASTU/TEQIP-III/Collaborative Research/2019/2474, Dated July 17, 2019.

## References


1. Dounia M, Dairi S, Djebbar Y (2014) Rainfall-rain off modeling using artificial neural network. ScienceDirect
2. Danladi A, Stephen M, Aliyu BM, Gaya GK, Silikwa NW, Machael Y (2018) Assessing the influence of weather parameters on rainfall to forecast river discharge based on short-term. Alexandria Eng J 57(2):1157–1162
3. Hasan N, Nath NC, Rasel RI (2015) A support vector regression model for forecasting rainfall. In: 2015 2nd international conference on electrical information and communication technologies (EICT), pp 554–559
4. Zhao S, Wang L (2010) The model of rainfall forecasting by support vector regression based on particle swarm optimization algorithms. In: Life system modeling and intelligent computing. Springer, Berlin, pp 110–119
5. Kashiwao T, Nakayama K, Ando S, Ikeda K, Lee M, Bahadori A (2017) A neural network-based local rainfall prediction system using meteorological data on the Internet: a case study using data from the Japan Meteorological Agency. Appl Soft Comput 56:317–330
6. Abhishek K, Singh MP, Ghosh S, Anand A (2012) Weather forecasting model using artificial neural network. Procedia Technol 4:311–318
7. Rasel RI, Sultana N, Meesad P (2017) An application of data mining and machine learning for weather forecasting. In: International conference on computing and information technology, pp 169–178
8. Ouyang Q, Lu W, Xin X, Zhang Y, Cheng W, Yu T (2016) Monthly rainfall forecasting using EEMD-SVR based on phase-space reconstruction. Water Resour Manage 30(7):2311–2325
9. He X, Guan H, Zhang X, Simmons CT (2014) A wavelet-based multiple linear regression model for forecasting monthly rainfall. Int J Climatol 34(6):1898–1912
10. Adhani G, Buono A, Faqih A (2013) Support vector regression modelling for rainfall prediction in dry season based on southern oscillation index and NINO3. 4. In: 2013 international conference on advanced computer science and information systems (ICACSIS), pp 315–320

# A Study on Classification Accuracy of Different Features in Identification of Parasite Eggs from Microscopic Images of Fecal Samples

**Kaushik Ray, Sarat Saharia, and Nityananda Sarma**

**Abstract** Selecting and extracting a robust feature set is very much important in automatic identification or classification of parasitic eggs in microscopic images. In this paper, we have explored different types of features and used in identification of the segmented microscopic images of three different types of parasitic eggs namely Ascaris Lumbricoides, Necator Americanus and Trichuris Trichiura. We have extracted four types of image moments as four different feature sets from the segmented gray-scale images of parasite eggs. We have also extracted several features from gray level pixel values and some texture features along with a few shape-based features to automatically identify the above mentioned types of parasite eggs. The performances of the different feature sets are examined using three different classifiers viz SVM, ANN and kNN. The highest classification accuracy of 96.5% is obtained by SVM using texture and shaped based features.

**Keywords** Feature · Moments · Texture · Pixel intensity · Ascaris · Necator · Trichuris · Parasite egg · Microscopic image · Classification · Identification

## 1 Introduction

Automatic identification of different types of parasite eggs from microscopic images is a challenging task due to the variations of size, shape and texture of the eggs. Many works reported in the past used image processing and machine learning techniques to develop systems that can effectively identify various species of parasite eggs from microscopic images. However, like any other pattern recognition task, selection and

K. Ray · S. Saharia (✉) · N. Sarma
Tezpur University, Tezpur, Assam, India
e-mail: sarat@tezu.ernet.in

K. Ray
e-mail: kray90@tezu.ernet.in

N. Sarma
e-mail: nitya@tezu.ernet.in



A. Maji et al. (eds.), *Proceedings of the International Conference on Computing and Communication Systems*, Lecture Notes in Networks and Systems 170,
https://doi.org/10.1007/978-981-33-4084-8_29

extraction of a robust feature set which can discriminate the parasite eggs of different species is very critical. Due to the image noise, presence of impurities and various objects that look like parasite eggs, it becomes a complex task to extract a robust set of features from microscopic images. Microscopic images prepared from the fecal samples are one of the most popular kinds of images that are used to analyze and detect parasite eggs. Each of these images may contain a single or multiple eggs of same or different types. Usually these images contain a lot of fecal impurities along with parasite eggs which makes the process of detection and identification of those eggs difficult. Therefore, a strong image analysis process is required to identify the most relevant features. Color of the images may vary and so color of the eggs. Size of the eggs vary depending on the zooming effect of camera, texture of the eggs may also change due to the presence of noise and sometimes eggs may fully or partially covered with impurities. These are some issues related to the microscopic images of parasite eggs that needs to be addressed before extracting a feature set. During the last two decades, researchers have introduced many different features for identification of various types of parasite eggs. But the same type of features may not be effective for identifying many different species of parasite eggs. Hence it is necessary to choose a robust feature set that can be use to classify a many of different types of parasite eggs from microscopic images.

In this paper, we have explored and experimented with some popular features like Hu's invariant moments [1], GLCM based texture features [2] and gray level pixel intensity based feature sets to classify eggs of Ascaris Lumbricoides, Necator Americanus and Trichuris Trichiura parasites in microscopic images of fecal matters of pigs. We have also used three different types of well-known image moments as feature sets and investigated their behavior in the classification of parasite eggs.

The remaining parts of this paper are divided in multiple sections as follows: Sect. 2 describes about the materials and methodology used in this paper, Sect. 3 states about the different types of features extracted in our work, Sect. 4 provides the classification results obtained from the experiments and lastly we draw the conclusion of our work in Sect. 5.

## 2 Materials and Methodology

We have used microscopic images of fecal samples of pigs that contain three types of parasite eggs namely Ascaris Lumbricoides, Necator Americanus and Trichuris Trichiura. A total of 1350 images are used where, 800 images mostly having the eggs of only Ascaris, 250 of Necator and 300 images of Trichuris. Most of these images contain more than a single parasite egg of same species along with fecal impurities or non-egg substances. Sample images of our data-set containing eggs of above mentioned parasites are shown in Fig. 1.

In order to identify all the eggs present in an image, several operations viz image pre-processing, segmentation, feature extraction form the segmented objects and classification are need to be performed. Here, we have used the segmentation method

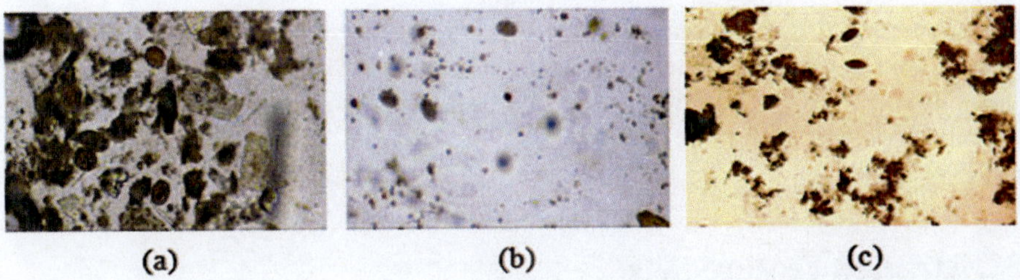

Fig. 1 Images containing parasite eggs. **a** Ascaris lumbricoides [3], **b** Necator americanus [3] and **c** Trichuris trichiura [4]

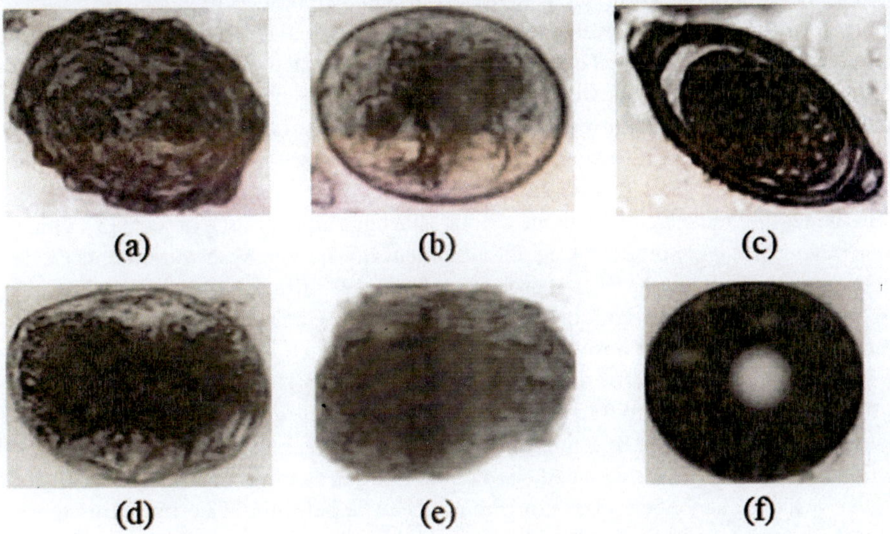

**Fig. 2** Segmented images of Parasite egg of **(a)** Ascaris lumbricoides, **(b)** Necator Americanus and **(c)** Trichuris Trichiura, **(d–f)** segmented non-egg objects

used in one of our previous work [3] that segments out several objects including the parasitic eggs as well as some non-egg objects from the original images in gray-scale. Among these segmented objects 1453 of them are Ascaris eggs, 348 Necator eggs, 384 Trichuris eggs and 1865 non-egg objects. Example of some of the segmented images of parasite eggs and non-egg objects are shown in Fig. 2. Before extraction of any kinds of feature from these segmented objects, we performed data augmentation by rotating the segmented images in 45°, 90°, 135°, 180° and 270°. From all the original and rotated segmented images, a total of 11362 images are selected for feature extraction. Among these images 3590 images are of Ascaris, 1808 images are of Necator eggs, 2239 images are of Trichuris eggs and 3725 are of non-egg objects. Different types of feature sets are extracted from these segmented gray scaled images and labeled each feature vector with its respective class of parasite egg or as non-egg object. After the feature extraction and labeling process, the feature matrices are fed to the classifier for training process that classify each feature vector (One single feature vector corresponds to a single segmented object) as the specific type parasite egg or as non-egg object.

## 3 Extracted Feature Sets

After analyzing the images of our dataset and carefully reviewing the different kind of features used in literature, we have extracted six different feature sets: four types of image moments, gray pixel intensity based features and texture based features from the segmented images. The work presented here is an extension of our previous work reported in [3]. Apart from the features used in our previous work, here we have added another moment based feature set and some new textural features. We have also increased the dimensions of the feature vectors used in [3] and analyze their performances. Hu's seven invariant moments [1] are extracted as a feature set from the segmented gray scale images of parasite eggs and non-egg objects [3]. Other three kinds of image moments: Legendre moments [5, 6], Chebyshev moments [7, 8] and Krawtchouk moments [9, 10] are also extracted as different feature sets. Earlier, we have calculated only 10th order of Legendre and Chebyshev moments. In this work, we have extracted and examined the performance of those two as well as of Krawtchouk moments for orders starting from 5 to 50. Apart from these image moments, we have also extracted texture features using gray level co-occurrence matrix (GLCM) described by Haralick [2]. Fourteen different types of texture features are extracted from the gray-scaled segmented images as introduced in [2]. Parasite eggs have a regular shape and this information can be helpful in discriminating them from the non-egg objects. Hence, along with the texture features, we have included three shape based measures such as: circularity, major axis and minor axis length to construct the feature vector. Pixel intensity based feature set is constructed by resizing the segmented gray scale images and dividing into multiple blocks of same sizes as described in [3]. Standard deviation, variance, maximum, minimum and mean pixel intensity values for each block are then calculated. A feature vector is constructed by concatenating these pixel based values of all the blocks of an image. Size of each block used in [3] is only of $30 \times 30$ pixels, where as in this work we have experimented with multiple sized blocks such as $10 \times 10$, $20 \times 20$, $30 \times 30$, $40 \times 40$ and $60 \times 60$ pixels.

## 4 Results and Discussion

To check the effectiveness of the extracted feature sets, we have trained each of the individual feature set with three different classifier viz. Artificial Neural Network (ANN), k-Nearest Neighbor (kNN) and Multiclass-Support Vector Machine (MC-

SVM). In our experiment with ANN, three hidden layers are used for all the cases. We have observed that, increase of hidden layers does not improve the classification accuracy significantly. To choose the value of k in kNN, we performed multiple experiments by randomly choosing $k$ values in the range of 1 to the square root of the number of training samples. From all the experiments we have observed that k value from 80 to 83 shows promising results for all six feature sets. Here, 10-fold cross validation method is used in all the training processes in order to validate the performance of the classifiers.

## 4.1 *Classification Using Different Image Moments*

Classification accuracies obtained from all the three classifiers using Hu's seven moments are shown in Table 1.

Tables 2, 3 and 4 respectively shows the results obtained by ANN, SVM and kNN using the feature sets produced by different orders of Legendre, Chebyshev and Krawtchouk moments.

Since the higher order moment generates high dimensional feature vector, it may impede the classification accuracy. Hence, here we have used the Principal Component Analysis (PCA) technique to reduce the feature dimension of $50^{th}$ order moments of Legendre, Chebyshev and Krawtchouk. These reduced feature sets are than trained with the classifiers. Training process is performed by keeping different numbers of PCA components starting from 50 to 700 for all the three types of moments and observed the obtained classification accuracies. Number of PCA

**Table 1** Classification accuracy of Hu's seven moments

| ANN | SVM | kNN |
|---|---|---|
| 88.6% | 89.4% | 83.3% |

**Table 2** Classification accuracy of Legendre, Chebyshev and Krawtchouk Moments with ANN

| Moment order | Legendre (%) | Chebyshev (%) | Krawtchouk (%) |
|---|---|---|---|
| 5 | 89.8 | 87.5 | 78.6 |
| 10 | 92.7 | 89.6 | 77.4 |
| 15 | 92.1 | 93.6 | 84.5 |
| 20 | 91.3 | 91.7 | 88.1 |
| 25 | 90.0 | 91.2 | 88.6 |
| 30 | 88.7 | 90.1 | 89.5 |
| 35 | 88.1 | 89.6 | 90.8 |
| 40 | 86.3 | 88.3 | 92.1 |
| 45 | 85.5 | 86.8 | 91.9 |
| 50 | 83.7 | 85.2 | 92.0 |

**Table 3** Classification accuracy of Legendre, Chebyshev and Krawtchouk moments with SVM

| Moment order | Legendre (%) | Chebyshev (%) | Krawtchouk (%) |
|---|---|---|---|
| 5 | 90.7 | 86.8 | 76.0 |
| 10 | 93.1 | 93.0 | 77.9 |
| 15 | 93.0 | 92.5 | 83.0 |
| 20 | 91.6 | 91.4 | 87.5 |
| 25 | 90.5 | 90.5 | 85.7 |
| 30 | 89.3 | 89.4 | 89.6 |
| 35 | 88.6 | 88.1 | 90.8 |
| 40 | 85.9 | 87.6 | 91.2 |
| 45 | 84.4 | 86.3 | 91.6 |
| 50 | 83.2 | 84.6 | 91.7 |

**Table 4** Classification accuracy of Legendre, Chebyshev and Krawtchouk moments with kNN

| Moment order | Legendre (%) | Chebyshev (%) | Krawtchouk (%) |
|---|---|---|---|
| 5 | 87.7 | 89.3 | 79.1 |
| 10 | 90.4 | 90.3 | 80.3 |
| 15 | 89.1 | 89.2 | 82.1 |
| 20 | 88.2 | 87.4 | 84.8 |
| 25 | 87.3 | 87.5 | 85.8 |
| 30 | 86.8 | 86.9 | 87.0 |
| 35 | 85.2 | 84.7 | 88.9 |
| 40 | 83.6 | 83.2 | 89.3 |
| 45 | 81.5 | 82.6 | 89.8 |
| 50 | 81.3 | 80.7 | 90.0 |

components for which the ANN, SVM and kNN shows best classification result are shown in Table 5.

## 4.2 *Classification Using Pixel Based Features*

Table 6 shows the classification accuracies obtained using pixel intensity/value based features extracted using different sizes of blocks as described in Sect. 3.

**Table 5** Classification accuracy of Legendre, Chebyshev and Krawtchouk moments with different numbers of PCA components

| Moment type | Classifier | No. of PCA components | Accuracy (%) |
|---|---|---|---|
| Legendre | ANN | 110 | 93.0 |
| | SVM | 80 | 92.8 |
| | kNN | 170 | 90.2 |
| Chebyshev | ANN | 150 | 92.8 |
| | SVM | 40 | 93.7 |
| | kNN | 130 | 90.5 |
| Krawtchouk | ANN | 130 | 91.7 |
| | SVM | 50 | 92.4 |
| | kNN | 220 | 89.0 |

## 4.3 *Classification Using Shape and Texture Based Features*

Classification accuracies of the classifiers obtained after training them using texture and shape based feature set are shown in Table 7.

From all the classification results obtained using different types of moments, we have observed that the classification accuracy increases from 5th to 10th order for Legendre and Chebyshev moments and then it gradually decreases with the increase of moment order. This may happen because these two types of moments keeps introducing noise with increase of order. On the other hand, accuracy increases with the increase of moment order for Krawtchouk moments. However, after reaching a certain order, accuracy doesn't increase significantly. We have obtained highest classification rate of 93.1% for Legendre moments with SVM, 93.6% for Chebyshev with ANN and

**Table 6** Classification accuracy using pixel intensity based features

| Block size | Size of single feature vector | SVM (%) | ANN (%) | kNN (%) |
|---|---|---|---|---|
| 10 × 10 | 720 | 95.4 | 94.8 | 92.4 |
| 20 × 20 | 180 | 95.2 | 94.3 | 92.7 |
| 30 × 30 | 80 | 94.3 | 93.6 | 91.6 |
| 40 × 40 | 45 | 93.3 | 92.7 | 91.3 |
| 60 × 60 | 20 | 88.8 | 86.5 | 89.4 |

**Table 7** Classification accuracy using texture and shape based features

| ANN (%) | SVM (%) | kNN (%) |
|---|---|---|
| 95.9 | 96.5 | 95.2 |

92.1% for Krawtchouk moments with ANN. Use of PCA with 50th order moments improved the results a little for some cases but it is not significant enough. In our experiment with Hu's seven invariant moments, it shows highest accuracy of 89.4% with SVM which is not very much satisfactory. Pixel intensity based features have shown some promising results for all the three classifiers. It shows highest accuracy rate of 95.4, 94.8 and 92.7% with ANN, SVM and kNN respectively. We have observed that the accuracy rate decreases with the increasing size of the blocks. This may happens because the smaller blocks increase the number of features in a feature vector and hence gives more granular information about intensity distribution of an image. With the texture and shape based feature set, classification results are again improved. All the three classifiers showed very promising results with this type of feature set but SVM produced the highest classification rate of 96.5%.

## 5 Conclusion and Future Work

The primary goal of our work was to explore the effective features that can be use to classify different species of parasite eggs. In our work, we have extracted four types of image moments as feature vector and experimented their performance using three different classifiers. PCA analysis is used to reduce the dimension of the feature vectors generated by higher order moments. Apart from image moments, we have extracted pixel intensity based features by dividing the segmented gray scale images into multiple blocks. Some textural features are also extracted and a feature vector is contracted by adding three shape based features along with these texture features. In our experiment, we trained three different classifiers ANN, SVM and kNN to evaluate the performance of each feature set in identification of Ascaris, Necator and Trichuris parasitic eggs. Form the experimented results, we have observed that shape and texture based features are the most suitable for classification of different parasite eggs. All the three classifiers shows promising results with this type of feature set. Apart form the texture features, image pixel intensity based features also shows a satisfactory result. Due to the unavailability of images, we have considered image of only three types of parasite eggs in this work. We believe that texture and pixel intensity based feature sets will also perform satisfactorily with other species of parasite eggs. Our future work will be experimentation of these feature sets with other different types of parasite egg's images and evaluate their performance. Classification by combining different types of feature set together is also a future work of our study.

**Acknowledgements** We are thankful to Dr. Nagappa S. Karabasanavar from KVAFSU, Karnataka and Raafat Salih Hadi, from University of Malaysia Pahang (Former) for providing us the microscopic images of parasite eggs for this research work.

## References


1. Ming-Kuei H (1962) Visual pattern recognition by moment invariants. IEEE Trans Inf Theory 8(2):179–187
2. Haralick RM, Shanmugam K, Dinstein I (1973) Textural features for image classification. IEEE Trans Syst Man Cybern 3(6):610–621
3. Ray K, Shil S, Saharia S, Sarma N, Karabasanavar NS (2020) Detection and identification of parasite eggs from microscopic images of fecal samples. In: Das A, Nayak J, Naik B, Pati S, Pelusi D (eds) Computational intelligence in pattern recognition, advances in intelligent systems and computing, vol 999. Springer, Singapore, pp 45–55. https://doi.org/10.1007/978-981-13-9042-5_5
4. Hadi RS, Khalidin IZ, Ghazali KH, Zeehaida M (2012) Human parasitic worm detection using image processing technique. In: Computer Applications and Industrial Electronics (ISCAIE). IEEE Symposium. Kota Kinabalu, pp 196–201
5. Shu H, Luo L, Bao X, Yu W (2000) An efficient method for computation of legendre moments. Graph Models 62(4): 237–262
6. Yang GY, Shua HZ, Toumoulinb C, Hanc GN, Luo LM (2006) Efficient Legendre moment computation for grey level images. Pattern Recogn 39(1):74–80
7. Yap PT, Raveendran P (2004) Image focus measure based on Chebyshev moments. IEEE Proc Vis Image Signal Process 151(2):128–136
8. Saharia S, Bora PK, Saikia DK (2004) A comparative study on discrete orthonormal Chebyshev moments and Legendre moments for representation of printed characters. In: 4th Indian conference on computer vision. In: Graphics and image processing. Allied Publishers, Kolkata, pp 491–496
9. Asli BHS, Flusser J (2014) Fast computation of Krawtchouk moments. Inf Sci 288:73–86
10. Yap PT, Raveendran P, Ong SH (2003) Image analysis by Krawtchouk moments. IEEE Trans Image Process 12(11):1367–1377

# Reconstruction of Contour Lines for Digitization of Contour Map

**Ashis Pradhan, Mohan P. Pradhan, and Sabna Sharma**

**Abstract** Contour map has contour lines that are significant in building a digital elevation model (DEM). During the digitization and preprocessing of contour maps, the contour line intersects with each other or breaks apart resulting in broken contour segments. These broken segments impose a greater risk while building DEM leading to a faulty model. The endpoints of the broken segments are matched and reconnected accurately and efficiently using the concept of minimum Euclidean distance and gradient direction. Elevation value is extracted from the map, and a digital elevation model is built in 3D space.

**Keywords** Digitization · Digital elevation model · Gradient · Euclidean distance · Contour maps

## 1 Introduction

One major feature in topographic sheet is a contour line that is represented by points at equal elevation from reference level. A map consisting of only contour lines can be generated which is known as contour map. A contour in a topographic sheet is highlighted by monochromatic brown color.

Contour lines are also called as isolines, and the relative spacing of the lines in the contour map indicates the relative slope of the surface. Thus, contours enable us to represent the mountains of varied heights, ocean or a lake depth and slope gradient on a two-dimensional surface. For steep places, the lines are close together, and the

A. Pradhan (✉)
Department of CSE, Sikkim Manipal Institute of Technology, SMU, Sikkim, India
e-mail: ashis.pradhan.2010@gmail.com

M. P. Pradhan
Department of Computer Application, Sikkim University, Sikkim, India
e-mail: mppradhan@cus.ac.in

S. Sharma
School of Information Technology, SRM University, Sikkim, India
e-mail: savna27@gmail.com

 
A. Maji et al. (eds.), *Proceedings of the International Conference on Computing and Communication Systems*, Lecture Notes in Networks and Systems 170,
https://doi.org/10.1007/978-981-33-4084-8_30

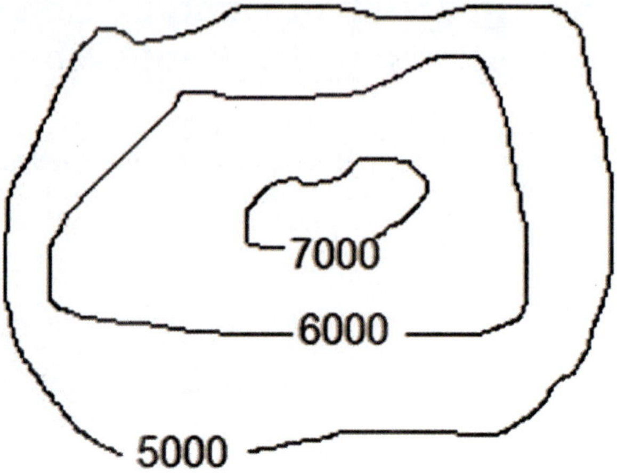


**Fig. 1** Sample contour map

elevation is changing rapidly while for the flat places, the elevation does not change much.

Two-dimensional surface imposes visualization problem for users; hence, a model has to be generated in 3D space. And, these contour lines act as significant aspect for digital elevation modeling (DEM) generation. The DEM data can then be processed to deliver a model in 3D space (Fig. 1).

After the preprocessing of contour maps, the closed contour lines in the map breaks resulting in the broken segments and unwanted pixels as noise which imposes a greater risk in building a DEM leading to a faulty model. Thus, a proper advanced strategy has to be used to reconnect the broken segments of the contour line with utmost accuracy and efficiency and to remove the unwanted noisy pixels.

The elevation values present in the map can be generated during the preprocessing phase or after the reconnection of contour lines phase is complete. The elevation value along with the generated DEM data can be processed further using different techniques and software to generate a proper DEM model (Fig. 2).

The generated DEM model can then be used to determine landslide possibility within a geographical region, to analyze changes within a topographic region within a certain period of time and has several other engineering and geographical applications. This paper deals with the digitization of contour maps, extraction of features such as elevation value from the contour maps, so that it can act as a basis for constructing a digital elevation model.

## 2 Literature Survey

The digitization of traditional map has always been a challenging issue but important. It is important to digitize a map, so that it can be easily stored in a computer and

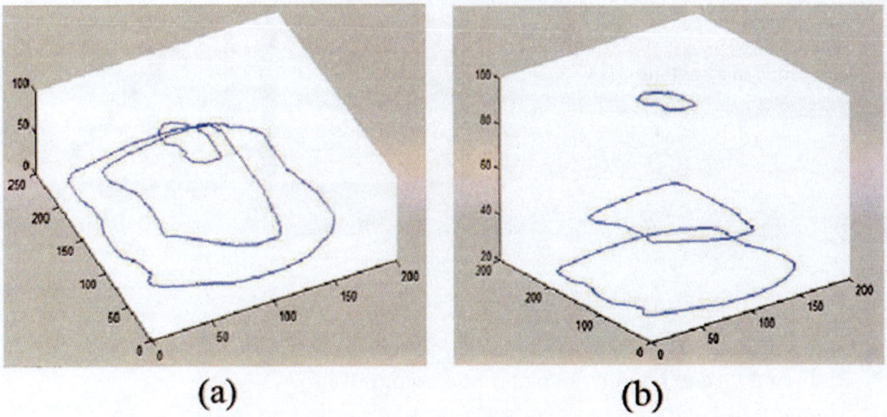

Fig. 2 **a** Sample DEM 1 (top view), **b** sample DEM 2 (side view)

editable. Since, it is editable, so it has many applications in the field of GIS application. The author Hancer in [1] highlighted the efficiency and accuracy of reconnection of broken contours. Consideration of both opposite as well as parabolic direction during endpoint matching has been highlighted. Also, it minimized the drawbacks of conventional methods for reconnecting of broken segments. Hence, research based on development of novel reconstruction strategy is of prime importance.

The author in [2] have proposed different knowledge-based efficient reconnection technique based on the concept of leech, water flow, and wiper. The use of local information, i.e., Euclidean distance between extreme points of the curve and their directions combined for the joint decision, was introduced by Ghircoias and Brad [3]. The author Samet [4] resolved and removed crossed points effectively by using masks. Li [5] highlighted location of text feature and identification of external feature incident on text feature.

## 3 Methodology

The following are the basic steps to be followed for digitization of contour maps into DEM:

(a) Image acquisition (Scanning the contour maps)
(b) Preprocessing

   i. Binarization
   ii. Noise removal
   iii. Thinning
   iv. Character localization and extraction
   v. Reconstruction of broken contour line

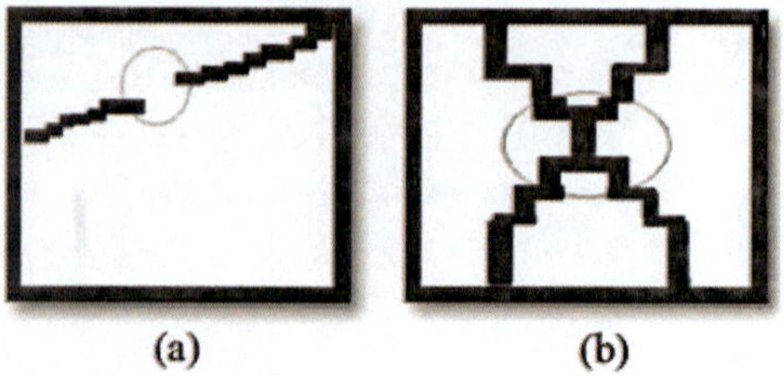


**Fig. 3** **a** Terminal points in a broken contour segment, **b** crossed points in a contour segment

vi. Identification of broken points
vii. Interpolating line between valid contour broken points.

(c) Refinement of extracted character and recognition
(d) Mapping of contour line as per recognized elevation point into 3D space.

The challenges faced while converting a contour map to a 3D model are:

1. **Identifying terminal and crossed points**: Removing crossed points and devising a suitable technique to reconnect the broken contour line segments (Fig. 3).
2. **Character extraction and recognition to generate the elevation value**: Some digits like 0 and 8 do not have horizontal and vertical lines and endpoints as well (Fig. 4).
3. **Recognizing characters that are intersected/crossed by contour lines** (Fig. 5)

To detect the crossed points, 3 by 3 mask is applied to the whole image and since a contour can move only in one direction, the pixels moving in two or more directions are deleted (Fig. 6).

To detect the endpoints, filters are used. A search window of 11 * 11 pixel is established to detect the endpoints in the region. To match the endpoint of a broken contour with the other endpoint, combination of minimum Euclidean distance and direction of the pixel computed using gradient operator is used, and the matching endpoints are reconnected through the cubic-hermite spline interpolation technique. Cubic-hermite spline is defined in the equation:

**Fig. 4** Zoning of digit 0

**Fig. 5** Elevation value crossed by a contour line

| | | |
|---|---|---|
| X | 1 | 1 |
| X | 0 | 1 |
| X | 1 | 1 |

| | | |
|---|---|---|
| 1 | 1 | X |
| 1 | 0 | X |
| 1 | 1 | X |

| | | |
|---|---|---|
| 1 | 1 | 1 |
| 1 | 0 | 1 |
| X | X | X |

| | | |
|---|---|---|
| X | X | X |
| 1 | 0 | 1 |
| 1 | 1 | 1 |

**Fig. 6** Mask applied to detect crossed points

$$p(t) = \left(2t^3 - 2t^2 + 1\right) p_0 + \left(t^3 - 2t^2 + t\right) m_0 + \left(-2t^3 - 2t^2\right) p_1 + \left(t^3 - t^2\right) m_1$$

where $p_0$ and $p_1$ are endpoint and $m_0$ and $m_1$ are tangents of endpoints.

The spline is a curve that is used to reconnect two or more points. It is implemented using inbuilt function $s = \text{spline}(x, y, xq)$ that returns a vector. This vector contains interpolated values that correspond to query point $xq$. The contents of $s$ are evaluated using cubic spline interpolation method. The used method adopts the concept of interpolating the pixel based on control point. The key advantage of same is that it retains the continuity of the broken contour. Spline selects slopes differently to make continuous representation, so that the resultant is smoother. The quality of result depends on data provided by smooth function.

Binary image is divided into nine zones, and the number of foreground (black) pixels in each zone is counted to find out the zone with specified foreground pixel value to detect a specific digit. For example, digit 0 has zero foreground pixel value in zone 5 (Fig. 7).

To label the contour lines and build DEM, the concept of point extrema, triangulation, and generation of height values using bilinear interpolation technique is used. The DEM can also be generated automatically using software such as ArcGIS.

To generate the DEM from contour maps, following steps are followed:

1. **Preprocessing**: To enhance raw images. The image preprocessing technique mainly deals with the thresholding and noise removal operations so as to perform

**Fig. 7** Feature extraction of digit 0

morphological operations and to implement general algorithm on the image acquired.

***Thresholding operation***: Used to extract useful information while reducing the background noise. Thresholding operation is performed using histogram shape-based method to extract only the essential information.

***Noise Removal operation***: Used to remove unwanted pixels as distortions. Median filter is used to remove the noise such as Gaussian noise and salt and pepper noise.

2. **Thinning operation:** Used to remove identified significant pixels from processed images for skeleton representation. Zhang-Suen's algorithm is used to convert the thick pixel image into single pixel thick image.
3. **Detection of endpoints:** Terminal points of the broken contour segments are detected using filters over a window of certain pixels.
4. **Matching of endpoints:** The detected endpoints are matched with each other using the concept of Euclidean distance and gradient direction to proceed with the process of reconstruction of contour line.
   The Euclidean distance of any two points ($x1$, $y1$) and ($x2$, $y2$) is defined as: $\mathrm{De} = \sqrt{(x1 - x2)^2 + (y1 - y2)^2}$
5. **Reconstruction of broken contour segments:** The matched endpoints are reconnected to reconstruct broken segments using cubic-hermite spline interpolation technique.
6. **Character extraction and recognition:** To identify the location of the text feature in the map and recognize it.
7. **Digital elevation model (DEM):** To generate 3D model from the DEM data in 3D space.

## 4 Results and Discussions

With the help of histogram information, thresholding operation was performed (Fig. 8).

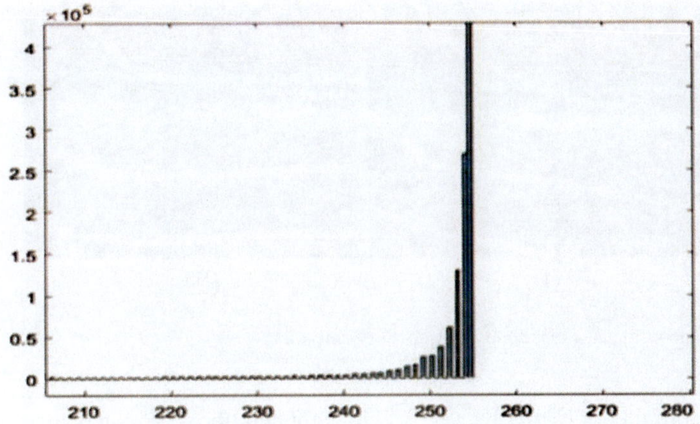


**Fig. 8** Histogram for the image read

Pseudo-code for Thresholding:

1. Read image as 'ip_image'.
2. Generate histogram values to find the spread of pixel values.
3. Compute the mid-point of the spread as m.
4. for i <- 1 to r
    4.1 for j<- 1 to c
        4.1.1 if ip_image (i,j) > m
            4.1.1.1 ip_image (i,j) = 0
        4.1.2 else
            4.1.2.1 ip_image (i,j) = 255

**Noise removal operation**: Salt and pepper noise generated after binarization was removed with the help of median filter by using the inbuilt function medfilt2(). Different colored pixels in the binary image as Gaussian noise was also removed using the inbuilt function medfilt2() (Fig. 9).

**Thinning operation**: The inbuilt morphological operator function bwmorph() produced unnecessary protrusions and to remove these protrusions, the thinned image required additional preprocessing. Instead of bwmorph morphological operator function, the thinning algorithm as proposed by Zhang-Suens was implemented and produced satisfactory results. The Zhang-Suen thinning algorithm is a two-pass algorithm, i.e., for each iteration, it performs two sets of checks to remove pixels from the image. The checks are devised, so that the first set removes from the southeast of the image, and the second set removes from the northwest (top left) corner.

$A(i, j)$ = number of transition from significant to in-significant pixel, in clockwise direction of the eight neighbors around significant pixel $(i, j)$, where the sequence starts and ends at the same neighbor.

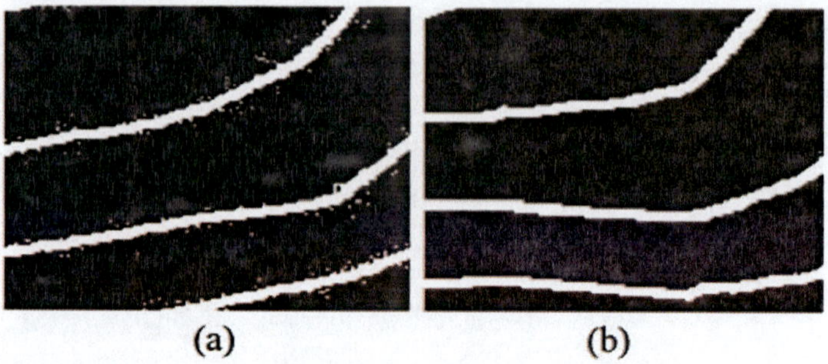

Fig. 9 **a** Image with salt noise, **b** after removal of salt noise

$B(i,j) =$ number of in-significant pixel in eight neighbors around pixel $(i,j)$. Mark such pixel for removal if it satisfies all of the following conditions:

**Pass 1**

1. The pixel is significant and has eight neighbors.
2. $2 < = B(i,j) < = 6$.
3. $A(i,j) = 1$.
4. At least one in north, east, and south neighbors is insignificant.
5. At least one in east, south, and west neighbors is insignificant.

**Pass 2**

1. The pixel is significant and has eight neighbors.
2. $2 < = B(i,j) < = 6$.
3. $A(i,j) = 1$.
4. At least one in north, east, and west neighbors is insignificant.
5. At least one in north, south, and west neighbors is insignificant (Fig. 10).

**Detection of endpoints**: Terminal points of the broken contour segments are detected. The terminal points are detected based on number of neighbors of $3 \times 3$ mask. The total number of neighbors should be exactly 1 (Fig. 11).

**Matching of endpoints**: The detected endpoints are matched with each other to facilitate the reconstruction process. The gradient features of the terminal points along *x*- and *y*-axis as *Gx* and *Gy* are evaluated using 'sobel' operator, and also the gradient direction is evaluated to help in the reconstruction purpose.

All detected terminal points are stored in database. If match is found, then reconnection is done using spline interpolation technique to generate the smooth curve. Any endpoint that does not match is saved and checked against other similar endpoints at the end. The starting point of broken segment of contour line in the counter-clockwise direction will have positive gradient direction while the endpoint of the broken segment will have negative gradient direction.

**Fig. 10** Resultant thinned image

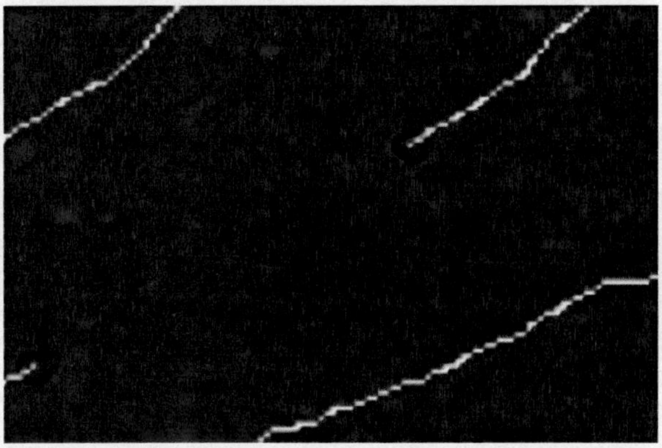

**Fig. 11** Detection of terminal points

*Case 1: When two endpoints are detected in the search space:* If the endpoints have opposite gradient direction, then the endpoints are matched, and the matched endpoints are reconnected to reconstruct the broken segments.

*Case 2: When the matching endpoint is not detected in the search space:* The endpoint is saved and is checked against other similar endpoints at the end.

An endpoint is selected, and its gradient direction is determined. If the determined gradient direction is positive, then the list of endpoints having negative gradient direction are probable matching endpoints, based on minimum Euclidean distance. Finally, the matching endpoints are then reconnected (Fig. 12).

**Reconstruction of broken contour segments**: The matched endpoints are reconnected using smooth polynomial functions called splines. For reconnection, cubic-hermite spline interpolation technique is used.

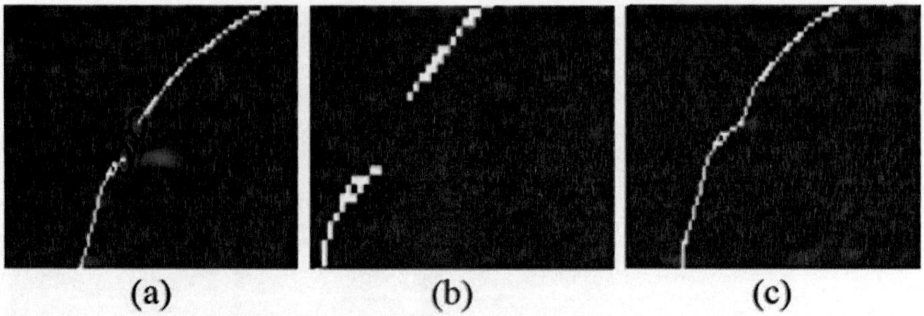


**Fig. 12** **a** Endpoints in a contour line, **b** interpolated using cubic-hermite spline, **c** endpoints reconnected by adding foreground pixel along the spline

## 5 Conclusions

This paper deals with the digitization of contour maps and to facilitate a input for digital elevation model generation. The input contour map was successfully digitized, and a threshold value was generated through the spread of pixels. The noise was then removed using filters, and the thinning operation was performed using Zhang-Suen algorithm. The preprocessing and thinning phase resulted in the broken contour segments. The endpoints of the broken contour segments were identified, and a search space was used to match the endpoints. The matching of endpoints was done using the concept of gradient direction and minimum Euclidean distance. Cubic-hermite spline interpolation technique was used to generate the path between the matching endpoints. Foreground pixels were added along the path to complete the reconstruction process. The reconstructed map was then used to generate a DEM model in 3D space. The use of histogram information helped to process the raw image sharply, and the implementation of remaining steps happened to be easier. The median filters were able to remove the noise, and the thick pixel image was converted into single pixel image using Zhang-Suen algorithm. The concept of Euclidean distance and gradient direction was able to match the six endpoints detected. The use of only two features for matching could cause problems while matching if many endpoints are detected. The character recognition step is tricky if the characters are overlapped with the contour lines. The concept of Euclidean distance and gradient direction is though applicable, some new features has to be taken into account while trying to match the endpoints for reconnection process. More preprocessing steps can be performed to sharpen the contour map read and to nullify the unwanted distortions as noise. The matching of endpoints can be made more accurate by considering other directional features like parabolic directions. The broken segments can be reconstructed by generating smooth curves using an efficient interpolation technique. The extraction and identification of elevation values from the contour map can also be done using efficient techniques. The development of an automatic technique for contour line reconstruction can also be considered as one of the future direction of this research work.

## References

1. Hancer E, Samet R, Karaboga D (2014) A hybrid method to the reconstruction of contour lines from scanned topographic maps. In: IEEE 23rd international symposium on industrial electronics (ISIE), Istanbul, pp 930–933. https://doi.org/10.1109/ISIE.2014.6864736
2. Pradhan MP, Ghose MK, Rai PS, Mukherjee N (2013) Knowledge based contour line reconnection techniques. Int J Comput Appl 65(9)
3. Ghircoias T, Brad R (2010) Contour lines extraction and reconstruction from topographic maps. UbiCC J 6
4. Samet R, AskerzadeAskerbeyli IN, Varol C (2010) An implementation of automatic contour line extraction from scanned digital topographic maps. Appl Comput Math 9(1):116–127
5. Li L, Nagy G, Samal A et al (2000) Integrated text and line-art extraction from a topographic map. IJDAR 2:177–185. https://doi.org/10.1007/PL00021524
6. Chaikin G (1974) An algorithm for high-speed curve generation. Comput Graph Image Process 3:346–349
7. Song J, Wang P, Miao Q, Liu R, Huang B (2017) The reconnection of contour lines from scanned color images of topographical maps based on GPU implementation. IEEE J Select Top Appl Earth Observ Remote Sens 10(2):400–408. https://doi.org/10.1109/JSTARS.2016.2580903
8. Joy KI (2002) Chaikin's algorithms for curves, visualization and graphics research group. Department of Computer Science, University of California, Davis. [Online] Available at: https://www.cs.unc.edu/~dm/UNC/COMP258/LECTURES/Chaikins-Algorithm.pdf. Accessed on 16/03/2020
9. Eikvil L, Aas K, Koren H (1995) Tools for interactive map conversion and vectorization. In: Proceedings of 3rd international conference on document analysis and recognition, Montreal, Quebec, Canada vol 2, pp 927–930. https://doi.org/10.1109/ICDAR.1995.602053
10. Gul S, Khan MF (2010) Automatic extraction of contour lines from topographic maps. In: International conference on digital image computing: techniques and applications. Sydney, NSW, pp 593–598. https://doi.org/10.1109/DICTA.2010.105
11. Shimada S, Maruyama K, Matsumoto A, Hiraki K (1995) Agent-based parallel recognition method of contour lines. In: Proceedings of the third international conference on document analysis and recognition. vol 1. ICDAR '95. IEEE Computer Society, USA, pp 154
12. Spinello S, Guitton P (2004) Contour line recognition from scanned topographic maps, pp 419–426
13. Xie X, Wunsche B (2010) Efficient contour line labelling for terrain modelling. In: Proceedings 33rd Australasian computer science conference (ACSC), Brisbane, Australia
14. Xu B, Chen J, Yao M (2016) Identification of contour lines from average-quality scanned topographic maps. Math Prob Eng 1–14. https://doi.org/10.1155/2016/3089690

# 2D Image Stitching to Generate a Film of Fired Bullets

**Barun Barua, Genevieve Chyrmang, Kangkana Bora, R. Suresh, and Lipi B. Mahanta**

**Abstract** In forensic science, the identification of firearms used in a crime scene can be performed by studying the individual and class characteristics of the striation marks left on the surface of the fired bullets and cartridge cases. Each firearm has unique characteristics based on their manufacturing, especially the gun barrel which contains spiral patterns bored into it. When fired it engraves the patterns on the surface of the bullet by which identification of the firearm is possible. Firearm identification is done by ballistic experts in the laboratory by comparing the unique characteristics of the specimen (bullets or cartridges) in the crime scene, which is a time consuming process. For this purpose, an automated firearm identification system is the demand of time. In doing so, the 3D structure of the bullets always challenges their study. In this paper, an attempt has been made to generate a continuous film of bullets using image stitching technique, which can represent their 3D structure in a single image. This method can be used by ballistic experts to study the striations in just one unbroken image.

**Keywords** Forensic Science · Firearm · Bullets · Ballistic Expert · Automated Firearm Identification · Image Stitching

---

B. Barua · G. Chyrmang
Assam Don Bosco University, Guwahati, India
e-mail: barunbarua.2012@gmail.com

G. Chyrmang
e-mail: chyrmanggenevieve@gmail.com

K. Bora (✉)
Cotton University, Guwahati, India
e-mail: kangkana.bora89@gmail.com

R. Suresh
Central Forensic Science Laboratory, Guwahat, Assam, India
e-mail: rsureshcfsl@rediffmail.com

L. B. Mahanta
Institute of Advanced Study in Science and Technology, Guwahati, India
e-mail: lipimahanta@yahoo.co.in

 
A. Maji et al. (eds.), *Proceedings of the International Conference on Computing and Communication Systems*, Lecture Notes in Networks and Systems 170,
https://doi.org/10.1007/978-981-33-4084-8_31

# 1 Introduction

In the domain of forensic science, firearm identification involves the examination of the individual and class characteristics of the striation marks present on the surface of cartridges cases or bullets that are recovered from a crime scene. These striation marks are considered as the 'fingerprints' [5] of a firearm, as they are unique to the firearms. When a firearm is fired, it leaves some unique characteristics marks on the surface of bullets and cartridge cases due to rifling, the firing pin, breach, and ejector. Rifling marks (striation marks) on the surface of the bullets will be emphasized in this paper. To improve the flight path and accuracy of the bullet, some spiral patterns are bored into the barrel of the firearm, this is called rifling. These spiral patterns are a collection of land and grooves, which are one of the characteristics of rifling. Grooves are designed to spin the bullet, this spin stabilizes the flight path of the bullet and improves accuracy. Lands are the area between grooves in the barrel of the firearm. The number of lands and grooves are equal. Grooves and Lands inside a barrel are illustrated in Fig. 1. The specification of the firearm used in the picture are Make: Standard factory made firearm, Type: Semi-automatic pistol, Calibre: 7.65 mm, No. of lands and grooves: five, Direction of twist: right, Ammunition: 7.65 × 17 mm SR. The spinning direction is another characteristic of rifling. When a bullet is fired from the firearm, the land and grooves are engraved on the surface of the bullet with a twist in a direction of clockwise or anticlockwise due to spinning. The impression of land and grooves is illustrated in Fig. 2. Some general class characteristics of rifling are the number of grooves and lands, and direction of twist of the grooves and lands on the surface of bullet, either clockwise or anticlockwise.

These general characteristics help ballistic experts to identify the firearm. Figure 2 illustrates the land of a bullet under a microscope of 50x magnification. Usually, the land of the bullet displays a higher number of striations as it is in closer contact with the gun barrel. When the lands of the bullet are seen under magnification of

**Fig. 1** Lands and Grooves inside the barrel of Firearm

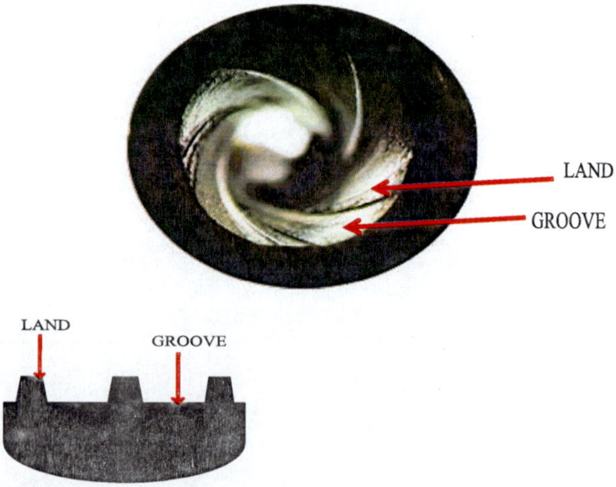

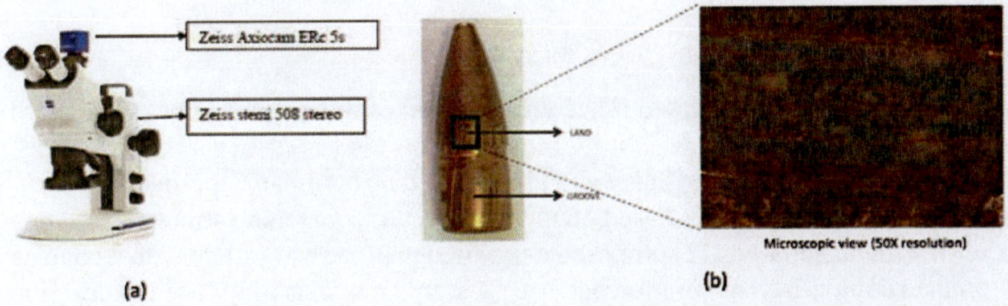


**Fig. 2** **a** Microscope (Zeiss stemi 508 stereo) and camera (Zeiss Axiocam ERc 5s) used of image acquisition, **b** Lands and Grooves on the surface of a bullet and its maginified view showing striation marks

a microscope, the striation marks can be observed, which help in the identification of firearm. In India, the current scenario in the ballistic examination is a traditional process, where the examination is done by ballistic experts, who closely examine the class and individual characteristics present on the bullet (especially land areas) in a laboratory using sophisticated equipments, employing measuring devices such as calipers. Stereo microscopes are used to determine the class and individual characteristics of the fired bullet. A comparison microscope is used for the examination of a fired bullet. The traditional process has a drawback because it is labor-intensive, time-consuming, and susceptible to human error. So, to overcome the drawbacks of the traditional process of ballistic examination an automated identification and comparison is the demand of time [5]. One of the challenges in image processing is the 3D structure of the bullets, due to which it is very difficult to get a single image for analysis. Image stitching is required because, the magnification with which the individual characteristics of striations are visible in the microscope are ensconced within a limited area/portion on the rifling of fired bullet. Hence, multiple images have to be acquired for one rifling mark say four images acquired. These four images are acquired in four separate files and belongs to one rifling mark. The sequence of image acquisition is carefully noted and marked in the digital saved file.These four images, if stitched suitably, then it is possible to view the complete rifling mark present on the fired bullet representing characteristic striations. These characteristic striations are most useful for forensic experts to determine the firearm through which suspected bullet has been fired through. Hence, this work has been taken up. With this motivation, this paper proposes to develop an automated image stitching system that will stitch the individually grabbed bullet images together and generate one unbroken image of lands of the surface of the bullet. This stitched image/film can be later used for comparison and bullet signature generation, using the power of computer vision and machine learning. Here the main focus is given on stitching land images of the bullet, as it can display a large number of striation marks. This automated stitching technique can help peer researchers to ease their work on ballistic imaging.

## 2 Related Works

Some existing works performed related to the topic of this paper can be categorized into the following groups -

*Identification of firearm*: Chu et al. [1] presented an automatic approach for identification of bullet signatures based on their consecutive matching striate (CMS) criteria. Heizmann and Leon [2] proposed a new image processing strategy that enables automated comparison of striation patterns. A signal model is introduced to describe the interesting features of forensically relevant striation marks. Heizmann [3] worked on an image processing method that uses an illumination series to robustly segment striation patterns from mainly isotropic background areas. Chu et al. [4] also presented an automated bullet identification system by selecting valid correlation areas. Leon and Beyerer [5] provided a reliable feature technique from the images using data fusion and a model-based abstraction is accomplished by adaptively projecting the image intensities of relevant grooves. Changmai et al. [6] proposed an automated technique for the identification and classification of bullet striations using image processing and machine learning techniques. Bigdel and Ebrahimi Moghaddam [7] focus on 2-D optical images using a multimodal technique and propose several distinct methods as its modalities. This proposed method uses a multimodal rule-based linear weighted fusion approach.

*Image stitching*: Brown and Lowe [8] presented a system of automated panorama image stitching using invariant local features of images and a probabilistic model to verify image matches. Adel et al. [9] performed a survey about the feature-based image stitching and the comparison between different algorithms for feature detection. Adel et al. [10] presented an image stitching system based on ORB and compensation blending.

## 3 Methodology

### *3.1 Overview of Proposed Work*

To achieve the goal of automated image stitching, the work is carried out in two phases: (1) Acquisition of microscopic images of bullets, (2) Designing automated image stitching technique. Above mentioned steps is illustrated in Fig. 3.

### *3.2 Acquisition of Microscopic Images*

For this study, six test-fired bullets have been provided by Central Forensic Science Laboratory (CFSL), Directorate of Forensic Science Services, Ministry of Home Affairs, Govt. of India, Guwahati, Assam. Out of these six bullets, four bullets are of

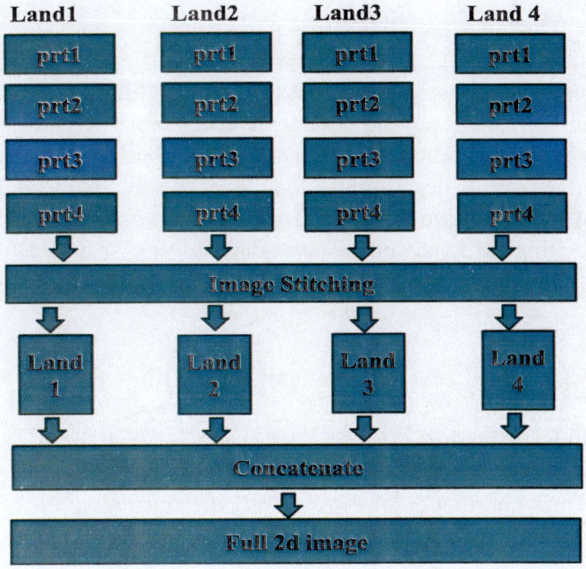


**Fig. 3** Overview of the propose work

5.56 mm caliber and two bullets are of 9 mm caliber. The Land images present on the Bullet are acquired in parts by a Stemi-508doc stereo microscope available at Central Computational and Numerical Studies Division (CCNS), Institute of Advanced Study in Science and Technology (IASST), Department of Science Technology, Govt. of India, Assam. A microscopy camera (model: Zeiss Axiocam ERc 5s) attached to the stereomicroscope is used to capture the required images. Land images are captured in parts because under magnification of microscope the full length of the land is cannot be seen. Tools and techniques used for Acquisition of images are:

- Zeiss stemi 508 stereo microscope (Fig. 2): The Zeiss Stemi 508 stereo microscope is apochromatic with an 8:1 zoom ratio providing excellent image contrast and color accuracy.
- Zeiss Axiocam ERc 5s (Fig. 2): It is a 5 megapixel CMOS camera attached to the microscope to capture an image.
- For documentation of image, a PC with an 8GB ram installed with Carl Zeiss Zen software is connected to the microscope camera through an HDMI cable. Carl Zeiss Zen software lets the user control the camera attached to the microscope through a PC.
- Magnification of 50x of the microscope is used to capture the images of the bullet. At this magnification, striation marks can be seen clearly. For the illumination, we have used two gooseneck lights at the angle of 45 degrees, one at clockwise direction and another one in an anticlockwise direction. Since the full length of the land of the bullet cannot be captured under magnification, it is captured in parts as shown in Fig. 5.

## 3.3 *Image Stitching*

Image stitching is the process of constructing a seamless image from the sequence of input images of the same scene taken in parts, as explained above, and with a little overlapping. The overlapping is bound to occur in case of a manual acquisition.This paper aims to design an automated image stitching technique to stitch the land images of bullet captured in parts. The overview of the image stitching technique proposed for this paper has been represented in the form of a block diagram in Fig. 4. The techniques that will be used for achieving image stitching for this paper are as follows.

### 3.3.1 Feature Detection and Matching

Standard Feature Detection and Matching Techniques of Image Processing have been used to achieve the proposed image stitching goal. Feature detection and matching is an important task in many computer vision applications, such as image retrieval, object detection, and more. Main components of feature detection and matching are:

- Detection: It is responsible for identifying the feature points.
- Description: It is a process of computing descriptors for each feature point. These feature points are sort of numerical 'fingerprint' that can be used to differentiate feature points of images. Descriptors are obtained in the form of binary vector for each feature point.
- Matching: Descriptors are compared across the images, to identify similar features. For two images we may get a set of pairs $(X_i, Y_i)$ $(X_i^{'}, Y_i^{'})$, where $(X_i, Y_i)$ is a feature in one image and $(X_i^{'}, Y_i^{'})$ its matching feature in the other image.

Some algorithms of feature detection and matching are [10]: Harris Corner, SIFT(Scale Invariant Feature Transform) [8], SURF (Speed Up Robust Feature), FAST (Features from Accelerated Segment Test), ORB (Oriented FAST and Rotated BRIEF) [10].

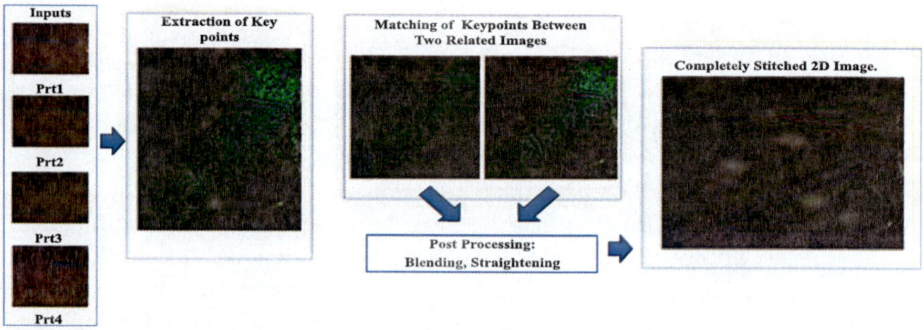


**Fig. 4** Block diagram of propose image stitching System

### 3.3.2 Homography Using Ransac (Random Sample Consensus)

Ransac Homography is used for increasing the accuracy of matching the correct key points (feature points) between two images. It is used to choose the best match between two images by separating inliers and outliers and it removes wrongly detected points [10].

### 3.3.3 Post Processing

Post-processing, like blending and straightening, must be performed to get a good quality image. For image blending, the image should be aligned with each other, so straightening of the image is important. Image blending is used to blend the pixels of the overlapping region of images to get a seamless stitched image.

## 4 Results

*Image Acquisition*: Land images of bullets are captured in parts using Stemi-508doc stereo microscope fitted with Zeiss Axiocam ERc 5s as in Fig. 5, of resolution 2560 × 1920 where rifling characteristics can be seen clearly. Figure 5 shows some images acquired with 5.56mm caliber bullet.

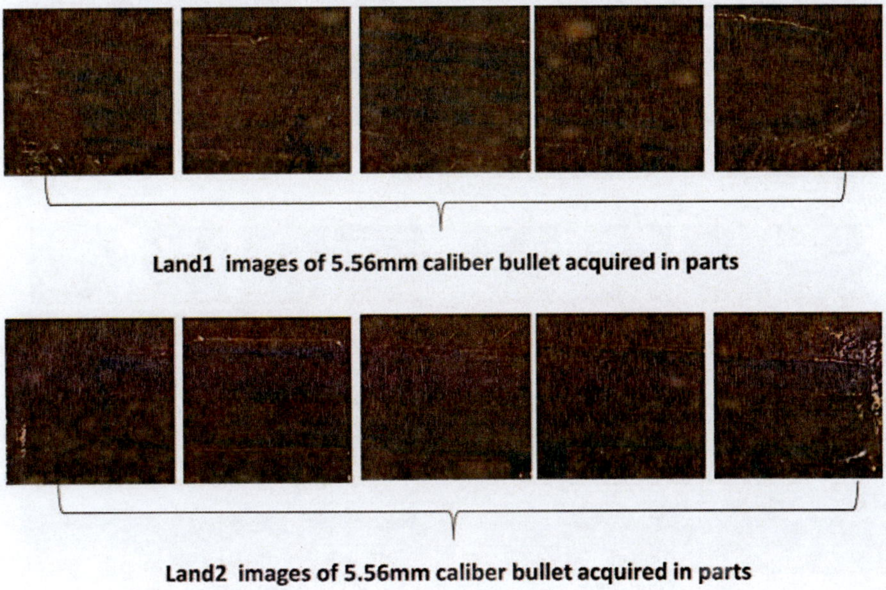


**Fig. 5** Land images of 5.56 mm caliber bullet acquired

*Feature Detection and Matching*: Feature Detection and Matching is an essential aspect of image stitching. For the proposed image stitching system in this paper, ORB feature detection technique is used. ORB stands for Oriented FAST and Rotated BRIEF. ORB is the combination of two techniques known as FAST and BRIEF. FAST is responsible for locating key points in an image and BRIEF is to generate descriptors from the obtained key points. The descriptors generated is used by the Matcher to find the best match between two images.

Brute-Force matching algorithm is used as a matcher here. Results after implementation is shown in Figs. 6 and 7. The results in Fig. 7 displays the completely generated film for two lands. These new completely magnified stitched images or film can be used for further analysis by the experts like striation segmentation etc. For this study, validation of the achieved results were done visually and approved by the ballistic expert involved in the study. However, the results which are promising enough requires to be validated with a larger sample size. The future scope involves automated validation by collecting more bullet samples to extend the study.

**Fig. 6** Images with the detected key points. Green dots represents the Key points

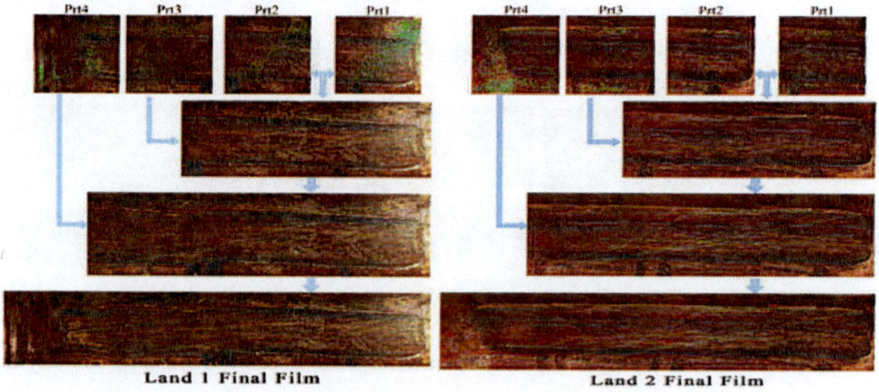


**Fig. 7** Final stitched image obtained after implementation of Matcher. The color dots represents the match found

## 5 Conclusion and Future Work

The automation of the firearm identification system will assist the forensic department by reducing the time required in analyzing and comparing. It will ease the work of ballistic experts and also increase the efficiency of identification of firearm. This paper emphases on the image stitching system which will be necessary to develop the automated firearm identification system. This image stitching technique generates a fully stitched 2D image of the bullet, which is very challenging due to the 3D structure of the bullets. This study can be considered as the first step towards the development of an automated firearm identification process. The future works include the segmentation of striation marks and generation of signature from the striation marks from the stitched images.

## References

1. Chu W, Thompson RM, Song J, Vorburger TV (2013) Automatic identification of bullet signatures based on consecutive matching striae (CMS) criteria. Forensic Sci Int 231(1–3):137–141
2. Heizmann M, Leon FP (2003) Imaging and analysis of forensic striation marks. Opt Eng 42(12):3423–3433
3. Heizmann M (2005) Model-based segmentation of striation marks. In: Fifth meeting of the ENFSI working group marks (SPTM), Stavern, Norway. http://ies.anthropomatik.kit.edu/english/publ.php
4. Chu W, Song J, Vorburger TV, Thompson R, Silver R (2011) Selecting valid correlation areas for automated bullet identification system based on striation detection. J Res Natl Inst Stand Technol 116(3):647–653
5. Leon FP, Beyerer J (1999) Automatic comparison of striation information on firearm bullets. In Intelligent robots and computer vision XVIII: algorithms, techniques, and active vision. Int Soc Opt Photon 3837:266–277
6. Changmai P, Bora K, Suresh R, Deb N, Mahanta LB (2019) On the study of automated identification of firearms through associated striations. In: 31st International symposium on ballistics, pp 1192–1203
7. Bigdeli S, Ebrahimi Moghaddam M (2019) A multimodal fusion approach for bullet identification systems. J Forensic Sci 64(3):741–753
8. Brown M, Lowe DG (2007) Automatic panoramic image stitching using invariantfeatures. Int J Comput Vis 74(1):59–73
9. Adel E, Elmogy M, Elbakry H (2014) Image stitching based on feature extraction techniques: a survey. Int J Comput Appl 99(6):1–8
10. Adel E, Elmogy M, Elbakry H (2015) Image stitching system based on ORB feature-based technique and compensation blending. Int J Adv Comput Sci Appl 6(9):55–60

# An Efficient Approach to Remove Specular Reflection from Non-Ideal Eye Image

Md. Amir Sohail, Chinmoy Ghosh, and Satyendranath Mondal

**Abstract** Iris recognition is very widely used in biometrics-based personal identification, due to the uniqueness of its texture pattern. Segmentation and feature extraction are two key stages of iris recognition system. The presence of noise may affect edge detection-based and active contour-based segmentation and always affects the feature extraction. Specular reflection is a noise due to the reflection of light on eye. Reflection removal is considered as a preprocessing stage in iris recognition. In this paper, we have presented an efficient approach for specular reflection removal to improve the iris recognition. It removes the reflection from eye images very efficiently, without disturbing its characteristics.

**Keywords** Iris texture · Specular reflection · Linear interpolation · Bilinear interpolation · Inward interpolation · Mean

## 1 Introduction

Iris-based biometric is a very secure authentication approach. Iris recognition is divided into four main steps: acquisition, segmentation, feature extraction, identification or verification [1]. Segmentation and feature extraction may not work properly if there is noise [2] such as specular reflection [3]. Specular reflection is a regular reflection of light on the eye surface. There are three types of reflections in eye

Md. A. Sohail (✉) · C. Ghosh
Department of Computer Science and Engineering, Jalpaiguri Government Engineering College, Jalpaiguri, West Bengal, India
e-mail: mas1953@cse.jgec.ac.in

C. Ghosh
e-mail: chinmoy.ghosh@cse.jgec.ac.in

S. Mondal
Department of Information Technology, Kalyani Government Engineering College, Kalyani, West Bengal, India
e-mail: satyen_kgec@rediffmail.com

 
A. Maji et al. (eds.), *Proceedings of the International Conference on Computing and Communication Systems*, Lecture Notes in Networks and Systems 170,
https://doi.org/10.1007/978-981-33-4084-8_32

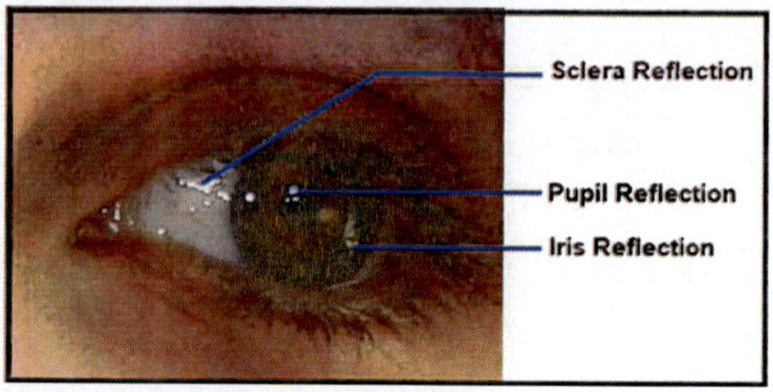


**Fig. 1** Different types of specular reflection

image: sclera reflection, iris reflection and pupil reflection (shown in Fig. 1). Sclera reflection is of less importance. But iris reflection significantly affects texture pattern which causes improper feature extraction, and pupil reflection may create problem during segmentation.

Daugman [4] proposed a revolutionary idea of iris recognition using an integro-differential operator that detects the inner and outer boundary of iris. After that, iris recognition has evolved a lot.

He et al. [5] have given a basic idea of specular reflection removal. They first used an adaptive threshold (top 5% brightest intensity) to produce a binary reflection map $R_{(X,Y)}$ of an image $I_{(X,Y)}$ and then applied bilinear interpolation to remove the reflection areas. In this article, authors have suggested a method to find four envelop points { $P^{\text{left}}_{(x_l,Y)}, P^{\text{right}}_{(x_r,Y)}, P^{\text{top}}_{(X,y_t)}, P^{\text{bottom}}_{(X,y_b)}$ } which are used to interpolate the reflection point $P_{(X,Y)}$. Those are the second nearest non-reflection pixels to the reflection area along four directions (left, right, top, bottom) of x-axis and y-axis, which can be calculated as follows (Eqs. 1–4):

$$x_l = \max\left\{x : \sum_{i=0}^{L-1} R_{(x+i,Y)} = 0, R_{(x+L,Y)} = 1, x < X\right\} \quad (1)$$

$$x_r = \min\left\{x : \sum_{i=0}^{L-1} R_{(x-i,Y)} = 0, R_{(x-L,Y)} = 1, x > X\right\} \quad (2)$$

$$y_t = \max\left\{y : \sum_{i=0}^{L-1} R_{(X,y+i)} = 0, R_{(X,y+L)} = 1, y < Y\right\} \quad (3)$$

$$y_b = \min\left\{y : \sum_{i=0}^{L-1} R_{(X,y-i)} = 0, R_{(X,y-L)} = 1, y > Y\right\} \quad (4)$$

where $L = 2$ controls the necessary separation (one pixel away) between the reflection area and non-reflection area. Hence, in any reflection area, all the points in same row (or column) will have same $P^{\text{left}}$ and $P^{\text{right}}$ (or same $P^{\text{top}}$ and $P^{\text{bottom}}$) envelop points. With the help of these envelop points, reflection pixel is then interpolated as shown in Eq. (5):

$$I(P) = \frac{I(P^{\text{left}})(x_{\text{right}} - X) + I(P^{\text{right}})(X - x_{\text{left}})}{2(x_{\text{right}} - x_{\text{left}})} + \frac{I(P^{\text{top}})(y_{\text{bottom}} - Y) + I(P^{\text{bottom}})(Y - y_{\text{top}})}{2(y_{\text{bottom}} - y_{\text{top}})} \tag{5}$$

Jarjes et al. [6] suggested another method to denoize the eye image. Authors suggested to make the complement of the image (replace each pixel by subtracting from maximum pixel value, i.e. 255) firstly. As the reflections are the brightest area, those would appear as dark holes in the complement image. Now, dark holes are filled based on connectivity of four-connected pixels. Then again complement is evaluated, and to reduce the sharpness, Gaussian smoothing filter is applied. It can successfully remove specular reflection from CASIA or MMU database images, but not quite a good approach for non-ideal eye images.

Aydi et al. [7] proposed a new corneal reflection removal method using extension of linear interpolation. At first, a reflection map is generated by using a top-hat operator and then applying suitable average threshold to isolate the reflection points. Then, dilation is applied to enlarge the reflection area. After that, linear interpolation function **f** is computed at pixel $P(x, y)$ with the help of four top (left, right) and down (left, right) pixels $\{ Q_{\text{TL}}, Q_{\text{TR}}, Q_{\text{DL}}, Q_{\text{DR}} \}$ (shown in Fig. 2) along $x$ and $y$ axes direction, as shown in the following Eqs. (6–8):

$$f(P_1) = \frac{x_2 - x}{x_2 - x_1} f(Q_{\text{DL}}) + \frac{x - x_1}{x_2 - x_1} f(Q_{\text{DR}}) \tag{6}$$

$$f(P_2) = \frac{x_2 - x}{x_2 - x_1} f(Q_{\text{TL}}) + \frac{x - x_1}{x_2 - x_1} f(Q_{\text{TR}}) \tag{7}$$

$$f(P) = \frac{y_2 - y}{y_2 - y_1} f(P_1) + \frac{y - y_1}{y_2 - y_1} f(P_2) \tag{8}$$

Jamaluddin et al. [8] has given two faster and memory-efficient approaches for reflection removal: pupil reflection and iris reflection removal approach. To remove pupil reflection, the contrast of the image is enhanced, and unsharp masking is applied to sharpen the texture. Then, pupil area is determined (generally threshold less than 40) to apply morphological closing to enlarge foreground boundary and shrink background boundary targeting to eliminate the holes (reflection points). Remaining holes are removed by flood-fill operation. Tolerable iris reflections are removed with intensity adjustment (set between 0.21 and 0.8), and intolerable iris reflection pixels are removed by inward interpolation where the reflection area is substituted with the mean of its neighbour pixels. It solves the discrete laplacian equation over the image without affecting the boundary pixels of reflection area. Reddi et al.

Fig. 2 Different points for interpolation scheme

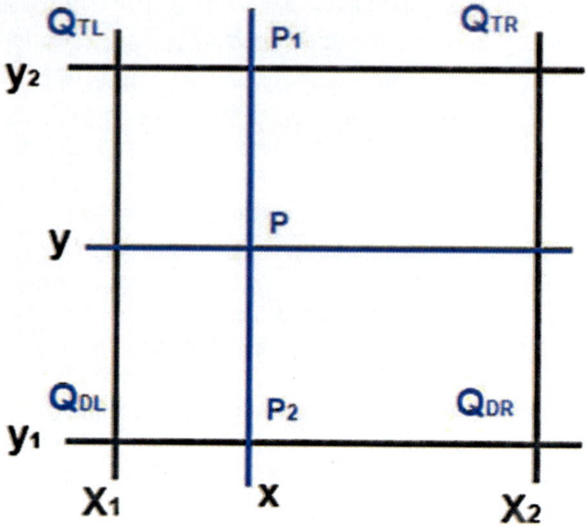


[9] used a threshold-based specular reflection removal approach using inpainting algorithm [10]. Wang et al. [11] had suggested reflection area filling approach using the inpainting method based on Navier–Stokes [12] equations.

## 2 Proposed Methodology

The reflection areas are the brightest pixels of the eye image. Mostly, reflection pixels have intensity greater than 240. Here, we have used threshold-based reflection detection, where pixels with intensity greater or equal to 230 belong to reflection area. Sometimes, reflection pixels nearer to the boundary of reflection area have slightly lower intensity than the pixels located towards the centre of a reflection spot. This is why, threshold-based reflection detection may exclude those outer pixels. To overcome this situation, we have proposed an algorithm **EnlargeReflectionArea** (1) that uses either $3 \times 3$ or $5 \times 5$ filters to enlarge the reflection area. This algorithm converts all the pixels of $3 \times 3$ or $5 \times 5$ search window into white (intensity = 255) if there exists any pixel with threshold $\geq 230$ under the window. It enlarges the reflection areas by at most 2 pixels and 4 pixels (outward), respectively. These filters are applied to all the three components of RGB input image. Then, we apply our reflection removal methods on this reflection-enlarged image. The reflection-enlarged image outputs are shown in Fig. 3.

We have suggested two novel and efficient reflection removal techniques here, using the concept of replacing reflection area pixels by the mean of its nearest non-reflection neighbour pixels. Depending upon how many direction the algorithm traverses, we have named them as: four-direction traverse approach and eight-direction traverse approach.

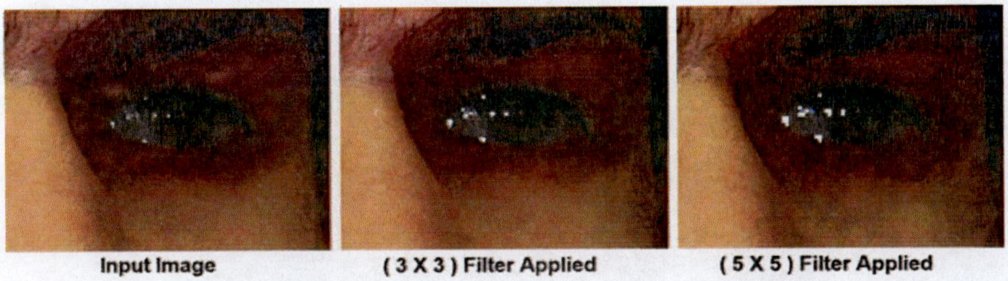


**Fig. 3** Reflection-enlarged images after applying (3 × 3) and (5 × 5) filters

**Algorithm 1** Enlarge Reflection Area Algorithm (Applying 3 × 3 Filter)

```
procedure EnlargeReflectionArea(I)                    ▷ I(1 : row, 1 : col) is input image
  Output ← I(1 : row, 1 : col)
  for r = 2; r <= (row − 1); r = (r + 2) do
    for c = 2; c <= (col − 1); c = (c + 2) do
      flag ← 0                                        ▷ Apply 3*3 window, search
      for i = r − 1; r <= (r + 1); r + + do
        for j = c − 1; c <= (c + 1); c + + do
          if I(i, j) ≥ 230 then                       ▷ Reflection Pixel Found
            flag ← 1
            break
          end if
        end for
      end for
      if flag == 1 then                               ▷ Make neighborhood: white
        for i = r − 1; r <= (r + 1); r + + do
          for j = c − 1; c <= (c + 1); c + + do
            Output(i, j) = 255
          end for
        end for
      end if
    end for
  end for
  return Output                                       ▷ The output image (reflection enlarged)
end procedure
```

## *2.1 Four-Direction Traverse Approach*

As the name of the method suggests, our algorithm traverses towards four directions (along left, right, top, bottom) from the reflection pixel $P(x, y)$ as shown in Fig. 4. During that traversal, we choose the nearest non-reflection pixels (i.e. chosen direction pixels) of the reflection pixel $P$. The chosen direction pixels $\{P_{\text{left}}(x_{\text{left}}, y), P_{\text{right}}(x_{\text{right}}, y), P_{\text{top}}(x, y_{\text{top}}), P_{\text{bottom}}(x, y_{\text{bottom}})\}$ are evaluated in the following Eqs. (9–12).

$$x_{\text{left}} = \max\{x_l : I(y, x_l) < 230, \forall 1 \leq x_l \leq (x - 1)\} \tag{9}$$

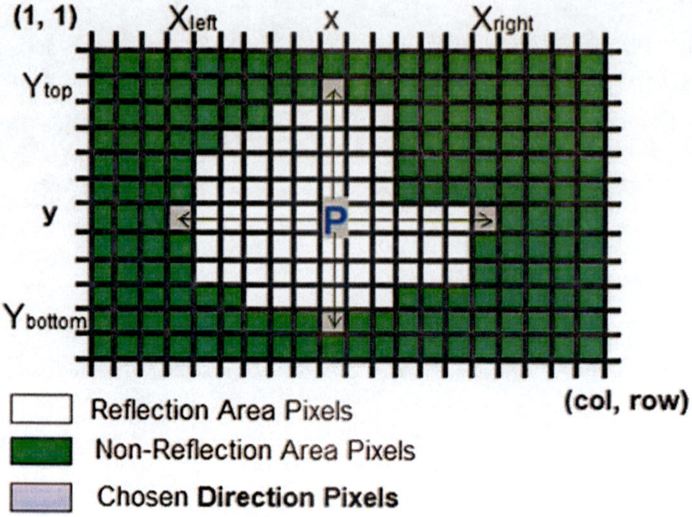


**Fig. 4** Four-direction traverse approach

$$x_{\text{right}} = \min\{x_r : I(y, x_r) < 230, \forall (x+1) \le x_r \le \text{col}\} \quad (10)$$

$$y_{\text{top}} = \max\{y_t : I(y_t, x) < 230, \forall 1 \le y_t \le (y-1)\} \quad (11)$$

$$y_{\text{bottom}} = \min\{y_b : I(y_b, x) < 230, \forall (y+1) \le y_b \le \text{row}\} \quad (12)$$

where $I(y, x)$ represents a pixel of the input image $I$ located at row '$y$' and column '$x$'. The image has 'row' number of rows and 'col' number of columns, i.e. $I$ is of dimension (col, row). And the pixel $P(x, y)$ is replaced by the mean of chosen direction pixels as shown in Eq. (13).

$$I(P) = \frac{I(P_{left}) + I(P_{\text{right}}) + I(P_{\text{top}}) + I(P_{\text{bottom}})}{4} \quad (13)$$

## 2.2 *Eight-Direction Traverse Approach*

The main logic of this approach is also similar to four-direction traverse approach. The only difference is that here, we traverse along eight directions (north, south, west, east, north-west, north-east, south-west, south-east) and find nearest non-reflection pixels in each direction (chosen direction pixels), shown in Fig. 5.

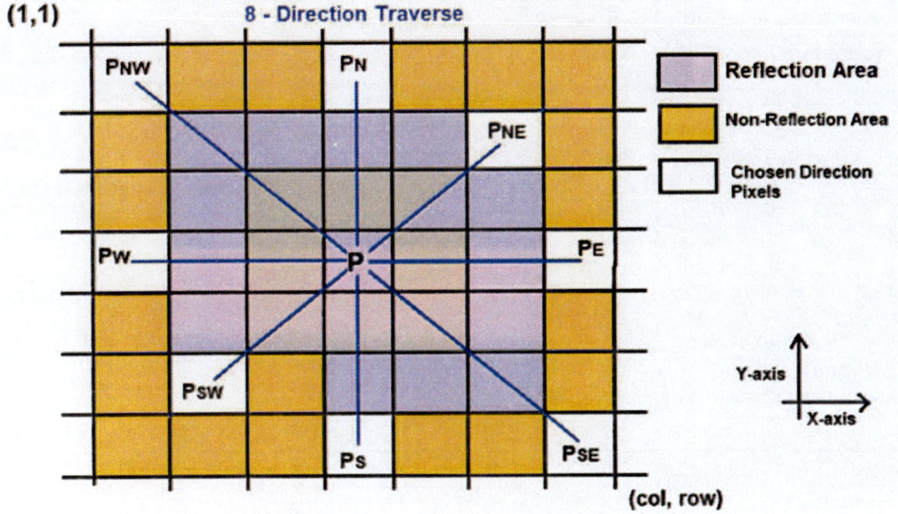


**Fig. 5** Eight-Direction traverse approach

For a pixel $P(x, y)$, the chosen direction pixels are: $P_N(x, y_\text{N})$, $P_\text{S}(x, y_\text{S})$, $P_\text{E}(x_\text{E}, y)$, $P_\text{W}(x_\text{W}, y)$, $P_\text{NE}(x_\text{NE}, y_\text{NE})$, $P_\text{NW}(x_\text{NW}, y_\text{NW})$, $P_\text{SE}(x_\text{SE}, y_\text{SE})$, $P_\text{SW}(x_\text{SW}, y_\text{SW})$. First four of them are computed as (Eqs. 14–17):

$$y_\text{N} = \max\{b : I(b, x) < 230, \forall 1 \leq b \leq (y - 1)\} \tag{14}$$

$$y_\text{S} = \min\{b : I(b, x) < 230, \forall (y + 1) \leq b \leq \text{row}\} \tag{15}$$

$$x_\text{W} = \max\{a : I(y, a) < 230, \forall 1 \leq a \leq (x - 1)\} \tag{16}$$

$$x_\text{E} = \min\{a : I(y, a) < 230, \forall (x + 1) \leq a \leq \text{col}\} \tag{17}$$

For image **I(1:row, 1:col)**, we can locate the pixels $P_\text{NW}$ and $P_\text{NE}$ using the **LocatePixel_NW** and **LocatePixel_NE** algorithms (Algorithms 2, 3). Similarly, we get the other two diagonal pixels $P_\text{SE}$, $P_\text{SW}$ with reverse algorithms. We now replace the reflection pixel $P(x, y)$ by the mean of its chosen direction pixels using the following Eq. (18):

$$I(P) = \frac{I(P_\text{N}) + I(P_\text{S}) + I(P_\text{E}) + I(P_\text{W}) + I(P_\text{NE}) + I(P_\text{NW}) + I(P_\text{SE}) + I(P_\text{SW})}{8} \tag{18}$$

For both of the two approaches discussed above, there could be a special case when the reflection area is attached to the image boundary. At that time, towards one or more direction, there might not be any non-reflection pixels. For those cases,

**Algorithm 2** Algorithm to find Pixel $P_{NW}(x_{NW}, y_{NW})$

```
procedure LocatePixel_NW(I)
    if pixelP(x, y) ∈ ReflectionArea then
        i ← (y − 1)                              ▷ Traverse North-West, outward direction
        j ← (x − 1)
        while i ≥ 1 && j ≥ 1 do
            if I(i, j) < 230 then                ▷ Non-reflection neighbor found
                y_NW ← (i)                       ▷ Get non-reflection pixel index
                x_NW ← (j) break
            end if
            i ← (i − 1)                          ▷ Next pixel in North-West direction
            j ← (j − 1)
        end while
    end if
end procedure
```

**Algorithm 3** Algorithm to find Pixel $P_{NE}(x_{NE}, y_{NE})$

```
procedure LocatePixel_NE(I)
    if pixelP(x, y) ∈ ReflectionArea then
        i ← (y − 1)                              ▷ Traverse North-East, outward direction
        j ← (x + 1)
        while i ≥ 1 && j ≤ col do
            if I(i, j) < 230 then                ▷ Non-reflection neighbor found
                y_NE ← (i)                       ▷ Get non-reflection pixel index
                x_NE ← (j) break
            end if
            i ← (i − 1)                          ▷ Next pixel in North-East direction
            j ← (j + 1)
        end while
    end if
end procedure
```

some directions are not considered (as no nearest non-reflection point exists) and mean is also calculated accordingly. That means, if we get non-reflection pixels in six directions successfully, then we sum up those six direction pixels and calculate the mean of those pixels only. It helps us to remove reflection along the image boundary area effectively.

## 3 Experiment Results

We conducted the experiments on an Intel Core i3 6th Gen laptop (2.00 GHz) with 8GB RAM using MATLAB R2016a environment, on UBIRIS-V2 [13] database images. For our experiment, we have chosen 1000 images (based on alphabetical order of Image Id) from the database as it contains almost 5800 images in total. The computation is very fast while processing grayscale image. But we performed every operation on three component matrices (i.e. red, green and blue) of RGB image, which takes 3X time than usual grayscale image processing. So, we take RGB images as input and our reflection removed images are also in RGB format.

Our approach has shown outstanding result with CASIA iris database images and for some other ideal iris database images. But our goal was to remove reflection from non-ideal eye images, hence we proceeded with non-ideal eye images of UBIRIS-V2 database, and the results are good enough. Some reflection removal examples are shown in Table 1. The second approach (Sect. 2.2) has given far better outcome than the first one (Sect. 2.1), as it considers non-reflection pixel search in eight directions and helps to acquire more similarity with its neighbourhood.

The results of our four-direction and eight-direction traverse approaches are shown in Table 1. It depicts how well our approaches remove specular reflection from pupil, iris and sclera region of eye image. The computation time is also given there. It is calculated based on the time taken by filtering, four-direction and eight-direction algorithm, for each image (using tic-toc MATLAB function).

**Table 1** Image outputs of different approaches

| Input | | 3X3 Filter & Output | | | 5X5 Filter & Output | | |
|---|---|---|---|---|---|---|---|
| Id | Input Image | (3x3) filter | 4-Dir (3x3) | 8-Dir (3x3) | (5x5) filter | 4-Dir (5x5) | 8-Dir (5x5) |
| (a) | 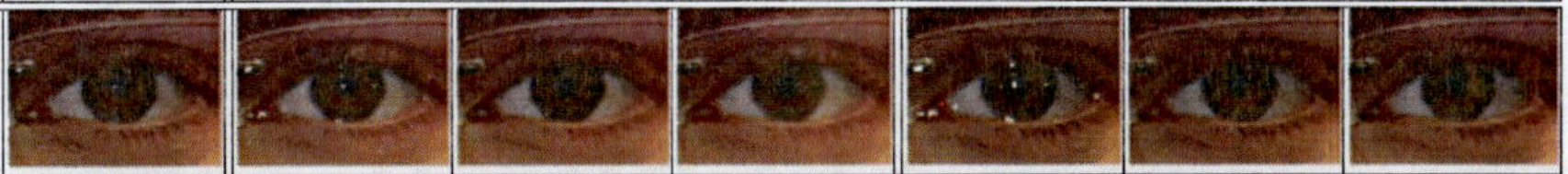 | | | | | | |
| Processing Time | | 0.043114 sec | | | 0.040483 sec | | |
| (b) |  | | | | | | |
| Processing Time | | 0.042365 sec | | | 0.037898 sec | | |
| (b) | 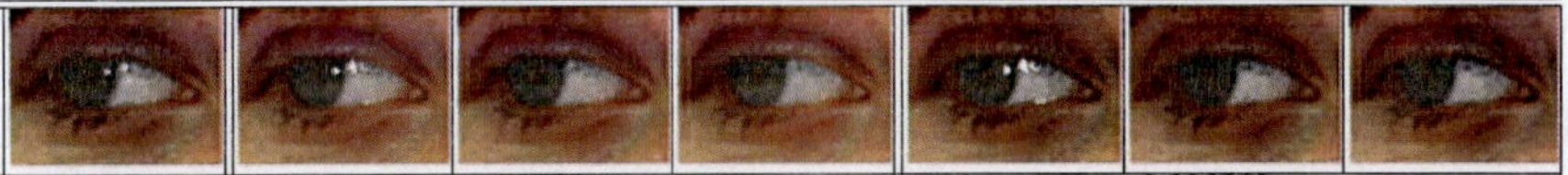 | | | | | | |
| Processing Time | | 0.040481 sec | | | 0.036308 sec | | |
| (d) | 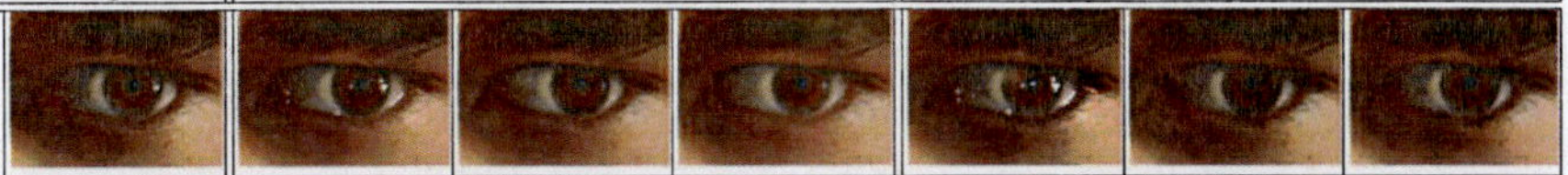 | | | | | | |
| Processing Time | | 0.046060 sec | | | 0.046948 sec | | |
| (e) | 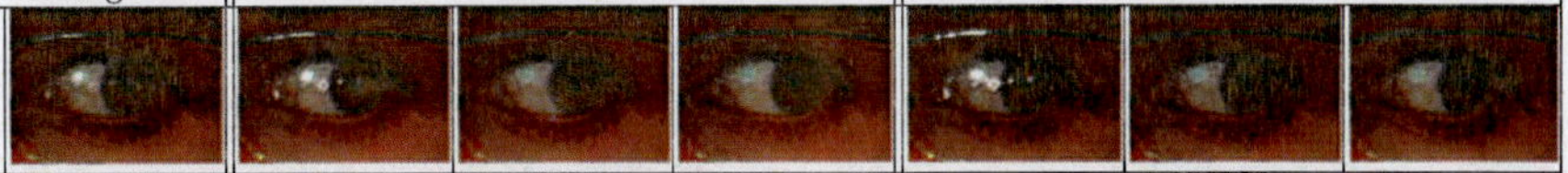 | | | | | | |
| Processing Time | | 0.045037 sec | | | 0.054706 sec | | |
| (f) | 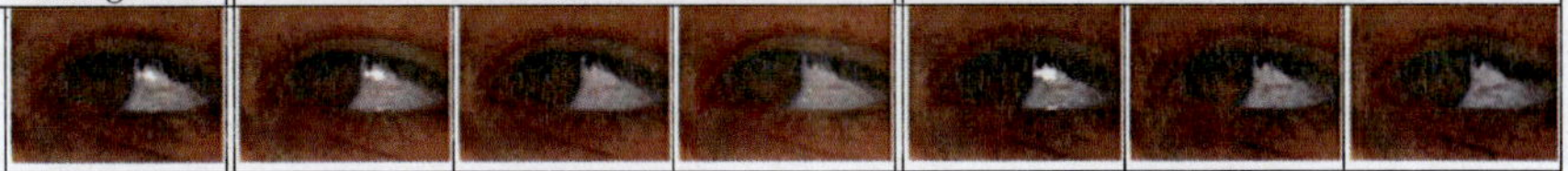 | | | | | | |
| Processing Time | | 0.042235 sec | | | 0.038798 sec | | |

(a, b)—Images with pupil reflection; (b) is a low brightness image. (c), (d)—Off-angle images with iris reflection; (d) has some low-intensity reflection area (properly captured by 5X5 filter). (e), (f)—Images with sclera reflection; (e) has reflection on eyeglass

We applied both 3 × 3 and 5 × 5 filters during four-direction and eight-direction traverse approach. For smaller reflection area, the difference was not noticeable, but for larger reflection areas, reflection area enlargement by 5 × 5 filter shows better performance. Both approaches have been compared basically on the basis of structural similarity index measurement (SSIM) with respect to the original input image, illustrated on Table 2. The SSIM index finds structural similarity between two images, and it is computed based on: luminance, contrast and structural term. The results clarify that our reflection removal approaches did not affect the original structure of the eye image.

To differentiate between the output of four-direction and eight-direction approaches, we calculated the SSIM index between their output images as well. The comparison is shown in Table 3. It indicates that outputs of these two approaches are little different by means of their structural similarity.

**Table 2** SSIM index (similarity to input image) Comparison

| Input Image ID | For 3 × 3 filter | | For 5 × 5 filter | |
|---|---|---|---|---|
| | Four-direction traverse | Eight-direction traverse | Four-direction traverse | Eight-direction traverse |
| a | 0.992860 | 0.992973 | 0.991310 | 0.991336 |
| b | 0.997672 | 0.997684 | 0.997139 | 0.997175 |
| c | 0.994445 | 0.994413 | 0.993395 | 0.993324 |
| d | 0.991357 | 0.991593 | 0.989135 | 0.989434 |
| e | 0.987160 | 0.986610 | 0.983556 | 0.982818 |
| f | 0.994977 | 0.994936 | 0.994481 | 0.994503 |

**Table 3** SSIM index comparison between output images

| Input Image ID | Output for 3 × 3 filter (four-direction traverse versus eight-direction traverse) | Output for 5 × 5 filter (four-direction traverse versus eight-direction traverse) |
|---|---|---|
| a | 0.9997 | 0.9996 |
| b | 0.9998 | 0.9997 |
| c | 0.9994 | 0.9993 |
| d | 0.9996 | 0.9993 |
| e | 0.9992 | 0.9982 |
| f | 0.9997 | 0.9996 |

## 4 Conclusion

We have proposed threshold-based reflection detection and two novel reflection removal techniques in this article. It has always been challenging to work with non-ideal iris database like UBIRIS. For non-ideal eye images, the reflection areas are not always the brightest or they are not captured in ideal scenario. Sometimes, in a reflection area, the pixels towards the boundary are less brighter than the pixels which are nearer to the centre. Though we have applied two types of filters to cover those less bright reflection areas using EnlargeReflectionArea function (Algorithm 1), sometimes it becomes harder to capture the whole less bright reflection area. Our approach has provided quite good results for UBIRIS database images as shown in Table 1. The proposed methods faced difficulty in removing specular reflection from eye images which have very poor illumination. So we are looking forward to a new strategy based on intensity or histogram adjustment where we can also modify those low brightness reflection areas without disturbing the property of original eye image.

## References


1. Sung H, Lim J, Park J-H, Lee Y (2004) Iris Recognition using collarette boundary localization. In: Proceedings of the 17th international conference on pattern recognition, vol 4
2. Shin KY, Nam GP, Jeong DS, Cho DH, Kang BJ, Park KR, Kim J (2012) New iris recognition method for noisy iris images. Pattern Recogn Lett 33(8)
3. Laughton MA, Warne DJ (2003) Electrical engineer's reference book, 16th edn. Elsevier, pp 21–22 (Chapter-lighting, subsection-glare)
4. Daugman J (2004) How iris recognition works. IEEE Trans Circ Syst Video Technol 14(6)
5. He Z, Tan T, Sun Z, Qui X (2009) Toward accurate and fast iris segmentation for iris biometrics. IEEE Trans Pattern Anal Mach Intell 31(1)
6. Jarjes AA, Wang K, Mohammed GJ (2010) Gvf snake-based method for accurate pupil contour detection. Inf Technol J 9(2):1653–1658
7. Aydi W, Masmoudi N, Kamoun L (2011) New Corneal reflection removal method used in iris recognition system. Int J Electr Comput Energetic Electron Commun Eng 5(5)
8. Jamaludin S, Zainal N, Zaki WMDW. The removal of specular reflection in noisy iris image. J Telecommun Electron Comput Eng 8(4). ISSN 2180-1843, eISSN 2289-8131
9. Reddi N, Rattani A, Derakshani R (2016) A robust scheme for iris segmentation in mobile environment. In: IEEE symposium on technologies for homeland security (HST). https://doi.org/10.1109/THS.2016.7568948
10. Chui CK (2009) An MRA approach to surface completion and image inpainting. Appl Comput Harmonic Anal 26(2). https://doi.org/10.1016/j.acha.2008.05.001
11. Wang N, Li Q, Abd El-Latif AA, Zhang T, Niu X (2014) Toward accurate localization and high recognition performance for noisy iris images. Multimedia Tools Appl 71(3) 1411–1430. https://doi.org/10.1007/s11042-012-1278-7
12. Bertalmio M, Bertozzi AL, Sapiro G (2001) Navier-Stokes, fluid dynamics, and image and video inpainting. In: Proceedings- IEEE computer society conference on computer vision and pattern recognition (CVPR), vol 1, Kauai, USA
13. Proenca H, Filipe S, Santos R, Oliveira J, Alexandre KA (2010) The UBIRIS.v2: a database of visible wavelength iris images captured on-the-move and at-a-distance. IEEE Trans Pattern Analy Mach Intell 32(8):1529–1535. ISSN 0162-8828, http://doi.ieeecomputersociety.org/10.1109/TPAMI.2009.66

# Dynamic Block Matching Algorithm (BMA) Scheduling Algorithm for Fast and Accurate Motion Estimation

**Hussain Ahmed Choudhury, Nidul Sinha, and Monjul Saikia**

**Abstract** The challenge of motion estimation (ME) process in the video compression (VC) technique is to balance the complexity, visual experience, and speed, which directly affects the efficiency of video compression. An abundant number of block matching algorithms (BMA) was designed only to achieve the above objective of ME. This paper presents a dynamic version of existing BMAs for ME, which uses the ideas and concepts of the three most effective BMAs, namely reduced three-step logarithmic search (RTSLS), simple and efficient search (SES), and adaptive rood pattern search (ARPS) reported till date. The proposed algorithm yields equivalent visual experience than that of full search (FS) but is several times faster than FS.

**Keywords** Block matching algorithm · Mean absolute difference · Sum of absolute difference · Three-step search

## 1 Introduction

The research on video compression is a very old area, but growing demand and the use of videos in all sectors of the current digital world never let the area go out of scope. The use of appropriate and efficient BMAs in the ME block is highly recommended to make the whole VC process efficient. Accordingly, Jain and Jain (1981) first put forward the concept of full search [1–3] BMA, which paves the way for all other developed algorithms. The BMAs usually follow some fixed initial search patterns,

H. A. Choudhury (✉)
Associate Professor, Department of IT, CVR College of Engineering, Hyderabad 501510, India
e-mail: jainhussain2018@gmail.com

N. Sinha
National Institute of Technology, Silchar, Assam 788010, India
e-mail: nidul.sinha@gmail.com

M. Saikia
North Eastern Regional Institute of Science and Technology, Nirjuli, Arunachal Pradesh 791109, India
e-mail: monjuls@gmail.com

 
A. Maji et al. (eds.), *Proceedings of the International Conference on Computing and Communication Systems*, Lecture Notes in Networks and Systems 170,
https://doi.org/10.1007/978-981-33-4084-8_33

and gradually, the search window and search patterns change based on different criteria, which are mostly the location of minimum block distortion measure (BDM). Gradually, different BMAs are developed, namely three-step search (TSS) [3–5], fast TSS [4], efficient TSS [6], cross-search algorithm (CSA) [7], new three-step search (NTSS) [8], four-step search (4SS) [9], diamond search (DS) [10–12], small cross-diamond search (SCDS) [13], reduced three-step logarithmic search (RTSLS) [14], etc. Few other BMAs are available which considers the correlation between the current MB and neighboring MBs like adaptive rood pattern search (ARPS) [15], inter-block correlation-based BMA [16], and correlation-based rood pattern search (CBRPS) [17, 18]. Out of the BMAs mentioned above, RTSLS, SES is found to be better than most of the BMAs, whereas ARPS was found to be one of the best BMA, which not only reduces the search point, i.e., computational complexity, but also maintained the visual representation of the compensated image. There exist many different types of videos, whether they are of different formats, whether they are of different resolution based on camera quality or speed of the content of the videos. To address the above issues, a new algorithm called dynamic BMA scheduling algorithm (DBMASA) is proposed in this paper which takes the advantages of RTSLS, SES, and ARPS and based on the type of movement and number of moving MB, allows the switching from one BMA to another BMA per-frame basis which may suit better keeping in mind the overall type of video to be processed.

This paper is organized into six sections, which are as follows: Sect. 2 explains the block-based ME technique in brief. The proposed methodology and its key terms used are elaborated in Sect. 3. The proposed BMA is explained in Sect. 4. Section 5 highlights the simulation results and analyzes them with the results of different BMAs. The comparative results with existing techniques are followed by the conclusion under Sect. 6.

## 2 Block-Based Motion Estimation (BBME)

The process of block matching is a searching practice that is realized through a comparison between the reference frame and the candidate frame. As pixel-level comparison may result in the wrong result due to spatial redundancy and high computations are involved, the block-level comparison is preferred. In block-level comparison, each frame is split into equal-sized macro blocks (MB), and then comparisons are made to find out the best matching MB in a reference frame assuming that all pixels of an MB undergo the same amount of translation. With the help of BMA, the difference between coordinate locations of MB in the target frame to that of anchor frame is found, which is known as Motion vector (MV), which in turn helps in the estimation of motion of MB. This process is referred to as block-based motion estimation (BBME or BME).

## 3 Proposed Methodology and Key Parameters

The proposed algorithm works in two stages. In the first phase, the frames are classified into nine categories based on motion vector indicator and percentage (%) of static MBs. Then, in the second phase, select the most appropriate algorithm out of three prevalent efficient algorithms, SES, RTSLS, and ARPS for motion estimation. The parameter '$\mu$' called as MV indicator or speed indicator, which is calculated by Eq. (1) and used to specify if the videos are slow, medium, fast, or very fast, etc. The '%' of static MBs denoted by '$\rho$' is calculated using Eq. (2), used to notify other sub-categories of frames under the top four categories as given in Table 1. The frames are classified based on '$\mu$' and '$\rho$' values, and accordingly, '$\mu_{avg}$' and '$\rho_{avg}$' are calculated using Eqs. (3) and (4) and used to classify the experimental videos based on rules listed in Table 1.

$$\mu = \frac{\sum_{i=1}^{n}\{\text{abs}(\text{MV}_x) + \text{abs}(\text{MV}_y)\}}{n} \tag{1}$$

$$\rho = \frac{\text{No. of static MB}}{\text{Total no. of MB per frame (n)}} \times 100\% \tag{2}$$

**Table 1** Classification rules for frame and videos

| S. No. | $\mu$ values | $\rho$ (in %) | Category | Remarks (types of videos) |
|---|---|---|---|---|
| 1 | $0 \leq \mu \leq 2$ | $0 < \rho \leq 40$ | SH | Slow video with high moving MB |
| 2 | | $40 < \rho \leq 70$ | SM | Slow video with moderate moving MB |
| 3 | | $70 < \rho \leq 100$ | SL | Slow video with low number of moving MB |
| 4 | $2 < \mu \leq 4$ | $0 < \rho \leq 40$ | MH | Medium speed video with high number of moving MB |
| 5 | | $40 < \rho \leq 70$ | MM | Medium speed video with moderate moving MB |
| 6 | | $70 < \rho \leq 100$ | ML | Medium speed video with low no. of moving MB |
| 7 | $4 < \mu \leq 6$ | $0 < \rho \leq 40$ | FH | Fast video with high number of moving MB |
| 8 | | $40 < \rho \leq 70$ | FM | Fast video with a moderate number of moving MB |
| 9 | | $70 < \rho \leq 100$ | FL | Fast video with low number of moving MB |
| 10 | $6 < \mu \leq \text{p}$ | $0 < \rho \leq 40$ | VFH | Very fast video with high number of moving MB |
| 11 | | $40 < \rho \leq 70$ | VFM | Very fast video with a moderate number of moving MB |
| 12 | | $70 < \rho \leq 100$ | VFL | Very fast video with low number of moving MB |

$$\mu_{\text{avg}} = \frac{\mu_1 + \mu_2 + \mu_3 + \cdots + \mu_n}{n} \quad (3)$$

$$\rho_{avg} = \frac{\rho_1 + \rho_2 + \rho_3 + \cdots + \rho_n}{n} \quad (4)$$

Based on the above table and criteria, the frames are first classified, and then, frame by frame, the algorithms will check the speed of an object and static MBs to choose the appropriate algorithm out of three efficient algorithms namely RTSLS, ARPS, and SES. Since the proposed algorithm uses the features of frames for selection of best-suited one, the algorithm is named as dynamic BMA scheduling algorithm (DBMSA).

## 4 Proposed Algorithm

The proposed DBMSA will start the implementation of the sequence of BMA and will try to explore results obtained by the combination of three most efficient algorithms namely reduced three-step logarithmic search (RTSLS), simple and efficient search (SES), and adaptive rood pattern search (ARPS). For our convenience, the first letter of each BMA mentioned above is to represent the sequence, i.e., RSA is used for the above sequence ("R" for RTSLS, "S" for SES, and "A" for ARPS) which are taken. Since the motion estimation is done by switching between three existing BMAs depending upon the frame criteria for which it also can be regarded as dynamic BMA scheduling algorithm. The algorithm has the following steps:

**Algorithm: Dynamic Block Matching Scheduling Algorithm**

| | |
|---|---|
| Step 1 | Set MB size = 8 × 8, window parameter ($p$) = ± 7. Apply ARPS on the first frame. Calculate the displacement parameter for the first frame ($\mu_1$) and % of static blocks ($\rho_1$) |
| Step 2: | (a) For each frame 1:29, carry out the following steps if $0 < \mu_1 < p$, else go to step 3. If the value of $0 < \mu_1 < = 2$, check the conditions (i) through (iii)<br>(i) If ($0 < \rho_i < = 40\%$): Choose the algorithm "RTSLS"<br>(ii) If ($40 < \rho_i < = 70\%$): Choose the algorithm "SES"<br>(iii) If ($70 < \rho_i < = 100\%$): Choose the algorithm "ARPS"<br>(b) If the value of $2 < \mu_1 < = 4$, check the conditions (i) through (iii) again<br>(iv) If the value of $4 < \mu 1 < = 6$, check the conditions (i) through (iii) again.<br>(v) If the value of $6 < \mu_1 < = p$, check the conditions (i) through (iii) again<br>Apply ARPS to the video |
| Step 3: | Frames are classified and calculate the average value of $\mu$ and $\rho$, i.e., '$\mu_{\text{avg}}$' and '$\rho_{\text{avg}}$,' and apply the rules mentioned in Table 1 to classify the videos |

## 5 Experimental Results

To bring out the merits and efficiency of the proposed algorithm, extensive experiments are carried out on a system equipped with Intel(R) Pentium(R) 3558U 1.70 GHz 4 GB RAM and 64-Bit Windows 10 operating system. The software used for carrying out the experiments is MATLAB R2013a. The simulations are done on standard video sequences namely Claire, Foreman, News, Football, Akiyo, Container, Bus, Cricket Ball, Salesman, Coastguard, etc., and they are classified by applying the rules stated in Table 1, and the classification of videos are given in Table 5.

Since all the videos are of different qualities and different types, the proposed algorithm is applied to 30 frames of each video type. The proposed algorithm is applied on different videos, and the detailed results obtained on "Coastguard.qcif" "carphone.qcif" are given in Table 2 which shows the MV displacement vector, % of moving MV. The computational complexity obtained by the proposed algorithm DBMASA represented by 'RSA' on different videos is given in Table 3, and the obtained PSNR values in Table 4.

Based on the results mentioned in Tables 3 and 4, the "$\mu_{avg}$" and "$\rho_{avg}$" are calculated, and the videos are classified in 12 different sub-categories under four types of videos, namely slow, medium, fast, and very fast as given in Table 5.

### *5.1 Comparative Analysis*

The results in terms of the number of comparisons required per MB are shown in Fig. 1 by proposed DBMASA(RSA) over the video "CricketBall.avi" and a few other existing pattern-based BMAs wherein it is evident that the proposed algorithm is better. The PSNR values for the reconstructed image for the same video sequence is also highlighted in Fig. 2. It has been observed that the proposed RSA algorithm performs better than most of the existing BMAs both in case of computational complexity and the quality of the reconstructed image.

## 6 Conclusion

The proposed BMA is dynamic in nature and schedules the three most efficient BMAs based on the properties of MBs in a frame and video as a whole. This DBMASA represented by RSA combines the advantages of RTSLS, SES, and ARPS to overcome the rigidity of existing BMA and provides the provision of selecting a BMA based on the properties of MB which was not available in no other reported work. Firstly, the proposed algorithm overcomes the constraint of existing BMA of not considering the video types and provides the facility to choose the BMAs per-frame basis. The

**Table 2** Proposed RSA results showing frame-wise switching and frame types (Coastguard. qcif)_RSA

| Parameters | Frames | | | | | | | | | |
|---|---|---|---|---|---|---|---|---|---|---|
| | F1 | F2 | F3 | F4 | F5 | F6 | F7 | F8 | F9 | F10 |
| Smax | 1.011 | 1.007 | 1.143 | 1.028 | 1.007 | 1.102 | 1.021 | 1.107 | 1.066 | 1.078 |
| Static% | 32.071 | 27.273 | 39.899 | 27.525 | 27.525 | 40.657 | 38.636 | 45.707 | 46.212 | 35.606 |
| F. type | SH | SH | SH | SH | SH | SM | SH | SM | SM | SH |
| Algorithm | ARPS | RTSLS | RTSLS | RTSLS | RTSLS | RTSLS | SES | RTSLS | SES | SES |
| Comp | 6.917 | 12.394 | 12.404 | 12.394 | 12.394 | 12.396 | 17.073 | 12.394 | 17.088 | 17.058 |
| PSNR | 34.206 | 35.581 | 33.600 | 36.167 | 35.052 | 32.595 | 34.821 | 32.531 | 34.117 | 32.891 |

**Table 3** Computational complexities obtained by proposed DBMASA (RSA) on different videos

| Videos | Frames | | | | | | | | | |
|---|---|---|---|---|---|---|---|---|---|---|
| | F1 | F2 | F3 | F4 | F5 | F6 | F7 | F8 | F9 | F10 |
| Cricket ball | 6.061 | 6.479 | 6.415 | 6.541 | 6.347 | 6.577 | 6.395 | 6.319 | 6.293 | 6.942 |
| Stefan | 11.042 | 12.706 | 12.711 | 12.713 | 12.698 | 12.702 | 12.704 | 12.711 | 12.698 | 12.705 |
| Claire | 7.586 | 16.359 | 16.745 | 6.740 | 7.346 | 16.682 | 6.823 | 7.409 | 16.735 | 6.694 |
| Cycling | 10.771 | 12.863 | 17.063 | 17.241 | 8.586 | 17.238 | 7.740 | 8.369 | 17.271 | 7.518 |
| Cricket Batting | 8.942 | 16.756 | 8.870 | 8.888 | 8.951 | 16.786 | 9.128 | 16.663 | 9.064 | 16.696 |
| Bus | 10.520 | 12.563 | 12.573 | 12.558 | 12.545 | 12.518 | 12.500 | 12.518 | 12.543 | 12.548 |

**Table 4** PSNR values obtained by the proposed DBMASA (RSA) on different Videos

| Videos | Frames | | | | | | | | | |
|---|---|---|---|---|---|---|---|---|---|---|
| | F1 | F2 | F3 | F4 | F5 | F6 | F7 | F8 | F9 | F10 |
| Cricket ball | 36.346 | 35.668 | 35.808 | 35.489 | 36.282 | 35.045 | 36.349 | 35.275 | 37.263 | 32.268 |
| Stefan | 25.378 | 22.701 | 21.984 | 21.346 | 20.608 | 19.986 | 19.658 | 19.561 | 19.411 | 19.504 |
| Claire | 41.405 | 36.963 | 39.605 | 42.696 | 38.834 | 42.556 | 40.176 | 39.514 | 39.174 | 40.046 |
| Cycling | 26.795 | 28.923 | 29.257 | 28.971 | 28.970 | 28.361 | 29.606 | 30.101 | 29.814 | 29.944 |
| Cricket batting | 39.766 | 41.289 | 41.651 | 41.540 | 41.485 | 40.952 | 39.366 | 39.962 | 40.010 | 38.317 |
| Bus | 21.364 | 19.769 | 20.481 | 20.329 | 18.728 | 20.081 | 20.244 | 20.270 | 20.722 | 20.789 |

**Table 5** Classification of videos based on the results obtained by proposed DBMASA(RSA)

| S. No. | Video sequence | Format | $\mu$ values | $\rho$ (in %) | Remark |
|---|---|---|---|---|---|
| 1 | Coastguard | Qcif | 1.058 | 28.224 | SH |
| 2 | Garden | Sif | 2.89 | 4.34 | MH |
| 3 | Carphone | Qcif | 2.12 | 45.682 | MM |
| 4 | Claire | Qcif | 4.19 | 76.01 | FL |
| 5 | Stefan | Sif | 3.78 | 31.41 | MH |
| 6 | Cricket ball | Avi | 4.903 | 82.6845 | FL |
| 7 | Cycling | Avi | 4.24 | 75.09 | FL |
| 8 | Crew | cif | 3.83 | 27.89 | MH |
| 9 | Cricket batting | avi | 7.34 | 67.564 | VFM |
| 10 | Balloon falling | avi | 6.14 | 13.91 | VFH |
| 11 | Takeoff | cif | 5.60 | 16.12 | FH |
| 12 | Bus | qcif | 4.105 | 8.55 | FH |

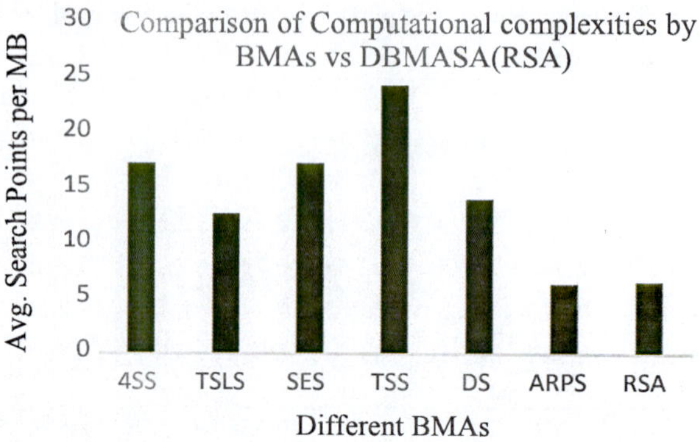


**Fig. 1** ASP for comparison by proposed DBMASA (RSA) versus all algorithm for "Cricket Ball.avi"

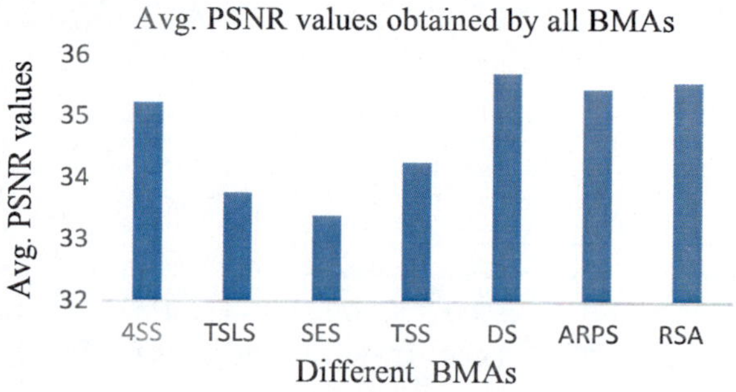


**Fig. 2** Avg. PSNR for first 30 frames by proposed DBMASA (RSA) versus all algorithm for Cricket Ball.avi

same algorithm is not applied to all frames as the movement of MB between a pair of frames is different from another pair of frames. Secondly, using the values of '$\mu_{avg,}$' the different videos are classified as slow, moderate, fast, and very fast type of video, and using the value of '$\rho_{avg,}$' the above videos are further classified in SH, SM, SL, MH, MM, ML, FH, FM, FL, VFH, VFM, and VFL. Lastly, the results have proved that the proposed RSA performs better than almost all the existing BMAs for the video sequences even when compared with ARPS, which was regarded as the best among the existing pattern-based BMA.

**Acknowledgements** The authors would like to thank the National Institute of Technology, Silchar, for providing enough scope and infrastructure to carry out my research work. The authors would also like to thank Ms. Nasima A. Laskar for her valuable suggestions and contributions.

## References

1. Jain J, Jain A (1981) Displacement measurement and its application in interframe image coding. IEEE Trans Commun 29(12):1799–1808
2. Chan MH, Yu YB, Constantinides AG (1990) Variable size block matching motion compensation with applications to video coding. IEE Proc I (Commun, Speech, Vision) 137(4):205–212

3. Koga T, Iinuma K, Hirano A, Iijima Y, Ishiguro T (1981) Motion-compensated interframe coding for video conferencing. Proc NTC81, pp C9.6.1–9.6.5
4. Kim JN, Choi TS (1998) A fast three-step search algorithm with minimum checking points using unimodal error surface assumption. IEEE Trans Consum Electron 44(3):638–648
5. Choudhury HA, Saikia M (2013) Comparative study of block matching algorithm for motion estimation. In: International conference on advanced computer science and information technology, Bhubaneswar, India. vol 17
6. Chau LP, Jing X (2003) Efficient three-step search algorithm for block motion estimation in video coding. In: 2003 IEEE international conference on acoustics, speech, and signal processing, 2003. Proceedings. (ICASSP'03). IEEE, vol 3, pp III-421
7. Ghanbari M (1990) The cross-search algorithm for motion estimation (image coding). IEEE Trans Commun 38(7):950–953
8. Li R, Zeng B, Liou ML (1994) A new three-step search algorithm for block motion estimation. IEEE Trans Circuits Syst Video Technol 4(4):438–442
9. Po LM, Ma WC (1996) A novel four-step search algorithm for fast block motion estimation. IEEE Trans Circuits Syst Video Technol 6(3):313–317
10. Zhu S, Ma K-K (1997) A new diamond search algorithm for fast block-matching motion estimation. In: 1997 proceedings international conference information, communications and signal processing (ICICS), vol 1, pp 292–296
11. Barjatya A (2004) Block matching algorithms for motion estimation. IEEE Trans Evol Comput 8(3):225–239
12. Choudhury HA, Saikia M (2015) Block matching algorithms for motion estimation: a performance-based study. In: Advances in communication and computing. Springer, New Delhi, pp 149–160
13. Cheung C-H, Po L-M (2002) A novel small-cross-diamond search algorithm for fast video coding and videoconferencing applications. In: Proceedings international conference on image processing, vol 1. IEEE, pp I–I
14. Choudhury HA, Saikia M (2014) Reduced three steps logarithmic search for motion estimation. In: International conference on information communication and embedded systems (ICICES2014). IEEE, pp 1–5
15. Nie Y, Ma KK (2002) Adaptive rood pattern search for fast block-matching motion estimation. IEEE Trans Image Process 11(12):1442–1449
16. Hsieh CH, Lu PC, Shyn JS, Lu EH (1990) Motion estimation algorithm using inter block correlation. Electron Lett 26(5):276–277
17. RSAi JC, Hsieh CH, Weng SK, Lai MF (1998) Block-matching motion estimation using correlation search algorithm. Signal Process Image Commun 13(2):119–133
18. Choudhury HA, Sinha D, Saikia M Correlation based rood pattern search (CBRPS) for motion estimation in video processing. J Intel Fuzzy Syst (Preprint) 1–11

# Analysis of Biomedical Image, Data and Signal

# BrainTalk—An EEG Motor Imagery BCI for Giving Non-speaking People a Voice: A Position Paper

**Upasana Talukdar and Dhanapati Shougrakpam**

**Abstract** Patients suffering from locked-in syndrome or disorders of consciousness cannot answer questions or express their thoughts, and thus making the daily interactions very challenging. In such cases, systems that can express their desires/views and thoughts without requiring any movement hold promise. Brain-Computer Interface is physiological computing where brain signals are used to identify the intent of a user. This paper presents the idea of implementing a system using Brain-Computer Interface (BCI) that can aid people suffering from disorders of consciousness or locked-in syndrome to communicate their ideas and views.

**Keywords** Locked-in-syndrome · Motor imagery · Language dictionary · Brain-computer-interface · Speech

## 1 Introduction

Motor neuron disorders such as Primary lateral sclerosis (PLS), amyotrophic lateral sclerosis (ALS), paralysis, stroke, and brain stem injuries upset the neural transmission and affects the motor neurons. This declines the muscle control and hinders the communication ability of the patients. As a result of which, they are incompetent to express their needs, desires, and views [1]. Thus, it is challenging for caregivers or the medical staff to evaluate their mental state, making their everyday interactions difficult [1]. Although such patients cannot give any voluntary response to commands, they can persue instructions, like imagination of movement [2]. To study these imagination of movements, neuroscience provides a sterling platform with

U. Talukdar (✉)
Department of Data Science and Analytics, Central University of Rajasthan, Ajmer, Rajasthan, India
e-mail: upasana@curaj.ac.in

D. Shougrakpam
Department of Linguistics, Central University of Rajasthan, Ajmer, Rajasthan, India
e-mail: dhanapati@curaj.ac.in



A. Maji et al. (eds.), *Proceedings of the International Conference on Computing and Communication Systems*, Lecture Notes in Networks and Systems 170,
https://doi.org/10.1007/978-981-33-4084-8_34

different neuroimaging methods that monitors the brain activities linked with task performance as these patients can't move, Hence, they may be able to understand and persue simple instructions although they do not demonstrate behavioural indicators of consciousness [3, 4].

Brain-Computer Interface (BCI) is a methodology where brain signals are used to control external devices. They are useful for the persons suffering from motor disabilities who need a system that enable them to function at the bedside without the requirement of any movement or vision. The history of BCI starts with Hans Berger's first introduction of Electroencephalogram (EEG) in 1929. Starting therefrom, research on BCI has been observed a fast growth. Brain signals can be acquired by means of *invasive* and *non-invasive* methods [5]. The *invasive* methods detect the electric fields produced within the cortex (e.g. spikes, local field potentials) while the *non-invasive* methods acquire brain signals from the scalp of the user (e.g. EEG, Magnetic resonance imaging (MRI) etc.). Besides, these two modes, another mode is *partially invasive* which detects the electric fields produced at the cortical surface (e.g. Electrocorticogram (EcoG)). Traditionally BCI is perceived as an actor, that aims at the users motor disabilities to aid them to connect with the outside world. Motor imagery (MI) BCI that transforms the imagination of movements to external commands, thus attained quite popularity in the field of rehabilitation robotics. MI is the mental process of simulating a given action without actually performing the movement. EEG is normally used for MI BCI because of its non-invasiveness, low cost, portability and high temporal resolution. They are appealing and have been studied for 40 years [1].

Apart from using BCI as an actor, BCI can also be used for other applications such as estimation/detection of cognitive states, remediation of depression, anxiety etc. Such systems have the potential to aid and enable people suffering from locked in syndrome (LIS) to answer questions using only their thoughts. Loss of verbal communication leads to loss of social interaction [6]. Under such situation, a caregiver or an attendant faces challenges in finding out the desires and needs of the patient. Henceforth, designing of BCIs to aid the patients suffering from LIS in communicating their desires/views holds promise. EEG would be the preferred choice. This paper presents the idea of using EEG-based motor imagery BCI to aid people in answering questions/expressing thoughts or giving voluntary response to commands.

## 2 BCI for Generating Speech

BCI can acquire and decodes brain signals that correspond to speech production [6]. A typical speech BCI is illustrated in Fig. 1 (picture adapted from Cooney et al. [6]). In the figure, the user imagines the sentence "I am thirsty!". The brain signals are acquired which are temporally aligned with each word followed by extraction of features and classification. The BCI generates the output: the display of the sentence "I am thirsty" on the monitor.

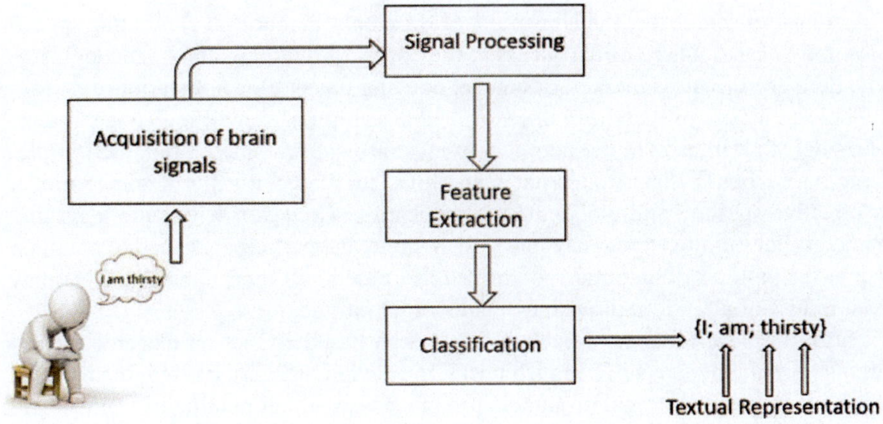


**Fig. 1** A typical speech BCI

Attempts to decryption of the content of languages directly from brain activity have gained notable success. Literature reports various works on the neural association of consonants as well as vowels [7], phonemes [8], syllables [9], and sentences [10].

Leuthardt et al. [11] presented a review on the scientific, clinical, and technical findings illustrating the motivation and effectiveness of speech-based BCIs. Brumberg et al. [12] presented a framework for silent speech based BCI to predict intended speech information directly from the neuronal activities associated with the production of speech. A group of San Francisco neuroscientists and engineers is developing a BCI that tries to figure out what people with vocal disabilities are trying to say by examining how they attempt to move their mouths. By analysing the neuronal activities that corresponds to the movement of the lips, tongue, jaw and larynx, a speech prosthetic is trying to give these people a voice.[1] Another biomedical device named as mindBEAGLE [1] is the first EEG based BCI system that gives people suffering from locked-in syndrome, the ability to answer yes-or-no questions using only their thoughts.

Chang et al. [13] tried to comprehend how the electrical activity in the brain that codes information, can generate speech and came up with a neural decoder that explicitly decodes the kinematic and sound representations encoded in human cortical activity and gives rise to audible speech. A team at Columbia University [14] portrayed that such a system could give rise to speech using neural activities from the auditory cortex of the brain. The work is still preliminary, and there are a lot of challenges.

---

[1] https://www.discovermagazine.com/mind/brain-machine-interface-could-give-voice-to-the-voiceless.

A computer scientist at the University of Bremen said that BCI generating speech is limited to cases where direct access to the cortex of the brain can be made. If it is to be done for the mass market, of course, opening the skull is not an option.[2]

The second issue is that the brains are not alike. This makes it difficult to generalize the models that transform the neural activities into speech. Oken et al. [15] implemented a keyboard that amalgamates an EEG classifier with the language model for effective spelling and eloquent communication. However, with such a system, characters per minute were fairly low [15]. Second, the participants need to maintain gaze at the laptop or computer monitor, so as to maintain their attention. This may result in loss of attention, drowsiness and mental fatigue.

Third, the development of such direct speech-based BCI is an extremely challenging. For, the people suffering from LIS, the design of BCI uses imagined speech which encompasses internal pronunciation of words without making movement and emitting sounds. Imagined speech has been least explored [6]. In designing BCIs based on imagined speech, researchers searching to differentiate word classes from brain activities should look at the effect of length of the word, frequency of the word, word repetition, as well as syntactic and semantic context [6]. A neurophysiological response is a complex concoction of the semantic and syntactic interactions of the given words [6].

Let us consider two scenarios of linguistics cues for lexical mapping. Taking into consideration the most significant speech of caregivers of locked-in syndrome patients, what one can deduct from is the most semantically paired "yes" and "no" questions. In English, the response could be either a Yes or No, which is not universally applicable. Here, language data from Manipuri, a language belonging to the Tibeto-Burman language family spoken in the north-eastern part of India is examined for the present study. Manipuri is a highly agglutinative language, where one word qualifies the status of a sentence. Let us observe the following data from Manipuri which can be used for task-elicited imagined speech by caretakers of patients with locked-in syndromes.

**Scenario 1: Lexical Mapping of Action Words**

1. cani bra 'Would you like/wish to eat?'
2. $t^h$akni bra 'Would you like/wish to drink?'
3. tumni bra 'Would you like/wish to sleep?'

From the above set of examples, it can be drawn that if the response is in affirmative, it can be stated by saying "hoi" (Yes) for all of the above (1–3) but if the response is negative, it cannot be indicated by just a single word as indicated below:

caningde, $t^h$akni de and tumni de for 1, 2 and 3 respectively. Here, we can see that

cani bra: 'Would you like/wish to eat?' is a question.
cani de: 'No.' is a response.

Here, it takes the form

[2]https://www.wired.com/story/machine-reads-your-mind-talks/.

a. ca-ni -bra eat-like/wish → INT[3]
b. ca-ni -de eat-like/wish → NEG.[4]

**Scenario 2: Lexical Mapping of the State of Mind or Emotions**

1. wabra 'Are you tired?'
2. nu aibra 'Are you fine or feeling better?'

if the response is in affirmative, it can be stated by saying "hoi" (Yes) for both (a) and (b) but if the response is negative, it cannot be indicated by just a single word as shown below:

Here, wabra: 'Are you tired?' is a Question.
wade 'No/I am not.' is a Response.
wa-bra :Tired → INT
wa-de : tired → NEG.

Similarly,

nu aibra 'Are you fine?' is a Question.
nu aite 'No/I am not fine.' is a Response.
nu ai-bra → INT
nu ai-te → NEG.

Here again, another challenge is how to map the negative morpheme te de, in which the phoneme /*t*/ and /*d*/ are alternates.

From scenario 1 and 2, we can draw that this may take the form of semantic or phonological priming.[5] The semantic categorisation leads to the distinction between words. Categorical differences between words can produce notably distinct neural activities, and this alteration may support classification [6]. Take for instance action words from English like eat, drink, sleep. They can have the effect on the neuronal activities associated with actually carrying out the activity.

In scenerio 2 (i) and (ii), the phonemes /*t*/ and /*d*/ in the negative morpheme **te** and **de** are in complementary distribution i.e., there is no change in the meaning. Varying speakers pronounced words differently. As for instance, "might" and "mite" are homophones. Similarly, "fees" and "fish" are homophones [16]. Varying speakers pronounced the word wade also as wate 'No/I am not (tired)', and this is the case of free variation i.e., the interchangeable relationship between two phones, in which the phones may substitute for one another in the same environment. It is substantial to understand the phenomena of imagined speech and individual speech processes [6]. This may lead to incorrect recognition of the letters and may give an incorrect predicted word.

This review claims for a framework of BCI that can lead towards a naturalistic form of communication.

---

[3]Interrogative.

[4]Negative.

[5]Phone: The smallest identifiable unit in speech.

# 3 BrainTalk—An EEG Based Motor Imagery BCI for Aiding People with LIS

Here, we present an idea to design a system BrainTalk by amalgamating the two tasks of BCI: (a) motor imagery and (b) generation of response to the commands. The aim is to enable people with vocal disabilities to reach to a level of independence for performing their daily living activities.

BCI has attained much popularity in recognizing MI tasks. So, this paper emphasizes on using the MI tasks to identify words. The idea is to maintain a language dictionary along with an EEG-based motor imagery BCI system. The proposed framework has the advantage over the aforesaid mentioned methods as the detection of MI is relatively easier as compared to that of speech imagery, hence it is believed that this approach would come up with higher performance and better prediction of words as compared to that of speech-based BCIs.

## 3.1 *Phase 1: Recognition of Motor Imagery Tasks*

Common spatial pattern (CSP) [17] would be employed for the extraction of MI EEG features. MI EEG data can be analysed by splitting the EEG trial into distinct temporal segments and choosing the optimal segment conveying the most insightful information based on a predefined criterion. ADSWIN [18], an adaptive temporal segmentation of EEG trial would be used here to obtain the optimal spatial-temporal patterns. To pick up the best features, KPLS-mRMR [19], a KPLS based feature selection method would be employed. Finally, a traditional classifier would be applied to classify and recognise the motor imagery tasks.

## 3.2 *Phase 2: Maintaining Language Dictionary*

Language dictionary is a database that lists the different motor imagery tasks and the associated words to be inferred by the system, i.e. if a person is imagining to eat (i.e., the motor imagery task is 'eat'), the word 'eat'/ 'I am hungry' would be stored corresponding to the MI task 'eat'. Similarly, for the motor imagery task-imagination of the movement of his/her right leg, the thoughts 'he wants to turn right' can be stored in the language dictionary.

### 3.3 *Phase 3: MI BCI for Generating Response to Questions*

The language dictionary would direct which motor imagery task would iden- tify which word/thoughts. Once a motor imagery task would be identified, BCI would look for the associated word stored in the dictionary and then would reflect that particular word on the monitor screen. As for instance, if a person is imagining the movement of his/her right leg, it would imply that he wants to turn right. The same would be reflected in the computer screen and the attendant would come to know the same.

### 3.4 *Envisaged Application Through a Scenario*

Figure 2 illustrates the working of BrainTalk. The figure shows a person imagining the movement of his right leg. The EEG signal would be recorded. MI BCI would process, analyze and recognise the MI task as the right leg movement imagery. The system would then look for that MI task in the language dictionary and would come up with the associated thought "Wants to turn right". This would be reflected on the computer screen enabling the attendant to come to know that the patient desires to turn right.

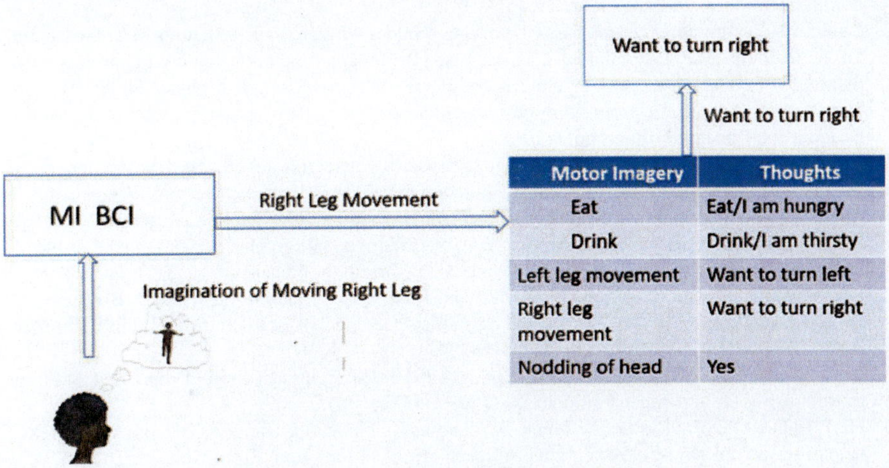


| Motor Imagery | Thoughts |
|---|---|
| Eat | Eat/I am hungry |
| Drink | Drink/I am thirsty |
| Left leg movement | Want to turn left |
| Right leg movement | Want to turn right |
| Nodding of head | Yes |

**Fig. 2** BCI for aiding people to communicate their ideas/views

## 4 Conclusion

Traditionally BCI has been perceived as an actor for controlling external devices through the classification of motor imagery tasks. This is an achievement in the area of rehabilitation robotics; and hence, the use of motor imagery opens up far more directions! Undoubtedly one of the greatest applications of BCI is to drive external devices through motor imagery, amalgamating motor imagery with the language/words could provide a better means to the people with motor disabilities. Such amalgamation ensures the user to communicate with their relatives by answering to their questions as well as expressing their thoughts and lead them to live a better life.

## References

1. Guger C, Allison B, Spataro R, La Bella V, Kammerhofer A, Guttmann F, von Oertzen TJ, Annen J, Laureys S, Heilinger A et al (2017) Mindbeagle'a new system for the assessment and communication with patients with disorders of consciousness and complete locked-in syndrom. In: 2017 IEEE international conference on systems, man, and cybernetics (SMC). IEEE, pp 3008–3013
2. Ashwal S, Cranford R, Bernat J et al (1994) The multi-society task force on PVS. Medical aspects of the persistent vegetative state. N Engl J Med 330, 1499–1508
3. Lugo ZR, Rodriguez J, Lechner A, Ortner R, Gantner IS, Laureys S, Noirhomme Q, Guger C (2014) A vibrotactile P300-based brain-computer interface for consciousness detection and communication. Clin EEG Neurosci 45(1):14–21
4. Lulé D, Noirhomme Q, Kleih SC, Chatelle C, Halder S, Demertzi A, Bruno MA, Gosseries O, Vanhaudenhuyse A, Schnakers C et al (2013) Probing command following in patients with disorders of consciousness using a brain-computer interface. Clin Neurophys 124(1):101–106
5. Tan D, Nijholt A (2010) Brain-computer interfaces and human-computer interaction. In: Brain Comput Interfaces. Springer, pp 3–19
6. Cooney C, Folli R, Coyle D (2018) Neurolinguistics research advancing development of a direct-speech brain-computer interface. IScience 8:103
7. Yoshimura N, Nishimoto A, Belkacem AN, Shin D, Kambara H, Hanakawa T, Koike Y (2016) Decoding of covert vowel articulation using electroencephalography cortical currents. Front Neurosci 10:175
8. Leuthardt EC, Gaona C, Sharma M, Szrama N, Roland J, Freudenberg Z, Solis J, Breshears J, Schalk G (2011) Using the electrocorticographic speech network to control a brain-computer interface in humans. J Neural Eng 8(3):036004
9. Deng S, Srinivasan R, Lappas T, D'Zmura M (2010) EEG classification of imagined syllable rhythm using Hilbert spectrum methods. J Neural Eng 7(4):046006
10. Herff C, Heger D, De Pesters A, Telaar D, Brunner P, Schalk G, Schultz T (2015) Brain-to-text: decoding spoken phrases from phone representations in the brain. Front Neurosci 9:217
11. Leuthardt EC, Cunningham J, Barbour D (2013) Towards a speech BCI using ECoG. In: Brain-computer interface research. Springer, pp 93–110
12. Brumberg JS, Nieto-Castanon A, Kennedy PR, Guenther FH (2010) Brain-computer interfaces for speech communication. Speech Commun 52(4):367–379
13. Anumanchipalli GK, Chartier J, Chang EF (2019) Speech synthesis from neural decoding of spoken sentences. Nature 568(7753):493
14. Akbari H, Khalighinejad B, Herrero JL, Mehta AD, Mesgarani N (2019) Towards reconstructing intelligible speech from the human auditory cortex. Sci Rep 9(1):874

15. Oken BS, Orhan U, Roark B, Erdogmus D, Fowler A, Mooney A, Peters B, Miller M, Fried-Oken MB (2014) Brain-computer interface with language model-electroencephalography fusion for locked-in syndrome. Neurorehabilit Neural Repair 28(4):387–394
16. Davenport M, Hannahs S (2005) Introducing phonetics and phonology. Oxford University Press
17. Blankertz B, Tomioka R, Lemm S, Kawanabe M, Muller KR (2008) Optimizing spatial filters for robust EEG single-trial analysis. IEEE Sig Process Mag 25(1):41–56
18. Talukdar U, Hazarika SM (2017) Designing optimal spatio-temporal filter for single trial EEG based BCI. In: 3rd international conference on advances in robotics (AIR). ACM
19. Talukdar U, Hazarika SM, Gan JQ (2018) A kernel partial least square based feature selection method. Pattern Recogn 83:91–106

# Familiarity Detection from EEG Signals Using Wavelet Transform and LSTM

**Vincent Kiplimo Ngetich**

**Abstract** The advancement in brain–computer interface technology has been influenced by the use of electroencephalography (EEG) signals. Electroencephalography (EEG) is a unique non-invasive technology used for capturing the electrical signals emitted by neurons inside the brain using electrodes placed at the scalp of the subject. Brain–computer interface systems does the recording, transforming, and extraction of meaningful features with the aim of "inferencing human thought" from signals. The identification of distinctive EEG features that relate to familiarity is not well explored in research so far. The potential of usability in criminal investigations, mental illness diagnosis, and EEG-based authentication is promising. The objective of this paper is to enhance research in this area by separating EEG signals into multiple components using stationary wavelet decomposition to classify familiarity and non-familiarity of phenomena. Detailed coefficients of stationary wavelet transform are useful in identifying patterns and events. To collect data, an experimental setup was done where subjects were exposed to a set of familiar–unfamiliar images for a definite amount of time. DC offset removal, filtering and independent component analysis were performed on the data before applying different mother wavelets to decompose it into different components which were then fed as input in a stacked LSTM deep learning model. An accuracy of 82.22% was achieved from training the model.

**Keywords** Electroencephalography · Wavelet transform · Long short-term memory

## 1 Introduction

Electroencephalography (EEG), a technique for electrophysiological monitoring is one of the most convenient ways of brain wave recording that has been in use for over four decades. EEG is useful in the development of various brain–computer interface

---

V. K. Ngetich (✉)
Department of Computer Science and Engineering, Maulana Abul Kalam Azad University of Technology, Kolkata, West Bengal, India
e-mail: vinkiplee@gmail.com



A. Maji et al. (eds.), *Proceedings of the International Conference on Computing and Communication Systems*, Lecture Notes in Networks and Systems 170,
https://doi.org/10.1007/978-981-33-4084-8_35

systems. Myriads of applications have been implemented using this technology. EEG-based BCI is currently being used by disabled people to communicate their interior motive to the outside world. Strategies of receiving neurofeedback have been implemented in BCI as a therapy tool in children with attention-deficit to boost their attention [1, 2]. Another application area of electroencephalography is in the lateralization of the sick in case of epileptic seizure attacks [3]. Since EEG signals are not strong enough, they are susceptible to interferences from artifacts; hence, proper pre-processing has to be done. EEG signal patterns differ according to factors like emotions, exercise, concentration among others. Analysis of the relationship between external phenomena and its corresponding electroencephalography (EEG) waves has become a subject of interesting research.

Decoding human thought is the ultimate goal of many brain–computer interface systems. Therefore, it is practically beneficial for disabled people for performing motor tasks like wheelchair movement, typing, cursor control, and even drone flying. Motor imagery domain deals with perceived movement [4, 5].

Several studies have been advanced in the area of analysis of familiarity/unfamiliarity in EEG signals. The majority of these studies has utilized evoked potentials. For instance, Sun et al. [2] employed a directed-lying method to distinguish between differentiation and classification ways that are utilized in identification of familiar faces by deeply looking at the event-related Potentials.

In [3], the authors did a classification of EEG signals obtained from the familiar/unfamiliar phenomena experimental setup by running it through a machine learning algorithm. Subjects clicked on the familiar phenomena while ignoring those that were unfamiliar. EEG data was then wavelet transformed and features extracted in the transform domain. Classification accuracy of 67% was obtained.

In [4], the authors utilized channel selection and machine learning methods to distinguish familiarity. In their research, the channels with high discriminative potential were decided computing pairwise distance. Next, the selected channel data acted as input to train a linear SVM machine classifier. Using the Kullback Leibler distance measure, a success rate of 72.67% was achieved.

The authors in [6] analyzed the effects of familiarity in the frequency domain. The EEG oscillation power was recorded from the theta and alpha bands. The author in [7] found out that EEG signals had time–frequency domain theta and gamma bands pointing toward the starting of a known phenomenon. Several researchers have used classification for segregating EEG signals as depicted in [8] where participants were shown visual stimuli made up of birds and cars and were required to point out their familiarity. This study achieved an accuracy of 59.6%.

Even though much progress has been achieved in this particular domain, the accuracies of the work already done is still low and needs to be improved. Traditional machine learning algorithms deal with very many limitations. One of the problems of traditional machine learning algorithms is that most of them are very open to significant errors, and it is tough to correct those errors because of the complexity of the algorithms.

This paper aims to enhance research in this area by separating EEG signals into multiple components using stationary wavelet decomposition to classify familiarity and non-familiarity of phenomena. The decomposed signal is then filtered and cleaned and then wavelet decomposition components were obtained which then acted as input to an LSTM model. This paper is organized this way. In Sect. 2, materials (EEG recording system and data, subjects, stimuli, and procedure) are presented. Section 3 deals with the methodology and then subsequently findings and discussion in Sect. 4. Section 5 is the conclusion of the paper.

## 2 Materials

### 2.1 *EEG Recording System and Data*

Eight electrodes of the "OpenBCI Ultracortex Mark IV" EEG cap were used for the recording of the EEG data. It uses eight electrodes with 10–20 International Electrode System. Also, it has one reference electrode and one bias electrode that can connect to each ear lobules where the disturbance is very low. The cap is using ADS1299 32-bit ADC chip for recording the analog EEG voltage and convert it to digital EEG which can be visualized in OpenBCI Graphical User Interface. The tools in GUI help to visualize the signal and have some simple signal pre-processing tools like applying notch filter and bandpass filters. The data was recorded by OpenBCI at a sampling rate of 256 Hz. For this particular experiment, we focused on only four electrodes, P7, P8, O1, and O2. It is because these four-electrode positions cover the visual-cortex portion of the brain [9]. Thus, we can analyze how the visual region is responding to the familiar–unfamiliar images separately.

### 2.2 *Subjects*

For the collection of data, we chose ten healthy subjects. We beforehand confirmed that all ten subjects have no health problems as well as no neurological disorders. All the selected subjects were in the age of 21–30 years. All have a normal or normal-to-corrected vision.

### 2.3 *Stimuli*

Two sets of images were utilized for the subjects, one set being that of familiar images and the other set being that of unfamiliar images. Each set is made up of a total of a hundred images. In the familiar set, we used the photos of prominent personalities in different domains of the world, the reason they were chosen was that those people are very familiar with the subjects hence easily distinguishable. In the unfamiliar set, we carefully chose 100 photos of people who are completely unknown to the subjects hence a total of 2000 images were used. In the training phase, we first showed the set of familiar photos to the subject, thereafter the set of unfamiliar images. There were five min relax time between two sets for each subject. EEG recording was done separately for two sets. After the training phase, we took the EEG data of subjects with testing phase images. In the testing phase, we randomly showed familiar and unfamiliar images and used a total of 200 images for each subject.

### 2.4 *Procedure*

As shown in Fig. 1, the participant was asked to sit in front of a liquid–crystal display (LCD) monitor placed at a distance of 65 cm from the subject ensuring that they get a clear vision of each stimulus. The room used for setting up the experiment was noise-free and overexposure to light was minimal. Each image was shown for 2 s with an inter-stimuli interval time of 2 s, where the subject got relaxed and got ready for the next image. The first 25 s were used as the baseline signal. Three trials of the experiment were performed and each trial comprised of a testing and training phase. In between each trial was a 20 min gap to make the subject relaxed and refreshed. Finally, the best data was used out of the three trials with less noise interference.

## 3 EEG Processing

The summary of the entire process is shown in Fig. 2.

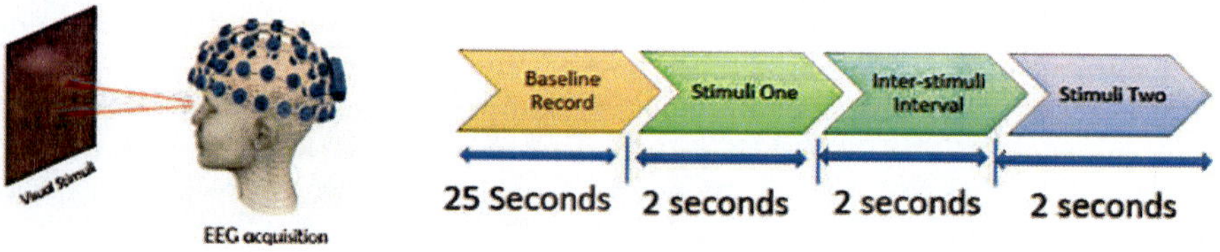


**Fig. 1** Experimental timing protocol

**Fig. 2** Diagram of the steps followed in familiarity detection

## 3.1 Filtering

Signal acquired by the EEG device always contains the powerline noise of high frequency and other unwanted physiological artifacts due to the effect of eye and muscle movements. DC offset and baseline correction of the recorded EEG signal were done to remove the unwanted drifts/offset. The main focus was in the alpha and theta bands of recorded EEG data; therefore, a band-pass filter of the range 4–13 Hz was applied, beginning with the lowpass and then highpass.

## 3.2 Independent Component Analysis

Independent component analysis (ICA) procedure removed the artifacts present in the filtered signal. The main function of ICA is to introduce different measures of non-gaussianity [10].

## 3.3 Model Architecture

After the removal of the artifacts using independent component analysis, the signal was decomposed into several subsequences using a discrete wavelet transform (DWT). The decomposed components were then fed into a stateful long short-term memory model which automatically learned the features of the EEG data. We used LSTM cell vanilla architecture as shown in Fig. 3. It consists of three gates, which are the input-gate, forget-gate, and an output-gate. It has a cell state which is the central part of the (LSTM) cell. All the equations related to the gate and cell state are mentioned in Eqs. (1–6) [11]. Equation (1) takes the inputs and previous output

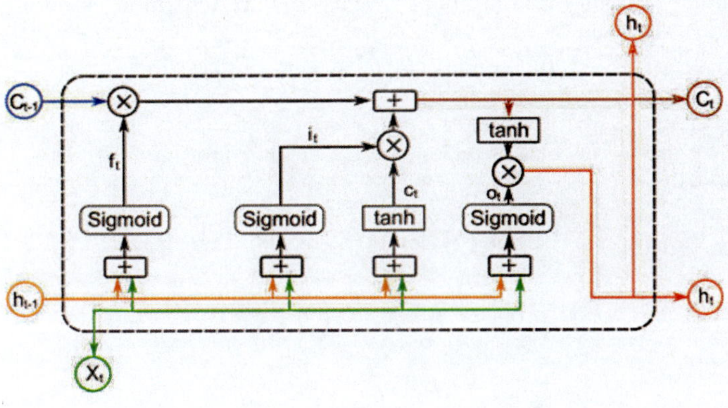


**Fig. 3** Long short-term memory cell architecture

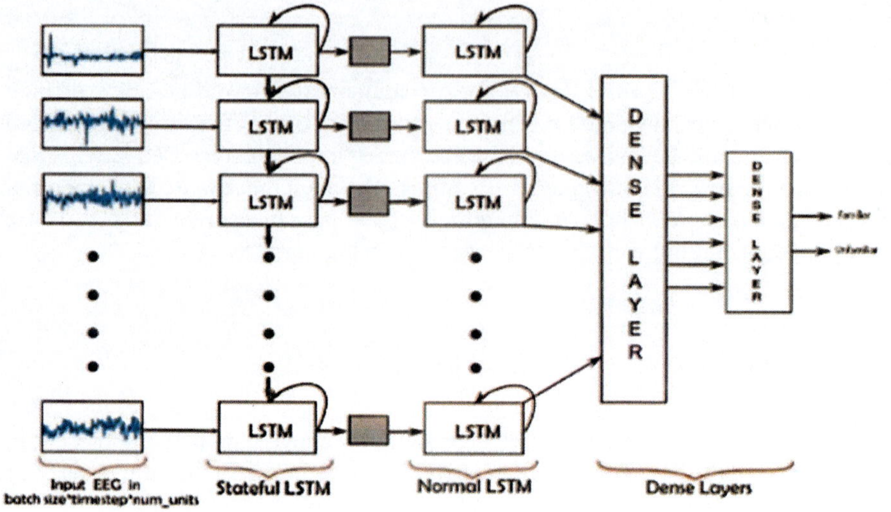


**Fig. 4** Stateful long short-term memory model

which gives the ability of what to forget or remember by the network with the help of a Sigmoid activation function, where $b_f$ is the bias and $W_f$ is the weight vector. New input along with the candidate layer output ($C_t$) and add this to the last cell state ($C_{t-1}$). Then the cell state will be updated using the $C_{t-1}$ and $C_t$. Finally, the cell will give the new output ($o_t$) and pass it through a tan$h$ activation layer, thus deciding the final output.

In the LSTM layer, we used six LSTM cell units in the first layer as shown in Fig. 4, with Relu as the activation layer. Usually, in stateful LSTM, the inputs are checked in batches, and the output of one batch is given to the next and the output from this is given to a normal LSTM. It is then further forwarded to a dense layer of six neurons and then to two neurons dense layer where we got the final classified output as either familiar or unfamiliar using a SoftMax activation function.

$$\left(f_t = \sigma\left(W_f \cdot \left[h_{t-1}, x_t\right] + b_f\right)\right) \tag{1}$$

$$\left(i_t = \sigma\left(W_i \cdot \left[h_{t-1}, x_t\right] + b_i\right)\right) \tag{2}$$

$$\left(C_t = \tan h\left(W_c \cdot \left[h_{t-1}, x_t\right] + b_c\right)\right) \tag{3}$$

$$\left(C_t = f_t * C_{t-1} + i_t * C\right) \tag{4}$$

$$\left(o_t = \sigma\left(W_o \cdot \left[h_{t-1}, x_t\right] + b_o\right)\right) \tag{5}$$

$$(h_t = o_t * \tan h(C_t)) \tag{6}$$

From the complete data, we used 70% of the data from each subject used for the training purpose of the models implemented and the rest 30% was used for testing purposes.

## 4 Findings and Discussions

The aim is to introduce effective deep learning techniques for the efficient classification of the signal data, which can pave a way to use the implemented methods further for familiarity detection. The confusion matrix of the LSTM model is shown in Fig. 5, where 1644 images were accurately classified from ten subject's data.

The Stateful LSTM model is providing an accuracy of 82.22%, which is an improvement from the accuracy rate of the SVM method in reference work [12], which got only 70.72% as shown in Table 1. Subject-wise classification accuracy is depicted in Fig. 6. The model was also able to generalize well given new input hence the minimal loss. The graphical plots of the model loss is shown in Fig. 7.

We calculated the precision, recall, and *f*-score of the model as shown in Table 2. The equations related to the evaluation matrix were given in Eqs. 7, 8, and 9.

$$\text{Precision} = \frac{\text{True Positives}}{\text{True Positives} + \text{False Positives}} \tag{7}$$

**Fig. 5** Confusion matrix for LSTM model

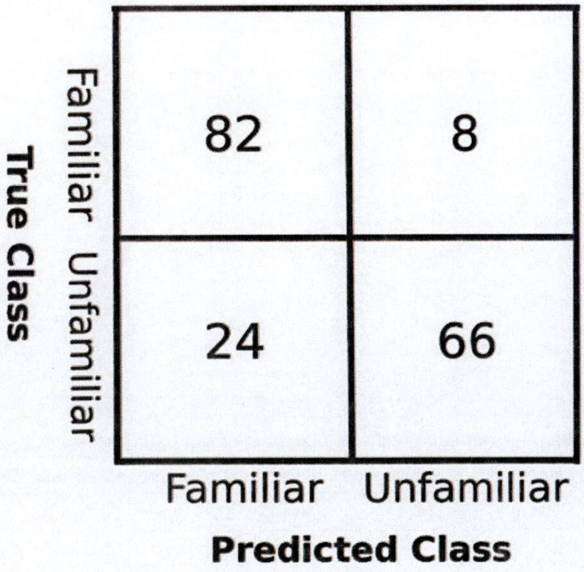

**Table 1** Comparison of proposed and existing method

| Model | Subjects | Peak value (familiar, unfamiliar) | Peak-peak value (familiar, unfamiliar) | Band power (familiar, unfamiliar) | Classifier accuracy (%) |
|---|---|---|---|---|---|
| SVM | 7 | 17.1762, 17.7488 | 1.1472, 1.1250 | 6.24E−06, 6.11E−06 | 70.72 |
| LSTM | 10 | 15.849, 16.397 | 4.0018, 3.5513 | 1.68E−13, 1.2138E−03 | 82.22 |

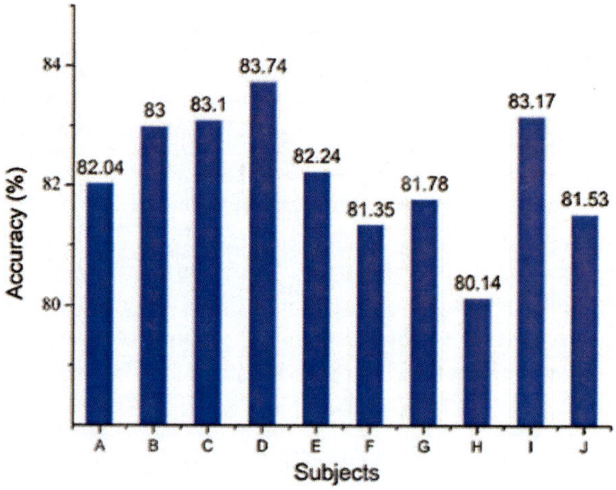


**Fig. 6** Subject-wise classification accuracy

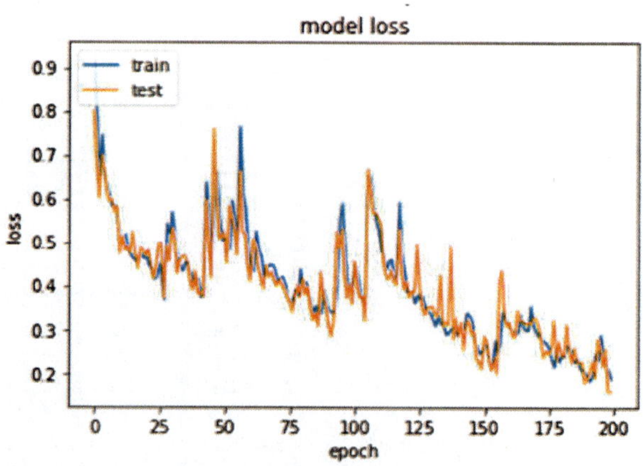


**Fig. 7** Plotted diagram of the model loss

**Table 2** Evaluation matrix parameters of the LSTM model for familiarity detection

| Model | Precision (%) | Recall (%) | $F_1$ score (%) | Accuracy rate (%) |
|---|---|---|---|---|
| LSTM | 91.11 | 77.35 | 83.66 | 82.21 |

$$\text{Recall} = \frac{\text{True Positives}}{\text{True Positives} + \text{False Negatives}} \tag{8}$$

$$F_1 = 2 * \frac{\text{Precision} * \text{Recall}}{\text{Precision} * \text{Recall}} \tag{9}$$

## 5 Conclusion

In this experiment, different familiar and unfamiliar images of persons were used as the visual stimuli and EEG signals were acquired while showing these stimuli to the subjects. We introduced a deep learning model for the classification purpose, where we use a Stateful LSTM network. The model achieved an accuracy rate of 82.21% which was more satisfactory than the reference models which got only 70.71%.

Also, the techniques reduced the burden of manual feature extraction done for traditional learning methods. Thus, there was no need to rely on specific features for the final detection because the feature combination cannot be generalized for the classification every time with different models but by using deep learning, the model only takes care of the features automatically.

## References

1. Shenjie S, Thomas KP, Smitha KG, Vinod AP (2014) Two player EEG-based neurofeedback ball game for attention enhancement. In: IEEE international conference on systems, man, and cybernetics (SMC), San Diego, pp 3150–3155
2. Thomas KP, Vinod AP, Guan C (2013) Enhancement of attention and cognitive skills using EEG based neurofeedback game. In: 2013 6th international IEEE/EMBS conference on neural engineering (NER). IEEE
3. Zandi AS, Javidan M, Dumont GA, Tafreshi R (2010) Automated real-time epileptic seizure detection in scalp EEG recordings using an algorithm based on wavelet packet transform. IEEE Trans Biomed Eng 57(7):1639–1651. https://doi.org/10.1109/TBME.2010.2046417
4. Blankertz B, Dornhege G, Krauledat M, Müller K-R, Kunzmann V, Losch F et al (2006) The Berlin brain–computer interface: EEG-based communication without subject training. IEEE Trans Neural Syst Rehabil Eng 14:147–152
5. Burgess A, Gruzelier J (2000) Short duration power changes in the EEG during recognition memory for words and faces. Psychophysiology 37(5):596–606
6. Kim S-P et al (2012) Modulation of theta phase synchronization in the human electroencephalogram during a recognition memory task. Neuroreport 23(11):637–641
7. Hanslmayr S, Staudigl T (2014) How brain oscillations form memories—a processing based perspective on oscillatory subsequent memory effects. Neuroimage 85:648–655
8. Ungureanu F, Lupu RG, Romila AR (2019) Experimental analysis of visual cortex activity—an EEG study. In: 2019 22nd international conference on control systems and computer science (CSCS), Bucharest, Romania, pp 223–227, https://doi.org/10.1109/CSCS.2019.00044
9. Hyvärinen A, Oja E (2000) Independent component analysis: algorithms and applications. Neural Netw 13(4–5):411–430

10. Hochreiter S, Schmidhuber J (1997) Long short-term memory. Neural Comput 9(8):1735–1780
11. Tan ZHE, Smitha KG, Vinod AP (2015) Detection of familiar and unfamiliar images using EEG-based brain-computer interface. In: 2015 IEEE international conference on systems, man, and cybernetics. IEEE, pp 3152–3157

# A Comparative Analysis on Denoising Techniques in Brain MRI and Cardiac Echo

**Imayanmosha Wahlang, Goutam Saha, and Arnab Kumar Maji**

**Abstract** Preprocessing plays a vital role in image for reducing noise and other unwanted data. There is a need for preprocessing in many fields, including medical imaging. Many of the medical images are contaminated with noise. In this paper, works related to denoising for brain magnetic resonance imaging (MRI) and cardiac echo have been studied and implemented. Numerous types of traditional filters were compared for this purpose using mean square error (MSE), peak signal-to-noise ratio (PSNR), and structural similarity index (SSIM).

**Keywords** MRI · Echo · Rician noise · Speckle noise · Median filtering · Gaussian filtering

## 1 Introduction

The human body consists of many biological systems that carry out specific functions inevitable for everyday life, such as the digestive system, nervous system, endocrine system, and so on [1]. There are various radiologies that can be used to visualize all those systems in the human body. The different radiology helps to perceive the structural and functional parts of an organ. The most common radiology used in most clinics and hospitals is electrocardiogram (ECG), echocardiogram, magnetic resonance imaging (MRI), ultrasonography (USG), electroencephalogram (EEG), positron emission tomography (PET), cardiac catheterization, computed tomogra-

---

I. Wahlang (✉) · G. Saha · A. K. Maji
North-Eastern Hill University, Shillong, Meghalaya 793022, India
e-mail: imayanwahlang@gmail.com

G. Saha
e-mail: dr.goutamsaha@gmail.com

A. K. Maji
e-mail: arnab.maji@gmail.com

 
A. Maji et al. (eds.), *Proceedings of the International Conference on Computing and Communication Systems*, Lecture Notes in Networks and Systems 170,
https://doi.org/10.1007/978-981-33-4084-8_36

phy(CT), etc. [2]. In this paper, we intend to deal with MRI imaging and echo as they are the most common radiologies applied in the brain and heart aspects, respectively.

Though MRI and echo help in visualizing the anatomy of brain and heart, respectively, there exists some unwanted portion that is not a part of the anatomy. Such type of unwanted region is noise or artifact. Almost all medical images are corrupted with noise. Noises vary with different types of imaging techniques. In the case of MRI imaging, Rician noise is most prominent [3–5], and in echo, speckle noise is most prominent [6]. Rician noise is a type of complex Gaussian noise. The presence of noise may mislead a physician and make him take the wrong decision.

Therefore, it is a must to preprocess the images by applying denoising techniques before a physician further evaluates it. Many such works are done manually where images are to be reread. Sometimes re-scanning is required, which takes time and creates many issues. However, automated methods came into existence and have helped in this process. Several methods have been used for the elimination and reduction of Rician noise in all types of images used in medical field. To identify artifact or noise is the biggest challenge. Some might not be unwanted data but are considered to be and vice versa. Therefore, to remove such distortion is one of the significant issues that will be taken into account. Figure 1 shows an MRI corrupted Rician noise image in (a) and echo image in (b). In this paper, the analysis and application of different filtering techniques have been compared. Some of the methods are already applied in the preprocessing of brain MRI images and echo images and some of which are newly included.

This paper differs from other existing paper where the result has also been provided with more number of data and methods. It also compares two types of imaging modalities, i.e., MRI and echo in the brain and heart, respectively. Although all methods could not be compared, few methods have been considered for our purpose.

A summary of different existing denoising techniques that has been used is covered in Sect. 2. Different methodologies that are compared are explained in brief in Sect. 3 with their advantages and disadvantages. Section 3 provides the experimental result analysis, and conclusion and future work are delivered in the last section.

(a)

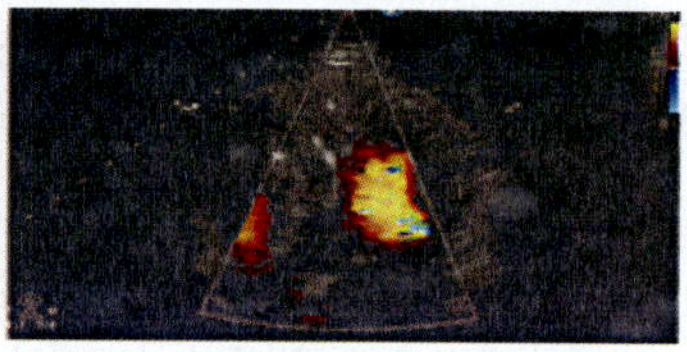

(b)

**Fig. 1** **a** Image showing an MRI modality and **b** image showing echo modality

## 2 Related Works

The brain is a sophisticated structure that consists of nerve cells [1]. Taking MRI as our imaging technique, methods used in preprocessing of this radiology, before further proceeding with clinical suggestions and treatment, are explained below. When noise removal is visualized, contrast and intensity enhancement always play a major role. More to this are techniques called filtering that can be used in obtaining a better output of image and can be even used for further processing if desired. Image smoothing is, therefore, useful in obtaining a softer visual effect and noise removal [7]. There can be various types of noise like speckle noise, Gaussian noise, Rayleigh, uniform noise, etc. [4]. In most imaging techniques, including MRI, speckle noise can be mostly found. Using median filtering [8, 9], it was found that it gives the most desirable result. In addition to median filtering, max and min filters were used in MR brain images [9], and a comparative study between them was given without experimental results. There are other techniques like discrete wavelet transform (DWT) for removal of noise, which has become increasingly popular [10]. They showed that using biorthogonal wavelet(bior 1.1 and bior 1.3), the better result could be achieved compared to any other type of wavelet. Other works include [11] where non-local means are used to reduce noise from images obtained from brain web. Considering Rician noise techniques that have been used is median filtering [12, 13], which proves to be the best in removing outliers, soft thresholding, which is done by Pallabi et al. [14] that gives a good result, adaptive filtering and histogram [15] and fuzzy filters [16, 17]. In [18], a comparison was done for removal of Gaussian noise using median filter, averaging filter, non-local means(NLM), Wiener filter, [19–21], and fuzzy filters where fuzzy filters were proved to be better than other filters. Other filters include are KSL filter [22], anisotropic diffusion filter [5], wavelet [4], Gaussian filter and anisotropic diffusion [23], bilateral filter [21], and block matching, and 3D filtering (BM3D) [24].

In the case of cardiac system, a work by Danilov et al. [25] shows four filtering techniques, out of which median filtering gives the best output compared to Kuan [26], ramponi, and sigma filter. Median filtering has been used in [25–28]. A paper using Gaussian filter can be seen in [28–30]. A comparison of median filter, Kuan filter, Lee filter, Frost filter, and SRAD filter was performed in [31, 32]. Other works include using singular value decomposition (SVD) [33], anisotropic diffusion [34], Wiener filter [28], proposed median and averaging technique [35], and isomap algorithm [36].

## 3 Methodologies

After a brief research on the techniques used in noise removal in MRI images, some of the methods that are used in this paper for comparison are median filtering, max filtering, bilateral filtering, mean filter, Gaussian filter, and Savitzky–Golay filter.

### 3.1 Median Filtering

It is one of the easiest and popular filtering techniques that is available and is useful for reducing salt and pepper noise in an image [8, 9, 26]. It is a nonlinear spatial filter where each pixel of the image is taken into account. The filter can be illustrated in this way; an image $X$ is taken, which is corrupted with noise where statistically, it corresponds to the 50th percentile. $X$ is taken in grayscale, and using Eq. (1), the median of an image is obtained, i.e., the middle pixel value of a window.

$$\text{Median}(i, j) = \text{Median}\, f(i, j) \tag{1}$$

with $i$ and $j$ being the pixel value of an image $X$.

### 3.2 Max Filtering

Max filter works precisely like that of a median filter where statistically, it corresponds to 100 percentile, not 50 percentile, as in the former case [37]. The max value of an image in a window can be obtained using Eq. (2). The basic operation is averaging, which is a linear operation. Such filter obeys the property of the superposition concept. It replaces the central pixel with the lightest one in the running window.

$$F(pi_x, pi_y) = \max g(s, t) \tag{2}$$

$pi_x, pi_y$ is the pixel value and $s, t$ the value of the sub image area.

### 3.3 Bilateral Filter

It is a nonlinear filter mainly used in CT images [9]. It is an edge-preserving filter used for smoothing noise in an image. Since the time it was discovered, its application has increased rapidly. The formulation is simple as weights are used. Each pixel is replaced by the weighted average of values from nearby pixels. Weights do not only depend on Euclidean distance but also radiometric distance [38]. The bilateral filter can be defined as [39]

$$\text{BF}[I]_p = \frac{1}{w_p} \sum_{q \epsilon s} G_{\sigma_s}(\| p - q \|) G_{\sigma_r}(|I_p - I_q|) I_q \tag{3}$$

$$W_p = \sum_{q \epsilon s} G_{\sigma_s}(\| p - q \|) G_{\sigma_r}(|I_p - I_q|) \tag{4}$$

### 3.4 Mean Filter

It is a linear filter that uses a simple sliding window. The central pixel is replaced by the average of the rest of the pixel values in a window. It is a spatial filter and works as a low pass filter. It can be calculated using Eq. (5) [37]:

$$\text{Mean} = \frac{1}{m} \sum g(s, t) \tag{5}$$

### 3.5 Gaussian Filter

It modifies the input signal by convolution with Gaussian function [40]. It is used to blur images and a smoothing operator. The Gaussian function is given in Eq. (6) [41]. It is assumed to have a zero mean distribution.

$$G(x) = \frac{1}{\sqrt{2\pi\sigma^2}} \mathrm{e}^{-\frac{x^2}{2\sigma^2}} \tag{6}$$

### 3.6 Savitzky–Golay (SG) Filter

It is a digital filter that uses the polynomial method. It is a low-pass filter for smoothing data [42]. SG filter finds the filter coefficient that preserves higher moments. The idea of the SG filter is to derive different filter coefficients. These coefficient preserves the higher moments. Polynomial of higher order is used (quadratic or quartic). The derivation can be seen in [42].

Table 1 provides the advantages and disadvantages of different methodologies based on the output obtained and how the filters work.

## 4 Result Analysis

### 4.1 Dataset

The dataset used for this purpose is obtained from Radiopaedia, NEIGRIHMS, and Hope Clinic. Here, four datasets have been used and compared using the existing techniques. The first dataset is from Radiopaedia with 320 images; the second dataset is from NEIGRIHMS having 222 images; the third dataset is doppler echo with 122 images and 2D echo with 163 images.

**Table 1** Advantages and disadvantages of median, max, bilateral, Gaussian, SG and mean filters

| Filtering technique | Advantages | Disadvantages |
|---|---|---|
| Median | Edges are intact. smooths white noise | Good for salt and pepper noise |
| Max | Easy method | Edges are lost,blurry image and important region are totally degraded. |
| Bilateral filter | Edge is preserved | Blurring of images |
| Gaussian filter | Reduces noise, effective and efficient | Reduces detail |
| SG filter | Area preserving | computation time proportional to window width |
| Mean filter | Noise is less apparent | Blurring of images |

## 4.2 Parameters

Different parameters used are mean square error (MSE), peak signal-to-noise ratio (PSNR), root mean square error (RMSE), and structural similarity index (SSIM) [8, 9, 13, 14, 22, 31]. MSE [13, 14] can be calculated as:

$$\mathrm{MSE} = \frac{1}{mn} \sum_{i=0}^{n-1} \sum_{j=0}^{n-1} [S_i, j - T_i, j]^2 \tag{7}$$

where $S$ and $T$ are input and output images of size $m \times n$. RMSE [31] is calculated using:

$$\mathrm{RMSE} = \sqrt{\mathrm{MSE}} \tag{8}$$

PSNR [13, 14] is calculated using:

$$PSNR = 10 \log_{10} \left( \frac{Max}{\sqrt{\mathrm{MSE}}} \right) \tag{9}$$

SSIM [31] can be calculated using:

$$\mathrm{SSIM} = \frac{1}{m} \sum \frac{(2\mu 1\mu 2 + C1)2\sigma_{1,2} + C2}{(\mu_1^2 + \mu_2^2 + C1)(\sigma_1^2 + \sigma_2^2 + C2)} \tag{10}$$

where $\mu$ is the mean $\sigma$ is the standard deviation and $\sigma_{1,2}$ is the covariance.

From Table 2, using the brain dataset, it could be observed that MSE and RMSE are better in the case of Gaussian filter for both datasets, and PSNR and SSIM are better in median filter in both datasets. From Table 3, using the heart dataset, the output shows that MSE and RMSE are better in the case of Gaussian filter, PSNR

**Table 2** Output of brain dataset

| Filtering techniques | Dataset 1 | | | | Dataset 2 | | | |
|---|---|---|---|---|---|---|---|---|
| | MSE | PSNR | RMSE | SSIM | MSE | PSNR | RMSE | SSIM |
| Median | 0.061 | 0.109 | 0.247 | 0.0029 | 0.345 | 0.131 | 0.58 | 0.0043 |
| Mean | 0.181 | 0.0115 | 0.0425 | 0.0017 | 0.183 | 0.0237 | 0.42 | 0.0026 |
| Max | 0.233 | 0.003 | 0.482 | 0.0013 | 0.212 | 0.030 | 0.46 | 0.0022 |
| Bilateral | 0.080 | 0.016 | 0.282 | 0.0026 | 0.099 | 0.037 | 0.31 | 0.0039 |
| Savitzky Golay | 0.098 | 0.104 | 0.313 | 0.0029 | 0.479 | 0.125 | 0.35 | 0.0042 |
| Gaussian | 0.056 | 0.0227 | 0.236 | 0.0028 | 0.0846 | 0.041 | 0.28 | 0.0042 |

**Table 3** Output of Heart dataset

| Filtering techniques | Dataset 1 | | | | Dataset 2 | | | |
|---|---|---|---|---|---|---|---|---|
| | MSE | PSNR | RMSE | SSIM | MSE | PSNR | RMSE | SSIM |
| Median | 0.1774 | 0.0815 | 0.421 | 0.0026 | 0.305 | 0.265 | 0.552 | 0.007 |
| Mean | 0.0792 | 0.0139 | 0.28 | 0.0019 | 0.327 | 0.0379 | 0.571 | 0.0052 |
| Max | 0.128 | 0.0138 | 0.357 | 0.0014 | 0.500 | 0.0305 | 0.707 | 0.0038 |
| Bilateral | 0.0398 | 0.0215 | 0.199 | 0.002 | 0.192 | 0.0523 | 0.43 | 0.007 |
| Savitzk–Golay | 0.1710 | 0.0822 | 0.41 | 0.0026 | 0.312 | 0.266 | 0.558 | 0.007 |
| Gaussian | 0.0252 | 0.0262 | 0.158 | 0.002 | 0.129 | 0.0639 | 0.359 | 0.0077 |

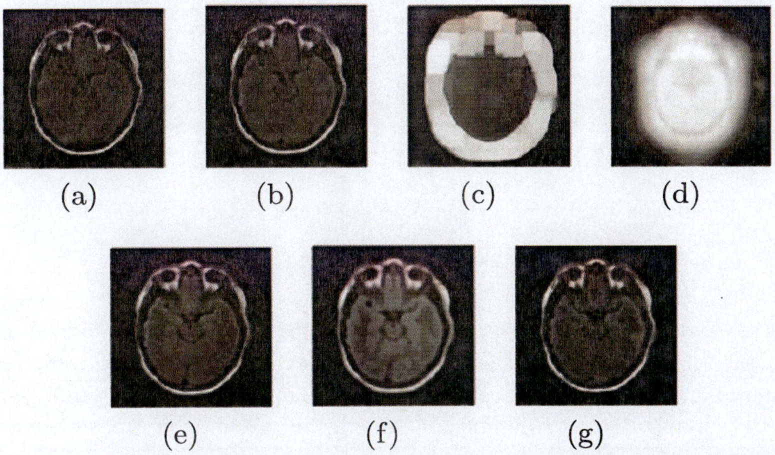

**Fig. 2** Output of dataset 1 with **a** input image **b** median filter **c** max filter **d** mean filter **e** SG filter **f** bilateral filter and **g** Gaussian filter

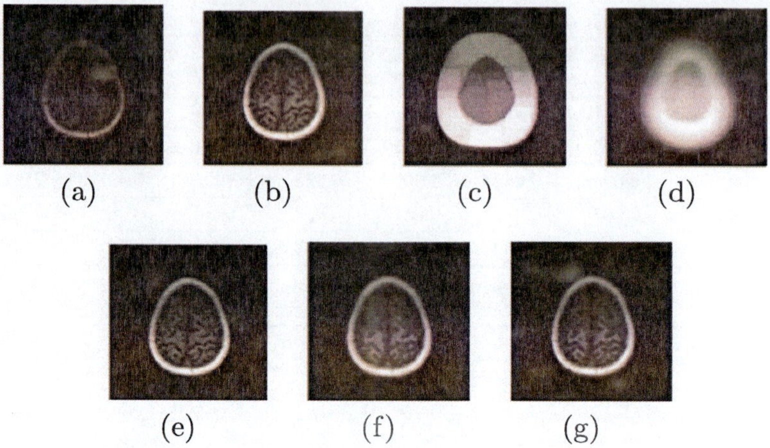


**Fig. 3** Output of dataset 2 with **a** input image **b** median filter **c** max filter **d** mean filter **e** SG filter **f** bilateral filter and **g** Gaussian filter

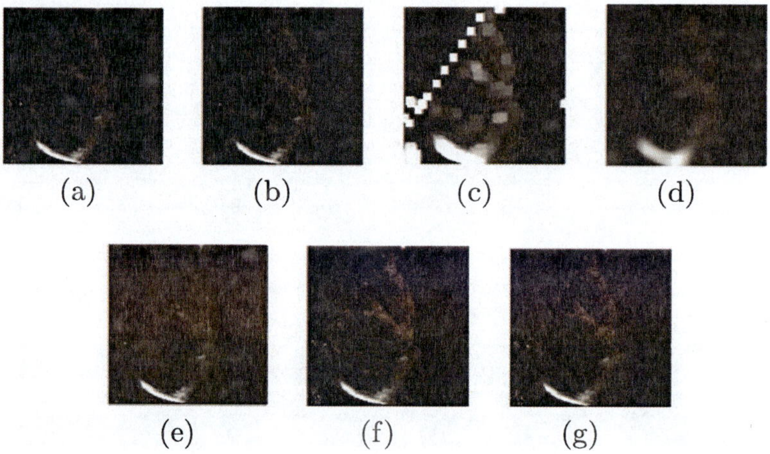


**Fig. 4** Output of dataset 3 with **a** input image **b** median filter **c** max filter **d** mean filter **e** SG filter **f** bilateral filter and **g** Gaussian filter

in the case of median and SG filters, and SSIM in the case of median, bilateral, SG and Gaussian in both datasets. All the datasets show almost the same results using performance metrics, which indicates that the median filter and Gaussian filters are most likely suitable in removing noise in both the brain and heart aspects compared to the other methods (Figs. 2, 3, 4 and 5).

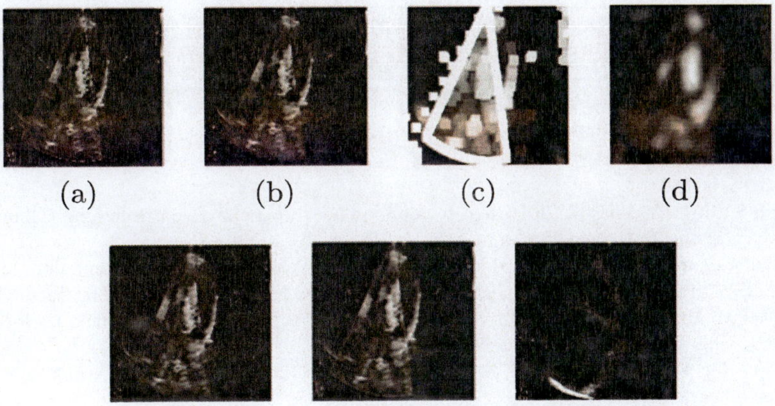


**Fig. 5** Output of dataset 4 with **a** input image **b** median filter **c** max filter **d** mean filter **e** SG filter **f** bilateral filter and **g** Gaussian filter

## 5 Conclusion and Future Work

Preprocessing is a trivial and challenging task in medical imaging. It has been investigated and reviewed by many researchers. In this paper, work based on denoising of brain MRI images and cardiac echo images was done. From the four datasets and using the existing filtering techniques, it was found that the median filter and Gaussian filter are better. Although only a few methods were used, this work can be further improved using more data and other methodologies (traditional methodologies, hybrid methodologies, advanced methodologies, etc.). As future work, work can be done using preprocessing and post-processing methods. It will help in the understanding of images in a better way and can give efficient results in segmentation and classification phases.

**Acknowledgements** A special thanks go to Dr. Pranjal Phukan of North Eastern Indira Gandhi Regional Institute of Health and Medical Sciences, Shillong, [43] for providing us with brain MRI data and Dr. Daljit Singh Sethi, Director of Hope clinic for providing and analyzing the echo data.

## References

1. European society of radiology. Brain watch. https://connect.myesr.org/course/pl-3-building-the-human-brain-molecular-logic-of-neural-circuit-formation/
2. https://medlineplus.gov/ency/article/007451.htm
3. Jangra S, Yadav S (2014) A review of Rician noise reduction in MRI images using wave atom transform. Int J Comput Sci Mob Comput 3:454–457
4. Sheshadri HS (2016) Denoising of Rician noise in magnitude MRI images using wavelet shrinkage and fusion method

5. Pal C, Das P, Chakrabarti A, Ghosh R (2017) Rician noise removal in magnitude MRI images using efficient anisotropic diffusion filtering. Int J Imag Syst Technol 27(3):248–264
6. Gifani P, Behnam H, Sani ZA (2014) Noise reduction of echocardiographic images based on temporal information. IEEE Trans Ultrason Ferroelectr Freq Control 61(4):620–630
7. Wang Y, Fan Y, Bhatt P, Davatzikos C (2010) High-dimensional pattern regression using machine learning: from medical images to continuous clinical variables. Neuroimage 50(4):1519–1535
8. Singh S, Singh Y, Vijay R (2016) An evaluation of features extraction from lung CT images for the classification stage of malignancy
9. Balasubramanian C, Sudha B (2014) Comparative study of de-noising, segmentation, feature extraction, classification techniques for medical images. Int J Innov Res Sci Eng Technol 3
10. Gabhel M, Hiradhar A (2014) Comparative analysis of various denoising techniques for MRI image using wavelet. Int J Sci Res (IJSR)
11. Manjon JV, Robles M, Thacker NA (2007) Multispectral MRI de-noising using non-local means. Proc MIUA 7:41–45
12. Patil MG, Galande MV, Kekan MV, Dange MK (2014) Sentiment analysis using support vector machine. Int J Innov Res Comput Commun Eng 2(1):2607–2612
13. Ali HM (2018) MRI medical image denoising by fundamental filters. In: High-resolution neuroimaging-basic physical principles and clinical applications. InTech, pp 111–124
14. Patil MPL, Raskar MV (2015) MRI images techniques. Reconstruction 4(2)
15. Andotra AS, Sharma S (2018) MRI image enhancement: optimized filtering mechanism for achieving high accuracy in diagnose process
16. Toh KKV, Isa NAM (2009) Noise adaptive fuzzy switching median filter for salt-and-pepper noise reduction. IEEE Sig Process Lett 17(3):281–284
17. Kwan BY, Kwan HK (2011) Impulse noise reduction in brain magnetic resonance imaging using fuzzy filters. World Acad Sci Eng Technol 60:1344–1347
18. Krishnan MH, Viswanathan R (2013) A new concept of reduction of Gaussian noise in images based on fuzzy logic. Appl Math Sci 7(12):595–602
19. Yang J, Fan J, Ai D, Zhou S, Tang S, Wang Y (2015) Brain MR image denoising for Rician noise using pre-smooth non-local means filter. Biomed Eng Online 14(1):2
20. Mathen SJ, George A (2014) Analysis of MRI enhancement techniques for contrast improvement and denoising. Int J Curr Eng Technol 4:3853–3860
21. Saladi S, Amutha Prabha N (2017) Analysis of denoising filters on MRI brain images. Int J Imag Syst Technol 27(3):201–208
22. Devasena CL, Hemalatha M (2011) Noise removal in magnetic resonance images using hybrid KSL filtering technique. Int J Comput Appl 27(8):1–4
23. Renukalatha S, Suresh KV (2016) Brain tumor analysis of Rician noise affected MRI images. Int J Comput Appl 141(14):0975–8887
24. Djurović I (2016) BM3D filter in salt-and-pepper noise removal. EURASIP J Image Video Process 2016(1):13
25. Danilov VV, Skirnevskiy IP, Gerget OM (2017) Segmentation of anatomical structures of the heart based on echocardiography. J Phys Conf Ser 803(1):012031
26. Sharma A, Jain S (2016) A review on noise reduction of echo cardiographic images based on temporal information. Int J Comput Appl 975:8887
27. Balaji GN, Subashini TS, Chidambaram N (2015) Automatic classification of cardiac views in echocardiogram using histogram and statistical features. Proc Comput Sci 46:1569–1576
28. Perumal S, Velmurugan T (2018) Preprocessing by contrast enhancement techniques for medical images. Int J Pure Appl Math 118(18):3681–3688
29. Elalfi A, Eisa M, Ahmed H (2013) Artificial neural networks in medical images for diagnosis heart valve diseases. Int J Comput Sci Issues (IJCSI) 10(5):83
30. Tulin IJ (2017) A comparative study of Gaussian noise removal methodologies for gray scale images. Int J Comput Appl 975:8887
31. Jaybhay J, Shastri R (2015) A study of speckle noise reduction filters. Signal Image Process Int J (SIPIJ) 6(3):71–80

32. Raj B, Sharma A, Kapoor K, Jyoti D (2016) A novel approach for the reduction of noise. Int J Adv Res Ideas Innov Technol 2(3):182–193
33. Bydder M, Du J (2006) Noise reduction in multiple-echo data sets using singular value decomposition. Magn Reson Imag 24(7):849–856
34. Yu Y, Acton ST (2002) Speckle reducing anisotropic diffusion. IEEE Trans Image Process 11(11):1260–1270
35. Yousuf MA, Nobi MN (2011) A new method to remove noise in magnetic resonance and ultrasound images. J Sci Res 3(1):81
36. Gifani P, Behnam H, Shalbaf A, Sani ZA (2011) Noise reduction of echocardiography images using Isomap algorithm. In: 2011 1st middle east conference on biomedical engineering. IEEE, pp 150–153
37. Gonzalez R, Woods R (2014) Digital image processing
38. http://everestkanto.com/kwbn/numpy-euclidean-distance-matrix.html
39. Paris S, Kornprobst P, Tumblin J, Durand F (2009) Bilateral filtering: theory and applications. Found Trends Comput Graph Vision 4(1):1–73
40. Foster I, Kesselman C (1999) The grid: blueprint for a new computing infrastructure. Morgan Kaufmann, San Francisco
41. Gaussian filtering. University of Auckland
42. Savitzky A, Golay MJE (1964) Analytical chemistry, vol 36, pp 1627–1639
43. Wahlang I, Sharma P, Nasreen SM, Maji AK, Saha G (2019) A comparative study on segmentation techniques for brain tumor MRI. In: Information and communication technology for competitive strategies pp 665–673

# A Novel Spatial FCM-Based Method for Brain MRI Image Segmentation in the Presence of Noise and Inhomogeneity

**Kamarujjaman, Mausumi Maitra, and Susanta Chakraborty**

**Abstract** Unsupervised classification of the brain MRI image into different tissue regions shows salient role to analyze the neuro-imaging and scientific research in the field of medicine and diagnosis. In this paper, an efficient unsupervised spatial fuzzy *c*-means technique is proposed for classification of distorted brain MRI image with different bias factors. The spatial information has been measured by analyzing three different neighbouring distances to suppress the noise and inhomogeneity as well as preserving the image details at the time of classification. The correlation among the neighbours have been decided by spatial information so that if the centers are itself affected with bias error, it would not affect the correlation. The qualitative and quantitative results are used to state the superiority of the proposed method from the considered methods of interest with reference to the partition coefficient and partition entropy.

**Keywords** Brain MRI · Neuro-imaging · Noise · Spatial information · Inhomogeneity · Fuzzy *c*-means · Segmentation · Classification · Gaussian distribution

---

Kamarujjaman
Department of Computer Application, Techno India University, Kolkata, West Bengal, India
e-mail: skmasum000@gmail.com

M. Maitra (✉)
Department of Information Technology, Govt. College of Engineering and Ceramic Technology, Kolkata, India
e-mail: mou1232005@yahoo.com

S. Chakraborty
Department of Computer Science and Technology, Indian Institute of Engineering Science and Technology, Shibpur, India
e-mail: susanta.chak@gmail.com



A. Maji et al. (eds.), *Proceedings of the International Conference on Computing and Communication Systems*, Lecture Notes in Networks and Systems 170,
https://doi.org/10.1007/978-981-33-4084-8_37

## 1 Introduction

Automatic computerized neuro-image analysis for medicine and diagnosis has been most essential requirement in the field of medicine and scientific research. The magnetic resonance imaging (MRI) image plays an momentous role to diagnose the human brain abnormalities [1]. So by analyzing the brain MRI image and based on the structural and functional activities, brain MRI image would be classified into three broader regions, which are White matter (WM), Gray matter (GM) and Ceribrospinal fluid (CSF) [2]. The vast and complex details of the brain MRI image needs to classify for better diagnosis and clinical practice [3, 4]. The computerized and unsupervised techniques have been come into scenario to deal with the MRI image details [1–4]. Different MRI image segmentation methods have been thoroughly compared in [5], and shows that the fuzzy based techniques [6] are more suitable techniques for the segmentation. The regular FCM [7] is not insensitive to noise and bias effects and in [8] it is shown that the FCM is not fulfilling the details preserving principle at the time of classification of brain MRI image. Some of those methods have improved the efficiency of the FCM with improved objective function to obtain the optimum classes [9, 10].

So, the methods reviewed above have been reported that the performance has been controlled and limited by the effect of artifacts and bias effects of the acquisition system. The robust discriminative clustering [11] was introduced to improve the performance over the previous works. The Spatial weighted information [12] has been proposed with especial care to the features of the scene to overcome the problems. But in the active existence of the artifacts and inhomogeneity in MRI, these methods also fail to classify the brain MRI image. In [13–15] authors tried to measure the spatial information and incorporate it to compute the fuzzy membership function in the presence of artifacts and inhomogeneity. The traditional FCM method has been improved considering the uncertainty of the classification as entropy in [16].

In this paper, an efficient unsupervised spatial fuzzy $c$-means technique has been presented for classification with the purpose of significantly improved intra-class correlation for brain tissues in the distorted brain MRI with different bias factors. The qualitative and quantitative outcomes have been compared with the state of the art and recently published FCM based methods for brain MRI. The results reveal that the proposed method is a step ahead in many scenarios for brain MRI with high noise and inhomogeneity. This paper has been concluded in Sect. 5, preceded by preliminaries in Sect. 2, proposed method in Sect. 3 and result and discussion in Sect. 4.

## 2 Validation Function

The partition coefficient, $V_{pc}$, interprets the fuzziness of the system [13, 17, 18] and defined as follows

$$V_{pc} = \frac{\sum_{i=1}^{C} \sum_{j=1}^{n} U_{ij}}{n} \tag{1}$$

The partition entropy, $V_{pe}$ [13, 17, 18] is expressed as:

$$V_{pe} = \frac{\sum_{i=1}^{C} \sum_{j=1}^{n} (U_{ij} * \log(U_{ij}))}{n} \quad (2)$$

where $U_{ij}$, $C$ and $n$ are membership value, no. of clusters and number of elements respectively.

## 3 Proposed Method

The proposed unsupervised spatial FCM method has been designed based on the spatial information to classify the brain MRI image $(X)$ in four distinct anatomical and homogeneous regions i.e. WM, GM, CSF and background. The segmented results in the form of partition coefficient and entropy and qualitative outcomes, it has been observed that presented method is well designed to segment the brain MRI images, when noise and bias effects is high in the MRI image as well. The local spatial information has been measured as Local membership function $(L_{ij} | \sum_{i=1}^{C} L_{ij} = 1)$ and spatial regions have been selected as dynamic windowing $(M_j)$ to avoid the unwanted padding. The number of clusters and size of the MRI image is defined as $C$ and $(nr * nc * nz)$ respectively. The proposed method has been simulated on brainweb dataset [19] of brain MRI image $(X)$, the image and dynamic mask have been defined as Eqs. (3) and (4) respectively.

$$X = \{x_j | j = 1, 2, \ldots (nr * nc * nz)\} \quad (3)$$

$$M_j = \{x_k | M_j \subset X, \forall k\} \quad (4)$$

Here, $x_k$ is considered to be the $k$th neighbor of $x_j$ and dimension of the mask $(M_j)$ is $(mr * mc)$. The initial parameters for the iterative proposed method have been collected by histogram peak associative rules [20]. The number of classes and their centers have been defined as $C$ and $V(V = \{v_i | i = 1, 2, \ldots C\})$ respectively. The proposed method has two membership functions i.e. global $(\mu_{ij})$ and local$(L_{ij})$. Now the global membership function is defined based on the Gaussian distribution of the image with respect to centers as mean of the distribution. The intensity based differences is sensitive to the noise and inhomogeneity of the images, to deal with this the proposed global membership is defined as a function of compliment Gaussian distribution of the image $(X)$ with respect to the centers $(v_i)$. So, the global membership function is defined as,

$$u_{ij} = \frac{1}{\sum_{l=1}^{C} \left(\frac{d_{ij}^2}{d_{lj}^2}\right)^{\frac{2}{(m-1)}}} \Bigg| d_{ij} = \left(1 - \mathrm{e}^{-\frac{\|x_j - v_i\|^2}{2\sigma_i}}\right) \quad (5)$$

where $d_{ij}$ is the Gaussian based distance and $\sigma_i^2$ is the standard deviation of the image with respect to centers. $\sigma_i^2$ is defined as,

$$\sigma_i = \sqrt{\frac{\sum_{j=1}^{(nr*nc)} ||x_j - v_i||^2}{(nr * nc)}} \tag{6}$$

The global membership may classify the MRI image but the effects of artifacts and bias error will miss lead the process of classification. To reduce the effect of artifacts on the resultant classes, here a local membership function has been explored based on the dynamic mask and three different distance parameters, which will be discussed in the remaining parts in this section. The local membership function $(L_{ij})$ has been defined as,

$$L_{ij} = \frac{1}{\sum_{l=1}^{C} \left(\frac{dm_{ij}^2}{dm_{lj}^2}\right)^{\frac{2}{(m-1)}}} \mid \sum_{i=1}^{C} L_{ij} = 1 \tag{7}$$

Here $dm_{ij}^2$ is the normalize distance $(\widetilde{dm}_{ij}^2)$ of the processing pixel from the centers in the spatial regions. The normalization has been done as,

$$dm_{ij}^2 = \left\{\frac{\widetilde{dm}_{ij}^2}{\sum_l^C \widetilde{dm}_{lj}^2}\right\} \tag{8}$$

The distance $(\widetilde{dm}_{ij}^2)$ has been computed based on the two different distance matrix in the spatial regions to detect and suppress the noise and inhomogeneity. $(\widetilde{dm}_{ij}^2))$ is defined as,

$$\widetilde{dm}_{ij}^2 = \left\{\sum_{\forall k} (DS_{ijk} * \widetilde{DS}_{ijk})\right\} \tag{9}$$

Here $DS_{ijk}$ and $\widetilde{DS}_{ijk}$ are the distances in the spatial regions. $DS_{ijk}$ and $\widetilde{DS}_{ijk}$ is defined in Eqs. (10) and (11) respectively,

$$DS_{ijk} = (d_{ijk} * ds_{jk}) \tag{10}$$

$$\widetilde{DS}_{ijk} = (\widetilde{d}_{ijk} * ds_{jk}) \tag{11}$$

Here the distance $DS_{ijk}$ is the inverse geometrical distance of the neighbours with respect to the processing pixel and defined as,

$$ds_{jk} = \left\{\frac{1}{\sqrt{||(\overrightarrow{ii, jj}) - (\overrightarrow{s, t})||^2}}\right\} \tag{12}$$

The $(\overrightarrow{ii, jj})$ and $(\overrightarrow{s, t})$ are the geometrical position of the neighbour and processing pixel respectively. The $d_{ijk}$ is the Euclidean distance of the neighbours from the class centers and defined as,

$$d_{ijk} = \left\{ \| x_j - V_i \|^2 \, | x_k \in M_j \wedge v_i \in V \right\} \tag{13}$$

The most important distance matrix $(\widetilde{d}_{ijk})$ has been used to measure the correlation among the neighbours. $(\widetilde{d}_{ijk})$ is defined as,

$$\widetilde{d}_{ijk} = \left\{ \| d_{ijk} - d_{ijc} \| \right\} \tag{14}$$

Here $d_{ijc}$ is the distance of the processing pixel $(x_j \in M_j)$ in reference with the center center $(v_i \in V)$.

The global membership and local membership functions are defined in Eqs. (5) and (7) respectively. The membership functions are combined to generate a new membership function $(U_{ij})$ to reduce the effect of noise and inhomogeneity from the global membership. The new function for membership estimation is employed to compute the optimum cluster centers and classes as well. The new membership function is defined as,

$$U_{ij} = \frac{u_{ij}^p * L_{ij}^q}{\sum_{i=1}^{C} \left( u_{ij}^p * L_{ij}^q \right)} \Big| \sum_{i=1}^{C} U_{ij} = 1 \tag{15}$$

where, the control variables ($p$ and $q$) are employed to govern the participation of the membership functions in new membership function. The optimum cluster centers are estimated upon the new membership function. The cluster centers are computed as,

$$V_{i\_\text{new}} = \frac{\sum_{j=1}^{(nr*nc)} (U_{ij}^m * x_{ij})}{\sum_{j=1}^{(nr*nc)} U_{ij}^m} | x_j \in X \tag{16}$$

The fuzzy based methods are iterative in nature. The optimality of the proposed unsupervised spatial FCM has been controlled by the objective function. The convergence of the objective function ensures that the optimal cluster centers have been reached. The objective function is defined in Eq. (17).

$$J_{\text{USFCM}} = \sum_{i=1}^{C} \sum_{\substack{j=1,\\ x_j \in X}}^{(nr*nc)} \left( u_{ij}^m * d_{ij}^2 \right) + \tau \sum_{i=1}^{C} \sum_{\substack{j=1,\\ x_j \in (M_j \cap X)}}^{(nr*nc)} \left( L_{ij}^m * \sum_{\forall k | x_k \in M_j} \left( \frac{\widetilde{DS}_{ijk}}{(mr * mc)} \right) \right) \tag{17}$$

Figure 1 explains the working principle of the proposed technique.

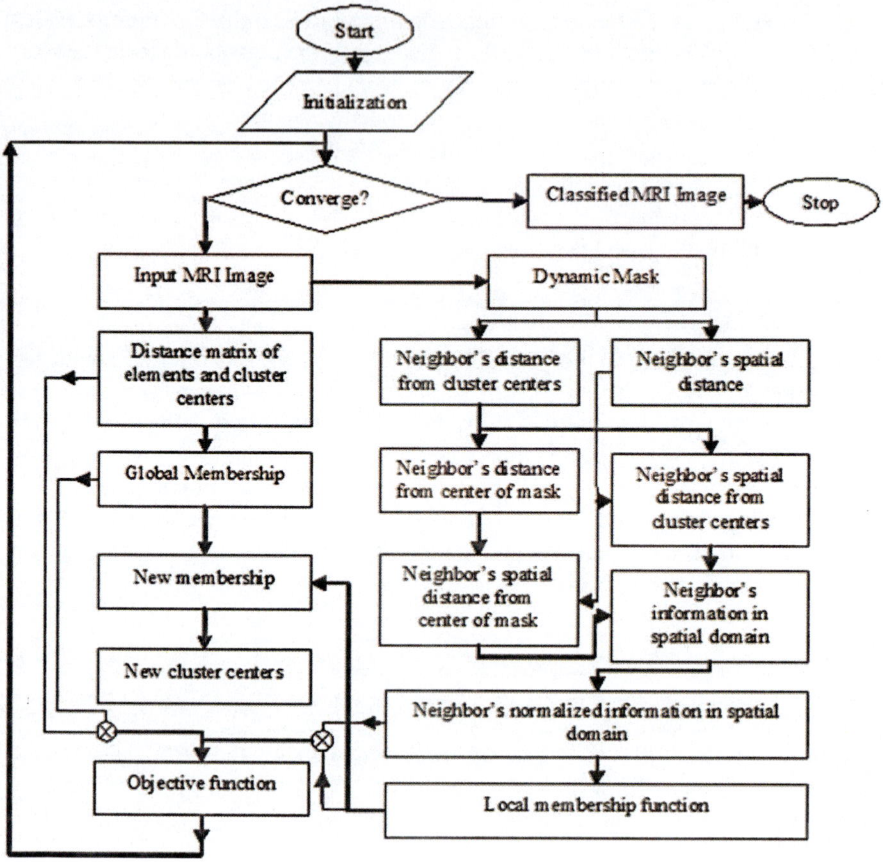


**Fig. 1** Working principle of the proposed technique

## 4 Results and Discussions

The presented method has been simulated and analyzed on the T1 weighted brain MRI image (resolutions 181 × 217 × 181 and size 1 × 1 × 1 mm), freely available in Brainweb [19] dataset of McGill University. The simulation has been done for slice 84 of brain MRI on the Matlab (v. 2016a) at windows 10 with 4GB RAM and intel core i3 processor. The slice 84 has been chosen from total 181 slices because of the details present in this slice are very complex and sensitive for brain MRI volume. The qualitative and quantitative outcomes of the proposed unsupervised spatial FCM have been used to check the superiority of the method compared to the state of the art methods and recently published FCM based methods, i.e. FCM [7], fuzzy *c*-means clustering with spatial information (sFCM) [13], adaptive spatial information-theoretic fuzzy clustering (ASIFC) [15], unsupervised modified spatial FCM (UMSFCM) [14] and entropy-based approach for fuzzy *c*-means (EFCM) [16].

**Table 1** Comparisons of validation functions for considered FCM based methods with proposed method

| Noise versus IIH | Validation function | FCM | sFCM | ASIFC | UMSFCM | EFCM | Proposed method |
|---|---|---|---|---|---|---|---|
| Noise 7% and IIH 20% | Vpc | 0.823 | 0.912 | 0.866 | 0.982 | 0.921 | 0.989 |
| | Vpe | 0.343 | 0.153 | 0.294 | 0.031 | 0.106 | 0.018 |
| Noise 7% and IIH 40% | Vpc | 0.823 | 0.910 | 0.844 | 0.981 | 0.921 | 0.989 |
| | Vpe | 0.343 | 0.141 | 0.309 | 0.032 | 0.089 | 0.018 |
| Noise 9% and IIH 20% | Vpc | 0.803 | 0.897 | 0.845 | 0.978 | 0.911 | 0.988 |
| | Vpe | 0.378 | 0.780 | 0.299 | 0.037 | 0.117 | 0.019 |
| Noise 9% and IIH 40% | Vpc | 0.803 | 0.897 | 0.843 | 0.976 | 0.906 | 0.987 |
| | Vpe | 0.378 | 0.180 | 0.290 | 0.039 | 0.119 | 0.019 |

All the methods have been simulated for brain MRI with combination of noise from 9 to 7% and intensity inhomogeneity (IIH) from 40 to 20%. The quantitative results have been measured with Vpc and Vpe. Table 1 depicts the comparative quantitative results for proposed method and considered FCM based methods and different noise versus inhomogeneity combinations. The qualitative results have been checked to know that the classification of the brain MRI image into WM, GM and CSF has not been effected by noise and inhomogeneity and details of the regions have been preserved after classification. The qualitative results have been shown in Fig. 2 for proposed and considered methods for brain MRI image with noise and inhomogeneity of 9 and 40% respectively. From Fig. 2, it might be clear that there is no sign of noise or bias effects in case of proposed method in the classified image. So, from Fig. 2 and Table 1, it can be concluded that the proposed unsupervised spatial FCM method performs well over the state of the art and recently published fuzzy based methods, FCM [7], sFCM [13], ASIFC [15], UMSFCM [14] and EFCM [16] in terms of cluster validation functions, as well as visual quality of the segmented results.

## 5 Conclusion

The proposed unsupervised spatial FCM method has been designed by giving importance of the local information in spatial domain, which is used to classify distorted MRI image with high noise and IIH, especially for brain. The purposes have been

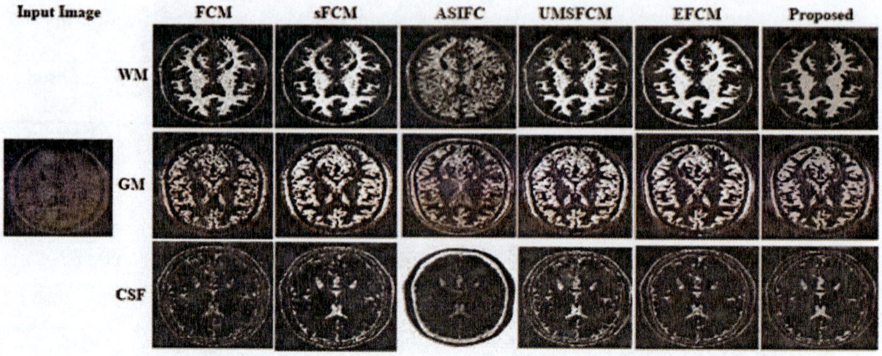


**Fig. 2** Qualitative results for T1 weighted brain MRI. 1st column represents original image (slice 84) with 9% noise and 40% inhomogeneity, 1st row contains WM, 2nd row contains GM and 3rd row contains CSF for FCM [7], sFCM [13], ASIFC [15], UMSFCM [14], EFCM [16] and proposed method

served by combining the spatial information and global information as membership function. The iterative proposed method has been stopped by converging objective function, which is defined based on global as well as spatial domain. The cluster validation and qualitative performance from results and discussion, it has been proved that the resultant clusters and clusters centers are optimized with the optimization of the objective function. The quantitative and qualitative results in Table 1 and Fig. 2 respectively explains that the proposed method is enhanced appreciably over FCM [7], sFCM [13], ASIFC [15], UMSFCM [14] and EFCM [16]. The proposed method has performed better compared to the existing methods in terms of qualitative results and efficiency. In future, this work will extend to classify and detect abnormality within brain MRI in 3D.

**Acknowledgements** The authors of this paper acknowledge the Indian Institute of Engineering Science and Technology, Shibpur and Govt. College of Engineering and Ceramic Technology, Kolkata and also Techno India University, WB for providing necessary supports. The authors are thankful to the editor and reviewers for their valuable remarks for the improvement in the paper.

## References

1. Simmons A, Tofts PS, Barker GJ, Arridge SR (1994) Sources of intensity nonuniformity in spin echo images at 1.5 T. Magn Reson Med 32(1):121–128
2. Kirchhoff B, Gordon B, Head D (2014) Prefrontal gray matter volume mediate age effects on memory strategies. Neuro Image 90(15):326–334
3. Clark MC, Hall LO, Goldgof DB, Clarke LP, Vethuizen RP, Silbiger MS (1994) MRI segmentation using fuzzy clustering techniques. IEEE Eng Med Biol Mag 13:730–742
4. Pitiot A, Toga AW, Thompson PM (2002) Adaptive elastic segmentation of brain MRI via shape-model-guided evolutionary programming. IEEE Trans Med Imag 21:910–923

5. Balafar MA, Ramli AR, Saripan MI, Mashohor S (2010) Review of brain MRI image segmentation methods. J Artif Intell 33(3):261–274
6. Pham DL, Xu C, Prince JL (2000) Current methods in medical image segmentation. Annu Rev Biomed Eng 2(1):315–327
7. Bezdek JC (1981) Pattern recognition with fuzzy objective function algorithms. Kluwer Academic Publishers, Norwell, MA, USA
8. Olabarriaga SD, Smeulders AW (2001) Interaction in the segmentation of medical images: a survey. Med Image Anal 5(2):127–142
9. Ahmed MN, Yamany SM, Mohamed N, Farag AA, Moriaty T (2002) A modified fuzzy *c*-means algorithm for bias field estimation and segmentation on MRI data. IEEE Trans Med Imaging 21(3):193–199
10. Nayak J, Naik B, Behera HS (2015) Fuzzy *c*-means (FCM) clustering algorithm: a decade review from 2000 to 2014. Comput Intell Data Min 2:133–149
11. Kong Y, Deng Y, Dai Q (2015) Discriminative clustering and feature selection for brain MRI segmentation. IEEE Sig Process Lett 22(5):573–577
12. Kang M, Kim JM (2014) Fuzzy *C*-means clustering with spatially weighted information for medical image segmentation. CIMSIVP. IEEE symposium. Orlando, FL, pp 1–8
13. Chuang KS, Tzeng HL, Chen S, Wu J, Chen TJ (2006) Fuzzy *c*-means clustering with spatial information for image segmentation. Comput Med Imaging Graph 30(1):9–15
14. Kamarujjaman, Chakraborty S, Maitra M (2016) An unsupervised modified spatial fuzzy *C*-mean method for segmentation of brain MR image. In: International conference on advances in computing, communications and informatics (ICACCI), Jaipur, pp 125–131
15. Wang Z, Song Q, Soh YC, Sim K (2013) An adaptive spatial information-theoretic fuzzy clustering algorithm for image segmentation. Comput Vis Image Underst 117(10):1412–1420
16. Kahali S, Sing JK, Saha PK (2019) A new entropy-based approach for fuzzy *c*-means clustering and its application to brain MR image segmentation. Soft Comput 23:10407–10414
17. Greenspan H, Ruf A, Goldberger J (2006) Constrained Gaussian mixture model framework for automatic segmentation of MR brain images. IEEE Trans Med Imag 25(9):1233–1245
18. Ahmed M, Yamany S, Mohamed N, Farag A, Moriarty T (2002) A modified fuzzy *C*-means algorithm for bias field estimation and segmentation of MRI data. IEEE Trans Med Imag 21:193–199
19. http://www.bic.mni.mcgill.ca/brainweb/
20. Namburu A, Samayamantula SK, Edara SR (2017) Generalised rough intuitionistic fuzzy *c*-means for magnetic resonance brain image segmentation. IET Image Process 11(9):777–785

# Melanoma Classification Through Transfer Learning by the Analysis of Skin Lesion Images

**Bunil Kumar Balabantaray, Rommel Chakravarty, Akash Kumar Panda, and Rajashree Nayak**

**Abstract** Malignant melanoma is a form of skin cancer that develops from melanocytic cells in the human body and can prove to be dangerous if not treated early. With the advent of artificial intelligence, deep learning approaches have been applied for the diagnosis of the disease which in turn shall help medical professionals in the line of treatment. In this paper, transfer learning is used by implementing the pre-trained VGG19 Net model. The model is pre-trained with ImageNet dataset. Certain layers of the pre-trained model are used for the analysis of the dataset of our interest by freezing rest of the convolutional layers. Here, skin lesions images from the ISBI2016 challenge (Gutman et al. in skin lesion analysis toward melanoma detection: a challenge at the international symposium on biomedical imaging, (Gutman D, Codella NCF, Celebi E, Helba B, Marchetti M, Mishra N, Halpern A (2016) skin lesion analysis toward melanoma detection: a challenge at the international symposium on biomedical imaging (ISBI) 2016, hosted by the International Skin Imaging Collaboration (ISIC), [1]) dataset are considered for the classification purpose. Proposed classification method records a commendable validation accuracy of 81.33% along with a testing accuracy of 86.67%, precision of 95.08%, recall of 82.25%, IoU of 78.26% and Dice score of 85.29%.

**Keywords** Skin lesions · Deep learning · VGG19 Net model · Convolutional neural networks · Transfer learning

---

B. K. Balabantaray (✉) · R. Chakravarty · A. K. Panda · R. Nayak
Computer Vision and Machine Intelligence Laboratory, National Institute of Technology Meghalaya, Shillong, Meghalaya 793003, India
e-mail: bunil@nitm.ac.in

R. Chakravarty
e-mail: rommelchac@gmail.com

A. K. Panda
e-mail: akpanda734@gmail.com

R. Nayak
e-mail: rnayak@nitm.ac.in

 
A. Maji et al. (eds.), *Proceedings of the International Conference on Computing and Communication Systems*, Lecture Notes in Networks and Systems 170,
https://doi.org/10.1007/978-981-33-4084-8_38

## 1 Introduction

Melanoma in dermatology is considered as one of the most severe forms of skin cancer which develops from the uncontrolled growth of skin (melanocytic, i.e., melanin producing) cells. Though the exact cause of the condition is unknown, the widely recognized cause is thought to be from the exposure of UV rays and other harmful light sources. Considered as the fifth most common form of cancer in the world, melanoma is projected to invade 108,420 patients and cause approximately 11,480 deaths alone in the USA in 2020 [2]. The traditional methods of melanoma detection include the ABCDE test [3] (asymmetric shape, irregular borders, change in color, diameter and evolution) which helps doctors to get a vivid idea about the onset of the disease in patients. A biopsy can be performed to confirm the disease, and staging is performed therein to make way for treatment. Figure 1 shows some of the sample malignant and non-malignant skin lesions images from ISBI2016 dataset [1].

These methods however are not accurate at all times as it becomes extremely difficult to classify melanoma clinically from other forms of skin cancers (like basal cell carcinoma, Bowen's disease) and other skin lesions due to similar parameters of analysis and classification. Looking incisively at the loopholes, researchers have started working avidly in the field of biomedical imaging to develop solutions for diagnosis. Different deep learning techniques have been applied in the past for classifying various forms of skin cancer and have yielded good results. But with exceptional rise in the condition, more consistent and transparent methods need to be found for successfully identifying melanoma.

In the following paper, we try to implement a method of classification which uses a custom CNN model for training and validating the skin lesion dataset and identifying the condition of melanoma with proper sensitivity. The custom CNN model used here is the Visual Geometry Group (VGG) Network with 19 specialized layers for learning (VGG19 Net model). The process of experimentation involves two steps: (i) pre-process the ISBI2016 dataset by sampling and augmentation (ii) train and validate the dataset with the pre-trained VGG19 Net model through transfer learning.

Accordingly, this paper has been dissected as follows: Sect. 2 gives a vivid description of the previous work performed in this domain and how our work shall improve

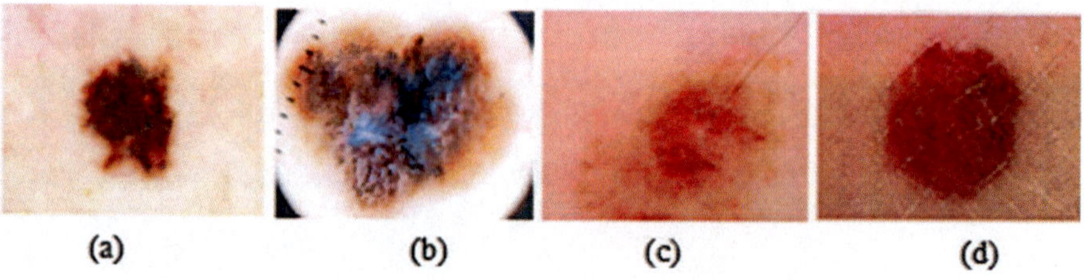


**Fig. 1** Images from ISBI2016 skin lesion dataset [1]. **a**, **b** malignant, **c**–**d** non-malignant

it; Sect. 3 describes the methodology used in the experiment; Sect. 4 reviews the results found from the experiment and Sect. 5 ends with a definite conclusion.

## 2 Literature Survey

Table 1 provides a brief idea of some previous works which have been very prominent in the field of skin cancer classification and image recognition.

## 3 Methodology

### *3.1 Dataset Description*

The dataset used for this experiment is the ISBI2016 challenge dataset which consists of 900 training and 379 testing images which are already labeled with two classes, i.e., benign and melanoma. Each image in the dataset has a dimension of 224 × 224 which is considered ideal for the VGG19 model.

### *3.2 Dataset Pre-processing*

**Sampling** First, the images in the dataset are clubbed in different classes according to the labels provided. Post this, the unbalanced dataset is balanced by taking the first 346 images from the benign class of training set and oversampling the malignant class to another 346 making a total of 692 training images which are balanced. This shall increase the training samples for the balanced dataset of an already relatively small dataset. The testing set is sampled by taking first 75 images from each class making a total of 150 images.

**Augmentation** Image augmentation helps to diversify the set of images without the need to collect new samples [11]. Accordingly, few methods of data augmentation have been applied in our dataset to increase the variety of images and prevent over fitting problem. Initially, half of the images from each class of training set are rotated toward the left and the other half are rotated toward the right. Saturation of images is slightly increased after this to enhance the images.

**Table. 1** Literature survey

| Work | Dataset | Method used | Observation |
|---|---|---|---|
| [4] | 129,450 clinical images which include 3374 dermoscopy images | Google Inception v3 CNN architecture with partitioning algorithm (CNN with PA) | CNN-PA is clearly seen outperforming with a three-way accuracy of 72.1 ± 0.9% and a nine-way accuracy of 55.4 ± 1.7% |
| [5] | ImageNet 2012 | ResNet-152, ResNet-101, ResNet-50, ResNet-34 C, ResNet- 34 B, BN-Inception, PReLU-net, VGG, Google-Net | The ResNet-152 Model has error rate % of 19.38 top-1 err. and 4.49 top-5 err. which is the lowest among all |
| [6] | ISBI 2016 Challenge dataset for skin lesion analysis | Three methods were the modified VGG16 ConvNet: M1-Training from Scratch; M2-ConvNet as feature extractor; M3-Fine-tuning the ConvNet | The proposed approach achieves promising results—most notably, a sensitivity value of 78.66% and a precision of 79.74% on the testing dataset which is observed in M3 |
| [7] | ISBI 2017 Challenge dataset for skin lesion analysis | ResNet similar CNN architecture is trained end to end. Parallely, hand-coded feature extraction was done which is followed by random forest classifier. The end result is the averaged output of both classifications. U-Net similar architecture is trained for segmentation purpose | The classification task achieves an accuracy of 80.3%, AUC score of 0.69 and a precision score of 0.81. For segmentation, we implement a convolutional–deconvolutional architecture and the segmentation model achieves a Dice coefficient of 73.5% |
| [8] | Dermatology database used in MED-NODE (2015) | A self-designed CNN architecture was designed for the purpose | The proposed model secured an accuracy of 81%, PPV of 75%, NPV of 86%, sensitivity of 81% and specificity of 80% |
| [9] | ISIC dataset | Model-1: Caffe CNN was used to learn the features. Model-2: Unsupervised feature learning was done by sparse coding and is then followed by non-linear SVMs. Finally, the models were combined by late fusion | The presented approach achieves an accuracy of 93.1% (94.9% sensitivity, and 92.8% specificity) for the first task and 73.9% accuracy (73.8% sensitivity, and 74.3% specificity) for the second task |

(continued)

**Table. 1** (continued)

| Work | Dataset | Method used | Observation |
|---|---|---|---|
| [10] | ISIC 2017 dataset | A deep learning framework consisting of two fully convolutional residual networks (FCRN) was proposed for task 1 and task 3. A straight-forward CNN is used for task 2 | Accuracies for the frameworks were observed to be 0.753 for task 1(lesion segmentation), 0.848 for task 2 (lesion dermoscopic feature extraction) and 0.912 for task 3 (lesion classification) |
| [Proposed work] | Training and validation images were taken from ISBI 2016 and test images from ISBI 2017 | Transfer learning using customized VGG19 model | Accuracy of 86.67%, recall of 82.85%, precision of 95.08%, IoU of 78.26% and Dice score of 85.29% were observed in the process |

## *3.3 CNN Architecture and Its Modification*

The custom CNN model used here is the VGG19 Net model, a variant of the Visual Geometry Group [12] which has been used widely in image classification problem and comes pre-trained with the ImageNet database, thereby being trained with a wide variety of features as the ImageNet dataset contains about 1000 different object classes. This model can accept input images of dimensions $224 \times 224$ and comes with a deep network consisting of 19 layers. It has five convolutional blocks with each block consisting of varying number of convolution layers with filters as described in Fig. 2. Each convolutional layer uses the rectified linear unit (ReLU) [13] activation function. Maxpooling layers are provided at the end of each block with strides of $2 \times 2$. The architecture concludes with a final block of fully connected layers consisting of three fully connected layers, i.e., one flattened layer and two dense layers.

At the end of the fully connected layer lies the activation function used to classify object classes. It is in the activation function that a major modification is used. Here, instead of using the softmax function which is used to perform a classification on 1000 object classes of the ImageNet database, we shall use the sigmoid function [14] which is more suitable for binary classification as our dataset has only two classes, i.e., benign and malignant. Typical architecture of a VGG19 model [15] with 19 specialized layers is depicted in Fig. 2.

## *3.4 Proposed Work*

The main challenge in our experiment is to train and validate a relatively small dataset with less diversity. Under such circumstances, the method of transfer learning is

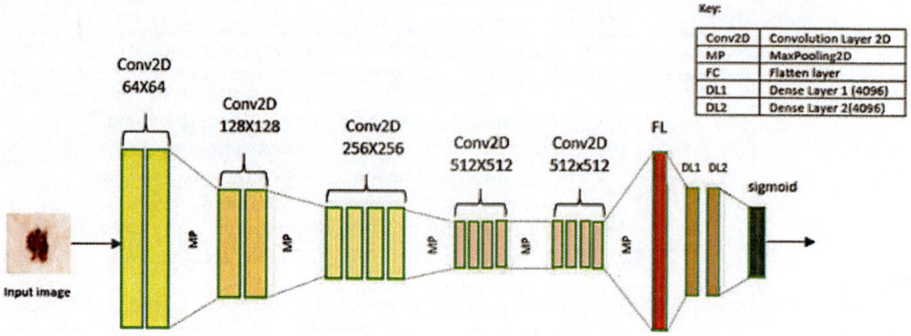


**Fig. 2** VGG19 model architecture with 19 specialized layers [15]

utilized, the concept of which lies in storing knowledge from one problem and using it to solve another problem. Here, the pre-processed dataset consisting of 692 training images is taken and validated against 150 testing images using the pre-trained VGG19 model which is laden with rich knowledge base which it had gathered from training the ImageNet dataset. While training the model with our dataset, all layers of the network are freezed apart from the fully connected layers as it is presumed that the model is sufficiently trained from before. Once the model is trained and validated, certain matrices are calculated to test the performance of the model which shall be discussed in the next section.

## 4 Experimental Results and Discussion

The entire code for the experiment was developed using Keras [16], a Python framework which is used commonly for developing deep learning models. Additionally, other libraries used consisted of scikit-learn, PIL, pandas and NumPy.

The experiment required the use of a computer server with dedicated graphical processing unit (GPU) as a result of which a machine with AMD Ryzen 5 3550H processor and NVidia GeForce GTX 1050 [17] graphics card was used.

### *4.1 Performance and Observation:*

Certain parameters used in the model:

- Epochs = 35; Batch size = 16; Optimizer = RMSProp [18]; Loss = binary cross entropy [19]

  Performance of the model on the pre-processed dataset is provided in Table 2. From the tabular data, it is noticed that the model provides successful results with

**Table. 2** Evaluation of VGG19 model

| Training loss | Training accuracy | Validation loss | Validation accuracy |
|---|---|---|---|
| 0.1033 | 0.9872 | 0.1170 | 0.8133 |

the training accuracy being recorded at an excellent 98.72% with a loss of 0.1033 and its corresponding validation accuracy at 81.33%.

In order to measure the performance of the model on random input images, a validation set of 150 images was created from ISBI2017 challenge dataset [20] consisting of similar images. Accordingly, several quality measures as described below are computed for quantifying the classification performance. Accuracy = number of correctly identified subjects out of total subjects

$$\text{Accuracy} = (\text{TP} + \text{TN})/(\text{TP} + \text{TN} + \text{FP} + \text{FN})$$

• Recall (Sensitivity) = fraction of true positives identified correctly

$$\text{Recall} = (\text{TP})/(\text{TP} + \text{FN})$$

• Precision = fraction of true negatives identified correctly

$$\text{Precision} = (\text{TN})/(\text{TN} + \text{FP})$$

• IoU (Jaccard Index) = used for measuring the similarity and diversity of sample sets

$$\text{IoU (Jaccard Index)} = (\text{TP})/(\text{TP} + \text{FP} + \text{FN})$$

• Dice score (or $F1$-score) = size of the overlap of the two segmentations divided by the total size of the two objects

$$\text{Dice-Score} = (2 * \text{TP})/(2 * \text{TP} + \text{FP} + \text{FN})$$

TP = true positive, TN = true negative, FP = false positive and FN = false negative.

Figure 3 demonstrates the plot of several performance measures of our proposed classifier. Figure 4 shows the TP, TN, FP and FN images resulted in the classification process.

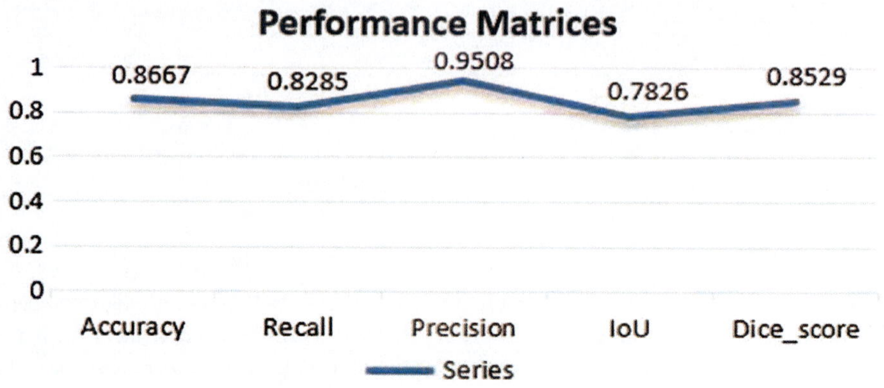


**Fig. 3** Performance measures of the proposed classifier

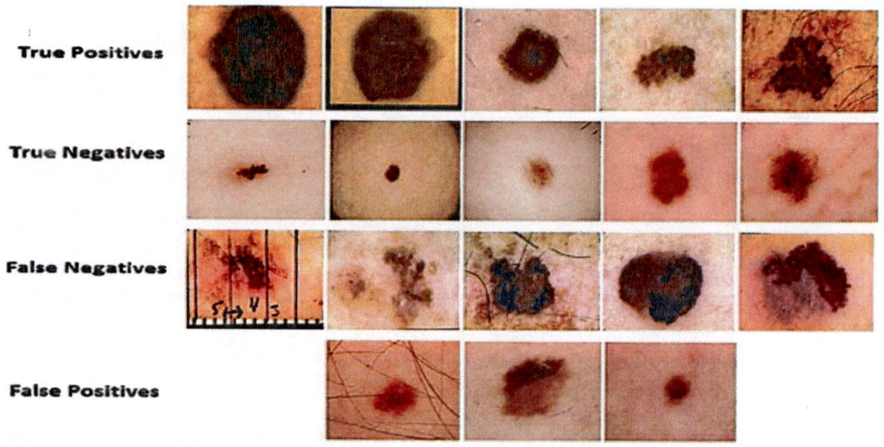


**Fig. 4** Prediction results by the classifier

## 4.2 Discussion

From the results of the experiment, it is clear that the concept of transfer learning has greatly helped in providing better performance matrices for the ConvNet variant. The use of 19 layers in the deep learning model makes VGG19 more effective than the VGG16 model. The problem of overfitting was minimized with the use of selected feature extraction by freezing certain layers of the VGG19 model. As the top four convolutional blocks of the model are freezed, computational time greatly decreases as we only need to backpropagate the gradient and update weights for the bottom layers. The augmentation of the dataset also contributes significantly to the performance as it diversifies the images from a relatively small dataset. Figure 5

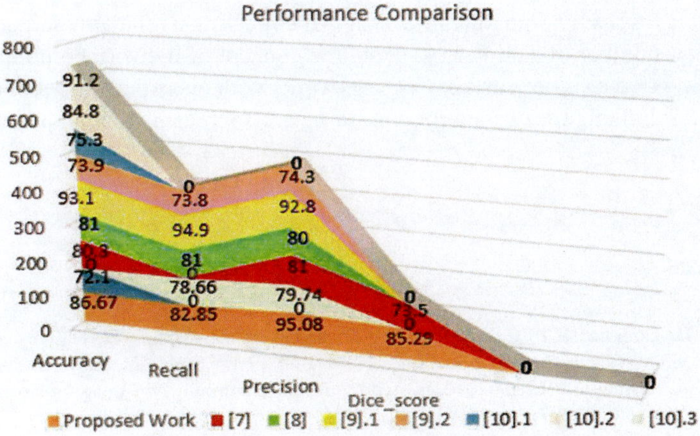


**Fig. 5** Performance comparison of our proposed work with related works

shows the performance comparison of our proposed classifier compared to existing related works in literature.

Also, it is observed that there is a substantial growth of up to 82.85% in the sensitivity which is greater than the average value of 82% achieved in the ISBI2016 challenge and better than the work of [8]. A validation accuracy of 81.33% is recorded which equals the accuracy achieved in the work of [6], and our testing accuracy is found to be better than the work of [7] and [8]. The proposed work even manages to better the precision of all existing works discussed in the paper by recording a commendable value of 95.08%. Below, we observe the relative performance of our work with the work of other authors who have significantly contributed to the field of melanoma classification. In the graph, [9] 0.1 and [9] 0.2 refer to the two tasks involved in the related paper [9] and similarly, [10] 0.1, [10] 0.2 and [10] 0.3 refer to the three individual tasks in the related paper [10].

## 5 Conclusion

This paper shows the ability of a custom deep learning model to classify melanoma from other skin lesions successfully which becomes rather difficult in case of human diagnosis. The VGG19 model with its relatively deeper neural network architecture forms a platform for better learning and training of relatively smaller datasets. The performance matrices derived from the work indicate the successful classification of melanoma from other skin lesions with commendable results. A validation accuracy of 81.33% demonstrates the performance of the VGG19 model which further records a testing accuracy of 86.67%, precision of 95.08%, recall of 82.25%, IoU of 78.26% and Dice score of 85.29% despite having a smaller number of sample images. This work forms a fundamental platform for our future work through which we shall try to achieve better results by doing the following: (i) collecting a relatively larger dataset with a greater number of samples (ii) using the concept of network combination by amalgamating two or more custom CNN networks with classifier for better classification (iii) minimizing the chances of overfitting and increasing overall performance matrices.

## References


1. Gutman D, Codella NCF, Celebi E, Helba B, Marchetti M, Mishra N, Halpern A (2016) Skin lesion analysis toward melanoma detection: a challenge at the international symposium on biomedical imaging (ISBI) 2016, hosted by the International Skin Imaging Collaboration (ISIC). arXiv:1605.01397v1 [cs.CV] 4 May 2016
2. Siegel RL, Miller KD, Jemal A (2020) Cancer statistics, 2020. CA Cancer J Clin 70:7–30
3. Melanoma—symptoms and causes. https://www.mayoclinic.org/diseases-conditions/melanoma/symptoms-causes/syc-20374884
4. Esteva A, Kuprel B, Novoa RA, Ko J, Swetter SM, Blau HM, Thrun S (2017) Dermatologist-level classification of skin cancer with deep neural networks. Nature 542:115
5. He K, Zhang X, Ren S, Sun J (2016) Deep residual learning for image recognition. In: The IEEE conference on computer vision and pattern recognition (CVPR), pp 770–778
6. Romero Lopez A, Giro-i-Nieto X, Burdick J, Marques O (2017) Skin lesion classification from dermoscopic images using deep learning techniques. In: 2017 13th IASTED international conference on biomedical engineering (BioMed), Innsbruck, Austria, pp 49–54
7. Singh V, Nwogu I (2018) Analyzing skin lesions in dermoscopy images using convolutional neural networks. In: 2018 IEEE international conference on systems, man, and cybernetics (SMC), Miyazaki, Japan, pp 4035–4040
8. Nasr-Esfahani E et al (2016) Melanoma detection by analysis of clinical images using convolutional neural network. In: 2016 38th annual international conference of the IEEE engineering in medicine and biology society (EMBC), Orlando, FL, pp 1373–1376
9. Codella N, Cai J, Abedini M, Garnavi R, Halpern A, Smith JR (2015) Deep learning, sparse coding, and SVM for melanoma recognition in dermoscopy images. https://doi.org/10.1007/978-3-319-24888-2
10. Li Y, Shen L (2018) Skin lesion analysis towards melanoma detection using deep learning network. Sensors 18:556. https://doi.org/10.3390/s18020556
11. Perez L, Wang J (2017) The effectiveness of data augmentation in image classification using deep learning. arXiv:1712.04621v1 [cs.CV] 13 Dec 2017
12. VGG16—convolutional network for classification and detection. https://neurohive.io/en/popular-networks/vgg16/
13. Agarap AFM (2019) Deep learning using rectified linear units (ReLU). arXiv:1803.08375v2 [cs.NE] 7 Feb 2019
14. Difference between sigmoid function and softmax function with code and graph. https://medium.com/aidevnepal/for-sigmoid-funcion-f7a5da78fec2
15. Ratul AR, Mozaffari MH, Lee W, Parimbelli E (2019) Skin lesions classification using deep learning based on dilated convolution. https://doi.org/10.1101/860700
16. Dentropy Manaswi NK (2018) Understanding and working with keras. In: Deep learning with applications using python. Apress, Berkeley, CA
17. Developer resources for deep learning and AI, NVIDIA. https://www.nvidia.com/en-us/deep-learning-ai/developer/
18. Overview of different optimizers in neural networks. https://medium.com/datadriveninvestor/overview-of-different-optimizers-for-neural-networks-e0ed119440c3

19. How to choose loss functions when training deep learning neural networks. https://machinelearningmastery.com/how-to-choose-loss-functions-when-training-deep-learning-neural-networks/
20. Codella N, Gutman D, Celebi ME, Helba B, Marchetti MA, Dusza S, Kalloo A, Liopyris K, Mishra N, Kittler H, Halpern A (2017) Skin lesion analysis toward melanoma detection: a challenge at the 2017 international symposium on biomedical imaging (ISBI), Hosted by the International Skin Imaging Collaboration (ISIC). arXiv:1710.05006 [cs.CV]

# Construction of Reduced PPI Network for Breast and Lung Cancer Using Random Walk with Restart

**Richa K. Makhijani**

**Abstract** Distinguished procedures inspired from nature have been applied to obtain remarkable performance in exhaustive search for solution. This has helped in solving many real-life problems such as computational biology for critical diseases, optimization problems, design, and many others. Random walk is one such strategy used as an intrinsic metaheuristic, nature-inspired algorithm. It is seen to be widely used in design of algorithms for analysis, interpretation, compaction, and summarization of large crucial protein–protein interaction (PPI) networks of complex diseases such as cancer. Several computational biology studies revolve around complex research domain of identifying candidate cancer genes, from numerous data sources. The large volumes of data belonging to different platforms pose a challenge in design of efficient bioinformatics pipelines. We provide and implement a computational approach for identification of candidate cancer genes in two most popular cancer conditions in men and women, lung and breast. The approach uses gene expression data from two efficient data platforms, microarray and RNA-seq. Genes jointly extracted from both data platforms in the two cancers are further analyzed at the protein level by constructing PPI network. The large networks produced are further reduced to significant genes by prioritization of genes by selecting hubs and then performing random walk with restart procedure. The results prove the relevance of the selected genes to cancer and thus help in producing compact PPI network which can be further studied for its functional characteristics.

**Keywords** PPI network · Cancer · Random walk · Summarization

## 1 Introduction

Interactions between biological molecules are widely being studied in bioinformatics research for various diseases. These interactions show association of genes or proteins at a physical or a functional level. Protein–protein interaction (PPI) networks are a

R. K. Makhijani (✉)
Indian Institute of Information Technology, Nagpur, India
e-mail: richa_makhijani@yahoo.co.in



A. Maji et al. (eds.), *Proceedings of the International Conference on Computing and Communication Systems*, Lecture Notes in Networks and Systems 170,
https://doi.org/10.1007/978-981-33-4084-8_39

very popular example of such associations between proteins which help in revealing the underlying biological processes in a cell. These data are derived from different clinical experiments or literature mining and stored in primary databases such as HPRD, STRING, DIP, and MINT. The topological as well as biological aspects of such networks contribute effectively to devise a solution to multiple problems at genetic level. Also, the quick access to a large PPI database of human proteins facilitates the analysis of such problems. The advancement in construction of such networks explores outstanding opportunities to unravel functional characteristics of gene in particular diseases, specifically cancer.

Random walk strategy is seen as a powerful approach that uses guilt-by-association principle and is applied to variety of problems in bioinformatics. The random walk approach was first developed to globally traverse biological networks in a topological manner. It simulates a network propagation strategy showing movement of a particle from one node to any randomly selected neighboring node in an iterative fashion [5–7, 10, 12, 17]. The restart idea was implemented for search engines on internet. The primary problem on which this approach was used was ranking of Web pages by modeling the behavior of a user. The hyperlinks and pages are considered as graphs, and topological structure is studied to find the most visited pages. The relevance of Web pages is then measured by the number of visits [1]. Another idea that emerged out of this approach was to force the particle to restart from a fixed node called seed and then measure the adjacency of this seed or set of seeds to all other nodes [13]. Since then, RWR emerged as the state-of-the-art guilt-by-association algorithm in computational biology dealing with networks. RWR methods are commonly applied to problems involving identification of significant genes and gene sets in different disease types. Authors in [8] used this strategy to identify novel genes in Meniere's disease (MD) using the known diseased genes as seed nodes. It was also used to find essentiality of proteins by integrating the topological and biological properties of PPI networks [6]. It was also used to computationally identify novel epigenetic factors at the genome level to completely understand the underlying mechanisms of epigenetic regulation [7].

It was initially applied, as commented above, to prioritize candidate disease genes. It was used to rank all the network nodes based on their adjacency to the known diseased nodes considered as seeds [8]. Moreover, RWR has also found applications in drug discovery by identification of drug–target interactions [5].

In this study, we therefore use this approach to predict common significant gene markers in the two most common cancer types, lung and breast. We first identify differentially expressed genes from microarray and RNA-seq data individually. We then estimate the joint results from this analysis which are used further to construct PPI network. The large PPI network is then reduced by selecting top 10 genes that are closest to the common hubs in the two networks. This is accomplished by the RWR approach. These genes thus characterize a reduced cancer network significantly enriched in cancer pathways and at the same time highlight the common hubs in two cancer conditions.

## 2 Materials and Methods

### 2.1 *Datasets Used*

Data used in this study includes expression of genes extracted from RNA-seq and microarray platforms, along with PPI data from STRING database [19]. The two most occurring cancers in men and women are chosen in this experiment, lung and breast. Gene Expression Omnibus (GEO) data repository was accessed for downloading gene expression data belonging to two platforms. The microarray dataset used in this study is GSE45827 (149/11) [2], GSE48984 (3/9) [20], GSE26910 (6/6) [14], GSE19804 (60/60) [11] and GSE10072 (57/50) [4]. The numbers in the brackets indicate the tumor/normal samples included. Also, GSE62944 (1621/164) [15] is the RNA-seq dataset that includes data for 24 cancers. The data for breast and lung cancer was extracted from these samples and analyzed.

### 2.2 *Identification of Significant Genes*

Methods using linear models have proven to be successful in efficient identification of differential expressed genes (DEGs). Hence, we used LIMMA that applies Bayes procedure and is improved with reference to increased power, less false positives, fast and easy execution [3, 18]. For normalization of values, *voom* method is implemented that performs a Locally Weighted Scatterplot Smoothing (LOWESS) regression. LIMMA provides benefit of using the similar modeling for RNA-seq data as that of microarrays. The parameters used for LIMMA were $P$-value $< 0.05$ with BH correction and log fold change $\geq 1$. When compared to other methods, LIMMA proved to be efficient [16]. LIMMA was seen to have equally better and comparable performance for some parameters when compared with others, even for RNA-seq data type. For all individual datasets of breast cancer, LIMMA was used to identify the DEGs. After obtaining genes for all datasets, the gene lists were merged by a union operation, hence forming the list of DEGs in breast cancer microarray data. Then, DEGs were extracted from RNA-seq data of breast cancer, and an intersection of gene lists from microarray and RNA-seq was conducted to select genes from these platforms. Similar procedure was conducted for lung cancer microarray and RNA-seq data.

### 2.3 *Computational Workflow for Identification of Top Genes in PPI Network*

From the previous step, we get the list of significant genes in breast and lung cancer. In the next step, we downloaded the PPI data for each of these genes from the

STRING database. Hence, we obtained two PPI datasets for the two cancers being studied. We read these datasets in R as an igraph object and identify the hub nodes which have large betweenness centrality and degree $\geq 10$. We used random walk with restart strategy (RWR) to further analyze the nodes of the networks with the aim to uncover their functional association. The strategy was originally used to exploit overall topological aspects of the networks. Many variations of this strategy have been proposed for improvements over the basic concept of random walks [9, 22]. We use RandomWalkRestartMH 1.3.0 R package for identification of top significant genes in the individual PPI networks of breast and lung cancer which tend to be functionally similar to each other [21].The algorithm produces top 10 genes along with the score which represents its closeness to the seeds of the network. We then create a network with the top scored genes (top 15 genes network). The networks are then visualized with the seeds highlighted.

## 3 Emperical Results

This section shows the results of our pipeline.

### *3.1 Formulation of DEGs in Breast and Lung Cancer*

The aim of identifying significant cancer markers in the two cancer types was accomplished by applying linear modeling technique to the two cancer datasets. Individual microarray datasets were analyzed to produce a DEG list for each cancer. Similarly, RNA-seq datasets were analyzed to produce DEG list for individual cancer. The results are elaborated in Table 1. Genes common to the two platforms were extracted by an intersection across the results of microarray and RNA-seq analysis. Using data from two platforms improves the reliability of the results.

### *3.2 Top Genes in PPI Network*

For the two DEG lists obtained in Table 1, we construct the corresponding PPI networks. This is accomplished by using STRING database for selecting their respective

**Table 1** Identified differentially expressed genes using data from two platforms

| Cancer type | Microarray DEGs | RNA-seq DEGs | Common to both |
|---|---|---|---|
| Breast | 2629 | 8639 | 1236 |
| Lung | 1080 | 7234 | 860 |

**Table 2** List of hub genes common in breast and lung cancer

| Id | Items_Details | Genes (DCN, CAV1, PDE2A, BMP2, ARRB1) |
|---|---|---|
| 0 | Cytokine–cytokine receptor interaction | BMP2 |
| 1 | Purine metabolism | PDE2A |
| 2 | MAPK signaling pathway | ARRB1 |
| 3 | Pathways in cancer | BMP2 |
| 4 | Chemokine signaling pathway | ARRB1 |
| 5 | Focal adhesion | CAV1 |
| 6 | Endocytosis | CAV1,ARRB1 |
| 7 | Phototransduction | ARRB1 |
| 8 | TGF-beta signaling pathway | DCN, BMP2 |
| 9 | Bacterial invasion of epithelial cells | CAV1 |
| 10 | Basal cell carcinoma | BMP2 |
| 11 | Hedgehog signaling pathway | BMP2 |
| 12 | Viral myocarditis | CAV1 |

proteins and their interaction partners. Thus, in the PPI network for breast cancer, there were 1001 nodes-6108 edges. Similarly, in the PPI network of lung cancer, 512 nodes-2609 edges were seen. The next step was to distinguish hub nodes from these networks with degree $\geq 10$ and large betweenness centrality values. This resulted in 78 hubs from breast cancer PPI network and 25 hubs from lung cancer PPI network. Among both the lists of hubs, five genes were found to be common. The list of these nodes is given in Table 2. The KEGG pathways for which these genes are enriched are given in Table 3.

For efficient characterization of candidate cancer genes functioning in the two cancer conditions, RWR was applied to the two PPI networks. These networks were treated as igraph objects in *R* for which the adjacency matrix was obtained. For starting the RWR, seeds were initialized as the five common hub genes(nodes) listed in Table 2. Random.Walk.Restart.Multiplex() function from RandomWalkRestartMH R package was then executed for the two individual PPI networks, keeping the seed lists same. Hence, the results produced top 10 genes which were close to the seeds for each PPI network. The results along with the score are illustrated in Table 3. We then create a reduced PPI network with the top 15 genes. This would give a global view of the candidate prioritized top genes in individual cancer conditions. The corresponding networks are shown in Fig. 1.

**Table 3** Top 10 genes in lung and breast cancer after RWR

| S. no. | Breast cancer | | Lung cancer | |
|---|---|---|---|---|
| | Node names | Score | Node names | Score |
| 1 | ALB | 0.004826647 | OGN | 0.006506405 |
| 2 | EGF | 0.004643741 | FGF2 | 0.005730947 |
| 3 | EGFR | 0.004623857 | JUN | 0.005165530 |
| 4 | BGN | 0.004579291 | IL1B | 0.005125314 |
| 5 | PRKAR1B | 0.003744878 | KDR | 0.004976911 |
| 6 | PRKAR1A | 0.003711779 | CXCL12 | 0.004778854 |
| 7 | FGF2 | 0.003442627 | EDN1 | 0.004684532 |
| 8 | OGN | 0.003408295 | RXFP1 | 0.004553666 |
| 9 | FMOD | 0.003319703 | ADCY9 | 0.004474954 |
| 10 | IGF1 | 0.003181494 | ECM2 | 0.004463404 |

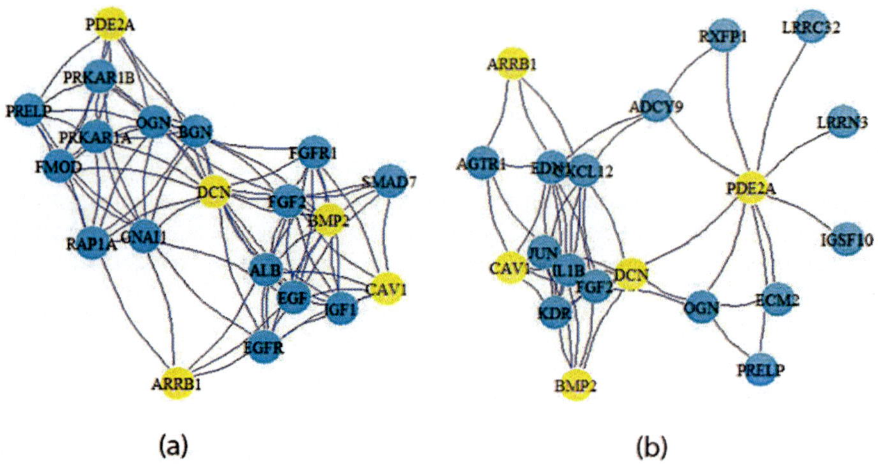


**Fig. 1** Compact PPI network of top genes with seeds in yellow **a** for breast cancer **b** for lung cancer

## 4 Discussion and Conclusion

Selection of diseased genes in cancer is a trivial research domain from the point of view of both molecular biologists and computer scientists. There is a high dependency of these researchers over each other, as the analysis of large data requires computational expertise, whereas generation of these data is to be done in wet laboratories. This has led to diverse data types and methodologies for selection of drug targets, specifically in complex deadly disease such as cancer. We propose an approach for detection of significant genes which may be markers in these cancers. The gene expression data from microarray and RNA-seq platforms helped in evaluating a reliable list of DEGs in the two cancers. Further analysis at the interactome level using

PPI networks resulted in characterization of significant hub nodes. It was interesting to get five genes, DCN, CAV1, PDE2A, BMP2, ARRB1, as the common hubs in both cancers, pointing toward the common functional profile at the molecular level. These nodes when taken as seeds for random walk with restart on the top genes produced a reduced PPI network for breast and lung cancer. This gives a clear insight of the roles of these hubs in individual cancers, as well as provides a global insight of their common functions in two cancer conditions. The KEGG pathways enriched by these genes include significant cancer-related pathways such as MAPK signaling pathway, pathways in cancer, basal cell carcinoma, and other. Hence, therapeutic targets for multiple cancers can be studied using this approach to unravel the complex functional characteristics at the genetic as well as interactome level.

**Acknowledgements** We greatly acknowledge the efforts of Dr. Hemant J. Purohit for his invaluable guidance and suggestions in carrying out this work.

## References

1. Brin S, Page L (1998) The anatomy of a large-scale hypertextual web search engine. Comput Netw ISDN Syst 30(1–7):107–117
2. Gruosso T, Mieulet V, Cardon M, Bourachot B, Kieffer Y, Devun F, Dubois T, Dutreix M, Vincent-Salomon A, Miller KM et al (2016) Chronic oxidative stress promotes H2AX protein degradation and enhances chemosensitivity in breast cancer patients. EMBO Mol Med 8(5):527–549
3. Jeanmougin M, De Reynies A, Marisa L, Paccard C, Nuel G, Guedj M (2010) Should we abandon the *t*-test in the analysis of gene expression microarray data: a comparison of variance modeling strategies. PLoS ONE 5(9):e12336
4. Landi MT, Dracheva T, Rotunno M, Figueroa JD, Liu H, Dasgupta A, Mann FE, Fukuoka J, Hames M, Bergen AW et al (2008) Gene expression signature of cigarette smoking and its role in lung adenocarcinoma development and survival. PLoS ONE 3(2):e1651
5. Lee I, Nam H (2018) Identification of drug-target interaction by a random walk with restart method on an interactome network. BMC Bioinformatics 19(8):208
6. Lei X, Yang X, Fujita H (2019) Random walk based method to identify essential proteins by integrating network topology and biological characteristics. Knowl Based Syst 167:53–67
7. Li J, Chen L, Wang S, Zhang Y, Kong X, Huang T, Cai YD (2018) A computational method using the random walk with restart algorithm for identifying novel epigenetic factors. Mol Genet Genomics 293(1):293–301
8. Li L, Wang Y, An L, Kong X, Huang T (2017) A network-based method using a random walk with restart algorithm and screening tests to identify novel genes associated with Menière's disease. PLoS ONE 12(8):e0182592
9. Li Y, Li J (2012) Disease gene identification by random walk on multigraphs merging heterogeneous genomic and phenotype data. BMC Gen 13(7):S27
10. Lovász L et al (1993) Random walks on graphs: a survey. In: Combinatorics, Paul Erdos is eighty, vol 2(1), pp 1–46
11. Lu TP, Tsai MH, Lee JM, Hsu CP, Chen PC, Lin CW, Shih JY, Yang PC, Hsiao CK, Lai LC et al (2010) Identification of a novel biomarker, SEMA5A, for non-small cell lung carcinoma in nonsmoking women. Cancer Epidemiol Prev Biomarkers 19(10):2590–2597
12. Niu YW, Wang GH, Yan GY, Chen X (2019) Integrating random walk and binary regression to identify novel miRNA-disease association. BMC Bioinformatics 20(1):59

13. Pan JY, Yang HJ, Faloutsos C, Duygulu P (2004) Automatic multimedia cross-modal correlation discovery. In: Proceedings of the tenth ACM SIGKDD international conference on knowledge discovery and data mining. ACM, pp 653–658
14. Planche A, Bacac M, Provero P, Fusco C, Delorenzi M, Stehle JC, Stamenkovic I (2011) Identification of prognostic molecular features in the reactive stroma of human breast and prostate cancer. PLoS ONE 6(5):e18640
15. Rahman M, Jackson LK, Johnson WE, Li DY, Bild AH, Piccolo SR (2015) Alternative preprocessing of RNA-sequencing data in the cancer genome atlas leads to improved analysis results. Bioinformatics 31(22):3666–3672
16. Rapaport F, Khanin R, Liang Y, Pirun M, Krek A, Zumbo P, Mason CE, Socci ND, Betel D (2013) Comprehensive evaluation of differential gene expression analysis methods for RNA-seq data. Genome Biol. 14(9):3158
17. Ren J, Shang L, Wang Q, Li J (2019) Ranking cancer proteins by integrating PPI network and protein expression profiles. BioMed Res Int 2019
18. Ritchie ME, Phipson B, Wu D, Hu Y, Law CW, Shi W, Smyth GK (2015) limma powers differential expression analyses for RNA-sequencing and microarray studies. Nucleic Acids Res 43(7):e47
19. Szklarczyk D, Franceschini A, Wyder S, Forslund K, Heller D, Huerta-Cepas J, Simonovic M, Roth A, Santos A, Tsafou KP et al (2014) String v10: protein-protein interaction networks, integrated over the tree of life. Nucleic Acids Res 43(D1):D447–D452
20. Timmerman LA, Holton T, Yuneva M, Louie RJ, Padró M, Daemen A, Hu M, Chan DA, Ethier SP, van't Veer LJ et al (2013) Glutamine sensitivity analysis identifies the xCT antiporter as a common triple-negative breast tumor therapeutic target. Cancer Cell 24(4):450–465
21. Valdeolivas A, Tichit L, Navarro C, Perrin S, Odelin G, Levy N, Cau P, Remy E, Baudot A (2017) Random walk with restart on multiplex and heterogeneous biological networks. bioRxiv, p 134734
22. Xie M, Xu Y, Zhang Y, Hwang T, Kuang R (2015) Network-based phenome-genome association prediction by bi-random walk. PLoS ONE 10(5):e0125138

# Analysis of Influence of Yoga-Asana on the Digestive Process Using Electrogastrograms

**Alagumariappan Paramasivam, D. Najumnissa Jamal, C. Emmanuel, K. B. Bhaskar, M. Mohit Jaisingh, and R. Kannan**

**Abstract** Human beings have the capability to do multi-tasking in an efficient way. Recent years, due to pressure, time constraints, and environmental conditions in the working environment, both physically and mentally, humans are subjected to high-stress level. Yoga plays a major role in controlling stress since it has no side effects with proper mentoring. In this work, the influence of Yoga-Asana on the digestive process is analyzed using electrogastrograms (EGG). Further, the fast Fourier transform (FFT) is used to analyze the frequency range of EGG signals acquired from an abnormal patient before and after Yoga. Results demonstrate that the fast Fourier transform of an electrogastrograms acquired from patient with digestive abnormality before Yoga is out of the EGG normal frequency range. Further, the FFT of an electrogastrograms acquired from patient with digestive abnormality after Yoga is in EGG normal frequency range. This work appears to be of highly significant since Yoga-Asana can improvise the human digestion process.

**Keywords** Yoga · Digestive disorders · Electrogastrograms · Fast Fourier transform

---

A. Paramasivam (✉) · D. Najumnissa Jamal · M. Mohit Jaisingh · R. Kannan
B. S. Abdur Rahman Crescent Institute of Science and Technology, Chennai 600048, India
e-mail: parama.ice@gmail.com

D. Najumnissa Jamal
e-mail: najumnissa.d@crescent.education

M. Mohit Jaisingh
e-mail: mohitmjaisingh@gmail.com

R. Kannan
e-mail: kannanraja962@gmail.com

K. B. Bhaskar
Adhi College of Engineering and Technology, Chennai 631605, India
e-mail: jaibhaskar15@gmail.com

C. Emmanuel
Global Hospitals and Health City, Perumbakkam, Chennai 600100, India
e-mail: cyem69@yahoo.co.in



A. Maji et al. (eds.), *Proceedings of the International Conference on Computing and Communication Systems*, Lecture Notes in Networks and Systems 170,
https://doi.org/10.1007/978-981-33-4084-8_40

## 1 Introduction

The stomach is a muscular organ located on the left side of the upper abdominal cavity whose physiological functions include storing, mixing, and digesting swallowed food and liquid [1]. Digestion is the process of breaking down complex substances into simpler components namely nutrients. These nutrients can be used for energy, growth, and cell repair. The stomach muscle controls the motility of the stomach using a rhythmic electrical activity which aids in the digestive process simultaneously. This noncontrolled rhythmic electrical activity is called gastric myoelectrical activity [1].

Abnormalities in the digestion process result in heartburn, nausea, vomiting, bleeding, bloating, diarrhea, constipation, etc. Abnormal digestion occurs due to factors like hereditary, climatic conditions, food habits, etc. Li et al. [2] have discussed the importance of the food digestion rate in the human body. Further, the authors have suggested many suitable solutions for maintaining a human digestion rate. Some of the major methods used for the diagnosis of digestive disorders are endoscopy, biopsy, EGG, etc. [3]. However, the endoscopy is one of the commonly followed methods used to diagnose digestive abnormalities, and it has major drawbacks such as very high recovery time and a semi-intrusive process.

Electrogastrography is the methodology used for the measurement of human gastricmyoelectrical activity. However, the EGG is a non-intrusive process and does not require any recovery time; it gains more importance in recent days [3–8]. Yoshida et al. [9] have analyzed the effect of placement of electrodes in different places of the stomach.

Yoga is a system of mental, physical, and spiritual practices that originate from ancient India [10]. Yoga helps improve health-related issues and health behavior; different Yoga practices are associated with specific health outcomes. Al Kafee et al. [1] have proposed a study on the impact of Yoga on curing digestive disorders. Ostermann et al. [10] have concluded that Yoga provides significant improvement for a patient having Anorexia nervosa and helps the patient to get relieved from lifelong medication. Also, the authors have concluded that Yoga has a major positive impact on the process of healing and improving digestive system disorders.

The information measures of bio-signals are highly useful features for diagnosis of digestive diseases. While reviewing several literatures, it is found that EGG is very simple and efficient for digestive health diagnosis and the study is proposed as an alternate method for diagnosis of digestive disorders.

The objective of this work is to analyze the influence of Yoga-Asana on digestive disorders using electrogastrograms.

# 2 Methodology

## 2.1 Proposed System

Figure 1 shows the overview of electrogastrogram acquisition device.

In this work, an electrogastrogram (EGG) acquisition system is designed and developed using Raspberry PI hardware. Further, a three-electrode system is used to acquire EGG signals from humans with digestive abnormalities. Since the amplitudes of EGG signals are in microvolts, the INA 12-based instrumentation amplifier circuit is utilized to amplify the acquired abnormal EGG signals before and after Yoga. Also, the amplified EGG signals are converted into digital using 10-bit analog-to-digital converter (ADC) namely MCP3008. The converted EGG signals of every individual patient are stored as individual.csv (comma-separated value) files.

The Yoga-Asana's like Balasana, Pavanamuktasana, Trikonasana, Ardha Matsyendrasana, etc., were selected by the Yoga practitioner for this study. Further, the Yoga therapy was performed and monitored under the guidance of certified trained Yoga practitioner. The acquisition protocols were explained clearly to the abnormal patients and the EGG signals from 10 abnormal patients were acquired before and after Yoga-Asana with their consent. Also, the questionnaire was obtained to observe previous history of abnormalities. The proposed study is approved by the Institutional Ethics Committee, Global Hospitals and Health City, Chennai, and proper clearance has been obtained. Also, the proposed EGG system was tested and calibrated against a standard EGG system.

Since, the amplitude and frequency of an electrogastrograms play a vital role in diagnosis. However, the amplitude of the EGG signals varies due to many factors such as pre-prandial and postprandial conditions, etc.; in this work, the fast Fourier transform is used to analyze the frequency of the acquired EGG signal. The FFT algorithm is programmed using Python software inside the Raspberry PI hardware. The frequency of the acquired electrogastrograms from patients before and after Yoga-Asana stored in a.csv file is analyzed using the FFT algorithm.

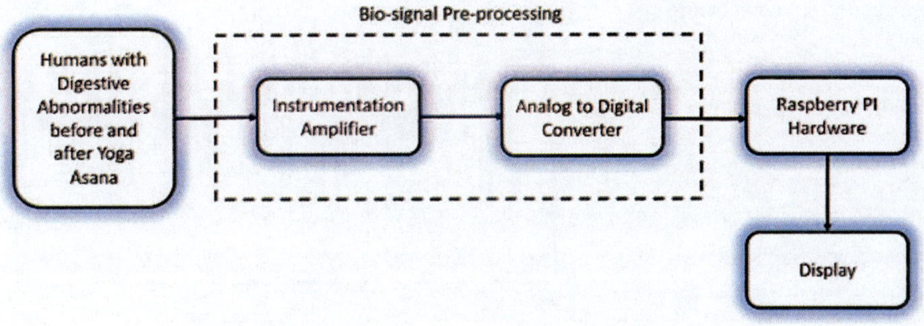


**Fig. 1** Overall block diagram of proposed electrogastrogram acquisition device

### 2.2 Analysis Using Fast Fourier Transform (FFT)

Generally, there are two ways to reckon a bio-signal. One is using time domain analysis and the other is using frequency domain analysis. In this work, the frequency domain analysis is used to represent the spectral entropy of the electrogastrograms. These spectral signals can be measured using a method called fast Fourier transform (FFT). The electrogastrogram is analyzed using FFT which is programmed using Python language inside the Raspberry PI controller. The FFT of an EGG signal can be calculated using Eq. 1.

$$F(x) = \sum_{n=0}^{N-1} e^{-2\pi j \frac{xn}{N}} y(n) \tag{1}$$

where $F(x)$ is the sum up to $N$ terms. When the frequency reaches its highest peak that frequency is considered as the fundamental frequency of the electrogastrograms.

## 3 Results and Discussion

Figure 2 shows a typical electrogastrograms measured from patient with digestive abnormality before Yoga. The EGG signal is taken for a period of 9 min in pre-prandial condition.

Figure 3 shows the FFT of an electrogastrogram measured from abnormal patient before Yoga. The normal frequency range of an EGG signal for a healthy person is 2.6–3.7 cycles per minute (CPM). It is seen that the FFT of an EGG signal acquired from patient with digestive abnormalities before yoga is not on the normal frequency range.

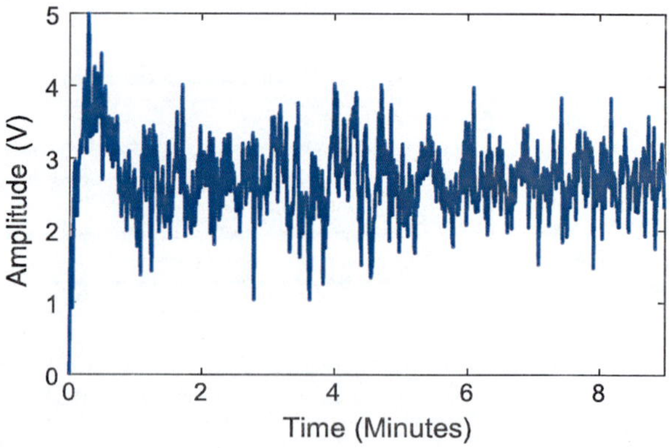


**Fig. 2** Typical electrogastrograms measured from abnormal patient before Yoga

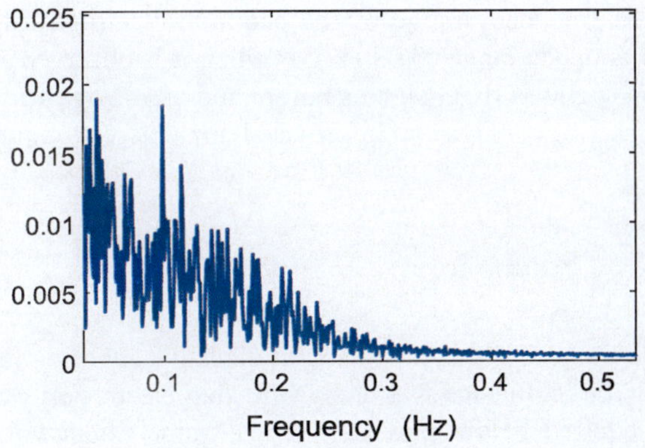


**Fig. 3** FFT of an electrogastrogram measured from abnormal patient before Yoga

Figure 4 shows a typical electrogastrogram measured from same patient with digestive abnormality after Yoga. Figure 5 shows the FFT of an electrogastrogram measured from abnormal patient after Yoga. Further, it is clearly demonstrated that

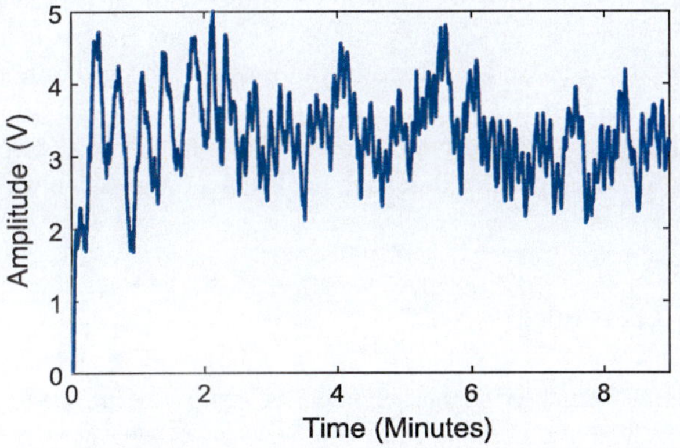


**Fig. 4** Typical electrogastrogram measured from abnormal patient after Yoga

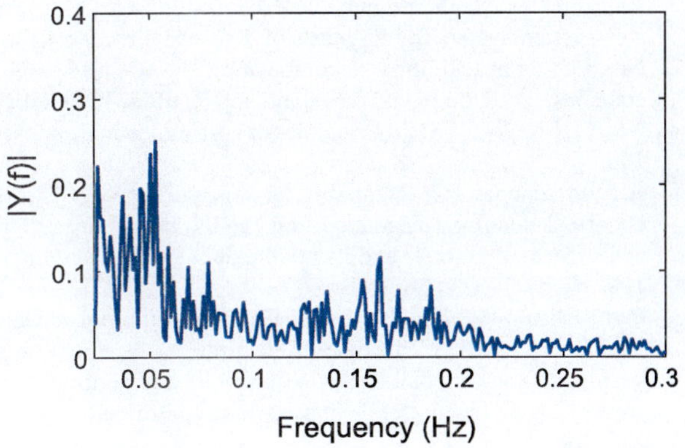


**Fig. 5** FFT of an EGG signal acquired from abnormal patient after Yoga

the frequency range of the acquired electrogastrogram is in normal EGG signal frequency range (0.05 Hz). Further, it is observed that there is a significant change in acquired EGG signals before and after Yoga which clearly demonstrates that the practice of Yoga can improvise human digestion process significantly.

## 4 Conclusion

In this work, the patients with digestive abnormalities are subjected to Yoga-Asana and its influence was analyzed using electrogastrograms. Further, the EGG signals acquired before and after Yoga from patient with digestive abnormalities were analyzed using FFT algorithm. Results demonstrate that the FFT of an electrogastrogram acquired before Yoga practice clearly shows that the frequency is not in normal EGG frequency range whereas the FFT of an EGG signal acquired after regular Yoga practice is in normal EGG frequency range (0.05 Hz). This work appears to be highly significant since EGG signals can be acquired during regular practice of Yoga and the improvement in the digestion process can be monitored. The acquisition of electrogastrograms from patient with fatty muscles is a very challenging task. Further, the signal interference due to other organs is high. However, the study has above discussed limitations, the proper filtering methods, and AI-based techniques can improve the efficacy of the proposed study. The future scope involves automated diagnosis of digestive disorders and their therapy. Also, the proposed study can be utilized to monitor digestive health in a progressive way.

## References


1. Al Kafee A, Eryiğit S, Akan A (2017) Electrogastrography in patients with diabetic gastroparesis. In: 2017 medical technologies national congress (TIPTEKNO). IEEE, pp 1–4
2. Li C, Yu W, Wu P, Chen XD (2019) Current in vitro digestion systems for understanding food digestion in human upper gastrointestinal tract. Trends Food Sci Technol 96:114–126
3. Rajagopal A, Alagumariappan P, Krishnamurthy K (2020) Development of an automated decision support system for diagnosis of digestive disorders using electrogastrograms: an approach based on empirical mode decomposition and K-means algorithm. In: Disruptive technology: concepts, methodologies, tools, and applications. IGI Global, pp 661–678
4. Keen T, Brooks C (2020) Principles of gastrointestinal endoscopy. Surgery (Oxford) 38(3):155–160
5. Alagumariappan P, Rajagopal A, Krishnamurthy K (2016) Complexity analysis on normal and abnormal electrogastrograms using Tsallis entropy. In: 3rd international electronic and flipped conference on entropy and its applications. Multidisciplinary Digital Publishing Institute, pp 1–6
6. Paramasivam A, Kamalanand K, Emmanuel C, Mahadevan B, Sundravadivelu K, Raman J, Jawahar PM (2018) Influence of electrode surface area on the fractal dimensions of electrogastrograms and fractal analysis of normal and abnormal digestion process. In: International conference on recent trends in electrical, control and communication (RTECC), Malaysia, pp 245–250

7. Alagumariappan P, Krishnamurthy K (2018) An approach based on information theory for selection of systems for efficient recording of electrogastrograms. In: Proceedings of the international conference on computing and communication systems. Springer, Singapore, pp 463–471
8. Paramasivam A, Kamalanand K, Sundravadivelu K, Emmanuel C, Rajinikanth V (2020) Feature extraction and genetic algorithm based feature selection for diagnosis of type-2 diabetes using Electrogastrograms. JMIR Biomed Eng
9. Yoshida M, Kitani Y, Inoue M (2000) Basic study for measurement of electrogastrography. In: Proceedings of the 22nd annual international conference of the IEEE engineering in medicine and biology society (Cat. No. 00CH37143), vol 3. IEEE, pp 1962–1963
10. Ostermann T, Vogel H, Starke C, Cramer H (2019) Effectiveness of yoga in eating disorders–a case report. Complem Therap Med 42:145–148

# Study the Cognitive Changes in Cerebral Palsy Children Employing Repetitive Transcranial Magnetic Stimulation and Neurofeedback Training

**Meena Gupta and Dinesh Bhatia**

**Abstract** Impairment of cognition is one of the biggest challenges for cerebral palsy (CP) survivors. Cognition could be defined as a mental process for acquiring knowledge and understanding the thoughts, experiences, and sensory process. The cognition process encompasses domains such as attention, knowledge, working memory, judgment, and logical reasoning. The study was conducted to evaluate the most effective therapy for understanding different cognitive changes in children with CP. The results obtained suggest that r-TMS is a result-oriented therapy for cognitive enhancement in the selected therapy groups. The cognitive brain training exercises were found to be the next most effective and cost-effective therapy for cognitive enhancement in brain-damaged children. These results could be employed for understanding the cognitive changes occurring in the children with CP for cognitive enhancement.

**Keywords** Cerebral palsy · Cognitive changes · Neuromodulation · Neurofeedback · Transcranial magnetic stimulation

## 1 Introduction

Cognition process involves the acquisition of knowledge by the understanding of thoughts, experiences, and senses. Different processes such as attention, knowledge, working memory, judgment, and reasoning are part of it [1]. Cerebral palsy is a disruption of the body's motor functions due to poor development of the motor areas of the brain that regulate, control, and coordinate movements to maintain a smooth and synchronized posture and movement of the body. This central disability

M. Gupta
Amity Institute of Physiotherapy, Amity University, Sector-125, Noida 201301, Uttar Pradesh, India
e-mail: mgupta9@amity.edu; meenaguptaphysio@gmail.com

D. Bhatia (✉)
Department of Biomedical Engineering, North Eastern Hill University, Shillong 793022, Meghalaya, India
e-mail: bhatiadinesh@rediffmail.com



A. Maji et al. (eds.), *Proceedings of the International Conference on Computing and Communication Systems*, Lecture Notes in Networks and Systems 170,
https://doi.org/10.1007/978-981-33-4084-8_41

is usually accompanied by other disorders such as epilepsy, speech deficits, hearing, vision, and intellectual disability. Performing tasks for cognitive changes in CP is one of the most difficult tasks that depend on neuromodulation. Children with CP have difficulty in any motor neuron function such as rolling, crawling, sitting, standing, and walking with many other associated problems such as poor understanding, intellectual disability, hearing impairment, and visual impairment. For hearing and visual impairment, several adaptive devices are available such as cochlear implant, spectacles with different intervention(s) along with therapeutically interventions such as visual exercises and auditory stimulation [2]. For children with CP, physical milestones can be set and achieved with the help of various intervention techniques such as surgical intervention, medical intervention, rhizotomy, physical therapy, occupational therapy, and conventional therapies. However, in the area of cognitive enhancement for CP children, both milestones and the methods through which they can be achieved are not well defined.

Presently, the neuromodulation technique is an emerging in medical as well as in the research field. It targets the nervous system to do restoration of functional performance, relief of pain and control of involuntary movements thereby improving neuroplasticity. Transcranial magnetic stimulation (TMS) and neurofeedback (NFB) both are examples of non-invasive stimulation techniques for neuromodulation [3]. In neuromodulation, neuromodulators directly stimulate the nervous system with electrical signals. A few examples can be stimulation of the spinal cord, peripheral nerve, and deep regions in the brain. In this study, we studied cognitive changes arising from different brain waves in CP children employing TMS [2] and NFB [3]. TMS directs a pulsed magnetic field into brain tissue non-invasively and helps to investigate cortical plasticity for treating several neurological disorders. The NFB trains the brain through self-regulation and helps in improving neuroplasticity by employing real-time display of brain activity. The aim of the present study was to evaluate the effects of rTMS and NFB in enhancing the cognitive changes in CP children for their cognitive impairment. Quite a few studies to date have been done in this area to understand cognitive changes in CP children by employing different therapeutical interventions.

## 2 Materials and Methods

### 2.1 *Participants*

Forty-five diagnosed cases of CP by child neurologists were recruited in this study. The study procedure was explained to these participants and written informed consent was collected from the parents/guardians of these subjects. Inclusion criteria for the study comprised of CP subjects with age group 5–15 years (age more than 5 years comes under concrete learning and child is able to understand the commands), willingness to participate in the study, both gender either male or female, seizure under

**Table 1** Demographic details of participants in different groups

| Participants' characteristics | Group A ($n$ = 15) | Group B ($n$ = 15) | Group C ($n$ = 15) |
|---|---|---|---|
| Mean age, years (SD) | 6.25 (4.06) | 8.26 (5.05) | 8.19 (5.08) |
| Mean height, cm (SD) | 112.93 (22.47) | 90.33 (45.39) | 90.27 (45.50) |
| Mean weight, kg (SD) | 17.27 (7) | 19.47(5.44) | 19.33 (5.39) |
| Sex; male: female | 8:07 | 11:04 | 9:06 |

antiepileptic control medicine and Gross Motor Function Classification System for cerebral palsy (GMFCS) level II and III. GMFCS level II and III children were included in the study so that the CP children can themselves voluntarily participate in the functional activities. Children with any metallic implant, not willing to participate in the study, development of an unstable medical condition or severe behavioral disturbance, refusal of therapy by caregiver or administration of any chosen therapy in the past three months, were excluded from the study. The demographic details of participants in different groups is shown in Table 1.

The selected participants were randomly divided into three groups namely group A ($n = 15$) as a control group who received only standard therapy, group B ($n = 15$) r-TMS therapy, and group C ($n = 15$) NFB therapy. Before starting the treatment for any group pre-assessment employing raw EEG sample and Vineland social maturity scale (VSMS) was recorded. In group B, r-TMS frequency of 10 Hz was administered to the study participants for 15 min duration for 20 days (5 days a week for 4 weeks), in group C; NFB training was provided to each participant for 30 min duration for 20 days (5 days a week for 4 weeks), and in group A only routine standard therapy was provided for 20 days.

## 2.2 *Stimulation Device and Process*

### 2.2.1 Transcranial Magnetic Stimulation

Transcranial magnetic stimulation (TMS) is administered to the motor cortex areas of the brain. The device consists therapeutic angulated coil in the figure of eight shape. The magnetic coil provides 4 T magnetic fields at the center (stimulation) of the coil which penetrates the cranium and the neighboring soft tissue brain region to produce excitatory modulation of the neurons for inducing muscle response.

### 2.2.2 Neurofeedback Brain Training

Neurofeedback or NFB is a type of biofeedback that measures the electrical activity of the brain waves and teaches self-regulation of brain functions [3]. It helps in improving the neuroplasticity of the brain. It has a capacity of neurons and neural networks to change their connections and functionality in response to new information, development, sensory stimulation, or damage [4]. It allows the neurons to adjust how they behave in response to repetitive experiences that change the environment by reshaping the brain. It based on phenomena of the neurobiological process of learning [4].

## 2.3 *Methodology*

Ethical clearance was provided by the Institutional Ethics Committee for Human Participants (IECHSP), NEHU, Shillong. In this study, the diagnosed cases of CP were randomly selected and divided into three groups such as group B ($n = 15$) r-TMS therapy, group C ($n = 15$) NFB therapy, and group A ($n = 15$) as a control group which received only standard therapy. Before recruiting participants, pre-assessment was done by pediatric neurologists by employing VSMS and raw EEG signals. Post-completion of the treatment for different groups, the post-assessment data of the same participants was recorded using similar outcome measures.

## 2.4 *Outcome Measures*

Vineland social maturity scale (VSMS) and raw electroencephalography (EEG) signal were used as outcome measures in this study. VSMS is used for the evaluation of cognitive and social skills from the age group upto fifteen years for the pediatric neurodevelopmental disease. It has a total of eight domains such as self-help groups, self-help eating, self-help dressing, occupation, socialization, communication, and locomotion. It is a descriptive-based assessment tool for evaluating cognitive, social skills, and communication abilities [4]. Another outcome measure employing raw EEG signals was recorded for the pre versus post-assessment of data and identifying the changes occurring during the therapy procedure.

## 2.5 *EEG Data Recording*

Raw EEG data was recorded using the BioTrace software for Nexus-10 (through technology Inc Russia). It has two-channel EEG recording setup and inbuilt bioamplifiers and filters. In the process of EEG recording, all the participants were instructed to wash their heads to remove oil and dust from their hair and the targeted area of the scalp was cleaned again by using an alcohol swab to reduce the risk of increased skin impedance. In a bipolar manner, button electrode was placed on, namely F3-F4, C3-C4, P3-P4, and O3-O4 regions of the scalp, and the reference electrode was placed on right mastoid region according to 10–20 international system [5] for EEG recording. After subject preparation, EEG was recorded for 15 min duration without the use of any sedatives during the awake condition of the participants as shown in Table 1.

## 3 Statistical Analysis

The data was analyzed using SPSS 20.0 (Armonk, NY, IBM Corp., USA) and Microsoft Excel 2010. A paired sample t-test was performed on each of the groups (A, B, and C) to determine statistical significance demonstrating the effectiveness of r-TMS and NFB training in the CP patients for their cognitive enhancement. The $p$-value of less than 0.05 was considered statistically significant.

## 4 Results

### *4.1 Sample Characteristics*

In each group, total of fifteen subjects was recruited with the ratio of male and female participants in each group as group A (9:6), group B (10:5), and group C (4:11), respectively. Similarly, according to the classification of CP type diplegic and hemiplegic, group A (6:9), group B (7:8), group C (6:9), and group D (4:11) comprised of children, respectively.

### *4.2 Vineland Social Maturity Scale (VSMS)*

The mean, minimum, maximum, standard deviation (SD), and median VSMS scores of pre- and post-treatment measures for different groups suggest, based on p-values, that each of the administered therapies incorporates a significant improvement. Further, based on this sample their order of effectiveness concerning $p$-values is B > C > A as shown in Table 2.

The paired $t$-test between and 95% confidence interval [CI] difference for pre- v/s post-assessment of VSMS scores in group A revealed significant differences ($t = -2.68$, $df = 14$, $p = 0.018$,[CI $-1.52$ to $-0.72$). The paired t-test between the pre- v/s post-assessment of VSMS scores in group B (r-TMS therapy) showed significant

**Table 2** Descriptive statistics of VSMS score for all the group (A, B, and C)

| Descriptive statistics of VSMS scores | | | | | | | | | |
|---|---|---|---|---|---|---|---|---|---|
| Groups | Min | | Max | | Median | | Mean ± SD | | p-value |
| | Pre | Post | Pre | Post | Pre | Post | Pre | Post | |
| A | 6.36 | 6.49 | 75.33 | 76 | 46.66 | 48 | 44.48 ± 22.71 | 45.33 ± 22.48 | 0.018 |
| B | 1.76 | 2.51 | 57 | 70.32 | 23.54 | 32.05 | 27.84 ± 19.68 | 29.12 ± 20.31 | 0.002 |
| C | 1.2 | 1.2 | 70.32 | 75.51 | 19.05 | 19.21 | 28.88 ± 26.53 | 29.86 ± 27.16 | 0.014 |

differences ($t = -3.83$, $df = 14$, $p = 0.002$, CI −1.70 to −0.59). The paired t-test between the pre- v/s post-assessment of VSMS scores in group C (NFB therapy) revealed significant differences ($t = -2.81$, $df = 14$, $p = 0.014$, CI −1.74 to −0.82). Therefore, it can be stated based on the above results that all groups showed a significant positive effect on the cognitive enhancement area of CP children.

Statistical analysis also revealed that the improvement in VSMS score was more in group B (r-TMS therapy), followed by group C (NFB) and group A (only standard therapy). The mean change score for VSMS shows a 1.28% improvement in group B, 1.17%, in group C and 0.85% improvement in group A. The result also suggested brain stimulation help in neuroplasticity that enhances learning.

### 4.3 EEG Data

The mean, minimum, maximum, standard deviation (SD), standard error of mean and median scores for pre- v/s post-alpha peak frequency measures for different groups and the pre- v/s post-alpha peak frequency for an individual subject for different groups (A, B, and C) were calculated. The paired t-test between the pre- v/s post-assessment of alpha peak frequency for groups A, B, and C shows statistical significant p-value, that is $p < 0.02$, $p < 0.029$ and $p < 0.012$, respectively. The mean change score for group A, B, and C show 0.47, 1.63, and 0.62 percent changes, respectively. The results suggest that brain stimulation by employing r-TMS and NFB show neuroplasticity but r-TMS shows more impact in learning for CP children as compared with NFB and standard therapy.

## 5 Discussion

CP is a common brain disorder seen in the pediatric population which is associated with many clinical symptoms such as motor delay, motor planning impairment, gait issue, epilepsy, speech impairment, and cognitive impairment. CP children who have cognitive impairment still show some cognitive functioning, but their cognition is impaired to a certain level [6]. In one of the studies by Decety [7] et al. 2012, it is suggested that the cognitive control and emotional responses are interplays in children and adults and it is not surprising that brain damage child behaviors would be closely interlinked with emotional and cognitive empathic response. Another study by Davidson [8] et al. 1992 showed that alpha waves exhibited greater frontal activation and a higher level of functioning in human beings. In their study, they showed a positive association between left-over-right frontal activation in affective perspective for cognitive empathy. In one more study on children who had speech impairment, the results showed normal cognition even though due to misarticulation problems these children are not able to express themselves freely which could be due to some cognitive issues during birth or at a later stage in their life [9]. Klimesch et al. [10]

in their study suggested that a wave in EEG has a close relationship for good performance and motor learning. Doufesh et al. [11] found in their study on EEG spectral analysis during prayers performed by Muslims show alpha wave indicating cortical inactivation [12] and relaxation in the human brain which is best observed during closed eye position. They also concluded that alpha wave increases during mediation and helps to make a person more decisive in a complicated situation. M. Gupta et al. in their study on spastic CP children to reduce muscle tightness employing r-TMS suggested that TMS is a highly effective tool in decreasing muscle tightness in spastic CP patients and also helps to improve cognitive changes in these children [13]. They also suggested r-TMS is a powerful tool that has a high capacity to neuromodulate brain signals for the formation of new pathways and affects neuroplasticity. Andel et al. 2006 suggested cognitive reserve theory, wherein the brain has capacity before starting the cognitive decline to support more amount of neuropathological changes and cognitive training delays that decline such as short-term memory and attention in dementia patients. Yoa Wang et al. 2016 in their work on NFB for cognitive impairment in autistic children found it to be a suitable tool for detecting and modulating neural plasticity due to its ability that alters the excitability of neural circuits and inducing functional recognition. One more study by Kouizer et al. 2010 found NFB as a new treatment option for autistic spectrum disorder (ASD), that helps the child in reducing hyperactivity and makes them calm while performing any routine activities and improves their concentration levels [14, 15].

## 6 Conclusion

The statistical analysis of data reveals that r-TMS therapy shows significant results as compared with (NFB) and standard therapy only. The results also suggest that r-TMS therapy shows a 1.28% improvement compared to NFB which shows a 0.98% improvement. We concluded from our findings that r-TMS is the most result-oriented therapy for cognitive enhancement in all selected therapy groups although being slightly expensive therapy treatment. Regarding cost-effectiveness treatment, we concluded that the most cost-effective therapy process to retrain the brain to learn new pathways [16] and neuroplasticity is r-TMS therapy with 1.28% improvement.

**Acknowledgements** The researchers acknowledge the equipment support employed for data collection by SEED, Division, Department of Science and Technology (DST), Government of India, New Delhi to North-Eastern Hill University Shillong, Meghalaya, 793022, India (Ref: SEED/TIDE/007/2013/G) during the above study. The authors also acknowledge the support of the entire team and participants at UDAAN, New Delhi in conducting this study.

## References


1. Blomberg O (2011) Concepts of cognition for cognitive engineering. Int J Aviat Psychol 21(1):85–104
2. Serruya MD, Kahana MJ (2008) Techniques and devices to restore cognition. Behav Brain Res 192(1):149–165
3. Gupta M, Bhatia D (2017) Effect of repetitive Transcranial magnetic stimulation on cognition in spastic cerebral palsy children. J Neurol Disord 5(2):1–2
4. Hilgenberg AM, Cardoso CC, Caldas FF (2015) Hearing rehabilitation in cerebral palsy: development of language and hearing after cochlear implantation. Braz J Otorhinolaryngol 81(3):240–247
5. Gupta M, Rajak BL, Bhatia D, Mukherjee A (2016) Effect of r-TMS over standard therapy in decreasing muscle tone of spastic cerebral palsy patient. J Med Eng Technol 40(4):210–216
6. Porrselvi AP, Shankar V (2017) Status of cognitive testing of adults in India. Ann Indian Acad Neurol 20(3):34–40
7. Decety J, Meidenbauer KL, Cowell JM (2017) The development of cognitive empathy and concern in preschool children: a Behavioral neuroscience investigation. Dev Sci 21(3):1–12
8. Gupta M, Bhatia D, Rajak BL (2017) Study of available intervention technique to improve cognitive function in cerebral palsy patients. Curr Neurobiol 8(1):51–59
9. Gupta M, Bhatia D, Sinha TK, Dogra S, Jha SK (2020) Investigation of cognitive changes in cerebral palsy children employing different integrated sensing techniques. Sensor Int 1(100016):1–21
10. Klimesch W (1996) Memory processes brain oscillation and EEG synchronization. Int J Psychophysiol 24(1):61–100
11. Doufesh H, Faisal T, Seang Lim K, Ibrahim F (2012) EEG spectral analysis on Muslim prayers. Appl Psychophysiol Biofeedback 37:11–18
12. Lefaucheur JP, Aleman A, Baeken C et al (2020) Evidence-based guidelines on the therapeutic use of repetitive transcranial magnetic stimulation. Clin Neurophysiol 131(2):474–528
13. Rajak BL, Gupta M, Bhatia D, Mukherjee A (2018) Repetitive transcranial magnetic stimulation improved gross and fine motor functions in spastic cerebral palsy children. Spectr Sci Technol 5(1):55–62
14. Rajak BL, Gupta M, Bhatia D, Mukherjee A (2019) Increasing number of therapy sessions of repetitive transcranial magnetic stimulation improves motor development by reducing muscle spasticity in cerebral palsy children. Ann Ind Acad Neurol 22(3):302–307
15. Gupta M, Rajak BL, Bhatia D, Mukherjee A (2019) Effect of repetitive transcranial magnetic stimulation on motor function and spasticity in spastic cerebral palsy. Int J Biomed Eng Technol 31(4):365–374
16. Gupta M, Bhatia D (2018) Evaluating the effect of repetitive transcranial magnetic stimulation in cerebral palsy children by employing electroencephalogram signals. Ann Ind Acad Neurol 21(4):280–284

# A Limelight on Human Gp130 and Its Deleterious Mutations: A Computational Sequence Level Approach for Hepatocellular Carcinomas

Shreya Bhattacharya, Debina Basu, Ritika Nandy, Arundhati Banerjee, and Sujay Ray

**Abstract** This present study centers hepatocellular adenomas (IHCAs) and its effect due to gp130 mutations based on evolutionary relationship as well as literature survey involving wet-laboratory experiments. Generally, IL6 family of cytokines interacts with their respective IL6 receptors and gp130, which in turn induces JAK/STAT pathway to activate pro-inflammatory STAT3 target genes. The activation of this signaling pathway poses a paramount impact in tumor development. Several mutations in gp130 were studied through evolutionary relatedness which aggravates the tumorigenesis. Mutations in IL6Rα binding domain region (positions: 131–218 of human gp130) from MSA were also studied. A point mutation: A200V in *Otolemur garnettii* showed the deleterious effect. Further, the deleterious effect upon proteins' functionality was also examined and seven mutations were responsible for it. V189G showed the highest destabilizing effect. Unstable conformation shifts were also observed in the proteins due to their respective mutations. Change in the transmembrane region, change in gp130s' ordered interface, loss of sulfation at Y186, lesser relative solvent accessibility, change in protein stability and metal binding properties were also observed in several mutations. These mutations might be responsible for

Shreya Bhattacharya, Debina Basu, Ritika Nandy, and Arundhati Banerjee are equally contributed to this work.

S. Bhattacharya
Department of Biosciences and Bioengineering, IIT Guwahati, Guwahati, Assam, India
e-mail: shreya199713@gmail.com

D. Basu · R. Nandy · S. Ray (✉)
Amity Institute of Biotechnology, Amity University, Kolkata, India
e-mail: raysujay@gmail.com

D. Basu
e-mail: debina.basu12@gmail.com

R. Nandy
e-mail: dyk.ritika19@gmail.com

A. Banerjee
Department of Biochemistry and Biophysics, University of Kalyani, Kalyani, West Bengal, India
e-mail: arundhati.92star@gmail.com

 
A. Maji et al. (eds.), *Proceedings of the International Conference on Computing and Communication Systems*, Lecture Notes in Networks and Systems 170,
https://doi.org/10.1007/978-981-33-4084-8_42

changing the activity of the IL6Rα binding domain region of human gp130 protein triggering hepatocellular adenomas. Therefore, targeting gp130 and its mutations would serve as an important therapeutic target to cause apoptosis of the tumor cells. Interaction studies within the three proteins for the respective mutations could also to be in the future pathway of research. It would also aim to instigate the drug discovery for the same.

**Keywords** Hepatocellular carcinomas · Gp130 · Mutations · Molecular changes · Deleterious effects · Destabilizing effect · Evolutionary relatedness

## 1 Introduction

Hepatocellular adenoma (IHCA) is a rare form of benign tumor in liver, evident mainly in females due to elevated levels of inflammatory proteins like C-reactive proteins (CRP) and serum amyloid A (SAA) [1, 2]. It is characterized by the presence of hepatocytes with cytoplasm containing large vacuoles that contain inflammatory by-products [1].

Interleukin 6 (IL-6) is a family of cytokines that activates the JAK-STAT cellular signaling pathway through gp130 interaction. It is the chief mediator for the differentiation factor of B-cell, inflammation and haematopoiesis [3]. They transduce the signals utilizing homodimeric glycoprotein 130 (gp130) receptors which is also called as Interleukin-6 receptor beta subunit (IL6ST) [4]. In general, IL-6 initially binds to Interleukin-6 receptor alpha subunit (also referred to as gp80 or IL6R) to form IL6/gp80 that further associates with the gp130 homodimer. This leads to the formation of a hexametric complex (two units each of IL-6, gp80 and gp130) which activates the signaling pathway of gp130 [4]. Activated gp130 induces JAK/STAT pathway to trigger activation of pro-inflammatory STAT3 target genes [4]. This activation of the JAK/STAT pathway has significant role in tumor development [4, 5]. It has been previously documented that mutations in IL6ST protein led to constitutive activation of this pathway that resulted in the development of adenomas [2, 4]. The constitutive activation of gp130 due to mutations resulted in STAT3 target genes getting activated every time, which is independent of IL6 binding to gp130 [2, 4].

Several experimental studies documented many deletions and point mutations in IL6ST protein that led to the activation of JAK/STAT signaling pathway leading to an acute inflammatory response in hepatocellular tumors [2, 6, 7]. Targeting IL6ST protein and its signaling pathway thus could serve as an important therapeutic target to inhibit tumor cell growth [8]. This made further studies on the molecular effects of different mutations on the gp130 protein important to further elucidate the mechanism of hepatocellular carcinoma. Detailed domain analysis and several mutations (through literature survey as well as evolutionary relatedness) were analyzed in this study. The exploration into the deleterious effects due to the mutations was the prime focus. Changes in the conformational fluctuations in the individual paramount mutations were also studied. The highly destabilizing effect through $\Delta\Delta G$ calculations in the mutants along with the detailed alterations in the molecular level of the function of the protein was examined in details. This would instigate to examine the contribution of these mutations in the further interaction process and drug targeting.

## 2 Materials and Methods

### 2.1 *Sequence and Domain Retrieval of Gp130 of Human as Well as Other Different Species*

The sequence of human gp130 protein was retrieved from the UniProt database, which was then subjected to Position-Specific Iterative Basic Local Alignment Search Tool (PSI-BLAST) [9] against the non-redundant database to identify the related gp130 protein in other organisms holding 100% query coverage and >75% sequence identity. The unnamed, hypothetical, isoforms, predicted, partial and putative proteins along with the redundancies from all the collected sequences were removed and the remaining unambiguous proteins were chosen for further analysis. It is essential to determine the domain position of a protein for functional analysis and structural prediction. The domain sequences of the gp130 protein in related species were then identified using Pfam protein family database [10].

### 2.2 *Identification of Mutations in the Human Gp130*

**Evolutionary Conservation Analysis for the Mutated Residues from MSA.** To identify the overall evolutionary conservation and the conservation in the regions of the mutation, multiple sequence alignment (MSA) was carried out through the implementation of Clustal Omega [11] software tool for MSA. The highly conserved regions in the alignment were identified while the mutated residues obtained from substitutions in MSA were also taken into account.

**Evolutionary Conservation Analysis for the Mutated Residues from Literature Survey.** In this study, we collected the mutagenesis data from the literature survey (as discussed in the Introduction section) [6–8] on the basis of previously referred point mutation in the amino acid sequence of human gp130. MSA data was utilized to determine whether the positions of these point mutations are conserved in other species.

### 2.3 Estimation of the Deleterious Effect upon Mutation in Human Gp130 Protein

Protein Variation Effect Analyzer (PROVEAN) was employed for the classifications of the mutations depending on the basis of their causative effect (deleterious effect or a neutral effect) on the functionality of the protein [12]. 'Default threshold' was considered to be −2.5, i.e., variants with a score equal to or below −2.5 were considered as 'deleterious,' whereas, variants with a score above −2.5 are considered to be 'neutral' [12]. MUpro [13] was then used to identify the changes in protein stability due to the point mutation between WT and mutant human gp130 proteins in terms of $\Delta\Delta G$ values, mainly by utilizing the support vector machine algorithm [13]. A positive $\Delta\Delta G$ value indicates enhanced stability, while a negative $\Delta\Delta G$ value indicates destabilizing effect of proteins because of the point mutations [13].

### 2.4 Comparison of the Secondary Structure of Human Gp130 with the Mutant Gp130 Proteins

The secondary structure of the wildtype (WT) and the mutant human gp130 proteins were elucidated using GOR4 server [14] and a comparative analysis of change in the content of helices, strands and sheets between WT and mutant proteins was performed to understand the structural effect of these mutations on the protein.

### 2.5 Analysis of Molecular Effects of Mutations

The effects of the deleterious mutations in the molecular level were analyzed using MutPred2 [15] server. This software helps in the probabilistic modeling of variant impact on more than 50 specific properties of protein structure and function that could be altered in point mutations [15]. This server thus helps to explore the molecular mechanisms of deleteriousness due to an amino acid substitution.

## 3 Results and Discussion

### 3.1 Sequence and Domain Retrieval of Gp130 in Human and Other Species

Human gp130 protein or Interleukin-6 receptor beta subunit protein (UniProt ID: P40189), encoded by the IL6ST gene was retrieved, which is 918 amino acids long

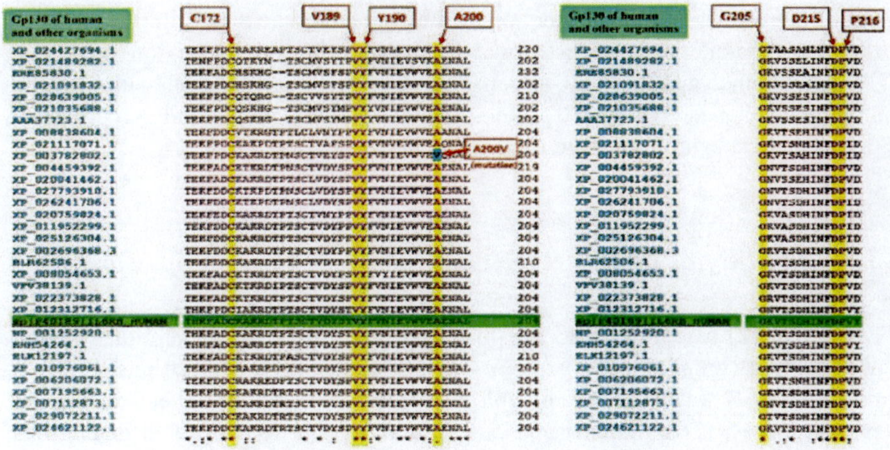


**Fig. 1** The part of MSA covering the positions of the deleterious mutations obtained from literature survey (yellow) and the deleterious mutation (A200V) obtained from MSA (cyan). The sequence of human gp130 is highlighted green

and has a molecular weight of 103.53 kDa. The Pfam server [10] identified four domains, namely (a) Lep receptor Ig (PF06328), an immunoglobulin-like ligand binding domain in position: 26–112, (b) Interleukin-6 receptor alpha chain binding domain (PF09240) that is the main domain responsible for binding of IL6/gp80 to gp130 protein to form the activated complex and (c) two sets of fibronectin type III domain (PF00041) in positions 223–311 and 518–605. Interleukin-6 receptor alpha chain binding domain is present in position 131–218 of human gp130.

To understand the evolutionary pattern of gp130, analysis of gp130 proteins from closely related species is essential. PSI-BLAST [9] with human gp130 revealed 32 unambiguous gp130 proteins from different species having >75% sequence identity and 100% query coverage. These sequences are homologous to human gp130 protein due to their high sequence identity and thus structural and functional similarity. These proteins were enlisted in Fig. 1 and were used to study the extent of conservation in residues of gp130, identify mutations that occur over the evolutionary period and finally to establish a phylogenetic relationship as described in later sections.

## 3.2 *Mutational Analysis of Gp130 Proteins*

Twelve point mutations activating gp130 (C172S, V189G, Y190G, D215G, P216H, V192I, V192F, V192T, S187A, N193S, G205R and I194F) and two natural variants (G148R and A200G) were retrieved from the literature survey [2, 6, 7]. The positions of these mutations were found to be present in the most conserved region of the MSA, thus any substitution in an amino acid residue in these regions by mutagenesis could disrupt protein's structure and function. Moreover, amino acid substitutions present in the conserved Interleukin-6 receptor alpha chain binding domain region (positions: 131–218 of human gp130) in MSA were also taken into consideration, which are the mutations that appeared in gp130 proteins over the evolutionary period. Accordingly, 23 amino acid substitutions were obtained from MSA result.

### 3.3 *Screening of the Most Deleterious Mutations of Gp130*

PROVEAN [12] was utilized to identify which mutations are deleterious for the functionality of human gp130 protein. The list of deleterious mutations screened from PROVEAN was tabulated in Table 1. Out of the mutations collected from the literature survey, six mutations and one natural variant demonstrated deleterious effect on the protein's functionality. Among 23 mutations identified from MSA data, one amino acid substitution: A200V demonstrated deleterious effect. This mutation was observed in *Otolemur garnettii* (the northern greater galago) having accession number: XP_003782802.1 as depicted in Fig. 1. The MUpro [13] results demonstrated a net destabilizing effect in all the deleterious mutations (Table 1) with a negative value of ΔΔG denoting that these mutations decreased the net protein stability. The point mutation: V189G has the highest destabilizing effect with ΔΔG value of –2.687 kcal/mol.

**Table. 1** List of deleterious mutations of human gp130 protein predicted by PROVEAN and its stability analysis by MUpro

| Variant | PROVEAN score | Prediction (cutoff = −2.5) | ΔΔG | Prediction (MUpro) |
|---|---|---|---|---|
| *Mutations obtained from literature survey* | | | | |
| C172S | −3.674 | Deleterious | −1.171 | Decrease in stability |
| V189G | −2.687 | Deleterious | −2.675 | Decrease in stability |
| Y190G | −2.665 | Deleterious | −1.712 | Decrease in stability |
| D215G | −4.711 | Deleterious | −1.994 | Decrease in stability |
| P216H | −6.048 | Deleterious | −0.937 | Decrease in stability |
| G205R | −7.718 | Deleterious | −1.557 | Decrease in stability |
| *Natural variants obtained from literature survey* | | | | |
| A200G | −3.751 | Deleterious | −1.237 | Decrease in stability |
| *Amino acid substitutions obtained from MSA* | | | | |
| A200V | −2.843 | Deleterious | −0.152 | Decrease in stability |

**Table. 2** Secondary structural alteration in human gp130 due to mutations

| Variants | Number of helices | Number of strands | Number of coils |
|---|---|---|---|
| WT | 138 | 245 | 535 |
| C172S | 136 | 245 | 537 |
| V189G | 138 | 247 | 533 |
| Y190G | 137 | 244 | 537 |
| A200G | 131 | 248 | 539 |
| G205R | 141 | 242 | 535 |
| D215G | 138 | 242 | 538 |
| P216H | 138 | 242 | 538 |
| A200V | 135 | 244 | 539 |

## 3.4 *Effect of Deleterious Mutations on Secondary Structure of Gp130 Protein*

The secondary structural propensities of WT and mutant gp130 proteins were elucidated using GOR4 [14] and the structural alterations due to mutation in the proteins were studied. The mutations were identified to cause minor alteration in the secondary structures as demonstrated in Table 2. Six out of eight mutations caused increase in the content of random coils in the protein, reducing the structured components like helices and strands. This correlated with the net destabilizing effect due to these mutations as observed in the previous section. The incorporation of Gly upon mutation in Y190G, A200G and D215G might be the driving force of net reduction of structured components into more flexible coils (Table 2). However, the mutation V189G was observed as an exemption where the substitution with glycine reduced coil content and increased structured strands in the protein. This mutation thus requires further investigation to understand the molecular basis of structural alteration. Additionally, the mutation: G205R, on the other hand, demonstrated a shift from highly structured strands to helices in the protein. The mutation A200V showed shifts from helices and strands to coils (Table 2). These observations revealed that these mutations cause a change in protein's structure that might serve as a reason for the deleterious nature of these mutations.

## 3.5 *Exploration into the Molecular Effects of the Mutations*

Furthermore, the way these mutations influence protein's structure and function in the molecular level was investigated. Significant changes in molecular level with *p*-value less than 0.05 were observed in five mutations: Y190G, G205R, D215G and A200G and A200V. All the five mutations have been observed to alter the transmembrane region of the protein. Mutations: Y190G and D215G altered the protein's ordered interface. The substitution: Y190G also caused the loss of sulfation at Y186. The substitution D215G was also found to be responsible for the reduction in relative solvent accessibility, changes in the stability and altered metal binding properties. Loss of relative solvent accessibility and altered metal binding was also observed in G205R mutation. This again confirms the deleterious impact of these mutations on the protein's activity.

Furthermore, we know that domains can function independently [16]. Mutations in IL6Rα binding domain region (positions: 131–218) of human gp130 were analyzed in this study. Deleterious effect of those mutations was studied as well. *Therefore, it could be inferred that mutations in such vital domain binding regions would affect the conformation and overall interaction pattern for the three protein interaction.* Thereby, in a way, it can cause a very high expression of pro-inflammatory STAT3 target genes in hepatocellular adenomas. This instigates the future study on the 3D changes in IL6-R binding pockets of gp130 that resulted in its constitutive activation.

Earlier studies showed that bazedoxifene obstructed the STAT3 phosphorylation in the cancer cell lines and thus serves as a novel efficient therapeutic agent [17, 18]. Even GP130 inhibitors like SC144 and evista were known to inhibit cell proliferation and glycolysis [17, 18]. In either of the situations, gp130 signaling pathway gets targeted for human cancer therapies like colorectal and pancreatic cancer [17, 18]. Further, in one our previous study, the participation of gp130 was studied for recovery of platelets especially for thrombocytopenia after chemotherapy [19]. It is accomplished through the interaction with IL11 protein and its receptor, where IL11 is also one of the members from IL6 cytokine family [19]. So, summing up previous wet-laboratory studies and present computational approach, gp130 can act as a competent drug target in humans.

## 4 Conclusion and Future Scope

This study mainly focuses on the hepatocellular adenoma (HCA) and its impact due to gp130 mutations based on evolutionary relationship as well as literature survey involving wet-laboratory experiments. Total of 14 point mutations were studied from early studies, and hereby, they were found to be in the highly conserved regions, which would thereby be responsible for the change in proteins' function. Additionally, mutations in Interleukin-6 receptor alpha chain binding domain region (positions: 131–218 of human gp130) from MSA were also studied. Accordingly, 23 amino acid substitutions were obtained through evolutionary relatedness among which A200V in *Otolemur garnettii* showed the deleterious effect. Further, the deleterious effect upon proteins' functionality was also examined and seven mutations were responsible for it. Mainly, V189G showed the highest destabilizing effect. Correlating the observations, the content of random coils in the protein got increased due to the mutations, reducing the highly structured components like helices and strands. The substitution by Gly upon mutation in 190th, 200th and 215th positions can serve for the unstable conformational shift. One exception was V189G, where the overall structured content increased although it caused overall destabilization in the protein. This requires further investigation to understand its molecular basis. Alanine in 200th position poses to be unstable upon either glycine or valine substitutions with deleterious impact. Noteworthy alterations in the molecular level through alteration in the transmembrane region with statistical significance ($p < 0.05$) were observed in Y190G, G205R, D215G and A200G and A200V. Among which 190th and 215th position also changed gp130s' ordered interface, loss of sulfation at Y186 and loss of relative solvent accessibility, altered stability and metal binding properties. G205R also showed loss of relative solvent accessibility and altered metal binding activity. These mutations might be the causative effect of alteration in the structure and activity of the IL-6 receptor alpha chain binding regions of human gp130 protein leading to its constitutive activation in hepatocellular adenomas.

Therefore, in gist, targeting gp130 and its mutations would serve as an important therapeutic target to cause apoptosis of the tumor cells. The changes in interaction pattern and further stability alterations in the molecular level through computational approaches due to mutation would serve novel of its kind for hepatocellular carcinoma. It would aim to instigate the drug discovery for the same.

**Acknowledgements** Authors are thankful to ICMR Project (ID: 2019-3336) for the funding of the project.

Puja Ghosh from Department of Biotechnology, Pondicherry University, Puducherry, is also acknowledged for her support. Authors are also grateful to the Department of Biotechnology, Amity University, Kolkata, for the cooperation. Authors are also grateful to the Department of Biosciences and Bioengineering, IIT, Guwahati, for the cooperation and support as well. Authors are also grateful to Dr. Rakhi Dasgupta from Department of Biochemistry and Biophysics, University of Kalyani, for the support.

**Conflicts of Interest** None.

## References

1. Rooks J (1979) Epidemiology of hepatocellular adenoma. The role of oral contraceptive use. JAMA: J Am Med Assoc 242:644–648
2. Rebouissou S, Amessou M, Couchy G, Poussin K, Imbeaud S, Pilati C, Izard T, Balabaud C, Bioulac-Sage P, Zucman-Rossi J (2008) Frequent in-frame somatic deletions activate gp130 in inflammatory hepatocellular tumours. Nature 457:200–204
3. Tanaka T, Narazaki M, Kishimoto T (2014) IL-6 in inflammation, immunity, and disease. Cold Spring Harbor Perspect Biol 6:a016295–a016295
4. Yu H, Pardoll D, Jove R (2009) STATs in cancer inflammation and immunity: a leading role for STAT3. Nat Rev Cancer 9:798–809
5. Tebbutt N, Giraud A, Inglese M, Jenkins B, Waring P, Clay F, Malki S, Alderman B, Grail D, Hollande F, Heath J, Ernst M (2002) Reciprocal regulation of gastrointestinal homeostasis by SHP2 and STAT-mediated trefoil gene activation in gp130 mutant mice. Nat Med 8:1089–1097
6. Poussin K, Pilati C, Couchy G, Calderaro J, Bioulac-Sage P, Bacq Y, Paradis V, Leteurtre E, Sturm N, Ramos J, Guettier C, Bardier-Dupas A, Boulai A, Wendum D, Selves J, Izard T, Nault J, Zucman-Rossi J (2013) Biochemical and functional analyses of gp130 mutants

unveil JAK1 as a novel therapeutic target in human inflammatory hepatocellular adenoma. OncoImmunology. 2:e27090
7. Schütt A, Zacharias M, Schneider N, Horn S, Grötzinger J, Rose-John S, Schmidt-Arras D (2013) gp130 activation is regulated by D2–D3 interdomain connectivity. Biochem J 450:487–496
8. Johnson D, O'Keefe R, Grandis J (2018) Targeting the IL-6/JAK/STAT3 signalling axis in cancer. Nat Rev Clin Oncol 15:234–248
9. Altschul S (1997) Gapped BLAST and PSI-BLAST: a new generation of protein database search programs. Nucleic Acids Res 25:3389–3402
10. El-Gebali S, Mistry J, Bateman A, Eddy S, Luciani A, Potter S, Qureshi M, Richardson L, Salazar G, Smart A, Sonnhammer E, Hirsh L, Paladin L, Piovesan D, Tosatto S, Finn R (2018) The Pfam protein families database in 2019. Nucl Acids Res 47:D427–D432
11. Sievers F, Higgins D (2017) Clustal Omega for making accurate alignments of many protein sequences. Protein Sci 27:135–145
12. Choi Y, Chan A (2015) PROVEAN web server: a tool to predict the functional effect of amino acid substitutions and indels. Bioinformatics 31:2745–2747
13. Cheng J, Randall A, Baldi P (2005) Prediction of protein stability changes for single-site mutations using support vector machines. Proteins: Struct Funct Bioinf 62:1125–1132
14. Garnier J, Gibrat J, Robson B (1996) GOR method for predicting protein secondary structure from amino acid sequence. Methods Enzymol 32:540–553
15. Li B, Krishnan V, Mort M, Xin F, Kamati K, Cooper D, Mooney S, Radivojac P (2009) Automated inference of molecular mechanisms of disease from amino acid substitutions. Bioinformatics 25:2744–2750
16. George RA et al (2002) An analysis of protein domain linkers: their classification and role in protein folding. Protein Eng 15(11):871–879
17. Xiang C et al (2019) Blocking IL-6/GP130 signaling inhibits cell viability/proliferation, glycolysis, and colony forming activity in human pancreatic cancer cells. Curr Cancer Drug Targets 19:417. https://doi.org/10.2174/1568009618666180430123939
18. Ling M et al (2019) Bazedoxifene as a novel GP130 inhibitor for colon cancer therapy. J Exp Clin Cancer Res 38:63
19. Banerjee A, Dasgupta R, Ray S (2019) Residual participation and thermodynamic stability due to molecular interactions in IL11, IL11Rα and Gp130 from *Homo sapiens:* an in silico outlook for IL11 as a therapeutic remedy. Int J Pept Res Ther 1–12. doi: https://doi.org/10.1007/s10989-019-09996-z

# Networks and Communication

# RCPDNPT:A Tool to Deploy a Post-disaster Network Under Resource-Constraint Scenarios

**Krishnandu Hazra, Swaraj Deep, Sujoy Saha, and Subrata Nandi**

**Abstract** In case of large scale natural disaster (like a cyclone, hurricane, etc.) conventional communication infrastructure (like cellular or internet) are damaged. In these scenarios deploying a temporary infrastructure for communication is of utmost importance. In this paper, we present a software tool to plan a rapidly deployable temporary opportunistic (DTN based) communication infrastructure for post-disaster scenarios under the constraint of the number of available deployable network resources. Such a tool may be used by the government disaster management authority to rapidly deploy a temporary communication infrastructure to aid the rescue relief operations in the aftermath of the disaster. Given the knowledge of disaster-affected area (through the Google Map or Open Street Map), the location of control center, the locations of shelter points, where the affected people can take shelter after the disaster and the pool of limited number of available deployable communication resources (like communication towers, data mules, etc.), the proposed tool could provide a near-optimal deployment plan to deploy the available resource pool to create a post-disaster communication infrastructure such that the end to end delay of the opportunistic network is minimized and having a decent delivery probability.

**Keywords** Post-disaster communication · Delay-tolerant networks · Resource constraint scenarios · Software tool

---

K. Hazra (✉) · S. Deep · S. Saha · S. Nandi
Department of CSE, National Institute of Technology Durgapur, Durgapur, India
e-mail: krishnandu91@gmail.com

S. Deep
e-mail: swarajdeep33445@gmail.com

S. Saha
e-mail: sujoy.ju@gmail.com

S. Nandi
e-mail: subrata.nandi@gmail.com



A. Maji et al. (eds.), *Proceedings of the International Conference on Computing and Communication Systems*, Lecture Notes in Networks and Systems 170,
https://doi.org/10.1007/978-981-33-4084-8_43

## 1 Introduction

Natural disasters are very well known catastrophic events that cause economic loss and destruction of human lives. There are numerous examples of such natural disasters like Tsunami in Japan in 2011, Nepal earthquake in 2015, Kerala flood in India in 2018, Cyclone Fani in India in 2019, etc. In such natural disasters, the conventional communication connectivity (i.e., Cellular or Internet) is often totally paralyzed or intermittent due to the damage of existing communication infrastructure. This is the prime reason for uncoordinated rescue/relief operations, which leads to a massive death toll and economic loss. Different stakeholders such as, police, fire team, army, medical teams, etc., which are the first responders in a post-disaster scenario, are not able to rescue the trapped victims properly due to the lack of information about the location of the trapped victims.

In such aforementioned scenarios, temporary communication infrastructure can provide an effective medium for communication until the backbone communication infrastructure is restored [4]. There are many works on designing post-disaster communication infrastructure like [1, 6, 7]. The work [1] described a robust infrastructure to share the internet in the post-disaster situation when conventional communication connectivity is intermittent. They have given a two-layered architecture, in the bottom layer the volunteers having a smartphone can create a mesh network using Bluetooth connectivity, and in the upper layer, there are three types robotic data carrier (UAV for areal communication, a movable ground communication unit and the submarine to cover the underwater communication. Note that, each of these 3 data carriers is equipped with two communication interfaces one is VHF goTenna[1] to communicate with each other in the upper layer and the other is Bluetooth to communicate with the lower layer.) to establish network connectivity on a large scale. In [6] the authors described an interim post-disaster communication infrastructure utilizing long-range communication towers and mobile data carriers equipped with a communication unit (i.e., data mule) when to meet a given delay constraint. The proposed communication infrastructure is capable of providing communication services to the roaming rescue/relief volunteers equipped with a smartphone. The authors in [7] described a post-disaster communication infrastructure based on a movable and deployable resource unit (MDRU), which is rapidly deployable, and has the capability of communication and information processing.

It is noteworthy to say that, the solutions like [1, 7] are effective for the developed countries, however not useful for under-developed of developing country like us. The main reason is the works utilize sophisticated and costly high-end set up (i.e., MDRU [7], high-end UAV with VHF goTenna interface [1], etc.) to deploy the network, which is not affordable to deploy in large scale in the under-developed of developing country due to the limited budgetary constraint. Again, the crucial element of deploying a post-disaster communication infrastructure is the network resources (like communication towers, data carriers equipped with a communication module, UAVs, etc.) which are often limited in case of underdeveloped countries due

[1]https://gotenna.com/.

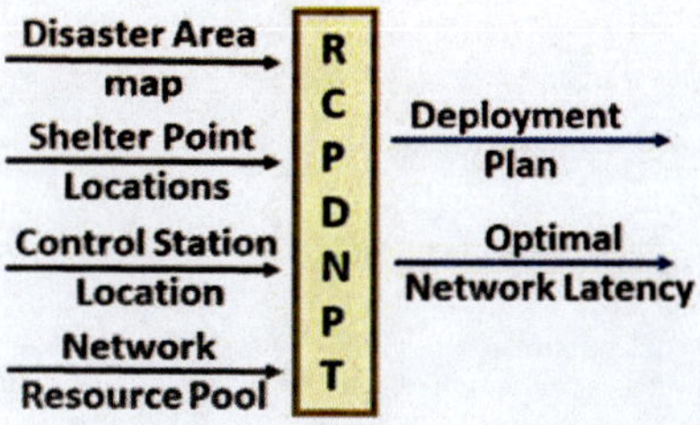


**Fig. 1** Overview of the tool (RCPDNPT): given inputs and predicted outputs

to budget limitation. However, none of the existing works [1, 6, 7] consider such resource constraint scenarios for deploying a post-disaster communication infrastructure, especially from the perspective of underdeveloped or developing countries. Such a resource constraint scenario is considered in our work [2] to deploy a post-disaster network.

In this paper, we present an interactive tool to deploy a post-disaster communication infrastructure under resource constraint scenarios, which meets the requirement of the under-developed or the developing countries. We termed this tool as RCPDNPT (**R**esource **C**onstraint **P**ost **D**isaster **N**etwork **P**lanning **T**ool). Given the limited set of communication resource pool,[2] disaster area map, location of shelter points (where the affected people take shelter after the disaster), location of the control station (from where the rescue operation are controlled) the tool (see Fig. 1) provide the suitable plan for deploying a delay tolerant network utilizing the available network resources such that the end to end network latency in between first responders/volunteers and the control station is minimized. The tool also simulates the predicted plan in ONE [3] simulator flawlessly and finally shows different performance parameters like delivery probability, average hop count, network latency, etc. This tool should be used by the government disaster management authorities or by NGOs in the under-developed or developing regions to deploy the post-disaster network.

## 2 Related Work

In [6], as mentioned earlier, the authors proposed delay-tolerant network architecture for post-disaster communication utilizing long-range communication tower, moving data carrier (data mule) equipped with a wireless interface, etc. to provide a communication medium in between the rescue volunteer and the control station. Note that, while deploying such infrastructure they consider a delay constraint to be achieved and assuming that there are plenty of network resources to deploy. However, such a thing is not possible in the case of the under-developed country (as mentioned above). In [5] a Google Map based user interface for planning the post-disaster network resource deployment is proposed, but again, they have not considered the

[2]Here we consider two types of communication resources, communication tower, movable data carrier with wireless communication module (data mule).

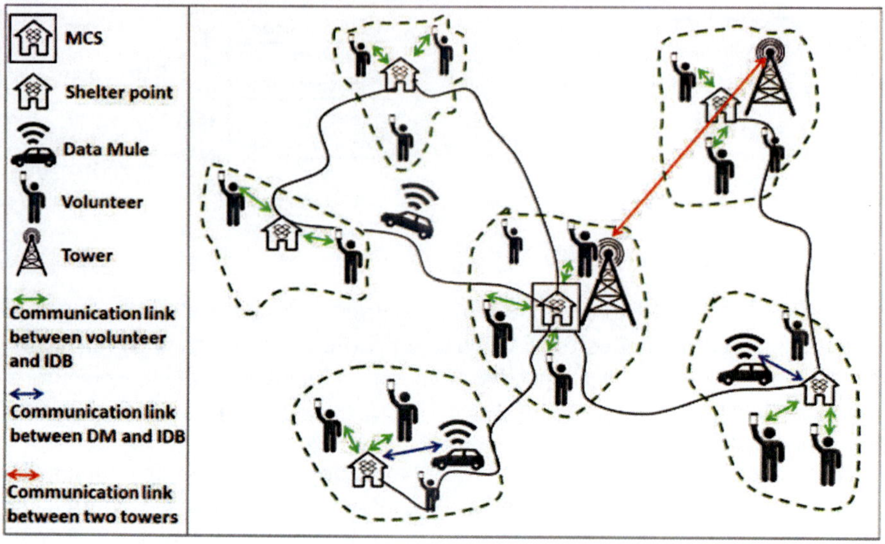


**Fig. 2** Overview of the resource constraint network architecture (Courtesy [2])

resource constraint. It is important to say that, deploying a limited set of network resources for creating a post-disaster network such that optimal network latency is achieved while maintaining nearly 100% delivery probability is a non-trivial problem. We consider such a problem in [2] and have proposed network architecture for post-disaster communication under resource constraint scenarios. This work is the building block of our proposed tool RCPDNPT. In [2] we have described a network architecture (see Fig. 2) to deploy an opportunistic Infrastructure. At the bottom layer, the rescue/relief volunteers (having smartphone) roaming around the shelter points to collect, create and delivered the information messages to the nearest shelter point in which there is an information dropbox, which can collect and transmit information. The data accumulated in the shelter point can be transported to the control station of the disaster area via long-range communication (LWC) towers and/or data mule (DM).

In this work, we have considered two types of network resource namely LWCs and DMs. Here we have provided the deployment plan of such limited resources, the trajectory of each DMs, etc. However, despite having this research, a user-friendly tool for network resource deployment is really helpful for disaster management authority to plan a temporary communication set up. To meet this gap we have designed the tool RCPDNPT.

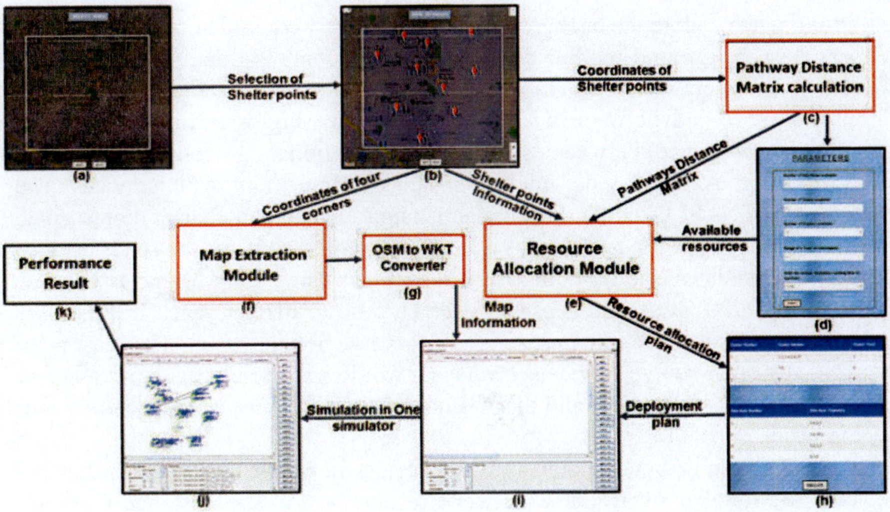


**Fig. 3** Flow diagram of RCPDNPT: **a** Disaster affected area; **b** selection of shelter points; **c** Calculation of pathway distance matrix using Bing map API; **d** Description of available network resources and parameters; **e** resource allocation module; **f** Extraction of disaster affected area map using OpenStreetMap API

## 3 Design Principle and Software Component

In Fig. 3, the flow diagram of the tool is described in details. Such tool provides an interactive and user-friendly interface for planning the deployment of available network resources in post-disaster scenarios. It can be used by the government disaster management authority or by NGOs to deploy a temporary communication infrastructure during the post disaster scenarios. Note that, to run this tool internet connectivity is needed to extract the map information (as discussed below). So, when the tool is used by the disaster management authority or NGOs possibly at the coordination center for planning the deployment of the available network resources, we can assume that there will be the provision for internet connectivity, and it is not an unrealistic assumption.

As shown in Fig. 3a, the tool provides an interface of Google map,[3] where we can select the affected area map. A disaster-affected area can be marked with an editable rectangle on the top of Google Maps. After the selection of a disaster-affected area, locations of the shelter points can be marked on the top of the map using a suitable marker (see Fig. 3b), here we use red marker to denote the shelter points. Note that, one of the shelter points can be treated as the control center of the disaster-affected area. Next, after annotating the affected area map and shelter point locations, the pathway distance matrix (see Fig. 3c) is calculated in background (note

[3]https://www.google.com/maps.

that, in Fig. 3 we denote the background modules as red-colored border like Fig. 3c, e, f, g). This background module (see Fig. 3c) represents the map and the shelter points as a complete weighted graph where the vertices are the shelter points and edges are the road-path between two shelter points. The weights are the road-distance (not Euclidean distance) between each pair of shelter points. To obtain the weights, Distance Matrix API[4] is used, which is hosted by Bing Maps API[5] services. The Distance Matrix API takes the latitude and longitude of each shelter points given on the map as a JSON file and returns a JSON file which contains the road-distance between every pair of shelter points. The reason for using Bing Maps[6] is because it is richer in information and is pseudo-free (125,000 calls per year). If the tool is deployed on a large scale, then an API key needs to be purchased. The reason for not using any open-source service is because it would not be rich (may give an error in the calculation of distance) and there is no open-source service that caters to our specific needs.

After calculation of such pathway distance matrix in the background, the tool will take the number of available network resources (i.e., Long-Range Communication(LWC) Towers and Data Mules) to be deployed, the range of LWC towers, the time to simulate the post-disaster network in ONE simulator, etc. as input (see Fig. 3d). Next, the tool invokes the network resource allocation module in the background feeding the shelter point information, pathway distance matrix[7] of the disaster-affected area and the available number of network resources as inputs. Note that, the algorithm for network resource allocation is taken from the paper [2]. The resource allocation module give proper deployment plan (see Fig. 3h) such as the set of shelter points where to place the available LWCs, allocation, and trajectory (also the depot) for each data mules to traverse a set of shelter points, etc., such that end to end network latency between volunteers and the control station is minimized while maintaining nearly cent percent delivery of the messages.

Next, the tool simulates the predicted deployment plan in the one simulator. The ONE simulator takes the disaster-affected area map in the .WKT[8] format only. However, we cannot extract the map information from the Google map directly. So, in this case, we use *OpenStreetMap*[9] for the extraction of the disaster area map. The map extraction module (see Fig. 3f) and the **OSM to WKT converter** module (see Fig. 3g) collaboratively run in background to extract the map in .WKT format. The map extraction module takes the four coordinates of the selected affected area in Google Map and gets the .OSM file from the Open Street Maps using a customized approach. The **OSM to WKT converter** converts the .OSM file to .WKT file and

[4]https://www.microsoft.com/en-us/maps/distance-matrix.

[5]https://www.microsoft.com/en-us/maps/choose-your-bing-maps-api.

[6]https://www.bing.com/maps.

[7]Note that, we have converted the JSON file of pathways distance matrix into a compatible input format of the resource allocation algorithm. For this we have used "json-simple-1.1.1.jar" library which is a JSON Parser and using this we parse the input and then use it in our program.

[8]https://maplarge.com/dataformats/wkt.

[9]https://www.openstreetmap.org/.

pass it to ONE simulator. The setting file of the ONE simulator is automatically written by the tool using a suitable approach to simulate the post-disaster network. After the completion of the simulation, the tool visualizes the performance report of the simulation for the deployment of the available resources. This report includes delivery probability, network latency, average hop-count, etc.

The proposed tool is very interactive to collect all the information about the disaster (i.e., the disaster-affected area, locations of the shelter points, etc.) and the available network resources and able to simulate the predicted network resource deployment plan seamlessly.

## 4 Result and Discussion

In this section, we discuss an overview of the tool. Figure 4a is the interface for taking the input for the disaster-affected area and shelter point locations. Figure 4b shows the user interface for collecting different input parameters like the number of available LWC towers, Data Mules, range of the towers, etc. Figure 4c shows the output of the resource allocation algorithm, which consists of the set of locations of the placement of LWC towers, trajectories of the data mules, etc. After that, the Fig. 4d describes the ONE simulator graphical user interface, which shows that the deployment plan is simulating in ONE simulator. Note that, in ONE simulator each of the given shelter points and LWC towers(associated with a certain shelter point) acts as a static DTN node and each data mule acts as moving DTN node which roams around a certain number of shelter points to collect the information messages from the volunteers (residing in shelter points) and thus forms a DTN-based post-disaster communication network. Finally, Fig. 4e shows the interface of the tool to visualize the performance report of the deployment of the DTN-based post-disaster network. Such a report contains network-related performance metrics like delivery probability, Network latency, Number of created messages within the simulation time, etc. Note that, this tool can take the disaster-affected area maximum up to 80 $km^2$. This restriction is due to the ONE simulator's limitation.

## 5 Conclusion and Future Scope

In this paper, we propose a tool for post-disaster scenarios to deploy a communication infrastructure utilizing the constrained set of network resource pool. Such a tool will be used by the government disaster management authority or NGOs to deploy an interim network infrastructure after a disaster. The objective of developing such a tool is to consider the resource constraint scenarios to deploy a post-disaster network, which often happens in the case of under-developed or developing countries. Such a tool will also be useful for the researchers in the field of post-disaster communica-

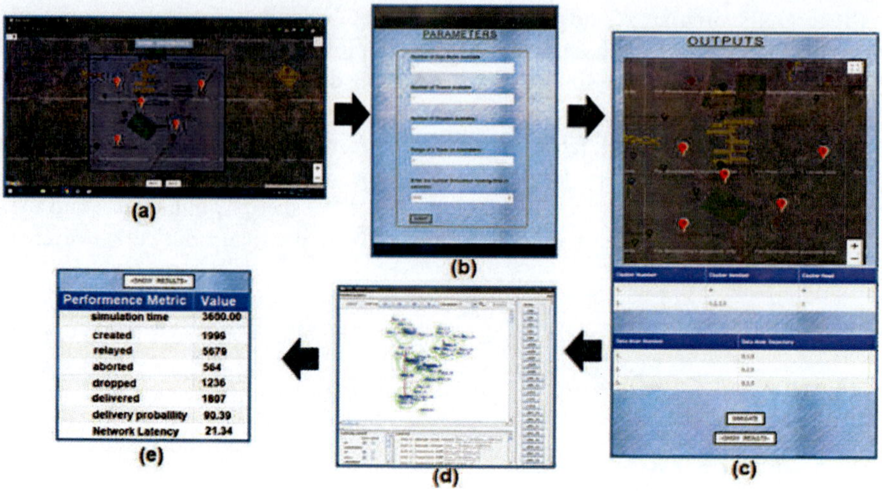


**Fig. 4** Snapshot of the various steps of RCPDNPT: **a** Interface for selecting the disaster affected are and shelter points; **b** Interface for inputting the available resource pool; **c** Output from resource allocation algorithm; **d** ONE simulator is simulating the disaster area network using deployment plan from the resource allocation algorithm; **e** Interface to show the performance report

tion, they can invoke their proposed resource allocation algorithms in this tool, and accordingly can validate the efficacy of their algorithms through simulation.

In the future, we will try to make this tool more interactive and general-purpose network planning and simulation tool. To do that, we will try to provide an API for the resource allocation algorithm so that users/researchers can use their proposed resource allocation algorithm to test for the efficiency of their algorithm.

**Acknowledgements** This work is an outcome of the R&D work undertaken in the project entitled "Developing ICT-based kiosks for post-disaster situational information management using opportunistic networking framework" funded by Department of Higher Education, Science and Technology and Biotechnology, Government of West Bengal, India.

## References

1. Ferranti L, D'Oro S, Bonati L, Demirors E, Cuomo F, Melodia T (2019) Hiro-net: self-organized robotic mesh networking for internet sharing in disaster scenarios. In: 2019 IEEE 20th international symposium on "a world of wireless, mobile and multimedia networks" (WoWMoM). IEEE, pp 1–9
2. Hazra K, Shah VK, Bilal M, Silvestri S, Das SK, Nandi S, Saha S (2019) A novel network architecture for resource-constrained post-disaster environments. In: 2019 11th international conference on communication systems and networks (COMSNETS). IEEE, pp 328–335
3. Keränen A, Ott J, Kärkkäinen T (2009) The one simulator for DTN protocol evaluation. In: Proceedings of the 2nd international conference on simulation tools and techniques, pp 1–10

4. Legendre F, Hossmann T, Sutton F, Plattner B (2011) 30 years of wireless ad hoc networking research: what about humanitarian and disaster relief solutions? what are we still missing? In: Proceedings of the 1st international conference on wireless technologies for humanitarian relief (ACWR'11), p 217
5. Saha S, Agarwal N, Dhanuka P, Nandi S (2013) Google map based user interface for network resource planning in post disaster management. In: Proceedings of the 3rd ACM symposium on computing for development, pp 1–2
6. Saha S, Nandi S, Paul PS, Shah VK, Roy A, Das SK (2015) Designing delay constrained hybrid ad hoc network infrastructure for post-disaster communication. Ad Hoc Netw 25:406–429
7. Sakano T, Fadlullah ZM, Ngo T, Nishiyama H, Nakazawa M, Adachi F, Kato N, Takahara A, Kumagai T, Kasahara H et al (2013) Disaster-resilient networking: a new vision based on movable and deployable resource units. IEEE Netw 27(4):40–46

# FoSDN: A Software-Defined Edge Computation for Resource Constraint Network

**Rohit Kumar Das, Fabiola Hazel Pohrmen, Arnab Kumar Maji, and Goutam Saha**

**Abstract** In recent years, edge computing has been widely used for the Internet of Things (IoT). The general prospect of edge computation is to provide computation near to the field of data generation. This helps to reduce delay and packet drop which are important parameter need to consider while designing an IoT network. The Software-Defined Network (SDN) is a convenient technology that can be used to perform computation at the edge level instead of cloud to enhance the network performance. The architectural advantage of SDN can help in providing a more robust solution for edge-based solutions in IoT networks. In this paper, an edge enabled SDN (FoSDN) architecture has been proposed which helps in reducing the delay, packet drop and latency by enabling computation at the edge of the network. The performance of the proposed FoSDN architecture was examined using the Mininet-WiFi simulator. The simulation results of FoSDN architecture were found to be satisfactory.

**Keywords** Edge computation · Internet of things · Software-defined network · Architecture · Reliability · Availability · Scalability

## 1 Introduction

The growth of IoT applications from the past few years is increasing at a high rate because of its useful features in day-to-day life and easy installation. The general operation of the IoT network is to collect data from the environment and process those

---

R. K. Das (✉) · F. H. Pohrmen · A. K. Maji · G. Saha
North-Eastern Hill University, Shillong, Meghalaya, India
e-mail: rohitdas.it.13@gmail.com

F. H. Pohrmen
e-mail: fhpohrmen@gmail.com

A. K. Maji
e-mail: arnab.maji@gmail.com

G. Saha
e-mail: dr.goutamsaha@gmail.com



A. Maji et al. (eds.), *Proceedings of the International Conference on Computing and Communication Systems*, Lecture Notes in Networks and Systems 170,
https://doi.org/10.1007/978-981-33-4084-8_44

data in the cloud. The data are transmitted to the cloud by the intermediate nodes such as sink nodes or gateway nodes. The user interacts with the system through some Application Program Interfaces (APIs). The user performs the necessary operation based on the gathered information.

The IoT network generally consists of a large number of nodes, which are resource constraints in nature. The processing power, storage and communication range of these nodes are limited which makes them reliable to other intermediate neighboring nodes. Some of the other major issues that the IoT network suffers from are availability, reliability, scalability, security and privacy [1]. The network designer has to take care of the above-mentioned issues while designing an IoT network. Most of the research work in this area focuses on the management, development and providing Quality of Services (QoS) of the network.

SDN technology recently has gained much interest because of its architectural advantage. It can handle the network more effectively with a powerful resource like the SDN controller. The basic principle of SDN is to separate the control and data plane from a device. The control plane is placed in the controller whereas the data plane simply follows rules to route a packet from the controller. The architectural advantage of SDN can help in reducing the congestion of data traffic at the network layer. It provides a programmable interface to the lower infrastructure layer and upper application layer which helps in the efficient management of the network [1].

The general SDN based IoT architecture consists of the Infrastructure layer, Network layer, Control layer and Application layer as shown in Fig. 1. The Infrastructure layer consists of a collection of IoT nodes and access points which are connected to the core network. The Network layer is the collection of OpenFlow switch which forms the core network for the SDN based IoT system. This OpenFlow switch acts as Data Plane which stores routing information in the flow table. The flow table is populated by flow entries from the SDN controller connected to it. The SDN controller resides in the Control layer which communicates with the Data plane using SouthBound Interface (SBI). From the Application layer, the various application or data analysis can be performed by interacting with the controller via NorthBound Interface (NBI) to manage various network operations [2, 3].

The SDN controller can have a network-wide view which is connected to the core network. The data are collected by the access points and transmitted to the controller via the core network whenever required. The distance between the SDN controller and the end nodes plays a significant role in network performance. As the packet needs to reach the controller via the access point and the core network, it increases the round trip time (RTT) of a packet to get processed. It also increases the packet drop possibility within the network. Moreover, the latency and bandwidth consumption will also increase which are not desirable for real-time based IoT applications.

One possible approach is to provide services and storage at the edge of the network. This is often regarded as Fog Computing or Mobile Edge Computing (MEC) for wireless network [4]. The Fog Computing works by providing servers for storage and processing the network information near to the field of data generation. This helps in reducing the latency and bandwidth consumption as computation is available on the edge of the network to that of cloud [5, 6]. In this paper, a Fog based SDN network

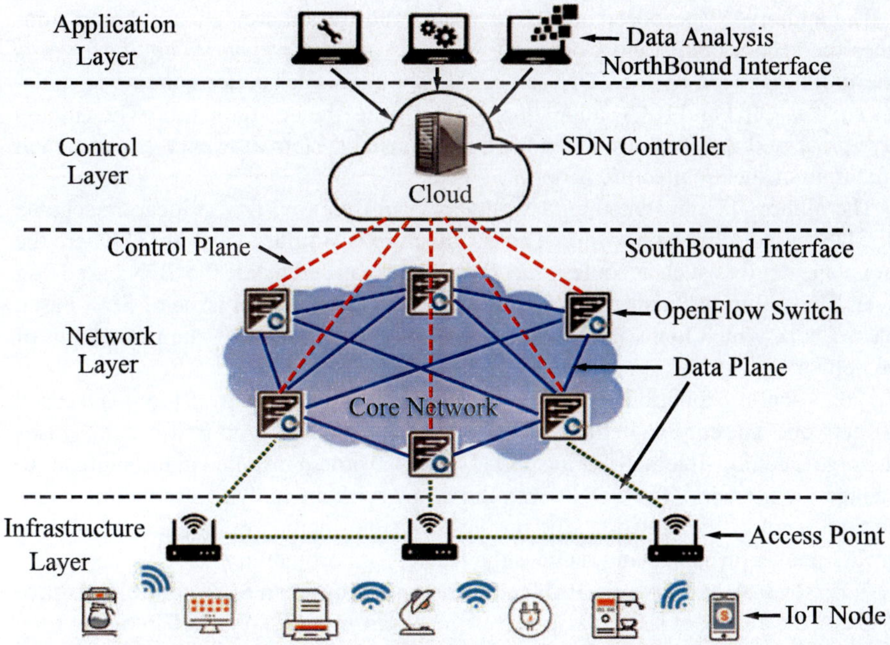


**Fig. 1** The general SDN based IoT architecture

(FoSDN) architecture has been proposed to provide computation at the edge of the network. The main contribution of the proposed FoSDN architecture is to increase the availability of services, reliability of the system and support scalability when required. These are achieved by introducing a Fog layer which consists of SDN controllers that are in direct contact with the OpenFlow based access points in the network. The FoSDN helps in decreasing the routing time taken by a packet which also helps in reducing the latency of the network.

The rest of the paper is organized as follows. Section 2 discuss various works related to the proposed FoSDN approach. Section 3 presents the FoSDN architecture for resource constraint network. Section 4 presents the experimental setup for FoSDN. The results are discussed in Sect. 5. Finally, Sect. 6 concludes the paper.

## 2 Related Work

The use of edge computing in SDN based IoT network can help to overcome the basic challenges of resource constraint networks. Several works have been reported so far in this domain. This section describes the works related to the proposed FoSDN architecture.

The authors [7] proposed an SDN-based orchestration architecture which combines the access points and OpenFlow switches in the Fog layer. The Fog layer is externally managed by the SDN controller. The controller collects the node statistics such as CPU and RAM allocation and performs the desirable task. The authors have presented different use cases for the proposed system, however, experimental validation of the architecture is not provided.

The authors [8] presented an assignment of differential flow allocation scheme for SDN based IoT/Fog network. The architecture is similar to that of [7] where the data plane devices such as routers and switches are placed inside the SDN based Fog layer. The author has addressed the dynamic flow management issue of SDN based IoT network which helps in reducing the latency and increasing the throughput of the network.

A data analytic functionality along the edge node is proposed by [9] for SDN based IoT network. The edge switch acts as a fog layer entity for the proposed system. It uses Message Queuing Telemetry Transport (MQTT) protocol installed in the switches to manage the network. The SDN controller along with an analytical model manages the QoS level of the network. The proposed system intends to increase the system throughput performance and reduces the packet loss probability.

A Personal Area Network (PAN) coordinator acting as an SDN controller is proposed by the author of [10]. It also serves the purpose of 6LoWPAN (IPv6 over Low -Power Wireless Personal Area Networks) gateway router. It placed at the edge of the 6LoWPAN based SDN network. The architecture is based on Network Functions Virtualization (NFV) where the PAN coordinator manages the flow rules to be used by the end nodes. It reduces the end-to-end delay of the overall network by the use of edge-based SD-NFV PAN coordinator. However, the use of a specialized type of resource constraint PAN coordinator restricts the scalability of the network.

An SDN based management IoT Management (SDIM) framework is proposed by [11] to manage IoT nodes in Wireless Sensor Network (WSN). Then SDIM intends to provide end-to-end communication and operational fault detection in the WSN network. It uses Multi-access Edge Computing (MEC) which is integrated with the cloud to facilitate multi-domain WSN topology. The SDN controller is connected to MEC which forms the edge network. The edge-based MECs are connected to the cloud which addresses the scalability issue by using the SDN controller to monitor the end nodes in the network.

From the above-listed literature, it is worth to note that most of the works provide a solution to a particular issue. The proposed FoSDN architecture uses the fog layer to increase the availability, reliability and scalability of the resource constraint network. It lowers the routing time and packet drop to increase availability and reliability. The use of the SDN controller in the Fog layer improves the scalability and also helps in reducing the latency of the network.

## 3 Proposed FoSDN Architecture

The proposed FoSDN architecture is shown in Fig. 2. It consists of a collection of OpenFlow switches which form the core network for the system. The SDN controllers are placed lower to the core network, i.e. in the Fog layer to manage the network more efficiently. The nodes from the Infrastructure layer are connected to the access points. The nodes communicate with each other via access points which are in direct contact with the SDN controllers placed in the Fog layer. The data are analyzed in the application layer which is stored in the cloud infrastructure.

The SDN controllers for the FoSDN architecture are directly accessible for the access points in the Infrastructure layer. The placement of the SDN controller in the Fog layer helps in reducing the average RTT of a packet. As the SDN controller is placed at the lower Fog layer, the time required by a packet to travel to the controller for further processing is much lesser when compared to the traditional SDN based IoT architecture. This increases the reliability and availability factor of the system. Depending upon the number of nodes in the network, the number of SDN controllers

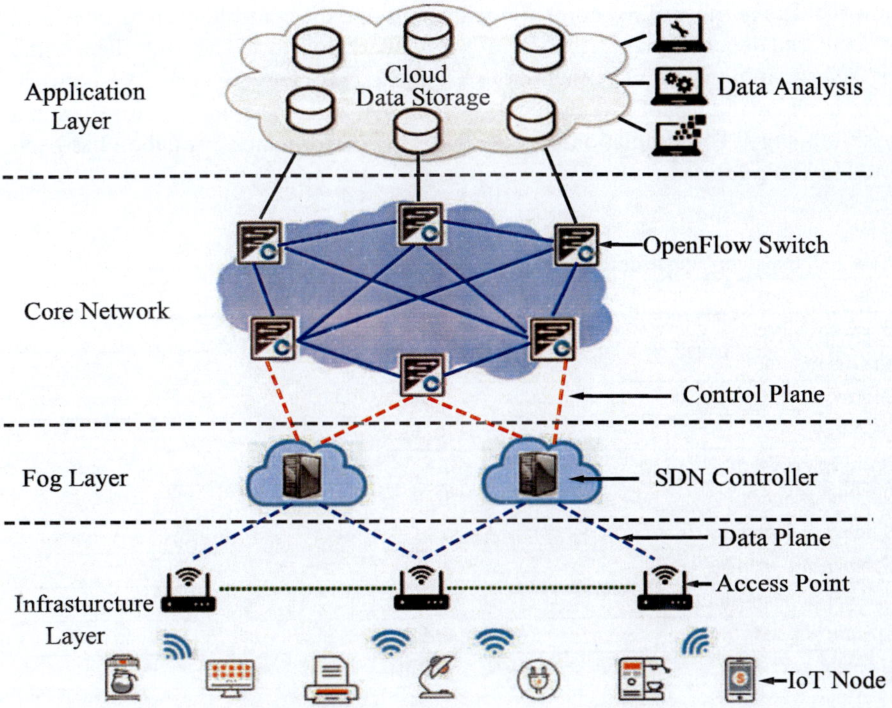


**Fig. 2** Proposed FoSDN architecture

can be increased which will sustain the scalability issue as well. The access points used in the FoSDN architecture are OpenFlow based access points. The routing path is stored in the flow table of the access point. Depending upon the requirement, the scalability of the system can be increased by using more number of SDN controller in the Fog layer.

## 4 Experimental Setup

The experimentation was performed in Dell Precision Workstation T5500 running Ubuntu LTS 18.04 with 8 GB RAM. To validate the performance of FoSDN, the Mininet-WiFi simulator [12] was used to simulate all the entities. It is a python based simulator that supports a large number of network operations.

A total of 100 IoT nodes were simulated and placed randomly in the network. The IoT nodes are connected to the access points which are directly accessible to the controller. For FoSDN, we have used two controllers in the fog layer. Each fog computing paradigm is enabled with a controller for the local management of the network. The number of fog computing elements and the controller can be increased or decreased depending upon the size of the network and cost. The controllers in the fog layer communicate with each other using the core network. The core network is simulated using OpenFlow switches, which are connected among themselves in a mesh topology. The simulation parameter for the performance evaluation has been listed in Table 1.

**Table 1** Simulation parameter

| Parameter | Value |
|---|---|
| Operating system | Linux (Ubuntu 18.04 LTS) |
| Simulator | Mininet-WiFi |
| Controller | OpenDaylight |
| Topology | Random |
| OpenFlow version | 1.5 |
| Number of nodes | 100 |
| Number of switch | 6 |
| Number of Fog layer | 2 |
| Number of controller | 2 |
| OpenFlow ports | 6655 |
| Time per test | 100 ms |

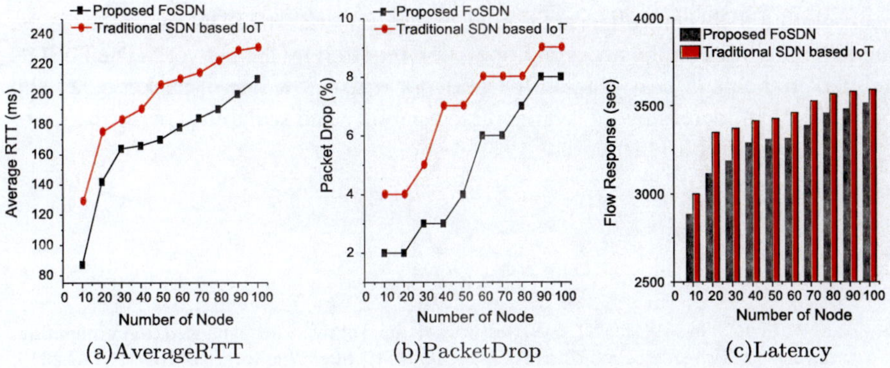


**Fig. 3** Performance comparison of proposed FoSDN with traditional SDN based IoT system

## 5 Result and Discussions

The experimentation was carried out in Dell Precision Workstations T5500 with 10 GB RAM. For the proposed FoSDN architecture, two OpenFlow based SDN Controllers (OpenDaylight) [13] were set up as Fog layer controllers. They are installed in two different physical systems which are in direct connectivity to the access points. The standard OpenFlow version 1.5 [14] protocol was used as a communication protocol between the nodes and the controller. To examine the performance of the FoSDN, 100 wireless nodes connected via access points were virtually created in the Mininet-WiFi simulator. The access points are connected to the nodes as well as to the Fog SDN controller in a mesh topology.

The FoSDN architecture was evaluated by measuring the average RTT of a packet, packet drop and latency. As shown in Fig. 3a, the FoSDN requires less time to process a packet in terms of communication when compared with the traditional SDN based IoT architecture. This is since the SDN controllers are placed on the edge of the network, hence it takes relatively lesser time to travel up to the controllers. This helps in increasing the availability of service factors for the network. The uses of the Fog layer for the SDN controller also help in reducing the packet drop as shown in Fig. 3b. The latency performance is indicated in Fig. 3c. The network suffers from a lower rate of latency as the network consumes lesser bandwidth to transmit a packet to the Fog layer SDN controllers.

## 6 Conclusion

The performance of any network is highly reliable in its computation platform. The inclusion of edge computation in IoT networks can greatly enhance its performance. The proposed FoSDN architecture enables edge computation at the Fog layer by incorporating SDN controllers at the edge of the network. The SDN controllers at the Fog layer provides all the necessary operation required by the network. The FoSDN architecture helps in providing an IoT network with a low rate of packet delay and loss. It helps in increasing the availability, reliability and scalability performance of resource constraint IoT network.

## References

1. Wang A, Zha Z, Guo Y, Chen S (2019) Software-defined networking enhanced edge computing: a network-centric survey. Proc IEEE 107(8):1500–1519. https://doi.org/10.1109/JPROC.2019.2924377
2. Kaur K, Garg S, Aujla GS, Kumar N, Rodrigues JJ, Guizani M (2018) Edge computing in the industrial internet of things environment: software-defined-networks-based edge-cloud interplay. IEEE Commun Mag 56(2):44–51. https://doi.org/10.1109/MCOM.2018.1700622
3. Das RK, Maji AK, Saha G (2019) Prospect of improving internet of things by incorporating software-defined network. In: Bera R, Sarkar S, Singh O, Saikia H (eds) Advances in communication, devices and networking, vol 537. Lecture notes in electrical engineering. Springer, Singapore, pp 537–544. https://doi.org/10.1007/978-981-13-3450-4_58
4. Chiang M, Zhang T (2016) Fog and IoT: an overview of research opportunities. IEEE Int Things J 3(6):854–864. https://doi.org/10.1109/JIOT.2016.2584538
5. Baktir AC, Ozgovde A, Ersoy C (2017) How can edge computing benefit from software-defined networking: a survey, use cases, and future directions. IEEE Commun Surv Tutor 19(4):2359–2391. https://doi.org/10.1109/COMST.2017.2717482
6. Sun X, Ansari N (2016) EdgeIoT: mobile edge computing for the internet of things. IEEE Commun Mag 54(12):22–29. https://doi.org/10.1109/MCOM.2016.1600492CM
7. Tomovic S, Yoshigoe K, Maljevic I, Radusinovic I (2017) Software-defined fog network architecture for IoT. Wirel Pers Commun 92(1):181–196. https://doi.org/10.1007/s11277-016-3845-0
8. Diro AA, Reda HT, Chilamkurti N (2018) Differential flow space allocation scheme in SDN based fog computing for IoT applications. J Am Intell Human Comput 1–11. https://doi.org/10.1007/s12652-017-0677-z
9. Xu Y, Mahendran V, Radhakrishnan S Towards (2016) SDN-based fog computing: MQTT broker virtualization for effective and reliable delivery. In: 8th international conference on communication systems and networks (COMSNETS). IEEE, Bangalore, pp 1–6. https://doi.org/10.1109/COMSNETS.2016.7439974
10. Al-Kaseem BR, Al-Dunainawi Y, Al-Raweshidy HS (2018) End-to-end delay enhancement in 6LoWPAN testbed using programmable network concepts. IEEE Int Things J 6(2):3070–3086. https://doi.org/10.1109/JIOT.2018.2879111
11. Mavromatis A, Colman-Meixner C, Silva AP, Vasilakos X, Nejabati R, Simeonidou D (2019) A software-defined IoT device management framework for edge and cloud computing. IEEE Int Things J 1–18. https://doi.org/10.1109/JIOT.2019.2949629
12. Fontes RR, Afzal S, Brito SH, Santos MA, Rothenberg CE (2015) Mininet-WiFi: emulating software-defined wireless networks. In: 11th international conference on network and service management (CNSM). IEEE, Barcelona, pp 384–389. https://doi.org/10.1109/CNSM.2015.7367387
13. Medved J, Varga R, Tkacik A, Gray K (2014) Opendaylight: towards a model-driven SDN controller architecture. In: Proceeding of IEEE international symposium on a world of wireless, mobile and multimedia networks. IEEE, Sydney, NSW, pp 1–6. https://doi.org/10.1109/WoWMoM.2014.6918985
14. Open networking foundation: OpenFlow-switch-v1.5.1. https://www.opennetworking.org/wp-content/uploads/2014/10/openflow-switch-v1.5.1.pdf

# Enabling Indirect Link Discovery Between SDN Switches

**Mir Wajahat Hussain and Diptendu Sinha Roy**

**Abstract** The removal of the control plane from a Software Defined Network (SDN) helps avoid flexibility issues that exist in the traditional networks thus enabling SDN to leverage more bandwidth utilization and hence renders it more suitable for big data analytics. A pertinent issue in network arises how the controller allocates network resources based on-demand/needs, viz. allocation of less delay path, optimal routing, less error path. The controller handles this issue by having global information of all the links in the network. SDN leverages protocols like link layer discovery protocol (LLDP) and broadcast domain discovery protocol (BDDP) for the identification of links in OpenFlow devices which are directly and indirectly connected. The detection of direct/indirect links between OpenFlow devices requires a considerable number of messages to be sent across every active interface of the switch thus puts this substantial load on the controller. This paper proposes an indirect link discovery (ILD) scheme which minimizes the number of messages required to fetch the links in OpenFlow devices by sending a single *Packet-Out* (*P-out*) across every OpenFlow switch (OFS), which circulates it to successive nodes. Experiments performed in Mininet shows the number of *P-out* messages and packet loss for fetching equivalent topology ILD shows a reduction in between 30.9–78.95% and 48.95–71.4% as compared to the state of art protocols with varied topologies.

**Keywords** Software-defined network · Link discovery · Controller · OpenFlow · Hybrid SDN

M. W. Hussain · D. Sinha Roy (✉)
Department of Computer Science and Engineering, National Institute of Technology, Meghalaya, Laitumkhrah, Shillong 793003, India
e-mail: diptendu.sr@nitm.ac.in

M. W. Hussain
e-mail: mir.wajahat.hussain@nitm.ac.in



A. Maji et al. (eds.), *Proceedings of the International Conference on Computing and Communication Systems*, Lecture Notes in Networks and Systems 170,
https://doi.org/10.1007/978-981-33-4084-8_45

# 1 Introduction

Managing traditional networks with prolific data generation from a multitude of devices is complicated and challenging [1]. This data generation from large devices that are connected to the Internet in IoT has to be analyzed for meaningful interpretation with big data analytics. Issues intrinsic to traditional networks like vendor-lock in, policy enforcement, proprietary protocols, with the surge in data generation and data analytics have paved the way to innovate and design a network that has features of programmability and manageability in it [2–4].

Software-defined networking (SDN) is a new network paradigm in which sets out apart control plane and forwarding plane [1, 5]. The interaction between control and the forward plane is done via well-defined *Southbound* API called OpenFlow. With the OpenFlow protocol, the controller can pull statistics inside all the devices in the network. The control plane is logically centralized and manages the forwarding inside the devices by installing rules based on specific needs [6]. Full deployment of SDN is far from reality on account of several issues, viz. considerable budgetary requirements, technical and business problems and hybrid SDN seems desired path [7, 8].

Discovering nodes and links inside the network is essential, for the controller to have dynamic control over the network. With the dynamic control over the network, the controller can allocate network resources on-demand [9]. These on-demand resource allocations include the finding optimal paths in a routing, giving a less delay path for movement of data for latency-sensitive flows, finding loopless topology in a network containing loops or a path with lesser error loss, etc. The controller handles these issues by having global link information of the devices [10].

Link discovery in SDN is done with the help of LLDP and BDDP protocols. LLDP and BDDP protocols are utilized to identify direct and indirect links in OFS [10, 11]. The mode of operation of the protocols (LLDP & BDDP) requires the controller to send the *P-out* events across every active interface of the switch, which puts a considerable strain on the controller as it is a continuous process [10]. Earlier works focused on the link discovery have either optimized the number of messages required by the controller to fetch the direct OpenFlow links or taking security issues into account while designing the protocol [10, 12–14].

This paper proposes a scheme called ILD, which gathers the link information in OFS which is either directly or indirectly connected. The mode of the operation of the scheme is that for every OFS, a single P-out event is sent. The OFS forwards this packet across every active port of the switch. Now if the OFS is connected by a legacy switch (LS), the packet will be forwarded across the outgoing interface or if the packet is received by the OFS it responds to the controller via a *Packet-In* (*P-in*) message. This *P-in* received by the controller helps it to find the link (directly/indirectly) in the OFS. Experiments performed in the paper that there is a reduction in the number of *P-out* messages, sent by the controller to fetch the links and packets lost to fetch the topology in OFS inside the network as compared to the state of art protocols with varied topologies.

The paper is organized as follows: Related works are described in Sect. 2. The proposed ILD scheme along with the frame format is presented in Sect. 3. Section 4 discusses the algorithmic description. Emulation environment and results discussion are described in Sect. 5. Conclusions are drawn in Sect. 5 and thus concludes the paper.

## 2 Related Work

Link discovery in SDN is the sole responsibility of the controller [10]. There have been several efforts to address the link discovery-related issues and are discussed as; Pakzad et al. [10] optimized the LLDP protocol by sending a single *P-out* message to every OFS. After the reception of the *P-out* message at the switch, the controller installed flow entry to handle the arrival of this message. Now with this new rule installed in the switch, the frame was forwarded on all the active ports of the switch by appending the MAC address of the interface. After the frame forwarded from the interface of the switch reaches nearby other OFS, it has a predefined rule via *P-in* to send it to the controller. Pakzad et al. [12] again modified their work by focusing on the optimization of the costly ternary content addressable memory (TCAM) of the OFS. The controller, in this case, sends the *P-out* along with the action list. This action list is utilized by the OFS to forward the frame and thus avoided the installation of setup rules inside the switch. In [11] with the usage of both LLDP and BDDP, multihop links are detected in OFS in a partial deployment (hybrid SDN) which has indirect connections within a single broadcast domain.

Ochoa et al. [15, 16] proposed an enhancement to the topology discovery mechanisms by incorporating both failure recovery and topology discovery. In work, they categorized nodes and their switch ports inside the network, based on the *topoRequest* and *echoReply* messages; subsequently, they were able to form a control tree thus being able to fetch the topology inside the network. They modified their work in [13] and profoundly discussed the necessary message types of the protocol proposed. These works removed possibility of an initial handshake between OFS and controller.

Azzouni et al. [13] pointed out the flaws in the LLDP protocol on account of its security and performance issues. They proposed a secure and efficient topology discovery protocol (sOFTDP) protocol that took security and performance into account to design the protocol and they delegated a part of the discovery mechanism from the controller to the switch. Besides finding the direct links between the switches for the topology discovery mechanism, the protocol also provides backup paths for handling failures. Nehra et al. [14] proposed a protocol secure and lightweight discovery protocol (SLDP), which addressed various attacks inside the protocol (flooding, security, and replay) and reduced the overhead on the controller. They proposed a new packet structure that is lightweight and requires fewer packets to fetch the topology and also addressed several issues on the security.

All of the above works have primarily focused on the detection of direct links between OpenFlow devices with the LLDP protocol. LLDP protocol is unable to detect the indirect links in OpenFlow devices [11]. However, [11] detects both the direct/indirect links in OFS but requires a large number of messages to fetch the link information. Although work in [10] is similar to our work, but [10] had only focused on optimizing the number of messages in a pure SDN with no consideration of a hybrid SDN. The proposed ILD scheme is promising as it identifies both direct and multihop links in OFS and requires only fewer messages to fetch topology.

## 3 Proposed Indirect Link Discovery Scheme

As pointed out the earlier majority of protocols used for link discovery requires considerable messages to be sent from the controller. Owing to larger *P-out* events to be sent from the controller limits the performance of the controller. Thus, this paper proposes a scheme that detects links in OFS in a partial deployment by limiting the number of *P-out* sent to an OFS to 1.

### 3.1 Illustrative Example

Figure 1 describes a network composed of OFS and LS in a partial deployment consisting of 4 switches. OFS $S_1$, $S_2$, and $S_3$ have a connection with the controller while LS $L_1$ has no connection with the controller.

The scheme operates by sending a *P-out* from the controller to an OFS $S_1$. With the reception of the *P-out* from the controller, OFS $S_1$ forwards this packet along with all the active interfaces of this switch based on the rules set up by the controller. On reception of the packet from switch $S_1$, switch $S_2$ sends a *P-in* to the controller. The reception of *P-in* results in the detection of the unidirectional link in OFS $S_1$ and $S_2$ and results in an update of the topology table in the controller.

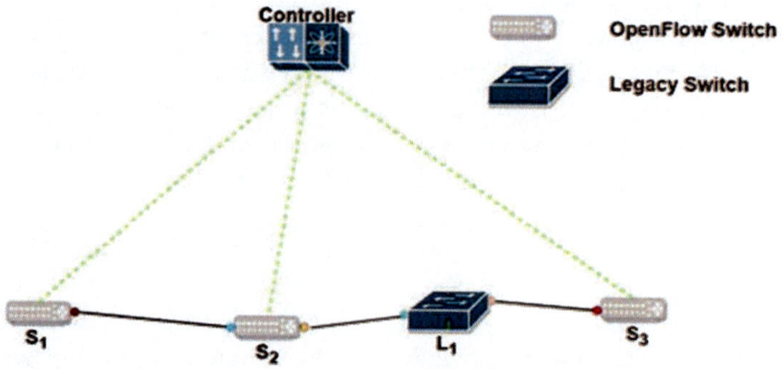


**Fig. 1** Network topology

Now the controller again sends a *P-out* event to an OFS $S_2$, forwards packet along with all the active interfaces. Packets are forwarded along with the link information and ultimately packets reach switches $S_1$ and $L_1$. OFS $S_1$ appends the metadata and replies to the controller via a *P-in*. The reception of *P-in* results in the detection of link in OFS $S_2$ and $S_1$, and subsequently, topology table is updated accordingly. LS $L_1$ is unable to handle this packet, so based on the destination address in the packet it forwards it without any addition inside. This forwarded packet is received by OFS $S_3$, which appends the link information and hence replies to the controller via a *P-in*. Thus, a link is detected in between OFS $S_2$ and $S_3$ by the controller. The same iterative procedure is used to detect a multihop link in OFS $S_3$ and $S_2$ (Figs. 2, 3, and 4).

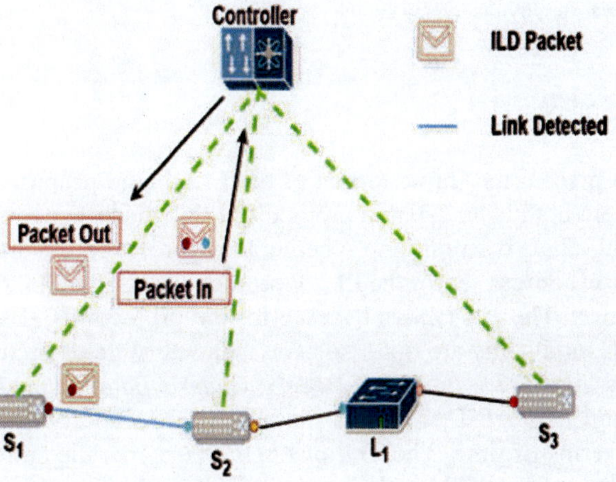


**Fig. 2** Packet-out sent by the controller

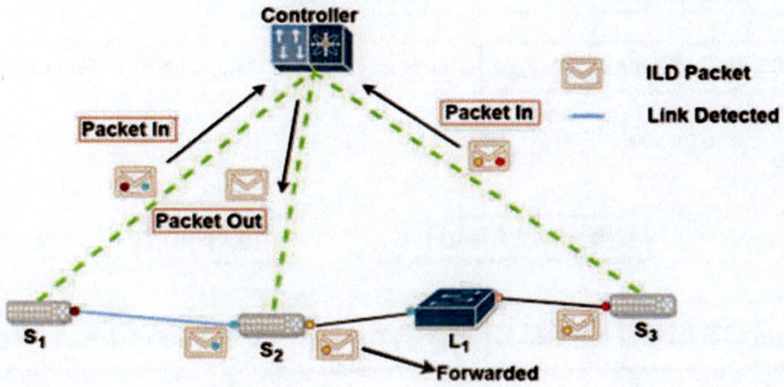


**Fig. 3** Packet-in received by the controller

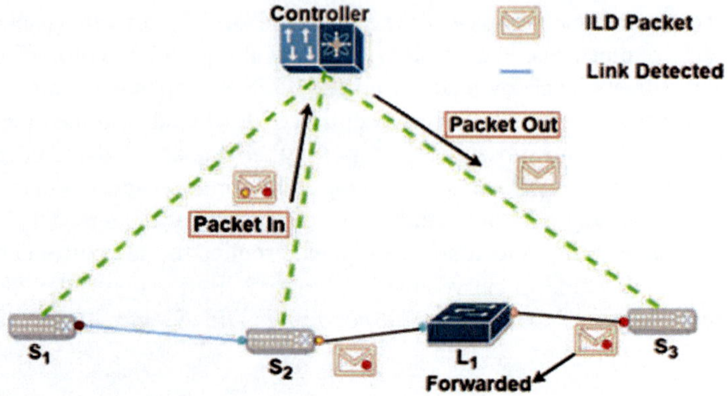


**Fig. 4** Packet forwarding by the legacy switch

## 3.2 *Packet Format*

Figures 5 and 6 present the frame format of the LLDP and proposed ILD scheme. LLDP and ILD are both layer 2-based protocol for link discovery. LLDP is a single-hop protocol while ILD is a multihop scheme; therefore, it detects multihop links.

The destination address set for the LLDP protocol is 00:80:c2:00:00:00, which is a multicast address. The *Eth type* is $0\times88$cc for the protocol [6]. LLDP frame has mandatory fields, and some are optional. The mandatory fields include *Chasis ID*, *Port ID*, *Time to live* (*TTL*), *End of LLDPDU*. *Chassis ID* and *port ID* identify the sender switch and its *port ID*. *TTL* is a measure of the validity of the packet and is quantified in terms of time. The *End of LLDPDU* marks the culmination of the LLDP frame. The optional field includes Optional *Type Length Value* (*TLV*).

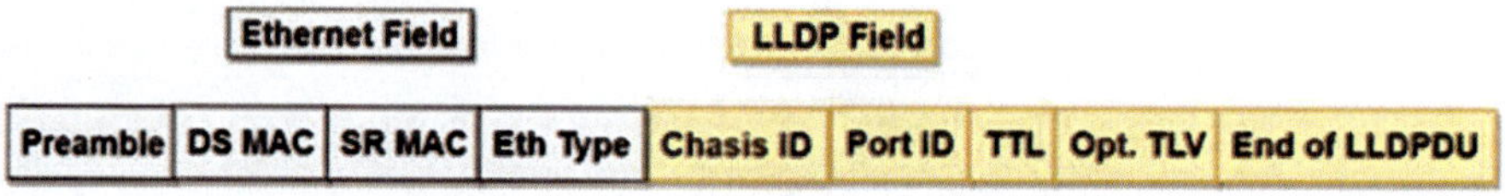


**Fig. 5** LLDP frame format

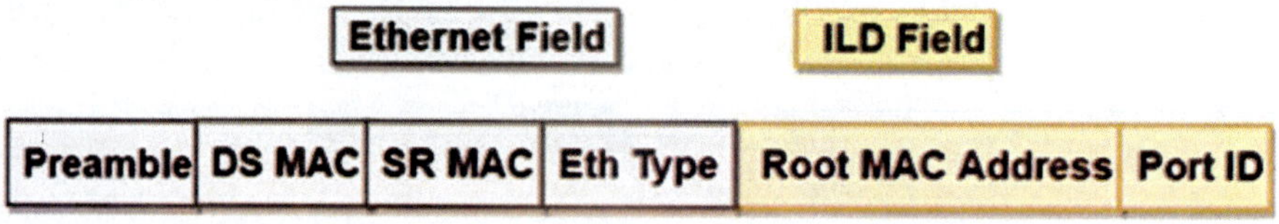


**Fig. 6** ILD frame format

The destination address set for the ILD is ff:ff:ff:ff:ff:ff, which is a broadcast address similar to the BDDP protocol. The *Eth type* is any custom type [11]. *Root MAC* address and *port ID* identify the interface and port of the switch, respectively.

```
I.    for all the packets received do
II.   if pkt.inport == CONTROLLER && pkt.etherType == ILD then // P-out
III.  for all OFS (S1) port do
IV.   pkt.srcMAC ← port.MACaddr
V.    pkt is sent on all the active ports of the switch
VI.   while (pkt received by from S1, Legacy switch (L1)) then
        forward pkt  //LS forwards the packet along with the outgoing interface
VII.  return pkt to the CONTROLLER  //OFS sends P-in
VIII. endfor
IX.   endif
X.    endfor
```

Algorithm 1: ILD algorithm

## 4 Algorithmic Description

The proposed algorithm operates by the reception of the ILD packet from the controller (line 1 and line 2). OFS updates the MAC address of the packet with the port address and sends a packet-in all the active ports of the switch (line 4 and line 5).

After the packet is sent from the switch $S_2$, receiving switch can either be an LS or OFS. If the recipient switch is an LS, it only forwards the packet to another outgoing interface (line 6) and ultimately reaches some OFS. The packet received by the OFS will reply to the controller via a *P-in* (line 7), and thus, links are identified.

## 5 Emulation Environment and Results Discussion

### *5.1 Emulation Environment*

Mininet is used for the evaluation of the proposed ILD scheme. Mininet is a widespread network emulator which creates virtual hosts, links, and switches in

**Table 1** Software used for the emulation

| Software | Version |
|---|---|
| Mininet | 2.3.0d5 |
| Ryu | 4.32 |
| OpenvSwitch | 2.9.2 |
| Python | 2.7 |
| Operating System | Ubuntu 18.04 LTS |

**Table 2** Topology specification used for the emulation

| S. No. | Topology | Switches | Ports |
|---|---|---|---|
| 1 | Linear | 50 | 198 |
| 2 | Tree, depth = 5, fanout = 2 | 31 | 60 |
| 3 | Fat-tree, $K = 2$ | 10 | 24 |

a network. The mininet is installed in a Ubuntu Operating System. The switch used is *Open vSwitch* a popular virtual software SDN based switch. Ryu is used as a controller which is written in python language. All experiments are done in a system comprised of Intel(R) Core(TM) i3-5005U CPU running at 2.00 GHz with 4 GB RAM. Table 1 describes the software specifications for the emulation required.

### *5.2 Topologies Used for the Experimentation*

The proposed ILD is tested for varied cases of topology like a tree, linear, and fat-tree. In each topology, we have OFS and LS. OFS is connected to the controller. LS has a fail mode attribute equivalent to standalone. The specifications of the topology described are described in Table 2. Inside each topology, at the lower layer in the edge level, each switch is connected to 2 hosts. The ports column mentioned in Table 2 describes only the switch-switch ports and excludes switch-host ports of the network. Each topology used has 2 LS. To limit the several possibilities of LS in the network terminal legacy condition is avoided because the packet from the OFS might be dropped by LS [5].

### *5.3 Performance Indicators*

To quantify the improvement of the proposed ILD scheme, we run a varied set of performance indicators. As pointed out earlier, the proposed scheme is able to discover direct/indirect links in OFS with less number of messages (*P-out*). So each specified topology is run and the number of *P-out* events is counted. Also since LLDP packets are consumed by the Legacy nodes, so packet loss becomes another key parameter to note the difference. The performance comparisons are drawn with the state of art protocols (LLDP and BDDP) and SLDP.

### 5.4 Results and Discussion

Figures 7 and 8 demonstrate the effect of the number of *P-out* sent and Packets lost sent by the controller to fetch the link discovery inside the network in varying topologies. In almost all cases, the state-of-the art protocols (LLDP + BDDP) requires a vast number of *P-out* and packets lost to fetch the same topology. This is due to the fact that LLDP sends the packet to all the interfaces of OFS. Also to fetch the indirect links in OFS, the controller has to send more packets to switches whose links have not been detected. SLDP requires less number to fetch the same topology, as the protocol does not send the *P-out* events to switches which are connected with the hosts which eventually also reduces the packets lost in that process. ILD performs better in all cases owing to sending a single *P-out* to every OFS. The *P-out* sent

**Fig. 7** Packet-out sent by controller versus varied topologies

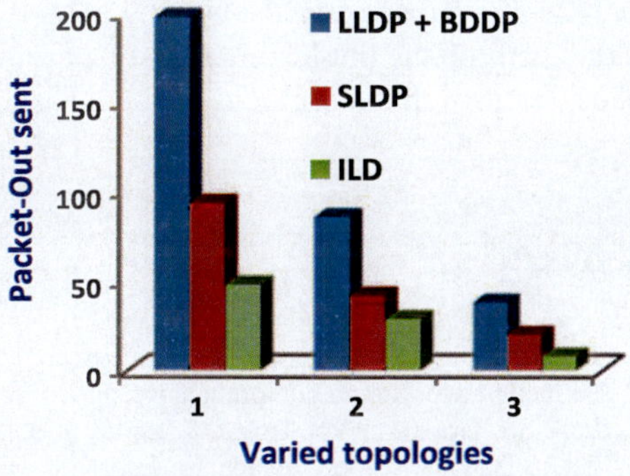


**Fig. 8** Packet-lost sent by controller versus varied topologies

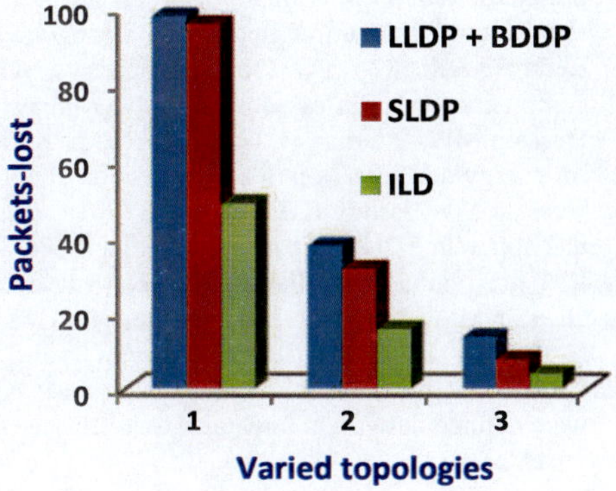

will be forwarded on all the interfaces. This sending of a single *P-out* to every OFS minimizes the packets lost to utmost by 1 for every OFS if it is connected to any number of hosts. The experiment shows that there is an reduction in *P-out* events of about 75.76–48.94%, 66.28–30.95%, and 78.95–60% in linear, tree, and fat-tree as compared to state-of-the art protocols (LLDP + BDDP) and SLDP. In packets lost, the reduction amounted between 50–48.95%, 57.89–50%, and 71.4–50% in linear, tree, and fat-tree as compared to state-of-the art protocols and SLDP with varied topologies.

## 6 Conclusion

In this paper, we have proposed an ILD scheme for fetching link information of the OFS inside the network. The packet format and the mode of operation are also discussed. ILD scheme operates by sending a single *P-out* event for every OFS. This received packet is disseminated to every interface and either by forwarding or replying back the packet reaches to the controller. Experiments performed shows that the number of *P-out* required for fetching topology is reduced amounted between 30.9–78.9% and packets lost by 48.95–71.4% in varied topologies. Future work will be the focus on the implementation of a protocol for discovering all links and adding robust features in the protocol to handle security aspects in a hybrid SDN.

## References


1. Kreutz D, Ramos F, Verissimo P, Rothenberg CE, Azodolmolky S, Uhlig S (2014) Software-defined networking: a comprehensive survey. arXiv preprint arXiv:1406.0440
2. Paik SS, Goswami RS, Roy DS, Reddy KH (2017, October) Intelligent data placement in heterogeneous hadoop cluster. In: International Conference on Next Generation Computing Technologies, pp. 568–579. Springer, Singapore
3. Ashu A, Hussain MW, Roy DS, Reddy HK (2019, December) Intelligent data compression policy for hadoop performance optimization. In: International Conference on Soft Computing and Pattern Recognition, pp. 80–89. Springer, Cham
4. Reddy KHK, Roy DS (2016) Dppacs: a novel data partitioning and placement aware computation scheduling scheme for data-intensive cloud applications. Comput J 59(1):64–82
5. Hussain MW, Pradhan B, Gao XZ, Reddy KHK, Roy DS (2020) Clonal selection algorithm for energy minimization in software defined networks. Applied Soft Comput 96:106617
6. Hussain MW, Reddy KHK, Roy DS (2019) Resource aware execution of speculated tasks in Hadoop with SDN. Int J Adv Sci Technol 28(13):72–84
7. Rathee S, Sinha Y, Haribabu K (2017) A survey: hybrid SDN. J Netw Comput Appl 100:35–55
8. Hussain MW, Reddy KHK, Rodrigues JJ, Roy DS (2020) An indirect controller-legacy switch forwarding scheme for link discovery in hybrid SDN. IEEE Syst J
9. Khan S, Gani A, Wahab AWA, Guizani M, Khan MK (2016) Topology discovery in software defined networks: threats, taxonomy, and state-of-the-art. IEEE Commun Surv Tutorials 19(1):303–324

10. Pakzad F, Portmann M, Tan WL, Indulska J (2014, December) Efficient topology discovery in software defined networks. In: 2014 8th international conference on signal processing and communication systems (ICSPCS), pp 1–8. IEEE
11. Floodlight Controller 2020. https://floodlight.atlassian.net/wiki/spaces/floodlightcontroller/pages/1343628/LinkDiscoveryManager+Dev
12. Pakzad F, Portmann M, Tan WL, Indulska J (2016) Efficient topology discovery in OpenFlow-based software defined networks. Comput Commun 77:52–61
13. Azzouni A, Boutaba R, Trang NTM, Pujolle G (2017) sOFTDP: Secure and efficient topology discovery protocol for SDN. arXiv preprint arXiv:1705.04527
14. Nehra A, Tripathi M, Gaur MS, Battula RB, Lal C (2019) SLDP: a secure and lightweight link discovery protocol for software defined networking. Comput Netw 150:102–116
15. Ochoa-Aday L, Cervelló-Pastor C, Fernández-Fernández A (2018) Self-healing topology discovery protocol for software-defined networks. IEEE Commun Lett 22(5):1070–1073
16. Ochoa-Aday L, Cervelló-Pastor C, Fernández-Fernández A (2019) ETDP: enhanced topology discovery protocol for software-defined networks. IEEE Access 7:23471–23487

# Scheduling Cloudlets to Improve Response Time Using CloudSim Simulator

**Lizia Sahkhar and Bunil Kumar Balabantaray**

**Abstract** The sustainability of any system is based on the performance and the service it gives to the users. The main objective of this work is to assess the performance of the cloud system in terms of its response time using CloudSim simulator. The performance is evaluated on a set of heterogenous cloudlets and the way the processing cores of the host are provision to the cloudlets and virtual machines (VMs) in the execution engine. The default scheduling algorithm used by the CloudSim simulator to allocate these cloudlets from the datacenter broker to the VM is based on the first come first serve (FCFS) scheduling algorithm. This paper proposed the shortest job first (SJF) algorithm to allocate the incoming cloudlets to VM in the ascending order of its length whose response time is evaluated when space-shared and time-shared provisioning policy is implemented. The experimental evaluation of the proposed algorithm shows better improvement when the order of length is considered.

**Keywords** CloudSim simulator · First come first serve · Shortest job first · Time shared · Space shared · Provisioning policy

## 1 Introduction

Cloud computing is a vast expanse of inter-connected, dynamic and shareable computer resources and services which operated on an on-demand and pay-as-you-go policy [1]. It has proved to be of great beneficial to the end-users and industry

L. Sahkhar (✉) · B. K. Balabantaray
Department of Computer Science and Engineering, National Institute of Technology Meghalaya (NITM), Shillong, India
e-mail: liziasahkhar@nitm.ac.in

B. K. Balabantaray
e-mail: bunil@nitm.ac.in

L. Sahkhar
Department of Computer Science, Lady Keane College, Shillong, India

 
A. Maji et al. (eds.), *Proceedings of the International Conference on Computing and Communication Systems*, Lecture Notes in Networks and Systems 170,
https://doi.org/10.1007/978-981-33-4084-8_46

through its cloud service module which includes platform, infrastructure and software as a service. These services are expected to be complete, efficient and guaranteed. Performance is the key to maintain the sustainability of the cloud system. Cloud performance can be determined by the functionality, improvements and contribution a cloud service contributed to the system as a whole [2]. It can be measured through parallel computing, scheduling and load balancing.

This paper focused on improving the response time of the cloud system by scheduling cloudlets using the SJF scheduling algorithm where the length of the cloudlets is considered while at the same time, provision over space-shared (SS) and time-shared (TS) provisioning policy.

This paper is organized as follows. Section 2 highlights the review of literature and related work, Section 3 draws out the importance of CloudSim simulator, Section 4 highlights the objective, theoretical description and algorithm of the proposed work, Section 5 covers the simulated experiment, analysis and results, Section 6 concludes the paper.

## 2 Review of Literature and Related Work

Scheduling is defined as a collection of policies that regulate the order of work performed by a cloud system [3]. Load balancing, on the other hand, is defined as a process of balancing the load in which no node of a system remains in idle state while other nodes are over utilized [4]. These properties help to achieve prioritization and efficient allocation of resources, improves resource utilization, processing time, response time and overall job execution [5].

Agarwal, Jain in [3], analyzes the performance of the execution time in a cloud system using the FCFS, round robin and generalized priority algorithm (GPA). GPA prioritized the task units according to their million instruction per second (MIPS) value in descending order. In this implementation, the generalized priority algorithm performs better than the FCFS and round robin algorithm.

Roy, Banerjee, Chowdhury, Biswas in [6] highlight their cloudlet allocation algorithm, the range wise busy-checking two-way balanced (RB2B), which improves the performance analysis of cloud parameters. This algorithm follows a three-phase process—the VM categorization phase, the two round busy-checking phase and cloudlet still not allocated (CSNA) phase. Certain balancing conditions are also specified to consistently balance the distribution of the cloudlet amidst the VMs such that a VM is considered suitable only when it is not busy and it satisfies such pre-conceived balancing conditions. The analysis shows great improvement in performance evaluation in terms of average turnaround time, waiting time, VM utilization rate as compared to max–min, min–min, conductance algorithm, round robin allocation and resource aware scheduling algorithm (RASA).

Chien et al. [7] in their work on scheduling and load balancing algorithm take into account the processing power of VM and length of assigned cloud tasks to estimate the end of service time in VMs. The algorithm uses all cases of VM's and Cloudlet's

Space-Shared and Time-Shared Scheduling Algorithm. Using this algorithm, the response time and processing time were determined which show improvement over the counter algorithm, the active monitoring load balancer algorithm.

Kumar et al. in [8] use optimization techniques such as cuckoo search algorithm, particle swarm optimization and artificial bee colony optimization on the resources of the VM to enhance the performance of cloud parameters such as execution of user's request, CPU utilization and remote storage of patient's digital information. The simulated analysis as shown in MATLAB and CloudSim shows improvement on the execution time and efficient handling and processing of real-time requests for the benefit of healthcare in the future.

Preethi, Jayavel in [9] introduces the weightage SJF scheduling algorithm with an aim to minimize any task from starving for resources for an extended period of time. The threshold value for the weightage queue is categorized in three groups—0 to 2000 MIPS for the first group, 2001–8000 MIPS for the second group and above 8000 MIPS for the third group. The analysis shows that when the weight is adjusted to the SJF scheduling algorithm, the completion time improves as compared to the FCFS algorithm.

## 3 CloudSim Simulator

A CloudSim simulator is proposed to conduct the experiments as it provides facilities to project various kinds of cloud entities such as hosts, virtual machines, cloudlets or tasks units, datacenters, datacenter broker and other services and utilities that can be configured appropriately, thus making it easier to analyze [10–12]. Cloudlet is also known as cloud task or application service [10], task unit [10–12] or job [7]. For the purpose of this work, cloudlet or task unit is used interchangeably.

### *3.1 CloudSim Architecture*

The Cloudsim architecture in Fig. 1 highlights the cloud information system (CIS), datacenter broker (DCB) and the datacenter which encapsulate the host, VM and the cloudlets. The DCB maintains a list of available cloudlets and VM and devised a way to send these cloudlets to the VM for execution. Once the request is being processed, the response is being send back to the DCB where the output is return to the console.

### *3.2 Provisioning Policy*

CloudSim implements SS and TS provisioning policy for scheduling VMs to host and scheduling tasks units to processing core of the VM. They are implemented

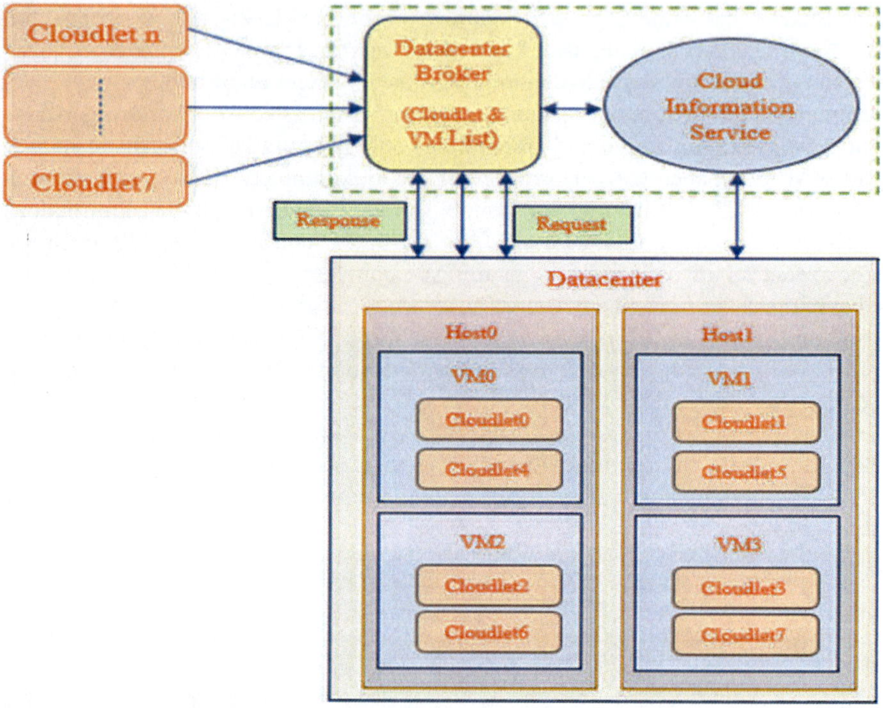


**Fig. 1** CloudSim architecture

by the VM Scheduler at the host level for the virtual machines and the Cloudlet Scheduler at the VM level for the cloudlets or tasks units [10–12]. Both SS and TS provision policy work around the area of how processing cores are assigned to VM or cloudlets. In SS mode, the processing core is wholly assigned to a VM or cloudlet entity hosted in the execution engine without any interruption from any other VM or cloudlet entity. In TS mode, time slices of the processing cores are assigned to a VM or cloudlet which can be multi-task or multi-switched during their life cycle of task execution. Four possible combinations of SS and TS provisioning policy can be projected side by side as shown in Table 1.

**Table 1** Combination of VM's and cloudlet's space-shared and time-shared provisioning policy

| Combination | VM scheduler (host level) | Cloudlet scheduler (VM level) |
|---|---|---|
| 1 | Space Shared (SS) | Space Shared (SS) |
| 2 | Time Shared (TS) | Space Shared (SS) |
| 3 | Space Shared (SS) | Time Shared (TS) |
| 4 | Time Shared (TS) | Time Shared (TS) |

### 3.3 *Effect of SS and TS Provisioning Policy for VMs and Cloudlets*

SS and TS scheduling policy in CloudSim have different effect on cloudlet execution [10, 11]. The effect takes place on the VM first which in turn effects the cloudlet or cloudlets residing on that VM. For example, if a VM is provisioned in the SS mode and cloudlet in TS mode, the processing core will be assigned to the VM first in SS manner, and within that domain, the cloudlet will have the processing core provisioned in a TS mode. Likewise, this effect will be projected in all other combinations. Let us look at this example, a host has two processing cores and hosted two VMs (VM1 and VM2), each of which requires two cores each. The VMs on the other hand, hosted eight tasks units, each of which requires one core each. Cloudlet 1–4 is hosted in VM1 and cloudlets 5–8 in VM2. Figure 2 illustrates the way the processing cores, VMs and cloudlets are distributed with regards to VM's and cloudlet's SS and TS scheduling policy.

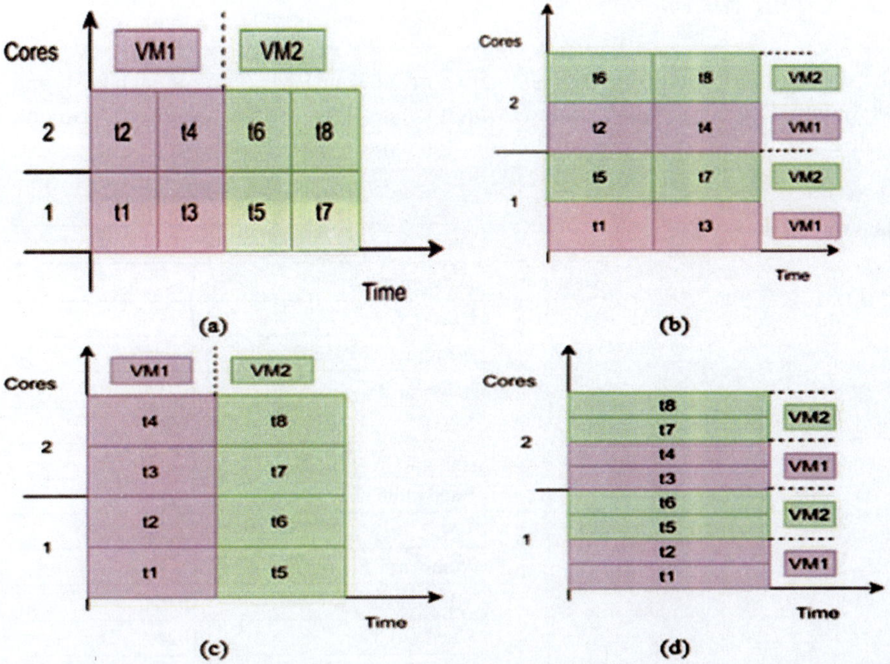


**Fig. 2** Effect of provisioning policy in context of cores, VM and cloudlet distribution when VM's and cloudlet's scheduling policy is: **a** SS-SS, **b** TS-SS, **c** SS-TS, **d** TS-TS

# 4 Objective, Description and Algorithm of the Proposed Work

## *4.1 Objective*

The default scheduling algorithm used by the CloudSim simulator to allocate the cloudlets from the datacenter broker to the VM is based on the FCFS algorithm. In this algorithm, the length of the cloudlets is not considered. Hence, this paper proposed the shortest job first (SJF) algorithm, where the cloudlets are arranged in the ascending order of the length. The main objective of this work is to assess the average response time using FCFS and SJF algorithm while the cloudlets and the VM are projected on all combinations of SS and TS provision policy as specified in Table 1.

## *4.2 Cloud Parameters*

The simulations include one datacenter, two hosts, four VMs (two for each host) and two hundred heterogenous cloudlets which is used as a dataset for the experiment. Table 2 listed all the cloud parameters for the experiment.

**Table 2** Cloud parameters

| Type | Parameters | Value |
|---|---|---|
| Datacenter | Datacenter | 1 |
| Host | Host | 2 |
| | Number of PE on Host | 2 |
| | MIPS of PE | 10,000 |
| | RAM | 4096 |
| | Storage (in mb) | 1,000,000 |
| | Bandwidth (BW) | 10,000 |
| VM | VM | 4 (2 per Host) |
| | Number of PE on VM | 1 |
| | MIPS of PE | 10,000 |
| | RAM | 1024 |
| | BW | 1024 |
| Cloudlet | Number of Cloudlet | 200 |
| | Cloudlet Length | 10,000–12,000 MI |
| | Number of PE requirement | 1 |

### 4.3 Response Time Calculation

The response time is calculated by Eq. 1 [7] as under:

$$\mathrm{RT} = e\mathrm{FT} - \mathrm{AT} + \mathrm{TD} \tag{1}$$

where RT signifies the response time, *e*FT the estimated finish time of request execution, AT the arrival time or submission time [14] of the cloudlet and TD the transmission delay, respectively [7, 13]. Since scheduling algorithms perform in the datacenter broker, the transmission delay can be ignored, that is TD = 0 [7].

When VM and cloudlet implement SS-SS policy and TS-SS policy, the estimated finish time *e*FT of the cloudlet *p* handled by the VM is specified by Eq. 2 [10]:

$$e\mathrm{FT} = e\mathrm{ST}(p) + \frac{\mathrm{tci}}{\mathrm{capacity} \times \mathrm{cores}(p)} \tag{2}$$

where *e*ST(*p*) is the estimated start time of the cloudlet *p*; tci is the total number of cloudlet's instructions; core(*p*) is the number of core or processing elements required by the cloudlet *p* and capacity is the overall processing capacity of a cloud's host [10].

The capacity of the host for the provisioning policy SS-SS and TS-SS with processing elements equal to *np* is given by Eq. 3 [10]:

$$\mathrm{capacity} = \sum_{i=1}^{np} \frac{\mathrm{cap}(i)}{\mathrm{np}} \tag{3}$$

where np signifies the number of core or processing element of host; cap(*i*) is the processing strength of each core elements [10]. As per cloud parameters, np is 2 and cap(*i*) is 10,000.

If the provisioning policy is SS-TS or TS-TS, the estimated finish time and the overall capacity of the host can be determined by the Eqs. 4 and 5 [10] as given below:

$$e\mathrm{FT} = \mathrm{ct} + \frac{\mathrm{tci}}{\mathrm{capacity} \times \mathrm{cores}(p)} \tag{4}$$

where ct is the current simulation time which is extracted by the clock () function of the CloudSim class [14].

$$\mathrm{Capacity} = \frac{\sum_{i=1}^{\mathrm{np}} \mathrm{cap}(i)}{\mathrm{Max}(\sum_{j=1}^{\alpha} \mathrm{cores}(j), \mathrm{np})} \tag{5}$$

where $\alpha$ is the total number of cloudlets that simultaneously multi-task inside the VM of a host [10] and core($i$) is the number of core or processing elements required by the cloudlet $i$.

### 4.4 SJF Algorithm

The SJF algorithm is incorporated in the datacenter broker whose aim is to arrange the cloudlet in the ascending order of its length.

*Step 1: Declare sortList As CloudletArrayList*
*Step 2: Declare tempList As CloudletArrayList*
*Step 3: Repeat Step 3 and 4 for all cloudlet as received from getCloud letList())*
*FOR (Cloudlet cloudlet : getCloudletList())*
*Step 4: Add cloudlet to tempList*
*END LOOP*
*Step 5: totalCloudlet = tempList.size()*
*Step 6: Repeat step 7 - 11 FOR k=0 TO totalCloudlet – 1*
*Step 7: SET smallestCloudlet = tempList.get (k)*
*//Assuming cloudlet at index 0 is the smallest cloudlet*
*Step 8: Repeat step 9-10 FOR (Cloudlet checkCloudlet : tempList)*
*//check cloudlet for each cloudlet instance in the temp list starting from index 1*
*Step 9: IF (smallestCloudletLength>checkCloudletLength)*
*Step 10: SET smallestCloudlet =CheckCloudlet*
*END IF*
*END LOOP*
*Step 11: ADD smallestCloudlet TO sortList*
*Step 12: REMOVE smallestCloudlet FROM tempList*
*END LOOP*

## 5 Simulated Experiments, Analysis and Results

The experimental evaluation and improvements of the SJF over FCFS algorithm on the average RT of a cloud system, using CloudSim simulator, are projected in Figs. 3, 4 and Table 3 respectively.

```
FCFS:Scheduling Cloudlets to VM
-> 1 : Current simulation time : 0.1: Broker: Sending cloudlet 0 Length:10419 to VM #0
-> 2 : Current simulation time : 0.1: Broker: Sending cloudlet 1 Length:10912 to VM #1
-> 3 : Current simulation time : 0.1: Broker: Sending cloudlet 2 Length:11558 to VM #2
-> 4 : Current simulation time : 0.1: Broker: Sending cloudlet 3 Length:10298 to VM #3
-> 5 : Current simulation time : 0.1: Broker: Sending cloudlet 4 Length:11887 to VM #0
-> 6 : Current simulation time : 0.1: Broker: Sending cloudlet 5 Length:11624 to VM #1
-> 7 : Current simulation time : 0.1: Broker: Sending cloudlet 6 Length:11902 to VM #2
-> 8 : Current simulation time : 0.1: Broker: Sending cloudlet 7 Length:10008 to VM #3
-> 9 : Current simulation time : 0.1: Broker: Sending cloudlet 8 Length:10235 to VM #0
-> 10 : Current simulation time : 0.1: Broker: Sending cloudlet 9 Length:11955 to VM #1
-> 11 : Current simulation time : 0.1: Broker: Sending cloudlet 10 Length:11851 to VM #2
-> 12 : Current simulation time : 0.1: Broker: Sending cloudlet 11 Length:11089 to VM #3
```

**Fig. 3** Scheduling cloudlets to VM in FCFS algorithm

```
SJF:Scheduling Cloudlets to VM
-> 1 : Current simulation time : 0.1: Broker: Sending cloudlet:160 Length:10006 to VM #:0
-> 2 : Current simulation time : 0.1: Broker: Sending cloudlet:93 Length:10032 to VM #:1
-> 3 : Current simulation time : 0.1: Broker: Sending cloudlet:58 Length:10044 to VM #:2
-> 4 : Current simulation time : 0.1: Broker: Sending cloudlet:104 Length:10053 to VM #:3
-> 5 : Current simulation time : 0.1: Broker: Sending cloudlet:4 Length:10054 to VM #:0
-> 6 : Current simulation time : 0.1: Broker: Sending cloudlet:40 Length:10056 to VM #:1
-> 7 : Current simulation time : 0.1: Broker: Sending cloudlet:172 Length:10065 to VM #:2
-> 8 : Current simulation time : 0.1: Broker: Sending cloudlet:56 Length:10069 to VM #:3
-> 9 : Current simulation time : 0.1: Broker: Sending cloudlet:193 Length:10073 to VM #:0
-> 10 : Current simulation time : 0.1: Broker: Sending cloudlet:174 Length:10083 to VM #:1
-> 11 : Current simulation time : 0.1: Broker: Sending cloudlet:189 Length:10083 to VM #:2
-> 12 : Current simulation time : 0.1: Broker: Sending cloudlet:67 Length:10096 to VM #:3
-> 13 : Current simulation time : 0.1: Broker: Sending cloudlet:20 Length:10100 to VM #:0
```

**Fig. 4** Scheduling cloudlets to VM in SJF algorithm in ascending order of the length

## *5.1 Scheduling Cloudlets to VM*

See Figs. 3 and 4.

## *5.2 Experiments Analysis and Results*

See Table 3.

**Table 3** Simulation result

| Case | VM provisioning policy | Cloudlet provisioning policy | Scheduling algorithm | Average RT (in secs) |
|---|---|---|---|---|
| Case I | SS | SS | FCFS | 28.61 |
| | | | SJF | 27.32 |
| Difference in the average RT with respect to capacity | | | | 1.29 |
| Case II | TS | SS | FCFS | 28.54 |
| | | | SJF | 27.22 |
| Difference in the average RT with respect to capacity | | | | 1.32 |
| Case III | SS | TS | FCFS | 54.05 |
| | | | SJF | 53.41 |
| Difference in the average RT with respect to capacity | | | | 0.64 |
| Case IV | TS | TS | FCFS | 53.86 |
| | | | SJF | 53.25 |
| Difference in the average RT with respect to capacity | | | | 0.61 |

### *5.3 Observation*

There is an overall improvement in the RT when the length of the cloudlets is considered, that is, when cloudlets are scheduled in SJF algorithm. The improvements can be seen in all cases of SS and TS scheduling policy with an increase of 1.29 s in SS-SS, 1.32 s in TS-SS, 0.64 s in SS-TS, and 0.61 s in TS-TS when the SJF algorithm is implemented as compared to FCFS.

## 6 Conclusion

To meet the demand of end-users and industry, cloud computing offers services such as virtualization of resources, multi-tenancy, system resilient, utility computing to end-users, business and industry. To maintain its stability, lot of research and development work are still going strong in this field. Finding better algorithm to improve response time or processing time, resource utilization or faster execution of the cloud system, to enhance user's experience and quality of service, are the future scope in this field. While looking at this perspective, energy-aware or power-aware computing is also the need of the hour which will transcends us toward green computing which is a great challenge at the moment.

## References


1. Mell P, Grance T (2011) The NIST definition of cloud computing. NIST Special Publication 800-145
2. Cloud Performance Management. https://www.techopedia.com/definition/30563/cloud-performance-management. Last accessed 23 July 2020
3. Agarwal D, Jain S (2014) Efficient optimal algorithm of task scheduling in cloud computing environment. In: arXiv preprint arXiv:1404.2076
4. Mondal A, Choudhury S, Islam S (2015) Performance analysis of VM scheduling algorithm of cloudsim in cloud computing. In: International Journal of Electronics & Communication Technology (IJECT), pp 49–53
5. Deepa T, Cheelu D (2017) A comparative study of static and dynamic load balancing algorithms in cloud computing. In: 2017 International Conference on Energy, Communication, Data Analytics and Soft Computing (ICECDS), pp 3375–3378
6. Roy S, Banerjee S, Chowdhury KR, Biswas U (2017) Development and analysis of a three phase cloudlet allocation algorithm. J King Saud Univ Comput Inf Sci 29:473–483
7. Chien NK, Son NH, Loc HD (2016) Load balancing algorithm based on estimating finish time of services in cloud computing. In: 2016 18th International Conference on Advanced Communication Technology (ICACT), pp 228–233
8. Kumar P, Silambarasan K (2019) Enhancing the performance of healthcare service in IoT and cloud using optimized techniques. IETE J Res 1–10
9. Preethi M, Jayavel K (2018) IOT based visualization of weightage based static task scheduling algorithm in datacentre. Int J Eng Technol (IJET) 439–443
10. Calheiros RN, Ranjan R, Beloglazov A, De Rose CA, Buyya R (2011) CloudSim: a toolkit for modeling and simulation of cloud computing environments and evaluation of resource provisioning algorithms. Softw Practice Exp 41:23–50
11. Buyya R, Ranjan R, Calheiros RN (2009) Modeling and simulation of scalable Cloud computing environments and the CloudSim toolkit: Challenges and opportunities. In: 2009 international conference on high performance computing and simulation, pp 1–11
12. Calheiros RN, Ranjan R, De Rose CA, Buyya R (2009) Cloudsim: a novel framework for modeling and simulation of cloud computing infrastructures and services. In: arXiv preprint arXiv:0903.2525
13. Phi NX, Hung TC (2017) Load balancing algorithm to improve response time on cloud computing. Int J Cloud Comput: Serv Archit 7:1–12
14. CloudSim 3.0 API (Application Programming Interface), The Cloud Computing and Distributed Systems (CLOUDS) Laboratory, The University of Melbourne. https://www.cloudbus.org/cloudsim/doc/api/index.html. Last accessed 1 July 2020
15. Suryateja PS (2016) A comparative analysis of cloud simulators. Int J Modern Educ Comput Sci 8:64
16. The Cloud Computing and Distributed System (CLOUDS) Laboratory, University of Melbourne. https://gridbus.cs.mu.oz.au/. Last accessed 11 Jan 2020
17. CloudSim Tutorial. https://www.cloudsimtutorials.online/. Last accessed 1 Jan 2020

# Design and Development of Time Synchronization System Based on GPS/IRNSS and PTP Protocol

**Shubham Kumar, Abhilasha Barla, G. Selvakumar, and Valluri Sarimela**

**Abstract** For highly stable and accurate time synchronization master system a highly stable and accurate clock is required. A multiple master system ends up in replication of hardware and increase in the cost of the system. Various defence applications, Intelligent Electronic Devices in transmission stations, GPS applications etc needs accurate timing information. An accurate time synchronization server needs a highly stable accurate clock, NMEA time code, 1-PPS (pulse per second) sync signal from GPS/IRNSS receiver and a core processor which processes these two information with high precision. Time information from different sources makes the system reliable but that increases the cost as well. Also, when the satellite signal is lost, different receivers have different time as they all drift with different speed from the reference signal. Thus, there is a need to reduce the cost and at the same time need to have one but highly accurate time server which not only provides stable and correct time but also functions in the same way when the signal from the satellite is lost. In this paper, we first describe our methodology of achieving high accuracy level of PTP grand-master clock and then a method to minimise the drift of the grand-master from the GPS/IRNSS time when the satellite signal is lost ….

**Keywords** Precision time protocol · Rubidium oscillator · IRNSS · GPS · NMEA · 1-PPS

S. Kumar (✉) · A. Barla · G. Selvakumar · V. Sarimela
Central Research Laboratory, Bharat Electronics Limited, Bengaluru, India
e-mail: shubhamkumar@bel.co.in

A. Barla
e-mail: abhilashabarla@bel.co.in

G. Selvakumar
e-mail: selvakumarg@bel.co.in

V. Sarimela
e-mail: vallurisarimela@bel.co.in



A. Maji et al. (eds.), *Proceedings of the International Conference on Computing and Communication Systems*, Lecture Notes in Networks and Systems 170,
https://doi.org/10.1007/978-981-33-4084-8_47

# 1 Introduction

Intelligent Electronic Devices (IED) systems, time distribution systems and various defence applications needs a highly accurate clock for critical requirements. A high precision time reference system is used for computer control system, power load mechanism, electric energy metering, MIS systems and scheduling systems [1]. Practically if one wants to calculate the distance of a location from the satellite, as the speed of light is known, we need only the time taken by the signal from point A to point B to calculate the accurate distance. Thus in defence application the time information becomes very critical. The time information at the time distribution server, needs to be highly accurate so that the clients receives accurate time. Even in absence of satellite signal, the master time server systems should be capable of providing as much accurate time as possible with least amount of drift from the main signal (GPS/IRNSS). In power industry, IEDs uses GPS receivers for accurate time because of easy availability and low cost. The IEDs take NMEA data and 1PPS signal from GPS receivers and there are a number of subsystems installed in a single power station [2, 3]. There are various concerns about the accuracy of GPS receivers and multiple instances of incorrect data from GPS receivers have been reported [4, 5]. Synchronization is severely affected if each subsystem is connected to separate GPS receiver [6]. Also we have observed that GPS receivers from different vendors/companies have different drift from actual 1-PPS. It is very rare that 1PPS from GPS receivers from two different suppliers is synchronous till micro and nano-second level. If at the source end itself there is so much drift then other devices in the system which rely on source, have more drift in time. This not only binds a company to purchase GPS receivers from one vendor, it increases the cost. Then there is a threat for GPS jamming which poses threat for synchronization [7]. High cost of GPS spoofers and multiple installations at each GPS device makes system more costly otherwise makes system more vulnerable [8]. Also ability of the GPS receivers to work in absence of GPS signal (1-PPS) adds another drawback for a complete reliable system.

Figure 1 shows the set-up of the system. A GPS/IRNSS receiver receives the signals from satellite and processes it. It generates the NMEA data, which has satellite information, time information and location information; and 1PPS. These two signals are then given to our developed hardware which then converts the time into seconds, and from 1PPS it generates the nanosecond values. These inputs are given to the PTP master for synchronizing the clocks. The output 1PPS of IRNSS receiver, out time synchronization hardware and the client's hardware are then kept to oscilloscope and the drift between the clocks is observed for accuracy. Most of the clients do not generate 1PPS and the accuracy of the PTP protocol is measured with the industrial grade instruments.

Guo and Crossley [9] shows multiple ways to achieve synchronization in different topologies and with different algorithms under multi GPS receiver system. In this paper, we mainly focus on the accuracy of the master node of the time reference system. We have done the work for both GPS and IRNSS receivers. The paper describes

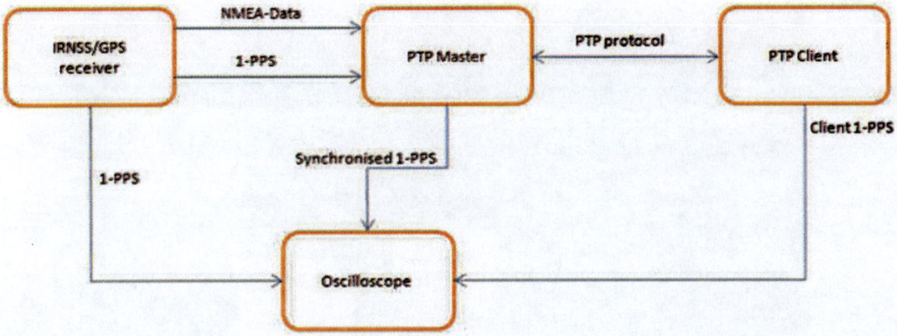


**Fig. 1** A set up of the system

a method to operate during the period when GPS signal is lost or when GPS/IRNSS services are blocked. The drift in the 1-PPS is measured for 48 h in the absence of GPS/IRNSS and accuracy till nanosecond level is obtained. The timing information is distributed to other units using IEEE 1588-2008 standard [10]. The accuracy of the 1PPS signal and PTP protocol timing information is measured using industrial grade measurement devices. Narula and Humphreys [11] proposes requirement of secure clock synchronization and how to secure IEEE 1588 protocol. This paper does not include the development of the IRNSS receiver which is indigenously developed at Bharat Electronics Limited. Many papers suggest that any GPS receiver should not be generating any data if the GPS signal is lost as it may lead to non-synchronization of the IED's in the system.

Section 2 describes a short difference for using PTP over NTP and the packet structure for timing information. Section 3 describes NMEA data format and need of 1PPS signal. Section 4 describes about the system hardware and how accuracy is achieved, Sect. 5 discusses a holdover mode which is used when satellite signal is not present. Section 5 describes some result and accuracy related data. Section 6 derives the conclusion.

## 2 PTP and NTP

For accurate timing, PTP is more reliable than NTP. NTP gives accuracy from milliseconds to sub microsecond level as it uses software time stamping where as PTP provides accuracy upto sub nanosecond level using hardware time stamping. In software time stamping, the time information is calculated in the software (at application level) where as in hardware time stamping mechanism, the moment an Ethernet packet is about to leave the system, the timing information is added in the packet and recalculated CRC is updated in the packet. In NTP the client ask for the time and master replies back with its time where as in PTP the master periodically announces its time and the client gets sync to that time. For timing purpose, the PTP and NTP

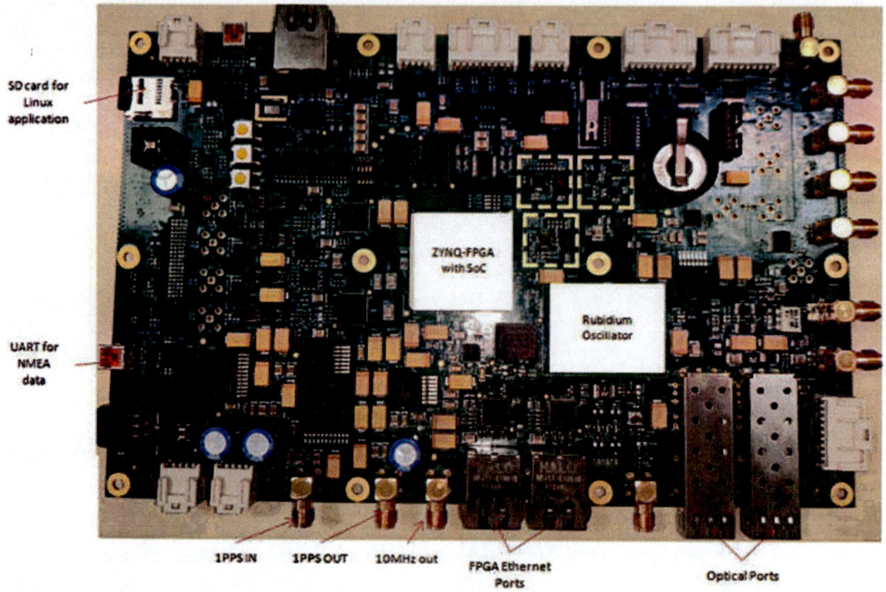


**Fig. 2** Developed hardware

packets both have two time fields one is seconds counter which is of 48 bits for PTP and 32 bit for NTP and another is nanosecond counter which is of 32 bits for both NTP and PTP. For more accuracy we have used PTP protocol for time synchronization.

Figure 2 it is the image of the developed hardware. There are multiple RF ports for different frequency inputs and it can be attached with another card for stacking. The two Ethernet ports are given, in case if one port fails the other port can take over as PTP master and the same for the optical ports. To utilize the most of the FPGA, a support for Linux applications is also given, so that in future other complex applications can be run on Linux.

## 3 NMEA Data

The IRNSS/GPS receiver give two outputs after processing satellite signals, one is the NMEA data [12] which contains the coordinates of the receiver, date, day, year, time and 1PPS (pulse per second) of 10% duty cycle (in general). This time is then converted to total number of seconds from epoch time (1 January 1970) and fed to the seconds field of PTP packet and the nanoseconds is generated from the 1PPS input. A counter in the hardware starts when the 1PPS is detected and the value of that counter than converted to corresponding nanoseconds and updated in the PTP packet just before it leaves from Ethernet port.

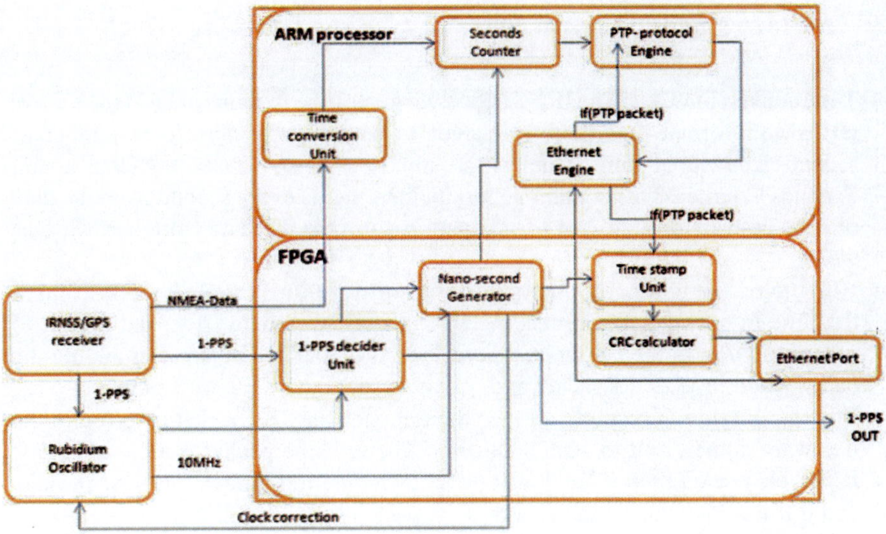


**Fig. 3** A set up of the system

Figure 3 describes the architecture of the hardware. The hardware consist of a Dual- core ARM FPGA SoC (System on Chip).

## *3.1 Hardware Section (FPGA): Coding Language Is VHDL*

- 1-PPS decider Unit: It will take the 1PPS from IRNSS receiver under normal circumstances, but if the system is under threat of jammer then it will take the 1PPS from rubidium oscillator.
- Rubidium Oscillator: It is a programmable oscillator. It has two outputs, one is disciplined 1PPS and another 10 MHz frequency signal. It takes a UART input through which one can modify the frequency output of the Oscillator.
- Nanosecond generator: It takes input as 1PPS and in the hardware it runs at a speed 250 MHz and generate time for nanosecond accuracy.
- Timestamp unit: It takes the nano-second time from hardware timestamp unit, and PTP packet from the arm core (software section) and updates the time in the PTP packet and gives the packet to the CRC calculator unit.
- CRC calculator: for normal Ethernet packets, the CRC calculator is constructed in the software section, but since PTP needs more accuracy, the nanoseconds time is undated in the hardware so CRC for the packet is recalculated and kept into the packet to be sent out from the Ethernet port.

### 3.2 Software Section: Arm System: Coding Language Is C

- Time conversion Unit: The IRNSS receiver gives time in terms of DDMMYYYY HHMMSS format thus this time needs to be converted into total number of seconds from epoch time (which is 1st January 1970) to be used in PTP protocol.
- Seconds Counter: This counter generally increments every second from the data of time conversion unit, but for syncing it can take the data from hardware as well.
- PTP protocol engine: This section of the software runs the PTP stack. It takes the time in seconds, constructs the PTP packet, calculate CRC and give it to timestamp unit in the hardware where CRC is calculated again after adding the nanosecond time in the packet and packet is sent out of Ethernet port.
- Ethernet Engine: it receives all the ethernet packets. The packet received in the hardware is then sent to Arm controller. There, if the packet is a normal ARP, ICMP etc packet then it handles it on its own creating replies for those packets and if it is a PTP packet then it gives the packet to PTP engine for processing.

## 4 Rubidium Oscillator Versus Normal Oscillator

A normal crystal oscillator used in the development was of 50 MHz with accuracy to 50 ppm. From mathematics point, this is very bad crystal for timing accuracy related hardware systems. 50 ppm means for every 1 million clock pulses there can be error of ±50 clock cycles. Thus at 50 MHz frequency, in 1 s there are 50 Mega clock cycles and every 1 million cycles can have error of 50 clock cycles that means in 1 s the error can reach upto $50 \times 50 = 2500$ clock cycles and at 50 MHz, 2500 clock cycles mean an error of 50 μs. This is the error which is mentioned by the manufacturer, which is bound to happen. Now if we accept this error at the master time server which is there to distribute time to other subsystems, then some additional error will be added to it by receivers and if they have a poor oscillator than master, the error introduced will be more. Therefore in our hardware we have introduced a highly stable oscillator, rubidium oscillator which is programmable and has much higher stability of 1ppb. This oscillator 10 MHz as output frequency and also its frequency can be changed through commands from UART port. Now, once a code is written in VHDL, the execution happens at one cycle of the clock of that process. To detect the 1-PPS rising edge, 1 clock cycle is needed. if we run the process at 10 MHz (as that of the rubidium oscillator, then to detect the 1-PPS it will take 0.1 μs, that means a delay of 0.1 μs is already introduced at the master end in just detection the pulse. Therefore we need a higher clock for 1PPS detection. Because of the software restrictions from Xilinx, we converted 10 MHz frequency input to 250 MHz. Now the detection of 1PPS will take 4 ns. Now, as soon as the 1PPS is detected by the FPGA, a counter 250 MHz will start running. Whenever the next rising edge of 1PPS arrives, the counter sets to 0 and again starts incrementing at every clock pulse of 250 MHz.

Whenever the PTP packets leaves the hardware, the corresponding nanosecond value is updated in the packet and packet is sent. In the code, we keep a separate count of 10 MHz pulses from the rubidium oscillator. The two rising edges of 1PPS signal from IRNSS receiver have an exact time separation of 1 s. That means, in every 1 s 10 M pulses 10 MHz oscillator should come. But there are chances that some extra or less number of clock pulses may be generated by the oscillator. For that purpose, we keep monitoring the counts 10 MHz rubidium oscillator, and depending on the counts generated by it, we send a command to rubidium oscillator to increase or decrease its frequency. In this way we ensure the 10 MHz frequency from the programmable rubidium oscillator, which is not possible with normal oscillators. This way, even if after some time if the frequency of the oscillator deviates due to any reason, it can be corrected. With numerous experimental trails, we found a suitable time of 400 s after which we send the command to oscillator and till 400 s, we monitor the frequency of the oscillator. Liu et al. [13] shows GPS correction methods and achieving nanosecond level accuracy based on L1 observations.

## 5 Algorithm for Clock Disciplining

An algorithm is developed for the hardware to work if IRNSS signal or GPS signal is not available. Then in that case, we take the past counts of 10 MHz rubidium oscillator and its number of counts for every second. Those values then used to make the algorithm learn about the frequency deviation of the rubidium oscillator. A continuous 20 min exposure to GPS signal is required for the algorithm to learn and function properly. As no oscillator is completely stable, the frequency of the oscillator drifts over time. That is why an algorithm is needed to keep track of the frequency drifts and then make corrections. After that even if the GPS signal is lost, the algorithm generates the 1PPS pulse from 10 MHz frequency output of Rubidium oscillator. And with 10 MHz counts, any change in the frequency of the Rubidium oscillator is made by UART command. The system provides a holdover mechanism as well, which has a very essential use, that is, if somebody is trying to spoof the GPS signal, as soon as this is detected, one can forcefully send the system to holdover mode in which it stops taking data from the GPS receiver and the 1PPS is generated from the output 10 MHz clock of Rubidium Oscillator. And with this input the time of the system is updated at seconds and nanoseconds level. This algorithm is tested for 48 h and has given accuracy of less than 20 ns drift from the original 1PPS pulse from IRNSS/GPS receiver.

Figure 4 shows the delay between the 1PPS from GPS receiver and the 1PPS pulse generated from the hardware by using a normal crystal oscillator. The delay in pulses is about 32 ns. That means with normal crystal the synchronization capacity of the system is 32 ns. Because normal crystal cannot generate stable clock and secondly, its frequency can't be controlled. Also in the absence of GPS signal, output from a normal crystal will slowly drift and that drift can't be corrected as well.

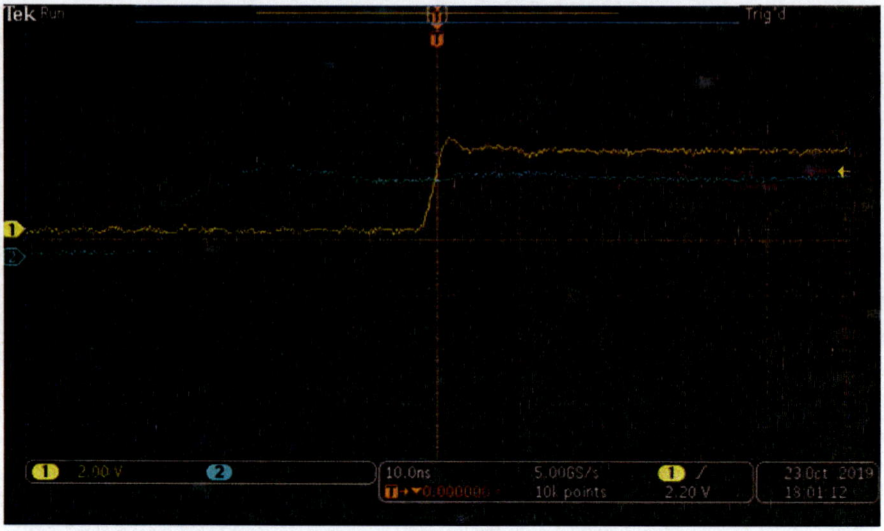


Fig. 4 1PPS from IRNSS receiver and synchronised 1PPS from hardware

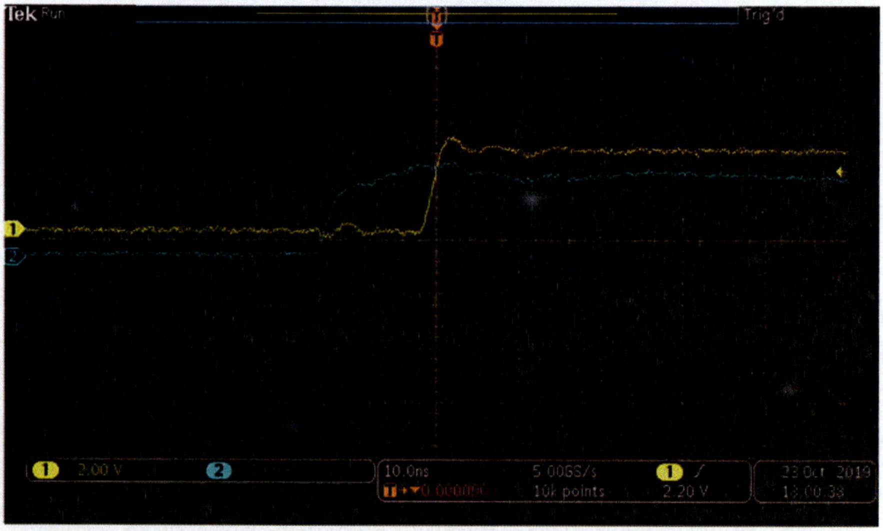


Fig. 5 1PPS from IRNSS receiver and synchronised 1PPS from hardware

Figure 5 shows the delay between the 1PPS from IRNSS receiver and the 1PPS pulse generated from the hardware by using a Rubidium Oscillator. The generated 1PPS is about 12 ns delayed from the IRNSS generated 1PPS pulse. This shows the importance of the rubidium oscillator, its output clock is not only highly stable but also it can be corrected if it drifts. Also in the absence if GPS signal, if the output 1PPS drifts by certain time, then that drift can be detected and can be corrected also.

## 6 Conclusion

This paper describes a high accurate stable timing system and a method for time synchronization even if GPS signal is not present. When the GPS signal comes back, the system aligns itself to the signal. This system not only gives accurate time but protection against GPS signal lost and GPS spoofing. It reduces the use of multiple GPS receivers at different substations as it uses the PTP protocol to synchronise the clients which provides accuracy up to nanoseconds level. Another important development can be the optimization of the algorithm and use of high frequency clock to reduce the errors.

## References

1. Southern E (1998) GPS synchronized current differential protection. Ph.D. dissertation. School of EEE, UMIST
2. Ingram DME, Schaub P, Campbell DA (2012) Use of precision time protocol to synchronize sampled-value process bus. IEEE Trans Instrum Meas 61(5):1173–1180 May
3. De Dominicis CM, Ferrari P, Flammini A, Rinaldi S, Quarantelli M (2011) On the use of IEEE 1588 in existing IEC 61850-based SASs: current behavior and future challenges. IEEE Trans Instrum Meas 60(9):3070–3081
4. An W, Nick T, Barron D, Bingham M, Hackett A (2012) A transmission utility's experience to date with feeder unit protection systems. In: Proceedings of the 11th international conference on developments in power systems protection (DPSP 2012). Birmingham, UK, 23–26 Apr 2012, pp 1–6
5. Xu C, Xiong H, He L, Li Z, Yang J (2013) The research of intelligent substation time synchronization system and the influence of its fault to relay protection. Energy Power Eng 5(4B):468–473
6. Baumgartner B, Riesch C, Schenk W (2014) GPS receiver vulnerabilities urban legends or sad, hard truth? In: Proceedings of the PAC world Americas conference 2014. Raleigh, USA, 22–25 Sept 2014, pp 1–20
7. Curry C (2014) SENTINEL project—report on GNSS vulnerabilities. Chronos Technology, UK [Online]. Available: www.chronos.co.uk/files/pdfs/gps/SENTINEL_Project_Report.pdf
8. Shepard DP, Humphreys TE, Fansler AA (2012) Evaluation of the vulnerability of phasor measurement units to GPS spoofing attacks. Int J Crit Infrastruct Protect 5(3–4):146–153
9. Guo H, Crossley P (2016) Design of a time synchronization system based on GPS and IEEE 1588 for transmission substations. IEEE Trans Power Deliv 32(4):2091–2100
10. IEEE standard for a precision clock synchronization protocol for networked measurement and control systems. IEEE Std 1588-2008, 24 Jul 2008

11. Narula L, Humphreys T (2018) Requirements for secure clock synchronization. IEEE J Sel Top Sig Process 12(4):749–762
12. https://www.sparkfun.com/datasheets/GPS/NMEA%20Reference%20Manual1.pdf
13. Liu W, Yuan H, Ge J (2019) Local-area nanosecond-accuracy time synchronisation based on GPS L1 observations. IET Radar Sonar Navig 13(5):824–829

# A Compact Dual-Band Reconfigurable Filtering Antenna Using PIN Diode

**Sangeeta Das and Pankaj Sarkar**

**Abstract** In this paper, a compact dual-band filtering antenna is presented. Initially a dual-band reconfigurable filter using integrated stepped impedance resonators (SIRs) is proposed to achieve the switching characteristics. The projected filter topology is implemented using four open-ended half-wavelength SIR to operate at 3.2 and 5.4 GHz. The PIN diodes are attached to the edges of the inner resonators which are used as switch ON–OFF elements. Finally, a dual-band monopole antenna is proposed using triangular radiating patch. Two different dimensions patch are constructed to radiate at 3.2 and 5.4 GHz. The filter and the antenna structures are cascaded together to realize the filtering antenna. The PIN diodes are biased properly to switch the radiation characteristics at the different operating bands. The composite structure is simulated and agreed well with the predicted response.

**Keywords** Filtering antenna · Reconfigurable filter · Stepped impedance resonators · PIN diode · Monopole antenna · Dual-band response

## 1 Introduction

With the growth of wireless communication in recent decade, the demand to realize antennas that are acute to tune the operating frequency in according to the requirement is increased rapidly. Moreover, using multiple-dedicated antennas to cater different wireless services that are sprinkled over wide frequency range increases the weight and the cost of a system. Furthermore, a significant amount of isolation among the antennas is very difficult to achieve. Reconfigurable filtering antennas are promising contenders for the forthcoming RF front end to minimize the number of antennas required for a particular system [1–3]. For the design of reconfigurable antennas,

S. Das · P. Sarkar (✉)
Electronics and Communication Engineering Department, School of Technology, North Eastern Hill University, Shillong 793022, India
e-mail: pankajsarkar111@gmail.com

S. Das
e-mail: sangdas273@gmail.com



A. Maji et al. (eds.), *Proceedings of the International Conference on Computing and Communication Systems*, Lecture Notes in Networks and Systems 170,
https://doi.org/10.1007/978-981-33-4084-8_48

the antenna designers are faced tremendous challenge mainly to achieve a good gain, stable radiation, and a good impedance match for different operating bands [4]. Electrically reconfigurable antennas use radio frequency microelectromechanical system (RF MEMS), PIN diodes, or varactors used as switching elements to incorporate reconfigurability and tunability in the antenna functionality. The turn on and off mechanism of these switching elements requires the assimilation of biasing lines in the radiating plane of the antenna. The interference of biasing lines and the radiating elements generates an undesirable resonances in the operating band of the antenna, and it also destroys the radiation pattern from the design requirements if the biasing lines are not designed properly [5, 6]. Considering the limitations, several researchers proposed optical switches to solve the limitations produced in the electrically switchable reconfigurable antennas [7]. Consequently, considering the present demand, wireless communication has observed exemplary growth in wireless local area network (WLAN), worldwide interoperability for microwave access (WiMAX), mobile phone, Bluetooth, global positioning system (GPS), etc. It is in the best interest and focus of designers and users if both antenna and BPF are implemented as a single compact module to achieve both filtering and radiation properties. Reconfigurable filter is a key component in reconfigurable filtering antenna design. Various reconfigurable filters are reported in [8, 9]. The integrated module of filter and antenna is referred to as filter antenna or filtering antenna [10]. In [11], a frequency agile monopole antenna using a reconfigurable stub coupled multiple-mode resonator (MMR) bandpass filter is reported. A circularly polarized reconfigurable filtering antenna is proposed in [12]. However, several drawbacks can still be observed in their designs, including an uncontrollable bandwidth (BW), too many switching elements, large circuit size, high return loss, spurious suppressions, not flexible for multiband operations, low gain, etc.

In this paper, a dual-band reconfigurable filtering antenna is offered. A new class of novel reconfigurable dual-band filter is presented consisting of two outer SIRs and two inner SIRs. Four different SIRs are assembled to achieve passband at 3.2 GHz and at 5.4 GHz. The reconfiguration from band pass to band stop filter is designed using PIN Diode switch ON–OFF topology. It enables easy switching from dual-band to single-band configurations used for WLAN and WiMAX operation. An antenna is implemented which is cascaded with the filter structure. Two triangular patches are utilized to radiate at WLAN and WiMAX band. Depending on the bias voltage given to the PIN diode, the proposed filtering antenna can be easily switched from dual-band to single-band configuration. The structure is implemented on FR4 substrate of height 1.0 mm. For the purpose of EM simulation, CST Microwave Studio is used.

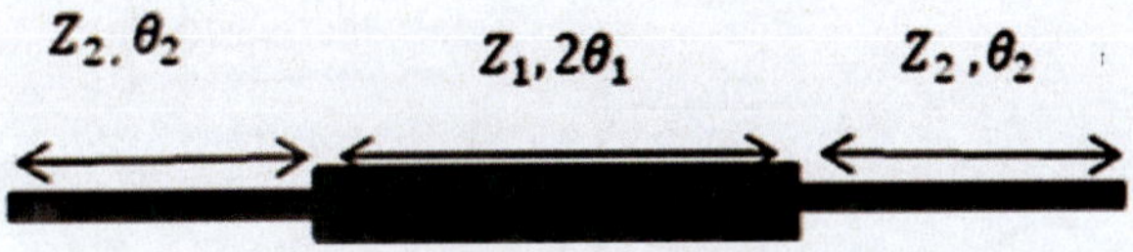


**Fig. 1** Structure of the stepped impedance resonators (SIRs)

# 2 Filter Design

## 2.1 *Resonance Properties of SIR*

The proposed filter is designed by using four half-wavelength open-ended SIRs. The configuration of a basic half-wavelength open-ended SIR is shown in Fig. 1.

The input admittance of the SIR is given below for the practical application when the length is the minimum for $\theta_1 = \theta_2 = \theta$ in [13].

$$Y_{\text{in}} = jY_2 \frac{2(1 + R_Z)\tan\theta(R_Z - \tan^2\theta)}{R_Z(1 + 2\tan^2\theta + \tan^4\theta) - 2(1 + R_Z^2)\tan^2\theta} \tag{1}$$

In Eq. (1) $R_z$ is the impedance ratio $Z_2/Z_1$. The corresponding solutions for fundamental resonance frequency ($f_0$) and first spurious resonance frequency ($f_1$) are given in Eq. (2) obtained by taking $Y_{\text{in}} = 0$.

$$\theta = \tan^{-1}\sqrt{R_Z} \tag{2}$$

The relation between the fundamental frequencies ($f_0$) and the first spurious frequency ($f_1$) is computed by Eq. (3). Rz is a pivotal parameter of a SIR that is responsible to control the position of spurious frequencies.

$$\frac{f_1}{f_0} = \frac{\pi}{2\tan^{-1}\sqrt{R_Z}} \tag{3}$$

## 2.2 *Reconfigurable Filter Design*

The proposed filter is designed using four $\frac{\lambda_g}{2}$ openended SIRs. The outer two SIRs, the Resonator 1 and the Resonator 2, are identical. Both the resonators are realized to operate at fundamental frequency ($f_{01}$) of 3.2 GHz and first spurious ($f_{02}$) of 5.4 GHz. To achieve the desired resonance, the impedance ratio of the SIR is calculated as 2.25 with the high and low impedance section values $Z_1 = 112\ \Omega$ and $Z_2 = 50\ \Omega$, respectively. Resonator 3 is realized to operate at the fundamental frequency ($f_{03}$) of 3.2 GHz. The impedance ratio is kept 0.62 to shift the first resonance ($f_{04}$) at 7.5 GHz. Resonator 4 is constructed to operate fundamental resonance of ($f_{05}$) 5.2 GHz and

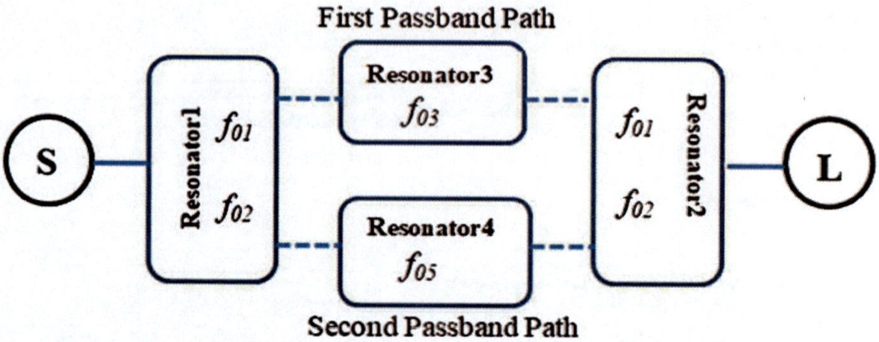


**Fig. 2** Coupling scheme for the proposed filter

spurious ($f_{06}$) at 12.5 GHz with an impedance ratio 0.60. Both the inner resonators are designed such a way that the spurious frequencies will not be coupled to the outer identical resonators. To comprehend the working mechanism, the coupling scheme is introduced in Fig. 2. It can be observed that the first passband is generated with the help of fundamental resonating frequency of resonator 1, resonator 2, and resonator 3. The second passband is implemented with the help of first spurious of resonator 1 and 2 and the fundamental frequency of resonator 4. The first spurious frequency of the resonator 3 and 4 are kept at higher side to accomplish wide upper stopband characteristics. The proposed filter is realized on a FR-4 substrate of relative dielectric constant 4.4 with loss tangent 0.015 and height 1 mm. For the purpose of EM simulation, Ansoft designer is used.

The layout of the proposed prototype is shown in Fig. 3. In order to switch on/off the first and the second passband, the PIN diodes are connected at the open end of the resonator 3 and resonator 4 to disperse the resonant frequency. The SMP1340-079LF pin diodes are connected at the open edge of resonator 3 and resonator 4 with the DC biasing circuits. The DC bias voltage is applied using a RFC of inductance 27nH. Two additional small microstriplines are embedded at the end of the PIN diode to construct vias. For the purpose of EM-simulations, the PIN diode is modeled with a series resistor of 2 Ω in ON state and a series capacitor of 0.3 pF in OFF state.

### 2.3 *Simulated Results for the Proposed Filter*

Reconfigurable operation is performed by changing biasing voltage across the PIN diode. The bias voltage $V_{dc}$ of 5 V is applied to turn on the PIN diode. The bias voltage must be removed to switch off the PIN diode. Four different characteristics can be obtained from the proposed filter. When both the PIN diodes are turned off the dual-band characteristics can be accomplished from the proposed filter. The frequency response of the filter is shown in Fig. 4a when $D_1$ and $D_2$ are off. The

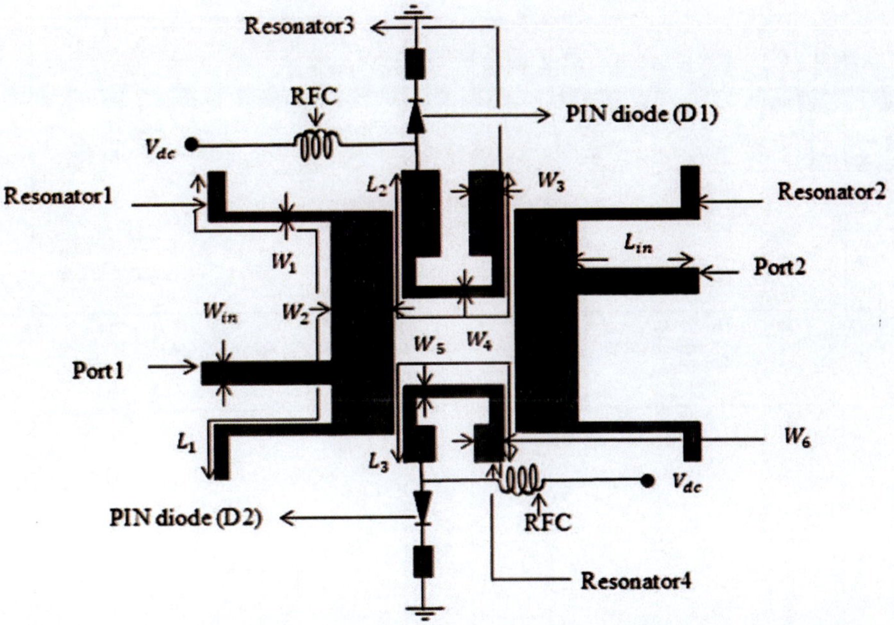


**Fig. 3** Layout of the dual-band reconfigurable bandpass filter with $L_1 = 32.4$, $L_2 = 22$, $L_3 = 11$, $L_{in} = 5$, Win = 1.8, $W_1 = 0.2$, $W_2 = 1.8$, $W_3 = 1.1$, $W_4 = 0.3$, $W_5 = 0.3$, $W_6 = 1.2$ (*all dimensions are in mm*)

passbands are centered at 3.25 GHz and 5.35 GHz with FBW 4.5 and 5.2%, respectively. The selectivity is satisfactory due to the present of transmission zeroes near to the passbands. Due to multipath coupling, the transmission zeroes are created. The insertion loss for both the bands is below 0.5 dB which signifies very low loss transmission. To accomplish the switching operation, the bias voltages are applied to the PIN diodes. To switch off the first passband, the bias voltage of 5 V is applied to the PIN diode $D_1$. The Resonator 3 will be connected to the ground as $D_1$ is on. As one of the edges of the resonator 3 is grounded, the resonating frequency of the resonator will changed to a different frequency. So the signal transmission for the frequency of 3.25 GHz is mitigated as illustrated in Fig. 4b. A transmission zero is observed at 3.25 GHz which stops the first passband with an attenuation level of 18.0 dB. The second passband is intact as the $D_2$ is off. So by turning on $D_1$, the proposed filter can be used as a single passband filter of passband resonating frequency at 5.35 GHz. Similarly when $D_1$ is off and $D_2$ is on, the proposed filter acts like a single passband filter with resonating frequency centered at 3.25 GHz. The characteristic is illustrated in Fig. 4c. It can be seen that the second passband is collapsed with an attenuation level of 17 dB. Finally, when both $D_1$ and $D_2$ are on, the both passbands are switched off with excellent attenuation level. The frequency response is plotted in Fig. 4d. So in this configuration, the filter can be utilized as an all stop filter.

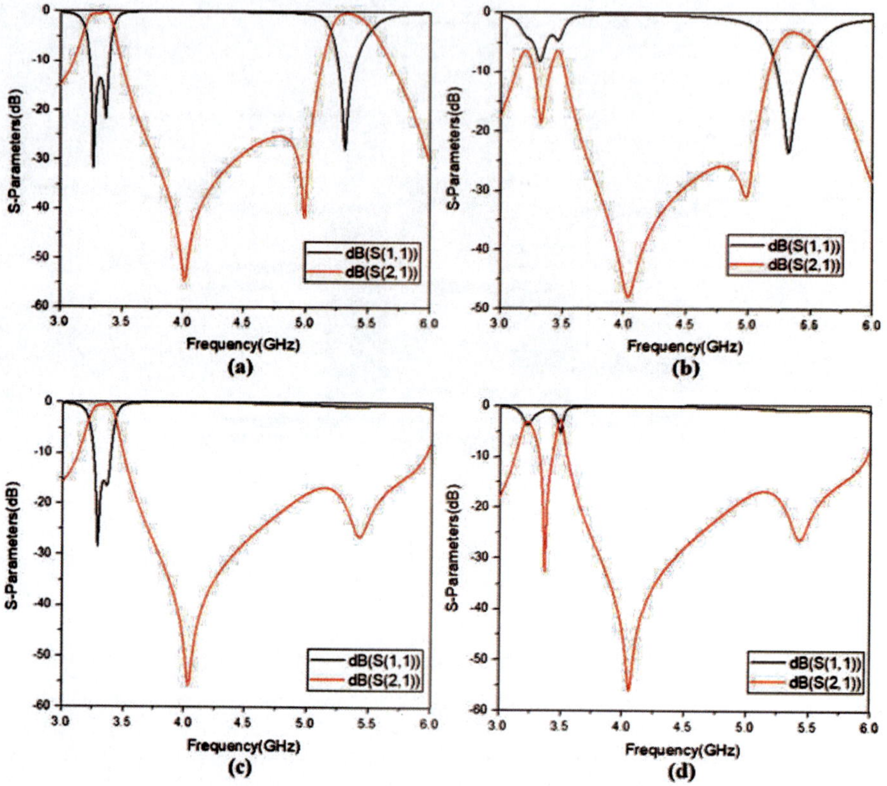


**Fig. 4** EM Simulated results of the suggested filter **a** when $D_1$ and $D_2$ are off, **b** when $D_1$ is on and $D_2$ is off. **c** when $D_1$ is off and $D_2$ is on. **d** when both $D_1$ and $D_2$ are on

## 3 Antenna Design

Figure 5 shows the proposed triangular monopole patch antenna. The resonant frequency can be calculated from [14]. The triangular patch has the side lengths of $L_1$ and $L_2$. The antenna is designed on the same FR4 substrate of height 1.0 mm. Two triangular patches are designed to resonate at 3.25 and 5.35 GHz. The small-size triangular patch is loaded over the large triangular patch. The radiating frequency can be controlled individually by changing the dimension of the patch structure. The radiating structure is excited by microstrip transmission line. The ground plane is tapered for proper impedance matching. The length $L_5$ is parametrically studied to obtain the best result. The proposed antenna structure is Em simulated and the S11 parameter is illustrated in Fig. 6a. It can be seen that the proposed antenna can be used as a dual-band antenna. The first band is centered at 3.25 GHz with fractional bandwidth (FBW) of 27%. The second band is little narrow compared to the first band. The center frequency of the second band is 5.4 GHz with radiating FBW of 5%.

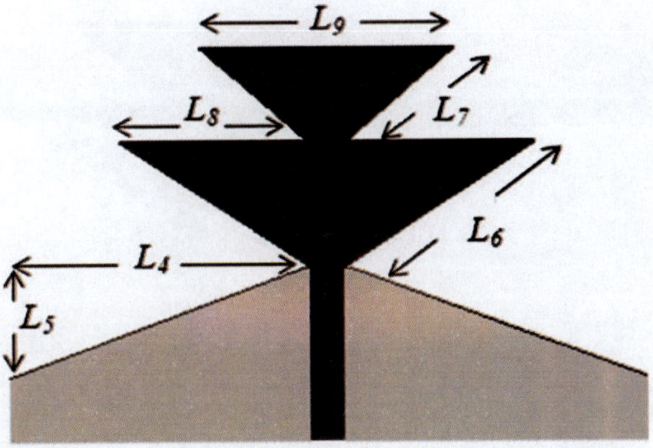


**Fig. 5** Proposed antenna structure with $L_4 = 20$, $L_5 = 7$, $L_6 = 14$, $L_7 = 9$, $L_8 = 11.8$, $L_9 = 16$ (*all dimensions are in mm*)

Figure 6b shows the gain of the antenna for different frequency. The variation of gain is from 1.9 to 5.0 dBi. For the first radiating band, the gain is around 3.0 dBi, whereas for second radiating band, the value is 5.0 dBi. Finally, the antenna structure is cascaded with the reconfigurable filter structure as shown in Fig. 7. The composite structure will provide the reconfigurable filtenna characteristics. The filter structure is embedded near to the antenna to minimize the space requirement. The PIN diodes which are connected to the inner resonator of the filter are biased externally through a RFC of inductance 27 nH. Another terminal of both the diodes $D_1$ and $D_2$ are connected to the metallic via.

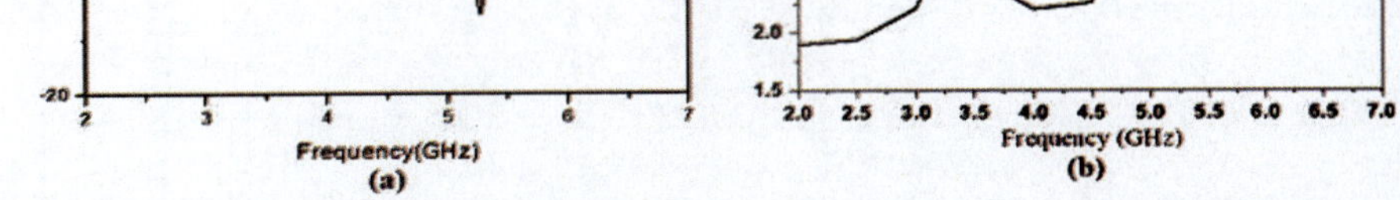


**Fig. 6** **a** The EM-Simulated response of the proposed Antenna. **b** Gain of the proposed antenna for different frequency

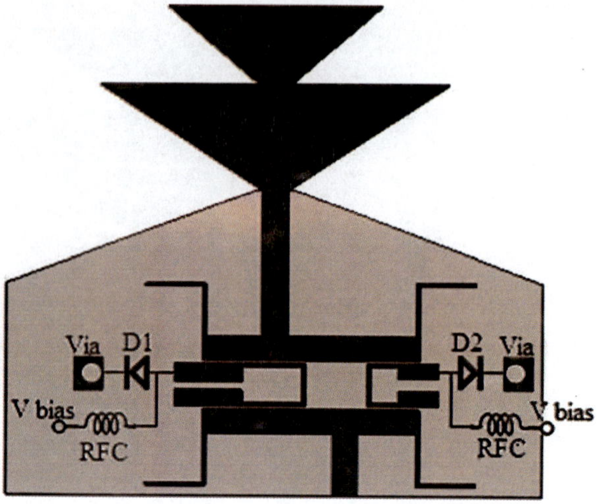


**Fig. 7** Layout of the proposed reconfigurable filtering antenna structure

## 4 Results and Discussion

The composite structure of the filter and the antenna is EM simulated to determine the frequency response of the proposed filtenna. The S11 parameter is plotted in Fig. 8. It is observed that the switching characteristics as discussed earlier are preserved. The return loss characteristics is excellent for different switching condition of the diode $D_1$ and $D_2$. By properly switching the PIN diodes, the antenna can be operated to no radiation, two different single-band and the dual-band states.

From the WiMAX and WLAN point of view, the antenna must have an omnidirectional radiation pattern. To determine the radio characteristics, the proposed filtenna structure is simulated at the center frequency of the operating band. Figure 9 depicts the radiation pattern at 3.25 and 5.35 GHz for the two principle planes of $\phi = 0^0$

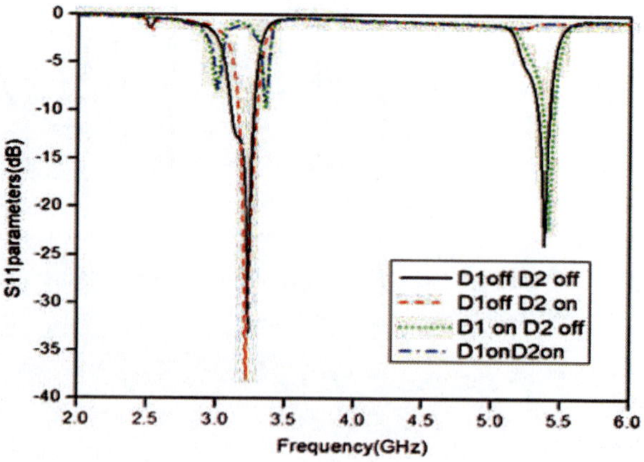


**Fig. 8** S11 parameters of the proposed filtenna for different switching condition of the diodes $D_1$ and $D_2$

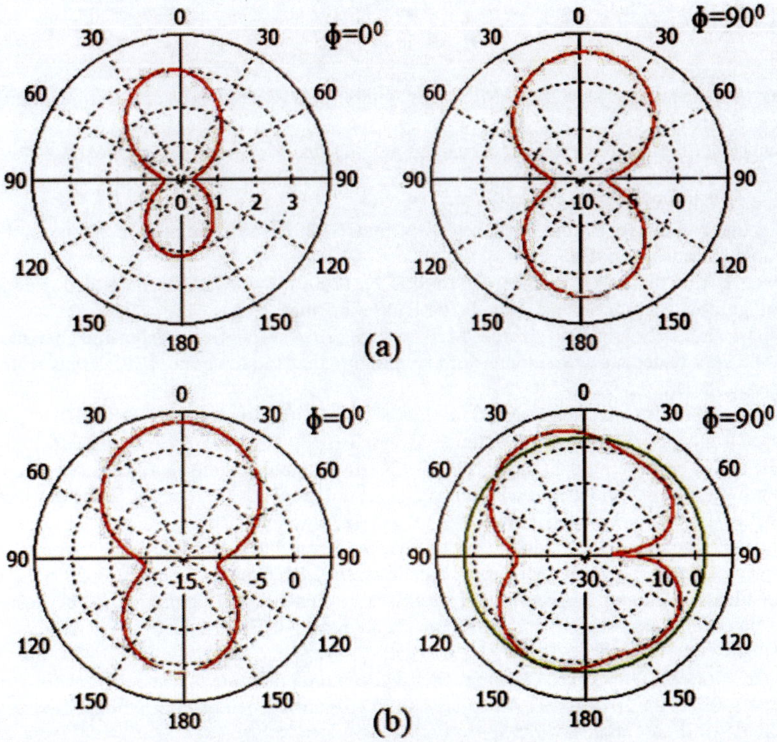


**Fig. 9** Radiation pattern of the filtenna for **a** 3.25 GHz, **b** 5.35 GHz

and $\phi = 90^0$. It can be perceived that an omnidirectional radiation is obtained with simulated gain at 3.25 GHz is 3.5 dBi and at 5.35 GHz is 5.08 dBi. The variation of the gain is very less over the corresponding bands.

## 5 Conclusion

This paper presents a compact dual-band reconfigurable filtering antenna using PIN diodes. Two triangular patches are introduced to design the antenna structure where the dimension of the individual patches is bringing the flexibility on the radiating frequency selection. The operating band of the proposed filtenna can be reconfigured to dual band, single band, and all stop characteristics. The passband and the stopband characteristics are excellent for all the configurations. Presence of multiple transmission zeroes at the non-radiating bands improves the selectivity of the filtenna structure. The size of the structure is also very compact which is around 48 × 40 mm$^2$. Thus, an easily switchable filtenna using two PIN diodes is obtained which can be used for modern multiband wireless communication systems.

## References


1. Hong JSG, Lancaster MJ (2004) Microstrip filters for RF/microwave applications, 2nd edn. Wiley, Hoboken
2. Balanis CA (2016) Antenna theory: analysis and design, 4th edn. Wiley, Hoboken
3. Tawk Y, Costantine J, Christodoulou C (2012) A varactor-based reconfigurable filtenna. IEEE Antennas Wirel Propag Lett 11:716–719
4. Costantine J, Tawk Y, Barbin SE, Christodoulou CG (2015) Reconfigurable antennas: design and applications. Proc IEEE 103(3):424–437
5. Cetiner BA, Crusats GR, Jofre L, Biyikli N (2009) RF MEMS integrated frequency reconfigurable annular slot antenna. IEEE Trans Antennas Propag 58(3):626–632
6. Perruisseau Carrier J, PardoCarrera P, Miskovsky P (2010) Modeling, design and characterization of a very wideband slot antenna with reconfigurable band rejection. IEEE Trans Antennas Propagat 58(7):2218–2226
7. Tawk Y, Albrecht AR, Hemmady S, Balakrishnan G, Christodoulou CG (2010) Optically pumped frequency reconfigurable antenna design. IEEE Antennas Wirel Propag Lett 9:280–283
8. Naglich EJ, Lee J, Peroulis D, Chappell WJ (2010) A tunable bandpass-to-bandstop reconfigurable filter with independent bandwidths and tunable response shape. In: IEEE transactions on microwave theory and techniques, vol 58, No 12, December 2010
9. Kingsly S, Kanagasabai M, Mohammed GNA, Subbaraj S (2018) Multi-band reconfigurable microwave filter using dual concentric resonators. Int J RF Comput-Aided Eng 28:e21290
10. Abunjaileh AI, Hunter IC, Kemp AH (2008) A circuit-theoretic approach to the design of quadruple-mode broadband microstrip patch antennas. IEEE Trans Microw Theory Tech 56(4):896–900
11. Qin PY, Guo YJ, Wei F (2014) Frequency agile monopole antenna using a reconfigurable bandpass filter. In: 2014 IEEE Antennas and Propagation Society International Symposium (APSURSI). IEEE, pp 1250–1251
12. Wen Z, Tang MC, Ziolkowski RW (2019) Band-and frequency-reconfigurable circularly polarised filtenna for cognitive radio applications. IET Microwaves Antennas Propag 13(7):1003–1008
13. Makimoto M, Yamashita S (1980) Bandpass filters using parallel coupled stripline stepped impedance resonators. IEEE Trans Microw Theor Tech 28(12):1413–1417
14. Kumprasert N, Kiranon W (1994) Simple and accurate formula for the resonant frequency of the equilateral triangular microstrip patch antenna. IEEE Trans Antennas Propag 42(8):1178–1179

# Varactor-Loaded Tunable SIW Filter with Wide Upper Stopband Characteristic

**Md. Atiqur Rahman and Pankaj Sarkar**

**Abstract** In this paper, varactor-loaded tunable substrate integrated waveguide (SIW)-based filter is presented. Initially, a SIW cavity is designed at higher frequency range. The stepped-impedance low-pass filter slots are embedded as a slot on the top of the cavity structure in order to suppress the higher order modes of the SIW cavity. The passband is centered 5.25 GHz to pass the WLAN band with excellent selectivity. The stopband characteristic of the proposed filter is satisfactory due to absence of higher order modes. The proposed structure is EM-simulated to verify the theoretical analysis.

**Keywords** Substrate integrated waveguide · Tunable passband · Cavity structure · Embedded low-pass filter · Higher order harmonics · Wide stopband

## 1 Introduction

Recently, the demand for low-cost, mass-producible, and high-performance subsystems is very high for modern microwave and millimeter-wave applications. Dealing with the radio coexistence of different high frequency bands and strong coupling between different subsystems basically brings tremendous challenge to the circuit designer. This concept is very crucial when designing the circuits such as resonators, highly selective filters, and voltage-controlled oscillators (VCOs) [1–3]. For high-performance passive circuits, conventional waveguide technology was a suitable choice [4]. But the heavy and bulky structures of the waveguide-based circuits are not suitable to integrate in active circuits. By the development of substrate integrated waveguide (SIW) technology, the problems associated with the conventional waveguide had been eliminated. The demand for SIW resonators has increased for

Md. A. Rahman · P. Sarkar (✉)
Department of Electronics and Communication Engineering, NEHU, Shillong 793022, India
e-mail: pankajsarkar111@gmail.com

Md. A. Rahman
e-mail: atiqurece@gmail.com



A. Maji et al. (eds.), *Proceedings of the International Conference on Computing and Communication Systems*, Lecture Notes in Networks and Systems 170,
https://doi.org/10.1007/978-981-33-4084-8_49

its low-cost, low loss, high-quality ($Q$) factor, compatible with planar printed circuit board (PCB) process [5–7]. Compared to the conventional microstrip resonators, the SIW resonators possess higher unloaded $Q$-factor (usually several hundred) [5] which facilitates high-selective filter design and low-phase noise VCO design.

As the demand for dynamic spectrum management increases, therefore the need for RF/microwave devices that can work in various bands is very high. One of the important parts in designing such systems that can cover more bands are microwave filters. Different solutions exist for such filters, such as filter banks, multiband filters, and tunable/reconfigurable filters. But filter banks and multiband filters occupy very large area which increases the weight of the system. Isolation and crosstalk problems are also associates with the filter banks. These limitations are overcomes by using tunable/reconfigurable filters. SIW technology is a good candidate for designing such filters due to its high-power handling, high linearity, narrow band, and highly selectivity. Various tuning methods have been investigated in [8–10]. However, in most of the methods, the problem is either with tuning range or low $Q$ for wider tuning range. In [8], a SIW cavity-backed antenna oscillator is proposed where a varactor diode is coupled for tuning. In [9], a metalized via post is introduced on top of the cavity resonator, where a varactor diode is loaded with the help of floating metal. The reactive loading effect changes by changing the bias voltage of the varactor diode and thereby the resonance frequency shifts. A mechanically controlled switching/tuning device has been proposed in [10], where a via post is connected to a screw, and the screw is connected to a flap. The screw is isolated by using ring slot from top wall. By changing the location of the flap at different angles $E$-field distribution varies, as a result the resonance frequency of the resonator changes.

In this paper, a varactor-loaded tunable narrow-band filter is implemented using SIW cavity and stepped-impedance, L-C ladder-type low-pass filter. Initially, a SIW cavity is constructed 5.3 GHz to miniaturize the cavity structure. To suppress the higher order harmonics, a low-pass filter using stepped-impedance resonator is implemented with cutoff frequency of 8 GHz. The low-pass filter is etched on the top of the cavity structure to restrict the higher order spurious up to 9 GHz. A varactor diode is then loaded to the proposed structure to achieve tuning of passband center frequency 4.6–5.3 GHz. The structure is designed on FR4 substrate of height 1.00 mm. For the purpose of EM simulation, Ansys HFSS is used.

## 2 Filter Design

**SIW cavity design**: The layout of the proposed SIW cavity is shown in Fig. 1. The cavity is implemented on FR 4 substrate of height 1 mm. The resonant frequency of the fundamental ($TE_{101}$) mode of the SIW cavity with length $L_{SIW}$ and width $W_{SIW}$ can be found by using

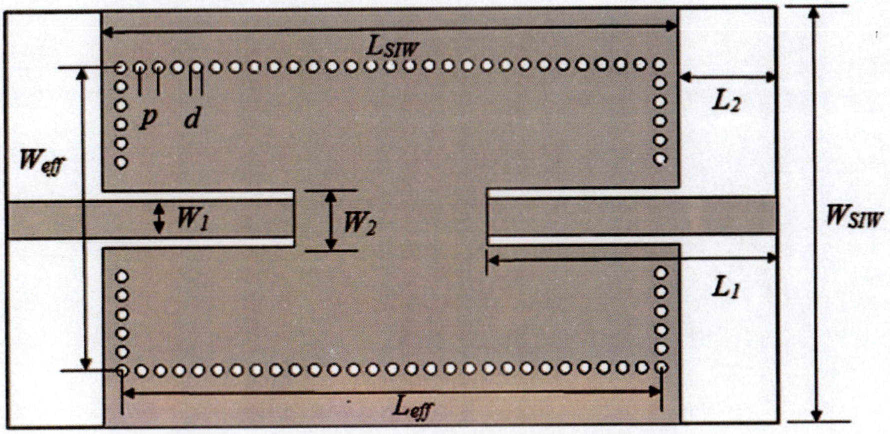


**Fig. 1** Layout of the SIW cavity with $L_{\text{eff}} = 30$, $L_{\text{SIW}} = 28$, $L_1 = 15$, $L_2 = 5$, $W_1 = 1.9$, $W_2 = 3$, $W_{\text{eff}} = 16$, $W_{\text{SIW}} = 14$, $d = 0.6$, $p = 1$ (all dimensions are in mm)

$$f_{101} = \frac{c}{\sqrt{2\ \epsilon_r\ \mu_r}}\sqrt{\left(\frac{1}{L_{\text{eff}}}\right)^2 + \left(\frac{1}{W_{\text{eff}}}\right)^2} \tag{1}$$

where $c$ is the speed of light in a vacuum and $\epsilon_r$ and $\mu_r$ are the relative permittivity and permeability of the substrate used for SIW cavity. The length and width of the SIW cavity can be found by using

$$W_{\text{eff}} = W_{\text{SIW}} - \frac{d^2}{0.95\ p} \tag{2}$$

$$L_{\text{eff}} = L_{\text{SIW}} - \frac{d^2}{0.95\ p} \tag{3}$$

where $W_{\text{eff}}$ and $L_{\text{eff}}$ are the effective width and length of the rectangular waveguide and $W_{\text{SIW}}$ and $L_{\text{SIW}}$ width and the length of the proposed SIW cavity, $d$ is the diameter of the via holes, and $p$ is the pitch between adjacent via holes.

The $S$ parameter response of the proposed SIW cavity is as shown in Fig. 2. It can be seen that the cavity operates at fundamental frequency of 5.3 GHz. The spurious bands which are present 7.1 and 9.1 GHz are affecting the upper stopband performance of the SIW cavity. For suppressing the higher order modes, a low-pass filter is designed to pass only the fundamental mode and to suppress the higher order modes. The layout of the proposed stepped-impedance low-pass microstrip filters is shown in Fig. 3. The structure is inherently a cascaded configuration of alternating high- and low-impedance transmission section. The high-impedance lines act as a series inductors, whereas the low-impedance lines act as a shunt capacitors. $Z_L$ and $Z_H$ denote the characteristic impedances of the low- and high-impedance lines,

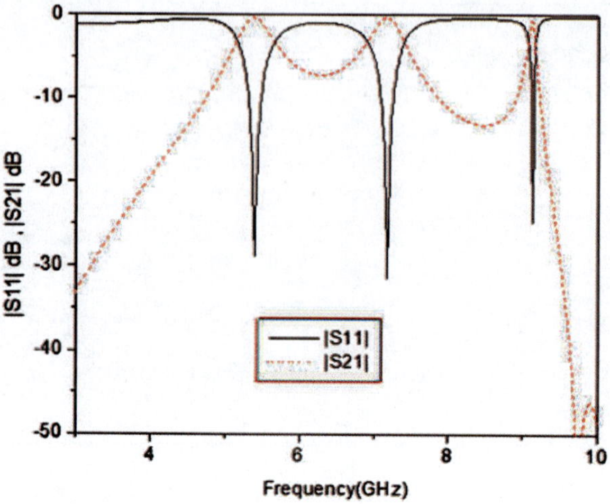


**Fig. 2** Simulated *S* parameter of the proposed SIW cavity

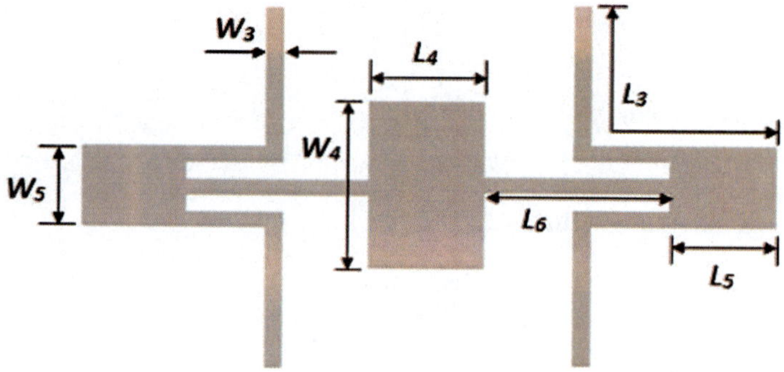


**Fig. 3** Layout of the stepped-impedance low-pass filter with $L_3 = 7.5$, $L_4 = 2$, $L_5 = 2.5$, $L_6 = 2$, $W_3 = 0.25$, $W_4 = 3.4$, $W_5 = 1.9$ (all dimensions are in mm)

respectively and $Z_0$ is the source impedance. The impedance value of high- and low-impedance section is chosen in such a way that $Z_L < Z_0 < Z_H$. The element values of the low-pass prototype filter are selected first which are usually normalized to make a source impedance $g_0 = 1$ and a cutoff frequency $\Omega_c = 1.0$. The elements values are further transferred to the L-C elements for the desired cutoff frequency and the desired source impedance. The proposed filter is designed on FR4 substrate of dielectric constant $\epsilon_r = 4.4$ and height of 1 mm. The order of the filter ($N$) is 3, and the cutoff frequency($f_c$) is 8 GHz. The high impedance ($Z_H$) and low impedance ($Z_L$) are taken as 100 Ω and 30 Ω, respectively. For proposed design work, to get maximally flat response in passband, we assume Butterworth approximation. The elemental values for ladder type circuits are $g_1 = 1.0$, $g_2 = 2.0$, $g_3 = 1.0$, $g_4 = 1.0$.

The value of inductors $L_i$ and capacitors $C_i$ is calculated from Eqs. (4) and (5).

$$L_i = \frac{Z_0 g_i}{2\pi f_c} \tag{4}$$

$$C_i = \frac{g_i}{Z_0 2\pi f_c} \tag{5}$$

The electrical length of the inductor and capacitor is calculated from Eqs. (6) and (7).

$$\beta l = \frac{L Z_0}{Z_H} \tag{6}$$

$$\beta l = \frac{C Z_L}{Z_0} \tag{7}$$

The simulated *S*-parameters are illustrated in Fig. 4. It can be seen that the proposed filter generates a passband up to 8 GHz. Throughout the passband, the $S11$ parameter is below −12.0 dB, and the $S21$ parameter is better than −0.2 dB. In order to suppress the spurious frequencies, the stepped-impedance low-pass filter is introduced at the top of the SIW cavity structure. The low-pass filter is etched on the top surface of the cavity. The layout of the proposed narrow-band filter is as shown in Fig. 5. This proposed filter basically allows the fundamental frequency of 5.3 GHz. The higher order modes of the cavity which are situated 7.1 and 9.1 GHz are restricted due to the presence of the embedded low-pass filter.

The simulated *S*-parameter of the proposed bandpass filter is shown in Fig. 6. It can be observed that the proposed structure generates a high selective response with center frequency of 5.2 GHz. It has the bandwidth of 100 MHz. Throughout the

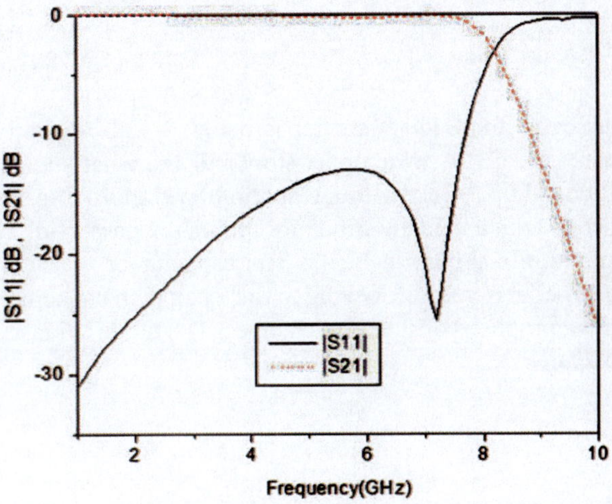


**Fig. 4** Simulated *S*-parameters of stepped-impedance low-pass filter

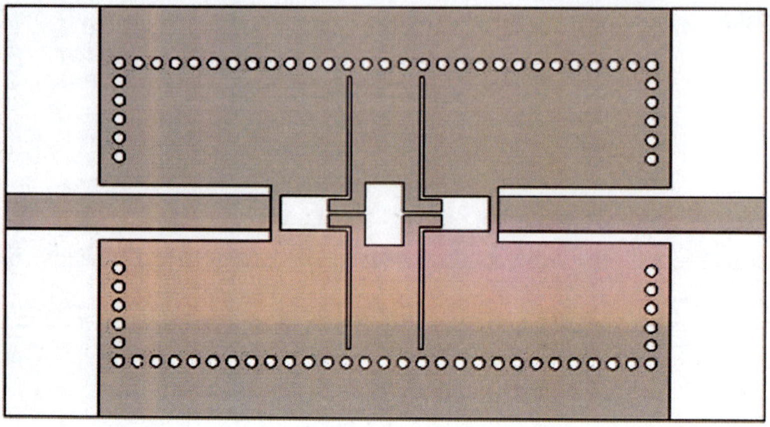

**Fig. 5** Layout of the proposed cavity with LPF etched at the top surface

**Fig. 6** Simulated *S*-parameter of the proposed cavity filter

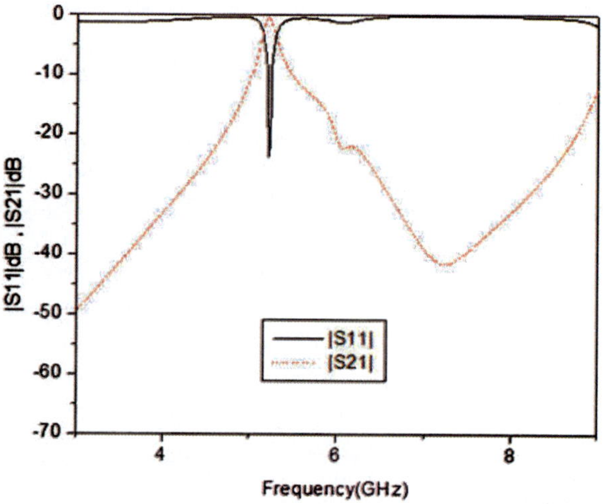


passband, the $S11$ parameter is below $-15.0$ dB, and the $S21$ parameter is better than $-0.2$ dB. A wide upper stopband characteristic can also be observed which is extended up 9 GHz with attenuation level of 15 dB. Now, the proposed structure is loaded with a varactor diode for tuning the passband. The proposed varactor-loaded structure is shown in Fig. 7. The capacitance of the varactor diode changes with external bias voltage. Owing to this change in capacitance, the resonating frequency of the proposed design also changes. The diode is biased externally with an inductor of 27 nH.

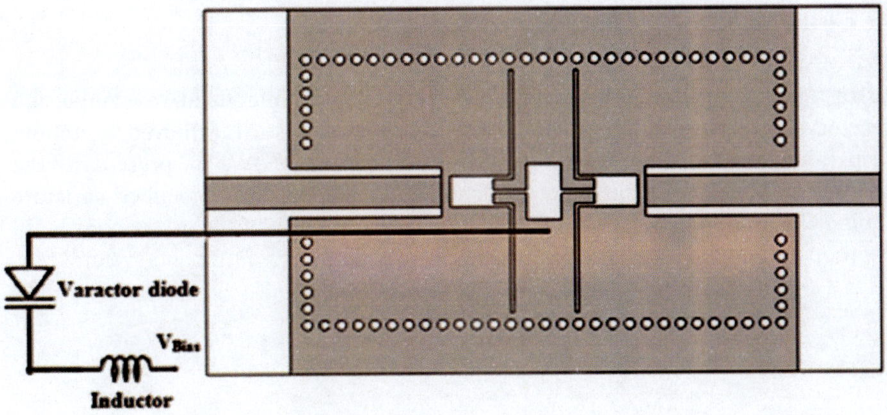


**Fig. 7** Layout of the proposed tunable SIW filter

## 3 Results and Discussion

The simulated *S*-parameter is illustrated in Fig. 8. It can be seen that the proposed filter generates a passband centered 5.2 GHz with no bias condition. Throughout the passband, the *S*11 parameter is below −12.0 dB, and the S21 parameter is better than −0.3 dB. The bias voltage of the varactor diode is varied to tune the center frequency of the filter. It can be observed that by increasing the reverse bias voltage, the passband center frequency can be 4.86–5.2 GHz. The proposed structure is tuned over a bandwidth of 340 MHz. The overall size of the proposed prototype is 40 × 28 $mm^2$.

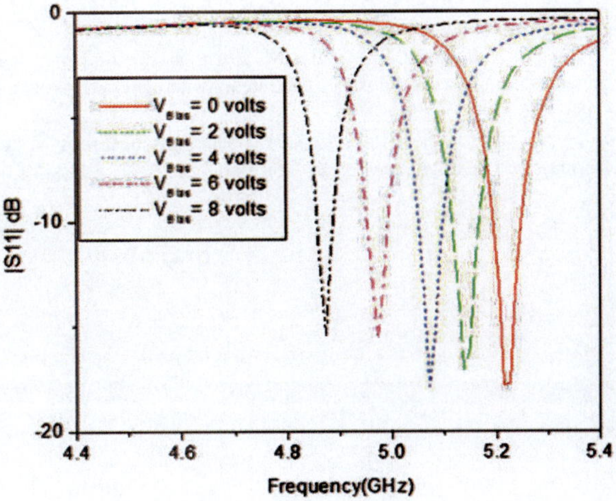


**Fig. 8** The *S*11 parameter of the proposed filter for different bias voltages

## 4 Conclusion

A varactor-loaded tunable filter is designed using substrate integrated waveguide, and varactor diode is presented here. Upper stopband characteristic is achieved by etching stepped-impedance low-pass filter on the top surface of the SIW. The presence of the low-pass filter improves the upper stopband performance. The embedded structure eliminates the additional space requirement. The selectivity of the passband is high. The proposed bandpass filter can be utilized for the modern multiband systems.

## References


1. Deslandes D, Wu K (2003) Single-substrate integration technique of planar circuits and waveguide filters. IEEE Trans Microw Theory Tech 51(2):593–596
2. Parameswaran A, Raghavan S (2017) Novel SIW dual mode band pass filter with high skirt selectivity. In: 2017 2nd international conference for convergence in technology (I2CT). IEEE 2017, pp 189–191
3. Almalkawi M, Westrick M, Devabhaktuni V, Alam M, Zhu L, Deng J (2011) Design of a dual-band dual-mode substrate integrated waveguide filter with symmetric transmission zeros. In: IEEE applied electromagnetics conference (AEMC), pp 1–3
4. Bozzi M, Georgiadis A, Wu K (2011) Review of substrate-integrated waveguide circuits and antennas. IET Microw Antennas Propag 5(8):909–920
5. Cassivi Y, Perregrini L, Wu K, Conciauro G, Low-cost and high-*q* millimeter-wave resonator using substrate integrated waveguide technique. In: 2002 32nd European microwave conference. IEEE 2002, pp 1–4
6. Cassivi Y, Wu K (2003) Low cost microwave oscillator using substrate integrated waveguide cavity. IEEE Microw Wirel Compon Lett 13(2):48–50
7. Georgiadis A, Via S, Collado A, Mira F (2009) Push-push oscillator design based on a substrate integrated waveguide (SIW) resonator. In: 2009 European microwave conference (EuMC). IEEE 2009, pp 1231–1234
8. Giuppi F, Georgiadis A, Collado A, Bozzi M, Perregrini L (2010) Tunable SIW cavity backed active antenna oscillator. Electron Lett 46(15):1053–1055
9. Sirci S, Martinez J, Taroncher M, Boria V (2011) Varactor-loaded continuously tunable SIW resonator for reconfigurable filter design. In: 2011 41st European microwave conference. IEEE, pp 436–439
10. Mira F, Mateu J, Collado C (2012) Mechanical tuning of substrate integrated waveguide resonators. IEEE Microw Wirel Compon Lett 22(9):447–449

# ASER of RQAM with MRC Receiver over Correlated Rayleigh Fading Channels

**Rajkishur Mudoi**

**Abstract** In this work, the average symbol error rate (ASER) of rectangular quadrature amplitude modulation (RQAM) technique has been investigated under Rayleigh fading channels, considering exponential correlation among the input antennas of a maximal ratio combining (MRC) diversity reception. Exponential correlation is noticed by the antennas placed in a linear arrangement at the receiver. The response of the ASER for various levels of diversity order, correlation coefficient as well as the quadrature to in-phase decision distance ratio has been observed.

**Keywords** ASER · MRC · RQAM · Rayleigh fading

## 1 Introduction

Quadrature amplitude modulation (QAM) is a very useful modulation technique in terms of higher spectral efficiency of wireless communications. QAM scheme can be classified as square QAM (SQAM), rectangular QAM (RQAM), and cross QAM (XQAM). RQAM is a common technique, since it has SQAM, QPSK, orthogonal BFSK, BPSK, and multilevel amplitude shift keying modulation techniques as certain examples [1]. Rayleigh fading is effective in highly built up urban environments. Rayleigh fading is relevant while there is no propagation of signals along a line of sight, between the base station (BS) and receiver. Rayleigh fading is a suitable distribution, when there are various objects in the surroundings to scatter the radio signals, prior to it appears at the receiver.

The performance of RQAM scheme for different fading environments has been reported in many studies. In [2], an ASER expression is developed for cooperative diversity systems, considering RQAM scheme under Rayleigh fading channels. The ASER expression is derived for RQAM technique using Nakagami-*m* fading

R. Mudoi (✉)
Department of Electronics and Communication Engineering, North-Eastern Hill University, Shillong 793022, India
e-mail: rajkishur@gmail.com



A. Maji et al. (eds.), *Proceedings of the International Conference on Computing and Communication Systems*, Lecture Notes in Networks and Systems 170,
https://doi.org/10.1007/978-981-33-4084-8_50

conditions in [3, 4]. ASER analysis of the RQAM is expressed for an MRC diversity reception under $L$ independent and identically distributed, Nakagami-$m$ fading conditions in [5]. In [6], the authors present the performance analysis of MRC reception over non-identically distributed $\eta - \mu$ fading environment. The ASEP, outage probability and the average channel capacity for MPSK, $M$-ary DPSK and RQAM modulation schemes are derived. MGF approach is utilized to find the expressions. The ASER of MRC receiver in $\eta - \mu$ and $k - \mu$ fading channels for QAM scheme is stated in [7]. In [8], the ASER of RQAM with XQAM schemes is analyzed using MGF in TWDP fading environment. ASER performance of coherent, DE-QPSK and $\pi/4$-QPSK schemes is also analyzed. ASER analysis of QAM schemes for cooperative AF dual-hop relaying system under Rayleigh fading channels is proposed in [9]. The expressions are derived using MGF-based approach with the MRC diversity scheme. Performance of a dual-hop decode-and-forward relaying system using QAM is explored in [10] under mixed $\eta - \mu$ and $k - \mu$ fading channels. The ASER expressions are derived for RQAM and XQAM schemes using MGF-based approach. ASER expressions for RQAM and XQAM under $k - \mu$ fading are presented in [11] by using two system models SISO and MIMO. In this paper, the ASER expression for an MRC receiver with RQAM scheme over exponentially correlated Rayleigh fading channels is presented. The PDF-based approach is implemented to obtain the expression of ASER.

## 2 Channel Model Description

The channel between the BS and mobile receiver is assumed to have Rayleigh fading distribution. An MRC receiver is considered to improve the quality of the downlink signals. In case of MRC receiver, the signals from all diversity antennas on the receiver are co-phased, intensified by a weight factor proportional to branch SNR and added jointly to enhance the output SNR. For an MRC receiver, instantaneous output SNR $\gamma$ is found by [12]

$$\gamma = \sum_{l=1}^{L} \gamma_l. \tag{1}$$

where $\gamma_l$ is the instantaneous SNR of $l$th branch and $L$ is the entire number of diversity antennas in the MRC receiver. The receiving antennas are positioned in a linear arrangement due to space constraints, and hence, the exponential correlation is observed among the received signals. The correlation coefficient between the $x$th and the $y$th received antennas is expressed as [12, 13]

$$\rho_{xy} = \rho^{|x-y|}; \quad x, y = 1, 2, \ldots, L. \tag{2}$$

Performing RV transformation, PDF of the output SNR, in case of an MRC diversity reception, can be shown as [13]

$$f_{\gamma}(\gamma) = \frac{1}{\Gamma\left(\frac{L^2}{r}\right)\left(\frac{r\overline{\gamma}}{L}\right)^{\frac{L^2}{r}}}\gamma^{\frac{L^2}{r}-1}\exp\left(\frac{-L\gamma}{r\overline{\gamma}}\right), \tag{3}$$

where $\overline{\gamma}$ is the average SNR in each input branch of MRC receiver, $r = L + \frac{2\rho}{1-\rho}\left(L - \frac{1-\rho^L}{1-\rho}\right)$, $\rho$ is the correlation among the input antennas of MRC receivers, and $\Gamma(.)$ is the Gamma function.

## 3 ASER Analysis

The ASER $P_e$ can be expressed as

$$P_e = \int_0^{\infty} P_e(e|\gamma) f_{\gamma}(\gamma)\mathrm{d}\gamma. \tag{4}$$

where $P_e(e|\gamma)$ is the conditional error probability (CEP) for AWGN channels and it can be specified for $M$-ary RQAM as [14]

$$P_e(e|\gamma) = 2\omega Q\left(h\sqrt{\gamma}\right) + 2\xi Q\left(k\sqrt{\gamma}\right) - 4\omega\xi Q\left(h\sqrt{\gamma}\right)Q\left(k\sqrt{\gamma}\right). \tag{5}$$

where $\omega = 1 - \left(\frac{1}{M_I}\right)$; $\xi = 1 - \left(\frac{1}{M_Q}\right)$; $M = M_I \times M_Q$; $Q(.)$ is Gaussian $Q$-function; $\beta = \frac{d_Q}{d_I}$; and $h = \sqrt{\frac{6}{(M_I^2-1)+(M_Q^2-1)\beta^2}}$; $k = \beta h$; $d_Q$ and $d_I$ are the quadrature and in-phase decision distance, respectively.

Using exponential functions, approximation of Gaussian $Q$-function, i.e., $Q(\delta) \approx \frac{1}{12}e^{-\frac{\delta^2}{2}} + \frac{1}{4}e^{-\frac{2\delta^2}{3}}$ [15, 16] in (5), $p(e|\gamma)$, is rewritten as

$$\begin{aligned} P_e(e|\gamma) = {} & 2\omega\left[\frac{1}{12}\mathrm{e}^{\frac{-h^2\gamma}{2}} + \frac{1}{4}\mathrm{e}^{\frac{-2h^2\gamma}{3}}\right] + 2\xi\left[\frac{1}{12}\mathrm{e}^{\frac{-k^2\gamma}{2}} + \frac{1}{4}\mathrm{e}^{\frac{-2k^2\gamma}{3}}\right] \\ & - 4\omega\xi\left[\frac{1}{144}\mathrm{e}^{-\left(\frac{h^2+k^2}{2}\right)\gamma} + \frac{1}{48}\mathrm{e}^{-\left(\frac{h^2}{2}+\frac{2k^2}{3}\right)\gamma}\right. \\ & \left. + \frac{1}{48}\mathrm{e}^{-\left(\frac{2h^2}{3}+\frac{k^2}{2}\right)\gamma} + \frac{1}{16}\mathrm{e}^{-\left(\frac{2h^2+2k^2}{3}\right)\gamma}\right]. \end{aligned} \tag{6}$$

Substituting Eqs. (3) and (6) into (4), we have

$$P_e = \frac{2}{\Gamma\left(\frac{L^2}{r}\right)\left(\frac{r\overline{\gamma}}{L}\right)^{\frac{L^2}{r}}} \left\{ \underbrace{\int_0^\infty \omega\gamma^{\frac{L^2}{r}-1} \exp\left(\frac{-L\gamma}{r\overline{\gamma}}\right)\left[\frac{1}{12}e^{\frac{-h^2\gamma}{2}} + \frac{1}{4}e^{\frac{-2h^2\gamma}{3}}\right] d\gamma}_{\Delta_1} \right.$$
$$+ \underbrace{\int_0^\infty \xi\gamma^{\frac{L^2}{r}-1} \exp\left(\frac{-L\gamma}{r\overline{\gamma}}\right)\left[\frac{1}{12}e^{\frac{-k^2\gamma}{2}} + \frac{1}{4}e^{\frac{-2k^2\gamma}{3}}\right] d\gamma}_{\Delta_2}$$
$$\left. -2\left[\underbrace{\int_0^\infty \omega\xi\gamma^{\frac{L^2}{r}-1} \exp\left(\frac{-L\gamma}{r\overline{\gamma}}\right)\left[\frac{1}{144}e^{-\left(\frac{h^2+k^2}{2}\right)\gamma} + \frac{1}{48}e^{-\left(\frac{h^2}{2}+\frac{2k^2}{3}\right)\gamma} + \frac{1}{48}e^{-\left(\frac{2h^2}{3}+\frac{k^2}{2}\right)\gamma} + \frac{1}{16}e^{-\left(\frac{2h^2+2k^2}{3}\right)\gamma}\right] d\gamma}_{\Delta_3}\right]\right\} \tag{7}$$

The ASER expression in (7) has three integral terms and putting $\Phi = \frac{2}{\Gamma\left(\frac{L^2}{r}\right)\left(\frac{r\overline{\gamma}}{L}\right)^{\frac{L^2}{r}}}$, it can be shown as

$$p_e = \sigma_1 + \sigma_2 - \sigma_3. \tag{8}$$

where $\sigma_1 = \Phi\Delta_1$; $\sigma_2 = \Phi\Delta_2$; $\sigma_3 = \Phi\Delta_3$ and consequently,

$$\sigma_1 = \frac{2\omega}{\Gamma\left(\frac{L^2}{r}\right)\left(\frac{r\overline{\gamma}}{L}\right)^{\frac{L^2}{r}}} \int_0^\infty \gamma^{\frac{L^2}{r}-1} \exp\left(\frac{-L\gamma}{r\overline{\gamma}}\right)\left[\frac{1}{12}e^{\frac{-h^2\gamma}{2}} + \frac{1}{4}e^{\frac{-2h^2\gamma}{3}}\right] d\gamma. \tag{9}$$

Applying [17, (3.381.4)] and after simplification, one can express (9) as,

$$\sigma_1 = \omega\left[\frac{1}{6}\left(1 + \frac{h^2 r\overline{\gamma}}{2L}\right)^{-\frac{L^2}{r}} + \frac{1}{2}\left(1 + \frac{2h^2 r\overline{\gamma}}{3L}\right)^{-\frac{L^2}{r}}\right]. \tag{10}$$

Similarly, the derivations for $\sigma_2$ and $\sigma_3$ can be shown as

$$\sigma_2 = \xi\left[\frac{1}{6}\left(1 + \frac{k^2 r\overline{\gamma}}{2L}\right)^{-\frac{L^2}{r}} + \frac{1}{2}\left(1 + \frac{2k^2 r\overline{\gamma}}{3L}\right)^{-\frac{L^2}{r}}\right], \tag{11}$$

and

$$\sigma_3 = \omega\xi\left[\frac{1}{36}\left(1 + \frac{r\overline{\gamma}}{L}\left(\frac{h^2+k^2}{2}\right)\right)^{-\frac{L^2}{r}} + \frac{1}{12}\left(1 + \frac{r\overline{\gamma}}{L}\left(\frac{h^2}{2} + \frac{2k^2}{3}\right)\right)^{-\frac{L^2}{r}}\right.$$

$$+\frac{1}{12}\left(1+\frac{r\overline{\gamma}}{L}\left(\frac{2h^2}{3}+\frac{k^2}{2}\right)\right)^{-\frac{L^2}{r}}+\frac{1}{4}\left(1+\frac{r\overline{\gamma}}{L}\left(\frac{2h^2+2k^2}{3}\right)\right)^{-\frac{L^2}{r}}\Bigg]. \quad (12)$$

## 4 Results and Discussions

The mathematical expression of ASER derived in Sect. 3 is numerically calculated. The results are arranged for different standards of $L$, i.e., the diversity order, correlation coefficient $\rho$, and decision distance ratio $\beta$. In Fig. 1, the ASER versus average SNR per branch is presented for $4 \times 2$ RQAM including $8 \times 4$ RQAM schemes with different $L$ values. In this figure, $\rho = 0.5$ and $\beta = 1$ are kept constant. It is seen from Fig. 1 that with the addition in the values of receiving antennas $L$ of the MRC receiver the ASER performance improves. In Fig. 2, the ASER performance versus average SNR is shown for certain values of $\beta = 1$, $L = 4$ and varying values of the correlation coefficient $\rho$. It is observed that with the improvement in the correlation coefficient $\rho$, the ASER performance of the receiver degrades for both $4 \times 2$ RQAM as well as $8 \times 4$ RQAM. In Fig. 3, the ASER versus average SNR is plotted for $4 \times 2$ RQAM as well as $8 \times 4$ RQAM schemes by changing $\beta$. From Fig. 3, it is seen that the ASER performance outperforms for $\beta = 1$ than the ASER performance with $\beta = 0.5$ and $\beta = 2$. For $\beta = 1$, quadrature phase distance $(d_Q)$ as well as in-phase distance $(d_I)$ are equal, hence the minimum ASER can be obtained from the channel.

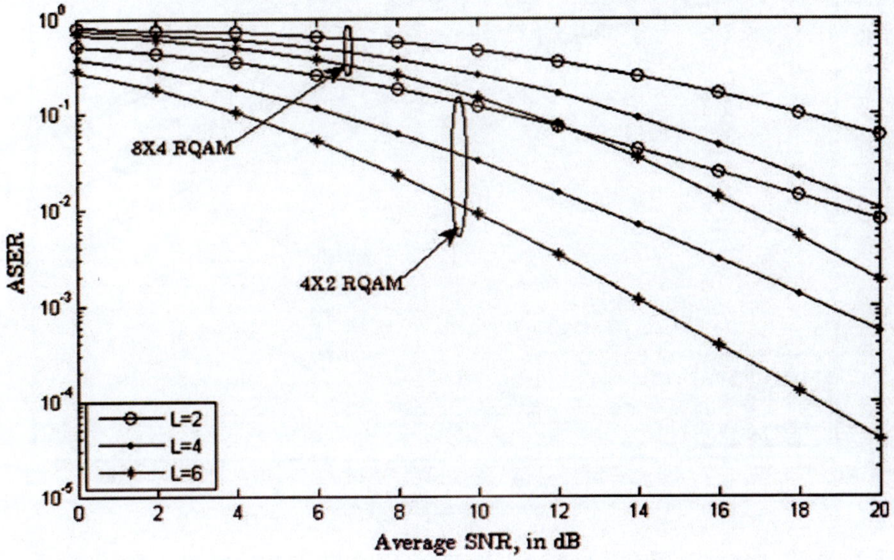


**Fig. 1** ASER vs. average SNR ($\overline{\gamma}$) of $4 \times 2$ RQAM and $8 \times 4$ RQAM with $\rho = 0.5$ and $\beta = 1$

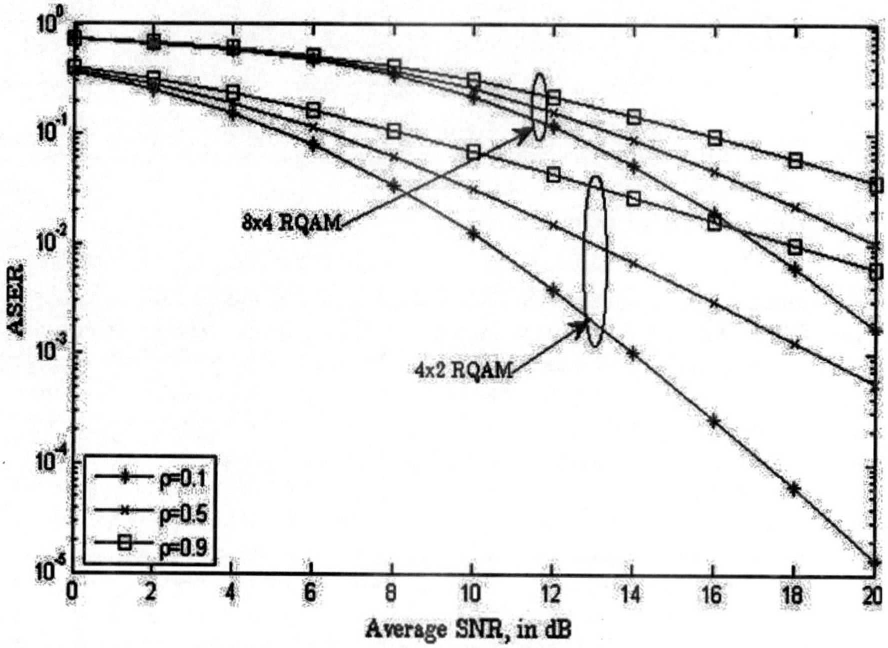


**Fig. 2** ASER vs. average SNR ($\overline{\gamma}$) of 4 × 2 RQAM and 8 × 4 RQAM with $\beta = 1$ and $L = 4$

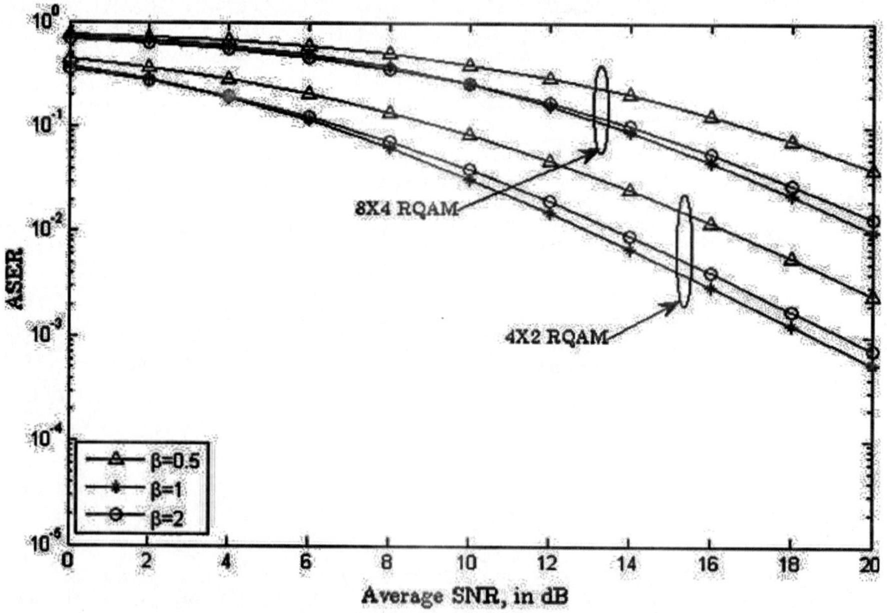


**Fig. 3** ASER vs. average SNR ($\overline{\gamma}$) of 4 × 2 RQAM and 8 × 4 RQAM with $\rho = 0.5$ and $L = 4$

In Fig. 3, $\rho = 0.5$ and $L = 4$ are constant values. It can be noticed from the figures that the ASER of $4 \times 2$ RQAM scheme outperforms the $8 \times 4$ RQAM scheme. Thus, as expected the ASER performance enhances with the increase in diversity order and deteriorates with increase in antenna correlations.

## 5 Conclusions

In the work, the ASER performance of an $L$-branch MRC receiver system with RQAM technique under Rayleigh fading channels is investigated. Closed form formulation is derived for ASER, which is obtainable in mathematical MATHEMATICA software. The determined results of the expression are plotted to illustrate the reaction of correlation and the number of receiving antennas on the receiver performance.

**Acknowledgements** The author would like to thank the anonymous reviewers and D. Nongrum for their valuable comments and suggestions.

## References

1. Proakis JG (2001) Digital communications, 4th edn. McGraw-Hill, New York
2. Dixit D, Sahu PR (2012a) Symbol error rate of rectangular QAM with best relay selection in cooperative systems over Rayleigh fading channels. IEEE Commun Lett 16(4):466–469
3. Kumar N, Bhatia V (2016) Exact ASER analysis of rectangular QAM in two-way relaying networks over Nakagami-$m$ fading channels. IEEE Wirel Commun Lett 5(5):548–551
4. Karagiannidis GK (2006) On the symbol error probability of general order rectangular QAM in Nakagami-$m$ fading. IEEE Commun Lett 10(11):745–747
5. Lei X, Fan P, Hao L (2007) Exact symbol error probability of general order rectangular QAM with MRC diversity reception over Nakagami-$m$ fading channels. IEEE Commun Lett 11(12):958–960
6. Peppas K (2011) Performance analysis of maximal ratio diversity receivers over Generalized fading channels. Adv Trends Wirel Commun 47–64
7. Dixit D, Sahu PR (2012b) Performance of $L$-Branch MRC receiver in $\eta$–$\mu$ and $k$–$\mu$ fading channels for QAM signals. IEEE Wirel Commun Lett 1(4):316–319
8. Dixit D, Sahu PR (2013) Performance of QAM signaling over TWDP fading channels. IEEE Trans Wireless Commun 12(4):1794–1799
9. Kumar N, Bhatia V, Dixit D (2017) Performance analysis of QAM in amplify-and-forward cooperative communication networks over Rayleigh fading channels. AEU Int J Electr Commun 72:86–94
10. Dixit D, Sahu PR (2017) Performance of dual-hop DF relaying systems with QAM schemes over mixed $\eta$–$\mu$ and $k$–$\mu$ fading channels. Trans Emerg Telecommun Technol 28(11)
11. Bilim M (2019) QAM signaling over $k$–$\mu$ shadowed fading channels. Phys Commun 34:261–271
12. Simon MK, Alouini M-S (2005) Digital communications over fading channels, 2nd edn. Wiley-Interscience, Hoboken

13. Aalo VA (1995) Performance of maximal-ratio diversity systems in a correlated Nakagami-m fading environment. IEEE Trans Commun 43(8):2360–2369
14. Beaulieu NC (2006) A useful integral for wireless communication theory and its application to rectangular signaling constellation error rates. IEEE Trans Commun 54(5):802–805
15. Bilim M, Kapucu N (2019) Average symbol error rate analysis of QAM schemes over millimeter wave fluctuating two-ray fading channels. IEEE Access 7:105746–105754
16. Chiani M, Dardari D, Simon MK (2003) New exponential bounds and approximations for the computation of error probability in fading channels. IEEE Trans Wireless Commun 2(4):840–845
17. Gradshteyn S, Ryzhik IM (2000) Table of integrals, series, products, 6th edn. Academic Press Inc., New York

# Dual Modulation of QPSK and OOK Using Silicon Microring Modulator

**Nitish Sinha, Sanjukta Bhowmik, Satyabrata Singha, and Bishanka Brata Bhowmik**

**Abstract** In this paper, we experimentally demonstrate an integrable and wavelength preserving dual-modulation format for the simultaneous generation of quadrature phase-shift keying(QPSK) signal over the on–off keying (OOK) signal. We have demonstrated the successful operation of dual modulation by using three microring modulators in two different configurations. The OOK and QPSK signals are fed with different bitrates. The performance of the modulator in the presence of amplified spontaneous emission (ASE) noise and phase noise has also been investigated.

**Keywords** Microring modulator · Microring resonator · Dual modulation

## 1 Introduction

Silicon photonics offers a promising platform in the past years for solving cost-effective and power-efficient optical communications [1]. Many researchers have implemented integrated lasers and photodetectors by examining heterogeneous epi growth of germanium on silicon [2]. Presently, many types of silicon microcavity modulators, like carrier-depletion microring modulators [3], carrier-injection silicon microring modulators [4], have been realized. Further, many research studies have been reported on QPSK modulation using silicon microring modulators [5–7]. In recent times, silicon-on-insulator-based devices show significant attention for implementing fabrication with PIC technologies [8–10]. Also, silicon in photon-

N. Sinha · S. Bhowmik (✉) · S. Singha · B. B. Bhowmik
Tripura University, Suryamaninagar, Tripura, India
e-mail: bhowmiksanjukta95@gmail.com

N. Sinha
e-mail: nitish.ns0176@gmail.com

S. Singha
e-mail: satyabrata.ece@tripurauniv.in

B. B. Bhowmik
e-mail: bishankabhowmik@tripurauniv.in



A. Maji et al. (eds.), *Proceedings of the International Conference on Computing and Communication Systems*, Lecture Notes in Networks and Systems 170,
https://doi.org/10.1007/978-981-33-4084-8_51

ics could provide a solid integration platform to diminish the volume of the optical element in significant level. Furthermore, silicon PIC has gathered much attention for implementing reliable integration with low cost [11, 12]. Furthermore, [13] presented electro-optic characteristics of the silicon, where spectral efficiency of optical signal in silicon microring has been shown. Microring resonator (MRR) has been presented to make add-drop multiplexer [14], logic gates [15], modulator [4, 16, 17] and also optical filters [18, 19]. Microring modulator (MRM) provides much attention by reason of small footprint, low power consumption using modulator particularly in silicon-based photonic integrated circuits. Many researchers have been reported electro-optic effect in MRR by implementing binary phase-shift keying (BPSK) and quadrature phase-shift keying (QPSK) [20–22]. The carrier dispersion effect has been reported using silicon as resonating form [4] and Mach–Zehnder modulator (MZM) configuration [23–25]. By applying electric voltage in silicon, the effect has also been demonstrated in a p-i-n diode [4, 24, 26, 27] and a PN diode [25, 27].

In recent years, silicon microring resonator-based devices have gathered much interest in integrated silicon photonics for its higher spectral efficiency (SE), optimum power utilization, small footprint and its compatibility with CMOS technology. One of the earliest works on microring modulators (MRM) is by Xu et al. The refractive index can be modulated by varying the concentration of electrons and holes in the waveguide [28]. Subsequently, phase modulators such as BPSK, DPSK [29] and intensity modulators [30] were implemented. Increasing demand for transmission capacity has led to higher modulation formats, and MRM modulators are implemented to generate QPSK, QAM, etc. [11, 22, 31]. But higher modulation formats are complicated to implement and are more susceptible to error in a noisy channel. 8PSK is another possible substitute out of all high spectral efficient optical multilevel modulation formats [31]. A more desirable and attractive technique for utilizing bandwidth and acquiring higher SE can be a dual-modulation technique. Sergio Pinna et al. proposed a transmission scheme for multiplexing two asynchronous amplitude (OOK) and phase (DQPSK) signals. The two independent signals were multiplexed into a single 8-APSK using Mach–Zehnder modulator (MZM) [32]. We propose an MRM-based dual modulation of two independent QPSK and OOK signals.

## 2 Device Characterization of Dual Modulation

### *2.1 Optical Phase Modulation in MRM*

Many researchers have reported the transfer characteristics of microring resonators (MRRs) many times elsewhere. The most common structure using a silicon optical modulator constitutes a ring associated with one straight waveguide. It behaves as a resonator because of its closed-loop structure, and light of a significant frequency builds up intensity within the structure. When the optical path length of the resonator becomes equal to an integer multiple of the guided waveguide as wave-

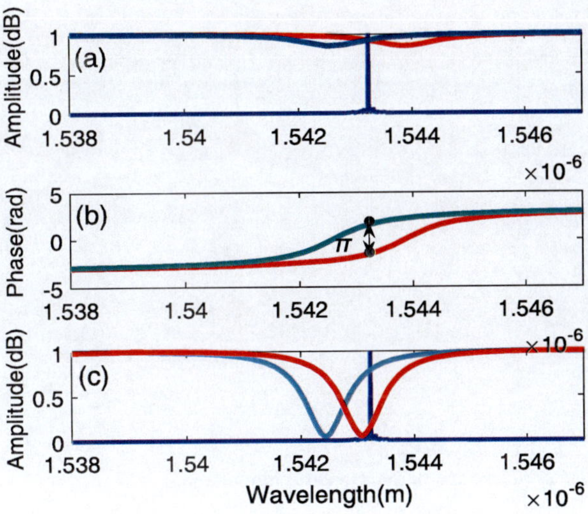


**Fig. 1** **a** QPSK amplitude response, **b** QPSK phase response, **c** OOK amplitude response

lengths, a resonance occurs. The resonance wavelength of the MRR is shifted by varying the free-carrier concentration, thereby changing the power transfer through the bus. Equation 1 depicted the transfer characteristics of MRM [11].

$$\frac{E_{\text{output}}}{E_{\text{input}}} = \frac{\sqrt{1-k} - \alpha e^{\frac{-j2\pi L}{\lambda}}}{1 - \alpha\sqrt{1-k}e^{\frac{-j2\pi L}{\lambda}}} \tag{1}$$

where $K$ is the cross-coupling coefficient between the ring and straight waveguide. $\alpha$ is the attenuation loss of the microring modulator. $L$ is the effective ring length, $\lambda$ is the operating wavelength of the particulate laser.

Figure 1 shows the amplitude and phase response of QPSK and amplitude response of OOK. In silicon microring modulator, $L' = \text{nL}$, where $L'$ is the effective path length of the ring which varies with change in the refractive index ($n$) and thereby causes the phase shifting of the optical signal into the ring.

## 2.2 *Dual Modulation with MRM*

Figure 2 shows the schematic of the proposed dual modulator. Two microring modulators are used in two different configurations, one as a dual-coupled ring for QPSK modulation in in-phase and $Q$-phase (IQ) configuration, and the other as a single-coupled ring for OOK signal generation. Here, we propose a $U$-shaped silicon microring modulator which significantly improves modulation accuracy greatly. Dual-coupled MRM is a $U$-shaped microring resonator device whose through the port and add port is connected. Dual-coupled ring provides a transfer function with

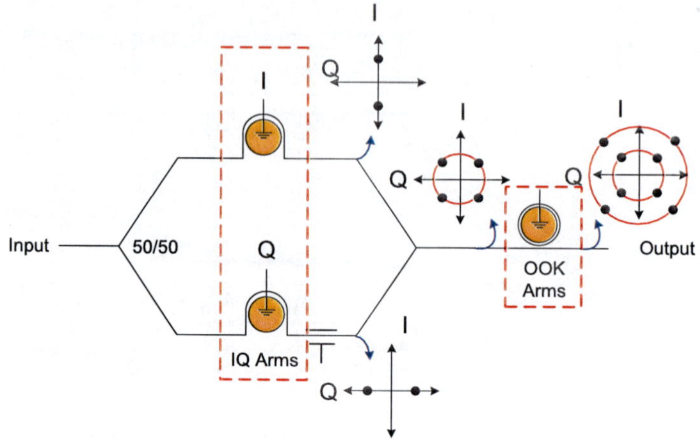


**Fig. 2** Schematic of the proposed modulator

lower free spectral range (FSR) and extinction ratio (ER) [33] of −0.8 dB as compared to the single-coupled ring of ER of −13 dB. MRMs with different ER ensure that the two QPSK and OOK signals are not affected by one another. A 50/50 coupler divides the input power equally to the two upper and lower MRMs. The two dual-coupled rings produce a QPSK signal, and the single-coupled ring then produces an 8-APSK signal. The bitrate of produced QPSK and OOK signal is 12.5 Gbps and 3.1250 Gbps respectively.

## 3 Simulation Results

For demodulation of QPSK signal, a hybrid demodulator is used, whereas for demodulating OOK signal, a photodiode is used. Figure 3 shows the simulation-based QPSK(Fig. 3a, b) and OOK (Fig. 3c) eye diagrams (using MATLAB). While trans-

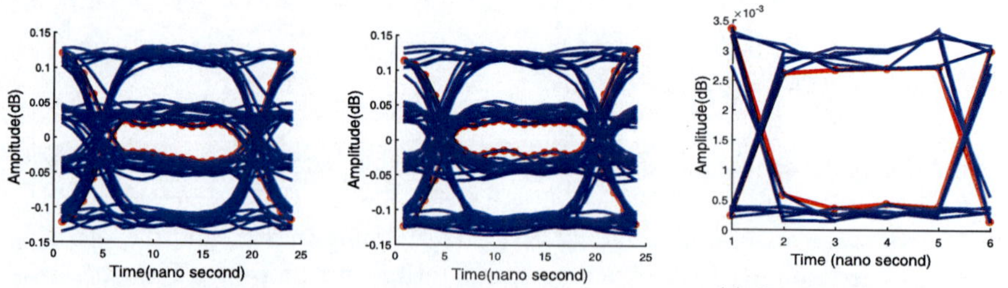


(a) I data (b) Q data (c) Simulated plots of the OOK eye opening

**Fig. 3** Simulated plots of the QPSK IQ-data and OOK eye diagram

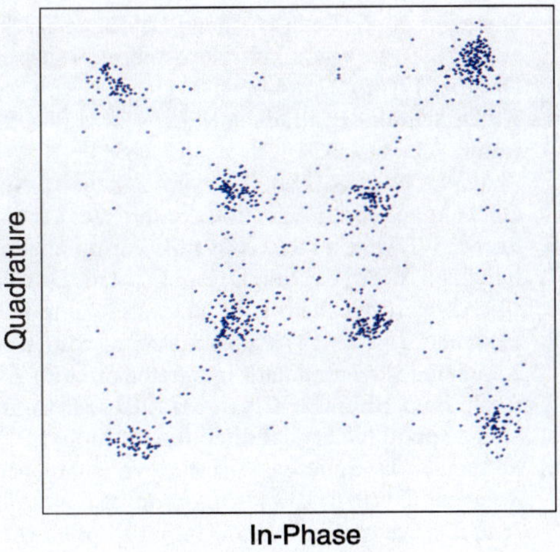


**Fig. 4** Constellation diagram with AWGN noise

mitting in optical fiber, a filter effect emerges in OOK signal which can be solved using DSP-MLSE.

Figure 4 shows a constellation diagram of the dual-modulation format, and this shows the presence of ASE noise and phase noise. In this proposed scheme, two modulation formats are present, i.e., QPSK and OOK.

From the result, it can be observed that the dual-modulation generated 25 Gbps signal, with a combination of QPSK signal and OOK signal. Figure 3a, b shows an eye diagram of the QPSK signal with *I* and *Q* data, and Fig. 3c is an OOK signal eye diagram.

## 4 Conclusion

We have proposed a dual-modulation scheme of QPSK over OOK using three microring modulators by applying a single carrier. We have presented a novel U-shaped microring configured in IQ manner. The proposed structure improves ER, thereby reducing the effect of noise. We have also investigated the effect of ASE noise and phase noise and found that the dual-modulation generated 25 Gbps signal, with the combination of QPSK signal and OOK signal. Here, the two modulations (QPSK, OOK) are asynchronous.

## References

1. Soref R (2006) The past, present, and future of silicon photonics. IEEE J Sel Top Quant Electron 12(6):1678–1687
2. Ahn D, Hong C, Liu J, Giziewicz W, Beals M, Kimerling LC, Michel J, Chen J, Kärtner FX (2007) High performance, waveguide integrated Ge photodetectors. Opt Exp 15(7):3916–3921

3. Dong P, Liao S, Feng D, Liang H, Zheng D, Shafiiha R, Kung C-C, Qian W, Li G, Zheng X et al (2009) Low $V_{pp}$, ultralow-energy, compact, high-speed silicon electro-optic modulator. Opt Exp 17(25):22484–22490
4. Xu Q, Schmidt B, Pradhan S, Lipson M (2005) All-optical control of light on a silicon chip. Nature 435:325–327
5. Zhang L, Wang R, Zou Z, Gray A, Olana L, Newell T, Webb D, Varangis P, Lester LF (2003) Optical fiber communication conference. Technical digest, paper no. FG2. Atlanta, GA
6. Sacher WD, Poon JKS (2009) Microring quadrature modulators. Opt Lett 34(24):3878–3880
7. Integlia RA, Yin L, Ding D, Pan DZ, Gill DM, Jiang W (2011) Parallel-coupled dual racetrack silicon micro-resonators for quadrature amplitude modulation. Opt Exp 19(16):14892–14902
8. Plouchart J-O (2011) Applications of SOI technologies to communication. In: 2011 IEEE compound semiconductor integrated circuit symposium (CSICS). IEEE, pp 1–4
9. Engin E, O'Brien JL, Cryan MJ (2012) Design and analysis of a gallium nitride-on-sapphire tunable photonic crystal directional coupler. JOSA B 29(6):1157–1164
10. Vermeulen D, Spuesens T, De Heyn P, Mechet P, Nötzel R, Verstuyft S, Van Thourhout D, Roelkens G (2010) III–V/silicon-on-insulator photonic integrated circuit for fiber-to-the-home central office transceivers in a point-to-point network configuration. In: 36th European conference and exhibition on optical communication. IEEE, pp 1–3
11. Bhowmik BB, Gupta S (2013) Silicon micro-ring resonator based 16-QAM modulator for WDM transmission systems. In: 2013 international conference on microwave and photonics (ICMAP). IEEE, pp 1–5
12. Bhowmik BB, Gupta S (2016) Wavelength division multiplexed 16-QAM generator using silicon micro ring modulator with preconditioned electronic input. J Opt 45(1):26–38
13. Soref RA (1987) Electrooptical effects in silicon. IEEE J Quant Electron 23(1):123–129
14. Xiao S, Khan MH, Shen H, Qi M (2008) Silicon-on-insulator microring add-drop filters with free spectral ranges over 30 nm. J Lightwave Technol 26(2):228–236
15. Kumar R, Liu L, Roelkens G, Geluk E-J, de Vries T, Regreny P, Van Thourhout D, Baets R, Morthier G (2010) 10 GHz all-optical gate on a silicon chip. In: Optical fiber communication conference. Optical Society of America, p JWA44
16. Xu Q, Manipatruni S, Schmidt B, Shakya J, Lipson M (2007) 12.5 Gbit/s carrier-injection-based silicon micro-ring silicon modulators. Opt Exp 15(2):430–436
17. Xi X, Li X, Xu H, Hu Y, Kang X, Li Z, Chu T, Yu J, Yu Y (2012) 44-Gb/s silicon microring modulators based on zigzag PN junctions. IEEE Photonics Technol Lett 24(19):1712–1714
18. Makino T, Gotoh T, Hasegawa R, Arakawa T, Kokubun Y (2011) Microring resonator wavelength tunable filter using five-layer asymmetric coupled quantum well. J Lightwave Technol 29(16):2387–2393
19. Rabus DG, Hamacher M, Heidrich H (2001) Active and passive microring resonator filter applications in GaInAsP/InP. In: Conference proceedings. 2001 international conference on indium phosphide and related materials. 13th IPRM (Cat. No. 01CH37198). IEEE, pp 477–480
20. Padmaraju K, Ophir N, Xu Q, Schmidt B, Shakya J, Manipatruni S, Lipson M, Bergman K (2011) Error-free transmission of DPSK at 5 Gb/s using a silicon microring modulator. In: European conference and exposition on optical communications. Optical Society of America, pp Th–12
21. Padmaraju K, Ophir N, Xu Q, Schmidt B, Shakya J, Manipatruni S, Lipson M, Bergman K (2012) Error-free transmission of microring-modulated BPSK. Opt Exp 20(8):8681–8688
22. Dong P, Xie C, Chen L, Fontaine NK, Chen Y (2012) Experimental demonstration of microring quadrature phase-shift keying modulators. Opt Lett 37(7):1178–1180
23. Liu A, Jones R, Liao L, Samara-Rubio D, Rubin D, Cohen O, Nicolaescu R, Paniccia M (2004) A high-speed silicon optical modulator based on a metal-oxide-semiconductor capacitor. Nature 427(6975):615–618
24. Gan F, Kartner FX (2005) High-speed electrical modulator in high-index-contrast (HIC) Si-waveguides. In: (CLEO). Conference on lasers and electro-optics, vol 1. IEEE, pp 104–106

25. Gunn C (2005). CMOS photonics/spl trade/—SOI learns a new trick. In: 2005 IEEE international SOI conference proceedings. IEEE, pp 7–13
26. Schmidt B, Xu Q, Shakya J, Manipatruni S, Lipson M (2007) Compact electro-optic modulator on silicon-on-insulator substrates using cavities with ultra-small modal volumes. Optics Exp 15(6):3140–3148
27. Zhou L, Poon AW (2006) Silicon electro-optic modulators using pin diodes embedded 10-micron-diameter microdisk resonators. Opt Exp 14(15):6851–6857
28. Xu Q, Schmidt B, Pradhan S, Lipson M (2005) Micrometre-scale silicon electro-optic modulator. Nature 435(7040):325–327
29. Zhang L, Yang J-Y, Song M, Li Y, Zhang B, Beausoleil RG, Willner AE (2007) Microring-based modulation and demodulation of DPSK signal. Opt Exp 15(18):11564–11569
30. Zhang L, Li Y, Yang J-Y, Song M, Beausoleil RG, Willner AE (2009) Silicon-based microring resonator modulators for intensity modulation. IEEE J Sel Top Quant Electron 16(1):149–158
31. Bhowmik S, Singha S, Bhowmik BB (2019) Optical 8PSK generation by cascading squared and rectangular QPSK using microring modulators. Fiber Integr Opt 1–7
32. Pinna S, Malacarne A, Lazzeri E, Bogoni A (2013) PPLN-based OOK and DQPSK optical grooming by amplitude and phase-signal multiplexing through pump depletion. Opt Lett 38(19):3870–3873
33. Li Z, Bai L, Li X, Gu E, Niu L, Zhang X (2018) U-shaped micro-ring graphene electro-optic modulator. Opt Commun 428:200–205

# Y-Arrangement Seven-Core Erbium-Doped Fiber Amplifier

**Balbindar Kaur and Subhash C. Arya**

**Abstract** A novel Y-arrangement seven-core erbium-doped fiber amplifier has been modeled and characterized in terms signal gain, noise figure (NF) cross talk, and mode analysis for C-band optical communication system. The optimum designing parameters used for the proposed model of Y-arrangement EDFA are size of the core ($R_{co}$) and cladding ($R_{ca}$)—10, 125 μm and pitch (Λ)—28 μm. A highest gain 42.25 dB with the lowest crosstalk of −42.07 dB has been obtained at 1550–1555 nm. The performance analysis of a Y-arrangement MC-EDFA is found to be in good agreement with available literature, and results are useful for the fabrication of MC-EDFA to meet explosive bandwidth and data rate demand in the area of a high-speed optical communication system.

**Keywords** Gain · Noise figure · Cross talk · Mode analysis · Multicore erbium-doped fiber amplifier

## 1 Introduction

At this present and highly advanced scenario, a requirement of ultra-high-speed of data transmission for longer distance communication is in great demand. Optical fiber is considered to be one of the modes of transmission to fit in this fastest-growing world of communication because of its high reliability, fast, and secured technology. Then for transmission of high capacity of data use of space division multiplexing (SDM) technique has drawn the attention of researcher as it offers 100 times higher capacity than single-mode fiber [1]. Two ways to attain SDM within an optical fiber are—multicore fiber (MCF) and few mode fiber (FMF) [2]. Later to cope up with the growing rate of capacity and long-distance transmission, various forms of

B. Kaur (✉) · S. C. Arya
Department of Electronics and Communication, North-Eastern Hill University, Shillong 793022, India
e-mail: balbindar27kaur@gmail.com

S. C. Arya
e-mail: aryasubh@yahoo.co.in



A. Maji et al. (eds.), *Proceedings of the International Conference on Computing and Communication Systems*, Lecture Notes in Networks and Systems 170,
https://doi.org/10.1007/978-981-33-4084-8_52

amplifiers have been used. Besides MCF, an amplifier with rare earth material named erbium is being suitably used that provides a high capacity and fastest direction to the optical fiber transmission which is referred to as multicore erbium-doped fiber amplifier (MC–EDFA). For incrementing the speediness, rate of data transfer and better amplification, MC-EDFA transmission methods tried different techniques such as—varying the number of cores, size of the core and cladding. Also used different structures/arrangement of cores based on coupled and uncoupled MCF such as—hexagonal structure, circular structure, square structure, 2 × 4 core layout, 1 × 4 core layout, equilateral triangular structure [3]. Different numbers of cores are being used in different types of core layout [3]. Most fibers to date have used a hexagonal structure of 7-cores for which the central core, with six nearest neighbors as opposed to three for cores in the outer ring. Along with hexagonal structured, circular, and square core layout, an equilateral triangular core layout consisting of 3-core microstructure fiber has been designed by Ryf et al. in the year 2012 [4]. In the years 2013 and 2015, Hayashi et al. [6, 7] and Nakanishi et al. [5] have designed a 32-core hexagonal structured fiber with a cladding diameter of 225 μm, $A_{eff}$ of 57 μm within a C- and L-band operating wavelengths. In 2016, Hayashi et al. [6] have used 8 cores circular structure to design an MCF suitable for O-Band short-reach optical interconnects and in the next year 2017 Hayashi et al. [7] have designed a coupled 4 core square structure fiber using a cladding diameter of 125 μm, $A_{eff}$ as 112 μm and a cut off wavelength of $\leq$1.47. Also, a 2 × 4 and 1 × 4 structure core layout has been designed by Hayashi et al. and Nagashima et al. in the year 2017 for short-reach communication [8, 9].

In this paper, a unique structure of a multicore erbium-doped fiber amplifier has been introduced using 7-cores, named as Y-structured. While designing the Y-structured MC-EDFA, core and cladding sizes are kept constant. The performance of Y-structured MC-EDFA has been analyzed by calculating the gain and the crosstalk through each of the core. In this analysis, wavelength of C-band range, i.e., 1525–1565 nm, is used throughout the study. Section 2 briefly describes the structure and the parameters used for designing the multicore erbium-doped Fiber amplifier. Section 3 discusses simulation and the obtained results of the Y-structured multicore erbium-doped fiber amplifier for performance analysis. Finally, we summarize the findings in Sect. 4.

## 2 Description of Y-Arrangement Core

This section briefly discussed the parameters used for designing the Y-arrangement comprised of seven cores inside a single cladding. Figure 1 shows the schematic diagram of the Y-structured core arrangement of MC-EDFA consisting of 7 cores. The size of the core ($R_{co}$) and the cladding ($R_{ca}$), pitch ($\Lambda$) distance between the cores are kept constant at 10, 125, and 28 μm, respectively, throughout the analysis for designing Y-arrangement 7-core EDFA [10].

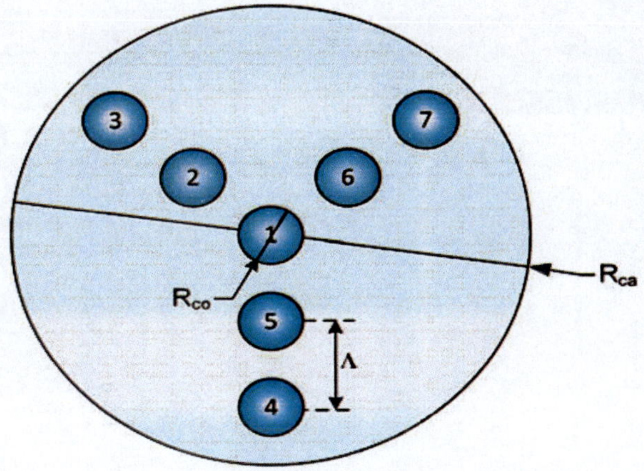


**Fig. 1** Cross section of 7-core Y-arrangement

## 3 Results and Discussions of 7-Cores Y-Arrangement EDFA

To model MC-EDFA, simulations are done in COMSOL Multiphysics to analyze different mode profiles over 7-cores Y-arrangement MC-EDFA. Table 1 shows the mode pattern for the proposed Y-arrangement core structure within C-band wavelength that solves the Maxwell equation [10]. The mode analysis of the designed fiber is carried out on C-band wavelength. For performance analysis, signal gain and noise figure (NF) have been calculated through each of the core for wavelength range 1525–1565 nm. Also, performance analysis includes measurement of crosstalk among the cores. The input signal power is kept constant at 0.001 W for Y-arrangement of core throughout the analysis. Further, the material is chosen in such a way that the concentration of the doped erbium remains constant at $7 \times 10^{24}/\text{m}^3$ throughout the analysis.

### 3.1 Description of Mode Analysis

This study involves the performance of a fiber using mode analysis. After performing the mode analysis, the mode patterns corresponding to the operating wavelength helps to understand the relation of electric and magnetic field for the light to propagate through the core. Table 1 shows the mode patterns for the wavelength range from 1525 to 1565 nm for Y-arrangement of cores. At 1525 nm wavelength, very low intensity and a dispersed mode of light are seen traveling inside the cores. Since core1 is the center core, so at low wavelength minimum amount of light is traveling in comparison with the rest of the cores. Then from 1530 to 1535 nm $LP_{02}$ mode is propagating through all the cores of Y-arrangement. At lowest wavelength either a very less intensity of light is traveling through the center core or a lower order of modes is propagating. As the wavelength increases, from 1540 to 1565 nm $LP_{12}$

**Table 1** Mode patterns for C-band wavelength, pitch = 28 μm for Y-arrangement of 7 cores EDFA

| Signal wavelength (nm) | Pitch-28 μm |
|---|---|
| 1525 | 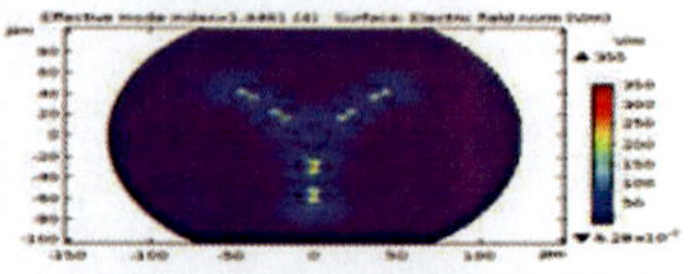 |
| 1530 | 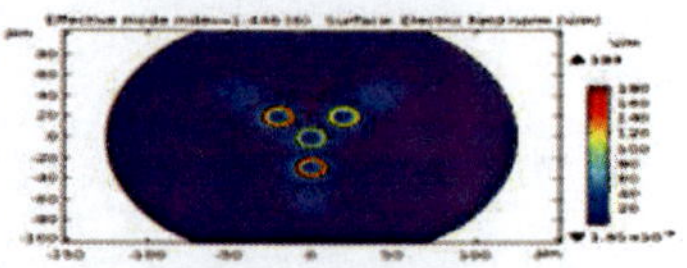 |
| 1535 | 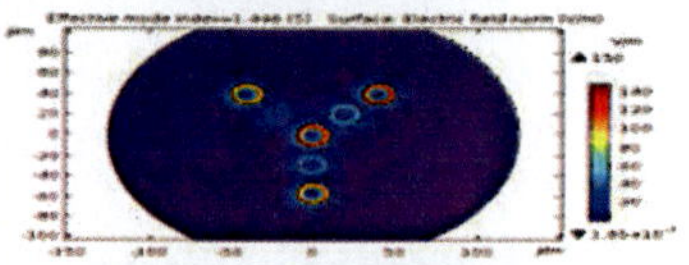 |
| 1540 | 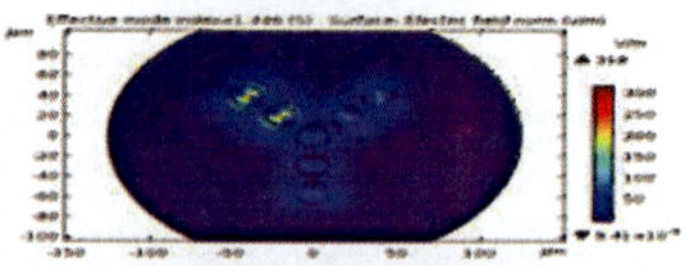 |
| 1545 | 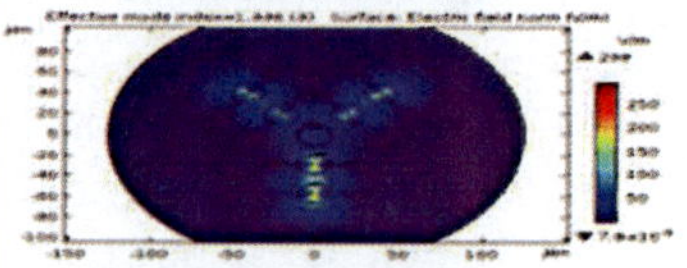 |
| 1550 | 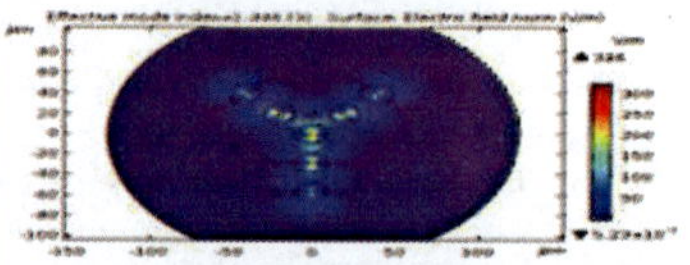 |
| 1555 | 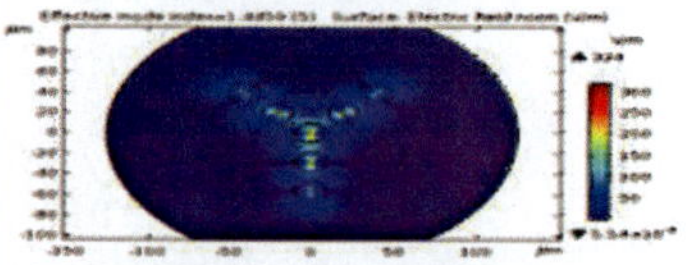 |
| 1560 | 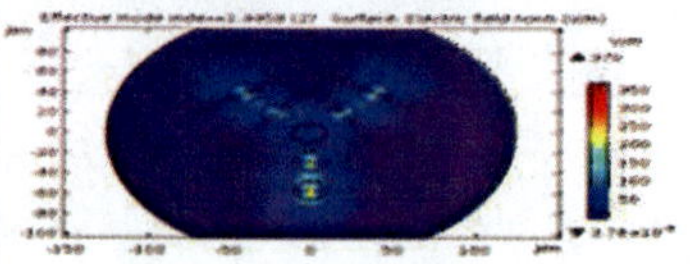 |

(continued)

**Table 1** (continued)

| Signal wavelength (nm) | Pitch-28 μm |
|---|---|
| 1565 | 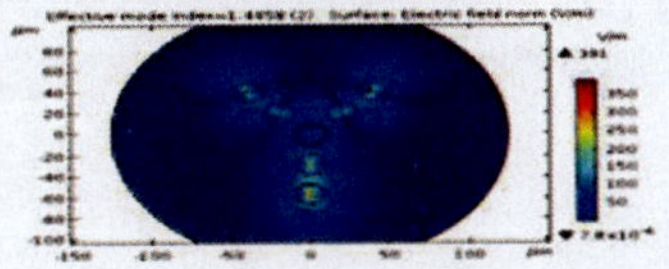 |

mode has been observed in the cores. From the mode profile pattern, it is observed that at 1550 and 1555 nm maximum intensity of $LP_{12}$ modes is equally distributed among all the cores of Y-arrangement. At 1560–1565 nm, intensity of $LP_{12}$ modes start dropping and mainly through core1 very low beam of light is seen propagating. Thus, we can be observed that for a different arrangement of the core, at 10 μm radius of the core, 125 μm cladding radius and pitch difference 28 μm, $LP_{12}$ mode found to be the dominating mode [10]. Along with the mode analysis, normalized frequency (V) is also calculated for wavelength range 1525–1565 nm to know the fraction of optical power bounded in the core in certain modes. For $R_{co} - 10$ μm, V number is within the range from 8.937 to 8.709. V' parameter which is considered as a governing parameter is higher than 2.405, so higher order of modes is traveling through the core.

## 3.2 *Performance Analysis of Y-Arrangement of MC-EDFA*

In this section, gain and noise figure (NF) for each of the cores are being calculated for the wavelength range from 1525 to 1565 nm. Figure 2 depicts the variation of gain and NF considering $R_{co} = 10$ μm, $R_{ca} = 125$ μm, and $\Lambda = 28$ μm. From the graph, it is clear that different core obtained different amounts of gain. Maximum gain peaks are obtained at 1530 and 1565 nm. But the highest gain of 42.25 dB is obtained at 1550–1555 nm. The minimum gain is attained at an operating wavelength of 1525 nm. Maximum gain is obtained when the incident frequency accompany with the frequency of the atomic transition, whereas gain decreases when incident frequency do not collaborate with the frequency of atomic transition Also, the NF obtained for $\Lambda = 28$ μm for Y-arrangement of the core as shown in Fig. 2 is within the range of 5.86–8.55 dB.

In addition to gain and NF, inter-core cross talk (IC-XT) at 1550 nm wavelength is also measured for Y-arrangement as shown in Fig. 3. To calculate the crosstalk among the neighboring cores is a crucial issue because low cross talk allows information to reach longer distances with minimum loss taking place among the adjacent cores during the propagation. The cross talk obtained for 28 μm pitch difference of the proposed Y –arrangement of MC-EDFA is −42.07 dB which is comparatively a low cross talk to avoid loss of information or coupling taking place among the cores within the fiber.

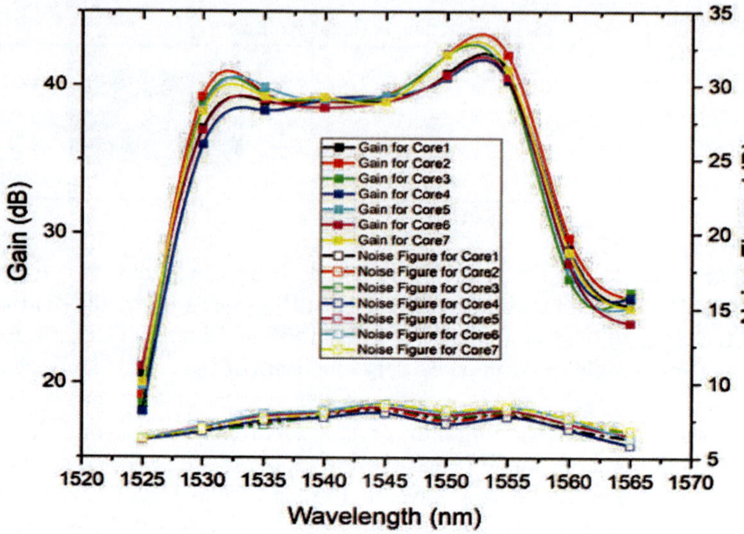


**Fig. 2** Gain and noise figure for pitch = 28 μm for Y-arrangement of core

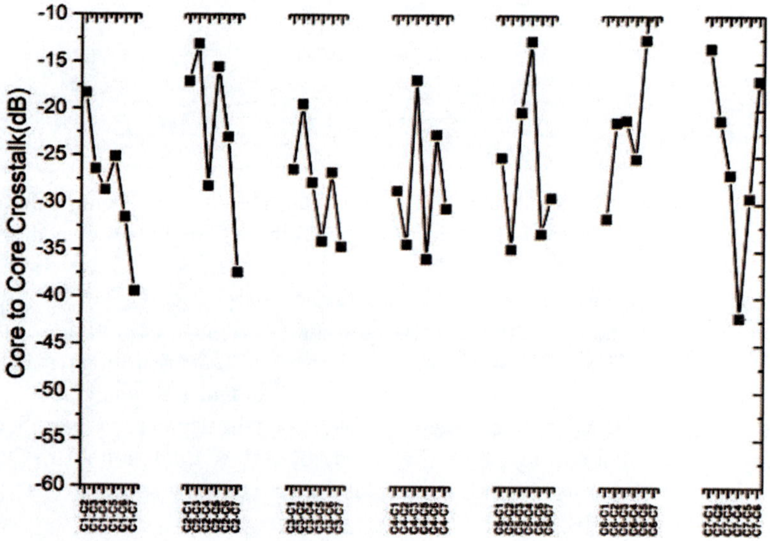


**Fig. 3** Inter-core cross talk for Y-arrangement of core

The IC-XT in dB is calculated by the following mathematical expression-

$$\mathrm{IC-XT\ (dB)} = 10\log_{10}\left( {P_{\mathrm{core\ A}}} / {(P_{\mathrm{core\ A-core\ B}})} \right)\Big|_{\mathrm{A}\neq\mathrm{B}} \quad (1)$$

Here, $P_{\text{coreA-coreB}}$ is the power coming from core A to core B. With reference to the considered geometry, if we consider core 1 as core A and $P_{\text{coreA}}$ is the power through core 1, than $P_{\text{coreB}}$ will be power transmitted through core2, core3, core4, core5, core6, and core7.

## 4 Conclusion

We proposed a unique structure of 7-cores MC-EDFA, named Y-arrangement at a wavelength range from 1525 to 1565 nm. We measured the performance of Y-arrangement of MC-EDFA in terms of gain, NF, cross talk and mode analysis. Based on the performance analysis, at 1550 and 1555 nm maximum intensity of $LP_{12}$ mode is observed and a maximum gain of 42.25 dB is obtained for Y-arrangement of the core. At 1550 nm, the lowest measured cross talk of around −42.07 dB has been obtained.

**Acknowledgements** We would like to acknowledge DST-SERB for supporting this work through financial aid by approving Project No. YSS/2015/000942 permitted to one of the authors. Also one the author acknowledges University Grant Commission (UGC) for granting fellowship under the scheme of Rajiv Gandhi National Fellowship. We also acknowledge North–Eastern Hill University for giving the infrastructural facilities to carry out this research work.

## References

1. Mizuno T, Takara H, Shibahara K, Sanoand A, Miyamoto Y (2016) Dense space division multiplexed transmission over multicore and multimode fiber for long-haul transport systems. J Lightwave Technol 34(6):0733–8724
2. Tulikumwenayo A (2013) Construction of multi-mode fibre modes using phase masks, M.S. Thesis, Department of Electrical Engineering, Rochester Institute of Technology
3. Hayashi T, NakanishiT (2018) Multicore optial fiber for next generation communication. SEI Tech Rev 86
4. Ryf R, Essiambre RJ, Gnauck AH, Randel S, Mestre MA, Schmidt C, Winzer PJ, Delbue R, Pupalaikis P, Sureka A, Hayashi T, Taru T, Sasaki T (2012) Space-division multiplexed transmission over 4200 km 3-core microstructured fiber. In: Optical fiber communication on proceedings
5. Nakanishi T, Hayashi T, Shimakawa O, Sasaki T (2015) Spatial-spectral-efficiency-enhanced multi-core fiber. In: Optical fiber communication on proceedings, OSA, Los Angeles California
6. Hayashi T, Nakanishi T, Hirashima K, Shimakawa O, Sato F, Koyama K, Furuya A, Murakami Y, Sasaki T (2016) 125-μm-cladding eight-core multi-core fiber realizing ultra-high-density cable suitable for O-Band short-reach optical interconnects. J Lightwave Technol 34(1):85–92

7. Hayashi T, Tamura Y, Hasegawa T, Taru T (2017) Record-low spatial mode dispersion and ultra-low loss coupled multi-core fiber for ultra-long-haul transmission. J Lightwave Technol 35(3):450–457
8. Hayashi T, Mekis A, Nakanishi T, Peterson M, Sahni S, Sun P, Freyling S, Armijo G, Sohn C, Foltz D, Pinguet T, Mack M, Kaneuchi Y, Shimakawa O, Morishima T, Sasaki T, Peter DD (2017) End-to-end multi-core fibre transmission link enabled by silicon photonics transceiver with grating coupler array. In: Optical communication on proceedings, Europe
9. Nagashima T, Toyokawa S, Hayashi T, Morishima T, Sakuma H, Nakanish T (2017) Multi-core fibre with concaved double-D shape cross section. In: European conference on optical communication, Gothenburg, Sweden
10. Kaur B, Arya SC (2020) A parametric study to create an optimal design of seven-cores hexagonal structured erbium-doped fiberamplifier. Opt Fib Technol 54:102081

# Symbol Error Rate of L-MRC Receiver in Correlated Non-homogeneous Non-line of Sight Environment

**Parag Moni Patangia, Niharika Barman, and Rupaban Subadar**

**Abstract** In this paper, we have derived a closed-form expression for the symbol error rate (SER) of the L-branch maximal ratio combiner (MRC) receiver over $\eta - \mu$ fading channel. Novel expression of moment generating function (MGF) of the correlated $\eta - \mu$ fading channel with constant correlation is calculated for the derivation of the SER expression. In the numerical analysis section, performance of the MRC receiver with respect to variation in different channel parameters along with the correlation is investigated. Monte Carlo simulation is used to verify the obtained results.

**Keywords** Correlated $\eta - \mu$ · Diversity receiver · Fading · MRC · Symbol error rate

## 1 Introduction

The diversity combining technique is used to mitigate the effect of fading arises due to the multipath nature of the received signals at the wireless receiver. MRC receiver gives optimum performance among all the diversity techniques. The multi-antenna system is used to achieve the spatial diversity in the receiver, in a practical scenario often the close spacing of the antennas introduces correlation between the received signals. Correlation degrades the advantages of diversity combiner.

The $\eta - \mu$ distribution exhibits an excellent fit to experimental fading channel measurements, and it is used to represent the small-scale variation of the fading signal, especially in a non-line of sight environment [8, 9]. As $\eta - \mu$ distribution

P. M. Patangia (✉) · N. Barman · R. Subadar
North Eastern Hill University, Shillong, India
e-mail: mail.parag19@gmail.com

N. Barman
e-mail: neyhabarman@gmail.com

R. Subadar
e-mail: rupaban.subadar@gmail.com



A. Maji et al. (eds.), *Proceedings of the International Conference on Computing and Communication Systems*, Lecture Notes in Networks and Systems 170,
https://doi.org/10.1007/978-981-33-4084-8_53

includes the Hoyt (Nakagami-$q$), Nakagami-$m$, Rayleigh, and one-sided Gaussian distributions as special cases, it is expected to be more flexible in matching experimental measurements for many physical propagation channels [8].

In [7], performance analysis of L-MRC receiver is carried out with equal correlation among the receiving signals. An expression of SER for generalized $\eta - \mu$ is derived in [5] for independent case. In [2], SER is calculated for correlated $\kappa - \mu$ shadowed fading channels for L-MRC receivers. We have seen that the SER for correlated $\eta - \mu$ environment is not calculated which motivates us for this work.

Remaining part of the paper is presented as per the following sequence.

Section 2 gives overview of fading the channel. The receiver is discussed in Sect. 3, and also MGF and SER expressions are calculated in this section. In Sect. 4, numerical results are discussed. The paper is concluded with Sect. 5.

## 2 Correlated $\eta - \mu$ Fading Channel

The channel is considered as non-homogeneous, non-line of sight represented by the $\eta - \mu$ statistical distribution presented in [9] as

$$\alpha^2 = \sum_{i=1}^{N} \left(x_i^2 + y_i^2\right), \tag{1}$$

where $\alpha$ is the $\eta - \mu$ distributed random variable (RV), $x_i$ and $y_i$ are independent Gaussian RVs with zero mean and variances ${\sigma_x}^2$ and ${\sigma_y}^2$, respectively. $N = 2\mu$ represents the cluster size of the distribution. The probability density function of $\alpha$ can be given as [9]

$$f_\alpha(\alpha) = \frac{4\sqrt{\pi}\mu^{\mu+\frac{1}{2}} h^{\mu} \alpha^{2\mu}}{\Gamma(\mu) H^{\mu-\frac{1}{2}} \Omega^{\mu+\frac{1}{2}}} \mathrm{e}^{-\frac{2\mu h}{\Omega}\alpha^2} I_{\mu-\frac{1}{2}} \left[\frac{2\mu H}{\Omega}\alpha^2\right], \tag{2}$$

where $\Omega = E\left[\alpha^2\right]$, $h \triangleq \left(2 + \eta^{-1} + \eta\right)/4$, $H \triangleq \left(\eta^{-1} - \eta\right)/4$, $I_v(\cdot)$ is the $v$th order modified Bessel function of the first kind. Parameter $\mu$ represents the cluster size of the distribution, whereas $\eta$ represents the power ratio or the correlation coefficient between the in-phase and the quadrature phase component in Format 1 and Format 2, respectively. However, mathematically one format can be converted to the other format [9]. The channel is slow and considered to be frequency non-selective.

## 3 The MRC Receiver

In the MRC diversity receiver, the signals received through different receiving antennas are co-phased, weighted with their respective SNR before algebraic addition at the combiner. The output of the combiner is processed by a suitable detector to extract the information received [7]. The output SNR is given by [6]

$$\gamma_{\text{mrc}} = \sum_{l=1}^{L} \gamma_l = \frac{E_b}{N_0} \left(\alpha_1^2 + \cdots + \alpha_L^2\right) \tag{3}$$

where $\gamma_l$ is the SNR of the $l$th antenna of the receiver.

### 3.1 PDF of the Output SNR

An expression of the PDF of the output SNR of the L-MRC receiver over correlated $\eta - \mu$ fading channel is given in ([7], 5)

$$\begin{aligned} f_{\gamma_{\text{mrc}}}(\gamma_{\text{mrc}}) = &\sum_{t_1=0}^{\infty} \sum_{t_2=0}^{\infty} \left(\frac{1-\rho}{1+(L-1)\rho}\right)^{2\mu} \left(\frac{\mu(1+\eta)}{\bar{\gamma}(1-\rho)\sqrt{\eta}}\right)^{2\mu L} \\ &\frac{(\mu)_{t_1} (\mu)_{t_2}}{t_1! t_2! \eta^{t_1} \Gamma(2\mu L + t_1 + t_2)} \left(\frac{\rho \mu L (1+\eta)}{\bar{\gamma}(1-\rho)(1+(L-1)\rho)}\right)^{t_1+t_2} \\ &\gamma^{(2\mu L + t_1 + t_2 - 1)} \mathrm{e}^{-\left(\frac{\mu(1+\eta)}{\bar{\gamma}(1-\rho)}\right)\gamma} \\ &{}_1F_1\left[\mu L + t_1, 2\mu L + t_1 + t_2, \frac{\mu(\eta^2-1)\gamma}{\bar{\gamma}(1-\rho)\eta}\right] \end{aligned} \tag{4}$$

where ${}_1F_1(\cdot, \cdot; \cdot)$ is the confluent hypergeometric function ([1], 13.1.2), $(.)_a$ is the Pochhammer's function ([1], 6.1.22).

### 3.2 MGF of the Output SNR

The MGF is calculated using

$$M_\gamma(s) = E_\gamma[\mathrm{e}^{-s\gamma}] \tag{5}$$

plugging (4) into the (5) and expressing the hypergeometric function in infinite series form, and the integration can be solved using ([3], 3.351). The final expression of the MGF is

$$M_\gamma(s) = \sum_{t_1=0}^{\infty}\sum_{t_2=0}^{\infty}\left(\frac{1-\rho}{1+(L-1)\rho}\right)^{2\mu}\left(\frac{\mu(1+\eta)}{\overline{\gamma}(1-\rho)\sqrt{\eta}}\right)^{2\mu L} \frac{(\mu)_{t_1}(\mu)_{t_2}}{t_1!t_2!\eta^{t_1}\Gamma(2\mu L+t_1+t_2)}\left(\frac{\rho\mu L(1+\eta)}{\overline{\gamma}(1-\rho)(1+(L-1)\rho)}\right)^{t_1+t_2} \sum_{k=0}^{\infty}\frac{(\mu L+t_1)_k\left(\frac{\mu(\eta^2-1)}{\overline{\gamma}(1-\rho)\eta}\right)^k}{(2\mu L+t_1+t_2)_k k!} (2\mu L+t_1+t_2+k-1)!\left(\frac{\mu(1+\eta)}{\overline{\gamma}(1-\rho)}+s\right)^{-(2\mu L+t_1+t_2+k)} \quad (6)$$

### 3.3 SER of the L-MRC Receiver

The SER of the MRC receiver can be calculated using the ([4], 10)

$$P_{\text{MPSK}} \approx \sum_{P=1}^{3}\beta_p M_\gamma(\alpha_p) \quad (7)$$

where $\beta_1 = \theta_M/(2\pi) - 1/6$, $\beta_2 = 1/4$, $\beta_3 = \theta_M/(2\pi) - 1/4$, $\alpha_1 = q\text{MPSK}$, $\alpha_2 = 4q\text{MPSK}/3$, $\alpha_3 = q\text{MPSK}/\sin^2(\theta_M)$, $q\text{MPSK} = \sin^2(\pi/M)$ and $\theta_M = (M-1)\pi/M$ [2] using the (6) in (7) we can get the SER of the L-MRC receiver for correlated $\eta - \mu$ fading channel as

$$P_{\text{MRC}} \approx \sum_{P=1}^{3}\beta_p\sum_{t_1=0}^{\infty}\sum_{t_2=0}^{\infty}\left(\frac{1-\rho}{1+(L-1)\rho}\right)^{2\mu}\left(\frac{\mu(1+\eta)}{\overline{\gamma}(1-\rho)\sqrt{\eta}}\right)^{2\mu L} \frac{(\mu)_{t_1}(\mu)_{t_2}}{t_1!t_2!\eta^{t_1}\Gamma(2\mu L+t_1+t_2)}\left(\frac{\rho\mu L(1+\eta)}{\overline{\gamma}(1-\rho)(1+(L-1)\rho)}\right)^{t_1+t_2} \sum_{k=0}^{\infty}\frac{(\mu L+t_1)_k\left(\frac{\mu(\eta^2-1)}{\overline{\gamma}(1-\rho)\eta}\right)^k}{(2\mu L+t_1+t_2)_k k!} (2\mu L+t_1+t_2+k-1)!\left(\frac{\mu(1+\eta)}{\overline{\gamma}(1-\rho)}+\alpha_p\right)^{-(2\mu L+t_1+t_2+k)} \quad (8)$$

## 4 Results and Discussion

This section covers the analysis of numerical results which are ploted in Figs. 1, 2 and 3 showing the effect of the variation of different channel and receiver parameters on the SER. It is visible from Fig. 1 that increase in correlation between the received signals causes an increase in error rate. The degrading effect of the correlation and

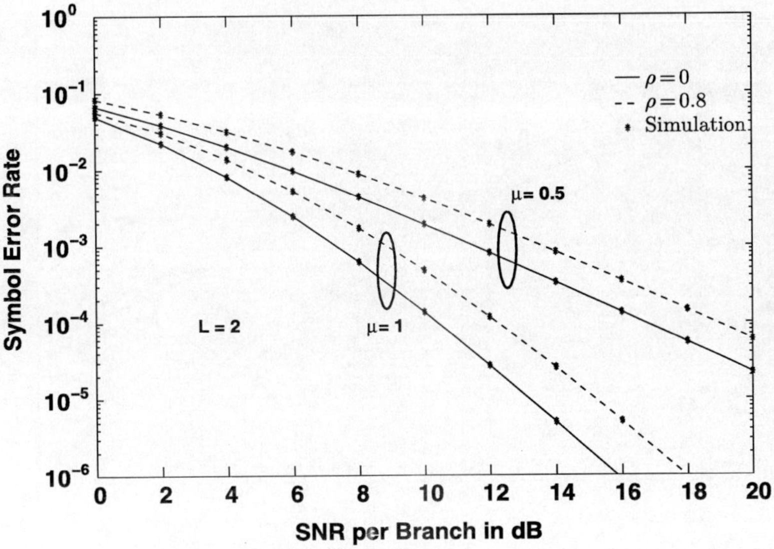


**Fig. 1** SER with variation of correlation

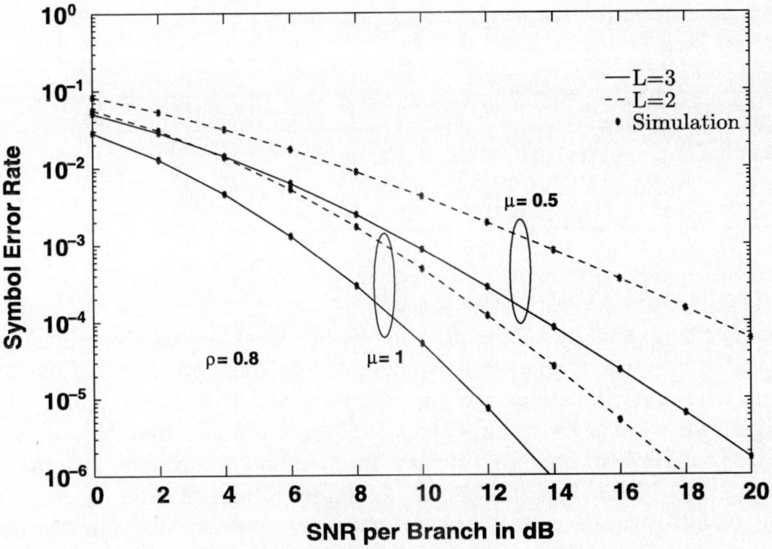


**Fig. 2** SER with variation of diversity branches

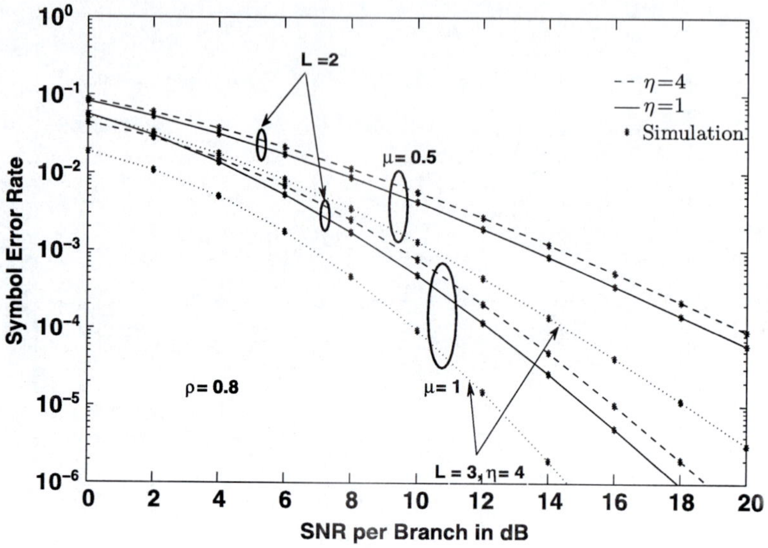


**Fig. 3** SER with variation of fading parameter

**Table 1** Number of terms ($T$) required for accuracy up to fifth place of decimal of SER with $\eta = 1$, $\mu = 0.5$, $\rho = 0.8$

| Avg. SNR | $\rho$ | $L = 2$ | $T$ | $L = 3$ | $T$ |
|---|---|---|---|---|---|
| 5 | 0 | 0.00964 | 4 | 0.001607 | 4 |
| 5 | 0.8 | 0.01716 | 8 | 0.00624 | 8 |
| 10 | 0 | 0.00190 | 2 | 0.00014 | 2 |
| 10 | 0.8 | 0.00190 | 5 | 0.00086 | 5 |

fading can be countered by adding diversity branches which can be verified by the results plotted in Fig. 2. In this work, we have demonstrated the result of the BPSK system only, however, the same expression can be used for higher-order modulations by changing the parameters appropriately. We can also notice that the lowest SER is obtained in the case of increased cluster size $\mu$ and no. of diversity antennas, i.e., when $\mu = 1$ and $L = 3$. Figure 3 depicts the effect of higher fading severity with variation of fading parameter $\eta$. As the derived expressions consist of infinite series, Table 1 is presented to put some light on the convergence of the expressions.

## 5 Conclusion

In this paper, we have derived an expression of performance parameter SER of L-MRC receiver in correlated $\eta - \mu$ environment with the help of MGF. The obtained expression is used to do the performance analysis of the L-MRC receiver with respect to different parameters. As we have considered the case of identical antennas with constant correlation, future work can be extended to the investigation of non-identical scenario.

## References


1. Abramowitz M, Stegun IA (1965) Handbook of mathematical functions: with formulas, graphs, and mathematical tables, vol 55. Courier Corporation
2. Bhatnagar MR (2014) On the sum of correlated squared $\kappa - \mu$ shadowed random variables and its application to performance analysis of MRC. IEEE Trans Veh Technol 64(6):2678–2684
3. Gradshteyn IS, Ryzhik IM (2014) Table of integrals, series, and products. Academic Press
4. McKay MR, Zanella A, Collings IB, Chiani M (2009) Error probability and SINR analysis of optimum combining in RICIAN fading. IEEE Trans Commun 57(3):676–687
5. Milisic M, Hamza M, Behlilovic N, Hadzialic M (2009) Symbol error probability analysis of l-branch maximal-ratio combiner for generalized eta-mu fading. In: VTC Spring 2009—IEEE 69th vehicular technology conference, Apr 2009, pp 1–5
6. Simon MK, Alouini MS (2005) Digital communication over fading channels, vol 95. Wiley
7. Subadar R, Sahu PR (2011) Performance of a L-MRC receiver over equally correlated $\eta - \mu$ fading channels. IEEE Trans Wirel Commun 10(5):1351–1355
8. Yacoub MD (2000) The $\eta - \mu$ distribution: a general fading distribution. In: 52nd vehicular technology conference, vol 2, Sept 2000, pp 872–877
9. Yacoub MD (2007) The $\kappa - \mu$ distribution and the $\eta - \mu$ distribution. IEEE Antennas Propag Mag 49(1):68–81

# Security and Cryptography

# Hybrid Framework for Android Malware Detection

**Pradeep Kumar Tiwari and T. Velayutham**

**Abstract** With the advent of affordable Android smartphones, the users have significantly increased to 85% of total market share (IDC in Smartphone Market Share, 2020, [1]). Given relaxed terms to become an android application developer, they are abusing the platform for user data collection through benign looking malicious applications. Google Play Protect does a scan to detect the authenticity and genuineness of the application, however fails to detect the intended use of such applications. We propose a framework to distinguish between a benign and malicious android application using comprehensive data flow analysis coupled with machine learning based malware classification. The system is complemented with the two techniques to detect benignness and maliciousness of an android application. We utilized the Androzoo and derbin dataset for building the model.

**Keywords** Android · Security · Malware analysis · Machine learning · Graph model · Android application

## 1 Introduction

The android applications offer and easy and cost-effective method of the running a business, government activities including awareness programs, subsidy programs, user friendly banking solutions etc. These utilities involve user's private information to be used in the process of application usage. This exposes user's private information at risk to the malicious applications on the smartphones. In order to identify such

P. K. Tiwari (✉) · T. Velayutham
Central Research Laboratory, Bharat Electronics Ltd., Bangalore 560013, India
e-mail: pradeepkumart@bel.co.in

T. Velayutham
e-mail: tvelayutham@bel.co.in

 
A. Maji et al. (eds.), *Proceedings of the International Conference on Computing and Communication Systems*, Lecture Notes in Networks and Systems 170,
https://doi.org/10.1007/978-981-33-4084-8_54

malicious applications, google has deployed a tool called Google Play Protect,[1] which analyzes the applications before getting published on the play store by the developers. However, it fails to protect applications changing their behavior after installation into user's phone. Also, sideloading [2] of the applications could also not be prevented using this technique.

We propose a system capable of distinguishing malicious applications from benign applications. The system could also indicate the potential of data leakage from an application based on the various use patterns from user. The proposed security framework of malware detection is a combination of two approaches using call graph pattern and static analysis of malwares using machine learning. The advantage of such system over traditional malware detection techniques includes:

- This approach combines static code analysis as well as call graph analysis of the application in a standard run time environment.
- This approach doesn't necessarily label an application as good or bad but it gives a similarity score with good as well as bad application. The higher value category indicates the inclination towards goodness and badness.
- This similarity value is then complemented with the other machine learning based approach to determine the category of the application.

The paper is organized starting from the background of instruments being using for the call graph generation, application data set collection techniques followed by the existing literature in the domain of call graph generation, malware detection in android, followed by the technique proposed for the security framework and finally it concludes with the results, future scope and conclusion.

## 2 Background

Soot is a fundamental intermediate representation in jimple format [3], a three-address code. As analysing a java bytecode for data flow is difficult because of the reason that implicit stack masks the flow of data. In jimple the data is stored in named variables making the local flow of data much more obvious.

The other literature [4–8] in the same domain has discovered static and dynamic ways to utilize the android features for malware classification. Call graph information is an inter-procedural feature of Soot provided by Spark. Spark over-approximates (Over-approximation means that any call which could occur in any execution of the program must appear in the call graph.) the set of calls that may occur at the run time. Call graphs edges connect sources (method, stmt) to targets (1st stmt in a method) (Fig. 1).

[1]Google Play Protect, https://www.android.com/play-protect/.

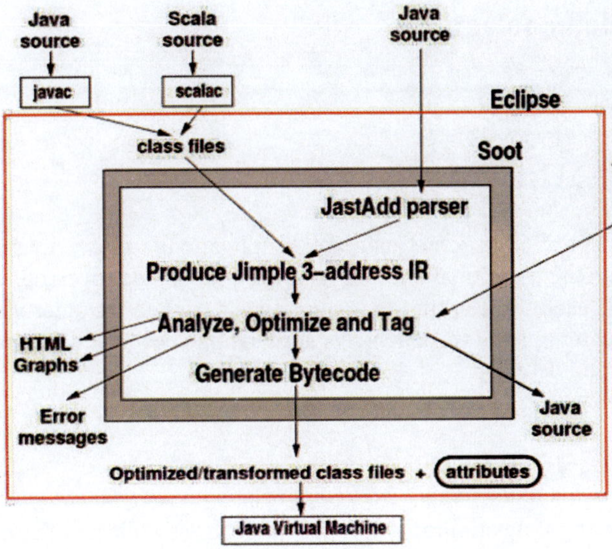


**Fig. 1** Overview of SOOT [3]

For Android Applications Soot can be used to

- Analyse (Call Graph Generation, Sources and Sinks analysis for a given application, Template-driven Intra-procedural data-flow analysis, Taint analysis in combination with FlowDroid [9])
- Instrument
- Optimize (can be used for Java optimization).

It takes input as 'apk file path' of the application and the related android platform, in case of android. Class files or java source files, in case of JAVA application and results in output as 'Android bytecode' (An intermediate representation). E.g. Call graphs pointed to by smali based procedures [10]. Soot results include:

- outputting transformed class files; (in bytecode format)
- outputting error messages; (violations of safety properties e.g trace matches of source sink for violations and reporting)
- generating HTML or graphs containing analysis results; or
- creating class files annotated with results obtained from program analysis.

# 3 Proposed Methodology

## 3.1 Datasets

### 3.1.1 Derbin Dataset [11, 12]

The dataset consists of 5560 different malicious applications from various sources. The dataset is rich in the traditional android features and includes even the obsolete APIs. The dataset is useful in classifying applications designed for older version of android. The shortcoming from this dataset is applications developed for recent OS will not be catered and detected.

### 3.1.2 Androzoo Dataset [13, 14]

AndroZoo is a growing collection of Android Applications collected from several sources, including the official Google Play app market [15]. It currently contains 5,642,879 different APKs, each of which has been (or will soon be) analysed by tens of different AntiVirus products to know which applications are detected as Malware. They provide this dataset to contribute to ongoing research efforts, as well as to enable new potential research topics on android apps.

Lists of APKs Big ($\geq$500 MB compressed) CSV file updated weekly, containing the following fields (not in that order): sha256, sha1, md5, apk size, dex size.

**dex date**: The date attached to the dex file inside the zip (sometimes invalid and manipulated).

**pkg name, vercode**: the name of the android Package and the version code (as reported in the manifest file). The pkg name might be unique inside one market (i.e. two apks with the same pkg name inside google play may have the same developer).

**vt detection, vt scan date**: The number of antivirus (AV) from VirusTotal (VT) that detected this apks as a malware on vt scan date (if available).

**Markets**: Separated list of the markets where we scarped this APK. The absence of a market does not mean that an APK was not published on this market. It means the status is not available.

**Malware Labels**: Euphony [16] is a command-line tool developers have developed to infer a single label per malicious application. The results are derived from VirusTotal reports. On the basis of that apks are catagorized into 2 catagories of malware and antimalware.

For the procedure of downloading the dataset from Androzoo we can refer to Fig. 2. The collection of dataset of APKs has been analysed by several anti-viruses to know which applications are detected as malware and which are secure to use. After downloading the dataset, we have a set of APKs labelled into 2 categories: **Malicious APKs** and **Benign APKs**.

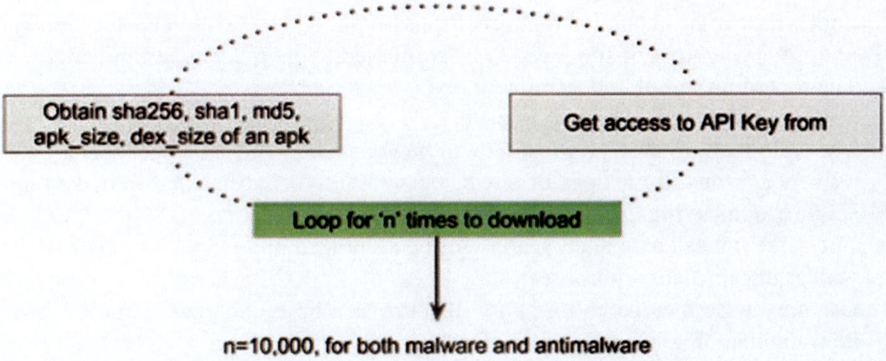


**Fig. 2** Procedure of dataset download

In the next step, the call graphs with respect to each APK are constructed using Soot, which is a Java analysis and optimisation framework provided by FlowDroid [9] (analysis tool for Android applications).

## 3.2 Categorization of Android Apps

The android apps are categorized in the following manner;

- The apps are decompiled using apktool [17], recursively.
- The AndroidManifest.xml file is analyzed to get parameters Activity Information, Intent Filters, Permissions, and few static attributes.
- With the meta information, the category of application is inferred. For example, an application having permissions LOCATION, INTERNET, DEIVCE ID, intents could be mapped to Travel or Social Media category.
- Some applications may belong to two different categories due to ambiguous meta information. Such applications are included in both such category.
- The categories included in the system are Location, Productivity, Travel, Gaming, e-Commerce, Social Media and Office.

## 3.3 Approach for the Malware Detection

The approach for malware detection has been divided into following steps:

1. The dataset of good and malicious android applications is stored.
2. The dataset need to be categorized into Location Based, Productivity Based, Lifestyle based, E-Commerce Based, Gaming, Office Based etc.

3. Soot will take head application directory as input and traverse in each category of apps. The output of soot need to be formatted to get the vertices and edges.
4. The record format of call graph will be
   *Seq No.—Timestamp—Edge Info—Edge Type—SrcNode—DstNode*
5. The call graph for each application will be generated in each category. The Call sequences from all the apps of one category will be concatenated to make one single graph having exhaustive list of edges.
6. The same concatenated call graph will be serialized and stored for security as well as encapsulation purposes.
7. Category wise merged call graph will serve as a base/reference graph for any new application to classify as good or bad.
8. When a new application is to be tested for information leakage, it will be fed to soot to give its call sequences. The sequence will then be weighed against the good as well as bad reference call graphs. The similarity score will be calculated with base/reference Genuine and malicious call sequences.
9. The inclination towards a category will flag the application accordingly for further analysis. Exact matching is also possible but will not give enough information to interpret.

### *3.4 Preprocessing and Analysis*

Preprocessing of the graphs include converting the graph into a set of edges [18]. The edges in the call graph, represent the attributes, android application takes to process the underlying function request. The type of edge determines the method invocation and frequency of such events in each case. Soot internally convert edges into following types:

**CLINIT**, also called as 'no invoke' or 'static' edge is formed when class object is initialized, which initializes static fields of a class.

**SPECIAL** edge is created, when class is being initialized in context of the activity of the application.

**VIRTUAL**, edge is formed when any non-static method of the class activity is called.

**STATIC** edge is created from the smali code of the class of application for static methods. These methods could be null as well.

**INTERFACE** edge is created, when java interfaces are invoked from the android application.

**FINALIZE** edge is created when the final block of the java code is invoked.

Various invocation methods include *virtual/special/static/interface* invoke methods. Figure 3 describes the output format of various edges and invocation methods.

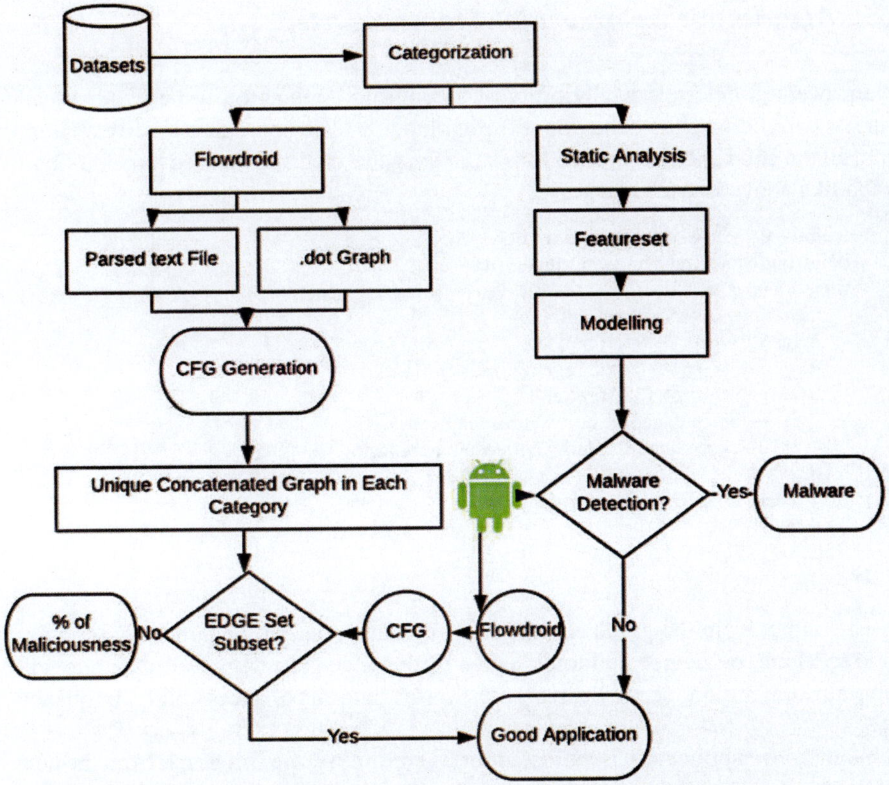


**Fig. 3** Malware classification technique, combining static analysis and flowdroid

```
FINALIZE edge: null in <java.lang.Object: void <init>()> ==>
SPECIAL edge: specialinvoke $r1.<java.lang.RuntimeException:
SPECIAL edge: specialinvoke $r1.<java.lang.RuntimeException:
SPECIAL edge: specialinvoke r0.<java.lang.Object: void <init>
SPECIAL edge: specialinvoke $r21.<com.android.insecurebank.Lc
<com.android.insecurebank.LoginScreen$3: void <init>(com.andr
VIRTUAL edge: virtualinvoke $r21.<com.android.insecurebank.Lc
```

**Fig. 4** Invoke methods and types of edge in sample soot output

For example the cumulative graph of the social network category applications have more concentration on class initialization-based edges. This indicates the implementation of such methods hasn't been implemented. In Fig. 4 the nodes are the source vertices with the initial function and edges contains the information like type of edge and the invocation method. The figure also indicates a sample output from an android application for the creation of graph.

## 3.5 Algorithmic Approach for Malware Detection

The procedure determines eligibility of an android application to be good or bad. Also, it will indicate the behavior of the application. The procedure involves: Graph from all the good and bad applications has been generated and merged into one single graph like above in each category.

```
procedure CLASS SIMILARITY CALCULATION
    buildAggregateGraphInEachCategory(dataset)
    for each app do
        (app)_c ← Category(metaInfo(app))
        EdgeSet(app) ← Soot(app)
        EdgeSet_c ← Soot(loadAggregateGraph(Cat_c))
        If(EdgeSet(app) ⊆ EdgeSet_c)
            P_b ← percentageSimilarityWithBenignClass
            P_m ← percentageSimilarityWithMaliciousClass
        Else
            checkInDifferentCategory(app)
    end for
    return P_b, P_m
end procedure
```

The output from the graph contains information about all the nodes, their neighbors, indegree, outdegree and total degree of the node. The degree of the node will determine the importance of the particular vertex in terms of accessibility, frequency etc.

When a new application is validated for its credibility, the call graph from the new application will be matched with the base graph from each category and percentage match score will be generated. The matching criteria for the new graph is it has to be subset of the base graph. The similarity score will determine the application's category based on inclination in the percentage value.

## 3.6 Machine Learning Based Model

We extracted following features from decompiled android applications for training the model (Table 1).

The above features are extracted and compiled from 800 bad applications and 200 good applications and the SVM Based model was incorporated to build the model. We didn't include dynamic features like CPU, Memory, Stack information from application as including these features was reducing the system's performance as well as it was deviating the system from intended objective.

**Table 1** The features from Android applications

| Category | Value |
|---|---|
| API | Android/telephony/telephony manager;<br>getDevice ID<br>Android/telephony/Telephony Manager;<br>getSubscriberid<br>... |
| Uses-feature | Android.Hardware.touchscreen<br>Android.Hardware.Camera<br>... |
| Intent/intent-filters | Android.intent.Action.BOOT_COMPLETED<br>android.Intent.Action.PHONE_STATE<br>... |

## 4 Experimental Results and Discussions

We gathered the malware applications from derbin dataset and benign applications were from androzoo dataset. From the set of 1000 applications, 80% were used to train the SVM Classifier and remaining 20% were used for the testing of the model. The following matrix shows the accuracy of the model since the applications were not systematically distributed (Table 2).

This result from the machine learning based model was complemented with the graph-based malware detection to enhance the framework's accuracy. These results were subjected to another test to bring a close classification. The results from the graph-based model's accuracy (Table 3).

The main advantage of this framework is that we can discretely identify the exact reasons of maliciousness and benignness in android applications. Though the unified system is under development, results are encouraging. The proposed technique, when compared over the existing techniques performs better (Table 4).

**Table 2** The confusion matrix

| | Actual class | | | Accuracy |
|---|---|---|---|---|
| | | Benign | Malware | |
| Expected class | Benign | 0.873 | 0.256 | 0.893 |
| | Malware | 0.180 | 0.918 | 0.941 |

**Table 3** Accuracy of the System on identifying the similarity between the classes

| Test sample size | Test class | Avg. similarity with benign class (%) | Avg. similarity with Malware class (%) |
|---|---|---|---|
| 1000 | Malware | 15–20 | 80–85 |
| 800 | Benign | 90–95 | 5–10 |

**Table 4** Comparison with existing techniques in graph and ML based approaches

| Technique | The key approach | Percentage accuracy (%) | False positives (%) |
|---|---|---|---|
| Permission based classification [17] | Machine learning | 85 | 15 |
| Subgraph2vec [18] | Graph based | 83 | 17 |
| HFAMD | ML + Graph | 94 | 6 |

## 5 Conclusion and Future Scope

We proposed a model for the detection of the android malware, which tries to analyze the static behavior of android applications in exhaustive manner through the call graphs and control flow graphs. The features of the call graph, selected for the training of the model was different from earlier approaches and found to be effective in detection of malwares. The approach, however can be complemented with signature-based approaches and LSTM [19] based techniques for better and effective results. The effectiveness of the system will also depend on the quality of dataset being used. Here, we have used derbin and androzoo dataset, which are somewhat matured. However, the derbin dataset is tending to become obsolete with new android OS releasing every year. A concrete methodology for malware detection from the above approaches in under evaluation. The accuracy of the same would affect and enhance with the integration of multiple malware detection approaches.

## References

1. Smartphone market share. https://www.idc.com/promo/smartphone-market-share/os. Accessed on 20 Jan 2020
2. Brady R (2018) App sideloading and cyber risk. ITNOW 60(4). https://doi.org/10.1016/0022-2836(81)90087-5
3. Bartel A et al (2012) Dexpler: converting android Dalvik bytecode to Jimple for static analysis with soot. In: Proceedings of the ACM SIGPLAN international workshop on state of the art in java program analysis
4. Yin H, Song D, Egele M, Kruegel C, Kirda E (2007) Panorama: capturing system-wide information flow for malware detection and analysis. In: Proceedings of the 14th ACM conference on computer and communications security. ACM, Oct 2007, pp 116–127
5. Gascon H, Yamaguchi F, Arp D, Rieck K (2013) Structural detection of android malware using embedded call graphs. In: Proceedings of the 2013 ACM workshop on artificial intelligence and security. ACM, Nov 2013, pp 45–54
6. Canfora G, Medvet E, Mercaldo F, Visaggio CA (2015) Detecting android malware using sequences of system calls. In: Proceedings of the 3rd international workshop on software development lifecycle for mobile. ACM, Aug 2015, pp 13–20
7. Rasthofer S, Arzt S, Bodden E (2014) A machine-learning approach for classifying and categorizing android sources and sinks. In: NDSS, Feb 2014

8. Tiwari PK et al (2019) Behaviour based authentication: a new login strategy for smartphones. In: 2019 second international conference on advanced computational and communication paradigms (ICACCP). IEEE
9. Arzt S et al (2016) Flowdroid: precise context, flow, field, object-sensitive and lifecycle-aware taint analysis for android apps. ACM Sigplan Not 49(6):259–269
10. Hoffmann J et al (2016) Slicing droids: program slicing for smali code. In: Proceedings of the 28th annual ACM symposium on applied computing
11. Arp D et al (2014) Drebin: effective and explainable detection of android malware in your pocket. Ndss. vol 14
12. The Drebin dataset. https://www.sec.cs.tu-bs.de/~danarp/drebin/index.html. Accessed on 20 Jan 2020
13. Allix K et al (2016) Androzoo: collecting millions of android apps for the research community. In: 2016 IEEE/ACM 13th working conference on mining software repositories (MSR). IEEE
14. AndroZoo (2020) A project of Université du Luxembourg. https://androzoo.uni.lu/. Accessed on 20 Jan 2020
15. Google Play Store. https://play.google.com/store?hl=en. Accessed on 20 Jan 2020
16. Malware labels. https://androzoo.uni.lu/labels. Accessed on 20 Jan 2020
17. Winsniewski R (2012) Android-apktool: a tool for reverse engineering android apk files
18. Narayanan A et al (2016) Subgraph2vec: learning distributed representations of rooted subgraphs from large graphs. arXiv preprint arXiv:1606.08928
19. Gers FA, Schmidhuber J, Cummins F (1999) Learning to forget: continual prediction with LSTM, pp 850–855

# A Publicly Verifiable and Cheating Resistant Secret Sharing Based on Public Key Cryptosystem and Hash Function

**Puja Sarkar and Amitava Nag**

**Abstract** Sensitive data kept in a single information-bearer is more vulnerable to attack. Due to a single point of failure, it could be easily accessible or modifiable by an unauthorized user. To provide security to sensitive data against illegal access or alteration, several secret sharing techniques have been proposed. In a $(k, n)$ threshold secret-sharing technique, $n$ shares of the secret has to be distributed to $n$ participants. And at the end, at least $k$ out of $n$ shares have to be collaborated to reveal the true secret. The beauty of this scheme is that collaboration of any shares less than $k$ will not be able to reveal any kind of information regarding the secret. But, what will be the scenario if the shareholders are dishonest or some attacker is trying to hamper the secrecy of the secret by providing wrong information? So, to verify the correctness of the shared information, the verifiable property is added to the secret sharing, which is known as Verifiable secret sharing (VSS). Secret verification and cheater recognition are the two most popular features of this mechanism. The first one verifies the correctness of the secret and the second one helps to identify the cheater who has provided the wrong share. In our recent work, a publicly verifiable secret sharing technique is proposed, which can recognize the malicious participants or dishonest coordinator. The resistance of our proposed scheme towards the different types of attacks has been shown in the security analysis in this paper.

**Keywords** Secret-sharing mechnism · Verifiable secret-sharing · Cheating detection · Cheater recognition

---

P. Sarkar (✉) · A. Nag
Department of Computer Science and Engineering,
Central Institute of Technology Kokrajhar, Btad, Assam, India
e-mail: p.sarkar@cit.ac.in

A. Nag
e-mail: amitava.nag@cit.ac.in



A. Maji et al. (eds.), *Proceedings of the International Conference on Computing and Communication Systems*, Lecture Notes in Networks and Systems 170,
https://doi.org/10.1007/978-981-33-4084-8_55

# 1 Introduction

In an unsafe cyber world, secret-sharing techniques play a vital role to provide safeguarding confidential information against modification and unauthorized access. Unlike cryptography and steganography, the secret sharing technique is free from the Single Point of Failure (SPOF) problem. Data Confidentiality as well as Data Availability both can be achieved by this technique at the same time. After the individual's proposal of Shamir [1] and Blakley [2] in 1979, the journey of the secret sharing technique has been started as a new research era in the information Security field. Shamir's [1] and Blakley's [2] concept of threshold $(k, n)$ secret sharing techniques are based on polynomial interpolation theorem proposed by Lagrange's and hyperplane geometry simultaneously. The TSS technique is mainly categorise into two phases. First one is the Sharing Phase, where the secret is coded into $n$ shares, and distributed to $n$ number of participants. And the second one is the reconstruction phase, where participants need to collaborate at least $k$ out of $n$ received shares to reconstruct the true secret as shown in Fig. 1. The beauty of this scheme lies in the following facts that the collaboration of any $(k - 1)$ or fewer shares reveals no information about the secret.

However, Shamir's SSS leaves a great security concern as it assumes that both the Coordinator and the participants are honest and they always submit the correct information throughout the transmission procedure, which may not be the case in real life [3]. The Verifiable SSS [4, 5] deal with this issue by providing features like cheating detection and cheater identification. VSS scheme fulfill all the security

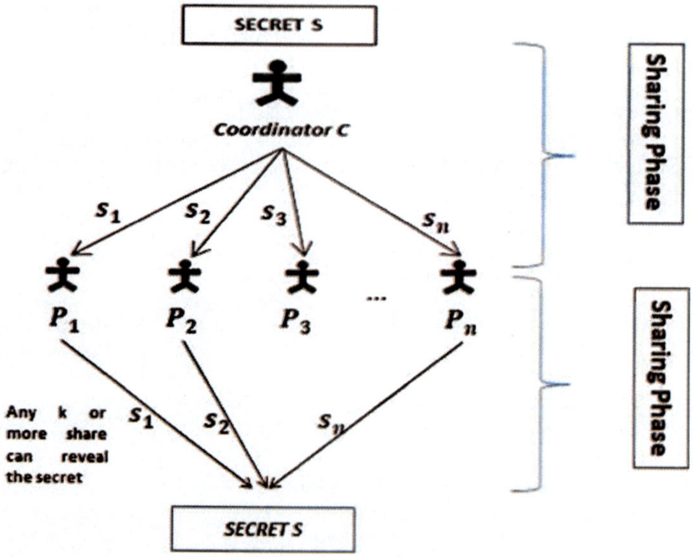


**Fig. 1** A model for $(k, n)$ threshold secret sharing scheme

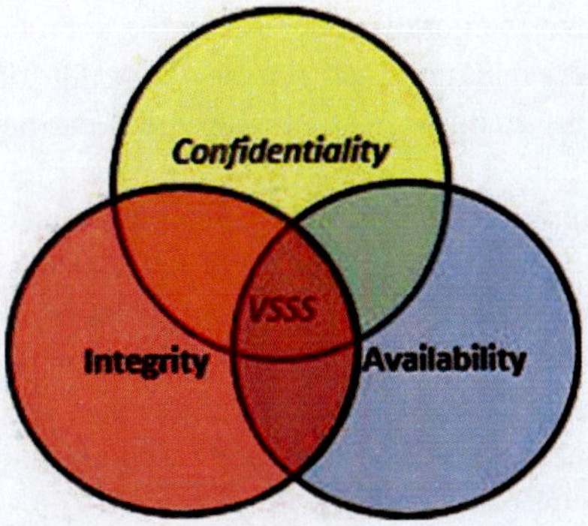


**Fig. 2** Modified security goals

goal as shown in Fig. 2. A Publicly Verifiable Secret Sharing (PVSS) mechanism [4, 6] is almost similar to VSS, with an additional property that the correctness of the distributed shares can be verified by any party or by any participants; hence verification is not restricted to the concerned participant receiving that particular share. We propose a PVSS scheme based on Shamir's secret sharing scheme which has the following features:

1. Our proposed scheme can able to point out the cheating by verifying the correctness of the shares.
2. It can also identify the cheater, no matter whether the cheater is the Coordinator or the participants.
3. Unlike Shamir's SSS, it doesn't demand for a secure channel to communicate.

The sequence of organizing the remaining part of the paper is as follows: the literature review is discussed in Sect. 2, followed by our proposed technique in Sect. 3. Some preliminary knowledge are discuss in Sect. 4. In Sect. 5, we did a performance analysis of our proposed scheme and compare it with other existing schemes. Finally, in Sect. 6, the conclusion is drawn along with the future scope.

## 2 Literature Review

Shamir's SSS [1] is not realistic as it assumed that both the coordinator and the participants are honest and the communication channel is secure, Which is not always true. To overcome this issue, the concept of VSS was proposed by Chor et al. [7]. Tompa et al. [8] proposed a more secure and efficient scheme as compared to Shamir's SSS [1] that its security does not depend on any hypothetical assumptions such as uncontrollability of computing number-theoretic functions. McEliece et al. [9] proposed a scheme, that is based on Reed-Solmon coding scheme, which provides more generalization and extensions of Shamir's traditional threshold SSS [1]. Harn et al. [10] presented an efficient and strong $(n, t, n)$ secret sharing, based on Benaloh's scheme, which is theoretically secure as like traditional Shanmir's SSS. Each share's size is equal to the size of the secret. Zhao et al. [11] presented a VSS for multiple secret data based on the RSA algorithm and can successfully detect cheating and can recognize the cheater also. Patil et al. [12] proposed a scheme for sharing secret image using image watermarking technique. At the time of distributing, the secret image is shared along with the watermarking image. Adhikari et al. [13] proposed a $(k, n)$ TSS scheme that can detect the rushing cheaters, who have submitted their forged share just after the submission of shares by some honest participants. Liu et al. [14] presented two different algorithms for cheating detection and cheating identification. For cheating detection, it demands $k$ or more user's participation. And cheater can be identified by the remaining participant, who has not taken part in the secret reconstruction. In PVSS, the shares are verified by the public, who are even part of the scheme [15, 16]. SSS can also be used for the image data, known as the Secret Image Sharing scheme. Multi-image secret sharing schemes with verifiable property are also available, which can transmit multiple secret images by sharing a single share to each participant [13, 17–19]. The concept of SSS are used in various fields like e-voting application [20], e-auction application [21], audio secret sharing [22] etc. some recent research work, the concept of secret sharing is used to address some problems of Blockchain technology such as low communication efficiency, low security issues, and tough recovery [23]. Alohali et al. [24] used the concept of SSS for secure group communication. As we have seen the diversity of secret sharing scheme in various area, so in such scenario the security issues also plays a vital role. Cheating detection and cheating identification become more important to protect our secret data from being damaged or lost, or falling into cheater's hands [5].

## 3 Preliminary Knowledge

As our scheme is mainly laid upon the concept of Shamir's SSS, so, first of all, we briefly discuss the said scheme below.

### 3.1 Shamir's $(k, n)$ Threshold Secret-Sharing Technique

Suppose we want to share a secret $S$ among a group of $n$ number of participants using $(k, n)$ TSS technique, where $k$ represents the threshold value that can be done in the following phases.

1. ***Initialization phase***: Here, the dealer have to choose any prime number $p$ and any set of $n$ numbers randomly from the range of 0 to $(p - 1)$ having values such as $x_1, x_2, x_3 \ldots, x_n$.
2. ***Share generation and distribution phase***: Here, the dealer secretly picks the coefficients $a_j \varepsilon \{0, 1, 2, \ldots, p - 1\}$ and build a $(k - 1)$ degree polynomial as $a(x) = \left(a_0 + \Sigma_{j=1}^{k-1} a_j x^j\right) \bmod p$ where $a_0 = S$, that is the secret number, and

evaluate the share by $y_i = a(x_i)$, where $1 \le i \le n$ and then simply distribute the share to the $n$ participants

3. ***Secret Reconstruction phase***: Using Lagrange's interpolation, the $k$ or more collaborating parties can reveal the confidential secret data as shown in Eq. (1)

$$S = a\,(0) = \left( \sum_{i=1}^{k} y_i \prod_{j=1, i \neq j}^{k} \frac{x_j}{x_j - x_i} \right) \bmod p \tag{1}$$

## 3.2 Public Key Mechanism: RSA Algorithm

In RSA algorithm [16], suppose we take the pair of public key as $(e, n)$ and secret key is $(d, n)$. Then, (Plain text)$^e$ mod $n =$ Cipher text and (Cipher text)$^d$ mod $n =$ Plain text The RSA algorithm is based on the mathematical fact that computation is easy in forward direction but almost impossible in reverse direction for very large numbers. That means it is computationally impossible to get back the secret key $d$ by knowing only $e$ and $n$ [16].

## 3.3 The Hash Function

In cryptography, hash function is defined as a one-directional function that generates unique strings of same length for all its inputs. It has the two most important properties. First, it is easy to calculate but hard to reverse the process and the second one is the Avalanche effect. SHA, MD4, etc. are some examples of Hash function.

# 4 The Proposed Scheme

Suppose the coordinator $C$ want to share a confidential data $S$ with a set of $n$ participants $P_1, P_2, P_3, \ldots, P_n$ satisfying the rules that only $k$ or more than $k$ participants can reconstruct the secret after combining their shares. The coordinator Chooses a

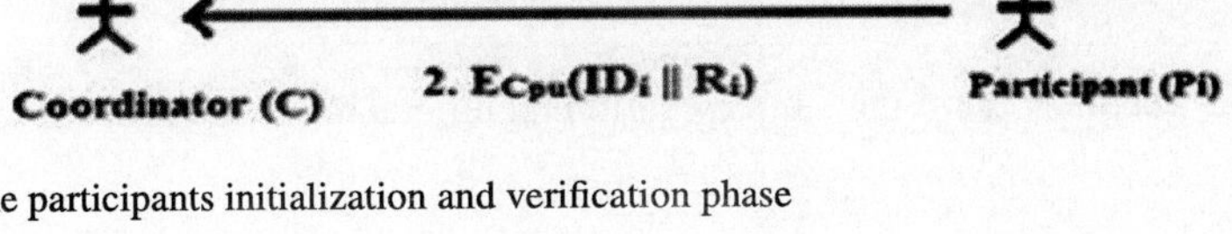


**Fig. 3** The participants initialization and verification phase

prime number $p$ and a threshold $k$ value and they agreed to work with some predefined algorithm such as RSA for encryption and decryption and SHA-2 for computing the hash values. The Coordinator having the public key-private key pair as $(C_p u, C_p r)$, publishes his public key honestly. Similarly the participants $P_i$ publishes their public key from the set of public key-private key pair as $(Pu, Pr)_i$ for $1 \leq i \leq n$. The proposed scheme is describe in the following phases:

1. ***Initialization and verification phase***: The coordinator $C$ generate the identification number $\text{ID}_i$ for each participants $P_i$ and send it back to the respective participant $P_i$ by encrypting it with the public key of $P_i$ as $E_{Pu}(\text{ID}_i)$, where $1 \leq i \leq n$. After performing decryption of the ID, each participants have to send their ID's to the coordinator $C$ concatenated with the secretly generated number $R_i$ encrypted by the public key of the coordinator $C$ as $E_{C_{Pu}}(\text{ID}_i \parallel R_i)$, which verifies the identity of the participants as shown in Fig. 3.
2. ***Share generation and distribution phase***: Here, the coordinator generates the $a_j \varepsilon \{0, 1, 2, \ldots, p-1\}$ and construct a $(k-1)$ degree polynomial as in Eq. (2).

$$Y_i = a(R_i) = \left(a_0 + \sum_{j=1}^{k-1} a_j R_i^j\right) \bmod p \quad (2)$$

where $a_0 = S$, that is the secret number, and evaluate the share for each of the participants as in Eq. (3)

$$\text{Share}_i = (R_i \oplus (R_i, Y_i)), \quad (3)$$

and then simply distribute the share to the respective $n$ participants using RSA, where $1 \leq i \leq n$.
3. ***Generating Hash values of the shares***: The coordinator calculates the hash values of secret $S$ along with all of the shares and stored them in a public bulletin as given in Eqs. (4) and (5).
4. ***Share Verification and Secret Reconstruction phase***: The participants can re verify the correctness of the received shares by computing the hash values and match it with the previously stored hash value in the public bulletin for the corresponding share. If both the hash value matches for all the collaborating parties combine to reconstruct the secret using Lagrange's interpolation theorem, and the correctness of the reconstructed secret is again verified by the hash values. If in any case, the information's authenticity mismatch, then it can detect the false information and also can identify the cheater. The reconstruction procedure is discussed as follows: First the participants perform the following XOR operation to retrieve $(R_i, Y_i)$

$$(R_i, Y_i) = (R_i \oplus \text{Share}_i) \quad (4)$$

and then perform Lagrange's interpolation technique to reconstruct the secret as

$$S = a(0) = \left( \sum_{i=1}^{k} Y_i \prod_{j=1, i \neq j}^{k} \frac{R_j}{R_j - R_i} \right) \bmod p \quad (5)$$

## 5 Performance and Security Analysis

We theoretically analyze the security of the proposed scheme to prove its robustness against the following kinds of attacks. Also, we did a comparative study of the proposed scheme with some existing schemes.

### 5.1 Proof of Robustness

The security of our presented technique is analyzed against the following kinds of attacks, presented in the form of theorems.

**Theorem 1** *The proposed scheme can identify the cheater, no matter whether it is a participant or the coordinator himself.*

***Proof*** In the secret recovery phase, if a participant $P_i$ gives a share $\text{Share}_i'$ instead of his actual received share $\text{Share}_i$. And try to cheat other. Then, anyone can check the correctness of his submitted share by simply comparing the hash value of $\text{Share}_i'$ with the corresponding hash value $\text{Share}_i$ stored in the public bulletin. If $h(\text{Share}_i') \neq h(\text{Share}_i)$, the the participant $P_i$ is identified as a cheater. Similarly, if the $h(\text{Shared Secret}) \neq h(\text{Reconstructed Secret})$ then, the Coordinator is marked as a cheater.

**Theorem 2** *An attacker is not able to reveal the secret by carrying out 'Man-in-the-middle attack'.*

***Proof*** Let us consider, the attacker A relay massages between the Coordinator and the participant $P_i$ and able to send an fake Initialization request as $E_{Pu}(ID_i')$ to $P_i$, then $P_i$ finds it authentic and generate a secret Random number $R_i$ and sent it as $E_{C_{Pu}}(ID_i' \parallel R_i)$. After fetching the messages of $P_i$ or the Coordinator C, the attacker is not able to decrypt the messages without knowing their Private keys. Thus, the fetch messages are useless to the attacker and he/she fails to reveal the actual secret.

**Theorem 3** *The Shares of the secret reveal no information about the secret.*

***Proof*** Here, the coordinator generates the shares for the participants $P_1, P_2, \ldots P_n$ using their corresponding random no.s $R_1, R_2, \ldots R_n$ applied to the polynomial interpolation theorem as shown in equation. It doesn't distribute the part of the secret, it distributes only the share of the secret, which doesn't reveal any information regarding the secret.

## 5.2 *Comparative Study with Some Existing Schemes*

In this subsection, our proposed scheme is compared with other schemes based on some parameters like cheating detection, cheater identification, about the verification process, and whether the scheme is publicly verifiable or not is as shown in Table 1 as follows.

In some comparable schemes, the cheating identification capability is lower than or equal to our proposed scheme. Some of these schemes are not able to detect cheater when, the no. of cheaters are greater than or equal to $k$ (threshold value). But our scheme is capable to detect cheating even if all the participants and the coordinator have submitted the wrong shares. It not only able to detect the wrong shares submitted for secret reconstruction, but also identify the cheater also.

# 6 Conclusion

In this paper, a new publicly verifiable secret sharing scheme is presented using the concept of public-key cryptosystem and hash function. We have added verifiable property into Shamir's $(k, n)$ threshold SSS, still maintaining the merits of it. It can

**Table 1** Comparative survey

| Properties | | | | | |
|---|---|---|---|---|---|
| Schemes | Cheating detection | Cheating identification | Publicly verifiable | Cheater identification capability | Encryption process used |
| Adhikari et al. [13] | Yes | Yes | Yes | $\left(\frac{k-1}{n}\right)$ | No |
| Liu et al. [14] | Yes | Yes | Yes | $\left(\frac{n-k+1}{2}\right)$ | Bivariate Polynomials |
| Tian et al. [25] | Yes | Yes | Yes | All the submitted share for reconstrution $(\geq k)$ | Diffie-Hellman |
| Zhao et al. [11] | Yes | Yes | Yes | All the submitted share for reconstrution | RSA |
| Hu et al. [26] | Yes | Yes | No | All the submitted share for reconstrution | AES |
| Mashhadi et al. [27] | Yes | Yes | Yes | All the submitted share for reconstrution | Bilinear pairing |
| The proposed scheme | Yes | Yes | Yes | All the submitted share for reconstruction | RSA |

detect cheating as well as can recognize the cheaters, no matter whether the cheater is a participant or the coordinator or some eavesdropper. The Shares are verified publicly, which means anyone can check the correctness of the submitted shared information as well as the reconstruction secret. As per the security concern, our proposed scheme is robust to some security threats as discussed. Verification for multiple secrets using SSS can be implemented in our future research work.

## References

1. Shamir A (1979) How to share a secret. Commun ACM 22(11):612–613
2. Blakley GR (1979) Safeguarding cryptographic keys. Proc Natl Comput Conf 48:313–317
3. Wu T-C, Wu T-S (1995) Cheating detection and cheater identification in secret sharing schemes. IEE Proc Comput Digit Tech 142(5):367–369
4. Harn L, Lin C (2009) Detection and identification of cheaters in $(t, n)$ secret sharing scheme. In: Designs, codes and cryptography
5. Liu Y, Zhang F, Zhang J (2016) Attacks to some verifiable multi-secret sharing schemes and two improved schemes. Inf Sci 329:524–539
6. Sarkar P, Sukumar N, Chowdhury MU (2017) Publicly verifiable secret sharing scheme in hierarchical settings using CLSC over IBC. In: International conference on applications and techniques in cyber security and intelligence. Edizioni della Normale, Cham
7. Chor B et al (1985) Verifiable secret sharing and achieving simultaneity in the presence of faults. In: 26th annual symposium on foundations of computer science (SFCS 1985). IEEE
8. Tompa M, Heather W (1989) How to share a secret with cheaters. J Cryptol 1(3):133–138
9. McEliece RJ, Sarwate DV (1981) On sharing secrets and Reed-Solomon codes. Commun ACM 24(9):583–584
10. Harn L, Lin C (2010) Strong $(n, t, n)$ verifiable secret sharing scheme. Inf Sci 180(16):3059–3064
11. Zhao J, Zhang J, Zhao R (2007) A practical verifiable multi-secret sharing scheme. Comput Stand Interf 29(1):138–141
12. Patil S, Deshmukh P (2014) Verifiable image secret sharing in matrix projection using watermarking. In: 2014 international conference on circuits. Systems, communication and information technology applications (CSCITA). IEEE
13. Adhikari, A et al (2016) Efficient threshold secret sharing schemes secure against rushing cheaters. In: International conference on information theoretic security. Springer, Cham
14. Liu Y et al (2018) Cheating identifiable secret sharing scheme using symmetric bivariate polynomial. Inf Sci 453:21–29
15. Jhanwar MP (2011) A practical (non-interactive) publicly verifiable secret sharing scheme. In: International conference on information security practice and experience. Springer, Berlin
16. Rui W, Chen J, Duan G (2011) A $k$-RSA algorithm. In: 2011 IEEE 3rd international conference on communication software and networks. IEEE
17. Nag A, Singh JP, Singh AK (2019) An efficient Boolean based multi-secret image sharing scheme. Multimedia Tools Appl 79:1–25
18. Chattopadhyay AK et al (2020) A verifiable multi-secret image sharing scheme using XOR operation and hash function. Multimedia Tools Appl 1–30
19. Liu Y-X, Sun Q-D (2018) Yang C-N (2018) $(k, n)$ secret image sharing scheme capable of cheating detection. EURASIP J Wirel Commun Netw 1:72
20. Schoenmakers B (1999) A simple publicly verifiable secret sharing scheme and its application to electronic voting. In: Annual international cryptology conference. Springer, Berlin
21. Larson M et al (2015) Secure auctions without an auctioneer via verifiable secret sharing. In: Proceedings of the 2015 workshop on privacy-aware mobile computing

22. Bharti SS, Gupta M, Agarwal S (2018) A novel approach for verifiable $(n, n)$ audio secret sharing scheme. Multimedia Tools Appl 77(19):25629–25657
23. Zheng W, Wang K, Wang F-Y (2020) Gan-based key secret-sharing scheme in blockchain. IEEE Trans Cybernet
24. Alohali BA et al (2018) A secure scheme for group communication of wireless IoT devices. In: 11th international symposium on communication systems. IEEE, Networks & Digital Signal Processing (CSNDSP)
25. Tian Y et al (2012) Publicly verifiable secret sharing schemes using bilinear pairings. Int J Netw Secur 14:142–148
26. Hu C et al (2017) A secure and verifiable access control scheme for big data storage in clouds. IEEE Trans Big Data 4(3):341–355
27. Mashhadi S (2017) Secure publicly verifiable and proactive secret sharing schemes with general access structure. Inf Sci 378:99–108

# Stand-alone Lightweight Security System for Native Language Using SPI Interface

**Navneet Kaur Brar, Jaskaranbeer Kaur, Ajay Kumar, Manu Bansal, and Alpana Agarwal**

**Abstract** In the defense, the native language is highly preferred among the soldiers for data communication. The data contains sensitive information which is susceptible to attacks. Therefore, cryptography algorithms are used to overcome these attacks. In this paper, a stand-alone lightweight security system has been designed for the native language using the PRESENT algorithm. In this algorithm, plaintext and key are input in the native Hindi language. This algorithm is configured on Basys 3 FPGA, and Arduino Uno has been used to display Hindi characters. The communication protocol Serial Peripheral Interface (SPI) is used between FPGA and Arduino. The PRESENT algorithm with the SPI communication protocols is programmed on Vivado 2016.2 software using VHDL language, and Arduino controller coding is done on IDE using C/C++ language. The results show that the proposed methodology consumes almost the same resources and power as in PRESENT algorithm.

**Keywords** Field programmable gate array (FPGA) · PRESENT Lightweight Block Cipher · Serial Peripheral Interface (SPI) · Arduino IDE (Integrated Development Environment)

---

N. K. Brar (✉) · J. Kaur · A. Kumar · M. Bansal · A. Agarwal
Electronics and Communication Engineering Department, Thapar Institute of Engineering and Technology, Patiala, Punjab, India
e-mail: navubrar1992@gmail.com

J. Kaur
e-mail: jsandhu228@gmail.com

A. Kumar
e-mail: erajay.kumar@thapar.edu

M. Bansal
e-mail: mbansal@thapar.edu

A. Agarwal
e-mail: alpana@thapar.edu

 
A. Maji et al. (eds.), *Proceedings of the International Conference on Computing and Communication Systems*, Lecture Notes in Networks and Systems 170,
https://doi.org/10.1007/978-981-33-4084-8_56

# 1 Introduction

In this digital era, security has become an important concern of the nations. The attacks such as eavesdropping, tampering, and modification have increased tremendously, be it in communication, internet banking, or in defense. Therefore, to prevent such situation, government has paid heed toward cryptography algorithms and started using such algorithms in defense and communicates the encrypted and decrypted data in their national language. The cryptography is basically encoding a normal message with the help of a key to an unreadable form, termed ciphertext i.e., decoded at the client side with a key and the initial message is attained as shown in Fig. 1 [1]. The key is categorized to symmetric and asymmetric algorithm where former uses the same key for encryption and decryption and the latter needs two keys, one for encryption, i.e., public key, the other for decryption called as private key [2]. Different algorithms have been designed such as AES, Blowfish, DES for securing the data [3]. But all these conventional algorithms had limitations such as large memory, area, and power consumption, whereas the lightweight cryptography algorithms, a sub-category of cryptography, require less resources [4, 5]. In the literature, number of lightweight cryptography algorithms proposed [6]. Selecting an appropriate algorithm from all these is a tedious job. According to the ISO/IEC 29,192-2, for data security, PRESENT algorithm is considered as highly secure algorithm and it is also accepted as standard for lightweight cryptography algorithms [7]. So, we have selected PRESENT algorithm for our work. PRESENT algorithm is a block cipher and based on substitution–permutation network. The algorithm contains 64-bit block size with two key variants 80 or 128 bit, and 32 rounds for data encryption.

This paper intends to design a stand-alone lightweight security system for native data communication. While communicating the confidential data, this hardware implementing the PRESENT lightweight algorithm aids people to protect their data form attackers. In India, majority of population communicate in Hindi. Therefore, we choose the plaintext and key to be in Hindi language. The PRESENT lightweight algorithm is tested on hardware using Basys 3 Artix-7 FPGA. In VHDL, the character generation for Hindi characters was a tedious job. Moreover, the character generation code increases the LUT consumption which would have increased the memory and power consumption. That is, the reason Arduino Uno is selected as a controller owing to its many libraries where displaying Hindi is quite convenient. For interface between Basys 3 and Arduino, Serial Peripheral Interface (SPI) is preferred over universal asynchronous receiver/transmitter (UART) [8] and inter-integrated circuit (I2C) [9]. As SPI works in full duplex mode, it sends data synchronously and is

**Fig. 1** Block diagram of encryption and decryption module

faster than asynchronous serial communication [10]. The paper contributes in the following manner:

- This technique provides security by encoding and decoding at the input and output side, respectively.
- Even after the hardware implementation, the resource and power utilization of the FPGA remains approximately same as SPI module occupies negligible area.
- The proposed methodology overcomes the problem of frequency mismatch between Basys3 and Arduino Uno.

The rest of the paper is classified in sections. In Sect. 2 the Present cipher algorithm, SPI is explained. Section 3 demonstrates the proposed methodology. Section 4 illustrates the results at software and hardware of the technique. Then, the conclusion is discussed in Sect. 5.

## 2 Related Terminology

In this proposed system, the PRESENT algorithm and SPI are the basic building blocks. The PRESENT lightweight cryptography algorithm is elaborated in Sect. 2.1. The Serial Peripheral Interface (SPI) modes and register are explained in Sect. 2.2.

### *2.1 PRESENT Lightweight Cryptography Algorithm*

PRESENT is lightweight symmetric block cipher having 64 bits block size and 80 bits or 128 bits key size [11]. It is a substitution–permutation network consisting of 31 rounds. The 32nd round is simply a key mixing operation. Figure 2 represents detailed approach of PRESENT block cipher.

Round operations performed in PRESENT cipher are described below [11]:

- *Add Round Key*: An XOR operation is performed in each round of PRESENT. It generates round key $K_m$ for $m = 1$ to 32, where $K_{32}$ is used to XOR the last data.

Consider the applied input is $d_{63} \ldots d_0$ and round key $K_m = k^m{}_{63} \ldots\ldots K^m{}_0$ for $m = 1$ to 32.

Add round key operation is explained in Eq. (1)

$$d_i = d_i \text{XOR} k_i^m \quad (\text{for } i = 0 \text{ to } 63) \tag{1}$$

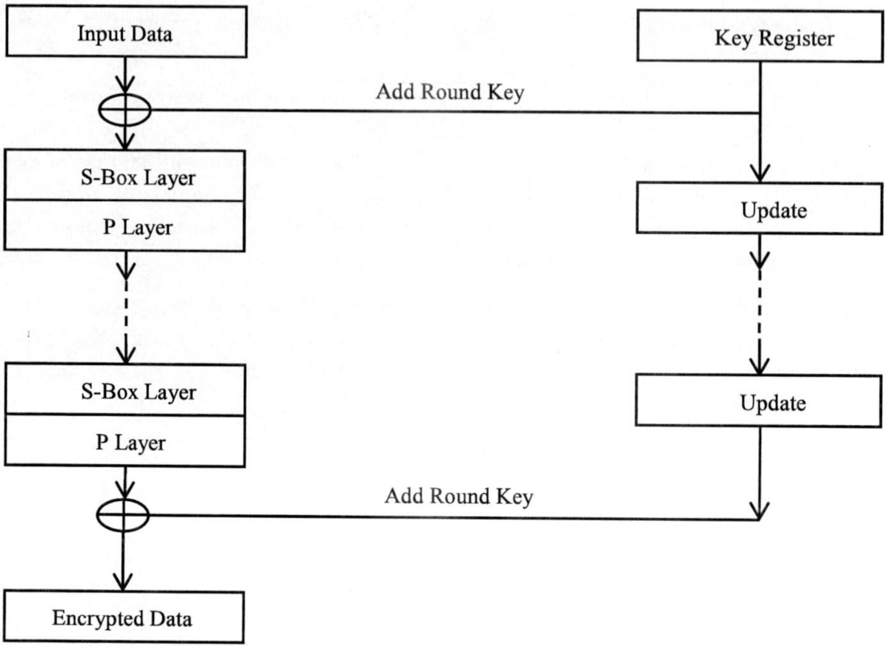


**Fig. 2** Algorithmic approach of PRESENT cipher [11]

- *S-box layer*: This layer performs nonlinear substitution where a single 4-bit box is applied 16 times in parallel for each round.

The current state of add round key $d_{63} \ldots d_0$ is considered as sixteen 4 bit words $f_{15} \ldots f_0$ which is provided by Eq. (2).

$$F_j = f_{(4*j)+3} \| s_{(4*j)+2} \| s_{(4*j)+1} \| s_{(4*j)} \quad (\text{for } j = 0 \text{ to } 15) \tag{2}$$

Each $s_j$ is substituted by corresponding fix value. Table 1 gives the hexadecimal values of PRESENT S-box.

- *P layer*: It creates confusion of data by performing linear bitwise permutation in which bit m of S-box output data is moved to bit position $X(m)$as given by Table 2.

**Table 1** PRESENT S-box [11]

| $f_j$ | 0 | 1 | 2 | 3 | 4 | 5 | 6 | 7 | 8 | 9 | A | B | C | D | E | F |
|---|---|---|---|---|---|---|---|---|---|---|---|---|---|---|---|---|
| $S[f_j]$ | C | 5 | 6 | B | 9 | 0 | A | D | 3 | E | F | 8 | 4 | 7 | 1 | 2 |

**Table 2** PRESENT P layer [11]

| $m$ | 0 | 1 | 2 | 3 | 4 | 5 | 6 | 7 | 8 | 9 | 10 | 11 | 12 | 13 | 14 | 15 |
|---|---|---|---|---|---|---|---|---|---|---|---|---|---|---|---|---|
| $X(m)$ | 0 | 16 | 32 | 48 | 1 | 17 | 33 | 49 | 2 | 18 | 34 | 50 | 3 | 19 | 35 | 51 |
| $m$ | 16 | 17 | 18 | 19 | 20 | 21 | 22 | 23 | 24 | 25 | 26 | 27 | 28 | 29 | 30 | 31 |
| $X(m)$ | 4 | 20 | 36 | 52 | 5 | 21 | 37 | 53 | 6 | 22 | 38 | 54 | 7 | 23 | 39 | 55 |
| $m$ | 32 | 33 | 34 | 35 | 36 | 37 | 38 | 39 | 40 | 41 | 42 | 43 | 44 | 45 | 46 | 47 |
| $X(m)$ | 8 | 24 | 40 | 56 | 9 | 25 | 41 | 57 | 10 | 26 | 42 | 58 | 11 | 27 | 43 | 59 |
| $m$ | 48 | 49 | 50 | 51 | 52 | 53 | 54 | 55 | 56 | 57 | 58 | 59 | 60 | 61 | 62 | 63 |
| $X(m)$ | 12 | 28 | 44 | 60 | 13 | 29 | 45 | 61 | 14 | 30 | 46 | 62 | 15 | 31 | 47 | 63 |

- *Key Scheduling*: PRESENT supports 80 bits and 128 bits keys which are stored in key register *N*. At each round, left most 64 bits of key register contents are considered as round key.
- *80 Bits Key*: Key register is updated as follows: [11]

1) $[N_{79}\ N_{78}\ N_{77} \ldots N_2\ N_1\ N_0] = [N_{18}\ N_{17}\ N_{16} \ldots N_{21}\ N_{20}\ N_{19}]$
2) $[N_{79}\ N_{78}\ N_{77}\ N_{76}] =$ S Box $[N_{79}\ N_{78}\ N_{77}\ N_{76}]$
3) $[N_{19}\ N_{18}\ N_{17}\ N_{16}\ N_{15}] = [N_{19}\ N_{18}\ N_{17}\ N_{16}\ N_{15}]$ XOR round counter

- *128 Bits Key*: Key register is updated as follows: [12]

1) $[N_{127}\ N_{126}\ N_{125} \ldots N_2\ N_1\ N_0] = [N_{66}\ N_{65}\ N_{64} \ldots N_{69}\ N_{68}\ N_{67}]$
2) $[N_{127}\ N_{126}\ N_{125}\ N_{124}] =$ S Box $[N_{127}\ N_{126}\ N_{125}\ N_{124}]$
3) $[N_{123}\ N_{122}\ N_{121}\ N_{120}] =$ S Box $[N_{123}\ N_{122}\ N_{121}\ N_{120}]$
4) $[N_{66}\ N_{65}\ N_{64}\ N_{63}\ N_{62}] = [N_{66}\ N_{65}\ N_{64}\ N_{63}\ N_{62}]$ XOR round counter

## 2.2 *Serial Peripheral Interface (SPI)*

SPI is a serial communication protocol that works in full duplex mode. In SPI communication, there is only one master and can be single or multiple slaves, which are configured by chip select in master. The SPI interface uses four pins generally called as Chip Select, Master Output Slave Input (MOSI), Master Input Slave Output (MISO), and Synchronous Clock (SCLK) as shown in Fig. 3 [10].

There are four modes in SPI depending on Clock Phase (CPHA) and Clock Polarity (CPOL) values. When the CPOL is 0, the leading edge is rising edge and lagging edge is falling edge and vice versa [13]. When CPHA is 0, data is received on rising edge and transmitted on falling edge and vice versa [13].

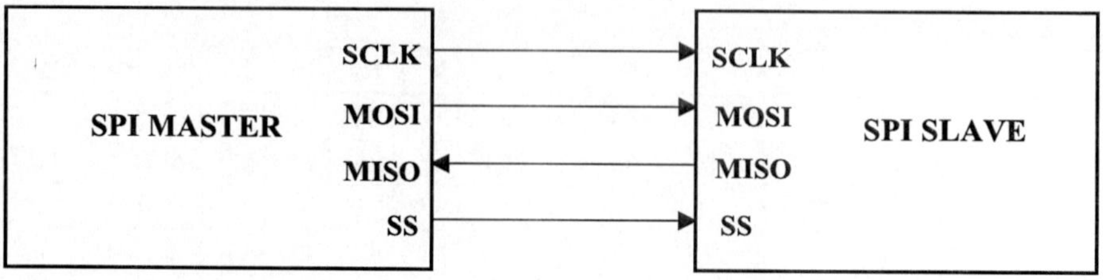


**Fig. 3** Master–slave configuration in SPI [13]

The SPI contains three internal registers named as Data Register (SPDR), Control Register (SPCR), and Status Register (SPSR) to configure the SPI transmission [13]. The SPI control register (SPCR) is of 8-bit and each bit has different functionality. The 8 bits are as follows starting from LSB SPR0, SPR1 (decide the clock rate), clock phase (CPHA), clock polarity (CPOL), MSTR (master–slave select-if '0′ slave is selected), DORD (data order-data send or received will be from LSB if 0 and MSB if 1), SPE (SPI enable-if 1 initialize the SPI operation), SPIE (interrupt enable) [11]. The SPI data register (SPDR) acts as buffer when data is transmitted from it and as shift register when the device receives data [13]. The data width can be 8 or 16 bits. The SPI status register (SPSR) mainly used bit SPI2X for setting the clock rate [13].

For SPI communication, initially, the frequency of slave is set to less than equal to twice of SCLK [13]. Then modes of CPOL, CPHA, and DORD are selected. In master, chip select is made low to configure the slave. After sending or receiving the data, chip select is made high. In slave, the SPIE is enabled so that data from MOSI can be transferred in SPDR and vice versa.

After rigorous research, we have found that none of the researcher has ever done PRESENT cipher hardware implementation using Arduino as controller and SPI as interface.

## 3 Proposed Methodology

The entire plan can be divided into three blocks: input from keypad, SPI communication interface of FPGA with Arduino, and the output is displayed on LCD. The algorithm of proposed technique is explained below:

**Algorithm:** Security System for Native Language Using SPI Interface
**Input :** Keypad (Hindi)
**Output :** LCD Display (Hindi)
**Method**
**Step 1:** Input Key and Plaintext (from Keypad)→ Buffer
**Step 2:** Control Signals *i.e.* Enc/Dec Select, Key Load, Data Load, Reset *via* Switches on Basys 3
**Step 3:** If ( Key Load = 1) then
 For i = 0 to 79
 Key = Input;
 If (Data Load = 1) then
 For i = 80 to 143;
 Plaintext = Input ;
 End;
/* Both key and Plaintext of 144 bits stored in Buffer*/
**Step 4:** Encryption/ Decryption:
 If ( Enc_Dec_Select = 1) then
 Encryption is done;
 Else
 Decryption is done;
/*Both encryption and decryption output is stored in Buffer*/
**Step 5:** Buffer Data → Arduino *via* SPI (Serial Peripheral Interface)

Figure 4 illustrates the proposed technique at block level and in the following steps each block is described in detail.

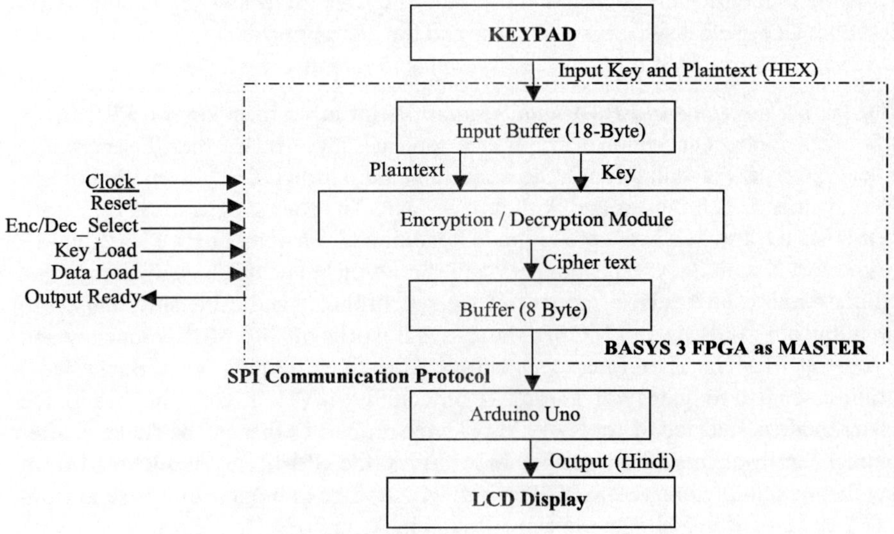


**Fig. 4** Design flow of proposed methodology

| | | | |
|---|---|---|---|
| 1 | 2 | 3 | A |
| 4 | 5 | 6 | B |
| 7 | 8 | 9 | C |
| 0 | E | F | D |

⟺

| | | | |
|---|---|---|---|
| म | आ | न | ट |
| क | ई | ल | प |
| त | फ | ह | र |
| ज़ | ख़ | उ | ड |

**Fig. 5** Lookup table of Hindi characters with HEX characters

**Step I**: *Input from Keypad*: The keypad contains Hindi characters and corresponding to each Hindi character there is Hex character in the lookup table as shown in Fig. 5. In Hindi language, there are total 52 characters but in this demonstration only 16 Hindi characters are shown as the Hex Keypad used has 16 Keys. In addition to it, if one wants to assign more than 16 characters, one can assign 3 values of Hindi characters corresponding to one value. In this manner, coding is done at the input side to provide security before the encryption is done. The input key and plaintext in Hindi are given from keypad, which is attached to one of the Basys 3 headers. The pressed keys correspond index values in HEX form. The index values are used to do coding of input data to secure the data at input side. Initially key load switch is made high, the key input of 80 bits is given, and after that, the key load switch is made low. Similarly, the data load switch is made high to store the next input as plaintext of 64 bits, which is later made low.

The input data as plaintext and key is stored in some buffer of 18 bytes in FPGA that is then send to encryption module, when Enc_Dec_Select switch is high. After that, Enc_Dec_Select switch is made low and the decryption is done. The decrypted data is again stored in another buffer of 8 bytes in memory as shown in Fig. 5.

**Step II**: *SPI interface of FPGA with Arduino*: After input from keypad PRESENT lightweight cipher encryption decryption is done in Basys 3 board, then the encrypted or decrypted data is send through SPI communication protocol. In this methodology, FPGA works as SPI master and Arduino as slave. The four signals of SPI, i.e., SS, MOSI, MISO, and SCLK are predefined in Arduino as pin number 10, 11, 12, and 13 are connected with Basys 3 header pins. SPI is nothing but using clock divider so that SPI master and slave device frequencies are synchronized and using shift register to communicate in MISO and MOSI. The Basys 3 works on 100 MHz frequency and Arduino at 16 MHz. Therefore, to synchronize the clock frequencies of Basys 3 and Arduino, we use frequency divider of 32-bit counter in VHDL code. In Vivado, the master mode is selected in Tools → Edit Device properties that can be changed after opening the implementation design. In Arduino, the SPI library is included from Sketch → Include library → SPI. Through MOSI, data in binary form is send from FPGA and stored in SPI data register (SPDR) in Arduino.

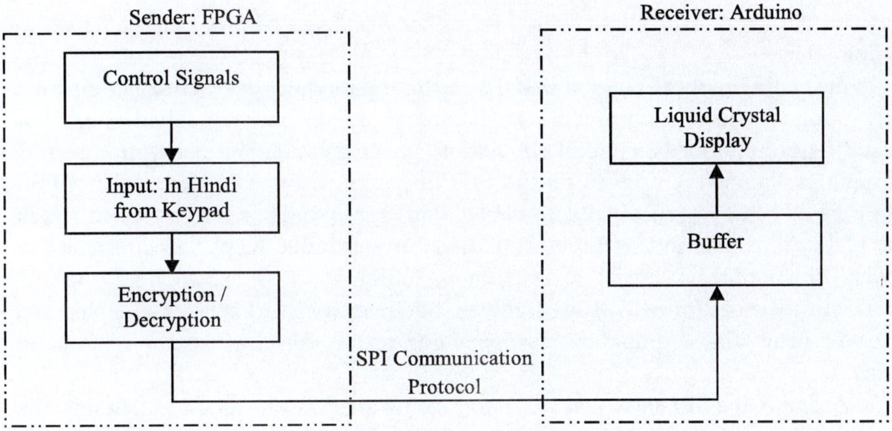


**Fig. 6** Flow diagram at sender and receiver end

**Step III**: *Output displayed on LCD*: The input data, i.e., plaintext is given by keypad and the character displayed on LCD connections with Arduino as per [14]. The liquid crystal library is included in Arduino code from Sketch → Include library. For regional languages, one can create characters on LCD character generator online [15] in form of Array (8 * 5). Then, in the Arduino code using *create.char()* command, one can create those characters. Corresponding to input Hindi characters, Hex index values lookup table is made that do decoding of the input data. Finally, the Hindi characters are displayed on LCD. The creation of the 16 characters lookup table occupies only 9% (2000) bytes of total memory.

The flow diagram of the combination of three steps is depicted in Fig. 6.

Therefore, in this section, the proposed technique is explained in detail. These three steps are pursued to implement the PRESENT lightweight cipher algorithm at hardware level.

## 4 Experimental Results

For implementing the technique at software level, the PRESENT lightweight cipher algorithm is coded in VHDL language on Vivado 2016.2 tool. The Arduino Uno coding is done in C/C++ language on integrated development environment (IDE) platform. To implement the technique at hardware level, the following components are used. Basys3 Xilinx Artix-7 FPGA Board (to burn the the Present algorithm); Arduino UNO board (controller); keypad (input); LCD Display (output).The SPI slave code and LCD interfacing are done on Arduino IDE.

Initially, in SPI master module, the counter of 32-bit is made to synchronize the clock frequency of Basys 3 [13] and Arduino. In shift register of master, the MOSI is given the decrypted output which is further displayed on the LCD attached to Arduino Uno.

In the coding part of Basys 3(VHDL) master top module; the PRESENT cipher's encryption and decryption module, keypad press and decoder are called as component. Therefore, after behavioral simulation, the encryption and decryption results of input as plaintext: $[\text{FFFF FFFF FFFF FFFF}]_{\text{Hex}}$ of 64 bits and Key: $[\text{FFFF FFFF FFFF FFFF FFFF}]_{\text{Hex}}$ of 80 bits is simulated in Figs. 7 and 8, respectively, in which the plaintext is referred as Input_Data and Key as Input_Key. The ciphertext in encryption module as Output_Data as shown in Fig. 7.

The test vectors for PRESENT lightweight cipher with SPI are implemented and after the behavioral simulation, corresponding results for ciphertext are shown in Table 3.

In Arduino, the SPI slave and LCD interfacing are done in same code. In this, the baud rate, i.e., a measure of the speed with which data is being communicated, is selected as 57,600 and SPI clock divider of 4 is done to synchronize both the devices data rate. After coding in the respective tools, the connections of FPGA and Arduino

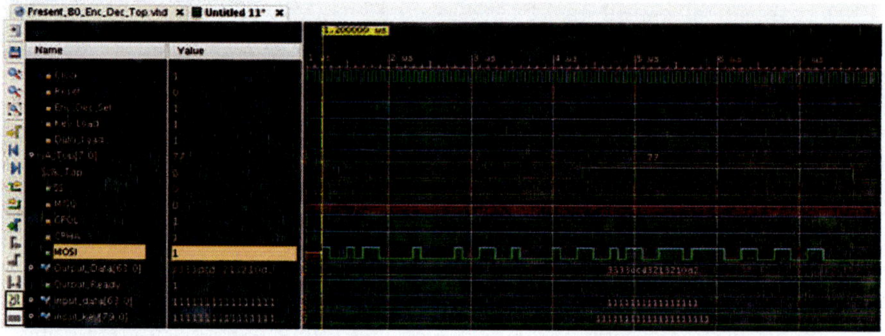


**Fig. 7** PRESENT encryption with SPI

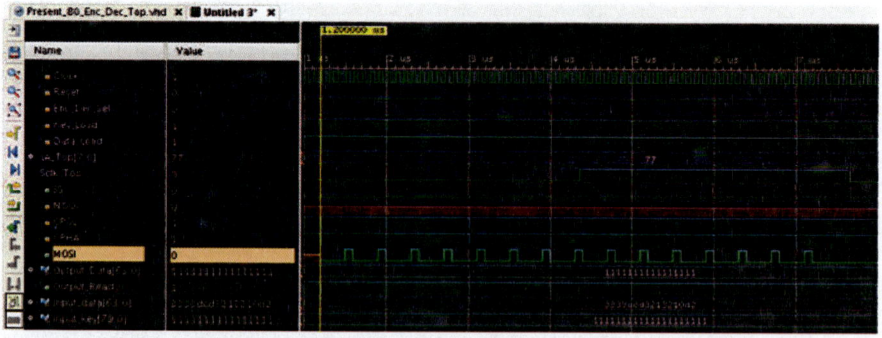


**Fig. 8** PRESENT decryption with SPI

**Table 3** Test vectors for PRESENT lightweight cipher

| Plaintext | Key | Ciphertext |
|---|---|---|
| $[0000\ 0000\ 0000\ 0000]_{Hex}$ | $[0000\ 0000\ 0000\ 0000\ 0000]_{Hex}$ | $[5579\ C138\ 7B22\ 8445]_{Hex}$ |
| $[0000\ 0000\ 0000\ 0000]_{Hex}$ | $[FFFF\ FFFF\ FFFF\ FFFF\ FFFF]_{Hex}$ | $[E72C\ 46C0\ F594\ 5049]_{Hex}$ |
| $[FFFF\ FFFF\ FFFF\ FFFF]_{Hex}$ | $[0000\ 0000\ 0000\ 0000\ 0000]_{Hex}$ | $[A112\ FFC7\ 2F68\ 417B]_{Hex}$ |
| $[0123\ 4567\ 89AB\ CDEF]_{Hex}$ | $[0123\ 4567\ 89AB\ CDEF\ 0123]_{Hex}$ | $[F8DD\ 5053\ 1D97\ 3BDE]_{Hex}$ |
| $[FEDC\ BA98\ 7654\ 3210]_{Hex}$ | $[FEDC\ BA98\ 7654\ 3210\ FEDC]_{Hex}$ | $[A490\ 536F\ 487F\ 5990]_{Hex}$ |

are done as discussed in Sect. 3. The setup apparatus of the proposed methodology is displayed in Figs. 9 and 10 showing encryption and decryption, respectively.

The comparison of parameters of PRESENT lightweight algorithm with and without SPI is shown in Table 4. The results show that area utilization which is

**Fig. 9** Setup apparatus of proposed methodology with encryption

**Fig. 10** Setup apparatus of proposed methodology with decryption

**Table 4** Comparison of hardware and software implementation parameters

| Parameters | PRESENT algorithm | PRESENT algorithm with SPI interface |
|---|---|---|
| Frequency | 100 MHz | 16 MHz |
| Slices | 693 | 757 |
| Flip-flop | 382 | 431 |
| Power consumption (mW) | 0.07 | 0.074 |

depicted in number of slices and flip-flops used is almost same. Moreover, the power consumption is approximately same in both the cases.

## 5 Conclusion

The paper demonstrated a stand-alone security system using PRESENT algorithm for native Hindi language. The PRESENT lightweight cipher is presented using Serial Peripheral Interface (SPI) communication interface that is implemented on Basys 3 Xilinx Artix-7 FPGA and Arduino Uno board. For input, keypad and a LCD for output display are required. The experimental results show that the hardware used does not utilize much of the resources as the power and LUT utilization are same at software and hardware level. With the help of this methodology, the common person can protect the data from intruders and use the encryption algorithm conveniently as the data communicated can be of any regional language. As far as the future work is concerned, an application-specific integrated circuit (ASIC) chip of this algorithm could be realized and in the proposed technique, Basys 3 FPGA could be replaced with ASIC chip. Finally, these algorithms ASIC could be inserted in mobiles or computers to prevent their data from malicious attackers.

**Acknowledgements** The authors are grateful to MeitY for the financial support through 'SMDP Chips to System Design' project. The authors also want to express their sincere gratitude toward the Director, Thapar Institute of Engineering and Technology, Patiala for his persistent support and encouragement.

## References

1. https://en.wikipedia.org/wiki/Cryptography. Accessed 19 Jan 2020
2. Forouzan BA, Mukhopadhyay D (2007) Cryptography & network security. Tata McGraw-Hill Education Pvt. Ltd., New Delhi
3. Thakur J, Kumar N (2011) DES, AES and Blowfish: symmetric key cryptography algorithms simulation based performance analysis. Int J Emerg Technol Adv Eng 1(2):6–12
4. Jana S, Bhaumik J, Maiti MK (2013) Survey on lightweight block cipher. Int J Soft Comp Eng (IJSCE) 5:183–187

5. Sallam S, Beheshti BD (2018) A survey on lightweight cryptographic algorithms. In: Proceedings of TENCON—IEEE Region 10 Conference
6. Mohd BJ, Hayajneh T, Vasilakos A (2015) A survey on lightweight block ciphers for low-resource devices: comparative study and open issues. J Network Comp App 58:73–93
7. ISO/IEC 29192-2 (2012) Information tech.-security techniques—part 2: block ciphers
8. Fang YY, Chen XJ (2011) Design and simulation of UART serial communication module based on VHDL. In: 3rd international workshop on intelligent systems and applications
9. Mankar J, Darode C, Trivedi K, Kanoje M, Shahare P (2014) Review of I2C Protocol. Int J Res Adv Tech 2:474–479
10. Shingare TD, Patil RT (2013) SPI implementation on FPGA. Int J Innov Tech Explor Eng (IJITEE) 2(2):7–9
11. Bogdanov A, Knudsen LR, Leander G, Paar C, Poschmann A, Robshaw MJB, Seurin Y, Vikkelsoe C (2007) PRESENT: an ultra-lightweight block cipher. In: International workshop on cryptographic hardware and embedded systems in Berlin, Heidelberg, pp 450–466
12. Hernandez-Castro JC, Peris-Lopez P, Aumasson JP (2012) On the key schedule strength of PRESENT. DPM SETOP Lect Notes Comput Sci 7122:253–263
13. Wooton C (2016) Samsung ARTIK reference. doi: https://doi.org/10.1007/978-1-4842-2322-2_21
14. https://reference.digilentinc.com/_media/reference/pmod/pmodclp/pmodclp_rm.pdf. Accessed 10 Dec 2019
15. https://maxpromer.github.io/LCD-Character-Creator/. Accessed 21 Dec 2019

# Design of Power Efficient SEC Orthogonal Latin Square (OLS) Codes

**Sayan Tripathi, Jhilam Jana, and Jaydeb Bhaumik**

**Abstract** Single event upset type of soft error caused by the ionizing radiation is a key constraint for reliable memory design. To correct soft errors in memories, single error correction (SEC) codes are generally used. However the use of multiple bits error correcting codes in memory is limited due to higher decoding complexity. Orthogonal Latin Square (OLS) codes are a type of one-step majority logic decodable (OS-MLD) codes which have been used for memory system. These codes require more number of parity check bits and larger area compare to Hamming and Hsiao codes, when error correction capability exceeds by two. In this paper, a scheme for $H$-matrix construction of SEC-OLS code has been introduced. The architectures of proposed and existing codes are synthesized on ASIC platform. The synthesis results show that proposed codec is power efficient (highest improvement 7.63%) compared to the existing schemes without increasing area.

**Keywords** Soft errors · Memories · Orthogonal latin square (OLS) code · SEC · ASIC

## 1 Introduction

Reliability of memory system is an emerging issue in recent years. Soft errors in memory are caused by alpha particle radiation, neutrons, protons. These errors are the main factors for determining the reliability of memory system [1–4]. Error correcting codes (ECCs) are used in memory to correct the errors. To protect memory, several single error correcting codes [5, 6], adjacent error correcting codes [7–10],

S. Tripathi (✉) · J. Jana · J. Bhaumik
Department of ETCE, Jadavpur University, Kolkata, India
e-mail: tripathysayan@gmail.com

J. Jana
e-mail: jhilamjana2014@gmail.com

J. Bhaumik
e-mail: bhaumik.jaydeb@gmail.com



A. Maji et al. (eds.), *Proceedings of the International Conference on Computing and Communication Systems*, Lecture Notes in Networks and Systems 170,
https://doi.org/10.1007/978-981-33-4084-8_57

and Reed Solomon (RS) code [11] have already been suggested. But several existing codes require higher area and delay that restrict their application in memory system. To diminish these impacts, one-step majority logic decodable (OS-MLD) codes which have lower delay and area with higher error correction capabilities are very effective. Different types of OS-MLD codes have been introduced for protection of memory cells such as Euclidean geometry (EG) codes, difference set (DS) codes. The Orthogonal Latin Square (OLS) code is a type of OS-MLD code and also has higher error correction capabilities. However, both EG and DS codes have a limited block sizes and error correction capabilities. Also OLS codes require more parity bits than other existing codes to protect memory against the same number of errors. Hsiao et al. have proposed a scheme for a multiple error correcting codes which can be constructed from OLS codes [12]. Reviriego et al. have described a method for extending the OLS codes to protect the larger data bits in memories. These codes have lower decoding complexity and delay. The different design and implementation of DEC OLS codes have also been described in [13]. Reviriego et al. have also introduced the single error correction OLS codes for protection of memories and caches [14–16]. Datta et al. have proposed a technique for an OLS code which can correct burst-errors in memories [17]. Also a construction scheme for double error correction codes has been proposed with the help of OS-MLD codes in [18]. Xiao et al. have described an optimized method of OLS codes which can correct single error and double adjacent errors. An optimization scheme has been developed using the cyclic shift algorithm in [19]. Lee at el. have proposed an error correction code in which the error correction capability changes according to the reliability requirement of the system [20]. To correct the double bit errors, the minimum Hamming distance ($d$min) value should be 5, but the $d$min value is 3 [20]. So, this code cannot correct all double bit errors. Also, Reviriego et al. [14] have proposed a (24, 16) modified error correction codes which satisfy only single error correction logic. It is found that this existing code [14] has $d$min value of 3. So, this code cannot correct triple adjacent error and double bit errors. But these existing codes require more power and delay. This paper aims to develop a power efficient SEC-OLS codes for protecting memory against soft errors. The proposed codes require lesser power and have simple decoding logic than other codes. The main contributions in this paper are as follows: (i) a new scheme to form the parity check matrix ($H$) for SEC-OLS code has been proposed. (ii) SEC-OLS codes with message length of 16-bit have been designed and implemented in ASIC platform and (iii) proposed codes are power efficient and have simpler decoding logic compared to existing designs. The rest of this paper is organized as follows. Section 2 provides basics of OLS codes. Section 3 presents proposed scheme. Section 4 contains ASIC based synthesis results and the paper is finally concluded in Sect. 5.

**Fig. 1** $H$-Matrix of (24, 16) SEC-OLS Code

$$H = \begin{bmatrix} 111100000000000010000000 \\ 000011110000000001000000 \\ 000000001111000000100000 \\ 000000000000111100010000 \\ 100010001000100000001000 \\ 010001000100010000000100 \\ 001000100010001000000010 \\ 000100010001000100000001 \end{bmatrix}$$

## 2 Basics of OLS Codes

Orthogonal Latin Square (OLS) codes are the type of one-step majority logic decodable codes. The theory of OLS codes is based on Latin Square (LS) [6]. A LS of size $m$ is a square matrix and elements are permutation of numbers $0, 1, \ldots, (m-1)$. Two LSs are orthogonal if the superimposed every pair of elements appears only once. The OLS codes contain $k = m^2$ data bits and $2tm$ check bits, where $m$ is an integer, and $t$ is the error correction capability of OLS codes. From $H$-matrix, it is observed that the error correction capability increases as the number of parity check bits increases. The $H$-matrix for (24, 16) SEC-OLS code ($t = 1$, $k = 16$) [14] that is derived from SEC-OLS codes is presented in Fig. 1.

The properties of OLS code that are observed from the $H$-matrix in Fig. 1 are in the following.

1. Each data bit is used to generate exactly two parity check bits.
2. A pair of data bits contributes not more than one of the parity check bits.

These two properties are necessary to employ OS-MLD algorithm in decoder.

## 3 Proposed Scheme

Proposed (24, 16) SEC-OLS code is a linear block code with parity check matrix ($H$). The $H$-matrix consists of 16 data columns and 8 parity columns. Proposed (24, 16) SEC-OLS code is presented in detail in the following subsection.

### *3.1 Proposed H-Matrix Construction Process*

In this section, the construction procedure of $H$-matrix for (24, 16) SEC-OLS code has been illustrated. The matrix of original code is restructured and proposed code is obtained by using circular left block shifting and parity bit shifting techniques. The proposed codes can correct only single error with an advantage in power consumption. This technique is described using Fig. 2, where vector size ($w$) is 4.

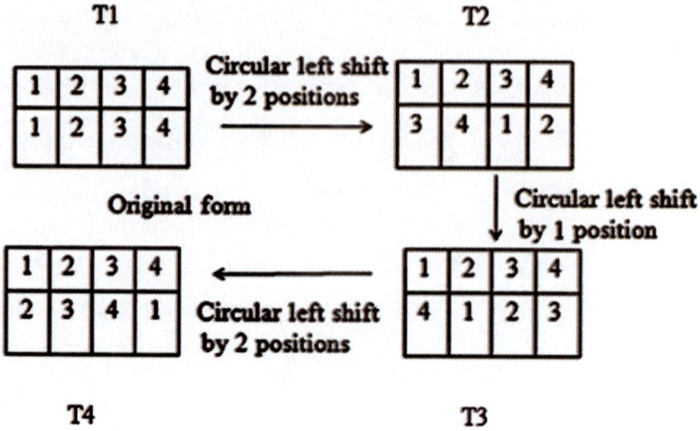


**Fig. 2** Proposed circular left block shifting algorithm

In Fig. 2, the number in the top row represents the bit position of blocks and the number in the bottom row is the position of data columns. In the bottom row of $T1$, (1, 2, 3, 4) represent that we have to place the 1st data column of 1st block, 1st data column of 2nd block, 1st data column of 3rd block and 1st data column of 4th block sequentially to obtain the block1 of proposed $H$-matrix. Similarly, in the bottom row of $T2$, (3, 4, 1, 2) represent that we have to place the 2nd data column of 3rd block, 2nd data column of 4th block, 2nd data column of 1st block and 2nd data column of 2nd block sequentially to obtain the block2 of proposed $H$-matrix. The other combinations of shifting vector ($w$) values are (4, 1, 2, 3) and (2, 3, 4, 1). Finally, the parity bits ($p_5$ to $p_8$) are appended after placing the data bits and rest of the parity bits ($p_1$ to $p_4$) are placed in the last of $H$-matrix. The construction of proposed $H$-matrix is done with the help of proposed circular left block shifting algorithm which is described in Fig. 3.

The proposed $H$-matrix for (24, 16) SEC-OLS is presented in Fig. 4.

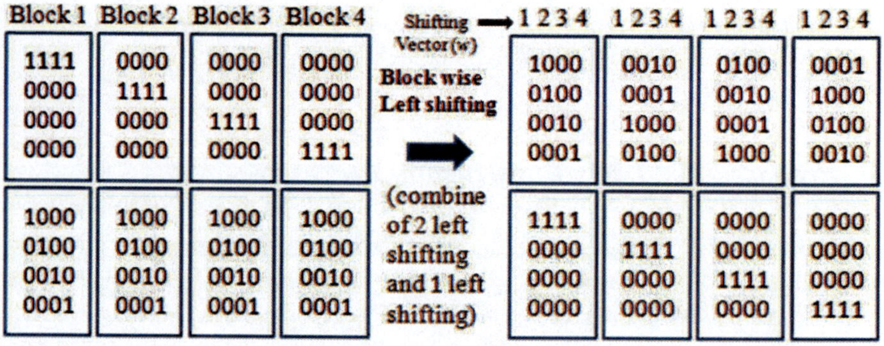


**Fig. 3** Proposed $H$-matrix for SEC-OLS code using circular left block shifting algorithm

**Fig. 4** Proposed $H$-matrix for (24, 16) SEC-OLS code

$$H = \begin{bmatrix} 100000100100000100001000 \\ 010000010010100000000100 \\ 001010000001010000000010 \\ 000101001000001000000001 \\ 111100000000000010000000 \\ 000011110000000001000000 \\ 000000001111000000100000 \\ 000000000000111100010000 \end{bmatrix}$$

## 3.2 Design of Proposed SEC-OLS Codec

Proposed SEC-OLS codec consists of an encoder and a decoder. In SEC-OLS codec, the encoder computes parity bits based on data bits. A codeword is formed with the combination of parity bits and data bits. All the parity bits are also stored along with the data bits in memory cells. The placement of data and parity bits in a codeword (c) of (24, 16) SEC-OLS code is presented in Fig. 5.

Equations for computing the parity bits for the proposed codec is as follows.

$$\begin{aligned} p_1 &= d_1 \oplus d_7 \oplus d_{10} \oplus d_{16} \\ p_2 &= d_2 \oplus d_8 \oplus d_{11} \oplus d_{13} \\ p_3 &= d_3 \oplus d_5 \oplus d_{12} \oplus d_{14} \\ p_4 &= d_4 \oplus d_6 \oplus d_9 \oplus d_{15} \\ p_5 &= d_1 \oplus d_2 \oplus d_3 \oplus d_4 \\ p_6 &= d_5 \oplus d_6 \oplus d_7 \oplus d_8 \\ p_7 &= d_9 \oplus d_{10} \oplus d_{11} \oplus d_{12} \\ p_8 &= d_{13} \oplus d_{14} \oplus d_{15} \oplus d_{16} \end{aligned} \tag{1}$$

whereas the decoder takes the stored codewords from the memory as input and produces corrected version of the stored data as output. The decoding scheme consists of syndrome ($S$) computation and error correction blocks. Error detection and correction is done using the syndrome values. The syndromes ($S$) are computed using Eq. (2).

$$S = r.H^T \tag{2}$$

| $c_1$ | $c_2$ | $c_3$ | $c_4$ | $c_5$ | $c_6$ | $c_7$ | $c_8$ | $c_9$ | $c_{10}$ | $c_{11}$ | $c_{12}$ | $c_{13}$ | $c_{14}$ | $c_{15}$ | $c_{16}$ | $c_{17}$ | $c_{18}$ | $c_{19}$ | $c_{20}$ | $c_{21}$ | $c_{22}$ | $c_{23}$ | $c_{24}$ |
|---|---|---|---|---|---|---|---|---|---|---|---|---|---|---|---|---|---|---|---|---|---|---|---|
| $d_1$ | $d_2$ | $d_3$ | $d_4$ | $d_5$ | $d_6$ | $d_7$ | $d_8$ | $d_9$ | $d_{10}$ | $d_{11}$ | $d_{12}$ | $d_{13}$ | $d_{14}$ | $d_{15}$ | $d_{16}$ | $p_5$ | $p_6$ | $p_7$ | $p_8$ | $p_1$ | $p_2$ | $p_3$ | $p_4$ |

**Fig. 5** Data-bit and parity-bit placement in a codeword of proposed (24,16) SEC-OLS code

**Table 1** Comparison of area and critical path delay of (24, 16) OLS codes

| OLS codec | Area overhead | | | | Critical path delay | | |
|---|---|---|---|---|---|---|---|
| | XOR2 | AND2 | OR2 | Equivalent NAND 2 | XOR2 | AND2 | OR2 |
| Lee et al. [20] | 72 | 16 | – | 320 | 8 | 1 | – |
| Reviriego et al. [14, 15] | 72 | 16 | – | 320 | 8 | 1 | – |
| Xiao et al. [19] | 72 | 109 | 31 | 599 | 8 | 7 | 2 |
| Reviriego et al. [14] | 72 | 16 | – | 320 | 8 | 1 | – |
| Proposed | 72 | 16 | – | 320 | 8 | 1 | – |

Error associated with each bit is detected and corrected in the error correction logic block. The error correction logic is used to correct the erroneous codeword and recover the original data. The error correction logic of first data bit of proposed (24, 16) SEC-OLS code is expressed as follows.

$$dc_1 = r_1 \oplus (s_1.s_5) \tag{3}$$

### 3.3 *Estimation of Logic Gates*

Proposed codec has been described using $H$-matrix presented in Fig. 4. Hardware requirements have been computed in terms of total number of equivalent NAND2 gates required for codec design. The area in terms of logic gates and critical path delay of proposed and existing codes have been presented in Table 1.

## 4 ASIC Synthesis Results

The proposed SEC-OLS codecs have been represented in verilog Hardware Description Language. Proposed and existing codes have been simulated and synthesized in ASIC platform using Cadence Genus synthesis solution (TSMC18) tool. In Table 2 the performance of existing and proposed codes have been presented in terms of area, delay, power, power area product (PAP), power delay product (PDP) and PADP (product of area, delay and power) for ASIC design. The proposed (24, 16) codec provides the maximum improvements of 7.63%, 11.38%, 27.31% and 30.26% in terms of power, PAP, PDP, and PADP respectively compared to the corresponding codec by Lee et al.

Table 2 ASIC synthesis results of (24, 16) OLS codes

| OLS codec | Area ($\mu m^2$) | Delay (ps) | Power ($\mu$W) | PAP ($\mu m^2$W) | PDP ($\mu$ W.ps) | PADP ($\mu m^2$W.ps) |
|---|---|---|---|---|---|---|
| Lee et al. [20] | 1969.22 | 386.7 | 254.63 | 501,422.49 | 98,465.421 | 193,900,076.34 |
| Reviriego et al. [14, 15] | 1889.39 | 304.3 | 237.39 | 448,522.29 | 72237.777 | 136,485,333.49 |
| Xiao et al. [19] | 1889.39 | 304.3 | 241.06 | 455,456.35 | 73,354.558 | 138,595,368.34 |
| Reviriego et al. [14] | 1889.39 | 304.3 | 247.37 | 467,378.40 | 75,274.691 | 142,223,248.43 |
| Proposed | 1889.39 | 304.3 | 235.18 | 444,346.74 | 71,565.274 | 135,214,713.04 |

## 5 Conclusion

This paper presents a power efficient (24, 16) SEC-OLS code. Using proposed algorithm, a $H$-matrix has been constructed. The proposed and existing codes have been designed and implemented on ASIC platform. ASIC-based synthesis results provide the highest 7.63% reduction in power, 11.38% reduction in PAP, 27.31% reduction in PDP and 30.26% reduction in PADP compared to existing codecs respectively. Comparisons of theoretical and synthesis results exhibit that the proposed SEC-OLS codec is more suitable for power efficient memory applications. In future the scheme can be extended for higher adjacent error correction capability codes.

**Acknowledgements** This research work has been performed under research grant received from RUSA 2.0 scheme of Jadavpur University. Authors would also like to thank to SMDP-C2SD Lab, Jadavpur University for Cadence tool.

## References


1. Li J, Reviriego P, Yi Xiao L, Wu H (2019) Protecting memories against soft errors: the case for customizable error correction codes. IEEE Trans Emerg Top Comput
2. Dixit A, Wood A (2011) The impact of new technology on soft error rates. In: 2011 IEEE international reliability physics symposium (IRPS). IEEE, pp 5B-4
3. Tripathi S, Jana J, Samanta J, Raj A, Ranjan D, Singh MP (2020) Design and evaluation of Neale-based multi-bit adjacent error-correcting codec for SRAM. In: Proceedings of the 2nd international conference on communication, devices and computing. Springer, Singapore, pp 259–268
4. Tripathi S, Jana J, Samanta J, Anand A, Kumar C, Raj G (2020) FPGA and ASIC implementation of SEC-DED-DAEC codes for SRAM applications. In: Proceedings of the 2nd international conference on communication, devices and computing. Springer, Singapore, pp 237–247
5. Hamming RW (1950) Error detecting and error correcting codes. Bell Syst Tech J 29(2):147–160
6. Hsiao MY (1970) A class of optimal minimum odd-weight-column SEC-DED codes. IBM J Res Dev 14(4):395–401

7. Dutta A, Touba NA (2007) Multiple bit upset tolerant memory using a selective cycle avoidance based SEC-DED-DAEC code. In VLSI test symposium, 2007. 25th IEEE. IEEE, pp 349–354
8. Reviriego P, Martínez J, Pontarelli S, Maestro JA (2014) A method to design SEC-DED-DAEC codes with optimized decoding. IEEE Trans Dev Mater Reliab 14(3):884–889
9. Samanta J, Tripathi S (2018) Comments on "A novel approach of error detection and correction for efficient energy in wireless networks". Multimedia Tools Appl 78(6):7579–7584
10. Tripathi S, Jana J, Samanta J, Bhaumik J (2019) Fast and power efficient SEC-DED and SEC-DED-DAEC codes in IoT based wireless sensor networks. In: TENCON 2019-2019 IEEE region 10 conference (TENCON). IEEE, pp 540–545
11. Pontarelli S, Reviriego P, Ottavi M, Maestro JA (2015) Low delay single symbol error correction codes based on reed solomon codes. IEEE Trans Comput 64(5):1497–1501
12. Hsiao MY, Bossen DC, Chien RT (1970) Orthogonal Latin square codes. IBM J Res Dev 14(4):390–394
13. Reviriego P, Pontarelli S, Sánchez-Macián A, Maestro JA (2013) A method to extend orthogonal Latin square codes. IEEE Trans Very Large Scale Integr (VLSI) Syst 22(7):1635–1639
14. Reviriego P, Liu S, Maestro JA, Lee S, Touba NA, Datta R (2013) Implementing triple adjacent error correction in double error correction orthogonal Latin squares codes. In: 2013 IEEE international symposium on defect and fault tolerance in VLSI and nanotechnology systems (DFTS). IEEE, pp 167–171
15. Reviriego P, Pontarelli S, Maestro JA (2012) Concurrent error detection for orthogonal latin squares encoders and syndrome computation. IEEE Trans Very Large Scale Integr (VLSI) Syst 21(12):2334–2338
16. Reviriego P, Pontarelli S, Evans A, Maestro JA (2014) A class of SEC-DED-DAEC codes derived from orthogonal latin square codes. IEEE Trans Very Large Scale Integr (vlsi) Syst 23(5):968–972
17. Datta R, Touba NA (2011) Generating burst-error correcting codes from orthogonal latin square codes—a graph theoretic approach. In: 2011 IEEE international symposium on defect and fault tolerance in VLSI and nanotechnology systems. IEEE, pp 367–373
18. Reviriego P, Liu S, Rottenstreich O, Lombardi F (2019) Two bit overlap: a class of double error correction one step majority logic decodable codes. IEEE Trans Comput 68(5):798–803
19. Xiao L, Li J, Li J, Guo J (2015) Hardened design based on advanced orthogonal Latin code against two adjacent multiple bit upsets (MBUs) in memories. In: Sixteenth international symposium on quality electronic design. IEEE, pp 485–489
20. Lee SE (2013) Adaptive error correction in orthogonal latin square codes for low-power, resilient on-chip interconnection network. Microelectron Reliab 53(3):509–511

# Game Theoretical Defense Mechanism Against Bandwidth Based DDoS Attacks

**Bhupender Kumar and Bubu Bhuyan**

**Abstract** DDoS attacks deplete network bandwidth and render the services unavailable to normal users. In spite of various qualitative and rule based defense mechanisms, the threat persists continuously because such mechanisms do not capture the incentive oriented attacker's intentions. Game theory promises to provide incentive based decision support system to defend against such attacks. The current work proposes a defense mechanism designed as two player zero sum finite horizon repeated game where decisions to allow or drop an incoming traffic at edge router are taken based on a flow-rate threshold. The flow-rate threshold is dynamically adjusted and updated based on maximum payoff obtained by the defender at a particular stage of the game. Payoff is computed by quantifying network conditions and parameters. The normal and attack traffic are modelled using normal and poisson distribution functions respectively. Using these functions, drop and pass probabilities of flows are computed. Further, costs and benefits in respect of the attacker and defender are computed based on the bandwidth utilized by the flows. Subsequently, payoff functions are designed for attacker and defender. Payoff functions are optimized to construct baseline optimal strategies. Nash Equilibrium is obtained using numerical computations which suggests an optimal flow-rate threshold to be set by the defender to secure the network against DDoS attacks round the clock.

**Keywords** Network security · Game theory · Distributed denial of service attack · Bandwidth exhaustion · Repeated game · Nash equilibrium

---

B. Kumar (✉) · B. Bhuyan
Department of Information Technology, North Eastern Hill University,
Umshing Mawkynroh, Shillong, Meghalaya 793002, India
e-mail: bhupinder.nayak@gmail.com

B. Bhuyan
e-mail: bbhuyan@nehu.ac.in



A. Maji et al. (eds.), *Proceedings of the International Conference on Computing and Communication Systems*, Lecture Notes in Networks and Systems 170,
https://doi.org/10.1007/978-981-33-4084-8_58

# 1 Introduction

DDoS can affect or destroy the networks of sensitive and vital installations of the nation such as defense, nuclear, railway, power, banking sectors and air traffic control systems. DDoS was known to the academic researchers since 1980 but first such incident was reported by Computer Incident Advisory Capability in 1999. Since then, numerous cases of major DDoS have occurred such as DDoS against commercial websites like Yahoo, Amazon, Ebay [1], Stuxnet [2], Brain Kreb, OVH and DNS system-Dyn which resulted in disruption of services provided by Netflix, Twitter and PayPal [3]. It causes an average loss to e-commerce approximately equals to $40,000 [4]. Attackers compromise machines by installing some malicious script, code or trojan horse through emails or freely downloadable file onto them. The installed code communicates back to its installer and acts as per the commands given directly to it by its installer. Commands can also be given indirectly through a Command & Control structure (C&C) using IRC or websites. Such compromised machines are then termed as Zombie or bots. Increased numbers of IoT devices have provided more interfaces to attackers to enter the network and compromise a huge number of IoT devices to work as zombies. 1 TB DDoS attack could be achieved using approximate 15 lakh of IoT devices against DNS system by Mirai botnet [5].

Perfect security against DDoS is still a challenging task due to high speed, complex, distributed and interdependent computer networks. Attackers are technically advanced and use sophisticated methodologies to remain undetected for a longer duration. Attacks are more organised, rational, dynamic and derived by economic incentives. Researchers have suggested various defense mechanisms against DDoS attacks based on packet filtering, packet marking, link testing, traffic validation, rate limiting, push-back and signature based schemes. These approaches can analyse the increased network traffic to detect the DDoS but fails to quantify the motives and intentions of the attacker and also suffer from the computational burden, scalability, memory issues and static rule based operations.

Researchers are using Data mining [6], Graph theory [7], Queuing theory [8], Artificial intelligence [9], Machine learning [10] and Game theory [11–14] as a methodology to suggest more realistic defense approaches against DDoS. Among them, Game theory [15] seems more promising because it provides a quantitative modelling of attack scenario and the corresponding intentions and goals of attacker behind launching such attack. It addresses the uncertainty about the actions of attacker and defender in a strictly competitive situation. The success probability of a DDoS attack depends upon the actions chosen by both the attacker and the defender. From game theory perspective, attacker and defender are called as players who are competing to capture more network bandwidth. To maximize the network bandwidth share, the attacker has to compromise enough number of machine. Compromising a machine incurs cost in terms of time and efforts. On the other hand, the defender competes to minimize the losses to legitimate flows and maximize the drop probabilities of the attack flow to make the attack costlier for attacker. It does so by setting flow-rate threshold. Since, rational players adhere to the best Course of Actions (CoA) to maximize the benefits and minimize the losses, a game is formulated between them. Inspired by this, a game theoretical defense mechanism against DDoS is proposed in this paper. We adopt and extend our earlier work carried out in [14] and modify the payoff functions of players by introducing finite horizon repeated game. Unlike in [11, 14], current work is based on repeated game and doesn't use honeypot due to its implementation cost born by defender. The main contribution of this work is given below.

- The work proposes two players, finite horizon, zero sum, repeated game over time $t = 1, 2, \cdots T$, where $T$ is finite. Repetition of the game over finite stages converges the game to accurate Nash Equilibrium strategies.

The rest of the paper continues with the discussion of relevant and related work in Sect. 2. Section 3 describes the network model, Sect. 4 presents the strategic game, Sect. 5 discusses the experimental results obtained using MATLAB and Sect. 6 concludes the paper with some ideas for future work.

## 2 Literature Survey

A static game theoretical defense mechanism against bandwidth consuming DDoS attacks is proposed in [11]. Normal probability distribution function is used to model the legitimate traffic and statistical techniques are used to calculate the ratio of bandwidth consumed by legitimate flows to number of lost legitimate flows. Decisions to drop, redirecting to honeypot or allow to pass to target machine are taken based on a flow-rate threshold using sigmoid function. Since, an attacker persistently sends flows with different rates, a static model is incompetent to address the issue of dynamic adjustment of flow-rate threshold. The situation could be modelled using mixed strategy and repeated game with incomplete information. A similar approach is suggested in [12] where the validation is done using NS-3. A good work on how to model and quantify attacker's intention behind launching a DoS attack is carried out in [13]. The authors discussed how to use game theory in modelling the objectives and strategies of attacker and defender in a DDoS scenario.

In our earlier work [14], we have proposed a game theoretical defense framework against bandwidth depleting DDoS attacks. The defense framework comprises two players with a finite action space. It is modelled as static and zero sum game. Realistic quantification of network parameters and attacker's intentions is done. Accordingly, the costs and benefits associated with attack and defense strategies are derived. Payoff functions are designed based on the bandwidth consumed by the players and associated costs. An optimum flow-rate threshold on incoming flows is suggested. Setting such an upper bound on rate of incoming flows is optimized based on the payoff obtained. Nash equilibrium is computed using numerical simulation in MATLAB. The proposed defense framework addresses only a single stage game which could further be analysed by modelling into a repeated game over $n$ number of stages for more refined results and more accurate Nash equilibrium.

Game theory is also used in designing defense frameworks against DoS attacks in wireless networks. Game theoretical models based on cooperation of players to achieve a common goal, where payoff of every player is maximum, are studied in [16–18]. The cooperation among the users is achieved by providing some incentives. The secrecy capacity of a channel is increased if the SINR is increased between an origin and destination path in a wireless network. Similarly, it is decreased if the channel has higher SINR. Exploiting such notion, friendly jammer is employed to degrade the secrecy capacity on a path between a user and eavesdropper so that eavesdropper is not able to flood the channel between normal users. Such a cooperative spectrum access [16] and cooperative jamming [17, 18] are induced by providing incentives. Interested readers can refer [19] for a detailed insight on using game theory to enhance network security.

# 3 Network Model

## 3.1 Network Topology

We consider generally used dumbbell shaped network topology as given in figure 1. The Bot-master has compromised $m$ number of machines and using them as attack nodes with an attack rate $r_i^a$ bps per $i$th node to launch DDoS against target server (1). Simultaneously, there exists $n$ number of normal or legitimate users with a flow rate $r_i^l$ bps per $i$th node, trying to access the server (1). Consequently, $0 < i \leq m$ for attack nodes, $0 < i \leq n$ for normal users, $\forall i \mapsto \Re^+$ and total number of nodes is $(m + n) = N$ and $N \in \mathbb{N}$. Through switches and network routers, all such nodes are connected to an Edge router (Gateway router) of a local area network inside which the target server (1) is placed. Edge router is connected to target server (1)

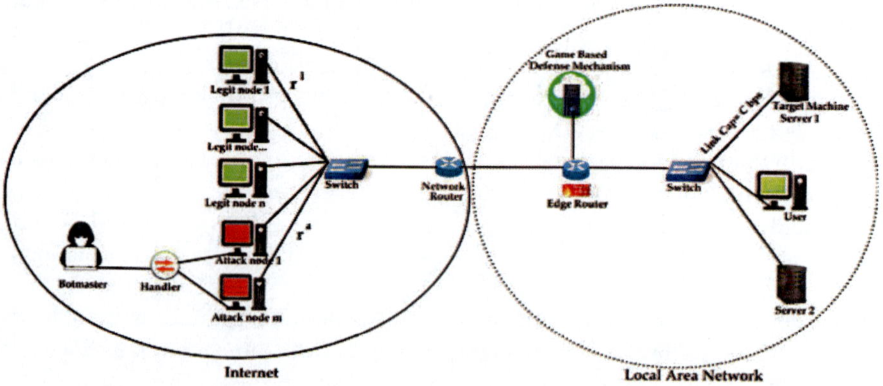


Fig. 1 Network topology

with a link capacity $C$ bps. Multiple flows can exists between one source-destination pair but for simplicity, these flows are aggregated and considered single flow per $i$th node. It results in a total flow, denoted by $f$, equals to $\sum_{i=1}^{i=m,n} mr_i^a + nr_i^l$ bps at the incoming interface of the Edge router. If the whole traffic received at incoming interface of the Edge router destined to target server (1) is higher than the link capacity between them, allowing the whole traffic to pass can result into congestion and eventually a successful DDoS attack. Hence, the Edge router either has to pass a flow to destination or simply drop it to avoid DDoS. Our Game Theoretical Defense Mechanism (GTDM) sets a flow-rate threshold $\tau$ and takes such decisions based on average available bandwidth $B_{aa}$ [21], total numbers of nodes $N$ and the link capacity $C$ between Edge router and destination server (1) as discussed in Sect. 4. The GTDM can be placed as a standalone device or hard-coded in Edge router itself.

### 3.2 Preliminaries

$B_{aa}$ is the average availability of bandwidth during different time slots of network operations. It is estimated over an interval of interest $\delta t$ which can vary between milliseconds to seconds. A user with administrative privileges to Edge router can compute $B_{aa}$ by using Simple Network Management Protocol queries. Some open source end to end bandwidth measurement tools such as *pathchar, pchar, nettimer, pathrate, pathload* [20] can also help to a user with no administrative privileges to a managed router. Since, instantaneous bandwidth utilization $B_{\text{inst}}$ at time $t$ is either 0 or 1, an average utilization of link's bandwidth $B_{\text{util}}$ from time $t = 1$ to $T$ gives $B_{aa}$ as given in Eq. (1).

$$B_{aa} = C \cdot \left[ 1 - \frac{1}{T} \int_{t=0}^{t=T} B_{\text{inst}}(x) d(x) \right] \tag{1}$$

To model the network traffic, flow-rates are considered as random variable. Flow rates of normal users are modelled using normal distribution with mean flow-rate $\mu$ as continuous random variable with $\sigma$ and $\sigma^2$ as standard deviation and variance respectively. Probabilities of normal flow-rates to be less then or equals to flow-rate threshold $\tau$ is computed using CDF of normal distribution as given in Eqs. (2) and 3. Similarly, attack flows are modelled using poisson distribution having flow-rates as discrete random variable. Poisson distribution is used in modelling attack flow rates is due to uncertain arrival time of such flows. We use $\lambda$ as the mean attack flow-rate. Using CDF of poisson distribution function, we compute the probabilities of attack flow-rates equal to or less than the flow-rate threshold $\tau$ as given in Eq. (4).

$$P(\mu = \tau) = \frac{1}{\sigma\sqrt{2\pi}} \mathrm{e}^{-\frac{1}{2}} \left( \frac{\tau - \mu}{\sigma^2} \right)^2 \tag{2}$$

$$\phi(\mu \leq \tau) = \int_{0}^{\tau} P(\mu = \tau)\mathrm{d}\tau \tag{3}$$

$$P(r_i^a \leq \tau) = \sum_{r_i^a=0}^{\tau} P(r_i^a \simeq \tau) \tag{4}$$

### 3.3 *Assumptions*

We assume that the attacker uses some tools to gather information about the current status of the link, its capacity and the value of flow-rate threshold set by GTDM. It can also infer these information using heuristic and historical knowledge. Using these inputs, it assigns the passing and dropping probabilities to its flows. Further, the link works in full duplex mode using full bandwidth for upstream and downstream of data. However, it doesn't always busy transmitting because a persistent and continuous transmission for an infinite duration needs infinite data either to send or receive. Finally, since 90% of internet traffic either follow TCP or is TCP friendly in nature, the mechanism is restricted to TCP or UDP flows till they behave as TCP friendly flows.

## 4 Game Formulation

The GTDM works based on two players, zero sum, repeated game over time $t = 1, 2, ..., T$. It is defined as $g = \{P, A, U\}$, where $P$ denotes players, $A$ denotes action set available to players and $U$ is the corresponding utility or payoff obtained. Two players include an attacker and the GTDM, which works as a defender. The action set of GTDM consists of setting optimum values of flow-rate threshold $\tau$ at different time slots of the game according to Eq. (5) so that maximum attack flows are dropped.

$$\tau = \frac{C}{N} \times \left[1 - \frac{1}{B_{aa}}\right] \tag{5}$$

GTDM is placed at Edge router and has an access to its information database. It gets feedback related to $C, N$ and $B_{aa}$ every $\delta t$ time of interval. Further, it directly observes the flow-rates $r_i$ of every flow received at incoming interface of the Edge router. It passes the value of $\tau$ to Edge router and instructs it to take decisions as mentioned below. Here, $r_i^a$ and $r_i^l$ are denoted by $r_i$ only.

- **if** $r_i \leq \tau$ Let the *i*th flow pass.
- **if** $r_i > \tau$ Drop the *i*th flow entirely.

Since we consider TCP or TCP friendly UDP flows, setting $\tau$ according to Eq. (5) gives every flow an equal and fair share of the network bandwidth. However, the value of $\tau$ at current time may not be optimal. To optimize it, GTDM computes payoff $U_d^t$ obtained by it at the current time $t$ according to Eq. (7). Furthermore, if $U_d^t$ at current time $t$ is higher than the payoff $U_d^{t-1}$ obtained at time $t-1$, GTDM marks the current value of $\tau$ as optimal and makes it as its baseline optimal strategy. However, if $U_d^t \leq U_d^{t-1}$, it immediately reverts back to previous value of $\tau$ and marks previous strategy as baseline optimal strategy. Setting $\tau$ strategically in this way leads a repeated play to Nash equilibrium or steady state of the game.

The attacker launches DDoS attack to consume higher bandwidth with lesser number of bots. Either of two results can be achieved by a DDoS attacker. One is a completely successful attack and second is a failed and futile attempt. For a successful attack, it has either to increase flow-rates $r_i^a$ per $i$th compromised machine or just to increase the number $m$ of compromised machines or bots to $S^{\text{bot}}$, where $S^{\text{bot}}$ denotes the size of botnet. Increasing $r_i^a$ is easy and cost effective but poses greater risk of detection and increases the drop probability of flows. Increasing botnet size $S^{\text{bot}}$ and splitting flow-rates among them lowers the detection risk. It also decreases the drop probability of attack flows. But compromising new machines and acquiring new bots involves a cost $\omega$ per bot in terms of time and efforts to exploit some vulnerability in new machines and installing malicious codes upon them. It makes the attack costlier. Thus, a rational attacker has to choose an optimal $r_i^a$ per $i$th flow along with an optimal $S^{\text{bot}}$. To prevent DDoS, GTDM exploits this notion of making the attack costly in terms of both attack flow-rates and botnet size by setting $\tau$ dynamically such that maximum attack flows are dropped.

## 4.1 Attack Costs, Benefits and Payoff Functions

Two parallel attack motives are considered as attack benefits to formulate the payoff function of the attacker. One is causing absolute impact on network bandwidth denoted by the ratio of bandwidth occupied by attack flows to the total network bandwidth. The bandwidth occupied $B_o^a$ by the $m'$ number of attack flows which successfully managed to pass is equals to $\sum_{i=1}^{m'}(r_i^a)$. Second is causing a relative impact on normal flows as well and denoted by the ratio of the bandwidth occupied by normal flows $B_o^l$ and total bandwidth required by such flows $B_w^l$. It depicts the gain of normal flows. Higher the ratio, lower the payoff of attacker. It seems rational because an attacker doesn't occupy maximum network bandwidth persistently over a period of time if there are no normal flows being transmitted by the link under consideration.

Similarly, two cost factors are considered to construct payoff function of the attacker. The effort made in compromising a machine in network and turning it into a bot by exploiting either its protocol vulnerability or some software flaws, is quantified as cost per bot and denoted by $\omega$ per bot. Similarly, if an attack flow is dropped by Edge router, an extra cost is incurred by the attacker in terms of efforts made to generate those flows. It is denoted by $\gamma$ per dropped flow. If $m'$ number of attack flows successfully managed to pass, the number of dropped attack flows are equal to $(m - m')$.

Based on the cost and benefit quantifications made above, the payoff function of attacker is constructed as given below.

$$U_a = \alpha\left(\frac{B_o^a}{C}\right) + \beta\left(1 - \frac{B_o^l}{B_w^l}\right) - (\omega S^{\text{bot}}) - ((m - m')\gamma) \tag{6}$$

Here, $\alpha$, $\beta$ are two weighing parameters to represent the rationality of the attacker.

Since the game is a zero sum game, defender's payoff function is negative to that of attacker's. It is represented as below.

$$U_d = -\left[\alpha\left(\frac{B_o^a}{C}\right) + \beta\left(1 - \frac{B_o^l}{B_w^l}\right) - (\omega S^{\text{bot}}) - ((m - m')\gamma)\right] \tag{7}$$

## 4.2 Baseline Strategy and Nash Equilibrium Analysis

The first two terms in the Eq. (6) are attack benefits and last two terms are attack costs. We denote the benefits by $x$ and costs by $y$ as given below.

$$x = \alpha\left(\frac{B_o^a}{C}\right) + \beta\left(1 - \frac{B_o^l}{B_w^l}\right) \tag{8}$$

$$y = (\omega S^{\text{bot}}) + ((m - m')\gamma) \tag{9}$$

The value of $x, y$ changes with change in network parameters like $\tau$, $B_{aa}$ and $N$ at every repeated stages of game. The attack payoff $U_a$ is positive at a stage only if total costs $y$ are less than the benefits $x$ of that stage, as depicted below.

$$U_a = \begin{cases} < 0 \ \forall \ x < y \\ = 0 \ \forall \ x = y \\ > 0 \ \forall \ x > y \end{cases}$$

The attacker tries to launch DDoS only if its payoff $U_a$ is positive at a stage of the game. The game is repeated over time $t$ such that $(1, 2, \ldots, t-1, t, t+1, \ldots T)$ number of times. Corresponding payoffs are calculated by the attacker and the game converges to an optimal baseline strategy as follow.

$$U_a^{\text{baseline}} = \begin{cases} U_a^t & \textit{if} \ \ U_a^t > U_a^{t-1} \ \text{(play again after } \delta t\text{)} \\ U_a^{t-1} & \textit{if} \ \ U_a^t < U_a^{t-1} \ \text{(play again immediately )} \\ U_a^{t+1} & \textit{if} \ \ U_a^t = U_a^{t-1} \ \text{(play again)} \end{cases}$$

A similar strategy is adopted by defender to mark its optimal baseline strategy and dynamically adjust the value of $\tau$ as discussed in Sect. 4. Thus, the Nash equilibrium strategies of the attacker and the GTDM as defender are denoted by the tuple $S^{\text{bot}*}, r^{a*}, \tau^*$, which satisfy the following.

$$U_a^{\text{baseline}}(S^{\text{bot}*}, r^{a*}, \tau^*) \geq U_a^s(S^{\text{bot}}, r^a, \tau^*) \ \forall \, r^a, S^{\text{bot}} \tag{10}$$

$$U_d^{\text{baseline}}(S^{\text{bot}*}, r^{a*}, \tau^*) \geq U_d(S_z^{\text{bot}*}, r^{a*}, \tau) \ \forall \, \tau \tag{11}$$

Nash equilibrium strategies suggest that the GTDM should set $\tau^*$ as flow-rate threshold to provide an optimum level of network security against DDoS no matter what is the botnet size or attack flow-rates adopted by the attacker. The attacker doesn't deviate from its Nash equilibrium strategies $S^{\text{bot}*}$ & $r^{a*}$ because it fetches a payoff which is always lesser than the Nash equilibrium strategies.

## 5 Experimental Results and Discussion

For experimental purpose and numerical simulation, MATLAB is used. We have taken initial values of network parameters as $n = 10$, $m = 10$, $\mu = 20$ bps, $\sigma = 5$, $\lambda = 30$ bps, $C = 1000$ bps, $B_{aa} = 20\%$, $\alpha = 0.5$, $\beta = 0.5$, $\omega = 0.01$ and $\gamma = 1.5\omega$. Based on the initial inputs at current stage of game, GTDM sets $\tau$ equals to 47.5 bps per flow. GTDM aggressively decreases $\tau$ if $B_{aa} < 10\%$. However, it behaves leniently if $B_{aa} \geq 10\%$ and sets $\tau$ equals to maximum share of a flow.

Stage:1 (at time $t_0$, Mean Attack Rate $\lambda = 30$ bps & $\tau = 47.5$ bps), $U_a^{t_0}$ is zero when $S^{\text{bot}} = 0$ as shown in Fig. 2. It starts increasing with an increase in $S^{\text{bot}} = 1$ to $S^{\text{bot}} = 11$ and significantly decreases thereafter. However, it remains positive till $S^{\text{bot}} \leq 18$. A further increase in the size of botnet fetches a negative payoff. The computation suggests that an attacker can achieve maximum payoff $U_a^{t_0} = 0.1378$ with $\lambda = 30$ by maintaining an optimal botnet size $S^{\text{bot}} = 11$ if $\tau = 47.5$ is best response strategy of GTDM. It can thus mark $S^{\text{bot}} = 11$ as its baseline optimal strategy. However, possibility exists for maximizing the attack payoff if the attacker chooses to adjust its mean attack flow-rates $\lambda$ beyond 30 bps as discussed next. Simultaneously, increasing botnet size means an increase in numbers of $m$ and $N$. It can also enforces GTDM to reset $\tau$ after $\delta t$ time to check if its payoff $U_d^{t_0+\delta t}$ at time $t_0 + \delta t$ can be maximized. Hence, game proceeds into next stage.

Stage:2..S (at time $t_1$, $S^{\text{bot}} = 11$ & $\tau = 47.5$ bps), with an optimal botnet size $S^{\text{bot}} = 11$, payoff of attacker still can be maximized at stage 2, time $t_1$ by adjusting its mean attack rate $\lambda$ per flow as shown in figure 3. The graph obtained shows that the payoff of attacker $U_a^{t_1} = 0.1394$ with a mean attack rate $\lambda = 25.5$. Since $U_a^{t_1} > U_a^{t_0}$, the baseline strategy of attacker is updated. It marks $S^{\text{bot}} = 11, \quad \lambda = 25.5$ bps as new optimal baseline strategy. Further, the game proceeds into next stage because the

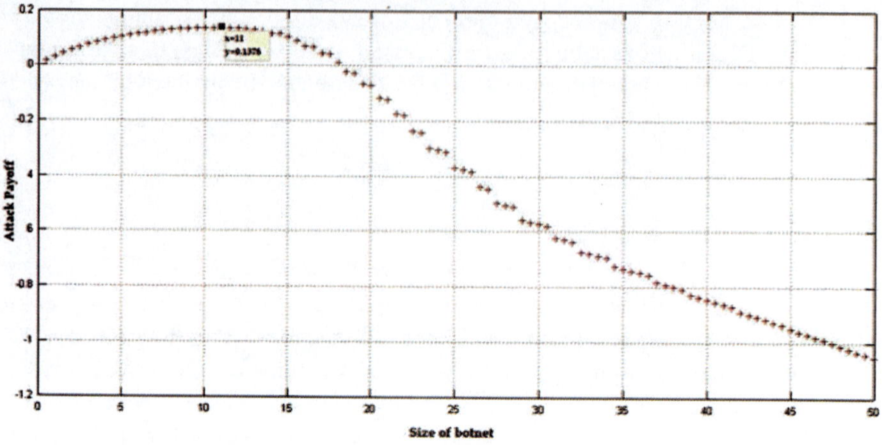


**Fig. 2** Botnet size versus attack payoff when $\lambda = 30$

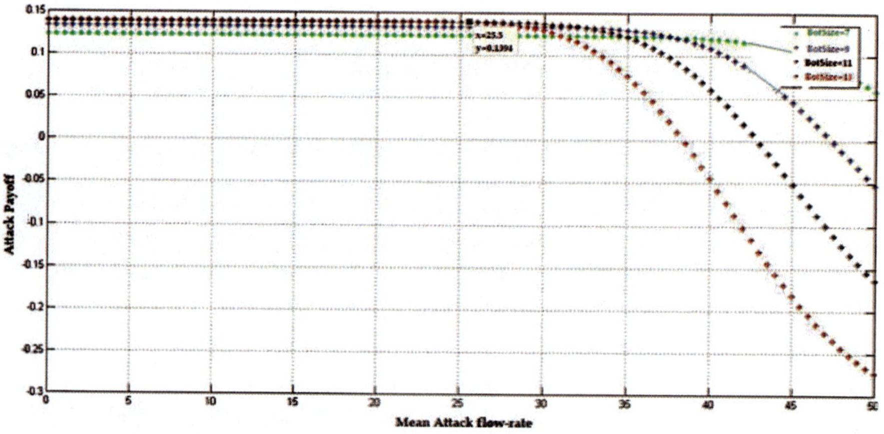


**Fig. 3** Attack payoff versus attack flow-rates when botnet size = 7, 9, 11 and 13

attacker can still achieve a better payoff equals to 0.1402 with $S^{\text{bot}} = 13$ with mean attack flow-rate $\lambda = 23.5$ bps. The game is played until time $t_T$ and converges into Nash equilibrium where an attacker does not get better payoff either by adjusting its botnet size or attack flow-rates.

Stage: Current (at Time $t_0 + \delta t$, $S^{\text{bot}} = 11$ and $\lambda = 25.5$ Bps), given the values of $S^{\text{bot}} = 11$ and mean attack flow-rate $\lambda = 25.5$bps as the optimal baseline strategy of attacker, setting $\tau = 47.5$ bps gives a negative payoff to GTDM as shown in Fig. 4. The payoff remains positive only until $\tau \geq 29.5$ bps. Thereafter, it goes to negative indicating loss of legitimate flows and depletion of network bandwidth by DDoS attack. Payoff of GTDM is maximum $U_d^{t_0+\delta t} = 0.1088$ at $\tau = 21$ bps. Hence,

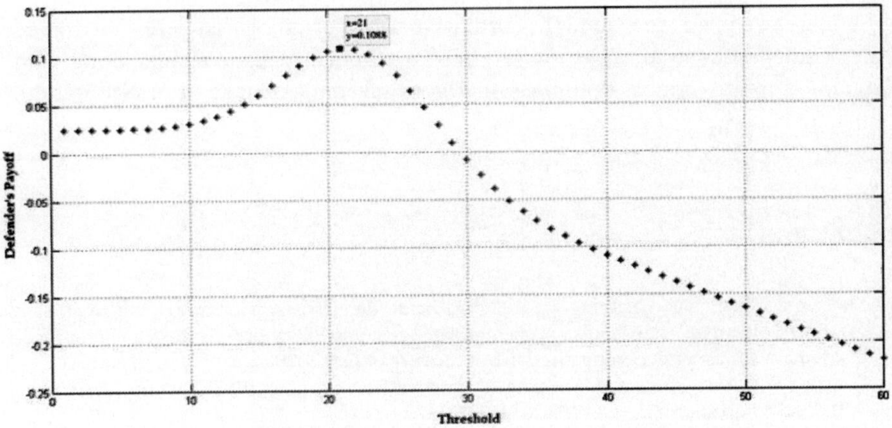


**Fig. 4** Threshold versus defender's payoff when BotSize = 11, $\lambda = 25.5$

after $\delta t$ time, GTDM sets $\tau = 21$ bps as its optimal baseline strategy and forces the attacker either to re-adjust $S^{\text{bot}}$ or to decrease the drop probability by decreasing $\lambda$. Such enforcement on attacker makes an attack costly as discussed in Sect. 4 and thus preventing DDoS.

## 6 Conclusion

DDoS is a severe security threat which renders the network services unavailable to normal and legitimate users. Security mechanisms suggested so far are based on ingress/egress filtering of the attack traffic, load balancing, traffic shaping, trace-back mechanisms and validating the incoming traffic. Such qualitative and rule based measures do not address the problem adequately and accurately due to technically evolved, rational and incentives oriented attackers.

We proposed a dynamically adaptive defense mechanism by quantifying incentives and network parameters based on game theory. Normal and attack traffics are modelled using probability distribution functions. We have used average available bandwidth, total numbers of flows and capacity of the link under consideration to compute a flow-rate threshold. The flow-rate threshold is dynamically updated over an interval by evaluating varying payoffs obtained at different stages of the two players, zero sum and repeated game. Attack costs and benefits are quantified and used to construct payoff function of the players. For experiments and numerical simulations, MATLAB is used. Baseline optimal strategies of the attacker and GTDM working as defender are suggested. Nash Equilibrium strategies are computed using the MATLAB analysis. The results obtained gives an insight to defender and helps him in putting an upper limit on incoming network traffic dynamically to prevent DDoS.

In future, we intend to extend the current work by designing non-zero sum, mixed strategy stochastic game to get more refined and much realistic results along with designing a multi variable optimization algorithm, implementation in NS-3 and a comparative analysis of existing works.

## References

1. Stephen J (2001) The changing face of distributed denial of service mitigation. Available via SANS Institute. https://www.sans.org/reading-room/whitepapers/threats/changing-face-distributed-denial-service-mitigation-462. Accessed 14 Feb 2020
2. Langner R (2011) Stuxnet: Dissecting a cyberwarfare weapon. IEEE Security & Privacy 9(3):49–51. https://doi.org/10.1109/MSP.2011.67
3. Corp Symantec (2017) Symanatec internet security threat report, vol 22. https://docs.broadcom.com/doc/istr-22-2017-enAccessed 14 Feb 2020
4. Innab N, Alamri A (2018) The impact of DDoS on e-commerce. In: IEEE 21st Saudi computer society national computer conference (NCC), Riyadh, Saudi Arabia, 25-26 Apr 2018, pp 1–4. https://doi.org/10.1109/NCG.2018.8593125
5. Said N B, Biondi F, Bontchev V et al (2018) Detection of Mirai by syntactic and behavioral analysis. In: IEEE 29th international symposium on software reliability engineering (ISSRE), Memphis, TN, USA, 15–18 Oct 2018, pp 224–235. https://doi.org/10.1109/ISSRE.2018.00032
6. Garg K, Chawla R (2011) Detection of DDoS attacks using data mining. Int J Comput Bus Res (IJCBR) 2(1)
7. Jaikumar P, Kak AC (2015) A graph-theoretic framework for isolating botnets in a network. Secur Commun Netw 8(16):2605–2623. https://doi.org/10.1002/sec.500
8. Singh S, Gyanchandani M (2010) Analysis of botnet behavior using queuing theory. Int J Comput Sci Commun 1(2):239-241. Available http://csjournals.com/IJCSC/PDF1-2/49.pdf. Accessed 07 Feb 2020
9. Zhang B, Zhang T, Yu Z (2017) DDoS detection and prevention based on artificial intelligence techniques. In: 3rd IEEE Int Conf on Computer and Communications (ICCC). p 1276-1280. https://doi.org/10.1109/CompComm.2017.8322748
10. He Z, Zhang T, Lee R B (2017) Machine learning based DDoS attack detection from source side in cloud. In: 4th international conference on cyber security and cloud computing (CSCloud), pp 114–120. https://doi.org/10.1109/CSCloud.2017.58
11. Bedi HS, Roy S, Shiva S (2011) Game theory based defense mechanism against DDoS attacks on TCP/TCP IP friendly flow. In: Proceedings of computational intelligence in cyber security (CICS), IEEE Symposium, pp 129–136. https://doi.org/10.1109/CICYBS.2011.5949407
12. Wu Q, Shiva S, Roy S et al (2010) On modeling and simulation of game theory-based defense mechanisms against DoS and DDoS attacks. In: Proceedings of 2010 spring simulation multi-conference, pp 1–8. https://doi.org/10.1145/1878537.1878703
13. Liu P, Zang W, Yu M (2005) Incentive based modelling and inference of attacker intent, objectives, and strategies. ACM Trans Inf Syst Secur (TISSEC) 8(1):78–118. https://doi.org/10.1145/1053283.1053288
14. Kumar B, Bhuyan B (2019) Using game theory to model DoS attack and defence. Sadhana 44(12):245. https://doi.org/10.1007/s12046-019-1228-4
15. Osborne MJ (2004) An introduction to game theory. Oxford University Press, London. ISBN 978-0-19-808610-9
16. Zheng G, Choo L, Wong K (2011) Optimal cooperative jamming to enhance physical layer security using relays. IEEE Trans Sig Process 59(3):1317–1322. https://doi.org/10.1109/TSP.2010.2092774

17. Yang J, Kim IM, Kim DI (2013) Optimal cooperative jamming for multiuser broadcast channel with multiple eavesdroppers. IEEE Trans Wirel Commun 12(6):2840–2852. https://doi.org/10.1109/TWC.2013.040413.120972
18. Zhang N, Lu N, Cheng N et al (2013) Cooperative spectrum access towards secure information transfer for CRNS. IEEE J Sel Areas Commun 31(11):2453–2464. https://doi.org/10.1109/JSAC.2013.131130
19. Manshaei MH, Zhu Q, Alpcan T et al (2011) Game theory meets network security and privacy. ACM Comput Surv 45(3). https://doi.org/10.1145/2480741.2480742
20. Prasad R, Constantinos D, Margaret M et al (2003) Bandwidth estimation: metrics, measurement techniques, and tools. IEEE Netw 17(6):27–35. https://doi.org/10.1109/MNET.2003.1248658
21. Antoniades D, Manos A, Papadogiannakis A et al (2006) Available bandwidth measurement as simple as running wget. In: Proceedings of 7th international conference on passive and active measurements (PAM), Adelaide, Australia, 30–31 March 2006, pp 61–70

# Handoff Authentication Through Diffie-Hellman Approach for Wireless Mesh Networks

**Amit Kumar Roy and Ajoy Kumar Khan**

**Abstract** Internet accessing through light-weight device becomes the most widely choice at present. Devices such as laptop and mobile become the most favourable choice for accessing the Internet anywhere and at anytime. However, an issue of handoff authentication arises as these devices migrates its location from its home MAP to foreign MAP. For handoff authentication, ticket-based schemes were mostly considered which allows the client to authenticate itself to the FMAP. These tickets should be exchanged in a secured manner among the mesh entities. Therefore, in this paper we had proposed a handoff authentication protocol with confidentiality of transfer ticket or identity ticket through Diffie-Hellman approach. Through experimental results, it is proved that our protocol achieves minimum handoff latency and minimum computational cost along with ticket confidentiality.

**Keywords** Handoff · Authentication · Confidentiality

## 1 Introduction

Wireless mesh networks (WMNs) due to its promising nature and advanced features as self-organized and self-healing capability in real life application, it ensures its massive role in today's environment [1]. Due to its promising nature it overcomes against certain issues which are limited in conventional networks such as LAN and MANET. The architecture of WMNs reflects its flexibility and reliability to provide continuous network service to the users. The architecture of WMNs consists of three components- Gateway router (GW), mesh router (MR) and mesh client (MC). Mesh routers plays an important role in WMNs as it formed the backbone of WMNs. The

A. K. Roy (✉)
National Institute of Technology Mizoram, Aizawl, Mizoram, India
e-mail: amitkroy12@gmail.com

A. K. Khan
Mizoram University, Aizawl, Mizoram, India
e-mail: ajoyiitg@gmail.com



A. Maji et al. (eds.), *Proceedings of the International Conference on Computing and Communication Systems*, Lecture Notes in Networks and Systems 170,
https://doi.org/10.1007/978-981-33-4084-8_59

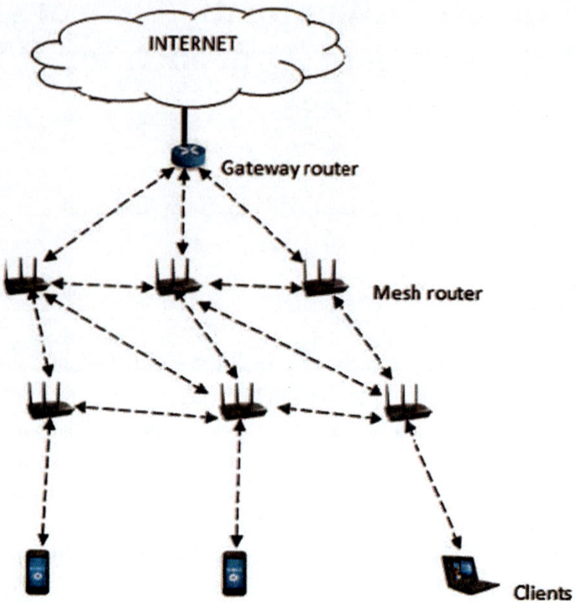


**Fig. 1** Architecture of WMNs

MR also known as mesh access point (MAP) is static in nature and forwards the request of a mesh client towards the gateway router (GW) for accessing the Internet [2–4]. As mesh clients are non-static and changes its position from home location to the foreign location, an authentication process is required to introduce for its legal access to the foreign location. Authentication allows the mobile clients to access the Internet in the foreign location. The authentication carried out during the migration of the mesh clients is known as handover. Many existing work had proposed in the past to allow the authentication during handover. Most of them were based on ticket-based authentication method [5, 6]. The existing protocols offer the authentication during handover with certain limitations and issues which had been discussed in Sect. 2.

In this paper, we have proposed a novel protocol to offer authentication during handover with confidentiality of the identity ticket shared between mesh entities. In our proposed protocol we have consider a Diffie-Hellman key exchange protocol [7] to preserve the confidentiality of a identity ticket/transfer ticket which is shared between the client and the foreign mesh access point (FMAP) (Fig. 1).

## 2 Related Work

In 2007, a secure proactive authentication protocol during handover had proposed for WLANs IEEE 802.11F which was an extended version of IEEE 802.11 that offers to access wireless point communications among multiple systems. When a client migrates from its home AP towards foreign AP, it sent a re-attachment request message to the foreign AP. Once the request message is received, the new AP further forwards the request message to the authentication server (AS) to verify the request message. If the request message is verified by the AS, than the AS generates the accept message to the new AP. On receipt of the accept message by the new AP, it allows the client to access the internet under its domain. However, the protocol suffers from authentication delay during handover as the access point (AP) had to verify the request message via AS each time the client migrates [8].

In 2010, a re-authentication service during handover had proposed for WLAN. At the initial stage of IEEE 802.11i full authentication process was established between the mobile station (STA) and the authentication server (AS) to mutually authenticate each other. After the mutual authentication, the AS generates the handover ticket (HT) list of the neighbor access point (AP) which was unicast to the STA. The STA stores HT list for future use. The HT allows the STA to re-authenticate itself during the handover process with the AP. As mobile stations (STA) are power constraint and offer with limited storage capacity, the storing of all the tickets in advance leads to overload in its constraint resources [9].

In 2011, a broadcast authentication process during handover had proposed that allows AS to authenticate the clients while maintaining every MAP. When a migration is about to start by a client, the AS generates the tickets and broadcast to each MAPs over a secure channel. Therefore, whenever a migration is done by a client from one MAP to another, it required re-authenticated by the AS. However, this protocol leads to massive latency time as AS lies many hops away from the client [10].

In 2013, a protocol had proposed to overcome from multi-hop authentication during handover. The protocol was based on one-hop authentication via login authentication protocol and handover authentication protocol. Various type of keys were generated to performed authentication among mesh entities. However, the key generated process consumed massive resources at the user side which was restricted with power constraint [11].

In 2014, a handoff authentication protocol had proposed for WMNs. The protocol was based on pre-distributed tickets to allow authentication between mesh entities. The tickets was generated by the AS and distributed among the mesh entities. The proposed protocol offers authentication during handoff with certain limitations such as authentication verification by AS in muti-hop fashion and pre-sharing of tickets to clients which causes burden to clients constraint resources [12].

In 2016, a fast handoff authentication method was proposed. The protocol requires to generate a master key shared between the client and the server and also generates group master key shared between the mesh router and the server for authentication. After the generation of both keys, the server generates the tickets based upon the keys generated and forwards the tickets towards the client and its corresponding mesh router. The ticket allows the client to proof its identity to the foreign mesh router during migration. However, it puts burden to the clients during key generation as it is restricted with power constraint [13] (Table 1).

**Table 1** Comparison table

| Protocol | Identity ticket generated by | Confidentiality | Authentication process |
|---|---|---|---|
| PKD [8] | AS | Yes | Multi-hop |
| Re-authentication [9] | AS | Yes | Multi-hop |
| PKD-TRSH [10] | AS | Yes | Multi-hop |
| LAP, HAP [11] | MAP | No | One-hop |
| Design ticket [12] | AP | Yes | One-hop |
| FHT [13] | AS | Yes | One-hop |
| Proposed | MAP | Yes | One-hop |

## 3 Ticket Types

In this paper, we consider four types of tickets to establish authentication between AS and mesh entities and also to establish authentication among mesh entities in WMNs.

1. Authentication between AS and MAP: To establish mutual authentication between AS and MAP through MAP ticket.
2. Authentication between AS and client: To establish mutual authentication between AS and client through client ticket.
3. Authentication between client and MAP: To establish mutual authentication between client and MAP through client ticket and MAP ticket at initial stage.
4. Authentication between client and FMAP: To establish authentication during handover between client and FMAP through identity ticket.

We assume that the mutual authentication had established between the AS and mesh entities and also mutual authentication had established among the mesh entities through the various tickets mentioned above. As the main aim presented in this paper is to provide confidentiality of the identity ticket shared between the client and FMAP during handover. Therefore, in this paper we have proposed a novel protocol which is based on Diffie-Hellman key exchange protocol to preserve the confidentiality of the identity ticket during handover. In our proposed protocol the client and MAP does not require any master key from the AS to preserve the identity ticket as it is required in the existing works. The identity ticket is generated by the MAP in our proposed protocol whereas in the existing works it is generated by the AS, Hence, the involvement of the AS is omitted in our proposed protocol during handover which results in efficient authentication during handover (Table 2).

**Table 2** Notation used

| Notation | Description |
|---|---|
| g | Generator |
| Mod p | Modulo prime |
| HMAP | Home mesh access point |
| FMAP | Foreign mesh access point |
| AS | Authentication server |
| PUB | Public key |
| CSK | Common secret key |
| HT | Handover ticket |
| ID | Identity |

## 4 Proposed Protocol

The proposed protocol employs Diffie-Hellman key exchange protocol between the HMAP and FMAP to generate a common secret key which is further used for encrypting and decrypting the identity ticket shared during handover. Once the mutual authentication is established between the client and the HMAP, the HMAP generates the identity ticket for its client during its roaming state. When the client wants to migrate towards the FMAP, the HMAP encrypts the identity ticket with the common secret key and sent to the roaming client. The client further forwards the encrypted identity ticket to the FMAP to proof its identity.
Step 1: Public key computed by HMAP

$$PUB_{\mathrm{HMAP}} = g^{a} mod p \tag{1}$$

where , a is the private key of HMAP.
Step 2: Public key computed by FMAP

$$PUB_{\mathrm{FMAP}} = g^{b} mod p \tag{2}$$

where b is the private key of FMAP.
Step 3: After step 1 and 2 both HMAP and FMAP exchange their public keys
Step 4: HMAP generates its common secret key

$$CSK_{\mathrm{HMAP}} = PUB_{\mathrm{FMAP}}^{a} mod p \tag{3}$$

Step 5: FMAP generates its common secret key

$$CSK_{\mathrm{FMAP}} = PUB_{\mathrm{HMAP}}^{b} mod p \tag{4}$$

Step 6: At Step 4 and Step 5 both parties compute the same secret keys

## 4.1 Steps of Proposed Protocol

Step 1. Input: Roaming client try to proof its authentication to foreign MAP during handover process.
Step 2. Output: Authentication successful via identity ticket along with the confidentiality of identity ticket to preserve its integrity.
3. Key generation process: Common secret key (CSK) is generated between HMAP and FMAP.
4. Generation of the identity ticket: HMAP generates the identity ticket for the roaming client. Identity ticket contains the ID of the client; ID of the HMAP and nonce to resists the duplication of the identity ticket.
5. HMAP encrypts the identity ticket with the common secret key shared with the FMAP as
$CSK_{\text{HMAP}}$ (Identity ticket) and send to the roaming client
6. Client forwards this encrypted identity ticket $CSK_{\text{HMAP}}$ (Identity ticket) to the FMAP.
7. On the receipt of this message, FMAP tries to decrypt the message $CSK_{\text{FMAP}}$ (Identity ticket)
If (decryption is successful, i.e, $CSK_{\text{HMAP}}$ (Identity ticket)) then
authentication is successful and client can access the Internet under FMAP
Else
If (decryption fails) then
authentication fails, because FMAP can know that the ticket has not generated by the relevant HMAP with whom FMAP had shared the common secret key.

# 5 Experimental Results

The implementation of our proposed work is carried out using NS-3. Two performance metrics had considered to evaluate the overall performance of our proposed work under different network size with 10 nodes, 20 nodes and 30 nodes.

## 5.1 Performance Metrics

The performance metrics of our proposed work is based on handoff delay and computational cost

1. Handoff delay occurs during the migration process when the client needs to authenticate itself to the FMAP. It is the time needed to authenticate the client during the migration process.
2. Computational cost is the time consumed during various security operations performed during handoff process (Fig. 2).

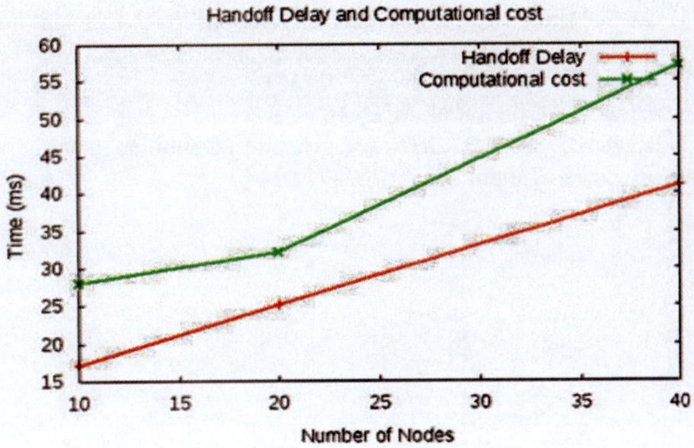


**Fig. 2** Handoff delay and computational cost

## 6 Conclusion

In this paper, the proposed protocol is designed to secure the confidentiality of the identity ticket during handoff operation. The confidentiality of the identity ticket is achieved through Diffie-Hellman approach. The experimental result shows that the proposed work offers handoff authentication with ticket confidentiality. The work in this paper could be further extended to achieve better performance as compared to other existing works.

## References


1. Akyildiz IF, Wang X (2005) A survey on wireless mesh networks. IEEE Commun Mag 9:S23–S30
2. Seyedzadegan M, Othman M, Ali BM, Subramaniam S (2011) Wireless mesh networks: WMN overview, WMN architecture. In: International conference on communication engineering and networks IPCSIT 19, p. 2. Singapore
3. Franklin AA, Murthy CSR (2007) An introduction to wireless mesh networks. In: Zhang Y et al (eds) Security in wireless mesh networks. CRC Press, Boca Raton
4. Sen J (2013) Security and privacy issues in wireless mesh networks: a survey. Wireless networks and security. Springer, Berlin, Heidelberg, pp 189–272
5. He D, Chan S, Guizani M (2015) Handover authentication for mobile networks: security and efficiency aspects. IEEE Netw 29(3):96–103
6. Wang K, Wang Y, Zeng D, Guo S (2017) An SDN-based architecture for next-generation wireless networks. IEEE Wirel Commun 24(1):25–31
7. Diffie W, Hellman M (1976) New directions in cryptography. IEEE Trans Inf Theory 22(6):644–654
8. Kassab, M., Bonnin, J.M., Guillouard, K.: Securing fast handover in WLANs: A ticket based proactive authentication scheme. In: Globecom Workshops, pp. 1–6. IEEE, New York (2007)
9. Li G, Chen X, Ma J (2010) A ticket-based re-authentication scheme for fast handover in wireless local area networks. In: 6th international conference on wireless communications networking and mobile computing (WiCOM). IEEE, Chengdu, China, pp 1–4
10. Li G, Ma J, Jiang Q, Chen X (2011) Anovel re-authentication scheme based on tickets in wireless local area networks. J Parallel Distrib Comput 71(7):906–914

11. Li C, Nguyen UT, Nguyen HL, Huda N (2013) Efficient authentication for fast handover in wireless mesh networks. Comput Secur 37:124–142
12. Xu L, He Y, Chen X, Huang X (2014) Ticket-based handoff authentication for wireless mesh networks. Comput Netw 73:185–194
13. Rathee G, Saini H (2016) A fast handoff technique in wireless mesh network (FHT for WMN). Procedia Comput Sci 79:722–728

# Multivariate Linkable Group Signature Scheme

**Satyam Omar and Sahadeo Padhye**

**Abstract** We propose a linkable group signature scheme based on Multivariate public-key cryptography. The security of the proposed signature scheme is based on the MQ problem. The security measures of a linkable group signature scheme have been apparently discussed, and correctness, signer anonymity, group leader traceability, linkability and unforgeability properties have been proved for the proposed scheme. Finally, the comparison between the already existing group signature in MPKC and our proposed scheme have been shown.

**Keywords** Multivariate PKC · MQ problem · IP problem · Group signature

## 1 Introduction

Chaum and Heyst [2] first time introduced the group signature in 1991. This group signature scheme was based on the RSA cryptosystem. In a group signature, the signing party is one of the people from a group of persons who can sign anonymously on behalf of the group. The group leader can disclose the identity of the real signer, and with given two signatures, except group leader, no one can distinguish that the same signer or different signer has signed both the documents.

There are various real-life applications of group signature as in e-commerce, tender filling, e-cash, e-auction, etc. Because of linkability property, our scheme is oriented to e-voting in companies and tender filling type scenarios where more than one signatures are not permitted. Due to the presence of linkability, with given two signatures, anyone can verify that someone has voted or filled the tender more than once.

S. Omar (✉) · S. Padhye (✉)
Department of Mathematics, Motilal Nehru National Institute of Technology Allahabad, Prayagraj 211004, India
e-mail: satyamomar840@gmail.com

S. Padhye
e-mail: sahadeomathrsu@gmail.com



A. Maji et al. (eds.), *Proceedings of the International Conference on Computing and Communication Systems*, Lecture Notes in Networks and Systems 170,
https://doi.org/10.1007/978-981-33-4084-8_60

Various group signature schemes have been given since 1991 on the factorization problem, DLP problem, etc., and many advancements have been made in those schemes according to their efficiency and security. But we know that quantum computers are the threat to all the public key systems based on factorization and DLP, so we have to develop the schemes that are secured against quantum computers. Many post-quantum techniques for digital signatures have been proposed, such as code-based schemes, hash-based schemes, lattice-based schemes, isogeny-based schemes, and multivariate schemes. Among all these quantum-resistant schemes, multivariate-based schemes are quite efficient and secured schemes that have a small size of signature in comparison to all other post-quantum techniques. Various multivariate signature schemes with additional functionalities, such as ring signature [9, 10], verifiable ring signature [7], proxy signature [11], blind signature [1], blind ring signature [3], and many more have been already proposed. In 2011, Yang et al. [4] introduced a Multivariate linkable group signature scheme [4], which was specially designed for e-voting only. But the signature size of the scheme given by Yang et al. [4] is very large. If there are $P_1$ number of variables and $P_2$ number of equations in the public keys and private keys of the group members, then the size of group signature in [4] is $(q + q * (P_1^2 + P_2^2) * 8 + P_2 * (P_1 + 1) * (P_1 + 2) * 8)$ bits, which is much larger. To reduce the signature size, we have proposed a multivariate linkable group signature with a very short signature size. In the scheme [4], the group leader himself can not disclose the identity of real signer without the help of the verifier, but in our scheme, the group leader can trace the real signer's identity without any help of the verifier.

## 2 Preliminaries

### 2.1 MQ Problem

Let $E = GF(p^r)$ be a finite field of order $p^r$; where $p$ is a prime number and $r$ is an integer. Then the quadratic system of equations

$$\begin{aligned} d_1(z_1, z_2, ..., z_n) &= \eta_1 \\ d_2(z_1, z_2, ..., z_n) &= \eta_2 \\ &\vdots \\ d_m(z_1, z_2, ..., z_n) &= \eta_m \end{aligned}$$

where, $d_j \in E[z_1, z_2, ..., z_n]$, $\eta_j \in E$, for all $j = 1, 2, ..., m$.

**Problem:** To find a vector $(s_1, ..., s_n) \in E^n$ that satisfies the equations $d_j(s_1, s_2, ..., s_n) = \eta_j$; for all $j$ with $1 \leq j \leq m$. This problem is called Multivariate Quadratic polynomial (MQ) Problem, which is NP-complete [8] even over the field $GF(2)$. The security of our proposed group signature scheme in Multivariate Public Key Cryptography is based on MQ Problem.

## 2.2 IP Problem

Patarin [5] discussed the IP Problem first time in 1996. Let $E$ be a finite field and $A$ is quadratic system of $P_2$ number of equations with $P_1$ number of variables. Now, let $\mu : E^{P_2} \rightarrow E^{P_2}$ and $\nu : E^{P_1} \rightarrow E^{P_1}$ be two random invertible affine transformations and

$$B = \mu \circ A \circ \nu$$

So, $B$ is also another quadratic system of $P_2$ number of equations with $P_1$ number of variables. Here, $A$ and $B$ are called isomorphic to each other and $(\mu, \nu)$ is called isomorphism between $A$ and $B$. Now, to find out random invertible affine transformations $\mu$ and $\nu$, for the given systems of quadratic equations $A$ and $B$, is called isomorphism of polynomials problem, i.e., IP problem.

## 2.3 Multivariate Signature Scheme

The general multivariate signature scheme works as follow:

**Key Generation:** Let $E$ be the finite base field, $\mu : E^{P_2} \rightarrow E^{P_2}$ and $\nu : E^{P_1} \rightarrow E^{P_1}$ be two invertible affine transformations, and $D : E^{P_1} \rightarrow E^{P_2}$ be a map named central map. Then public key is constructed as

$$\bar{D} = \mu \circ D \circ \nu = (\bar{d}_1(z_1, z_2..., z_{P_1}), \bar{d}_2(z_1, z_2..., z_{P_1}), ..., \bar{d}_{P_2}(z_1, z_2..., z_{P_1})),$$

where $\bar{d}_j \in E[z_1, z_2..., z_{P_1}]$, for all $j = 1, 2, ..., P_2$.
The private key is $(\mu, \nu, D)$, and the public key is the base field $E$ and the public map $\bar{D} : E^{P_1} \rightarrow E^{P_2}$ i.e. $(\bar{d}_1, \bar{d}_2, ..., \bar{d}_{P_2})$.

**Signature Generation:** Let the signer has the document $M \in E^{P_2}$, to sign on $M$ he computes as follows

$$\bar{D}^{-1}(M) = (\mu \circ D \circ \nu)^{-1}(M) = \nu^{-1} \circ D^{-1} \circ \mu^{-1}(M) = (z_1^{'}, z_2^{'}, ..., z_{P_1}^{'})$$

$(z_1^{'}, z_2^{'}, ..., z_{P_1}^{'}) \in E^{P_1}$ is the signature on the document $M$.

**Verification:** To verify the signature $(z_1^{'}, z_2^{'}, ..., z_{P_1}^{'})$ on the document $M$, the verifier checks the validity of

$$\bar{D}(z_1^{'}, z_2^{'}, ..., z_{P_1}^{'}) = M.$$

If equality holds, then accepts the signature; otherwise rejects it.
The main thing in multivariate signature scheme is to choose central map $D$ such as $D$ can be inverted easily but after hiding $D$ with $\mu$ and $\nu$, to invert $\bar{D}$ is impossible.

# 3 Multivariate Linkable Group Signature Scheme

In this section, we have proposed a multivariate linkable group signature scheme and all the properties associated with linkable group signature have been proved explicitly.

## 3.1 Key Generation

Let $E$ be the finite field. There is a group of $w$ members named $(U_0, U_1, ..., U_{w-1})$ and let $A$ is the group leader. The public key of each user is $\bar{D}_j$ and private key of each user is randomly chosen $(\mu_j, D_j, \nu_j)$, for all $j = 0, 1, ..., w-1$.
Where $\bar{D}_j = \mu_j \circ D_j \circ \nu_j$, for all $j = 0, 1, ..., w-1$.
Let $H : \{0, 1\}^* \rightarrow E^{P_1}$ be a collision-resistant cryptographically secured hash function. $A$ selects random numbers $u_j \in E^{P_1}$, for all $j = 0, 1, ..., w-1$, and gives it to each user respectively. So, definitely, each $u_j$ is known to $A$ and $j^{th}$ user only. In Table 1, we have shown this key generation process.

Moreover, the group members public keys are denoted as $G = (\bar{D}_0, \bar{D}_1, ..., \bar{D}_{w-1})$.

## 3.2 Group Signature Generation

Let among all the $U_i's$, $U_a$ is the real signer for a document $M$. Now $U_a$ proceeds as follows-

1. He randomly chooses $u \in E^{P_1}$ and computes

$$C_{a+1(\text{mod } w)} = H(G, M, \bar{D}_a(u) + \bar{D}_a(u_a)).$$

2. Now he chooses $k_j \in E^{P_1}$ randomly for all $j = a+1, a+2, ..., w-1, 0, 1, ..., a-1$, and computes

**Table 1** Key generation table

| Member | Public key | Private k | Group leader information |
|---|---|---|---|
| $U_0$ | $\bar{D}_0$ | $\mu_0, D_0, \nu_0$ | $u_0$ |
| $U_1$ | $\bar{D}_1$ | $\mu_1, D_1, \nu_1$ | $u_1$ |
| $\vdots$ | $\vdots$ | $\vdots$ | $\vdots$ |
| $U_{w-1}$ | $\bar{D}_{w-1}$ | $\mu_{w-1}, D_{w-1}, \nu_{w-1}$ | $u_{w-1}$ |

$$C_{j+1(\text{mod } w)} = H(G, M, \bar{D}_j(C_j) + \bar{D}_j(k_j))$$

for all $j = a+1, a+2, ..., w-1, 0, 1, ..., a-1$. Also, computes

$$k_a = \bar{D}_a^{-1}(\bar{D}_a(u) + \bar{D}_a(u_a) - \bar{D}_a(C_a)) \tag{1}$$

3. Group signature is a $(w+2)$-tuple $\sigma = (C_0, k_0, k_1, ..., k_a, ..., k_{w-1}, \bar{D}_a(u))$.

## 3.3 Group Signature Verification

The verifier has the document $M$ and the signature on this document $\sigma = (C_0, k_0, k_1, ..., k_a, ..., k_{w-1}, \bar{D}_a(u))$ for the group $G = (\bar{D}_0, \bar{D}_1, ..., \bar{D}_{w-1})$.
Now the verifier computes

$$C_{j+1} = H(G, M, \bar{D}_j(C_j) + \bar{D}_j(k_j)) \tag{2}$$

for all $j = 0, 1, 2, ..., w-1$.
Finally, the verifier checks that

$$C_0 = C_w \tag{3}$$

The signature is verified if the equality holds; otherwise rejected.

## 3.4 Correctness

According to the signing process, by Eq. (1)

$$k_a = \bar{D}_a^{-1}(\bar{D}_a(u) + \bar{D}_a(u_a) - \bar{D}_a(C_a))$$
$$\bar{D}_a(k_a) + \bar{D}_a(C_a) = \bar{D}_a(u) + \bar{D}_a(u_a)$$

So,

$$C_{j+1(\text{mod } w)} = H(G, M, \bar{D}_j(C_j) + \bar{D}_j(k_j))$$

is true for all $j = 0, 1, ..., w-1$.
Specially, for $j = w-1$, to compute $C_{w-1+1(\text{mod } w)} = C_0 = H(G, M, \bar{D}_{w-1}(C_{w-1}) + \bar{D}_{w-1}(k_{w-1}))$.
When the verifier will compute $C_w$, he will put $j = w-1$ in Eq. (2) as $C_{w-1+1} = C_w = H(G, M, \bar{D}_{w-1}(C_{w-1}) + \bar{D}_{w-1}(k_{w-1}))$
Cleary, we can see here that $C_0 = C_w$.

### 3.5 Signer Anonymity

In the group signature $\sigma = (C_0, k_0, k_1, ..., k_a, ..., k_{w-1}, \bar{D}_a(u))$, any component is not leaking any information about the signer's identity. We can see this as

1. All the $k_j's$ except $k_a$ are randomly selected in $E^{P_1}$, and $k_a$ is computed as

$$k_a = \bar{D}_a^{-1}(\bar{D}_a(u) + \bar{D}_a(u_a) - \bar{D}_a(C_a)).$$

   We can see here that value of $k_a$ depends on $u$ which has been randomly selected by the signer.
   So, clearly all the $k_j's$ are randomly distributed in $E^{P_1}$.
2. The value of $C_0$ depends on the $k_{w-1}$ which is randomly selected in $E^{P_1}$. So $C_0$ is also randomly distributed in $E^{P_1}$.
3. The value of $\bar{D}_a(u)$ also depends on $u$ which is randomly selected so the value of $\bar{D}_a(u)$ is also randomly distributed in $E^{P_1}$. Moreover, to compute the value of $u$ with $\bar{D}_a(u)$ given is equivalent to solve MQ problem, so randomness of scheme is preserved even after disclosing $\bar{D}_a(u)$.
   So, all the components of the group signature $\sigma$ are randomly and uniformly distributed in $E^{P_1}$. Hence, they are leaking no information about the signer.

### 3.6 Group Leader Traceability

The group leader $A$ checks that for which user $U_j$ for all $j = 0, 1, ..., w-1$, the equation

$$\bar{D}_j(k_j) + \bar{D}_j(C_j) - \bar{D}_j(u_j) = \bar{D}_a(u) \tag{4}$$

is true. By Eq. (1) we can see that this is only true for the user $U_a$.
Moreover, no other person from the group or from outside can verify that $U_a$ is the original signer because in Eq. (4), the values $u_j's$ are known only to the group leader.

### 3.7 Linkability

From Eq. (1), we have

$$\bar{D}_a(k_a) + \bar{D}_a(C_a) - \bar{D}_a(u) = \bar{D}_a(u_a). \tag{5}$$

The right-hand side (RHS) of the Eq. (5) is fixed because group leader information $u_a$ and public key $\bar{D}_a$ of the user $U_a$ are always fixed, and the left hand side (LHS) of the Eq. (5) can be computed by anyone. Whenever $U_a$ will sign again even with another random number $u' \in E^{P_1}$, Eq. (5) will look like

$$\bar{D}_a(k_a^{'}) + \bar{D}_a(C_a^{'}) - \bar{D}_a(u^{'}) = \bar{D}_a(u_a). \tag{6}$$

We can see here that the RHS of the Eq. (6) is still similar to the Eq. (5). Hence, in all the signatures constructed by the user $U_a$, the RHS of the Eq. (5) will always be the same.

## 3.8 Security Analysis

**Theorem 1** *Proposed group signature is unforgeable.*

***Proof*** We will show that the group leader, group member, or any other person outside the group can not sign a message on behalf of a particular signer.
Let the original signer is $U_a$ and any person, including group leader and group member, already has a group signature on a message $M$ signed by $U_a$, and now he wants to sign a new message $M^{'}$ on behalf of $U_a$. Let the forger has successfully signed on the message $M^{'}$. Let the group signature on the message $M$ is $\sigma = (C_0, k_0, k_1, ..., k_a, ..., k_{w-1}, \bar{D}_a(u))$ and group signature on the message $M^{'}$ is $\sigma^{'} = (C_0^{'}, k_0^{'}, k_1^{'}, ..., k_a^{'}, ..., k_{w-1}^{'}, \bar{D}_a(u^{'}))$ for another random number $u^{'}$. Now, we will see how the forger would have signed the message $M^{'}$.
The forger will start from computing $C^{'}_{a+1(\text{mod } w)} = H(G, M^{'}, \bar{D}_a(u_a) + \bar{D}_a(u^{'}))$, where $G$ is publicly known and $M^{'}$ is known to the forger. The forger can compute the value of $\bar{D}_a(u_a)$ with the help of the Eq. (1) as

$$k_a = \bar{D}_a^{-1}(\bar{D}_a(u) + \bar{D}_a(u_a) - \bar{D}_a(C_a))$$

$$\bar{D}_a(k_a) = \bar{D}_a(u) + \bar{D}_a(u_a) - \bar{D}_a(C_a)$$

$$\bar{D}_a(u_a) = \bar{D}_a(k_a) + \bar{D}_a(C_a) - \bar{D}_a(u) \tag{7}$$

So, it is easy to compute $\bar{D}_a(u_a)$ because the public key $\bar{D}_a$ of the signer $U_a$, and $k_a, C_a, \bar{D}_a(u)$ all are known to everyone who has the group signature $\sigma$ on the message $M$. Then, he will choose $k_j^{'}$, for all $j = a+1, a+2, ..., w-1, 0, 1, ..., a-1$, and he will compute

$$C^{'}_{j+1(\text{mod } w)} = H(G, M^{'}, \bar{D}_j(C_j^{'}) + \bar{D}_j(k_j^{'}))$$

for all $j = a+1, a+2, ..., w-1, 0, 1, ..., a-1$. But, when he will go to compute $k_a^{'}$, he will have to compute

$$k_a^{'} = \bar{D}^{-1}(\bar{D}_a(u^{'}) + \bar{D}_a(u_a) - \bar{D}_a(C_a^{'})) \tag{8}$$

and this is MQ problem which is intractable to solve. So, the forger can not generate $k_a^{'}$, hence, he can not sign the message $M^{'}$.

Moreover, if the forger wants that he will use the value of $k_a$ in place of $k_a^{'}$ then the Eq. (3) in verification process will not be satisfied because in the group signature $\sigma$, the value of $k_a$ depends on $C_a$, which depends on the message $M$. While in case of verification process of $\sigma^{'}$, value of $C_j^{'}$, for all $j = 0, 1, ..., w - 1$ depends on the message $M^{'}$.

The group leader has the only advantage over any other forger is that he knows all the secret values $u_j$, for all $j = 0, 1, ..., w - 1$. But in signing process, there is a use of $\bar{D}_a(u_a)$ instead of $u_a$, and $\bar{D}_a(u_a)$ can be obtained by anyone with given signature $\sigma$ as given in Eq. (1). So, there will be no extra advantage for the group leader by knowing all the $u_j^{'}s$.

### 3.9 Comparison

As we discussed earlier, Yang et al. [4] proposed a multivariate group signature scheme in 2011, specifically for e-voting only. It is the only multivariate group signature scheme presented in the literature. In this scheme, the group leader can not trace the original signer without the help of the verifier. In our proposed scheme group leader can trace the original signer without any help of the verifier. The security of [4] is based on IP problem while the security of our proposed signature scheme is based on the hardness of MQ problem.

Let we use the base field $E = GF(2^8)$, and there are $P_1$ number of variables and $P_2$ number of equations in the public keys and private keys of group members. Then the size of group signature in [4] is $((q/8) + q * (P_1^2 + P_2^2) + P_2 * (P_1 + 1) * (P_1 + 2))$ bytes. On the other hand, the size of the group signature in our proposed scheme is $((w + 1) * P_1 + P_2)$ bytes, which is a short signature in comparison to [4]. In particular, if we let $P_1 = 37$ and $P_2 = 27$ as in the Rainbow signature scheme [6], then the size of the group signature given in [4] is near about 71.86 *kb* for $q = 16$ as recommended in [5]. On the other hand, the size of our proposed group signature is $(37 * w + 64)$ bytes. We can see here that the size of the group signature given in [4] does not depend on the number of members in the group. At the same time, the size of the group signature in our proposed scheme depends on $w$, i.e., the number of group members. Now if we compare $37 * w + 64 = 73,584$ (size of signature in [4]) with the above mentioned parameter, we get $37 * w = 73,520 \Rightarrow w = 1987.02$ Hence, if we take group size 1987 then the size of our proposed group signature will be nearly equal to the group signature size of [4]. Particularly, if there are 100 group members in any group, then the size of our proposed group signature is 3.675 kb, which is a comparatively small size signature.

Now, according to the computation cost, we compare our scheme with the scheme [4]. Here, we compare both the schemes by taking $w$ as group size. In Table 2, we have compared both the linkable group signatures' computation costs on the basis of the number of operations performed in signing and verification which are

**Table 2** Comparison table

| Operation | Scheme [4] | | Our scheme | |
|---|---|---|---|---|
| | Signing | Verification | Signing | Verification |
| Hash | 1 | 1 | $w$ | $w$ |
| PKE | $4 * w + q$ | $2 * w + q$ | $2 * w + 1$ | $2 * w$ |
| SKE | 0 | 0 | 1 | 0 |
| Mat.Mult. | $8 * w + q$ | $2 * w$ | 0 | 0 |
| Mat.Inv. | $2 * w + q$ | $2 * w$ | 0 | 0 |

hash computation (Hash), public key evaluation (PKE), secret key evaluation (SKE), matrix multiplication (Mat.Mult.) and matrix inversion (Mat.Inv.).

In Table 2, we can see that our scheme is much efficient as per the computation cost of both the schemes.

## 4 Conclusion

A multivariate linkable group signature scheme has been proposed in this article. Security of the proposed scheme is based on the intractability of MQ problem. This scheme satisfies all the properties of a secure linkable group signature scheme. So, with a short signature size, it can be a convenient signature scheme that is secured against quantum computers.

## References


1. Cham D, van Heyst E (1991) Group signatures. Advances in cryptology- EUROCRYPT '91. Springer, pp 257–265
2. Mohamed MSE, Petzoldt A (2017) Ring rainbow—an efficient multivariate ring signature scheme. In: Joye M, Nitaj A (eds) Progress in cryptology—AFRICACRYPT 2017. AFRICACRYPT 2017, LNCS, vol 10239. Springer, pp 3–20
3. Wang S, Ma R, Zhang Y, Wang X (2011) Ring signature scheme based on multivariate public key cryptosystems. Comput Math Appl 62:3973–3979
4. Wang LL (2013) A verifiable ring signature scheme based on multivariate public-key cryptosystem. Appl Mech Mater 380–384:1899–1902 (Trans Tech Publications, Switzerland)
5. Tang S, Xu LL (2013) Towards provable secure proxy signature scheme based on Isomorphism of polynomials 30:91–97. Elsevier
6. Petzoldt A, Szepieniec A, Mohamed MSE (2017) A practical multivariate blind signature scheme. In: Kiayias A (eds) Financial cryptography and data security. FC 2017, vol 10322 LNCS. Springer, pp 437–454
7. Hoang Duong D, Susilo W, Tran H (2019) A multivariate blind ring signature scheme . Compu J. https://doi.org/10.1093/comjnl/bxz128

8. Yang G, Tang S, Yang L (2011) A novel group signature scheme based on MPKC. In: Bao F, Weng J (eds) Information security practice and experience. ISPEC 2011. Lecture notes in computer science, vol 6672. Springer, Berlin, Heidelberg, pp 181–195
9. Garey Michael R, Johnson David S (1991) Computers and intractability. A guide to the theory of NP-completeness. W.H Freeman
10. Patarin J (1996) Hidden field equations (HFE) and isomorphism of polynomials (IP0: two new families of asymmetric algorithms). In: Maurer U (ed), Eurocrypt'96, LNCS, vol 1070. Springer, pp 33–48. Extended version: http://www.minrank.org/hfe.pdf
11. Ding J, Schmidt D (2005) Rainbow, a new multivariable polynomial signature scheme. In: Ioannidis J, Keromytis A, Yung M (eds) Applied cryptography and network security. ACNS 2005. LNCS, vol 3531. Springer, Berlin, Heidelberg, pp 164–175

# Performance Evaluation of Zero Knowledge and HMAC-Based Authentication in Fog-Based Internet of Things

**Leki Chom Thungon, Ravinandan, and Md. Iftekhar Hussain**

**Abstract** A new computing model, called the Internet of Things (IoT), is surfacing and has grasped attention in various researches and industries in the past few years. The cloud-based applications where delays and jitters cannot be tolerated and need real-time decisions. For such applications, edge computing or fog computing plays a critical role by providing cloud-applications near the end-devices. With the increasing number of entities such as fog, security challenges also increases. In this work, we study different authentication challenges faced by a fog computing enabled IoT network. We analyze some of the selected authentication schemes like Zero-knowledge based Authentication protocol for entity authentication and HMAC authentication for message authentication. Performance results are compared for both types of protocols.

**Keywords** Authentication · Fog-computing · IoT · Lightweight · Security

## 1 Introduction

The Internet of Things, IoT, consist of entities or objects directly or indirectly available in the network. The security challenge is the primary and major factor for IoT success. A smart security solution is essential to encounter these pivotal security provocations. Fog computing is been widely used for better performance of traditional IoT networks. However, security challenges also increases due to such new entities in the network.

L. C. Thungon (✉) · Ravinandan · Md. I. Hussain
Department of Information Technology, North Eastern Hill University,
Shillong, Meghalaya 793022, India
e-mail: lekichom@gmail.com

Ravinandan
e-mail: ravinandan.bhadwaj@gmail.com

Md. I. Hussain
e-mail: ihussain@nehu.ac.in



A. Maji et al. (eds.), *Proceedings of the International Conference on Computing and Communication Systems*, Lecture Notes in Networks and Systems 170,
https://doi.org/10.1007/978-981-33-4084-8_61

Fogging or Fog computing is a decentralized computing where data, computation, storage, and applications are distributed logically and efficiently in the cloud and the devices. This computing method has its benefits when the cloud is closer to data where it has been created. Fogging improves efficiency and reduces the data to be processed, stored and analyzed by the cloud. This improves the efficiency and security of the cloud. Some of the fog computing examples are smart grid system, smart city, intelligent transportation, and software-defined networks [10]. Fog computing provides services to IoT users such as low latency, decentralization, mobility support, real or near real-time data processing. But, due to security threat in fog such as identity authentication, access control, etc. there is a need for a lightweight cryptographic algorithm to provide an authentication scheme which can support resource-constrained devices and fog node. In this paper, we have simulated some authentication protocol on fog based IoT network with help of Contiki-3.0 operating system using the Cooja simulator to support such tiny devices and analyze the performance of the algorithm for fog based IoT networks.

The rest of our paper is organised with the background survey of fog network architecture as described in Sect. 2. This section also gives a brief description about the existing authentication algorithms. This paper concludes with the Sect. 3, where the memory consumption and execution time for fog-based applications are evaluated.

## 2 Background

This section focuses on the architecture of fog network and discusses each and every layer of it.

### *2.1 Fog Network Architecture*

The architecture of the fog network consist of device, fog and cloud layer as shown in the Fig. 1. These layers have been arranged according to their computing and storage capability. Each layer can be enhanced with various entities making it more flexible and scalable. In addition, these layers are connected with authorities via. direct or indirect links so that in any vulnerability, concerned authorities can handle the situation easily [18]. The device layer has both mobile and fixed IoT devices. The mobile devices such as fitness trackers, smart clothes and cameras, etc., are owned and wearable and smart mobile devices are smartphones, smartwatches, smart glasses, vehicles. On the other hand, the fixed devices (sensors and FID tags) are used prior to deployment for predefined tasks. For example, a smart city will consist of both fixed and mobile IoT devices installed, which can collect data and can communicate with each other [16].

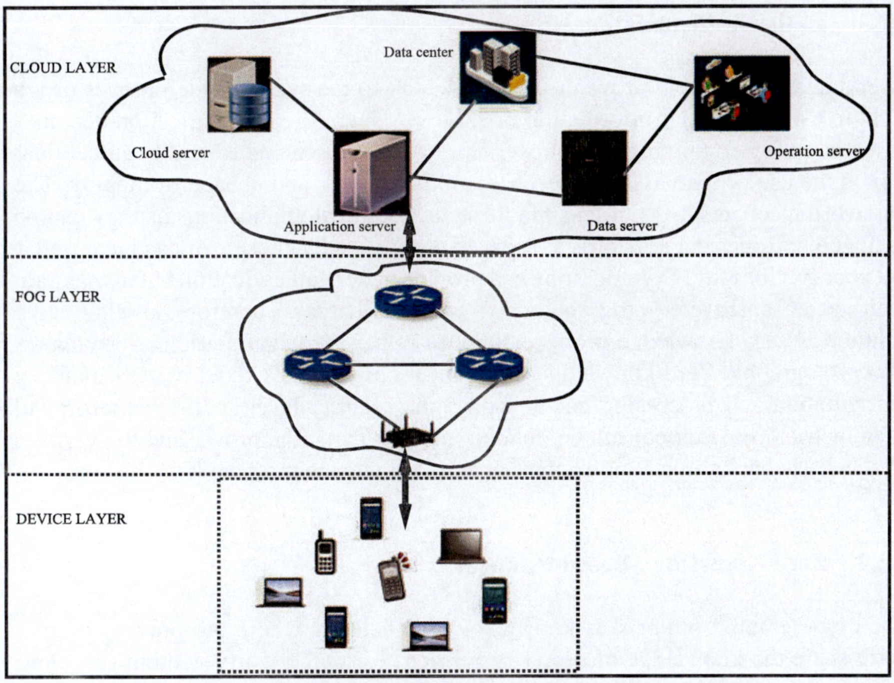


**Fig. 1** Architectural view of fog computing for IoT applications

The fog layer has fog nodes like switches, routers, bridges, gateways, and base stations with local servers, embedded servers, mobile phones, etc. Fog nodes can be situated anywhere based on the applications and are sequentially distributed both in IoT devices and cloud servers. A fog node can offer service without a cloud server like in decentralized vehicular navigation [9], smart traffic light system, etc. Further, it can collect real-time data from the IoT device and sends these data summaries to the cloud located at a higher level. The cloud layer in Fig. 1 has storage space and resources which can be accessed by the user anytime and from anywhere, when the devices are connected to the Internet [4].

## 2.2 Existing Authentication Algorithm

A brief description of some of the existing authentication algorithms are discussed as below.

### 2.2.1 Entity Authentication

User authentication can prove the user's identity to the system which proves user's right to use a particular information or services. Therefore, authentication becomes essential in fog computing. The most commonly used scheme for entity authentication is the use of the password mechanism. It is easier and cheaper to deploy. The password mechanism is not feasible in security-critical applications as they can be easily compromised. Passwords are easy to guess if badly chosen, or can be reused if not changed for long. Cryptography can provide better authentication but to maintain their secret sensitive cryptographic keys becomes harder. Therefore, an alternative solution is zero-knowledge proof techniques rather than symmetric and asymmetric cryptography. Zero-knowledge authentication protocols requires less memory consumption and processing power with same security level as of symmetric and asymmetric. The protocol can be divided into two parts: the prover and the verifier, which states background and content of zero-knowledge protocol [6].

### 2.2.2 Zero Knowledge-Based Protocol

The cryptographic protocol is known for its efficiency where the prover i.e., one party share the knowledge of secret to verifier i.e, another party without revealing the secret [7]. By doing this, both the verifier and eavesdropper cannot know the secret and cannot persuade the third party about the secret [14, 17]. The prover authenticates itself as legitimate node and the verifier authenticates the prover by challenge questions.

### 2.2.3 Fiat-Shamir Protocol

Fiat-Shamir is the identity-based identification system. The method enables a party to authenticate its identity to another. Both of the parties agrees on n, where $n = p \times q$, p, and q are two large prime numbers known only to both the parties. Then for every user it generates public key $v_1, v_2, \ldots, v_k$ and private key $s_1, s_2, \ldots, s_k$ such that $s_i = v_i^{-1} (\bmod n)$. To generate private keys computes values $v'_j = f(I, j)$, where $j$ = 1, 2, ..., $N$, and $I$, $j$ means concatenation of $I$ with string $j$, where $I$ contains information of user and a pseudo-random function f that maps random strings with range $[0, n)$. For public key, $k$ values of $v'_j$ is selected which has a square root (modulo $n$). The selected values become $v_1, v_2, \ldots, v_k$.

### 2.2.4 Guillou-Quisquater Protocol

The Guillou-Quisquater identification protocol strength is the computational complexity of $v$th power modulo $n$ [1–3, 8]. The protocol consists of a single round and is described as given in the below equation:

$$B^v J (\bmod n) = 1, \quad \text{where} \quad J = \text{Red}(I)$$

where, $B$ is unique authentication number, $n$ is composite number, $v$ is exponent number known to verifier, and $J$ is shadowed identity of device. Redundancy Rule (Red) is a published function which constructs $J$ from $I$. Red Computes $I$ to obtain $J$, where $I$ is half shorter than $n$.

#### 2.2.5 Message Authentication

A Message Authentication Code (MAC) or tag authenticates a message. MAC verifies whether the message has come from the stated sender or has been modified or not. It can protect the message's data integrity and authenticity by detecting changes in the message content. MAC mechanisms based on cryptographic hash functions are called HMAC. HMAC can be combined with any cryptographic hash functions like MD5, SHA1, etc. HMAC uses a secret key to verify and calculate message authentication values [19]. Examples of HMAC hash functions are SHA-1, MD5, SHA-256, etc. These hash functions are denoted by HMAC-SHA1, HMAC-MD5, HMAC-SHA256 [15].

#### 2.2.6 HMAC Using SHA-1, SHA-256, MD5 [15, 19]

Let us assume $H$ to be MD5 or SHA-1. Input of $H$ can be any length and output is of $l$ bit, where $l = 128$ for MD5, $l = 160$ for SHA-1, $l = 256$ for SHA-256, etc. Let 'Text' be the data where MAC is to be applied. Let '$K$' be message authentication secret key of the communicating parties. $K \leq 64$ bytes, size of hash block. If $K < 64$ bytes, zeros are appended to make it 64 bytes. We further define ipad and opad as: ipad $= 0 \times 36$ byte for 64 times and opad $= 0 \times 5\text{C}$ byte for 64 times. The function HMAC takes the key $K$ and *Text*, and produces:

$$\text{HMAC}_k(\text{Text}) = H(K \oplus \text{opad}, H(K \oplus \text{ipad}, \text{Text}))$$

The key should be at least l bit. The randomness of the key is inversely proportional to the key length. HMAC produces 80 bits output for the hash function. The cost of authenticating a stream text is similar with hashing that stream when text gets larger.

# 3 Performance Analysis

## *3.1 The Experimental Setup*

Our experimental setup approach consist of Contiki-3.0 Operating System on Cooja Simulator [11, 13]. The fog computing layer is achieved by using RPL border router of Contiki-3.0. This layer collects data from the packet payloads received from sensors. Sensor nodes are deployed using Sky-Websense [5, 12]. A node-js server on a computer is connected to a border router at one end and cloud platform to another end. It fetches sensor values and performs computation like authentication on received data and after validating forwards to the cloud [12]. The cloud computing layer runs on a desktop as a server system and stores sensor values in its database.

## 3.2 Experimental Result

We have deployed the required environment and implemented on zero knowledge based Authentication and HMAC (Message Authentication) as shown in the Figs. 2 and 3 respectively.

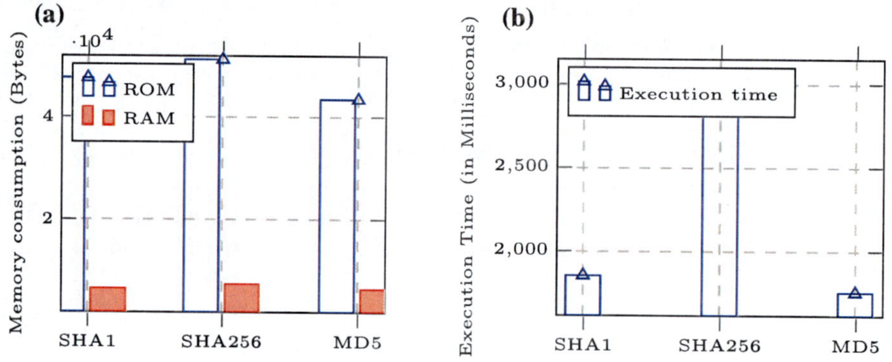


**Fig. 2** **a** HMAC algorithms memory consumption, **b** execution time of HMAC algorithms

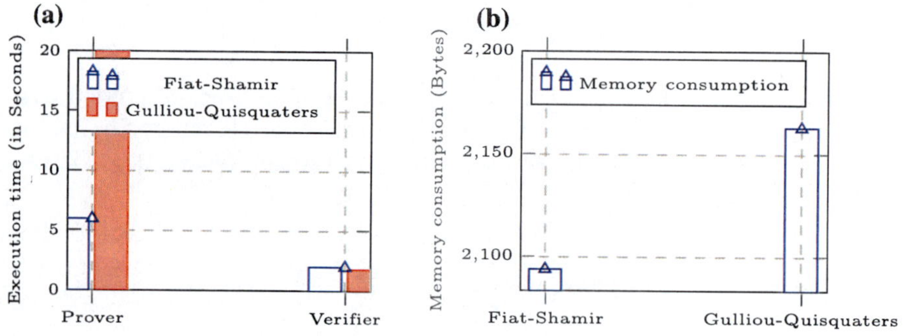


**Fig. 3** **a** Execution time of Fiat-Shamir and Gulliou-Quisquaters, **b** memory consumption of Fiat-Shamir and Gulliou-Quisquaters

### 3.2.1 Experimental Result of Zero-Knowledge Based Algorithms

### 3.2.2 Execution Time

Experiment results show that the Guillou-Quisquater's identification scheme is better in terms of security but takes more computational time as shown in Fig. 3. On another hand, the Fiat-Shamir simple identification scheme provides less security but performs faster than Guillou-Quisquater's authentication algorithm.

The memory consumption of Guillou-Quisquater's identification scheme is higher then of Fiat-Shamir as in Fig. 3. Guillou-Quisquater scheme store array $c$ value i.e., a challenge is given by the verifier and challenge is solved using array values. On another hand, in Fiat-Shamir algorithm challenge is 1 bit, i.e., $c$ is 1 bit, therefore, memory consumption in Fiat-Shamir is less as compared to Guillou-Quisquater.

### 3.2.3 Experimental Result of HMAC Algorithms

### 3.2.4 Execution Time

The experiment shows that the execution time for HMAC-SHA1, HMAC-SHA256 and HMAC-MD5. We can see that the execution time of HMAC-MD5 is lesser compared to another HMAC algorithm as shown in Fig. 2. The key size of HMAC-MD5 is 16 bit, HMAC-SHA1 key size is 20 bit and HMAC-SHA256 key size is 32 bit. Therefore, HMAC-MD5 has less round of computation to perform because of less key size.

In the above Fig. 2, we calculate the memory consumption of the HMAC-algorithm. We observe that HMAC-MD5 consumes less memory as compared to HMAC-SHA1 and HMAC-SHA256. This is because the output hash size of HMAC-MD5 is 126 bit where the output hash size of HMAC-SHA1 is 160 bit and HMAC-SHA256 is 256 bit. So the node has to calculate and store that hash output and then send to fog node with the original value that's why memory consumption of HMAC-MD5 is less as compared to another algorithm because of less size of the output.

## 4 Conclusion

We have successfully implemented the zero knowledge-based authentication protocols for entity authentication and HMAC protocols for message authentication in Contiki-3.0 operating system using cooja simulator. From the experiment, it is observed that the Guillou-Quisquater's identification scheme is better in terms of security but takes more computation time. On another hand, the Fiat-Shamir Simple identification scheme provides less security but performs faster then Guillou-Quisquater's. Moreover, message authentication HMAC-MD5 is faster then HMAC-SHA1 and HMAC-SHA256 and memory consumption of HMAC-MD5 is also less. But, because of smaller key size and smaller hash size HMAC-MD5 are more susceptible to collision attack. HMAC-SHA256 is a better option against collision attack but execution time and memory consumption is also more and is difficult to implement on resource-constrained devices.

## References

1. Bao F, Deng RH, Mao W (1998) Efficient and practical fair exchange protocols with off-line TTP. In: 1998 IEEE symposium on security and privacy, 1998. Proceedings, pp. 77–85. IEEE (1998)
2. Bengio S, Brassard G, Desmedt YG, Goutier C, Quisquater JJ (1991) Secure implementation of identification systems. J Cryptol 4(3):175–183
3. Boneh D, DeMillo RA, Lipton RJ (1997) On the importance of checking cryptographic protocols for faults. In: International conference on the theory and applications of cryptographic techniques. Springer, pp 37–51
4. Computing F (2015) The Internet of Things: extend the cloud to where the things are. In: Cisco white paper
5. Da Xu L, He W, Li S (2014) Internet of things in industries: a survey. IEEE Trans ind inf 10(4):2233–2243
6. De Santis A, Persiano G (1992) Communication efficient zero-knowledge proofs of knowledge. In: Annual symposium on theoretical aspects of computer science. Springer, pp 449–460
7. Goldwasser S, Micali S, Rackoff C (1989) The knowledge complexity of interactive proof systems. SIAM J Comput 18(1):186–208
8. Guillou LC, Quisquater JJ (1990) A "paradoxical" identity-based signature scheme resulting from zero-knowledge. In: Proceedings on advances in cryptology. Springer, pp 216–231
9. Huang C, Lu R, Choo KKR (2017) Vehicular fog computing: architecture, use case, and security and forensic challenges. IEEE Commun Mag 55(11):105–111
10. Klas GI (2015) Fog computing and mobile edge cloud gain momentum open fog consortium, ETSI MEC and cloudlets. Google Scholar
11. Kugler P, Nordhus P, Eskofier B (2013) Shimmer, Cooja and Contiki: A new toolset for the simulation of on-node signal processing algorithms. In: IEEE International conference on body sensor networks, pp 1–6. https://doi.org/10.1109/BSN.2013.6575497
12. Lavassani M, Forsström S, Jennehag U, Zhang T (2018) Combining fog computing with sensor mote machine learning for industrial IoT. Sensors 18(5):1532
13. Osterlind F, Dunkels A, Eriksson J, Finne N, Voigt T (2006) Cross-level sensor network simulation with Cooja. In: 31st IEEE conference on local computer networks, pp 641–648. https://doi.org/10.1109/LCN.2006.322172
14. Patel A, Kalajdzic K, Golafshan L, Taghavi M (2011) Design and implementation of a zero-knowledge authentication framework for java card. Int J Inf Secur Privacy (IJISP) 5(3):1–18
15. Sang-Su L, Han, Kim J-N (2005) An implementation of 3DES and HMAC-MD5 in Intel IXP 2400. In: The 7th International conference on advanced communication technology, ICACT, vol 1, pp 369–371. https://doi.org/10.1109/ICACT.2005.245875
16. Sehgal VK, Patrick A, Soni A, Rajput L (2015) Smart human security framework using internet of things, cloud and fog computing. In: Intelligent distributed computing. Springer, pp 251–263
17. Stinson DR, Wu J (2007) A zero-knowledge identification and key agreement protocol. IACR Cryptol ePrint Archive 2007:116

18. Vaquero LM, Rodero-Merino L (2014) Finding your way in the fog: towards a comprehensive definition of fog computing. ACM SIGCOMM computer communication review 44(5):27–32
19. Yiakoumis I, Papadonikolakis M, Michail H, Kakarountas AP, Goutis CE (2005) Efficient small-sized implementation of the keyed-hash message authentication code. In: EUROCON—The international conference on "Computer as a Tool", vol 2, pp 1875–1878. https://doi.org/10.1109/EURCON.2005.1630347

# Systems and Computing

# Finite Dimensional Representation of Fractional-Order System

**Jaydeep Swarnakar**

**Abstract** Fractional-order system (FOS) is a well-known topic at present among the researchers in numerous fields of science and engineering. Band-limited interpretation of infinite dimensional FOS is the most crucial thing to engage it practically for the several applications pertaining to control and signal processing. In this paper, a class of FOS has been represented within a limited band using the Oustaloup approximation method. The entire FOS has been modeled through approximating the individual fractional power term belonging to it. Finally, the frequency response of the approximated FOS has been compared with the frequency response of its finite dimensional counterpart as a function of changing the order of approximation of individual fractional power term. Simulation results have been provided to exhibit the efficacy of the approximation scheme.

**Keywords** Fractional-order system · Fractional power term · Oustaloup approximation

## 1 Introduction

Fractional-order system (FOS) has been studied in various disciplines of science and engineering at the present time, although the theory of fractional calculus is really very old one. The first mention regarding the fractional-order derivative was a discussion between two great mathematicians namely L'Hospital and Leibniz over three hundred years back [1]. Fractional-order systems are modeled by fractional-order differential equations or by fractional-order transfer functions [2, 3]. The main problem of analyzing the behavior of FOS was to get a solution for the fractional-order differential equation. Recently, many methods have come up to overcome this problem. The Laplace transform of the fractional-order differential equation results the fractional-order transfer function. For example, the Laplace transform of $\frac{d^{\alpha}y(t)}{dt^{\alpha}}$

J. Swarnakar (✉)
Department of Electronics and Communication Engineering, School of Technology, North-Eastern Hill University, Shillong 793022, Meghalaya, India
e-mail: jaydeepswarnakar@gmail.com

 
A. Maji et al. (eds.), *Proceedings of the International Conference on Computing and Communication Systems*, Lecture Notes in Networks and Systems 170,
https://doi.org/10.1007/978-981-33-4084-8_62

becomes $s^\alpha Y(s)$, where the fractional power term $s^\alpha$ is known as fractional-order operator, which is either fractional-order differentiator/integrator for $0 < \alpha < 1$ and $-1 < \alpha < 0$, respectively. Obviously, the term $s^\alpha$ becomes the backbone of the FOS. But unfortunately, this fractional power term $s^\alpha$ is an infinite dimensional filter. Therefore, the approximation of $s^\alpha$ becomes absolutely vital to represent the FOS within a finite dimension for practical interest. Many methods have been developed to obtain the finite dimensional representation of $s^\alpha$ [4–6]. Out of them, Oustaloup approximation method seems to be very popular one which has been employed for realization of fractional order operator in many occasions [7–13]. Therefore, this method has been capitalized to approximate a class of FOS in this presented work. The work is organized in two stages. In the first stage, the individual fractional power terms of the concerned FOS are approximated by using the Oustaloup method. In the second stage, the overall approximation of the FOS is obtained by combining the individual approximants. The ultimate focus of the paper is to analyze the impact of changing the order of approximation on the frequency responses of the approximated FOS. The entire work contains four sections. Section 1 introduces the FOS. Section 2 describes the approximation of the FOS employing Oustaloup method. Section 3 confers the simulation results. Section 4 concludes the work.

## 2 Approximation Method

Let, the FOS is assumed as follows:

$$F(s) = \frac{K}{\eta_2 s^{\alpha_2} + \eta_1 s^{\alpha_1} + \eta_0} \tag{1}$$

Equation (1) contains three coefficients ($\eta_2$, $\eta_1$, and $\eta_0$) along with two fractional power terms $s^{\alpha_2}$ and $s^{\alpha_1}$, respectively. At the first stage, these two terms are to be approximated by their equivalent integer-order counterparts. The generalized Oustaloup method [14] has been applied to approximate both the terms. The Oustaloup method mathematically computes the $N^{th}$ order approximation of $s^\alpha$ within a band $(\omega_\psi, \omega_\xi)$ as given below:

$$s^\alpha \approx \beta \prod_{m=1}^{N} \frac{s + \omega_\gamma^/}{s + \omega_\gamma} \tag{2}$$

where $\omega_\psi$ and $\omega_\xi$ are the low frequency and the high frequency, $\beta = \omega_\xi^\alpha$, $\omega_d = \sqrt{\omega_\xi / \omega_\psi}$, $\omega_\gamma^/ = \omega_\psi \omega_d^{(2m-1-\alpha)/N}$, and $\omega_\gamma = \omega_\psi \omega_d^{(2m-1+\alpha)/N}$.

It is noticeable that the different order of approximations of $s^{\alpha_2}$ and $s^{\alpha_1}$ can be obtained by taking different values of $N$. Therefore, replacing those approximants into Eq. (1), the original FOS can be approximated to produce the different integer order models of $F(s)$ as a function of $N$. It is obvious from Eq. (1) that the order of the resultant integer-order models of the FOS will have the order of $2N+1$. Finally, the frequency responses obtained from different integer order models are analyzed and compared with the ideal FOS as a function of order of approximation, i.e., $N$.

## 3 Simulation and Results

The following fractional-order model is considered by taking $\eta_2 = 14994$, $\eta_1 = 6009.5$, $\eta_0 = 1.69$, $\alpha_2 = 1.31$, and $\alpha_1 = 0.97$ as shown below [15]:

$$F(s) = \frac{1}{14994s^{1.31} + 6009.5s^{0.97} + 1.69} \tag{3}$$

The fractional power terms are $s^{1.31} = s \times s^{0.31}$ and $s^{0.97}$, respectively. The approximation of $s^{0.31}$ and $s^{0.97}$ are computed using Eq. (1) considering $\omega_\psi = 10^{-5}$ rad/s, $\omega_\xi = 10^5$ rad/s and taking three different values of $N$ ($N = 3, 5, 7$) as given in Tables 1 and 2.

**Table 1** Oustaloup approximations of $s^{0.31}$ taking different values of $N$

| $N$ | Equivalent integer-order transfer function |
|---|---|
| 3 | $\frac{35.48s^3+23270s^2+7083s+1}{s^3+7083s^2+23270s+35.48}$ |
| 5 | $\frac{35.48s^5+175500s^4+8598000s^3+4211000s^2+20620s+1}{s^5+20620s^4+4211000s^3+8598000s^2+175500s+35.48}$ |
| 7 | $\frac{35.48s^7+427400s^6+1.85\times10^8s^5+2.98\times10^9s^4+1.79\times10^9s^3+4.007\times10^7s^2+33390s+1}{s^7+33390s^6+4.007\times10^7s^5+1.79\times10^9s^4+2.98\times10^9s^3+1.85\times10^8s^2+427400s+35.48}$ |

**Table 2** Oustaloup approximations of $s^{0.97}$ taking different values of $N$

| $N$ | Equivalent integer-order transfer function |
|---|---|
| 3 | $\frac{70790s^3+3689000s^2+89170s+1}{s^3+89170s^2+3689000s+70790}$ |
| 5 | $\frac{70790s^5+7.662\times10^7s^4+8.211\times10^8s^3+8.798\times10^7s^2+94270s+1}{s^5+94270s^4+8.798\times10^7s^3+8.211\times10^8s^2+7.662\times10^7s+70790}$ |
| 7 | $\frac{70790s^7+2.88\times10^8s^6+4.21\times10^{10}s^5+2.291\times10^{11}s^4+4.646\times10^{10}s^3+3.513\times10^8s^2+98870s+1}{s^7+98870s^6+3.513\times10^8s^5+4.646\times10^{10}s^4+2.291\times10^{11}s^3+4.21\times10^{10}s^2+2.88\times10^8s+70790}$ |

Using Tables 1 and 2, the original FOS, i.e., $F(s)$, is approximated to the corresponding finite dimensional model $F_N(s)(N = 3, 5, 7)$ as shown in Eqs. (4), (5), and (6). The order of $F_N(s)$ becomes $2N + 1$, i.e., 7th, 11th, and 15th by taking $N = 3, 5$ and 7, respectively.

$$F_3(s) = \frac{\begin{matrix} s^6 + 96250s^5 + 6.353 \times 10^8 s^4 + 2.82 \times 10^{10} s^3 \\ +8.635 \times 10^{10} s^2 + 1.779 \times 10^9 s + 2.512 \times 10^6 \end{matrix}}{\begin{matrix} 532000s^7 + 4.821 \times 10^{10} s^6 + 3.611 \times 10^{13} s^5 + 1.464 \times 10^{15} s^4 \\ +9.362 \times 10^{14} s^3 + 2.098 \times 10^{13} s^2 + 2.322 \times 10^{10} s + 4.458 \times 10^6 \end{matrix}} \tag{4}$$

$$F_5(s) = \frac{\begin{matrix} s^{10} + 1.149 \times 10^5 s^9 + 2.036 \times 10^9 s^8 + 2.212 \times 10^{12} s^7 \\ +3.883 \times 10^{14} s^6 + 4.216 \times 10^{15} s^5 + 7.398 \times 10^{15} s^4 \\ +8.033 \times 10^{14} s^3 + 1.409 \times 10^{13} s^2 + 1.515 \times 10^{10} s+ \\ +2.512 \times 10^6 \end{matrix}}{\begin{matrix} 5.32 \times 10^5 s^{11} + 5.321 \times 10^{10} s^{10} + 3.043 \times 10^{14} s^9 + 2.555 \times 10^{17} s^8 \\ +1.555 \times 10^{19} s^7 + 1.364 \times 10^{20} s^6 + 1.065 \times 10^{20} s^5 + 1.053 \times 10^{19} s^4 \\ +1.274 \times 10^{17} s^3 + 1.651 \times 10^{14} s^2 + 4.781 \times 10^{10} s + 4.458 \times 10^6 \end{matrix}} \tag{5}$$

$$F_7(s) = \frac{\begin{matrix} s^{14} + 1.323 \times 10^5 s^{13} + 3.693 \times 10^9 s^{12} + 1.574 \times 10^{13} s^{11} \\ +1.581 \times 10^{16} s^{10} + 2.499 \times 10^{18} s^9 + 9.34 \times 10^{19} s^8 + 5.503 \times 10^{20} s^7 \\ +7.667 \times 10^{20} s^6 + 1.684 \times 10^{20} s^5 + 8.743 \times 10^{18} s^4 + 7.148 \times 10^{16} s^3 \\ +1.377 \times 10^{14} s^2 + 4.047 \times 10^{10} s + 2.512 \times 10^6 \end{matrix}}{\begin{matrix} 5.32 \times 10^5 s^{15} + 5.943 \times 10^{10} s^{14} + 8.391 \times 10^{14} s^{13} + 2.625 \times 10^{18} s^{12} \\ 1.355 \times 10^{21} s^{11} + 1.593 \times 10^{23} s^{10} + 3.235 \times 10^{24} s^9 + 1.483 \times 10^{25} s^8 \\ +1.271 \times 10^{25} s^7 + 2.372 \times 10^{24} s^6 + 9.195 \times 10^{22} s^5 + 7.224 \times 10^{20} s^4 \\ +1.33 \times 10^{18} s^3 + 6.023 \times 10^{14} s^2 + 9.311 \times 10^{10} s + 4.458 \times 10^6 \end{matrix}} \tag{6}$$

The frequency responses of $F_N(s)(N = 3, 5, 7)$ and $F(s)$ have been plotted together in Fig. 1. The magnitude approximation is worthwhile in each cases. However, the phase approximation reveals betterment as the order of approximation is enhanced. The errors are greater in the high frequency ranges as compared to the low frequency ranges. The errors have been plotted in Fig. 2. It is clear that the magnitude error and the phase error have been reduced as the order of approximation ($N$) goes high. A brief summary on resulting errors have been provided in Table 3. The data of Table 3 clearly suggest that both the magnitude error and phase error have been significantly reduced by considering the higher-order approximations. This further implicates that the $F_N(s)$ represents the underlying $F(s)$ in much better way in terms of frequency response characteristics as $N$ goes high.

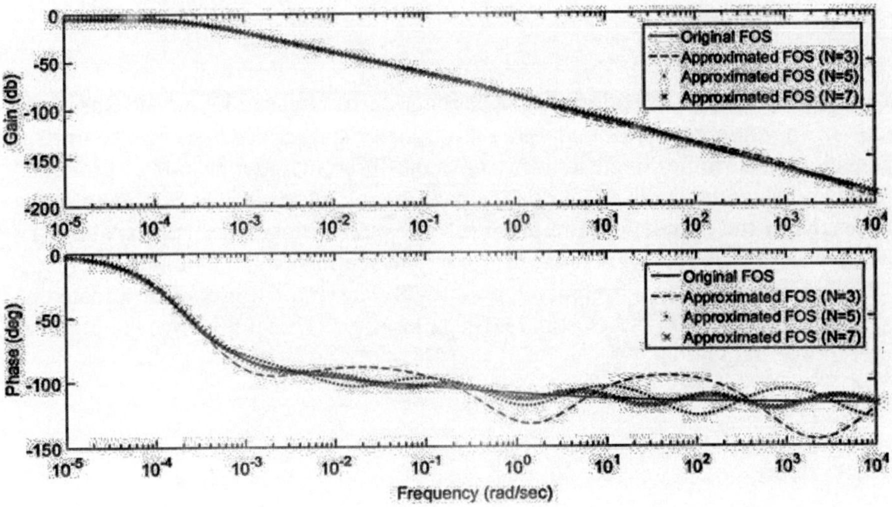


**Fig. 1** Frequency responses of $F_N(s)(N = 3, 5, 7)$ along with original FOS

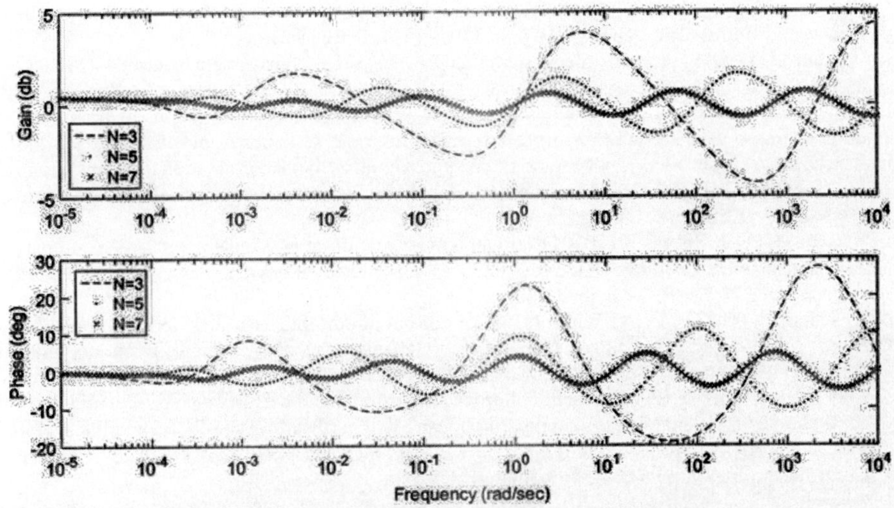


**Fig. 2** Comparison of errors taking different order of approximations of FOS

**Table 3** Comparison of errors as by taking three different orders of approximation of $F(s)$

| Order of approximation of individual fractional power term ($N$) | Maximum absolute magnitude error (dB) | Maximum absolute phase error ($^{o}$) |
|---|---|---|
| 3 | 4.4 | 27.7 |
| 5 | 1.65 | 11 |
| 7 | 0.7 | 4.6 |

## 4 Conclusions

In this paper, the approximation of FOS has been performed employing the Oustaloup method. The entire modeling undergoes through two stages. The first stage comprises of developing the finite dimensional model of the different fractional power terms, and the second stage finds the finite dimensional models of the whole FOS employing the results of the first stage. The primary endeavor of this paper is to analyze the effectiveness of the approximations through representing each fractional power term of the concerned FOS by different integer-orders. The simulation results validate that the approximation of FOS has been reached closer to the ideal FOS by taking higher order of approximations.

## References

1. Das S (2008) Functional fractional calculus for system identification and controls. Springer, Berlin
2. Chen YQ, Petras I, Xue D (2009) Fractional-order control—a tutorial. In: Proceeding of the American Control Conference (ACC), pp 1397–1411, IEEE, St. Louis, USA
3. Podlubny I (1999) Fractional order systems and $PI\lambda D\mu$. IEEE Trans Autom Control 44(1):208–214
4. Vinagre BM, Podlubny I, Hernandes A, Feliu V (2000) Some approximations of fractional-order operators used in control theory and applications. Fract Calculus Appl Anal 3(3):231–248
5. Xue D, Zhao C, Chen YQ (2006) A modified approximation method of fractional-order system. In: Proceedings of the International Conference on Mechatronics and Automation (ICMA), pp 1043–1048, IEEE, Luoyang, China
6. Khanra M, Pal J, Biswas K (2010) Rational approximation of fractional operator—a comparative study. In: Proceedings of the international conference on power, control and embedded systems, pp 1–5, IEEE, Allahabad, India
7. Oustaloup A (1991) CRONE control: Robust control of non-integer order. Hermes, Paris
8. Monje CA, Vinagre CA, Feliu V, Chen YQ (2008) Tuning and auto-tuning of fractional order controllers for industry applications. Control Eng Pract 16(7):798–812
9. Luo Y, Li H, Chen Y (2009) Fractional order proportional and derivative controller synthesis for a class of fractional order systems: tuning rule and hardware-in-the-loop experiment. In: Proceedings of the 48th IEEE conference on decision and control held jointly with 28th Chinese Control Conference, pp 5460–5465, Shanghai, China
10. Swarnakar J, Sungoh W (2020) Rational approximation of fractional-order system with multiple fractional powered terms—a comparative study. In: Pandian A, Ntalianis K, Palanisamy R (eds) Intelligent computing, information and control systems. Advances in intelligent systems and computing, vol 1039. Springer, Cham, pp 30–37
11. Swarnakar J, Adhikary B (2020) Band limited realization of fractional-order proportional integral controller for a class of finite dimensional system. In: Pandian A, Ntalianis K, Palanisamy R (eds) Intelligent computing, information and control systems. Advances in intelligent systems and computing, vol 1039. Springer, Cham, pp 70–77
12. Swarnakar J, Sarkar P, Singh LJ (2019) Rational approximation methods for a class of fractional-order SISO System in delta domain. In: Bera R, Sarkar SK, Singh OP, Saikia H (eds) Lecture Notes in Electrical Engineering, vol 537. Springer, Singapore, pp 395–402
13. Swarnakar J, Sarkar P, Singh LJ (2018) Realization of fractional-order operator in complex domains—a comparative study. In: Bera R, Sarkar S, Chakraborty S (eds) Lecture Notes in Electrical Engineering, vol 462. Springer, Singapore, pp 711–718

14. Sheng H, Chen YQ, Qiu TS (2011) Fractional processes and fractional-order signal processing. Springer, London
15. Narang A, Shah SL, Chen T (2010) Tuning of fractional PI controllers for fractional order system models with and without time delays. In: Proceedings of the American Control Conference, pp 6674–6679, IEEE, Baltimore, MD, USA (2010)

# Synthesis of Ternary Reversible Circuits based on Permutation Cycles

**P. Mercy Nesa Rani**

**Abstract** The main motivation to use ternary logic systems is that the amount of information per circuit line is higher as compared to the corresponding binary logic representation, thereby leading to more compact circuit realizations. This is particularly attractive for quantum computing as qutrits are expensive resources and minimizing their number is one of the main objectives during synthesis process. Hence, the present work exploits the ternary logic concept and synthesizes the ternary reversible circuits using a cycle-based approach. A permutation representing the ternary reversible specification of a given function in given as an input. The permutation cycles are obtained from the input and then it is factored into smaller length cycles viz., 3-cycles and 2-cycles. The simpler cycles are then represented with the ternary reversible gates. The gate library such as Ternary Not gate, Multi-polarity Ternary Feynman Gate ($MPT_F$) and Multi-polarity Ternary Toffoli Gate ($MPT_T$) is used to synthesize 3-cycles and 2-cycles. These gates are then decomposed into elementary ternary quantum gates. The experimental evaluation indicates that there is an improvement in the present approach i.e 49 and 19% in terms of quantum cost and gate count as compared with the published works.

**Keywords** Ternary reversible logic circuit · Permutation · Qutrits · Ternary reversible gates

## 1 Introduction

Recently, there is an increased research works on reversible circuits. The cascade of reversible gates present in a reversible circuit usually carry out some objective mapping between inputs and outputs. They are also structurally different from conventional circuits, in the sense that fanout and feedback connections are not permitted. Many researchers have further decomposed the reversible gates into elementary

P. Mercy Nesa Rani (✉)
College of Post-Graduate Studies in Agricultural Sciences, Umiam, Meghalaya 793103, India
e-mail: mercyranip@gmail.com



A. Maji et al. (eds.), *Proceedings of the International Conference on Computing and Communication Systems*, Lecture Notes in Networks and Systems 170,
https://doi.org/10.1007/978-981-33-4084-8_63

quantum gates, which are considered to be the primitive operations of a quantum computer. In a binary quantum system expressed as a quantum circuit, the number of circuit lines represent the basic unit of information being processed (called *quantum bits* or *qubits*). *Qutrit* is the term used to represent three-valued quantum system. For a quantum multi-valued logic (MVL) system, information unit is represented as *qudit*. As opposed to conventional computing, where the gates represent hardware resources and the total computation time is determined by the gate delays, the concept is somewhat different in reversible or quantum computing. Here, the circuit lines representing quantum bits are the hardware resources, and the gates represent the operations that are carried out in sequence of time. During the process of synthesis, various cost metrics have been used by researchers to compare the quality of the generated gate netlists, viz., number of reversible gates,number of quantum gates (called *quantum cost*), number of additional circuit lines required (called *ancilla lines*), etc.

A number of researchers have developed various synthesis techniques for reversible logic circuits. Much more works have been completed for the binary reversible circuits [1–3], but less number of works are reported in ternary reversible circuits [4–7]. Hence, the present work addresses the ternary reversible circuit synthesis using cycle based approach. The remaining of the research article has been formulated as follows. Section 2 discusses about the fundamentals of ternary quantum logic and various kinds of ternary reversible gates. Section 3 presents the permutation based synthesis approach and Sect. 4 shows the experimental results which is followed by conclusion in Sect. 5.

## 2 Background

In this section, we briefly discuss about the basics of ternary quantum logic and the various ternary reversible gates used in the ternary reversible logic synthesis process.

### 2.1 *Ternary Quantum Logic*

In this subsection, we discuss the representation of decimal numbers in the ternary logic system, and also the matrix representation in 3-dimensional Hilbert space.

**Definition 2.11** Consider an $n$-digit ternary representation of a number $D = (t_{n-1}\ t_{n-2}\ \ldots\ t_1\ t_0)$. The **decimal representation** of the number $D$ can be written as

$$D = t_{n-1} \times 3^{n-1} + t_{n-2} \times 3^{n-2} + \cdots t_1 \times 3^1 + t_0 \times 3^0$$

*Example 1* Consider a 3-digit ternary number 0 1 0. The decimal equivalent of the number is $0 \times 3^2 + 1 \times 3^1 + 0 \times 3^0 = 3$.

The manipulation of qutrit has been carried out in ternary quantum systems and it has three possible basis states $|0\rangle$, $|1\rangle$, and $|2\rangle$. Thus, it forms the basic unit for ternary quantum system. The state of a qutrit can be expressed as a linear combination of the basis states as:

$$|\alpha\rangle = \phi|0\rangle + \pi|1\rangle + \varphi|2\rangle \tag{1}$$

where $\phi$, $\pi$ and $\varphi$ are complex numbers and denote the probability amplitudes of the basic states as below. $|\phi|^2 + |\pi|^2 + |\varphi|^2 = 1$.

The three basis states $|0\rangle$, $|1\rangle$, $|2\rangle$ may be denoted by vectors as shown below.

$$|0\rangle = \begin{bmatrix} 1 \\ 0 \\ 0 \end{bmatrix} \quad |1\rangle = \begin{bmatrix} 0 \\ 1 \\ 0 \end{bmatrix} \quad |2\rangle = \begin{bmatrix} 0 \\ 0 \\ 1 \end{bmatrix}$$

It may be noted that the main advantage of Multi-Valued Logic (MVL) systems is the logarithmic reduction in the number of quantum digits (qudits) as compared to qubits in the binary domain. For a Hilbert space of $N$ dimension, the number of qubits required by a binary quantum system is given by:

$$n_2 = \log_2 N \tag{2}$$

Similarly, in an $m$-valued quantum system the number of qudits required is given by:

$$\begin{aligned} n_m &= \log_m N \\ &= (log_2 N)/(\log_2 m) \\ &= n_2/\log_2 m \end{aligned} \tag{3}$$

For ternary quantum system, $m = 3$, which results in 63% reduction in quantum register width, as

$$(1/\log_2 3) = (1/1.58) * 100 = 63\%$$

## 2.2 Ternary Reversible Gates

In this subsection, we briefly explain about the different kinds of gates used for the synthesis process.

**Definition 2.21 Multi-Polarity Ternary Feynman Gate** has two inputs and two outputs. An input $I_1$ is connected to the gate control and another input $I_2$ is connected to the gate target. It triggers according to Z-transformation when the value of $I_1$ is in the any one of the states such as $|0\rangle$, $|1\rangle$, or $|2\rangle$.

A schematic representation of $MPT_F$ and its truth table has been given in Fig. 1.

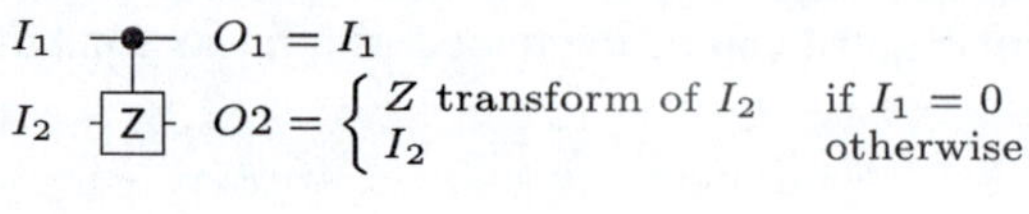


(a)

| $I_1$ | $I_2$ | $O_1$ | $O_2$ |
|---|---|---|---|
| 0 | 0 | 0 | 1 |
| 0 | 1 | 0 | 2 |
| 0 | 2 | 0 | 0 |
| 1 | 0 | 1 | 0 |
| 1 | 1 | 1 | 1 |
| 1 | 2 | 1 | 2 |
| 2 | 0 | 2 | 0 |
| 2 | 1 | 2 | 1 |
| 2 | 2 | 2 | 2 |

(b)

**Fig. 1** MPT$_F$ **a** Illustrative Picture, **b** truth table for $Z = [+1]$

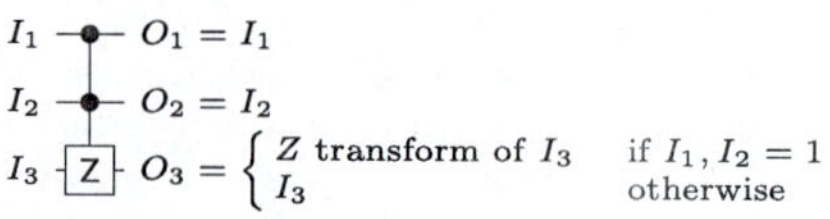


(a)

| $I_1$ | $I_2$ | $I_3$ | $O_1$ | $O_2$ | $O_3$ |
|---|---|---|---|---|---|
| 0 | 0 | $X$ | 0 | 0 | $X$ |
| 0 | 1 | $X$ | 0 | 1 | $X$ |
| 0 | 2 | $X$ | 0 | 2 | $X$ |
| 1 | 0 | $X$ | 1 | 0 | $X$ |
| 1 | 1 | 0 | 1 | 1 | 1 |
| 1 | 1 | 1 | 1 | 1 | 2 |
| 1 | 1 | 2 | 1 | 1 | 0 |
| 1 | 2 | $X$ | 1 | 2 | $X$ |
| 2 | 0 | $X$ | 2 | 0 | $X$ |
| 2 | 1 | $X$ | 2 | 1 | $X$ |
| 2 | 2 | $X$ | 2 | 2 | $X$ |

(b)

**Fig. 2** MPT$_T$ **a** Illustrative picture, **b** truth table when $Z = [+1]$ where $X \in \{0, 1, 2\}$

**Definition 2.22** The **Multi-Polarity Ternary Toffoli Gate** has three inputs and three outputs, with two inputs connected to gate controls and one input connected to the gate target. It triggers the target $I_3$ according to Z-transformation when the values of $I_1$ and $I_2$ are in the combination of the states such as $|0\rangle$, $|1\rangle$, or $|2\rangle$.

A schematic representation of MPT$_T$ and its truth table is given in Fig. 2.

## 3 Proposed Permutation Based Approach

We now present an approach that synthesizes the permutation cycles. Initially, essential transformations have been carried out to change the binary benchmarks given in RevLib [10] into an equivalent ternary specification with the help of reversible embedding and is reported in [8]. The permutation with longer cycles is obtained from the ternary reversible specification. The permutations which are having larger cycles are decomposed in terms of smaller length cycles viz., 3-cycles and 2-cycles. Then, a gate library such as $T_N$, $\mathrm{MPT_F}$ and $\mathrm{MPT_T}$ with target operations $\{+1, +2\}$ is used to synthesize every 3-cycles. In the similar fashion, a gate library viz., ($\mathrm{MPT_F}$) and ($\mathrm{MPT_T}$) with Z-transformation $\{01, 02, 12\}$ as target operation is used to synthesize each 2-cycle. Let us look into an example to synthesize a 3-cycle and 2-cycle separately as follows:

*Example 2* Let us take a 3-cycle (10, 14, 17) and the equivalent matrix representation is

$$\begin{pmatrix} 1, 0, 1 \\ 1, 1, 2 \\ 1, 2, 2 \end{pmatrix}$$

Initially, we have to check for the presence of three distinct values viz., 0,1,2 in any of the columns in the matrix. If it is present, fix that column as it is and do not apply any gate operations in that particular column. The number of 1's count is greater than the number of 2's count in the matrix, hence gates $N_\mathrm{T}$ and $\mathrm{MPT_T}$ have been utilized to modify the first and third columns of the present matrix to 1:

$$\begin{pmatrix} 1, 0, 1 \\ 1, 1, 2 \\ 1, 2, 2 \end{pmatrix} \xrightarrow{N_3^2} \begin{pmatrix} 1, 0, 0 \\ 1, 1, 1 \\ 1, 2, 1 \end{pmatrix} \xrightarrow{\mathrm{MPT}_{3[+1]}10} \begin{pmatrix} 1, 0, 1 \\ 1, 1, 1 \\ 1, 2, 1 \end{pmatrix}$$

Here, $N_3^2$ denotes the use of $N_\mathrm{T}$ to the third qutrit (suffix) two times (superscript). Similarly, $\mathrm{MPT}_{3[+1]}10$ indicates a $\mathrm{MPT_T}$ gate with target on third qutrit (suffix) with Z transformation [+1] and fired on the two control lines by the value 10.

In the matrix, the second column contains three distinct values, viz. 0, 1, 2. Hence, $N_\mathrm{T}$ is applied to the second column ($N_2$) to get the matrix:

$$\begin{pmatrix} 1, 1, 1 \\ 1, 2, 1 \\ 1, 0, 1 \end{pmatrix}$$

In order to comeback to the original state, Let us apply the gates previously used in reverse direction leaving the last gate applied.

$$\begin{pmatrix} 1, 1, 1 \\ 1, 2, 1 \\ 1, 0, 1 \end{pmatrix} \xrightarrow{\mathrm{MPT}_{3[+2]}10} \begin{pmatrix} 1, 1, 1 \\ 1, 2, 1 \\ 1, 0, 0 \end{pmatrix} \xrightarrow{N_3} \begin{pmatrix} 1, 1, 2 \\ 1, 2, 2 \\ 1, 0, 1 \end{pmatrix}$$

Figure 3 depicts the final netlist for the 3-cycle (10, 14, 17).

Thus, we obtain a netlist which consists of gate library viz., multi-polarity ternary Toffoli gate and multi-polarity ternary Feynman gate. The gates in the gate library

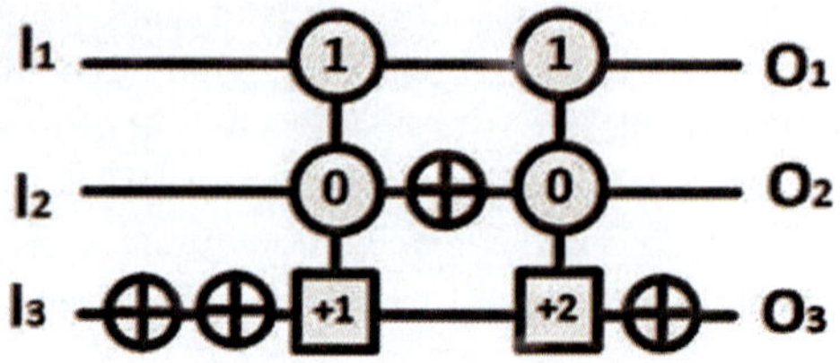


**Fig. 3** Generated netlist for 3-cycle synthesis

must be decomposed in terms of quantum gates such as M-S gate and ternary Shift gate which helps in the calculation of the quantum cost. The decomposition procedure is similar to the one which is presented in [8].

## 4 Experiment Results

The proposed approach is programmed in C++, and it is evaluated using ternary benchmarks. Initially, the binary benchmarks available in RevLib [10] are taken into consideration and the ternary reversible benchmarks are obtained using binary benchmarks by applying the above mentioned method. The hardware used for the execution of proposed method is an Intel i5 processor having 2.20 GHz clock with 4 GB Random Access Memory (RAM) and the operating system used is Ubuntu version 16.04. The results obtained by the cycle-based approach using gate library are shown in Table 1. Here, benchmark indicates the name of the ternary benchmark, inputs denotes the number of inputs for the corresponding benchmark. ($N_T$), ($MPT_F$), ($MPT_T$), (GC) (QC) denote the ($N_T$) count, $MPT_F$ count, ($MPT_T$) count, number of ternary reversible gates used and the number of ternary elementary gates used respectively. The run time indicates the total execution time of each benchmarks in seconds. Then, gate count (GC) and quantum cost (QC) calculated using the approach of [9] has been reported. Finally, comparison has been made between GC and QC of the proposed method with the method by [9] and the % improvement is presented.

It can be seen from Table 1, that the present method results in 49 and 19% reduction in terms of both QC and GC over [9]. It is observed that, the present method synthesizes the benchmarks up to 13 inputs. This method significantly outperforms the method of [9] with respect to quantum cost. It may be noted that only two benchmark circuits, take longer time viz., 2.33 and 2.18 s for the execution and the rest of the benchmark circuits take a fraction of a second for the execution.

## 5 Conclusion

In the current work, the synthesis of ternary benchmarks using a permutation approach is presented. Initially, the ternary reversible specification is represented as permutation cycles and the cycles are further decomposed into 2-cycles and 3-cycles

**Table 1** Cycle based synthesis results for ternary benchmarks

| Benchmark | Inputs | Proposed | | | | | | Rani et al. [9] | | Improvement(%) | |
|---|---|---|---|---|---|---|---|---|---|---|---|
| | | $N_T$ | $MPT_F$ | $MPT_T$ | GC | QC | Time (s) | GC | QC | GC | QC |
| 4_49_7 | 3 | 40 | 0 | 34 | 74 | 330 | 0.000 | 104 | 558 | 28.84 | 40.86 |
| graycode6_11 | 5 | 271 | 0 | 194 | 465 | 1805 | 0.004 | 637 | 4005 | 2.35 | 25.74 |
| half_adder | 3 | 13 | 0 | 22 | 35 | 179 | 0.000 | 88 | 524 | 60.22 | 65.83 |
| ham7_29 | 6 | 685 | 0 | 484 | 1169 | 4753 | 0.009 | 1541 | 9733 | 31.82 | 51.16 |
| mod5adder_66 | 5 | 118 | 0 | 82 | 200 | 760 | 0.002 | 319 | 1859 | 37.3 | 59.11 |
| mod5d1_16 | 4 | 58 | 0 | 46 | 104 | 380 | 0.001 | 237 | 1219 | 56.11 | 68.82 |
| mod5d2_17 | 4 | 130 | 0 | 98 | 228 | 924 | 0.001 | 340 | 1946 | 32.9 | 52.51 |
| one-two-three_27 | 3 | 25 | 0 | 30 | 55 | 267 | 0.000 | 101 | 565 | 45.54 | 52.74 |
| rd32_19 | 3 | 34 | 0 | 32 | 66 | 306 | 0.000 | 119 | 667 | 44.53 | 54.12 |
| mini-alu_84 | 5 | 118 | 0 | 88 | 206 | 854 | 0.002 | 294 | 1836 | 29.93 | 53.48 |
| 4gt13_25 | 4 | 13 | 0 | 14 | 27 | 135 | 0.000 | 47 | 309 | 42.55 | 56.31 |
| 4gt5_21 | 6 | 118 | 0 | 76 | 194 | 762 | 0.002 | 226 | 1530 | 14.15 | 50.19 |
| 4mod5_8 | 5 | 43 | 0 | 22 | 65 | 237 | 0.000 | 73 | 409 | 12.30 | 42.05 |
| alu_9 | 7 | 223 | 0 | 112 | 335 | 1111 | 0.003 | 348 | 2184 | 3.88 | 49.13 |
| 5xp1_90 | 10 | 2800 | 0 | 1346 | 4146 | 14186 | 0.049 | 3663 | 22979 | −11.64 | 38.26 |
| rd73_69 | 12 | 3859 | 0 | 1390 | 5249 | 15305 | 0.153 | 3539 | 21985 | −32.57 | 30.38 |
| sqn_203 | 10 | 1981 | 0 | 806 | 2787 | 8795 | 0.035 | 1358 | 13681 | −51.27 | 35.71 |
| sqrt8_205 | 13 | 8650 | 0 | 3268 | 11918 | 34158 | 0.418 | 8274 | 52658 | −30.57 | 35.13 |
| ham_3_28 | 2 | 1 | 6 | 0 | 7 | 7 | 0.000 | – | – | – | – |
| 4gt10_22 | 5 | 34 | 0 | 37 | 71 | 367 | 0.001 | – | – | – | – |
| 4gt12_24 | 4 | 19 | 0 | 25 | 44 | 226 | 0.000 | – | – | – | – |
| 4gt4_20 | 6 | 118 | 0 | 79 | 197 | 793 | 0.008 | – | – | – | – |
| 4mod7_26 | 4 | 37 | 0 | 55 | 92 | 490 | 0.001 | – | – | – | – |
| dist_144 | 12 | 7369 | 0 | 2803 | 10172 | 29192 | 0.218 | – | – | – | – |
| ham7_29 | 6 | 685 | 0 | 484 | 1169 | 4753 | 0.009 | – | – | – | – |
| majority_176 | 8 | 364 | 0 | 173 | 537 | 1743 | 0.006 | – | – | – | – |
| mlp4_184 | 12 | 6448 | 0 | 2515 | 8963 | 26647 | 0.209 | – | – | – | – |
| plus127mod8192_78 | 12 | 106600 | 0 | 57007 | 163607 | 532903 | 2.336 | – | – | – | – |
| plus63mod4096_79 | 11 | 46720 | 0 | 24091 | 70811 | 235347 | 0.946 | – | – | – | – |
| plus63mod8192_80 | 12 | 103531 | 0 | 50327 | 153858 | 480498 | 2.186 | – | – | – | – |
| radd_193 | 13 | 8347 | 0 | 3132 | 11479 | 34559 | 0.383 | – | – | – | – |
| rd53_68 | 7 | 469 | 0 | 231 | 700 | 2374 | 0.011 | – | – | – | – |
| root_197 | 12 | 7558 | 0 | 2854 | 10412 | 30292 | 0.220 | – | – | – | – |
| sqr6_204 | 9 | 1255 | 0 | 579 | 1834 | 6020 | 0.031 | – | – | – | – |
| squar5_206 | 5 | 199 | 0 | 167 | 366 | 1462 | 0.003 | – | – | – | – |

[a]—Stands for not reported

respectively. Then, these simpler cycles are realized with the help of ternary reversible gates. The gate count (GC) has been calculated using the number of ternary reversible gates used in the synthesis process. The quantum cost for each of the ternary benchmarks has been reported using the number of quantum gates such as M-S gate and ternary shift gates needed for the synthesis. There is an improvement over the existing work with respect to quantum cost and gate count of about 49% and 19% respectively. The maximum run time for largest circuit is observed as 2.33 s. In future, optimization in the synthesis approach can be worked out to enhance the presented results.

## References

1. Chattopadhyay A, Pal N, Majumder S (2014) Ancilla-quantum cost trade-off during reversible logic synthesis using exclusive sum-of-products. Emerging Technol 1–9. arXiv:1405.6073
2. Datta K, Rathi G, Sengupta I, Rahaman H (2012) Synthesis of reversible circuits using heuristic search method. In: Proceedings of the international conference on VLSI Design, Hyderabad, India, pp 328–333
3. Datta K, Rathi G, Sengupta I, Rahaman H (2015) Exact synthesis of reversible circuits using A* algorithm. J Inst Eng (Ser B) 96(2):121–130 June
4. Haghparast M, Wille R, Monfared AT (2017) Towards quantum reversible ternary coded decimal adder. Quant Inf Process 16(11):284/1–25
5. Khan M, Rice JE (2016) Ternary max-min algebra for representation of reversible logic functions. In: Proceedings of the international symposium on circuits and aystems (ISCAS), Montreal, Canada, pp 1670–1673
6. Kole A, Rani PMN, Datta K, Sengupta I, Drechsler R (2017) Exact synthesis of ternary reversible functions using ternary Toffoli gates. In: Proceedings of the 47th international symposium on multiple-valued logic (ISMVL), Novi Sad, Serbia, pp 179–184
7. Monfared AT, Haghparast M, Datta K (2019) Quantum computing using single photons and the zeno effect. Int J Theor Phys 4:1–16
8. Rani PMN, Kole A, Datta K (2019) A ternary decision diagram (TDD)-based synthesis approach for ternary logic circuits. J Inst Eng (India): Ser B 6:1–13
9. Rani PMN, Kole A, Datta K, Chakrabarty A (2016) Realization of ternary reversible circuits using improved gate library. In: Proceedings of the 6th international conference on advances in computing and Communications, Cochin, India, pp 153–160.
10. Wille R, Große D, Teuber L, Dueck GW, Drechsler R (2008) RevLib: an online resource for reversible functions and reversible circuits. In: Proceedings of the international symposium on multiple-valued logic, Texas, USA, pp 220–225

# Ultra Power Efficient Melior Quantum Multiplier with Reduced Ancilla and Garbage Outputs

R. Sai Roshan, Shaik Nawaz, Akshitha Vuppala, and J. V. R. Ravindra

**Abstract** Reversible Logic is an emerging field of research which finds its applications in low power computing, Nanotechnology and Quantum Computing. Reversible circuits should have one to one mapping i.e. one input can have only one output so that input vectors can be realized using output vectors. Reversible Circuits require Ancilla(constant inputs) and Garbage Outputs to retain reversibility. An efficient Reversible Circuit can be designed by optimizing their performance parameters. In this paper a $4 \times 4$ **Melior** Quantum Multiplier has been proposed which consists of an optimized Partial Product Generation and Multi-Operand Addition using primitive Quantum gates to reduce the count of Ancilla and Garbage Outputs. This proposed multiplier shows an improvement of 21.73% and 18.18% reduction of Ancilla and Garbage Outputs respectively. This multiplier has been implemented in Cadence Virtuoso with average power dissipation of 106.79 nW at 45 nm technology node and used in the implementation of a Linear Phase FIR filter with an average power dissipation of 456.1 nW.

**Keywords** Quantum computing · Reversible logic · Ancilla · Garbage outputs

R. Sai Roshan · S. Nawaz · A. Vuppala · J. V. R. Ravindra (✉)
Department of Electronics and Communication Engineering, Center for Advanced Computing and Research Laboratory (C-ACRL), Vardhaman College of Engineering, Shamshabad, Telangana, India
e-mail: jayanthi@ieee.org

R. Sai Roshan
e-mail: sairoshan1412@gmail.com

S. Nawaz
e-mail: shaiknawaz787@gmail.com

A. Vuppala
e-mail: akshithavuppala98@gmail.com



A. Maji et al. (eds.), *Proceedings of the International Conference on Computing and Communication Systems*, Lecture Notes in Networks and Systems 170,
https://doi.org/10.1007/978-981-33-4084-8_64

## 1 Introduction

A logic is said to be reversible if there exists one to one mapping between the inputs and outputs. Quantum Computing is a high-level abstraction of reversible logic and Quantum Circuits are logically reversible. Every reversible gate has a Quantum equivalent matrix which is unitary and is used to implement computation algorithms and these gates are also named as Quantum gates. A Quantum circuit is an interconnection of Quantum gates and they perform a specific function using qubits. A Qubit is a basic unit of information in the Quantum Computer, capable of existing in more than one state. The computation is made lossless as the count of inputs and outputs are equal and the amount of energy dissipated is reduced. In irreversible circuits there is loss of information and energy is dissipated as per Landauer's Principle [1]. The performance metrics of any Quantum Circuit are Ancilla and Garbage Outputs. Ancilla is the constant input that can be either 0 or 1. Garbage Outputs are those that are neither primary outputs nor the outputs that are used for further computation. Ancilla and garbage outputs should be less in an efficient quantum circuit design.

## 2 Literature Survey

Several Multiplier designs were proposed for an efficient design, a reversible full adder with reduced hardware complexity is used in [2]. Ripple Quantum Adders perform parallel addition of partial products in $N \times N$ Quantum multiplication based on binary tree algorithm in[3].

Digital compressors have less delay for carry propagation. Using this as an advantage, Nagamani et al. [7], a reversible design of Wallace multiplier using 4:2 compressors speeding up the process of multiplication. The authors of the paper [4] have proposed a reversible multiplier design using BVF gate which reduces the fan out by 50% and also optimises the cost and size of the circuit. BVF gate is used for copying the operand bits of the multiplier. Eshack et al. [8] proposed a design for low power multiplier using vedic mathematics and reversible gates with reduced delay and increased throughput. In [9], authors have proposed an efficient multiplication algorithm by modifying conventional algorthms and implemented it in two ways using peres gate in one design and toffoli gate in other design. In [10], authors have proposed a low power reversible multiplier using HNG gate. Anamika et al. [11] explained various reversible logic gates, their implementations and performance measures and advantages in design of various adders and subtractors. A low power consuming array multiplier is designed using multiplexer based full adder and reversible half adder in [12]. Quatum Computing also finds its several applications in nanotechnology. Table 1 summarizes the various multiplier designs proposed by various researchers.

**Table 1** Taxonomy of reversible multiplier designs

| Previous work | Ancilla | Garbage outputs | Published year |
|---|---|---|---|
| Moallem et al. [2] | 20 | 36 | 2010 |
| Thapliyal et al. [3] | 23 | 22 | 2014 |
| Bhagyalakshmi and Venkatesha [4] | 52 | 52 | 2010 |
| Shams and Navi [5] | 56 | 44 | 2008 |
| Pankaj et al. [6] | 25 | 32 | 2016 |

## 3 Proposed Melior Quantum Multiplier

This work includes three contributions, they are listed below:

1. Effective usage of target line and control lines in Partial Product Generation reduced the ancilla.
2. The count of full adders has been reduced to limit the Garbage Outputs.
3. Implementations of low power 4 × 4 reversible multiplier using pass transistor logic.

The process of 4 × 4 multiplication is shown in Fig. 1.

Multiplication operation is divided into two operations:

1. Partial Product Generation
2. Multi Operand Addition.

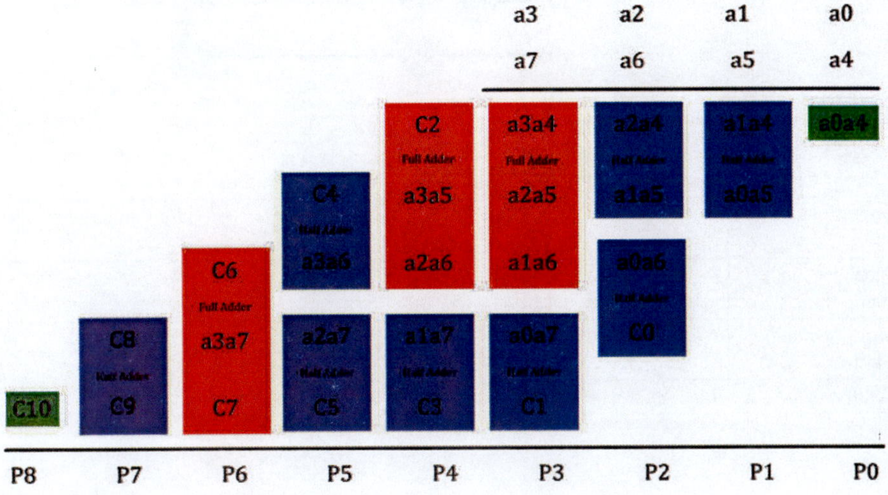


**Fig. 1** Multiplication operation

### 3.1 Partial Product Generation

The 4 × 4 **Melior** Quantum Multiplier produces 16 partial products. Sixteen Toffoli gates are used to generate these partial products. The efficient usage of Target Line by including them in the control line of other gates to reduced the number of ancilla of the **Melior** Quantum Multiplier.

### 3.2 Multi Operand Addition

In Multi Operand Addition, the outputs are generated using full and half adders, they have been realized using basic Quantum gates like $C^2$NOT and CNOT gates as illustrated in Fig. 2.

The count of $C^2$NOT, CNOT gate, half and full adders in both the operations and in the entire design has been tabulated in Table 2. The 4 × 4 **Melior** Quantum Multiplier has been implemented using IBM Q Experience and it is observed that the count of Ancilla and Garbage Outputs are 18 as shown in Fig. 3, the output P8 has been eliminated by considering all the possible combinations of inputs.

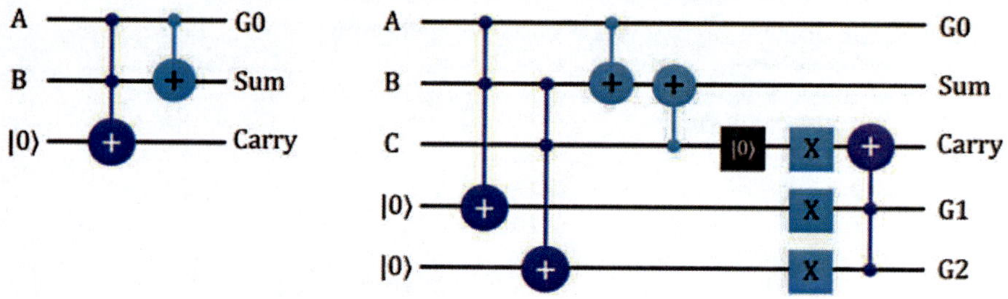


**Fig. 2** Quantum half adder and quantum full adder

**Table 2** Count of quantum gates and adders in **Melior** quantum multiplier

| Operations | No. of CNOT gates | No. of $C^2$NOT gates | Half adders | Full adders |
|---|---|---|---|---|
| Partial product generation | – | 16 | – | – |
| Multi operand addition | 14 | 17 | 8 | 3 |
| Total | 14 | 33 | 8 | 3 |

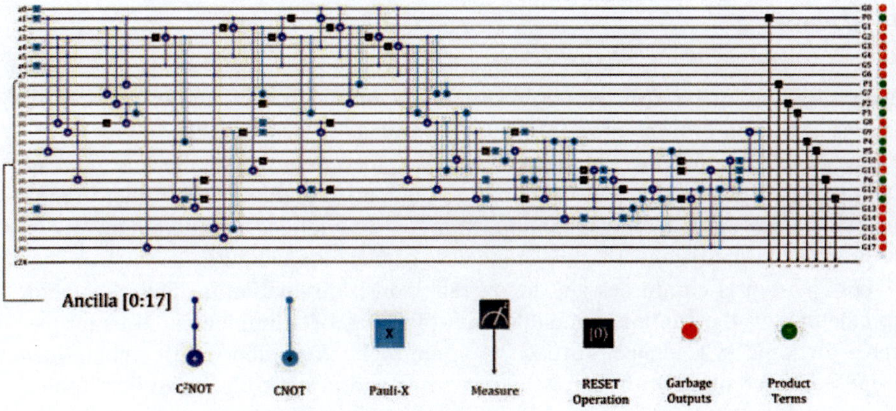


**Fig. 3** 4 × 4 **Melior** quantum multiplier

## *3.3 Power Efficient Linear Phase FIR Filter Using Melior Quantum Multiplier*

An FIR filter is said to be in linear phase if the coefficients are symmetrical about the center coefficient , so the count of multipliers is reduced in a linear phase filter. The shape of input signals is preserved by linear phase FIR filter which finds its applications in signal processing and demodulation, radar signal processing and audio processing. Due to its wide range of applications, it is essential that a linear phase filter should be fast and should dissipate less amount of power. The **Melior** reversible multiplier serves this purpose. In the third-order linear phase FIR filter, there are two multipliers that are replaced with the **Melior** Quantum Multiplier. The block diagram of the FIR filter for the below equation is shown in Fig. 4:

$$C_0 + C_1 \times Z^{-1} + C_1 \times Z^{-2} + C_0 \times Z^{-3} \quad (3.3.1)$$

**Fig. 4** Linear phase FIR filter with the **Melior** quantum multiplier

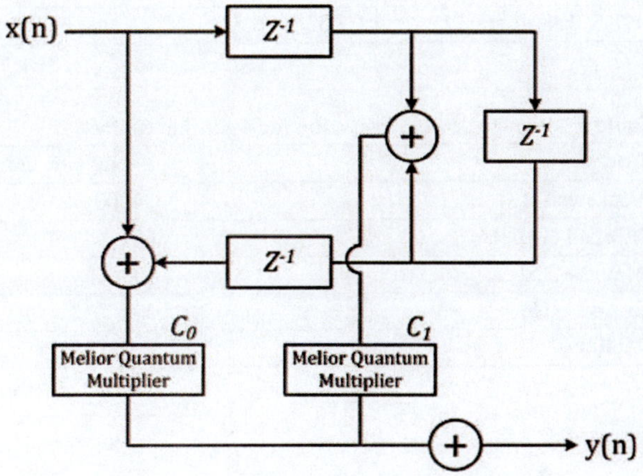

## 4 Results

A comparative analysis of the count of Ancilla and Garbage Outputs between the **Melior** Quantum Multiplier and the related works is shown in Table 3.

For an $N \times N$ **Melior** Quantum Multiplier, the count of Ancilla and Garbage Outputs were found to be $\frac{N(2N+1)}{2}$ and their comparison with the multiplier in [3] is given in Table 4. The **Melior** Quantum Multiplier has noticed a reduction of 21.73% in the case of Ancilla and 18.18% in the case of Garbage Outputs.

For any digital circuit design, the average power dissipation is the major factor. To calculate it, the **Melior** reversible multiplier has been implemented using pass transistor logic in Cadence Virtuoso at 45nm technology node. Researchers have proposed different algorithms for the reversible and irreversible multiplier. Table 5 compares the power of the existing design with the $4 \times 4$ **Melior** reversible multiplier.

**Table 3** Ancilla and Garbage outputs for $4 \times 4$ **Melior** quantum multiplier

| Comparison | [2] | [3] | [4] | [8] | [5] | Melior | Improvement (%) | | | | |
|---|---|---|---|---|---|---|---|---|---|---|---|
| | | | | | | | [2] | [3] | [4] | [8] | [5] |
| Ancilla | 20 | 23 | 52 | 58 | 56 | 18 | 10 | 21 | 65 | 68 | 68 |
| Garbage outputs | 36 | 22 | 52 | 46 | 44 | 18 | 50 | 18 | 65 | 61 | 59 |

**Table 4** Ancilla and Garbage outputs for $N \times N$ **Melior** quantum multiplier

| Size of multiplier | Thapliyal [3] | | Melior | | Efficiency (%) | |
|---|---|---|---|---|---|---|
| | Ancilla | Garbage outputs | Ancilla | Garbage outputs | Ancilla | Garbage outputs |
| 4 × 4 | 23 | 22 | 18 | 18 | 21.73 | 18.18 |
| 8 × 8 | 83 | 81 | 68 | 68 | 18.07 | 16.04 |
| 16 × 16 | 303 | 300 | 264 | 264 | 12.87 | 12 |
| 32 × 32 | 1135 | 1131 | 1040 | 1040 | 8.37 | 8.04 |
| 64 × 64 | 4351 | 4346 | 4128 | 4128 | 5.12 | 5.01 |
| 128 × 128 | 16959 | 16953 | 16448 | 16448 | 3.07 | 2.97 |

**Table 5** Average power dissipation for $4 \times 4$ multipliers

| Comparison | Average power dissipation (μ W) |
|---|---|
| Venkatesh [13] | 14000 |
| Meenali [14] | 694 |
| Mathew [15] | 433 |
| Srinivas [16] | 3.4 |
| **Melior** | 0.106 |

IBM Q Experience has been used to implement the linear phase FIR filter. The type of the multiplier and parallel adder varies according to the order of the filter and its coefficients. To obtain the average power dissipation, third-order FIR filter has been implemented in Cadence Virtuoso at 45 nm technology node with an average power dissipation of 456.1 nW.

## 5 Conclusion

The applications of Digital signal processing are digital filtering, digital communications and spectral analysis. Multiplier is the integral component of these systems. In this paper, a **Melior** Quantum Multiplier has been proposed to reduce Ancilla and Garbage Outputs. The **Melior** Quantum Multiplier is used in the implementation of the linear phase FIR filter of order 3. The third order has been implemented due to limitation in the usage of Q-Lines in IBM Q Experience. The **Melior** Quantum Multiplier and FIR filter are implemented in IBM Q Experience and Cadence Virtuoso to obtain the average power dissipation. Both the designs are power efficient and the count of Ancilla and Garbage Outputs have been reduced.

## References

1. Sangeetha P, Kumari P (2020) An implementation of KFDD based reversible logic using IBM-Q experience. In: 2020 7th International conference on computing for sustainable global development (INDIACom), New Delhi, India, pp 162–167
2. Ehsanpour M, Moallem P, Vafaei A (2010) Design of a novel reversible multiplier circuit using modified full adder. In: 2010 International conference on computer design and applications, Qinhuangdao, pp V3-230–V3-234
3. Kotiyal S, Thapliyal H, Ranganathan N (2014) Circuit for reversible quantum multiplier based on binary tree optimizing Ancilla and Garbage bits. In: 2014 27th International conference on VLSI design and 2014 13th International conference on embedded systems, Mumbai, pp 545–550
4. Bhagyalakshmi HR, Venkatesha MK (2010) An improved design of a multiplier using reversible logic gates. Int J Eng Sci Technol 2(8):3838–3845
5. Shams MHM, Navi K (2008) Novel reversible multiplier circuit in nanotechnology. World Appl Sci J 3:806–810
6. Pankaj NR, Venugopal P, Mortha P (2016) Design of quantum cost efficient reversible multiplier using Reed-Muller expressions. Int J Comput Sci Math 7(3):221–228
7. Nagamani AN, Kumar SS, Agrawal VL (2017) Design of garbage free reversible multiplier for low power applications. In: 2017 4th International conference on power, control and embedded systems (ICPCES), Allahabad, pp 1–4
8. Eshack A, Krishnakumar S (2019) Speed and power efficient reversible logic based vedic multiplier. In: 2019 International conference on recent advances in energy-efficient computing and communication (ICRAECC), Nagercoil, India, pp 1–5
9. Afrin S, Shihab F, Sworna ZT (2019) Two novel design approaches for optimized reversible multiplier circuit. In: 2019 IEEE international WIE conference on electrical and computer engineering (WIECON-ECE), Bangalore, India, pp 1–4

10. Radha N, Maheswari M (2018) High speed efficient multiplier design using reversible gates. In: 2018 International conference on computer communication and informatics (ICCCI), Coimbatore, pp 1–4
11. Anamika Bhardwaj R (2018) Reversible logic gates and its performances. In: 2018 2nd International conference on inventive systems and control (ICISC), Coimbatore, pp 226–231
12. Yugandhar K, Raja VG, Tejkumar M, Siva D (2018) High performance array multiplier using reversible logic structure. In: 2018 International conference on current trends towards converging technologies (ICCTCT), Coimbatore, pp 1–5
13. Venkatesh A, Rathan N, Saranya R (2017) Design of power efficient 4 × 4 multiplier based on various power optimizing techniques. Asian J Appl Sci Technol (AJAST) 1:115–119
14. Janveja M, Neeranjan V (2017) High performance wallace tree multiplier using improved adder. ICTACT J Microelectron 3:370–374
15. Mathew K, Asha Latha S, Ravi T, Logashanmugam E (2013) Design and analysis of an array multiplier using an area efficient full adder cell in 32 nm CMOS technology. Int J Eng Sci (IJES) 2:8–16
16. Srinivas E, Sharath Babu N, Sreenivasa Raju G (2019) Design of low power and high speed 4 × 4 multiplier using modified column bypassing scheme for DSP applications. Int J Recent Technol Eng (IJRTE) 8:643–647

# Is There Any Spooky Action at a Distance?

**Kisalaya Chakrabarti**

**Abstract** In his later years, Albert Einstein brilliantly criticized quantum physics particularly quantum entanglement which illustrates the ability of particles to share a combined state while in separation arbitrarily far away and grappled with oddity he termed this phenomenon as "spooky action at a distance". Lately, this ambiguity has intensified in various forms of non-locality. However in recent times, scientists have verified the reality of this phenomenon over ever greater distances, even from Earth to a satellite in space. In many cases, it violates the principle of special theory of relativity where the information travels faster than light, which is a mystery so far. In this paper, the author has proposed a probable mechanism of quantum entanglement which does not break the assumption of special theory of relativity and concept of time latency is also introduced.

**Keywords** Standing wave · Time latency · Distributed locality · Special theory of relativity

## 1 Introduction

In the year of 1935, Einstein, Podolsky, and Rosen wrote a paper [1] by which one can easily conclude that a quantum event which takes place at one part of the universe can immediately affect another event far away at arbitrary distance if the two events are entangled. Einstein termed this phenomenon as "spooky action at a distance" because it involves faster than light communication. Later this paradox was termed as "EPR paradox". Recently, many researchers have done this experiment [2, 3] and found something which seems apparently spooky.

K. Chakrabarti (✉)
Department of Electronics and Communication Engineering, Haldia Institute of Technology, Haldia 721657, WB, India
e-mail: kisalayac@gmail.com

 
A. Maji et al. (eds.), *Proceedings of the International Conference on Computing and Communication Systems*, Lecture Notes in Networks and Systems 170,
https://doi.org/10.1007/978-981-33-4084-8_65

## 1.1 Basics of Quantum Entanglement

All fundamentals particles have "spins" which manifest as angular momentum and orientation in space. When two particles spontaneously formed out of a photon, they will have opposite spins as the total angular momentum of that system containing of these particles is constant. However, it is impossible to predict which particular direction they are spinning until they are measured but it is well known that they are spinning opposite to each other. So, their wave functions are entangled as these particles have no well defined spin at all. Measurement of spin of one of them without delay tells us the spin of its entangled partner arbitrarily far away from another. It is mention worthy that the direction of spin is defined according to their spin axis. To measure the direction of spin, it is required to choose an axis and align it with the axis of its measuring device. Having done that it is observed that the spin always align itself with the selected measurement axis of the measuring device. As for example, if it is chosen to measure vertically the spin will turn out to be up or down. The important observation is that measurement procedure compels the alignment of the measured particle and according to the Copenhagen interpretation any measurement of one particle will collapse the entangled wave function which affects the spin of its entangled partner immediately. Einstein contemplated that these particles may have some hidden information for spooky action at a distance which he called local hidden variables. Later, Bell's experiment discards this idea of local hidden variables which directs about which spin they will manifest in different directions while measurement.

Now the fundamental question arises. If there are no local hidden variables then how come the influence of one particle that could theoretically be transmitted to its entangled partner instantaneously at any distance and even back in time which violates the causality principle.

## 1.2 Theory and Discussion

Let us consider two entangled particles located at points "*A*" and "*B*" separated by an arbitrary distance "*d*" and having superimposed states between them $\psi_s$. As because of the two particles are entangled with each other and there is no net transfer of energy in forward or backward direction, this combined wave function $\psi_S$ is composed of two $\psi$ states, $\psi_1$ and $\psi_2$, where $\psi_1$ is the forward traveling wave moving from "*A*" to "*B*" and a backward traveling wave $\psi_2$ moving from "*B*" to "*A*". Here, it is worth mentioning that for the time being the directions of the spin of both particles are not taken into account on the propagation of forward or backward moving waves, so we can express the superimposed wave function in the light of Schrodinger equation,

$$\psi_S = \psi_1 + \psi_2 = \left(\psi_{01}e^{\frac{iPx}{\hbar}} + \psi_{02}e^{\frac{-iPx}{\hbar}}\right) \tag{1}$$

As it is mentioned earlier, no net transfer of energy in any direction and it is also true at their boundary point "*A*" and "*B*", therefore

$$\psi_{02} = -\psi_{01} \tag{2}$$

Without considering spin effect in terms of directions, Eq. (1) can be written as

$$\psi_S = \left(\psi_{01}\mathrm{e}^{\frac{iPx}{\hbar}} - \psi_{01}\mathrm{e}^{\frac{-iPx}{\hbar}}\right) \tag{3}$$

Now, considering the direction of spin of the particle located at point "*A*" on the propagation of forward traveling wave $\psi_1$. In order to measure the spin direction, we need to choose an axis to align with the axis of the measuring device. Now aligning the axis of the measurement device located at point "*A*" in the direction of the forward traveling wave $\psi_1$ or in the forward "$+x$" direction which makes the spin of the particle located at "*A*" pointed in the "Right" direction and eventually we denote this observed spin of the particle at "*A*" by unit vector "$\hat{x}$" and introducing this factor to "$\psi_1$" we obtained the modified wave function expressed as:

$$\psi_1 = \psi_{01}\mathrm{e}^{\frac{iPx}{\hbar}}\hat{x} \tag{4}$$

From the conservation of angular momentum, we can denote the observed spin of the particle at "*B*" by unit vector "$-\hat{x}$" and introducing this factor to "$\psi_2$" we obtained the modified wave function expressed as:

$$\psi_2 = \psi_{02}\mathrm{e}^{\frac{-iPx}{\hbar}}(-\hat{x}) = -\psi_{02}\mathrm{e}^{\frac{-iPx}{\hbar}}\hat{x} \tag{5}$$

Therefore, after introducing the spin effect in Eq. (3), the final equation for combined wave function $\psi_S$ becomes:

$$\psi_S = \left(\psi_{01}\mathrm{e}^{\frac{iPx}{\hbar}} + \psi_{01}\mathrm{e}^{\frac{-iPx}{\hbar}}\right)\hat{x} = 2\psi_{01}\cos\left(\frac{Px}{\hbar}\right)\hat{x} \tag{6}$$

Considering the time evolution of the combined wave function, Eq. (6) is modified and can be written as follows:

$$\psi_S(t) = \mathrm{Re}(\psi_S \mathrm{e}^{i\omega t}) = 2\psi_{01}\cos\left(\frac{Px}{\hbar}\right)\cos(\omega t)\hat{x} \tag{7}$$

where $\omega$ is the angular velocity of $\psi_S(t)$, given by $E/\hbar$.

From Eq. (7) one can noticed that, this time evolved combined wave function $\psi_S(t)$ is a one-dimensional standing wave in $\sigma_x$ basis. In this case, there is no net transmission of energy in either direction. The energy density in the medium of this standing wave, having two particles at arbitrarily far distance, is proportional to the square of the amplitude of the time evolved combined wave function $\psi_S(t)$ and is given by

$4\psi_{01}^2 \cos^2\left(\frac{Px}{\hbar}\right)$ which varies from antinodes and zero at nodes. In three-dimensional case, considering the entire Pauli matrix basis, the time evolved combined wave function $\psi_S(x, y, z, t)$ looks like "3D Standing waves" or as rotating "Doughnuts" expressed by the formula $\sum_{\theta}^{2\pi} \psi_S(t)$. This phenomenon is recently established by an experiment by Sillanpää et al. [4] where an unusual state of entanglement persisted for half an hour.

By observing the nature of time evolved combined wave function $\psi_S(t)$, the position of nodes can be calculated by putting the spatial component of Eq. (7) to zero or $\cos\left(\frac{Px}{\hbar}\right) = 0$. This tells us the nodes are located at $\pm x = \lambda/4, 3\lambda/4, 5\lambda/4\ldots$; where $\lambda = \mathrm{h}/P$ with null energy density. It is quite obvious that as the energy density is zero, there will be no angular momentum associated at these points. Similarly, it can be inferred that there are two antinodes having with opposite spins residing at two terminal ends $\pm x = 0$, where the two entangled particles are physically present. The subsequent antinodes are located at $\pm x = \lambda/2, \lambda, 3\lambda/2, 2\lambda$ and so on. As mentioned earlier, the energy density is maximum at these points; consequently the angular momentum is also maximum at these particular points distributed periodically and evenly in the entire span of distance "$d$" between the points "$A$" and "$B$". This phenomenon can be described as "distributed locality". In this context, it is worthy to mention that the spin directions of the entangled particles are always opposite to each other and the time evolution has instantaneous effect on their relative phase irrespective of the distance between them.

For the entangled particles of having mass "$m$", the total mass of the system shall be 2 m.

According to the Klein Gordon equation $\psi_s = \mathrm{e}^{-i\omega t}\mathrm{e}^{i(k_x x + k_y y + k_z z)} = \mathrm{e}^{ik_\mu X^\mu}$ which gives rise to

$$\left\{-\omega^2 + k_x^2 + k_y^2 + k_z^2 + (2m)^2\right\}\psi_s = 0 \tag{8}$$

$$\text{or,} \quad \omega = \pm\sqrt{k_x^2 + k_y^2 + k_z^2 + (2m)^2} \tag{9}$$

where "$m$" is the mass of one quantum mechanical particle, energy is expressed by "$\omega$" and momentum "$k$".

These are Higgs like particles, having the

$$\text{Action} = \int -(2mc^2 + g\psi_s)\sqrt{1 - \frac{v^2}{c^2}}\mathrm{d}t \tag{10}$$

$$\text{gives rise to,} \quad \text{Action} = \int -(2mc^2 + g\psi_s)\left(1 - \frac{v^2}{2c^2}\right)\mathrm{d}t \tag{11}$$

where $g$ is the coupling constant and given by $g = 2e$ for electron positron pair. As $\psi_s \neq 0$ implies that the region of space associated with $\psi_s$ has some fixed value, then effectively this particle has distributed mass of $-\left(2mc^2 + g\psi_s(x, t)\right)$. Total action all together includes the action for the field and the action for the particles. Therefore,

$(-g\psi_s(x, t))$ is the effect of field on the particle when the particles are at rest state. In quantum mechanics, fields and particles are the same things. It creates a mass of the particles and the presence of particles will react back on the field and creates new fields. It is a chain process where particle affects fields and fields create new fields like scalar particles.

When the entangled particles consist of a system whose combined field is standing still, as in the case stated earlier for standing wave, implies $\psi_s$ is time independent or does not change with time.

Thus, the wave equation takes the form $\nabla^2\psi_s = g\delta^3(X)$, where $\delta^3(X)$ is the charge density and $X = x, y, z$. For one-dimensional case, it will be $\nabla^2\psi_s = g\delta(x)$. This charge density is the root cause for periodic field density which is manifested as nodes and antinodes. As usual this field density is maximum at antinodes and zero at nodes. This is termed as "distributed locality" of fields in space.

### 1.2.1 Time Latency After Establishment of Entangled Pairs

This most important aspect in the context of entanglement is time latency. It is the amount of time which is needed to establish the entanglement [5] after generation of entangled pairs.

Supposing, an entangled pair is created and then it is separated by a distance "$d$" in the "$x$" direction with a relative velocity "$v_x$". Let the time required for the evolution of the standing wave given by the combined wave function "$\psi_s$" is "$t$". Therefore, to establish a standing wave located at distance "$d$", the time required is "$t$" $= d/v_x$. Now if we displace any particle with the same velocity "$v_x$" which is still in entangled state with each other by an arbitrary small distance $\Delta d$, corresponds to the distance between them is "$d + \Delta d$", where $\Delta d$ is the minimum distance for the restoration of standing waves. As the standing wave is stationary and not traveling in any of the direction, so when one of the particles is moving, it is dragging the standing wave with it. The additional time $\Delta t$ required for establishing dragged version of standing wave is $\Delta t = \Delta d/v_x$. This $\Delta t$ is called the "time latency" of the entangled particles. In terms of proper time, time latency is given by $\Delta\tau_p$, which is an invariant quantity and does not change with any observant is given by

$$\Delta\tau_p = \left(\frac{\Delta d}{v_x}\right)\left(\sqrt{1 - \frac{v_x^2}{c^2}}\right) \tag{12}$$

Time latency is the minimum time which is needed to readjust the entangled particles to go back to its original standing wave condition.

So, if the particle is moving with velocity $v_x$ with the time latency $\Delta\tau_p$, in one-dimensional case, the modified wave equation will be:

$$-\frac{1}{c^2}\frac{\partial^2\psi_s}{\partial\tau^2}\left(1 - \frac{v_x^2}{c^2}\right) + \nabla^2\psi_s = g\delta(x - \Delta\tau_p v_x) \tag{13}$$

Now, when we are measuring the spin of any particle in any particular direction, suppose in $x$-direction which means the spin of the particle is forcibly aligning with the direction of the measurement, therefore $k_y$ and $k_z$ are equal to zero as because there is no measurable angular momentum in the $y$ and $z$ directions.

When the equipment of measurement aligned to $x$-direction, Eq. (9) becomes, $\omega = \pm\sqrt{k_x^2 + (2m)^2}$. So after aligning with a particular axis, only that component of angular momentum exists. This component will evolve with time in a fashion so that the directions of angular momentum of the two end particles will always remain opposite to each other.

### 1.2.2 Stationary Phase Approximation

Now, to understand the evolution of the quantum entanglement with time, we have to understand the stationary phase approximation of the combined wave function of the entangled particles as a whole which is given by combined wave function $\psi_S(t)$ depicted in Eq. (6).

Supposing, we are taking an integral given by:

$$J = \lim_{\Omega \to 0} \int_{-\infty}^{\infty} \mathrm{e}^{i\psi_S(t)/\Omega} \mathrm{d}t \tag{14}$$

Equation (14) can be expanded as follows:

$$J \approx \sum\nolimits_n \int_{-\infty}^{\infty} \mathrm{e}^{i\left\{\frac{\psi_S(t_0)}{\Omega} + \frac{\frac{d\psi_S(t_0)}{\mathrm{d}t}(t-t_0)}{\Omega} + \frac{\frac{d^2\psi_S(t_0)}{\mathrm{d}t^2}(t-t_0)^2}{2\Omega}\right\}} \mathrm{d}t \approx \mathrm{e}^{i\frac{\psi_S(t_0)}{\Omega}} \sqrt{\frac{2\Omega\pi}{i\frac{d^2\psi_S(t_0)}{\mathrm{d}t^2}}} \tag{15}$$

Here in Eq. (14) in order to introduce some scale to it, a book-keeping constant "$\Omega$" is forcefully inducted with $\Omega \to 0$ to arrive at a proper estimate after the aforementioned integration. If we consider the real part of $\psi_S(t)$, this is oscillating between 0 to +1 with periodic antinodes and nodes as mentioned earlier which is distributed and remain stationary with time and space in between the two entangled particles. Now at the point of antinodes, where the $\psi_S(t)$ is at extreme value which gives $\frac{\mathrm{d}\psi_S(t)}{\mathrm{d}t} = 0$ and "$J$" receives a nonzero contribution because the oscillatory character slows down locally. Now, if we expand the time around an arbitrary time instant "$t_0$" and putting the condition; $\frac{\mathrm{d}\psi_S(t_0)}{\mathrm{d}t} = 0$, we can re-establish Eq. (15) as follows:

$$J \approx \sum\nolimits_n \int_{-\infty}^{\infty} \mathrm{e}^{i\left\{\frac{\psi_S(t_0)}{\Omega} + \frac{\frac{d^2\psi_S(t_0)}{\mathrm{d}t^2}(t-t_0)^2}{2\Omega}\right\}} \mathrm{d}t \approx \mathrm{e}^{i\frac{\psi_S(t_0)}{\Omega}} \sqrt{\frac{2\Omega\pi}{i\frac{d^2\psi_S(t_0)}{\mathrm{d}t^2}}} \tag{16}$$

For reasons of convention, one should change "$i$" to "$-i$" inside the square root part of Eq. (16), following the convention of Gaussian integration formally. Mathematically of course the integral does not exist; we pretend the imaginary unit "$i$" is a real positive constant. Now as in the path integral case, we get nonzero contributions accurately where

$$\delta\, S[\psi_S(t), p(t), t] = 0 \tag{17}$$

which is the condition for the functional of classical trajectory "$S$" and is measured in the unit of $\hbar$, $p(t)$ is the momentum of the entangled system so in any physical domain where "$S$" is much bigger than $\hbar$, then any smallest variation of the path from the classical trajectory makes $\psi_S(t)$ wiggle out so fast that it will contribute nothing to the "$J$". The classical limit is $\frac{S}{\hbar} \gg 1$.

As it is clear that $J = \psi_S(t)$, squaring and comparing Eq. (16) with Eq. (7) and considering the real part, we obtain

$$\frac{\Omega\pi\, \sin\left(\frac{2\psi_S(t_0)}{\Omega}\right)}{2\psi_{01}^2 \cos^2(\omega t)} = \frac{d^2\psi_S(t_0)}{dt^2} \cos^2\left(\frac{Px}{\hbar}\right) \tag{18}$$

Equation (18) indicates that that acceleration on superimposed wave function at any arbitrary time instant "$t_0$" denoted by $\psi_S(t_0)$ will always correlates with change in momentum in space in a co-sinusoidal fashion associated with its intensity change ($\psi_{01}^2$) in an inverse manner.

### 1.2.3 Field Distribution Pattern of Superimposed Wave Function $\psi_S(t_0)$ at Any arbitrary time instant

Putting the average value of $\cos^2\left(\frac{Px}{\hbar}\right) = \frac{1}{2}$ at $x$ direction, Eq. (18) can be written as

$$\frac{\Omega\pi\, \sin\left(\frac{2\psi_S(t_0)}{\Omega}\right)}{\psi_{01}^2 \cos^2(\omega t)} = \frac{d^2\psi_S(t_0)}{dt^2} \tag{19}$$

Now, Eq. (19) can be rewritten as;

$$\frac{d^2\psi_S(t_0)}{dt^2} = \kappa \frac{\sin(\psi_S(t_0))}{\cos^2(\omega t)} \tag{20}$$

where $\kappa = \frac{\Omega\pi}{\psi_{01}^2}$ = constant with book-keeping constant "$\Omega$" for the sake of arithmetical simplicity which is nothing but a scaling factor.

Now solving the differential Eq. (20) we can derive the solution of $\psi_S(t_0)$ by double integration

$$\psi_S(t_0) = -\kappa[\sin(\psi_S(t_0))\ln|\cos(\omega t)|]/\omega^2 \quad (21)$$

Now plotting the function $\psi_S(t_0)$ from Eq. (21) for three particular cases shown below depicts how the instantaneous value of function $\psi_S(t_0)$ evolves with $\omega$.

**Case I:**

Putting the value of $\kappa = 10$, $\omega = \pi$ and plotting the values of superimposed wave function $\psi_S(t_0)$ field distribution with time shown in Fig. 1, one can observe that periodic value of $\psi_S(t_0)$ is equally likely in the interval of −20 to 20 time units and beyond as this pattern continues to $t = \infty$ in both sides. The envelop of the superimposed wave function is slightly oscillatory in nature. This value of $\psi_S(t_0)$ lies between ±2.5, which is quiet expected but we do not know which value it will take on measurement and it depends only at the external conditions of the measurement of that time instant.

**Case II:**

Now, putting the value of $\kappa = 10$, $\omega = 1.66\pi$, "$\omega$" is slightly higher than the previous case for observing the effect of angular distribution in superimposed wave function by plotting the values of $\psi_S(t_0)$ field distribution with time shown in Fig. 2, one can observe that periodic value of $\psi_S(t_0)$ is equally likely as in previous case in

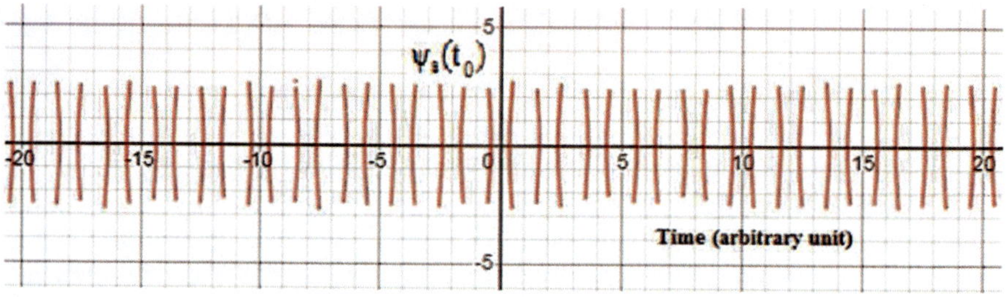


**Fig. 1** Superimposed wave function $\psi_S(t_0)$ field distribution with time at $\omega = \pi$

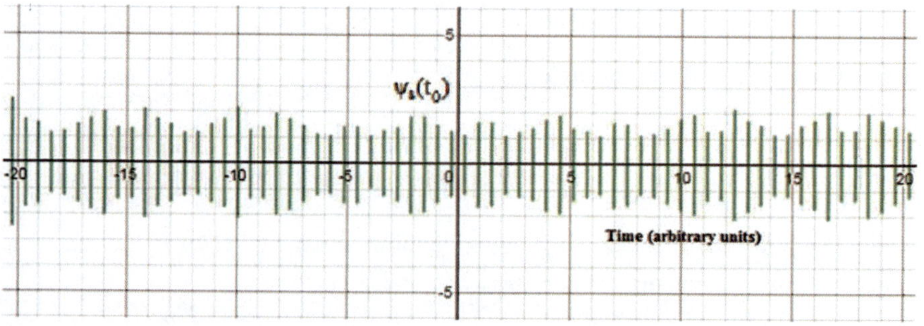


**Fig. 2** Superimposed wave function $\psi_S(t_0)$ field distribution with time at $\omega = 1.66\pi$

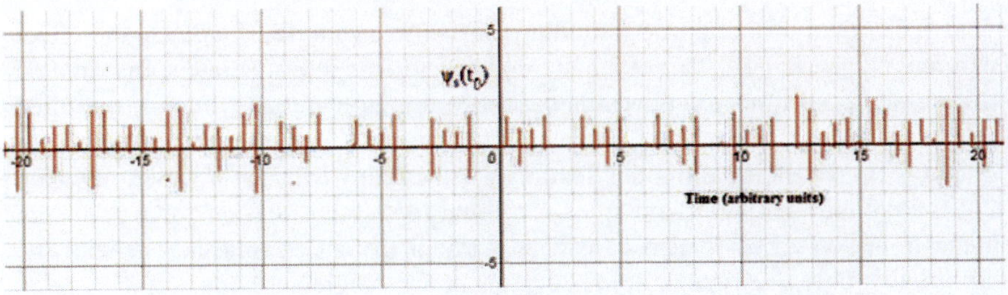


**Fig. 3** Superimposed wave function $\psi_S(t_0)$ field distribution with time at $\omega = 1.90\pi$

the interval of −20 to 20 time units and beyond (not seen in Fig. 2). In this case, the envelop of the superimposed wave function is quiet oscillatory in nature. This value of $\psi_S(t_0)$ lies between ±2, which is quiet expected and the value it take depends only at the external conditions of the measurement taken at that particular instant.

**Case III**:

Now, putting the value of $\kappa = 10$, $\omega = 1.90\pi$, "$\omega$" is still slightly higher than the second case and observing the effect of angular distribution in superimposed wave function and plotting the values of $\psi_S(t_0)$ field distribution with time shown in Fig. 3, one can notice that in a grave contrast to the precedent cases the value of $\psi_S(t_0)$ is not consistent in the interval of −20 to 20 time units and beyond (not seen in Fig. 3). The maximum values of $\psi_S(t_0)$ lies between ±2, but pattern of the superimposed wave function is erratic in nature.

The reason for these erratic natures in the values of $\psi_S(t_0)$ is due to superposition of two different amplitudes of forward moving wave "$\psi_{01}$" with the reflected backward moving waves "$\psi_{02}$". So $|\psi_{01}| \neq |\psi_{02}|$ at this particular angular frequency $\omega = 1.90\pi$, where "$\omega$" is a scalar measure of rotation rate.

From these graphs, one conclude for some particular angular frequency "$\omega$", the instantaneous value of superimposed wave function $\psi_S(t_0)$ at time $t_0$ may not remain periodic but the phenomenon of superposition of two $\psi$ states, $\psi_1$ and $\psi_2$ between two entangled particles located at points "*A*" and "*B*" separated by an arbitrary distance "*d*" remains intact.

## 2 Effect of Temperature and Tiny External Magnetic Field on the Intermediate Spins of Superimposed Wave Function

Phase transitions of intermediate spins of superimposed wave function $\psi_S(t_0)$ occur with sudden change in temperature. It is quiet natural that all the spins are lined up in a particular direction in the space at low temperature close to absolute zero but at the room temperature or at any sufficiently high temperature they are oriented 50%

up and 50% down of that particular alignment. Also, we have to analyze the effect of external magnetic field "*H*" of small amplitude on the intermediate spins situated at the antinodes sections of superimposed wave function $\psi_S(t_0)$, when it is applied to one of the entangled particles, supposing for this case located at point "*A*".

Suppose, at the entire antinodes sections (without any external magnetic applied at "*A*"), a number of total of up spins are equal to "*n*" and that of down spins are counted to be "*m*". Let us also consider, at initial instant of time $n > m$ with the condition that only one type of spin is present at any single antinode.

Let us consider that the numbers of antinodes are equal to "*N*". Therefore, the total numbers of spins equal to the total numbers of antinodes present in the superimposed wave function $\psi_S(t_0)$ are given by the following:

$$N = n + m \tag{22}$$

Now, if we applied an external tiny magnetic field "*H*" at "*A*" with "$\mu$" magnetic moment, which is directly related to the strength of the external magnetic field. Also, if we consider two groups associated with the types of spins "*m*" and "*n*", respectively, the energy "*E*" associated with the number of configurations out of these two groups can be written as:

$$E = (n - m)\mu H \tag{23}$$

The number of states associated with the above configurations is given by $N!/(n!m!)$ and the partition function "*z*" is given by the following expression:

$$z = \sum_{nm} \mathrm{e}^{-\beta(n-m)\mu H} = \sum_{n} \frac{N!}{n!(N-n)!} \mathrm{e}^{-\beta(n-m)\mu H} = 2^N (\cosh(\beta\mu H))^N \tag{24}$$

Now, we are going to define a quantity named as "spin factor" given by "*S*" which will depict the relative percentage of "up spins" to the "down spins" in superimposed wave function $\psi_S(t_0)$. Spin factor is positive if there are more "up spins" than "down spins". Therefore, by definition:

$$S = \langle (n - m) \rangle / N \tag{25}$$

which gives,

$$\langle E \rangle = N S \mu H \tag{26}$$

From Eq. (25), it can be inferred that at 0° K, when $n = 0$ and $m = N$, (all the spins are directed downwards) then $S = -1$; while at $\infty$° K, when $n = m$, entails $S = 0$. Therefore, average energy $\langle E \rangle$, when derived from the definition we obtain:

$$\langle E \rangle = -\frac{\partial \log z}{\partial \beta} = -\frac{N \sinh(\beta\mu H)}{\cos(\beta\mu H)} \mu H \tag{27}$$

Comparing Eqs. (26) and (27), we obtain the value of "S" as follows:

$$S = -\tanh(\beta\mu H) \tag{28}$$

From the established theory, we know that $\beta = 1/\mathrm{T}$. So, when the value of the temperature "$T$" is very small and it approaches to "absolute zero" which means the value of β is very large, then we will obtain the value as $S = -1$, as mentioned earlier, which also implies that $\tanh(\beta\mu H) = 1$. In this above condition, all the spins at the antinodes will be "down spins". Similarly, when $\beta = 1/T = 0$ or $T \to \infty$, the value of "spin factor" will be zero. When we change the direction of external magnetic field "$H$" opposite to the previous one, we shall get $S = \tanh(\beta\mu H)$ which corresponds to $S = 1$, that means all the spins at the antinodes will flip to "up spins" from the "down spins". When there is no external magnetic field, the spin of any antinodes is due to his neighbor contribution, so we found the direction of spin in any antinodes is not discrete, rather random, the neighboring antinode has also random spin direction. Therefore, the energy is always associated with pairs of "antinodal spins". We shall denote the "up spin" by $\sigma = +1$ and "down spin" by $\sigma = -1$ and also worth mentioning that, as the spins of the neighbors are randomly distributed, we have to consider the average of the direction of spins, which means likely to be oriented parallels or anti-parallels. Let us denote, average spin by $\bar{\sigma}$. We observed from the above discussions, when the temperature "$T$" is close to absolute zero, all the spins are directed in one particular direction, either up or down as it has only two stable ground states, in one-dimensional distribution in space but when the temperature is high, the average spins directions are directed randomly. So at any intermediate temperatures of the two extremes, we can conclude that the average spins directions are also directed either upward or downward, but in reality it is random. Therefore, in this one-dimensional system oriented in space, there is no phase transition or there is no sudden change in spin properties (directions) if the temperature is varied suddenly. Now, we will analyze the effect of external tiny Magnetic field applied at "$A$" oriented in a particular direction, likely at "up" or "down" direction while we are going to detect the spin of the entangled particle situated at "$A$", having done this, an extra energy "$O$" is introduced in the system of two entangled particles. From the **mean field approximation theory** [6] (self-consistent field theory), with the introduction extra energy "$O$" we can rewrite $\langle E\rangle$ with any combinatory configuration of "up spins" and "down spins" of the system as:

$$\langle E\rangle = -2j\bar{\sigma}\sigma - O\sigma = -2(j\bar{\sigma} + O)\sigma \tag{29}$$

where $j = \mu\mathrm{H}$ represents the total angular momentum of the system. From, Eq. (29), by solving the partition function "$z$" for the given energy $\langle E\rangle$, it can be easily obtained the value of $\bar{\sigma}$ with some mathematical calculations as:

$$\bar{\sigma} = \tanh[(2j\bar{\sigma} + \mathrm{O})\beta] \tag{30}$$

Rewriting Eq. (30) we obtained:

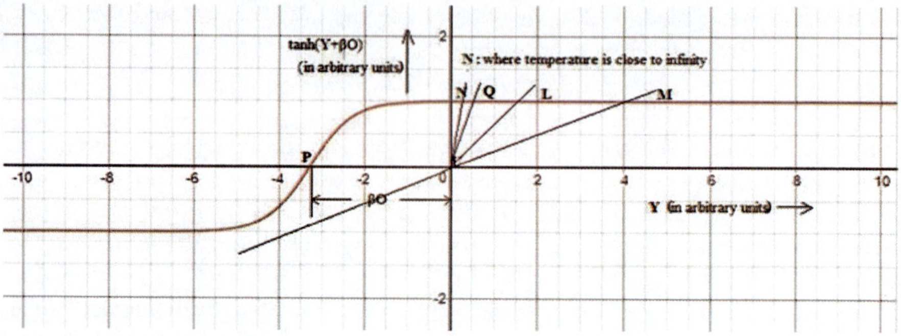


**Fig. 4** Relationship between $\bar{\sigma}$ versus $Y$

$$\frac{\mathrm{TY}}{2j} = \tanh(\mathrm{Y} + \beta\mathrm{O}) \tag{31}$$

where $Y = 2\beta j\bar{\sigma}$ and $\beta = 1/T$. Figure 4 shows the relationship between the functions $\tanh(Y + \beta O)$ and $Y$. Here, it is observed that the hyperbolic tangent (tanh) curve crosses the horizontal axis at the point "$P$" situated at a distance "$-\beta\mathrm{O}$" units from the origin point "$U$". Now, if we draw a tangent at the point "$P$", the slope of the tangent will be 1, which indicates $\frac{T}{2j} = 1$; which lies on the left side of the origin "$U$". Therefore, it is obvious that, all the points lying at the right (+ve) side of abscissa have the values $Y \neq 0 \Rightarrow \bar{\sigma} = +1$.

It can be observed that when the temperature is very high (depicted in Fig. 4 as point "$N$") and is approaching to infinity, i.e., $T \to \infty$ gives $\bar{\sigma} = +1$. The same value $\bar{\sigma} = +1$ is also observed for very low temperature also (depicted in Fig. 4 as point "$M$") and any intermediate temperatures (depicted in Fig. 4 as point "$Q$" and "$L$"). It seems that it has a long—range memory, pointing all in the same directions (up) immediately.

This phenomenon is called "Phase Transitions in 1-Dimension" space contrary to Prof. Ising [7] findings and which tells us, in one dimension at any temperature there is an average spin factorization that permeates the complete system. From Fig. 4, we can also notice that, there are no solutions for $\bar{\sigma} = -1$. For a tiniest magnetic field "$O$", the difference of energy is so enormous with millions of millions light years distance value that all the spins at antinodes will orient themselves at the up directions in no time. Consequently, comparing Eqs. (29) with (21), it can be inferred that:

$$\langle E\rangle = \Lambda[\psi_S(t_0)]^2 \tag{32}$$

where "Λ" is "**spin-wave function coupling factor**".

Therefore, this information of spin direction permeates immediately from the entangled particle position "$A$" to another entangled particle position "$B$" with billion light years worth and due to the standing wave pattern of superimposed wave function

$\psi_S(t_0)$ (already discussed in Sect. 1.2 of this article above). Therefore, $\bar{\sigma}$ on entangled particle situated on "$B$" orients to $\bar{\sigma} = -1$ immediately with the alignment of $\bar{\sigma} = +1$ with the entangled particle situated on "$A$", considering the "time latency" after establishment of entangled pairs is negligible.

## 3 Conclusion

For any entangled pair of particles which is separated by an arbitrary distance in nature sharing a combined state has its angular momentum conserved and has intensified its forms of non-locality which is verified over ever greater distances from Earth to a satellite in space. This does not violate the principle of special theory of relativity because this combined states manifest itself as a standing waves distributed over space between two entangled end particles which are evolved as a whole whatever the distance between them. However, a minimum time which is needed to readjust the entangled particles to go back to its original standing wave condition if the distance between them is altered is called time latency.

In a nutshell, the paper depicts that when two entangled particles are kept apart at any distance, a three-dimensional standing wave is formed comprising of many superimposed combined wave functions $\sum_{\theta}^{2\pi} \psi_S(t)$ distributed around the co-axis between them and the amplitudes of superimposed wave functions $\psi_S(t_0)$ at any time instant may differ in erratic manner with different angular frequency as discussed earlier. When someone measures the "spin" of any entangled particle of the entangled pair at any particular direction or in any particular orientation all the superimposed combined wave functions $\sum_{\theta}^{2\pi} \psi_S(t)$, except that (which is associated with that particularly directed combined wave function $\psi_S(t)$) collapse, so that one can instantaneously know the spin of the other entangled particle of the pair at any distance apart. It is worth mentioning that the corresponding phases of the intermediate fields of the standing wave are analogous to the spin directions of the entangled particles. Also, very important phenomenon is "phase transitions" will be seen in one-dimensional superimposed combined wave functions $\sum_{\theta}^{2\pi} \psi_S(t)$ subjected to external tiny magnetic field.

An important observation we can view from the discussions made in Sect. 2 is that when a magnetic field is applied to one of the entangled particle, whatsoever tiny it may be, then the change of temperature has no effect in the change of orientations of the spins in the intermediate antinodes.

## References


1. Albert E et al (1935) Can quantum-mechanical description of physical reality be considered complete? Phys Rev 47:777–780
2. Riedinger R et al (2018) Remote quantum entanglement between two micromechanical oscillators. Nature 556:473–477
3. Yin J et al (2017) Satellite-based entanglement distribution over 1200 kilometers. Science 356(6343):1140–1144
4. Sillanpää MA et al (2018) Stabilized entanglement of massive mechanical oscillators. Nature 556:478–482
5. Chakrabarti K (2020) Realization of original quantum entanglement state from mixing of four entangled quantum states. In: Castillo O, Jana D, Giri D, Ahmed A (eds) Recent advances in intelligent information systems and applied mathematics. ICITAM 2019. Studies in Computational Intelligence, vol 863. Springer, Cham
6. Chaikin PM, Lubensky TC (2007) Principles of condensed matter physics, 4th edn. Cambridge University Press, Cambridge. ISBN 978-0-521-79450-3
7. Ising E (1925) Beitrag zur Theorie des Ferromagnetismus. Z Phys 31:253–258

# Evolvable Hardware State of the Art

**Kamlendra Chandra, Amit Prakash Jagtap, and Sangeeta Srivastava**

**Abstract** Nature has always inspired traditional solutions to think differently. Modern applications require adaptive and fault-tolerant systems. Bio-inspired computing provides an option. Evolvable hardware (EHW) is a method where hardware is designed to adapt automatically by using optimization algorithms called evolutionary algorithms (EAs). Evolvable hardware has the potential to provide solution for complex real-world applications when compared with growing artificial intelligence (AI). This paper introduces concept of evolutionary algorithms and building blocks of evolvable hardware. Evolvable hardware has many applications, one of them being classifier systems discussed in detail. The concept of evolvable hardware came into world 23 years ago; from then onward, it has given various promising areas, but the challenges posed have stopped evolvable hardware to become competent with traditional solutions. This paper also discusses new advanced platforms and frameworks which opens a new horizon for evolvable hardware.

**Keywords** Evolvable hardware (EHW) · Evolutionary algorithms (EAs) · Genetic algorithms (GAs) · Fitness function · Category detection module (CDM) · Cartesian genetic programming (CGP) · Embedded Cartesian genetic programming (ECGP)

## 1 Introduction

Evolvable hardware is a combination of reconfigurable hardware, evolutionary algorithm and problem domain [1, 2]. Evolvable hardware (EHW) is a hardware that can change its architecture and behavior dynamically by continuously interacting with

K. Chandra (✉) · A. P. Jagtap · S. Srivastava
Central Research Laboratory, Bharat Electronics Limited, Bangalore, India
e-mail: kamlendrachandra@bel.co.in

A. P. Jagtap
e-mail: jagtapamitparakash@bel.co.in

S. Srivastava
e-mail: sangeetasrivastava@bel.co.in

 
A. Maji et al. (eds.), *Proceedings of the International Conference on Computing and Communication Systems*, Lecture Notes in Networks and Systems 170,
https://doi.org/10.1007/978-981-33-4084-8_66

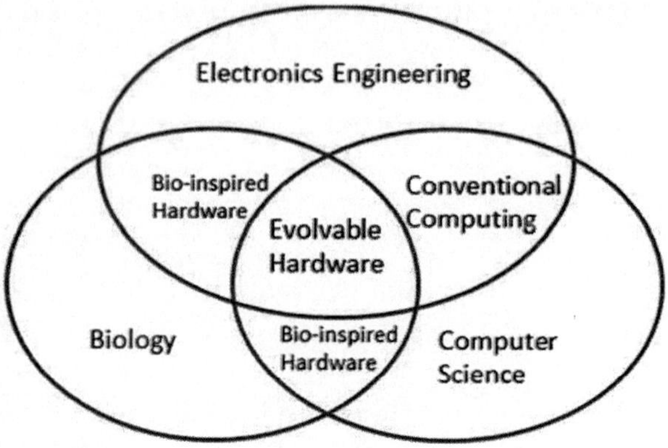


**Fig. 1** Evolvable hardware intersection of sciences [6]

its environment [3]. Evolvable hardware is having an adaptive nature and has the potential to find optimal solutions and would be excellent in unknown environment [4]. In evolvable hardware, the device has to adapt its behavior in a dynamic environment and find appropriate, optimal solutions over duration of the mission [5]. As seen from Fig. 1, evolvable hardware is inspired by biology and traditional computing.

## 2 Technology Implementation

As EHW is usually implemented on programmable devices, such as field-programmable gate arrays (FPGAs). The advanced development in FPGAs has recreated growing opportunities in the field of evolvable hardware. The circuits evolved from evolutionary algorithms need to be mapped to reconfigurable hardware at runtime. Since there are no commercial tools available to support dynamic reconfiguration suitable for EHW, presents a major challenge. Also, the FPGA vendors have proprietary tools with confidential ways to configure the FPGA internals, which user does not have direct access. This applies some restrictions on FPGAs to be used as an important entity for EHW. An alternative approach to online reconfigurability on FPGAs is the virtual reconfigurable circuit (VRC) method [7]. This method helps in providing virtual reconfigurability in FPGAs by changing register values and multiplexer control signals in the user circuit. The method provides a technique for fast reconfiguration and is applicable to all SRAM-based FPGAs. The evolvable hardware is categorized into extrinsic, intrinsic, unconstrained, constrained evolvable hardware and mixtrinsic based on evolution. Extrinsic EHW or off-chip simulates evolution by software and only downloads the best configuration to hardware in each generation , i.e., the hardware is only reconfigured once [8, 9]. Intrinsic EHW or on-chip evolution simulates evolution directly in its hardware, i.e., every chromosome will be used to reconfigure the hardware. The EHW will be reconfigured to validate the configuration, same number of times as the population size in each generation

[1, 4, 9]. The evolution which exploits physical and chemical properties of chip and environmental conditions is known as unconstrained evolution. Constrained evolution whether intrinsic or extrinsic evolution, gives the same results by ensuring that the representation or fitness function restricts the evolution of strange or practically non-realizable circuits [2]. In mixtrinsic evolution, some population is evaluated extrinsically and some intrinsically [10]. Extrinsic and intrinsic are the evolutionary approaches mostly discussed in the literature. Figure 2 shows extrinsic and intrinsic evolvable hardware flow. In extrinsic EHW, the evolution is done offline using model or simulator. In intrinsic EHW, the evolution is done in real hardware. Thus, intrinsic EHW is more accurate, fast and best suitable for real-time online EHW.

Figure 3 shows the block diagram of the evolvable hardware approach.

Where a configurable device is configured using a best-fit chromosome, decoded as output of genetic engine. The fitness unit evaluates the fitness of every chromosome

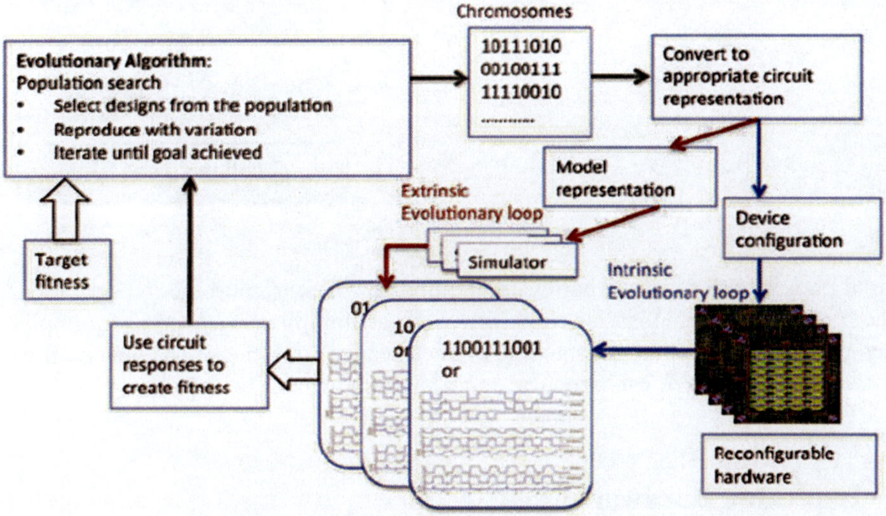


**Fig. 2** Extrinsic and intrinsic hardware loops [11]

**Fig. 3** Block diagram of the evolvable hardware approach [12]

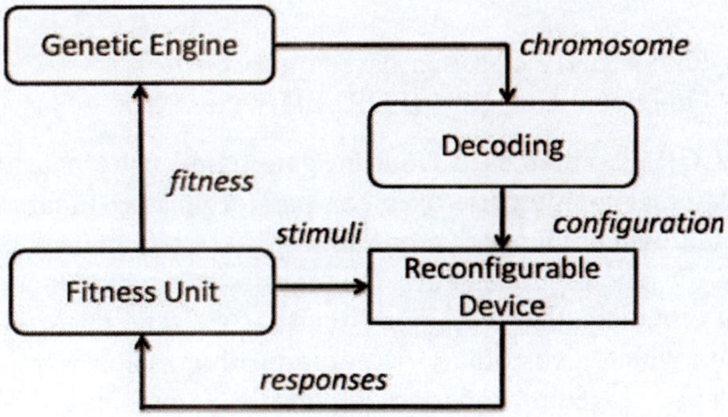

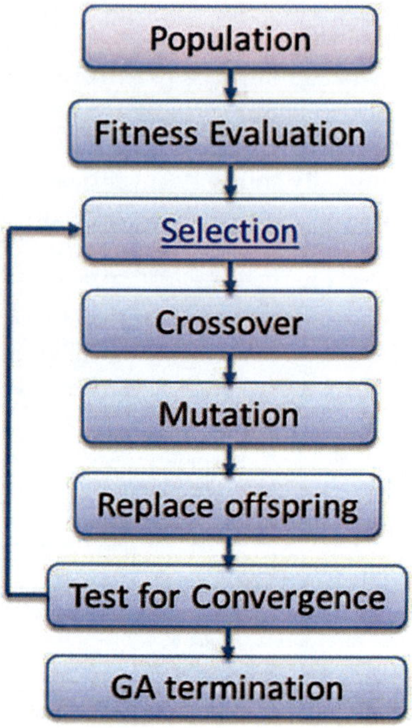


**Fig. 4** Generic flow of genetic algorithm

based on response from the hardware and updates the chromosome in population. The fitness mainly emphasizes on objectives to be minimized (resources for implementation, evolution time) or maximize functionality, robustness. The generic flow of genetic algorithm or evolutionary algorithm is shown in Fig. 4.

## 3 Evolvable Hardware as a Classifier

In the literatures related to EHW as a classifier, the two main classifier architectures are discussed.

- First is EHW classifier based on ECGP
- Functional Unit Row (FUR) EHW classifier.

ECGP classifier used Boolean gate-based pattern recognition circuit. FUR EHW classifier architecture uses comparators taking binary encoded numbers as input. Multiple comparators represent a pattern recognition circuit, fetching into a single AND gate. Figure 5 shows the high-level structure, and the input is given to number of circuits called classifier circuits (CC). The classifier circuits consist of a function which gives binary output indicating match or mismatch. Number of CCs are clubbed together, and their outputs are counted by a counter giving the total number

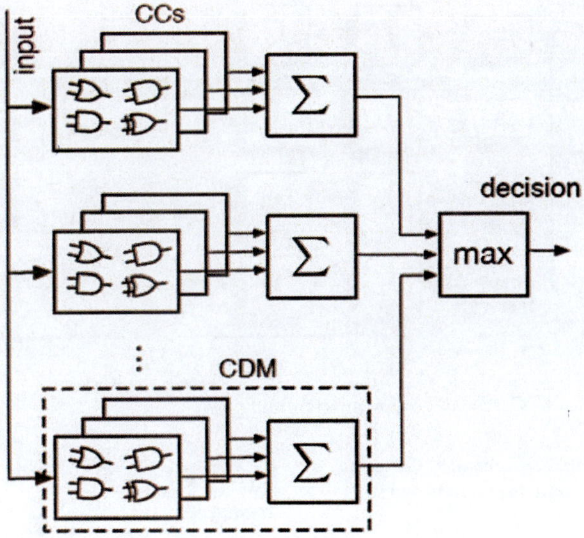


**Fig. 5** EHW classifier architecture [16]

of matches. A set of CCs and the counter are denoted as category detection module (CDM). CDM output having highest number of matches is selected for classifier decision. The category with lower index wins in case of a tie. The EHW-based classifier architecture is based on a variant of the Embedded Cartesian genetic programming (ECGP) model. ECGP is an extension of the Cartesian genetic programming (CGP) model [13] in FPGA. Figure 6 shows a model of Cartesian genetic programming (CGP); the model consists of $n_c \times n_r$ combinational logic blocks, $n_i$ primary inputs and $n_o$ primary outputs. A logic block has $n_n$ inputs and implements one out of $n_f$ different logic functions of these inputs. ECGP is a variant of CGP, and it consists of row $n_r = 1$ and column $n_c = 1$. It is having automatic definition and reuse of sub-functions (modules). Modules are defined as compositions of primitive nodes as shown in Fig. 7. The second classifier approach is the Functional Unit Row (FUR) EHW architecture which is specially used for classification tasks and online evolution. The classifier architecture is implemented as a circuit whose behavior and connections can be controlled by configuration registers, similar to VRC approach [14–16] to facilitate online evolution. Figure 8 shows block diagram of evolvable hardware Classifier, where each block is zoomed to show its internals. From Fig. 8, it can be seen that array of category detection module (CDM) forms an evolvable hardware classifier system. Each system consists of functional units. The representation of configuration bit string has been shown [14]. Configuration bit string will be used to evolve the connections and functions of the functional units.

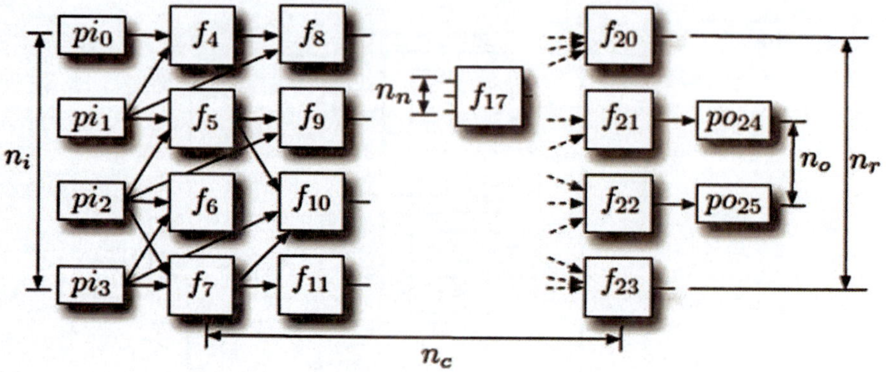


**Fig. 6** Cartesian genetic programming [16]

**Fig. 7** Embedded Cartesian genetic programming [16]

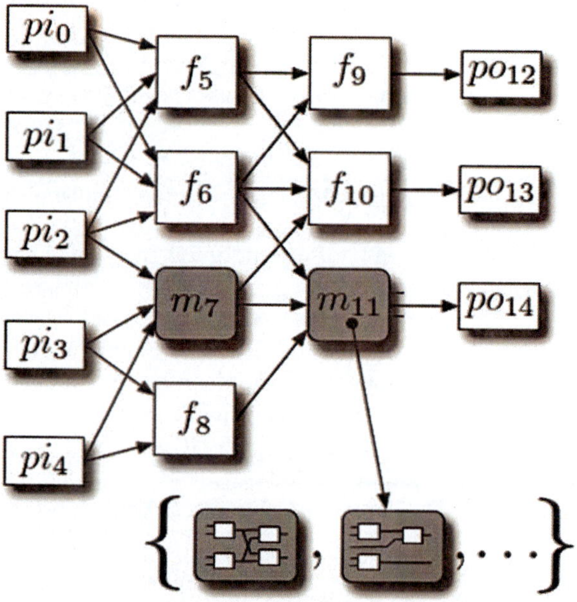


## 4 Promising Areas and Challenges

EHW can be very promising in some areas [2].

- EHW can perform better than some traditional solutions known so far [17].
- EHW can be used to calibrate or improve already operational traditional solutions.
- Designing solutions in areas where resources are constrained. Like solution in cases where space needed for implementation is high than the space available physically. Designing a circuit with 50-80% resources required by a traditional solution.

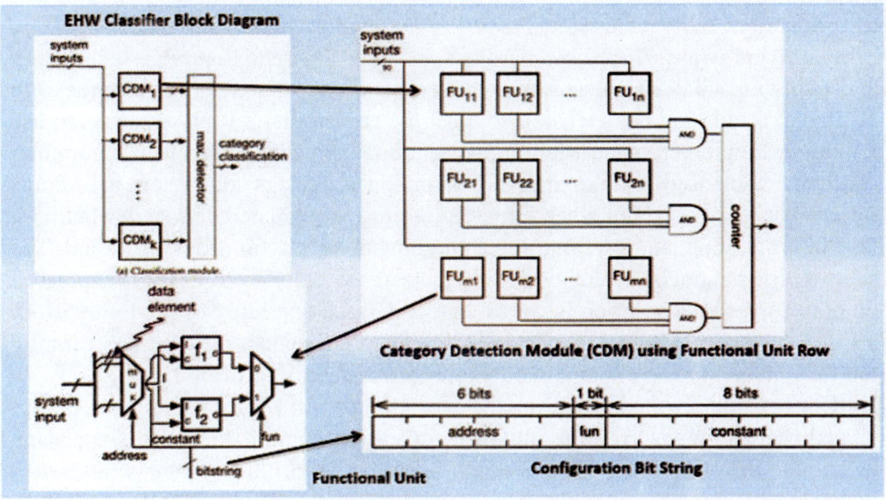


**Fig. 8** Evolvable hardware classifier [14]

– Evolution is a new way to discover solutions where our knowledge is less. It can give solutions in areas which are not in the scope of conventional engineering methods. EA can also utilize chemical and physical properties of a platform. It can be used to design efficient digital and analog circuits. It can also evolve mechanical designs such as antennas and structure of robot.

Challenges of Evolvable Hardware.

- **Scalability** is the major challenge of evolvable hardware. It is practically impossible to evolve a complex circuit from scratch. Scalability is the term which is widely misunderstood. Scaling up structures alone without a corresponding scale in functionality is not a scalability challenge. Scalability in EHW can be defined as use of evolutionary techniques to achieve complex functions without corresponding scaling up of resources. There can be two ways to build larger systems [18]

  – To explore general scalable principle of design.
  – To produce building blocks which are efficient and large as possible.

  The major approach which can be used to overcome scaling problem is functional-level evolution [19], incremental evolution [20] and embryonic approach.
- **Robustness of Evolved Circuits** Property of evolved circuits depends on the reconfigurable hardware (RH) and the environment in which it is evolved. But if the same evolved solution is applied in different environments, the functioning is questionable. Portability of solutions is trouble of designers. In the literature, requirement of robustness has been integrated into fitness function to overcome portability problem [2].
- **Evaluation Criteria or Measurements** Typically, in traditional electronics, terms such as functional correctness, area and power cost are general metric applied

to evaluate viability of circuit. Power aware evolution has started to appear in evolvable hardware. Evolvable hardware is basically reconfigurable technology and evolving small circuits. But it makes no sense that if more resources are available, traditional design can also meet the requirement. It makes sense to use EA only when search space is large and functional objective is not clear. A common challenge addressed in International Roadmap for devices and system is of fault-tolerant systems [21]. But when circuit reliability is evaluated, it is seen whether it is 100% functional and says nothing about how much circuit is dysfunctional. The same metric cannot be evaluated to EHW, because reliability of EHW is expressed on how corrected a solution is, across faults defined explicitly or randomly. Both are designed with different motive in mind and comparison is difficult. So common metric is needed for comparison of traditional and evolved design.

- **No Conceptual Evolution** During the evolution of combinational circuits, circuit is evaluated by measuring the output for all input combinations. This approach is suitable only for small circuits as evolutionary algorithm evaluates thousands of candidate solution. The situation becomes complicated where evaluation is required for intermediate circuit states. This is feasible only for small circuits. Circuit behavior can only be evolved, and its behavior can be evaluated. The fitness function can evaluate only final circuit. It cannot evaluate intermediate or internal circuit states because it cannot see the internal circuit structure. There is no conceptual evolution. The evolvable circuit design has the following objectives [2].
  - Behavioral requirements-Desired behavior in a large system in which the circuit is embedded or working independently (e.g., interested in behavior of system instead of input/output relation of constituting modules)
  - Nonbehavioral requirement based on resource availability and environmental conditions under which it must operate (e.g., power dissipation, current, voltage, temperature, humidity and pressure).

## 5 Dynamic, Real-Time Evolvable Hardware

The evolvable hardware favors where problem definition is clear and fitness function is static. Different fitness function during different stages of evolution also falls under the category of static EHW, for example, prosthetic-hand controller design. But in real-time system, the problem definition is not clear, and the fitness function is dynamic. The ultimate goal is to provide solution which is autonomous and continuously adapting. There is stimulus provided for a change showing fitness drop. But the real-time system cannot be made offline for configuration, unless it is very critical, for example, traffic control systems where images are taken in different environments having different noises in different times which cannot be predicted when the system is designed. Because the system has to work in real time and operate continuously 2-Reconfigurable Hardware strategy can be applied as shown in Fig. 9.

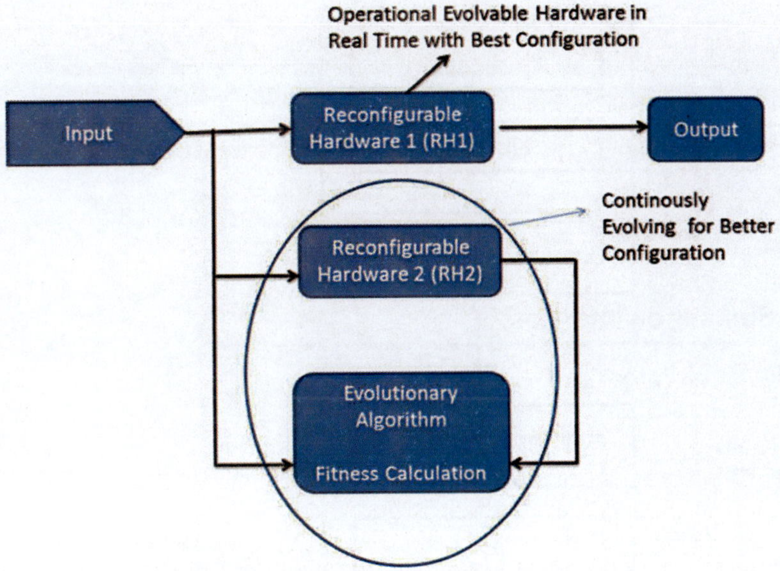


**Fig. 9** 2-Reconfigurable Hardware (RH) strategy

The reconfigurable hardware (RH1) contains the best configuration evolved so far. The reconfigurable hardware (RH2) is evolving continuously because it has to adjust with problem specification or changing fitness function. When something interesting is evolved in RH2 which is better than configuration running in RH1, it is uploaded to RH1. The system should be able to adapt continuously.

## 6 Future Platforms for Evolvable Hardware

Evolvable hardware is basically driven by reconfigurable hardware in both analog and digital domain. Field programmable gate arrays (FPGAs) in digital domain, field-programmable analog arrays (FPAAs) and field-programmable transistor arrays (FPTAs) in analog domain [1]. There are various heterogeneous computing frameworks, and reconfigurable platforms introduced recently can present a better prospect to support fast intrinsic evolution. The Intel Programmable Acceleration Cards are a family of FPGA board available to be used with the Intel Acceleration Stack for Intel Xeon CPU. It makes development and deploy of workloads easier in Intel FPGAs. The Acceleration Stack is a collection of software, firmware and tools that allows both software and RTL developers to take advantage of the power of Intel FPGAs. Computationally, intensive tasks are uploaded to FPGA, and the acceleration platform frees the Intel Xeon processor for other critical processing tasks. The Intel PAC with Intel Arria 10 GX FPGA, an accelerator card, connects to the Intel Xeon processor

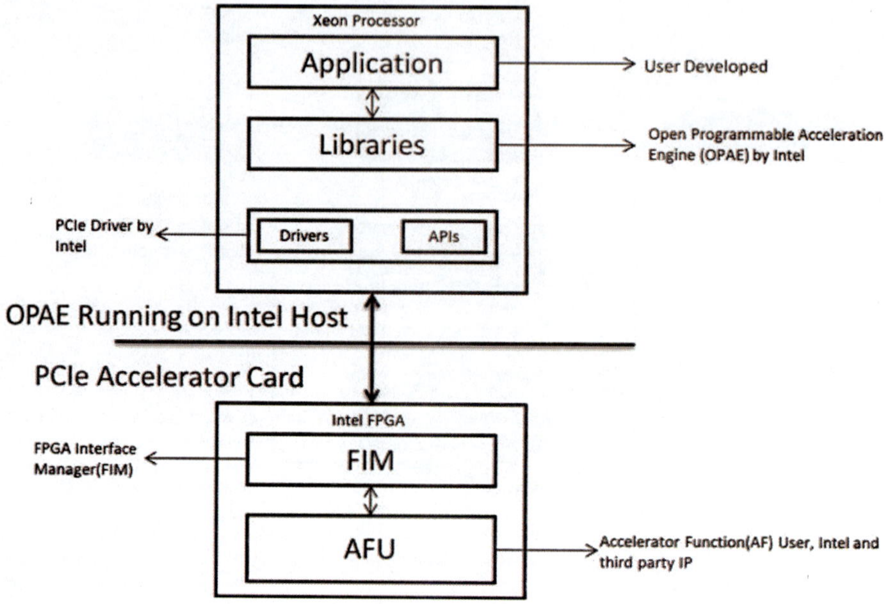


**Fig. 10** Block diagram of PAC card with Xeon processor [22]

through the peripheral component interconnect (PCIe) interface on the motherboard. Figure 10 shows the block diagram of programmable accelerator card (PAC) with Xeon processor [22]. The programmable accelerator card(PAC) is connected on PCIe with Host PC. PAC card has capability of reconfiguration over PCIe. The PAC card boots with blue bitstream. A partial reconfiguration portion Accelerator Functional Unit (AFU) is programmable using green bit stream, which is generated dynamically by user using tool chain as shown in Fig. 11. The available PAC card architecture is close to the required evolvable hardware architecture, where Acceleration Functional Unit (AFU) can hold the portion of VRC, ECGP and FUR. Similarly, Alveo cards from Xilinx can be used to accelerate computationally intensive applications such as machine learning, video processing and evolutionary algorithms [23]. Intel Distribution of OpenVINO toolkit for heterogeneous computing accelerates development of high-performance computer vision and deep learning interference into vision/AI applications. It works across platforms like FPGA, graphics processing unit (GPU), video processing unit (VPU), etc., by sharing workload to accelerate the performance. It has prebuilt models which can be used to bench mark evolved hardware or evolutionary algorithms.

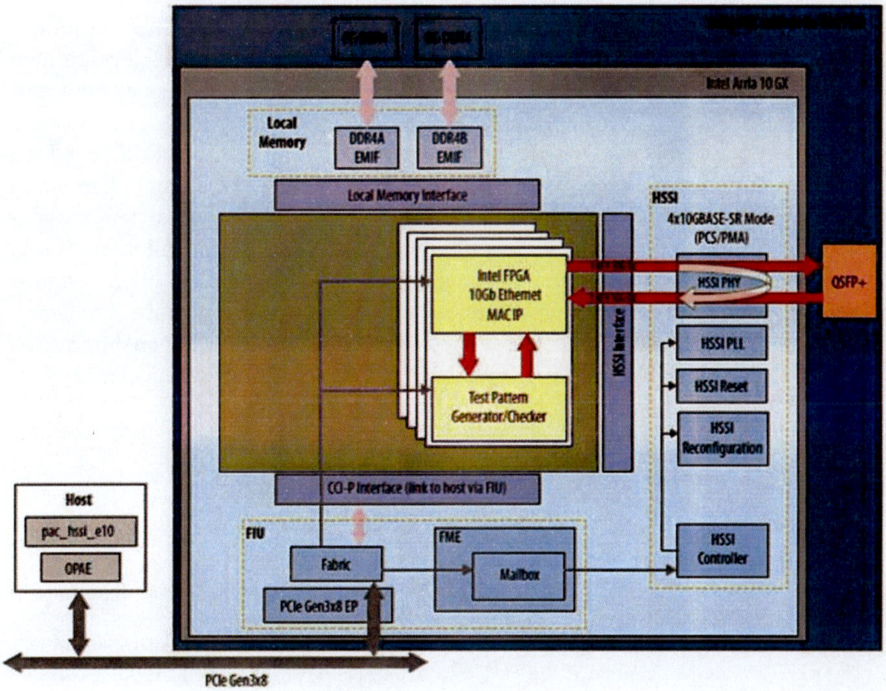


Fig. 11 Red bit stream and green bit stream AFU [24]

## 7 Explainability

We are in age of machine learning (ML) and artificial intelligence (AI) applications. There is a widespread adoption of ML model in various field providing tremendous benefits, but there effectiveness is limited because they are not able to explain its decision to users. Classification algorithms used presently are black-box algorithms, and they cannot give explanation for their answers, and cannot be applied to mission critical systems like firecontrol and target detection [25]. This gives advantage to evolvable hardware which is readable and explainable [26].

## 8 Conclusion

It has been seen from various literatures that there are two aspects of embedding intelligence in a system. The traditional way is through software, which involves mainly artificial intelligence. The other part of the system 'hardware' was normally untouched. The concept of evolvable hardware includes custom field-programmable transistor arrays (FPTAs), field-programmable analog arrays (FPAAs) and FPGAs.

Earlier, the small circuit development was done on these devices using evolutionary algorithms to get the required functionality. However, not much development was seen in the field of evolvable hardware due to insufficient access to the internal configuration details from programmable device manufacturers. In recent years, researcher overcame this with the idea of 'virtual reconfigurable circuit (VRC)' which gave complete control in reconfiguration features required for evolvable hardware. Based on VRC concept, many researchers came with evolvable hardware applications. Some of these applications are evolvable prosthetic arm controller, evolvable character recognition for autonomous vehicles, multi-target tracking video systems, self-reconfigurable system for adaptive image processing and many more. The accelerator hardwares and frameworks introduced like PAC card, alveo, OpenVino which supports heterogeneous computing, opens option for designers to accelerate their evolutionary algorithms and benchmark their evolved hardware.

## References

1. Haddow Pauline C, Tyrrell Andy M (2011) Challenges of evolvable hardware: past, present and the path to a promising future. Gen Program Evol Mach 12(3):183–215 Sep
2. Sekanina L (2004) Evolvable components: from theory to hardware implementations. Springer
3. Higuchi T, Iwata M, Kajitani I, Iba H, Hirao Y, Furuya T, Manderick B (1996) Evolvable hardware and its application to pattern recognition and fault-tolerant systems. In: Sanchez E, Tomassini M (eds) Towards evolvable hardware. Springer, Berlin, Heidelberg, pp 118–135
4. Wei L, Lijunb C (2019) Literature review of evolutionary hardware application based on Nios II and neural network. J Phys: Conf Ser 1176:032013
5. Lohn JD, Hornby GS (2006) Evolvable hardware: using evolutionary computation to design and optimize hardware systems. Comp Intell Mag 1(1):19–27 November
6. Tim G, Peter B (2002) On evolvable hardware, vol 101, Aug 2002
7. Sekanina L (2007) Virtual reconfigurable circuits for real-world applications of evolvable hardware. pp 186–197, Oct 2007
8. Dr. Hugo de Garis (1994) An artificial brain: Atr's cam-brain project aims to build/evolve an artificial brain with a million neural net modules inside a trillion cell cellular automata machine. New Gener Comput 12:215–221
9. Tørresen J (1997) Evolvable hardware-A short introduction. In: ICONIP
10. Stoica A, Zebulum RS, Keymeulen D (2000) Mixtrinsic evolution. In: Proceedings of the third international conference on evolvable systems: from biology to hardware, ICES'00. Springer, London, UK, pp 208–217
11. Trefzer M, Tyrrell A (2015) Evolvable hardware practice to applications, Jan 2015
12. Sekanina L (2009) Evolvable hardware: from applications to implications for the theory of computation. pp 24–36, Sept 2009
13. Vasicek Z (2018) Bridging the gap between evolvable hardware and industry using Cartesian genetic programming, pp 39–55, Jan 2018
14. Oscar G, Kyrre G, Jim T (2017) Comparing three online evolvable hardware implementations of a classification system. Genetic Program Evol Mach 10
15. Glette K, Torresen J, Kaufmann P, Platzner M (2008) A comparison of evolvable hardware architectures for classification tasks, vol 5216, pp 22–33, Sept 2008
16. Kaufmann P, Glette K, Gruber T, Platzner M, Torresen J, Sick B (2013) Classification of electromyographic signals: comparing evolvable hardware to conventional classifiers. IEEE Trans Evol Comput

17. Miller J, Job D, Vassilev V (2000) Principles in the evolutionary design of digital circuits-Part i. Gen Program Evol Mach 1:7–35
18. Miller J, Job D, Vassilev V (2000) Principles in the evolutionary design of digital circuits-Part ii. Gen Program Evol Mach 1:7–35
19. Higuchi T, Murakawa M, Iwata M, Kajitani I, Liu WX, Salami M (1996) Evolvable hardware at function level. In: Proceedings of 1997 IEEE international conference on evolutionary computation (ICEC'97), pp 187–192
20. Tørresen J (1998) A divide-and-conquer approach to evolvable hardware. In: ICES
21. Moore M (2017) International roadmap for devices and systems
22. Intel acceleration stack quick start guide for intel® Programmable acceleration card with intel® Arria® 10 GX FPGA, 5 Aug 2019
23. Getting started with Alveo Data Center accelerator cards, 12 Feb 2019
24. 10 Gbps ethernet accelerator functional unit design example user guide, 30 Apr 2019
25. Ribeiro MT, Singh S, Guestrin C (2016) "Why should i trust you?": explaining the predictions of any classifier. In: Proceedings of the 22nd ACM SIGKDD international conference on knowledge discovery and data mining, KDD'16. ACM, New York, NY, USA, pp 1135–1144
26. Iwata M, Kajitani I, Yamada H, Iba H, Higuchi T (1996) A pattern recognition system using evolvable hardware. In: Proceedings of the 4th international conference on parallel problem solving from nature, PPSN IV. Springer, London, UK, pp 761–770

# An Effective Comparative Study Among the Different Geometry of Electrodes in Performing the Tasks in Digital Microfluidic Biochips

**Riya Majumder, Amartya Dutta, Debasis Dhal, and Rajat Kumar Pal**

**Abstract** "Digital microfluidic biochip (DMFB)" emanates as one of the most promising technologies in the discipline of "Lab-on-a-Chip (LoC)" environment, which encompasses several application areas like biology, electronics, chemistry, etc., by replacing the conventional laboratory equipment. The geometry of digital microfluidic biochips is realized on a 2D grid with regular three-, four- and six-sided polygons. In this paper, we address the structural and behavioural aspects of the architectures as mentioned earlier and throwing sight to the enhancement of two primary modular operations such as droplet routing and mixing. Furthermore, a complete and tangible comparative study is performed efficiently based on these three design methodologies.

**Keywords** Digital microfluidic biochip · Droplet routing · Hexagonal electrode DMFB · Triangular electrode-based biochip

R. Majumder
Department of Computer Science and Engineering, Supreme Knowledge Foundation Group of Institutions, Chandannagar, Hooghly 712139, West Bengal, India
e-mail: riyamanai@gmail.com

A. Dutta (✉)
Department of Computer Science and Engineering, B. P. Poddar Institute of Management and Technology, Kolkata 700052, West Bengal, India
e-mail: dutta.kultu@gmail.com

D. Dhal · R. K. Pal
Department of Computer Science and Engineering, University of Calcutta, JD-2, Sector-III, Salt Lake, Kolkata 700106, West Bengal, India
e-mail: debasisdhal06@gmail.com

R. K. Pal
e-mail: pal.rajatk@gmail.com



A. Maji et al. (eds.), *Proceedings of the International Conference on Computing and Communication Systems*, Lecture Notes in Networks and Systems 170,
https://doi.org/10.1007/978-981-33-4084-8_67

## 1 Introduction

Microfluidics-based biochips cover many regions in biochemistry and biomedical sciences, and as a foremost accurate outcome, it performs the correct bioassay operations, which is the way of getting revolutions on the Lab-on-a-Chip (LoC) [1, 2] technology. Numerous applications like enzymatic analysis, immunoassays, various proteomic analysis, DNA sequencing, amplifications, blood testing for clinical diagnostics, environmental toxicity monitoring, etc., are the massive working fields of biochips. One of the popular categories of microfluidic biochips depends upon continuous fluid flow which uses either micro-pumps and micro-valves or some electrical techniques, such as electro-kinetics, to manage continuous fluid [1]. An alternative design outlook is formed based on "digital microfluidics", that follows the principles of electro-wetting-on-dielectric (EWOD) [1–3]. This digital-based system precisely handles minuscule (order of microlitre and nanolitre) biological droplet samples, yielding 100% or close to 100% accurate results [1, 2]. Compared to prosaic laboratory experiments, the new digital-based technology offers various improvements such as highly scalable system, lower design cost due to the smaller volume of samples, higher sensitivity and less erroneous arrangement [1–3].

Controlled digital microfluidic biochip (DMFB) can be formed using 2D grid structures; there can be regular three-, four- and six-sided polygons (triangle, square and hexagon) providing repeatability, uniform area division and useable for chip array [4]. Research works have generally focussed on square electrodes in digitally handled biochips technology, yet it is also possible to regulate the droplets through triangular [5] and hexagonal electrodes [6]. In this paper, we study the structural and behavioural aspects of these three representations of the cell shapes. An overall comparative analysis is performed based on square, triangular and hexagonal electrodes with respect to the bioassay completion time.

## 2 Problem Formulation

Formal DMFBs with square electrodes offer more accelerating and favoured design in the microfluidic realm [1]. In this existing technology, the droplet follows Manhattan routing, keeping four adjacent directions available. Another contemplation in this technology is biochip with equilateral triangular electrodes [5], which permits the opportunity of droplet movements with 0° and 60° phase shift in place of 0° and 90°. A further alteration takes a shift from the architectural point of view, which constitutes the chip array by regular hexagonal-shaped electrode [6]. The proposed mechanisms enhance the flexibility of droplet movements. As a result, a droplet can forward only in horizontal and diagonal directions addressing the furtherance in routing and mixing, moreover advancement in the whole bioassay scheme [6].

In this context, we have analysed a relative study among the three design structures described above and concentrated on the principal issues that are incompatible for conducting a bioassay, like size of the electrodes, fluidic constraints and pin constraints to handle the droplet on the chip. Finally, based on the experimental results, the affirmation on these three miniature architectonics has been focussed.

## 3 Design Methodology

In this module, we are going to talk about the design methodologies of three different electrode cell shapes. Keeping the side of the square electrodes' edge as a reference, a relationship among the edges of the triangular electrode and hexagonal electrode is achieved by considering some underlying assumptions; these are (*i*) the droplet volume is kept identical for all types of electrodes, (*ii*) the droplet portion outward of a polygonal side is also assumed the same and (*iii*) the voltage for activation and deactivation of electrodes is kept constant.

A. *Estimation of the portion of a droplet outside of the square electrode (x)*:

Figure 1a illustrates the top view of a droplet having radius $r$ placed on a charged square electrode, where the maximum radial portion of the remaining droplet exterior to the edges of the activated electrode is $x$. Let the diagonal and the edge of a square electrode be $d$ and $a$, respectively. So, $d = \sqrt{2}a$, and the droplet radius, $r = d/2 = a/\sqrt{2}$. Now, $a/2 + x = r = a/\sqrt{2}$; hence, $x = 0.207a$ [5], [6].

B. *Estimation of the edge of an equilateral triangular electrode (b)*:

From Fig. 1b, it is evident that $r = (1/3)h + x$, where $h$ is the height of the triangle with side $b$ and the droplet's radial spread exterior to the side of the charged triangular electrode is $x$. Hence, $h = \sqrt{3}b/2$. Thus, $r = (1/3) \times \sqrt{3}b/2 + 0.207a = b/(2\sqrt{3}) + 0.207a$. As the radii of the droplets on a square and a triangular electrode array must be same, we have, $a/\sqrt{2} = b/(2\sqrt{3}) + 0.207a$, and thus by solving we get, $b = 1.73a = \sqrt{3}a$ [5].

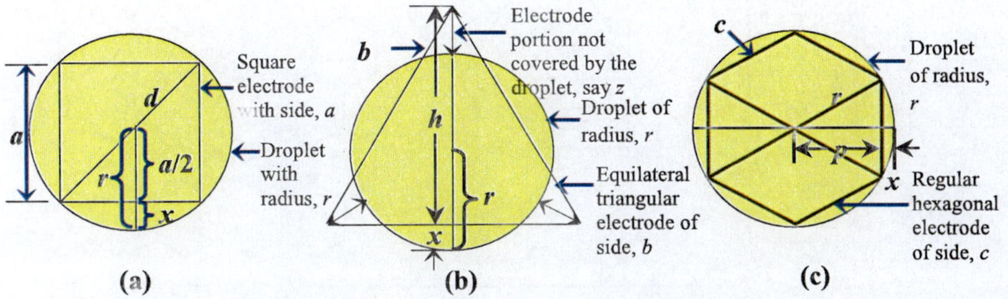


**Fig. 1** Top view of a droplet placed on an activated, **a** square electrode with side $a$, **b** equilateral triangular electrode with side $b$ and **c** regular hexagonal electrode with side $c$

C. *Estimation of the edge of a regular hexagonal electrode (c)*:

Here again, according to our supposition, the droplet radius is $r$ and $c$ is the edge of a regular hexagonal-shaped electrode (see Fig. 1c). Now, as a regular hexagon is the combination of six equilateral triangles, so it can be eventually concluded that the radius, $r$, of the droplet is same as the side of the hexagonal electrode, $c$. Thus, $p^2 + (r/2)^2 = r^2 \Rightarrow p = \sqrt{3}r/2$. By the above assumptions, $p + 0.207a = a/\sqrt{2}$, where $r = a/\sqrt{2}$. Hence, it can be deduced that, $p = 0.5a = \sqrt{3}r/2$. Now, as $c = r$, the relationship between the edges of regular hexagonal electrodes and the edges of square electrodes is $c = 0.58a$ [6], which reveals the fact that the edge of a hexagonal electrode is geometrically smaller than that of a square electrode by an amount of 42%. From the above computations, it can be concluded that (the edge of an equilateral triangular-shaped electrode) > (the edge of a conventional square-shaped electrode) > (the edge of a regular hexagonal-shaped electrode).

## 4 Constraints for Secure Droplet Routing

Fluidic constraints maintain a safe distance between two active droplets and control the secure movements of the droplets [1, 3]. Throughout the array during transportation, any unintentional mixing except merging is prevented by the fluidic constraints. The primary basis for maintaining fluidic constraints on the electrode array is that the two droplets say, $D_i$ and $D_j$ should not be adjacent to each other at time instant $t$ as well as at time instant $t + 1$. Generally, the fluidic constraints are classified into two types, (i) static fluidic constraints and (ii) the time-varying dynamic fluidic constraints. Several positions of droplets by maintaining the fluidic constraints are depicted in Fig. 2. To demonstrate the fluidic constraints, we consider the patterned array of the square, triangular and hexagonal electrodes, respectively, where $X_i(t)$

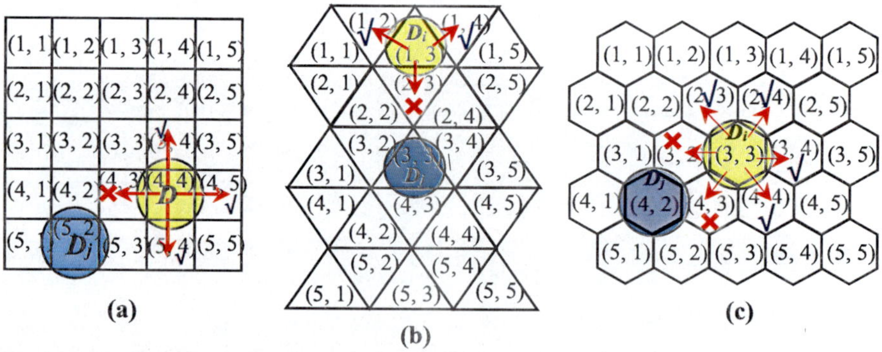


**Fig. 2** Allowable and restricted movements of droplets by maintaining fluidic constraints on **a** square electrode array, **b** triangular electrode array and **c** hexagonal electrode array. "√" refers to allowable movements and "**x**" refers to unacceptable movements of droplets

is the position of the $i$th droplet along the $x$-direction and $Y_i(t)$ is the same along $y$-direction at time $t$.

### 4.1 Fluidic Constraints on Square Electrode Array

(a) **Static fluidic constraints**

At time instant $t$, $|X_i(t) - X_j(t)| \geq 2$ and $|Y_i(t) - Y_j(t)| \geq 2$. Now, similarly, at time instant $t + 1$, the above conditions reform into $|X_i(t + 1) - X_j(t + 1)| \geq 2$ and $|Y_i(t + 1) - Y_j(t + 1)| \geq 2$.

(b) **Dynamic fluidic constraints**

In time-varying scenarios, while one droplet is moving and another is fixed with respect to time, the conditions can be stated as $|X_i(t + 1) - X_j(t)| \geq 2$ and $|Y_i(t + 1) - Y_j(t)| \geq 2$. Likewise, $|X_i(t) - X_j(t + 1)| \geq 2$ and $|Y_i(t) - Y_j(t + 1)| \geq 2$ are also true to reassure safe routing. Figure 2a portrays the allowable and restricted droplet movements on the square electrode array keeping the minimum safe separation of two electrodes.

### 4.2 Fluidic Constraints in an Equilateral Triangular Electrode Array

In the illustration of triangular electrode-based array, the fluidic constraints can be viewed into two ways with respect to $x$- and $y$-coordinates [5]. Considering the nature of the triangular electrodes, the fluidic constraints for $x$- and $y$-coordinates in static and dynamic time are as shown below.

(a) **Fluidic constraints for $x$-coordinate**

(i) *Static fluidic constraints at time instant t and t + 1:*
$|X_i(t) - X_j(t)| \geq 2$ and $|X_i(t + 1) - X_j(t + 1)| \geq 2$.

(ii) *Dynamic fluidic constraints*:

$|X_i(t) - X_j(t + 1)| \geq 2$ and similarly $|X_i(t + 1) - X_j(t)| \geq 2$; thus, the minimum distance along the $x$-coordinates of droplet $D_i$ at time instant $t + 1$ and droplet $D_j$ at time instant $t$ must be of two electrodes, and vice versa.

(b) **Fluidic constraints for $y$-coordinate**

The equilateral triangular electrodes are classified into two types, depending upon the restriction of droplet movement along $y$-axis [5].

(i) *Static fluidic constraints at time instant t and t + 1*:

$|Y_i(t) - Y_j(t)| \geq 2$ and $|Y_i(t+1) - Y_j(t+1)| \geq 2$ are the static fluidic constraints for droplets $D_i$ and $D_j$ along vertical directions.

(ii) *Dynamic fluidic constraints*:

$|Y_i(t) - Y_j(t+1)| \geq 2$. Similarly, $|Y_i(t+1) - Y_j(t)| \geq 2$. Therefore, finally, it can be concluded that the minimum distance between two droplets positioned along the same vertical column must be of two electrodes, for avoiding unwanted mixing in both static and dynamic scenarios (see Fig. 2b).

### 4.3 Fluidic Constraints on Hexagonal Electrode Array

(a) **Static fluidic constraints**

$|X_i(t) - X_j(t)| \geq 2$, $|Y_i(t) - Y_j(t)| \geq 2$, and ($|X_i(t) - X_j(t)| \geq 1$ && $|Y_i(t) - Y_j(t)| \geq 1$), while at time instant $t + 1$, the inequalities are reformed as $|X_i(t+1) - X_j(t+1)| \geq 2$, $|Y_i(t+1) - Y_j(t+1)| \geq 2$, and ($|X_i(t+1) - X_j(t+1)| \geq 1$ && $|Y_i(t+1) - Y_j(t+1)| \geq 1$).

(b) **Dynamic fluidic constraints**

If droplet $D_i$ changes its position at time instant $t + 1$ and droplet $D_j$ is static, then fluidic constraints must satisfy $|X_i(t+1) - X_j(t)| \geq 2$, $|Y_i(t+1) - Y_j(t)| \geq 2$, and ($|X_i(t+1) - X_j(t)| \geq 1$ && $|Y_i(t+1) - Y_j(t)| \geq 1$) (see Fig. 2c).

## 5 Pin Constrained Chip Design

### 5.1 Pin Constrained Chip Design in a Square Electrode Array

Pin count reduction is a critical requirement for an efficient biochip as it reduces voltage consumption. In direct addressing, a minimum number of control pins, say $K$, is needed to move a single droplet throughout the 2D square electrode array without any droplet interference [1]; it is known as a pin assignment problem that is NP-complete [1, 3] and can be visualized as a graph colouring problem. For a square electrode array, the value of $K$ is 5 according to the "Connect-5" algorithm [1]. By this algorithm, a three-layer printed circuit board (PCB) is required to connect the set of pins individually, and each layer comprises at most two non-intersecting wire layouts (see Fig. 3a).

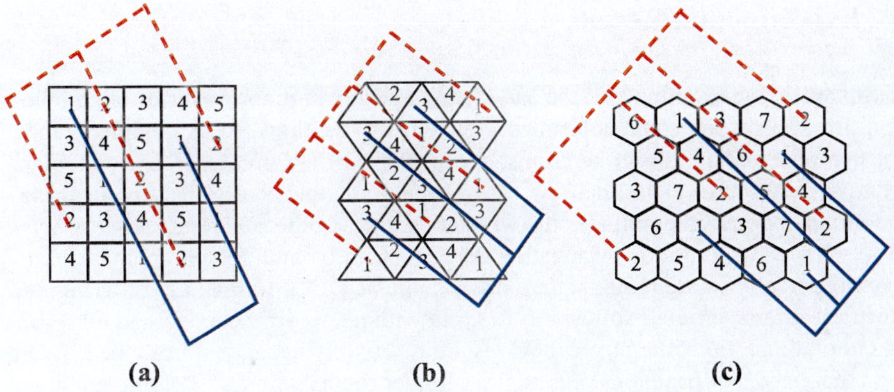

Fig. 3 Connections through electrical wires: **a** A 5 × 5 square electrode array designed by "Connect-5" algorithm with wire connections for control electrodes 1 and 2. **b** A 5 × 5 triangular electrode array is designed by "CTNT-4" algorithm with electrical wire connections for control electrodes 1 and 3. **c** A 5 × 5 hexagonal electrode array is covered by "CBPS" algorithm with wire representations for control electrodes 1 and 2

## *5.2 Pin Constrained Chip Design Problem in TEDMB*

In triangular electrode array (TEDMB), the minimization of control pins is possible by the connect three neighbouring triangles by 4 pins (CTNT-4) algorithm [5]. From the idea of a graph colouring problem, it can be said that the graph dual of a 2D TEDMB array has degree three, and thus cannot be coloured by less than four colours. In a TEDMB array, the electrodes are electrically connected by a two-layer PCB for avoiding any electrical hazards that may arise due to intersection of wires (see Fig. 3b).

## *5.3 Pin Constrained Chip Design Problem in HDMFB*

From a designer's point of view, hexagonal electrode-based DMFB (HDMFB) is a cluster-based organization [6]. A cluster is assembled with seven electrodes, forming a "honeycomb" shape of size $N_C = 7$, where $N_C$ is the number of electrodes belonging to the cluster [6]. According to the "Cluster-Based Pin Sharing (CBPS)" algorithm [6], the essential control pins become seven; that is, the value of $K$ is seven for avoiding the droplet interference in a single cluster. Furthermore, to evade any electrical hazard, a four-layer PCB is necessary for connecting all electrodes in HDMFB (see Fig. 3c).

## 6 Experimental Results

In this work, we examine glucose and lactate assay operations, in terms of completion times, based on some colorimetric enzymatic reactions [6]. A complete assay consists of several modular operations, like dispensing of sample and reagent droplets (routing from source to destination), mixing operations and detections. Among these, the mixing operation dominates the whole outcome. In the square electrode arrays, droplet movements follow Manhattan distance with 0° and 90° phase change and the mixer modules are either square or rectangular [7–9]. In the case of triangular electrode arrays, droplets follow *TZ_Routing* with phase difference 0° and 60° [10], and at the same time, the mixer modules are essentially hexagonal in shape [5]. On the other hand, for hexagonal electrodes, droplet routes with *HZ_Routing* maintain 0° and 60° phase change [6, 10], and here the mixer modules in effect form clusters [5].

The whole 2D array is first partitioned by "Time Division Pin Sharing (TDPS)" algorithm [1], and then control pins are assigned to it by using Connect-5, CTNT-4 and CBPS algorithms for square, triangular and hexagonal electrode arrays, respectively. A scheduled bioassay, illustrated in Fig. 4a, is simulated on the 15 × 15 array composed with the three types of electrodes using the abovementioned routing algorithms. Figure 4b represents the coordinate positions of the sample and reagent reservoirs, positions of mixer modules and detection sites on the 2D 15 × 15 associated patterned array. For this particular bioassay, we have used mixer modules of size 3 × 5, 2 × 7 and 3 × 5 for the square, triangular and hexagonal array of electrodes, respectively, and the total time of mixing is taken from the corresponding mixer library [5–7].

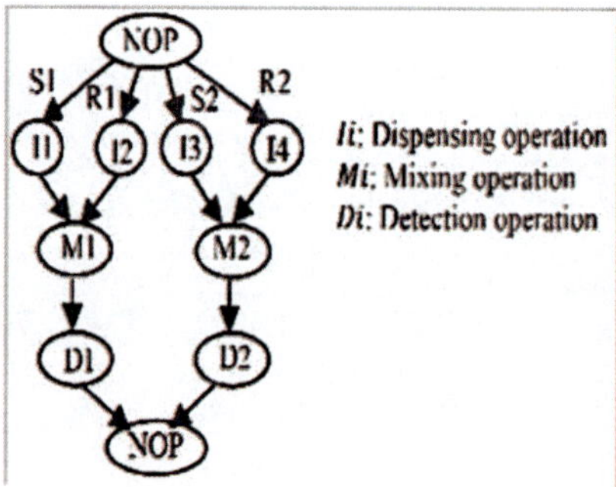


(a)

| Reservoirs/ Module | Positions on 15×15 2D array |
|---|---|
| $S1$ | (3,1) |
| $S2$ | (3,15) |
| $R1$ | (13,1) |
| $R2$ | (13,15) |
| $M1/M2$ | (7,6) |
| $D1$ | (2,6) |
| $D2$ | (2,10) |

(b)

**Fig. 4** Multiplexed bioassay: **a** A sequence graph, where $S_i$, $R_i$, $I_i$, $M_i$, $D_i$, NOP are the samples, reagents, allotting, mixing, detection and no operations, respectively. **b** Coordinate positions of samples, reagents, mixers and detection sites in 15 × 15 arrays with square-, triangular- and hexagonal-shaped electrodes

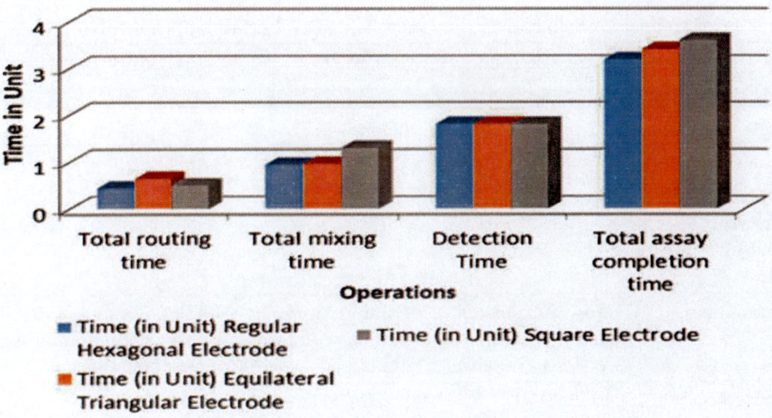


**Fig. 5** Performance metrics for square, triangular and hexagonal electrode array

In our experiment, we assume droplet speed to be 25 cm/s [2], and the required routing time to move from one cell to its neighbour to be 0.01 s [6]. The experiments as mentioned earlier are executed on a system with Intel Core *i*5 CPU (frequency 2.40 GHz) having 4 GB RAM in JAVA platform. Figure 5 depicts the execution of the three types of arrays where hexagonal electrodes may be assumed as the best alternative.

## 7 Conclusion

In this paper, we have focussed on better alterations over the regular square electrodes in the field of digital microfluidic biochip (DMFB). The shape of the electrodes has changed to hexagonal with a lot of positive features after its square and triangular counterparts. The droplet trace became optimized due to horizontal and diagonal routing paths in HDMFB, and at the same time, the zigzag movements of droplets are achieved in TEDMB and HDMFB as well. A relative comparison over several features for square DMFB, TEDMB and HDMFB has been presented in Table 1 based upon some structural and behavioural aspects and facets to think over the issues further.

**Table 1** A brief comparative study over the square DMFB, TEDMB and HDMFB

| Features | Square electrode DMFB | TEDMB | HDMFB |
|---|---|---|---|
| Droplet radius | *r* | *r* | *r* |
| Side of an electrode | *a* | 1.73*a* | 0.58*a* |
| Geometrical space between two neighbouring electrodes | *a*/2 | *a*/2 | *a*/2 |
| Droplet motion | Follows Manhattan distance with four directions available | Follows zigzag motion with three directions available | Follows more flexible zigzag motion through six directions |
| Radial portion of droplets outside a charged electrode | 0.207*a* | 0.207*a* | 0.207*a* |
| Minimum number of control pins required | 5 | 4 | 7 |
| Number of PCB layers needed | 3 | 2 | 4 |
| The shape of the mixer module | Square- or rectangular-shaped depending upon the size of the module | Cluster-based hexagonal-shaped | Cluster-based mixer zone centring an electrode |

## References


1. Chakrabarty K, Xu T (2010) Digital microfluidic biochips: design automation and optimization. CRC Press, Taylor and Francis Group, Boca Raton, FL, USA
2. Fair RB (2007) Digital microfluidics: is a true lab-on-a-chip possible? Microfluid Nanofluid 3:245–281
3. Su F, Chakrabarty K, Fair RB (2006) Microfluidics-based biochips: technology issues, implementation platforms, and design-automation challenges. IEEE Trans Comput Aided Des Integr Circuits Syst 25(2):211–223
4. Schneider L, Keszocze O, Stoppe J, Drechsler R (2017) Effects of cell shapes on the routability of digital microfluidic biochips. In: Proceedings of the international conference on design, automation and test in Europe, pp 1631–1634
5. Datta P, Dutta A, Majumder R, Chakraborty A, Dhal D, Pal RK (2017) Enhancement of mixing operation through new movement strategies in digital microfluidic biochip. In: Proceedings of the international conference on devices of integrated circuit, Kalyani, India, pp 611–614
6. Dutta A, Majumder R, Dhal D, Pal RK (2019) Structural and behavioural facets of digital microfluidic biochips with hexagonal electrode-based array. In: Proceedings of the 32nd international conference on VLSI design, New Delhi, India, pp 239–244
7. Maftei E, Pop P, Madsen J (2012) Droplet-aware module-based synthesis for fault-tolerant digital microfluidic biochips. In: Proceedings of the symposium on design, test, integration, and packaging of MEMS/MOEMS, ©EDA Publishing, Cannes, France
8. Maftei E (2011) Synthesis of digital microfluidic biochips with reconfigurable operation execution, Ph.D. Thesis, Technical University of Denmark, Kongens Lyngby, Denmark

9. Paik P, Pamula VK, Fair RB (2003) Rapid droplet mixers for digital microfluidic systems. Lab Chip 3–4:253–259
10. Dutta A, Majumder R, Dhal D, Pal RK (2019) Structural modelling, design automation, and a generalised routing technique for digital microfluidic biochip with hexagonal electrodes. In: Proceedings of the IEEE region 10 symposium (TENSYMP), Kolkata, India, pp 337–342

# A Novel Search Tree-Based 3D Sudoku Solver

**Sunanda Jana, Manjarini Mallik, Arnab Kumar Maji, and Rajat Kumar Pal**

**Abstract** Sudoku matrix is a unique Latin square-based mathematical structure, which may have several dimensions like 2D, 3D, etc. This structure has several applications in different domains of computer science. So, solving Sudoku and finding all its possible solutions are very important. Since 3D Sudoku has one extra dimension over 2D, it may give an edge to different applications of this puzzle game. Therefore, in this work, an effort has been made to explore all possible solutions of 3D Sudoku matrix. We have developed a permuted backtracking-based methodology to solve different 3D Sudoku instances. This method can solve an $n \times n \times n$ Sudoku having $n^2$ faces and provide all possible solutions for a valid Sudoku instance.

**Keywords** Sudoku puzzle · Cell · Grid · Minigrid · Minicube · Backtracking

## 1 Introduction

Sudoku is a logic-based puzzle. It may have dimensions either 2D or 3D. 2D Sudoku is represented by $n \times n$, 3D Sudoku is represented by $n \times n \times n$, where $n$ must be a perfect square integer. The logic for 2D Sudoku is such that, each row, each column, and each minigrid contains integers 1 through $n$ without repetition. Some of the popular possible sizes of Sudoku are $4 \times 4$ grid having $2 \times 2$ minigrids, $9 \times 9$

S. Jana (✉)
Department of Computer Science and Engineering, Haldia Institute of Technology, Haldia, India
e-mail: sunanda.only1@gmail.com

M. Mallik · R. K. Pal
Department of Computer Science and Engineering, University of Calcutta, Kolkata, India
e-mail: manjarini.mallik@gmail.com

R. K. Pal
e-mail: pal.rajatk@gmail.com

A. K. Maji
Department of Information Technology, North Eastern Hill University, Shillong, India
e-mail: arnab.maji@gmail.com



A. Maji et al. (eds.), *Proceedings of the International Conference on Computing and Communication Systems*, Lecture Notes in Networks and Systems 170,
https://doi.org/10.1007/978-981-33-4084-8_68

**Fig. 1** **a** An instance of 2D Sudoku. **b** Solution of the instance

| | | | | | | | | |
|---|---|---|---|---|---|---|---|---|
| | 2 | | 6 | | 8 | | | |
| 5 | 8 | | | | 9 | 7 | | |
| | | | | 4 | | | | |
| 3 | 7 | | | | | | | |
| 6 | | | | | | 5 | | 4 |
| | | 8 | | | | | 1 | 3 |
| | | | | 2 | | | | |
| | | 9 | 8 | | | | 3 | 6 |
| | | | 3 | | 6 | | 9 | |

(a)

| | | | | | | | | |
|---|---|---|---|---|---|---|---|---|
| 1 | 2 | 3 | 6 | 7 | 8 | 9 | 4 | 5 |
| 5 | 8 | 4 | 2 | 3 | 9 | 7 | 6 | 1 |
| 9 | 6 | 7 | 1 | 4 | 5 | 3 | 2 | 8 |
| 3 | 7 | 2 | 4 | 6 | 1 | 5 | 8 | 9 |
| 6 | 9 | 1 | 5 | 8 | 3 | 2 | 7 | 4 |
| 4 | 5 | 8 | 7 | 9 | 2 | 6 | 1 | 3 |
| 8 | 3 | 6 | 9 | 2 | 4 | 1 | 5 | 7 |
| 2 | 1 | 9 | 8 | 5 | 7 | 4 | 3 | 6 |
| 7 | 4 | 5 | 3 | 1 | 6 | 8 | 9 | 2 |

(b)

grid having $3 \times 3$ minigrids (the most common type of Sudoku, generally found in different magazines and newspapers), $16 \times 16$ grid having $4 \times 4$ minigrids (known as Super Sudoku [1]), and $25 \times 25$ grid having $5 \times 5$ minigrids (known as Giant Sudoku [2]). There are also some other types of Sudoku, like three-dimensional Sudoku [3] (invented by Dion Church), Godoku [4], or Wordoku [5] (puzzle with alphabetical placement).

## *1.1 Two-Dimensional Sudoku*

The Sudoku structure we are familiar with is a two-dimensional grid structure in which each constraint of Sudoku is applied in each row, column, and minigrid (each $\sqrt{n} \times \sqrt{n}$ square represented by bold outline). Generally, we find a $9 \times 9$ 2D Sudoku puzzle format in different newspapers. An example of such $9 \times 9$ Sudoku puzzle is shown in Fig. 1 along with its solution. However, it is also possible to extend this puzzle with one more dimension. The result will be a solid cubic structure filled with $m \times m \times m$ elements, where $m$ is a perfect square integer.

## *1.2 Three-Dimensional Sudoku*

The three-dimensional Sudoku is defined with a solid cubic structure with $m \times m \times m$ elements, where $m$ is a perfect square integer. The idea is such that, if we place nine two-dimensional Sudokus with grid size $9 \times 9$ each, one above another, by combining $i$-th row of each of them, we will get a 2D Sudoku and also by combining $i$-th column of each of them, we will get a 2D Sudoku, where $1 \leq i \leq 9$. So, in the three-dimensional Sudoku structure of size $9 \times 9 \times 9$, there will be $3^3$, i.e. 27 $9 \times 9$ two-dimensional Sudoku grids involved. An example of $9 \times 9 \times 9$ Sudoku solution [6] is shown in Fig. 2. There are nine grids in the figure which represent the two-dimensional Sudoku grids lying one above another. Grid 1 is the topmost layer, grid 2 lies just below grid 1, grid 3 lies just below grid 2, and so on. Thus, grid 9

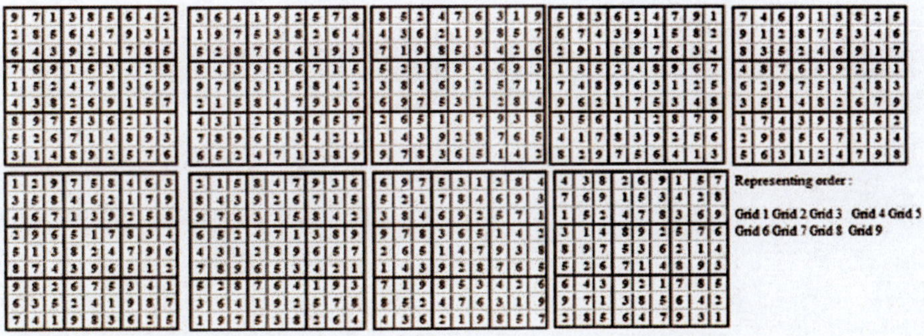


**Fig. 2** 2D layers of a 3D Sudoku solution lying top to bottom from grid 1 to grid 9 [6]

is the layer at the bottom. Note that each cell $[i, j]$ of each grid contains a different value. This is because by combining each $[i, j]$ cell of each grid, we get a row of another 2D Sudoku in the third dimension. For example, cell [1, 1] of grid 1 through grid 9 contains values 9, 3, 8, 5, 7, 1, 2, 6, 4, respectively. Hence, all nine elements are there, and they form a column in the third dimension.

There are $3^3$, i.e. 27, minicubes in the 3D $9 \times 9 \times 9$ Sudoku, each of size $3 \times 3 \times 3$. In each minicube, there are $3^2$, i.e. 9 numbers of 2D minigrid each of size $3 \times 3$. Each minicube has six neighbour minicubes, two in the $x$-direction, two in the $y$-direction, and two in the $z$-direction. It means that to compute the value for a blank cell in a minicube, we must consider the already placed values of its neighbour minicubes in the required position. The connectivity graph of minicubes is shown in Fig. 3. Each minicube is connected by an edge with its six neighbour minicubes. To elaborate more, the minicube *M1* with its neighbours is shown in Fig. 4.

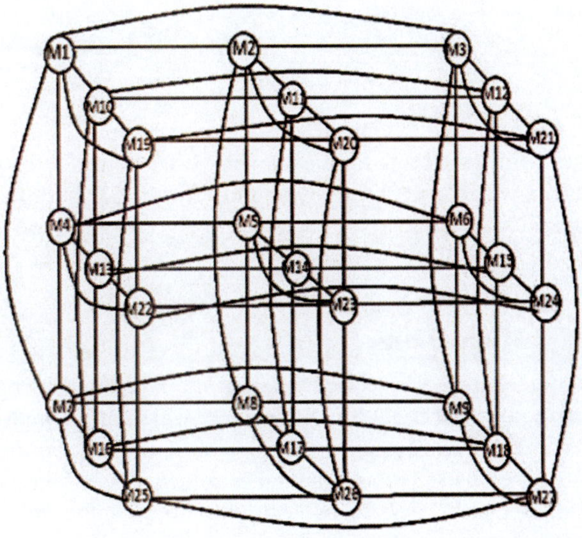


**Fig. 3** Connectivity graph for 3D Sudoku that represents relationship among the minicubes

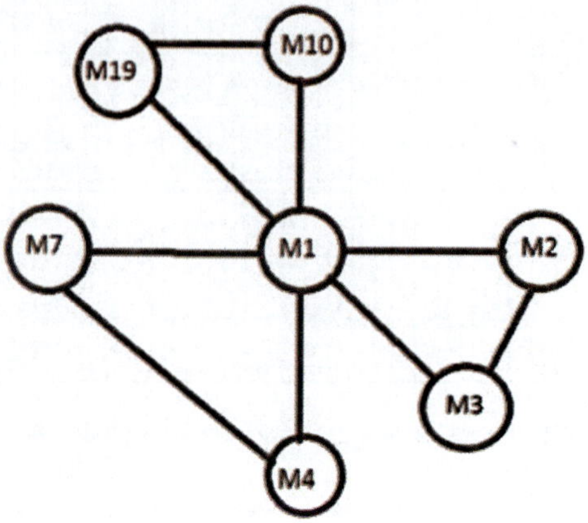


**Fig. 4** Connectivity graph for minicube *M1*

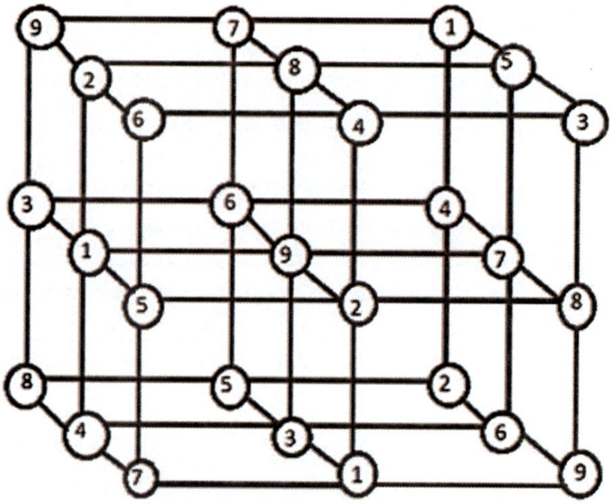

**Fig. 5** Minicube of position *M1*

Consider the example 3D Sudoku represented in Fig. 2. The grid 1, grid 2, and grid 3 form the three topmost layers. If we consider the first minigrids of these three layers, we get the first minicube *M1* for the 3D Sudoku, as has been represented in Fig. 5.

## 1.3 Motivation

There are several 2D Sudoku solvers which are completely guess-based and hence extremely time-consuming. In 2014, an entirely guess-free Sudoku solving algorithm was developed [7] that guarantees to generate all possible solutions for a valid 2D Sudoku instance. So this algorithm is very useful and efficient in the fields such as cryptography where 2D Sudoku with more than one solution is appreciated. However, the problem is the number of elements in 2D Sudoku is limited and we have to increase the grid size for application in encryption of huge data. It is expected if we can do the same for 3D Sudoku and use the huge structure for handling enormous data.

We hope that the 3D Sudoku solving algorithm can also be applied to resolve some setting and forecast-related quandary like aircraft scheduling problem. In this paper, we develop a brute force backtracking-based algorithm for 3D Sudoku that guarantees to generate all possible solutions for a given valid Sudoku instance. Till now, the researchers have developed a very few logic techniques to solve 3D Sudoku puzzles.

A valid Sudoku instance either in 2D or 3D may have more than one solution. Neither there is a constraint that a Sudoku problem instance must have only one unique solution [8] nor the existing rules force any possible unique solution of a Sudoku instance to be rational [9]. Some researchers have clearly proved the Another Solution Possible completeness (ASP-completeness ) of the Sudoku problem [10]. According to [11], there should be at least 17 clues to get a unique solution of a 2D Sudoku instance although it is also possible to get more than one solution with 17 or more clues. However, 2D Sudoku instances with single or multiple solutions have been applied in many fields such as in artificial intelligence, cryptography to generate one-time password (OTP) [12], encrypting SMS [13], steganography [14], biometric template encryption [15], digital watermarking [16], image–video encryption [17], DNA computing [18], visual cryptography [19], image authentication [20], and in many other areas. Extending the Sudoku structure to one more dimension will definitely make such applications possible to reach superior level.

## 2 Proposed Work

No such extensive work has been done in the field of 3D Sudoku. In our proposed method, brute force algorithm is used for solving 3D Sudoku puzzles using backtracking. A 3D Sudoku is fed as input, and a 3D array works as an associated array to store all the elements of that Sudoku. This algorithm uses linked list structure to store all the computed paths, either complete or incomplete, to find the solutions.

### *2.1 Physical Storage Structure of the 3D Sudoku Instance*

Table 1 shows the structure in which the 3D Sudoku of Fig. 2 is stored in the 3D array. Row 1 represents all the elements of grid 1, row 2 represents all the elements of grid 2, and so on. Each cell refers to the nine elements of the third dimension.

**Table 1** Storage structure of 3D Sudoku in 3D array

| | | | | | | | | |
|---|---|---|---|---|---|---|---|---|
| 971385642 | 285647931 | 643921785 | 769153428 | 152478369 | 438269157 | 897536214 | 526714893 | 314892576 |
| 364192578 | 197538264 | 528764193 | 843926715 | 976315842 | 215847936 | 431289657 | 789653421 | 652471389 |
| 852476319 | 436219857 | 719853426 | 521784693 | 384692571 | 697531284 | 265147938 | 143928765 | 978365142 |
| 583624791 | 674391582 | 291587634 | 135248967 | 748963125 | 962175348 | 356412879 | 417839256 | 829756413 |
| 746913825 | 912875346 | 835246917 | 487639251 | 629751483 | 351482679 | 174398562 | 298567134 | 563124798 |
| 129758463 | 358462179 | 467139258 | 296517834 | 513824796 | 874396512 | 982675341 | 635241987 | 741983625 |
| 215847936 | 843926715 | 976315842 | 652471389 | 431289657 | 789653421 | 528764193 | 364192578 | 197538264 |
| 697531284 | 521784693 | 384692571 | 978365142 | 265147938 | 143928765 | 719853426 | 852476319 | 436219857 |
| 438269157 | 769153428 | 152478369 | 314892576 | 897536214 | 526714893 | 643921785 | 971385642 | 285647931 |

Each cell represents the third dimension. The red-coloured elements in this table form minicube *M1*, which is shown in Fig. 5

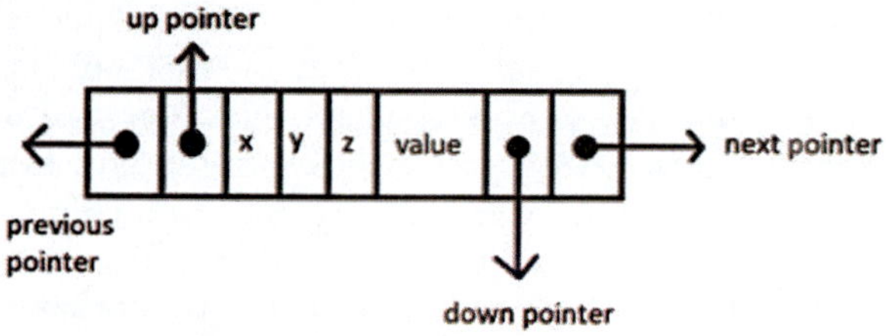


**Fig. 6** Node structure to store the blank cell value

### 2.1.1 Data Structure to Store the Computed Paths

A tree is generated using a linked list which is very important as it stores all possible solution paths, either complete or incomplete. For each value of a blank cell, a new node is created that stores row number *x*, column number *y*, position number *z* in cell [*x*, *y*], a computed value for the cell, a down pointer to point the next node for the same blank cell with a different computed value, a next pointer to point a node for another blank cell, an up pointer to point the previous node with a different value for the same blank cell, and a previous pointer to point the previous node for a different blank cell. Such a node structure is depicted in Fig. 6.

The linked list structure is represented in Fig. 7.

## 2.2 *Proposed Algorithm*

The algorithm stores the input 3D Sudoku puzzle in a 3D array. Then, it considers all blank cells one by one and computes probable value(s) for a current blank cell, temporarily places each of these values into the blank cell, and finds the possible values for the next blank cells. At some point, we may not find any possible value for the current blank cell. It implies that this combination of values in the previous and current blank cells does not lead to a solution. When we get certain possible values for each of the blank cells from first to last, we get a complete solution. In this way, a

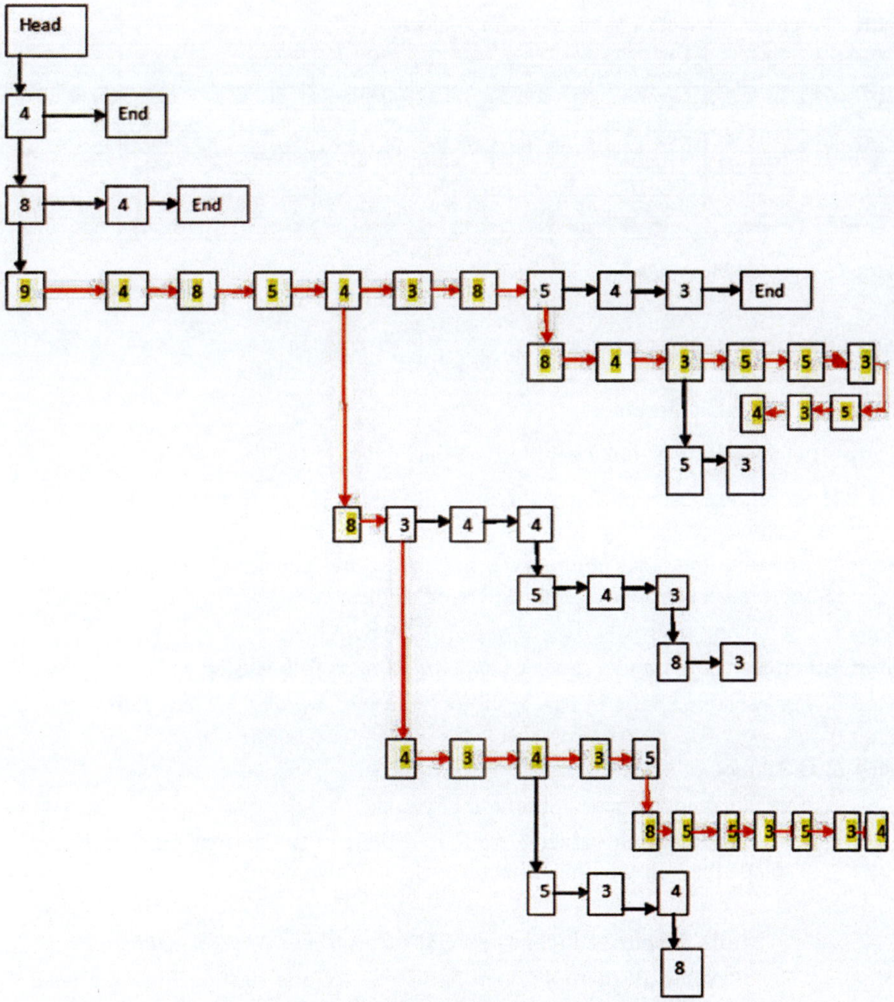


Fig. 7 Linked list structure to store solution paths for the given puzzle in Table 2

tree structure is formed using backtracking, and one or more solutions, if they exist, are obtained. A tree structure with the help of linked lists is shown in Fig. 7, to store the solution paths for the puzzle given in Table 2.

**Table 2** An input 3D Sudoku puzzle stored in a 3D array

| | | | | | | | | |
|---|---|---|---|---|---|---|---|---|
| 971385642 | 285647931 | 643921785 | 760153020 | 102478369 | 000269157 | 897536214 | 526714893 | 314892576 |
| 364192578 | 197538264 | 528764193 | 000926710 | 976315842 | 210847906 | 431289657 | 789653421 | 652471389 |
| 852476319 | 436219857 | 719853426 | 021784690 | 380692571 | 697531284 | 265147938 | 143928765 | 978365142 |
| 583624791 | 674391582 | 291587634 | 135248967 | 748963125 | 962175348 | 356412879 | 417839256 | 829756413 |
| 746913825 | 912875346 | 835246917 | 487639251 | 629751483 | 351482679 | 174398562 | 298567134 | 563124798 |
| 129758463 | 358462179 | 467139258 | 296517834 | 513824796 | 874396512 | 982675341 | 635241987 | 741983625 |
| 215847936 | 843926715 | 976315842 | 652471389 | 431289657 | 789653421 | 528764193 | 364192578 | 197538264 |
| 697531284 | 521784693 | 384692571 | 978365142 | 265147938 | 143928765 | 719853426 | 852476319 | 436219857 |
| 438269157 | 769153428 | 152478369 | 314892576 | 897536214 | 526714893 | 643921785 | 971385642 | 285647931 |

Zeros (0) represent blank cells

**Algorithm: 3D Sudoku Solver**

**Input:** A 3D Sudoku instance.

**Output:** One or more Sudoku solutions, if they exist.

**Step 1:** Store the 3D input *Sudoku_Matrix* puzzle in 3D array Sudoku. Place 0 in all the blank cells.

**Step 2:** Initialize the head node of the linked list and generate valid permutation *Search_Tree* considering the head node as the root.

**Step 3:** For each row $x$ in *Sudoku_Matrix*, follow Step 4.

**Step 4:** For each column $y$ in row $x$ of *Sudoku_Matrix*, follow Step 5.

**Step 5:** For each element $z$ in row $x$ and column $y$ of *Sudoku_Matrix*, follow Step 6.

**Step 6:** If *Sudoku_Matrix*$[x, y, z] = 0$

If the head node does not point to any node:

Check which value(s) from 1 to 9 fit in the current blank cell and place them in the linked list through the down pointer.

Else

Store the puzzle in temporary 3D array *Temp_Sudoku_Matrix*.

Traverse all possible open paths one by one starting from the head node and fill up all blank cells of *Temp_Sudoku_Matrix*.

Compute possible values for the current blank cell.

If found:

Create a new node for each possible value and connect it in the linked list through the down pointer. This list is connected to the end of that path through the next pointer of the last node.

Else

Create and append an end node to the end of the path.

**Step 7:** Initialize a 2D matrix, *Path*.

**Step 8:** Traverse each path of *Search_Tree* and store in each row of the matrix, *Path*.

**Table 3** Experimental results on puzzles having blank cells more on the same minicube

| Instances | # of blank cells | # of solution paths | Time in seconds | # of recursions |
|---|---|---|---|---|
| 1 | 15 | 1 | 0.28042841 | 137 |
| 1 | 20 | 1 | 0.287789583 | 209 |
| 1 | 29 | 2 | 0.046853065 | 533 |
| 1 | 35 | 4 | 0.093747854 | 2469 |
| 1 | 36 | 6 | 0.156235933 | 3807 |
| 1 | 37 | 16 | 0.249979973 | 6298 |
| 1 | 38 | 48 | 0.624959469 | 15419 |
| 1 | 47 | 96 | 81.21485853 | 136790 |
| 2 | 15 | 1 | 0.015622139 | 42 |
| 2 | 16 | 2 | 0.031252146 | 114 |
| 2 | 22 | 4 | 0.062496185 | 376 |
| 2 | 25 | 8 | 0.07812047 | 1011 |
| 2 | 26 | 16 | 0.140615702 | 1879 |
| 2 | 27 | 32 | 0.171845198 | 3350 |
| 2 | 29 | 96 | 0.609352112 | 15627 |
| 2 | 30 | 576 | 2.508261919 | 46902 |
| 2 | 31 | 1728 | 30.56493735 | 87490 |

## 3 Experimental Results

Table 3 includes the experimental results for solving 3D Sudoku puzzles. We have experimented the algorithm on two instances, each with different number of blank cells in different positions. Cells from the same minicube or related positions are made blank; hence, the number of solution paths increases drastically.

The performance of the algorithm is quite satisfactory for easy puzzles but depends more on the positions of blank cells and less on the number of the blank cells. Hence, we cannot measure the complexity in terms of the blank cells or in terms of the total number of elements. As this is a very first step towards designing an algorithm for solving 3D Sudoku, this algorithm may lack in performing efficiently for all cases; however, it is capable to furnish satisfactory performance for many cases.

## 4 Conclusion and Future Scopes

The proposed algorithm works well for easy 3D Sudoku puzzles; but in future, we like to update it for solving different difficulty levels and types of 3D Sudoku instances. Some techniques like lone ranger, twin, etc., also can be applied in that process to reduce the initial probable values for each blank cell and improve its performance.

## References

1. http://www.en.wikipedia.org/wiki/SuperSudoku
2. http://www.en.wikipedia.org/wiki/GiantSudoku
3. http://www.sudoku.org.uk/PDF/Dion_Cube.pdf
4. http://www.goduku.com
5. http://www.wordoku.biz
6. http://www.menneske.no/sudoku3d
7. Maji AK, Pal RK (2014) Sudoku solver using minigrid based backtracking. In: Proceedings of the IEEE international advance computing conference (IACC), pp 36–44. Gurgaon, India
8. http://sudopedia.enjoySudoku.com
9. http://www.mathoverflow.net/questions/27361/do-actual-Sudoku-puzzles-have-a-unique-rational-solution
10. Yato T, Seta T (2003) Complexity and completeness of finding another solution and its application to puzzles. IEICE Trans Fundam Electron Commun Comput Sci 86(5):1052–1060
11. McGuire G, Tugemann B, Civario G (2012) There is No 16—clue sudoku: solving the sudoku minimum number of clues problem, MS Thesis, School of Mathematical Sciences, University College Dublin
12. Okagbue HI, Omogbadegun ZO, Olajide FA, Opanuga AA (2015) On some suggested applications of sudoku in information systems security. Asian J Inform Technol 14(4):117–121
13. Shirali-Shahreza MH, Shirali-Shahreza M (2008) Steganography in SMS by Sudoku Puzzle. In: Proceedings of the IEEE/ACS international conference on computer systems and applications (ICCSA), pp 844–847
14. Maji AK, Roy S, Pal RK (2014) A novel steganographic scheme using Sudoku. In: Proceedings of the IEEE international conference on electrical information and communication technology (ICEICT 2013), Bangladesh, pp 116–121
15. Maji AK, Pal RK (2014) A novel biometric template encryption scheme using Sudoku. Appl Comput Secur Syst 2:109–128
16. Naini PM, Fakhraie SM, Avanaki AN (2010) Sudoku bit arrangement for combined demosaicking and watermarking in Digital Camera. In: Proceedings of the second international conference on advances in databases, knowledge, and data applications (ICADKDA), pp 41–44
17. Bourbakis N, Shujun L (2004) Data-image-video encryption. IEEE Potentials 23(3):28–34
18. Erlich Y, Chang K, Gordon A, Ronen R, Navon O, Rooks M, Hannon GJ (2009) DNA Sudoku: harnessing high-throughput sequencing for multiplexed specimen analysis. Genome Res J 19(7):1243–1253
19. Chang C-C, Lin P-Y, Wang Z-H, Li M (2010) A Sudoku based secret image sharing scheme with reversibility. J Commun 5(1):5–12
20. Wu W-C, Ren G-R (2009) A new approach to image authentication using chaotic map and Sudoku Puzzle. In: Proceedings of the fifth international conference on intelligent information hiding and multimedia signal processing (ICIIHMSP), pp 628–631

# Canonical Representation of Causal Parameters and Graph–Theoretic Approach for Identifying At-Risk Students

**Arnab Das, Saurav Malla, Nirmalya Mukherjee, and Suman Deb**

**Abstract** Classroom is a physical territory where students and teacher meet for certain time to exchange the subject specific knowledge through various interactions. This study specifically focuses on students of higher classes (undergraduate and beyond), where the self-paced attitude of students learning is highly dominant. In recent past, enormous increase in affordable technology and learning alternatives has added more issues to classroom where it is challenging for a teacher to have consistent presence and identify the learning progress of students. Positive correlation between learning outcome in terms of exam score and attendance is an established fact. Here in this article, the authors have tried to show the correlation between attendance and all other formative parameters contributing to the summative score of the students with a graph theoretic approach to identify academically at-risk students. Authors have also established a logical background for analysing the cumulative performance of a student taking attendance and required formative parameters into consideration. For the empirical observation and validating the claim, the group of engineering undergraduate students ($N = 189$) data is taken into consideration.

**Keywords** Canonical parameter representation · Decision tree · Academic correlation · Formative score

---

A. Das (✉) · S. Malla · N. Mukherjee · S. Deb
Department of Computer Science and Engineering, National Institute of Technology, Agartala, India
e-mail: dasarnab693@gmail.com

S. Malla
e-mail: sauravmalla4@gmail.com

N. Mukherjee
e-mail: mjeenirmalya005@gmail.com

S. Deb
e-mail: sumandebcs@gmail.com

 
A. Maji et al. (eds.), *Proceedings of the International Conference on Computing and Communication Systems*, Lecture Notes in Networks and Systems 170,
https://doi.org/10.1007/978-981-33-4084-8_69

## 1 Introduction

In a pedagogical teaching–learning environment, a proper procedure of teaching–learning is followed. In such an environment to make the classroom learning more productive, it is very much important to identify the causal parameter and at-risk students. So, it is a concern for a teacher to make their students more productive academically by identifying and working on at-risk students, as well students must be aware also to make their performance better by identifying what are the causal parameters affecting their performance concerning the class. To make a student more productive, a teacher always wants to know what is the academic status of any student with respect to others in the peer group. One such assumption may be a teacher wants to check if a student is performing more than the mean performance of the class. A student also wants to know whether all the class he or she is attending is productive or not. So, the authors have tried to figure out the solution to both the problems. Boolean expressions for two different sets of students have been derived, by which one can check the status of a student performing better than the mean performance of the class or not. And the graph–theoretic approach has been applied to identify the at-risk students by analysing the performance whether it leads to a zero or one. Leading to zero in leaf node means the student is not performing better than his/her peer group, and he or she is at risk and if his or her result leads to one in the leaf node, then performance is satisfactory.

## 2 Background

Omran and Saleh [1] studied different factors that are affecting the students' summative score. They focused on five different factors including the way teachers teach in class, proficiency in communicating in English language, language gauging, the pressure or tension exerted on students in academics and student's self-perception. However, their study was only focused on the feedback given by the students from a survey. They did not deal with how the feedback was correlating with the real marks of the student. Pudaruth et al. [2] have studied the correlation between attendance and total marks of computer science and engineering branch of Mauritius University. They have studied the declining performance of computer science and engineering students. They have shown how attendance is affecting the result of the students [3]. They have collected different data from different classes of different subjects and shown the correlation between the total marks and attendance. Harlen and Crick [4] have studied the extensive use of aggregate or collective scores from examinations. They claimed that students as well as the teachers can escalate their efforts based on the rewards and results of the tests. So, summative assessment is necessary for both acquiring knowledge and developing skills. Fadelelmoula [5] analysed how mandatory attendance is affecting the student's total marks. The work analysed the student's performance of four different courses. There it was also analysed the mean attendance and mean performance of the whole class. However, only a total mark out of 30 was considered for performance. The work did not analyse the parameters contributing to the total marks. Lina and Chen [6] studied the effect of cumulative class attendance in students' performance, and they found a positive result of the effect of cumulative attendance in students' performance [7]. Their study was based on the individual data they collected. Cohn and Johnson [8] tried to answer some questions related to attendance and students' performance [9]. They have used a sample of 347 students. Their main focus of the study was to examine whether attendance is affecting students' performance in Principle of Economics class [10]. They proposed four hypotheses for their work and found some major positive results for the same. However, their study was only focused on a particular subject. Purcell [11] did a linear regression analysis on the students' performance and attendance. Strong correlation between attendance and students' performance was found [12]. He showed that each 10% increase in the class presence increases the final performance by 3%. Khan et al. [13] studied the impact of class attendance on academic performance. They made two groups—one having attendance percentage greater than 75 and another having attendance percentage less than 75 [14]. They argued that their result validates the statement that attending lectures has a direct impact on the academic result of the students [15]. Kirby and Mcelroy [16] studied the interdependence between attendance and the result of the students. The authors found that not only class attendance has an impact on students' grades but the tutorial grades have more impact on it. Nyamapfene [17] studied second-year electronics students' performance where attendance was not mandatory [18]. The author showed that attendance has a strong linkage with the students' performance. The author found that there is no relation between the pre-university result and the current result. Also, the author found no gender bias in students' attendance and performance [19].

## 3 Methodology and Evaluation

The Academic details for the session 2017–18 and 2018–19 of four sections were collected. The total sample size of 2017–18 was 89 and 2018–19 was 100. The attendance percentage and the mean of each parameter were calculated. The truth table was made for each session based on the mean of all parameters of each class. If the value of any parameter of a student is greater than the mean value of that parameter, then the corresponding logic of that student was set as TRUE, otherwise FALSE. The truth table has four different independent variables named attendance (A), internal marks (I), mid-term marks (M) and end-term marks (E). So, the K-Map will have 16 entries for different combinations of these variables. So, for a dataset of 189 students, it is certain that the truth value of more than one student will overlap for the dependent variable final marks (Q). So, this thing may give two conditions. Let's say for a certain student the combination of independent variables is $A = 0, I = 1, M = 0, E = 1$, the final result may have the truth value 1 for that student. Another student may have the same combination of truth values for independent variables, but the truth value of the main variable maybe 0. So, the most overlapping truth values in the following section mean the greatest number of students for a certain combination of independent variables leading to a particular result and least overlapping truth values in the following section means a minority of students for that certain combination of independent variables leads to another result. $Q_1, Q_2, Q_3, Q_4$ in the following section are four different instances of the final marks ($Q$) for two different datasets.

### 3.1 Evaluation of Boolean Expression

Cumulative scores out of 100 were calculated using the standard distribution of marks. The distribution consists of 20% of the internal marks, 30% of the mid-semester exam marks and 50% of the end semester exam marks. To observe the relationship between formative parameters with total marks in binary logic, the independent variables were identified. In this work, four independent variables have been considered which are Attendance Percentage, Internal Marks, Mid-Term Marks and End-Term Marks. Total marks have been considered as a dependent variable. Since the number of independents was four, Karnaugh map was used to evaluate the final Boolean expression. Hence, the number of categories of students in each session will be 24 = 16. So, for more than sixteen students either the logical value should overlap or vitiate from values True, False. The authors made two Karnaugh maps, on the basis of the data set prepared, for example, a batch consisting of 89 students if 80 students get a combination of the attendance, internal marks, mid-semester exam marks and end semester exam marks as 1101, respectively, resulting to the cumulative score of 1 and the rest 9 students getting the combination of variable as 1101 resulting to a cumulative score of 0. So, for these two types of data, two Karnaugh maps were plotted for each data set.

**Karnaugh Maps for Session 2017–18** (Figs. 1 and 2).

**Karnaugh Maps for Session 2018–19** (Figs. 3 and 4).

The Boolean expression derived from Fig. 1 is:

$$Q_1 = ME + IM + IE \tag{1}$$

The K-Map in Fig. 1 has been formed according to the truth table prepared from the data of 89 students of session 2017–'18 in one subject. Here, it can be seen from equation $Q1$ that the summative score is not directly dependent on the attendance of a student. It is dependent on the other three causal parameters—internal marks, mid-term and end-term marks.

The Boolean expression derived from Fig. 2 is:

$$Q_2 = E\left(A' + AM\right) + I\left(E + A'M\right) \tag{2}$$

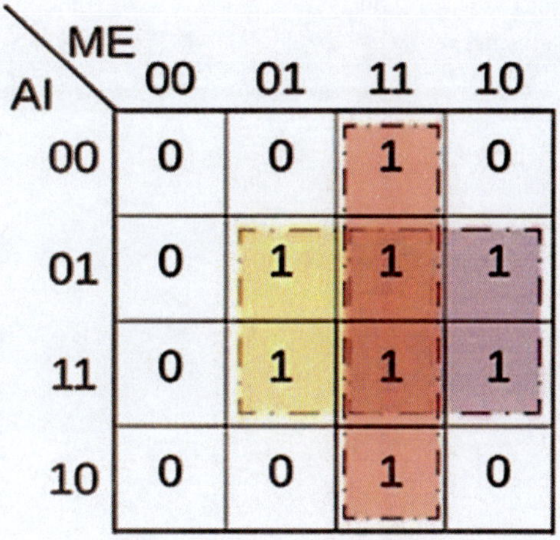


**Fig. 1** Most overlapping truth values

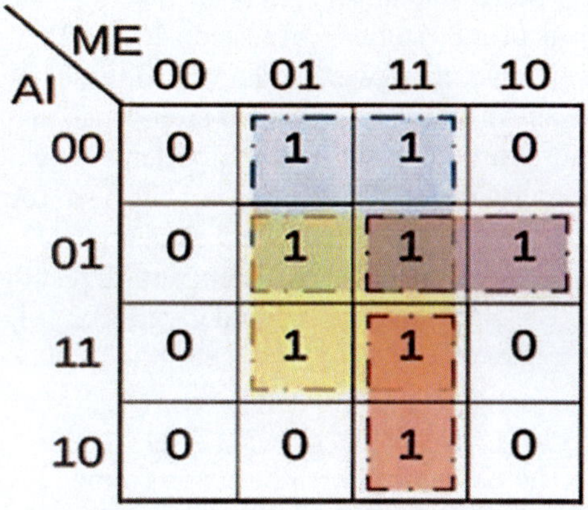


**Fig. 2** Least overlapping truth values

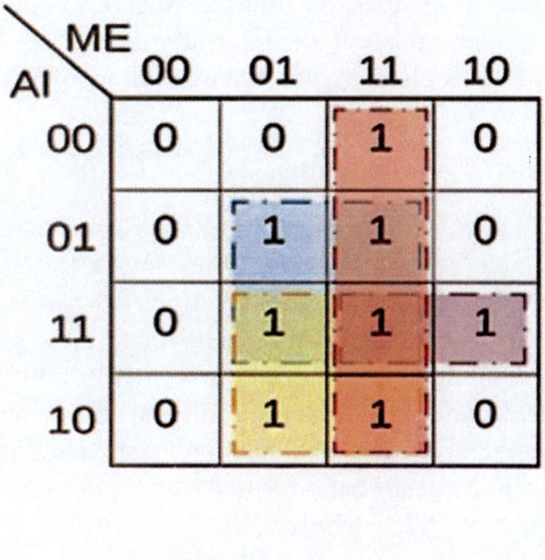


**Fig. 3** Most overlapping truth values

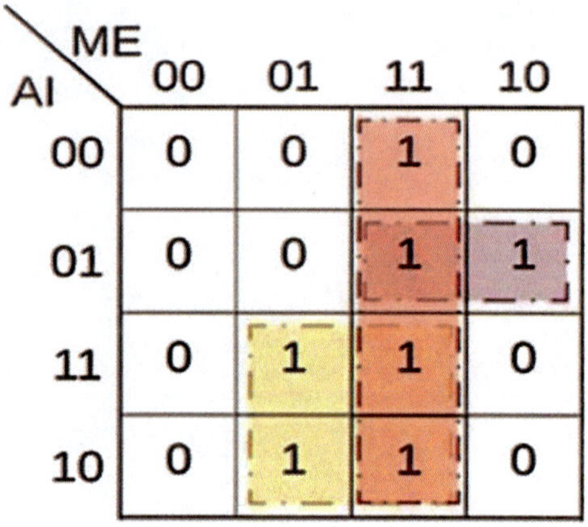


**Fig. 4** Least overlapping truth values

The K-Map in Fig. 2 has been formed according to the truth table prepared from the data of 89 students of session 2017–'18 in the same subject as that of Fig. 1. In that truth table, there were some variations. There are few students who did not perform well even after having a good attendance percentage and scoring good in internal and mid-term exams, while some students have done well only having a good score in the end-term exam. So, this K-Map has been formed by taking these two variations, and it can be seen from the deduced equation Q2 that the final score is partially dependent on attendance and fully dependent on rest three causal parameters.

The Boolean expression derived from Fig. 3 is:

$$Q_3 = E(A + M) + I(E + AM) \quad (3)$$

The K-map in Fig. 3 has been formed according to the truth table prepared from the data of 100 students of session 2018-'19 in one subject. Here, it can be seen from equation $Q_3$ that the summative score is dependent on all four causal parameters: attendance percentage, internal marks, mid-term and end-term marks.

The Boolean expression derived from Fig. 4 is:

$$Q_4 = E(A + M) + A'IM \quad (4)$$

The K-Map in Fig. 4 has been formed according to the truth table prepared from the data of 100 students of session 2018–'19 in the same subject as that of Fig. 3. In that truth table, some variations have been found. There are few students who did not perform well even after having a good attendance percentage and scoring good in internal and mid-term exams, while some students have done well only having a good score in the internal and end-term exams. There are also some students who scored well in the final exam by only having good internal and mid-term scores. So, this K-Map has been formed by taking these three variations, and from the deduced equation $Q_4$ it can be observed that the final score is partially dependent on attendance and fully dependent on rest three causal parameters.

## 4 Graph–Theoretic Approach

A graph is a pair G(V, E) of sets satisfying $E \subset V \times V$, where V is the set of vertices and E is the set of edges connecting the vertices. In this paper, the different causal parameters have been taken as vertices and the edges.

Several methods have been developed for expressing and managing Boolean functions. The methods based on truth table, sum-of-products or K-Maps are not so sufficient because the size of representing the arguments of these functions is generally $2^n$ or more. This exponential size in the function representation is a major drawback. A simple function doing the complement could be of large size which is not so feasible. Again, since none of these functions are canonical, so each and every function may be represented in different ways. Because of these drawbacks, the authors in this paper have opted to represent the causal parameters using Binary Search Tree as cited by Akers [20] in his paper. The main motive is to represent the digital functions pictorially in terms of a "tree structure", which will show us how to determine the output values based on some input. The authors have assumed the four causal parameters as the nodes of the trees and depending on the binary input given to these four parameters the tree will find the binary value of the respective Boolean functions. The Boolean expressions have been derived on the basis of the mean value of each parameter. So, if the inputs to the trees are given for a particular student, it will show whether the summative score of that student is below or above the mean value. This is one instance of representing the trees. Different levels can also be set for prediction instead of mean value, i.e. the performance level can be set according to our needs. There are four parameters: attendance (A), internal marks (I), mid-term marks (M) and end-term marks (E).

These decision trees have been represented by following the protocols given by Mythili and Shanavas [21] in his paper. Each non-terminal node represents one of the four causal parameters. One can choose the root node as any one of the causal parameters. The leaves represent the possible values the Boolean expressions can take on different values of the causal parameters. For example, in Fig. 5, if $M = 0, I = 1, E = 1, A = 0$, the final result will be 1. In Fig. 6, if $M = 1, I = 1, E = 0, A = 0$, the final result will be 1. In Fig. 7, if $M = 0, I = 1, E = 1, A = 0$, the final result will be 1. Similarly, in Fig. 8, if $M = 0, I = 1, E = 1, A = 0$, the final result will be 0.

## 5 Conclusion

The advantage of this binary tree representation is that it will yield the output in the time equal to the height of the tree which is less than the time if we calculate in naive approach. To validate the binary trees, the authors have taken a sample of some students. For the session 2017–'18, there are few students who did not perform well even after having a good attendance percentage and scoring good in internal and mid-term exams, while some students have done well only having good

**Fig. 5** Graphical representation of $Q_1$

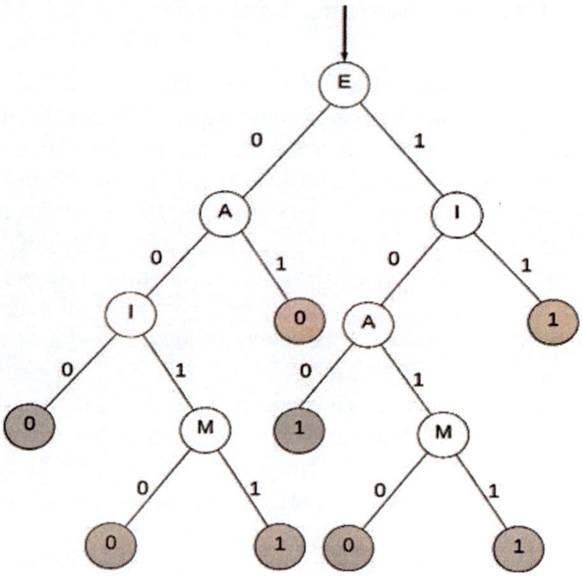


**Fig. 6** Graphical representation of $Q_2$

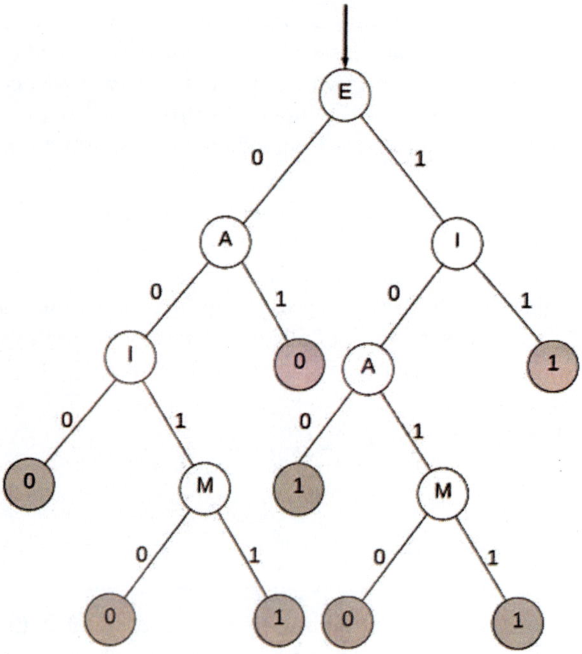

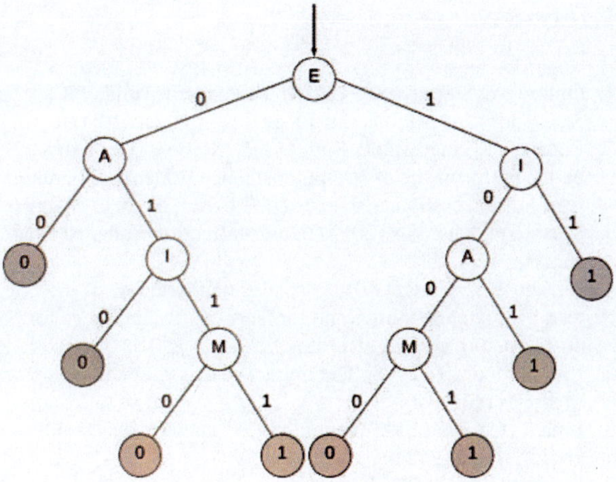


**Fig. 7** Graphical representation of $Q_3$

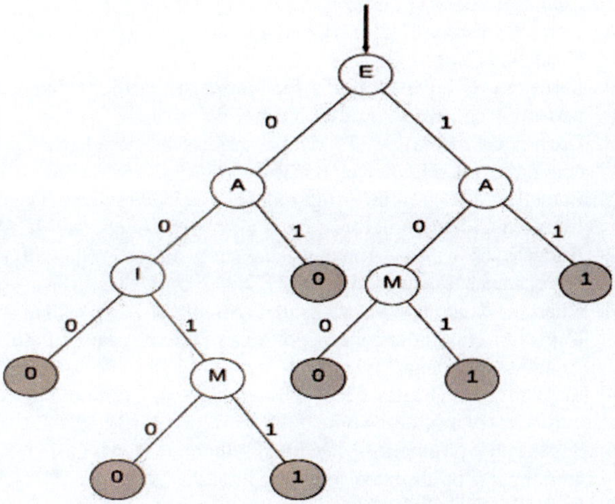


**Fig. 8** Graphical representation of $Q_4$

scores in the end-term exam. So, by taking these two variations, we have deduced equation $Q_2$ alongside the equation $Q_1$. Again, in the session 2018–'19, there are few students who did not perform well even after having a good attendance percentage and scoring good in internal and mid-term exams, while some students have done well only having a good score in the internal and end-term exams. There are also some students who scored well in the final exam by only having good internal and mid-term scores. So, by taking these three variations, we have deduced equation $Q_4$ alongside equation $Q_3$. Since four parameters have been taken to show the effect of the summative score, the whole class can be divided into $2^4 = 16$ groups. Each student of the class will fall under any one of the groups. This division will help both the student and teacher to have a perception of the status of the student, whether the student is doing well or the student is at risk.

## References


1. Omran A, Saleh MSH (2019) Factors affecting the academic performance of students at Academics Institutions. Int J Res Rev Appl Sci 40(1)
2. Pudaruth S, Nagowah L, Sungkur R, Moloo R, Chiniah A (2013) The effect of class attendance on the performance of computer science students. In: 2nd international conference on machine learning and computer science (IMLCS' 2013) August 25–26, Kuala Lumpur (Malaysia)
3. Mushtaq I, Khan SN (2012) Factors affecting students' academic performance. Glob J Manage Bus Res 12(1)
4. Harlen W, Crick RD (2002) A systematic review of the impact of summative assessment and tests on students' motivation for learning. In: Evidence for policy and practice information and co-ordinating centre, Institute of Education, University of London
5. Fadelelmoula T (2018) The impact of class attendance on student performance. Int Res J Med Medic Sci 6(2):47–49
6. Lina T, Chen J (2006) Cumulative class attendance and exam performance. Appl Econ Lett 2006(13):937–942
7. Credé M, Roch SG, Kieszczynka UM (2010) Class attendance in college: a meta-analytic review of the relationship of class attendance with grades and student characteristics. Rev Educ Res 80(2):272–295
8. Cohn E, Johnson E (2007) Class attendance and performance in principles of economics. Educ Econ 14(2):211–233
9. Devadoss S, Foltz J (1996) Evaluation of factors influencing student class attendance and performance. Am J Agric Econ 78(3):499–507
10. Durden GC, Ellis LV (1995) The effects of attendance on student learning in principles of economics. Am Econ Rev 85(2)
11. Purcell P (2007) Engineering student attendance at lectures: effect on examination performance. In: International conference on engineering education—ICEE 2007
12. Rodgers JR (2001) A panel-data study of the effect of student attendance on university performance. Austr J Educ 45(3):284–295
13. Khan H, Khattak AM, Mahsud I, Munir A, Ali S, Khan MH, Saleem M, Shah SH (2003) Impact of class attendance upon examination results of students in basic medical sciences. J Ayub Med College 15(2)
14. Stanca L (2010) The effects of attendance on academic performance: panel data evidence for introductory microeconomics. J Econ Educ 37(3):251–266
15. Alghamdi A, Yamani A, Khalil A, Albarkati B, Alrehili O, Salih M (2016) Prevalence, causes and impacts of absenteeism among medical students at UQU. Education 6(1):9–12
16. Kirby A, Mcelroy B (2003) The effect of attendance on grade for first-year economics students in University College Cork. Econ Soc Rev 34(3):311–326 (Winter)
17. Nyamapfene A (2010) Does class attendance still matter? Eng Educ 5(1):64–74
18. Hijazi ST, Naqvi SMMR (2006) Factors affecting students' performance: a case of private colleges. Bangladesh e-J Sociol 3(1)
19. Ajiboye JO, Tella A (2006) Class attendance and gender effects on Undergraduate Students' achievement in a Social Studies Course in Botswana. Essay Educ 18:1–14
20. Akers SB (1978) Binary decision diagrams. IEEE Trans Comput C-27(6):509–516
21. Mythili MS, Shanavas ARM (2014) An Analysis of students' performance using classification algorithms. IOSR J Comput Eng (IOSR-JCE) 16(1), Ver. III. e-ISSN: 2278-0661, p-ISSN: 2278-8727

# Author Index




A. Maji et al. (eds.), *Proceedings of the International Conference on Computing and Communication Systems*, Lecture Notes in Networks and Systems 170,
https://doi.org/10.1007/978-981-33-4084-8